Islamic Geometric Patterns

Jay Bonner

Islamic Geometric Patterns

Their Historical Development and Traditional Methods of Construction

with a chapter on the use of computer algorithms to generate Islamic geometric patterns by Craig Kaplan

 Springer

Jay Bonner
Bonner Design Consultancy
Santa Fe, New Mexico, USA

With contributions by
Craig Kaplan
University of Waterloo
Waterloo, Ontario, Canada

ISBN 978-1-4939-7921-9   ISBN 978-1-4419-0217-7   (eBook)
DOI 10.1007/978-1-4419-0217-7

This Springer imprint is published by Springer Nature
The registered company is Springer Science+Business Media LLC
The registered company address is: 233 Spring Street, New York, NY 10013, U.S.A.

*For my wife, Shireen, without whose support this book would never
have seen the light of day.*

Jay Bonner

*For Reza Sarhangi, who convinced me to cross a bridge
and never look back.*

Craig Kaplan

Foreword

For over eleven centuries, Islamic artists and architectural designers have developed extraordinary skills to produce many amazing forms of decoration. This superbly illustrated book, by Jay Bonner (with a final chapter by Craig Kaplan), is a wonderfully comprehensive and deeply thoughtful account of these highly distinctive designs. I feel sure that this book will represent a landmark in the study of Islamic design.

These remarkable artworks take various forms, but they are almost always of an entirely nonrepresentational character, only a very few showing indications of recognizable natural objects, such as leaves or flowers. Yet there is a distinctive beauty in these designs, most of these patterns being of a particular geometrical character, demonstrating a keen and subtle knowledge and interest in geometry and a profound skill in using geometrical motifs to produce some incredibly intricate patterns. These would normally be repetitive, in a planar arrangement, though sometimes bent to cover a curved surface, such as for a dome or ceiling.

They gain some of their beauty from the intricacy and ingenuity of the designs, which frequently involve sophisticated geometrical ideas in unexpected ways. These artists had clearly developed some significant understanding of mathematical notions that were not properly developed by professional mathematicians until the early twentieth century, when the plane Euclidean symmetry groups ("wallpaper symmetries") were finally classified into 17 distinct types. Representations of these different symmetries are copiously exhibited in Islamic patterns and, remarkably, all 17 are to be found in the single location of the Alhambra Palace in Granada, Spain, constructed more than half a millennium earlier than the time that these different symmetries were explicitly distinguished by mathematicians. In the early 1930s, these Alhambra designs had a seminal inspirational influence on the extraordinary work of the well-known Dutch artist M.C. Escher.

Yet, there is much more than the illustration of different Euclidean symmetries in the ingenious and intricate use of geometry in these Islamic designs. There is no doubt about the skill and artistry that have gone into the production of these creations. There is much, also, which is of mathematical interest in the local symmetries that have been frequently incorporated in this way, where we find symmetrical star-shaped regions of all sorts of improbable-seeming symmetry, such as combinations of stellar shapes with 13-fold and 9-fold symmetry, which are quite ruled out as overall symmetries of crystal-like arrangements, for which only 2-fold, 3-fold, 4-fold, and 6-fold symmetries are allowed.

The question has sometimes been raised, as to whether these ancient Islamic geometers might have discovered and made use of quasisymmetrical features, such as the fivefold ones that I found myself in the mid-1970s. These would give rise to patterns that "almost" (in a technical sense) repeat, but which never quite do so, and the fivefold symmetry is likewise a feature that "almost" holds but not quite. When I found such designs that arise naturally as the only arrangements in which one can assemble a pair of "jig-saw" shapes, I was particularly struck by the simplicity of these particular shapes, which could indeed force such fivefold quasisymmetric patterns. I thought it not unlikely that things of this nature might well have been made use of by the Islamic geometric artists of antiquity. Indeed, various later researchers have claimed that there is indeed evidence that some ancient Islamic buildings might well employ such quasisymmetric features. In this book, Jay Bonner argues otherwise, finding strong indications that the desire for exact periodicity was an overriding driving motive for the Islamic patterns, and that there is no evidence that quasisymmetric features were ever made use of.

It is, of course, still possible that convincing evidence might eventually come to light that the ideas of quasisymmetry did play a role for some of the ancient Islamic designers. However, I am inclined to agree with Jay Bonner that strict periodicity seems to have played such a key role for these ancient geometric artists that quasisymmetric considerations would be unlikely. It is, to me, quite extraordinary how some of these Islamic designers were able to fit such improbable symmetries as 13-fold symmetric stars combined with 9-fold ones so as to create a design with such a natural-looking elegance, and a periodicity that appears at a distance not much greater than the extent of these stellar shapes themselves. Mathematical quasisymmetric patterns with, say 11-fold or 13-fold quasisymmetry, can now be produced in computer-generated pictures, but it would only be at extremely remote places where local regions with this symmetry would be evident. If one is looking for beauty of design, with such regions of, say, 11-fold or 13-fold symmetry, then the ancient Islamic designs win hands down!

Roger Penrose
Emeritus Rouse Ball Professor of Mathematics
Mathematical Institute
University of Oxford

Preface and Acknowledgements

More than with any other cultural heritage, the work of Muslim artists over the centuries reveals the visual beauty that is inherent to geometry. Drawing inspiration from geometry led to an abundance of aesthetic innovations within the tradition of Islamic geometric design that were inexorably associated with methodological practices. Of particular importance was the discovery that geometric patterns could be extracted from underlying polygonal tessellations, herein referred to as the *polygonal technique*. Over time, and in the hands of skilled and dedicated practitioners, this design methodology engendered the extraordinary breadth of diversity that characterizes this artistic tradition. As architectural ornament, Quranic illumination, and its more limited use within the applied arts, geometric designs were enthusiastically embraced by succeeding Muslim cultures, and along with calligraphy and the floral idiom, became an integral aspect of Islamic aesthetics. As such, its role within Islamic art as a whole is paramount. Yet scholarship of this vast subject remains underrepresented in two key areas: historical development and design methodology. Related to this relative lack of attention is the surprising absence of a comprehensive approach to categorizing the range of patterns within this tradition. Additionally, while the exceptional achievements of past geometric artists serve as a source of inspiration for many contemporary artists, designers, craftspeople, and architects, the loss of methodological knowledge associated with this artistic tradition has substantially thwarted those with an interest in incorporating such designs into their work. In consideration of these gaps this book has several intentions: to provide a greater understanding of the developmental history of this remarkable artistic tradition; to emphasize a more nuanced attribution of the geometric and design diversity that is a hallmark of this distinctive form of art; and to provide a detailed elucidation of the methodological practices that are responsible for the diversity, beauty, and longevity of this artistic discipline. What is more, it is my hope that the focus upon design methodology, including the use of computer algorithms, will empower those with a sincere and dedicated interest in applying the creative processes of the past to their own original works.

The combined focus upon historical development and design methodology within this work requires an approach to the subject of Islamic geometric patterns that is both chronological and analytical. As such, there is an inevitable repetition of information concerning geometric characteristics and the dates and location of specific examples when bridging between these two approaches. This is necessary to clarify historical, geometric, and methodological context throughout the book, and I request the reader's patience and indulgence when encountering such repetition. As a practical consideration, all dates accord with the Gregorian calendar rather than either the Islamic or Persian calendars. In organizing the terminology employed throughout this work I have erred on the side of clarity and simplicity and have generally tried to avoid being overly technical. Wherever applicable, unless there is a compelling reason to do otherwise, I have adopted prior terminology. The glossary provides brief definitions of many of the terms used throughout this work, including those that are of foreign origin (Arabic, Persian, Urdu, Turkish, or Spanish), those that are technical and associated with the science of tiling and geometry, and those that pertain specifically to design methodology. In this latter category, much of the nomenclature is of my own invention. This is due to the fact that many significant features of this artistic discipline have not been previously identified as such and are therefore without name or title. I argue that the *polygonal technique* was the principle design methodology employed by Muslim geometric artists. This technique has been referred to variously as the *Hankin method* (in deference to Ernest Hanbury Hankin who first identified the historical use of this methodology), or the *PIC method* (polygons-in-contact). I prefer the term *polygonal technique* for its simplicity and descriptive accuracy. In previous publications I have referred to the polygonal mechanism that characterizes the polygonal technique as *subgrids*. However, in the interests of descriptive clarity, in this book I refer to this important methodological feature as the *underlying generative tessellation*, or alternatively as the *underlying polygonal tessellation*. The polygonal technique was employed in two very different modalities: systematically and nonsystematically. My identification of patterns as being either *systematic* or *nonsystematic* results from there being no such previous differentiation by prior scholars. My descriptive titles for the five historical design systems that employ repetitive modules to create geometric designs stem from their not having been identified as systems, *per se*, by other specialists prior to my work. I have titled these as the *system of regular polygons*, the *fourfold system A*, the *fourfold system B*, the *fivefold system*, and the *sevenfold system*. My classification of the four historical varieties of dual-level design has changed slightly from my 2003 account of this discipline (in which I had only identified three types) and results from there being no previous differentiation within the published literature. Rather than employing a descriptive title I have simply used the more prosaic terms *Type A*, *Type B*, *Type C*, and *Type D*. Similarly, my names for the four principle pattern families that are ubiquitous to this tradition stem from the absence of prior identifying classifications from previous sources in the English language. These are descriptively named the *acute*, *median*, *obtuse*, and *two-point* pattern families. In writing about a tradition that encompasses many distinct cultures with separate languages and artistic terms, I have chosen to refrain from employing terms that are specific to select Muslim cultures in writing about this discipline more generally. For example, despite being in common usage, I do not use the Farsi word *girih*, meaning "knot," when referring to geometric designs. I have sought to keep my geometric terminology as nontechnical as possible, while following convention to maintain clarity. For the most part my terminology corresponds with Craig Kaplan's in Chap. 4. He prefers the term *translational unit* for what I typically refer to as *repeat unit*. Similarly, he employs the phrase *template tiling* for what I refer to as the *underlying generative tessellation*, or alternatively as the *underlying polygonal tessellation*.

Chapter 1 chronicles the historical development of Islamic geometric patterns from initial influences and early manifestations through to full maturity. In identifying broad stylistic trends, geometric characteristics, and diverse varieties of design, I have referenced multiple individual pattern examples from successive Muslim cultures. My choice of examples reflects the chronological development from simplicity to complexity and will frequently build upon the familiar to introduce the less well known. This choice of historical examples cannot help but be subjective, but every attempt has been made in aligning my aesthetic preferences and value judgments with impartial historical significance. Similarly, the many photographs included within this chapter provide a sense of the broad aesthetic diversity contained within this tradition. In discussing the contributions of successive Muslim dynasties, the general structure within each section flows from the more basic patterns to more complex designs. Emphasis is always placed upon innovations that occurred during a given epoch, as these were primary vehicles for the advancement of this ornamental tradition. Considerable attention is given to the development of dual-level designs with self-similar characteristics. This was the last great outpouring of creative innovation, and despite the relatively small number of examples, their beauty and geometric ingenuity place them into a highly significant category of their own. Attention is also given to the application of geometric patterns to non-Euclidean surfaces of domes and domical niche hoods. There are two varieties of this form of Islamic geometric ornament: those that utilize radial gore segments as their repetitive schema and those that employ polyhedra as their repetitive device. The latter are far less common, and almost all of the significant historical examples are included within this study. To a very limited extent, each new dynasty is placed into a brief historical context that primarily describes their rise to power, and what set them apart from their predecessors. This will be redundant to historians, but many readers may benefit from the placement of the geometric idiom within a broader cultural and political milieu—however briefly outlined. Attention has been given to the relatively few examples of Islamic geometric design that were created for non-Muslim clients and in some cases by non-Muslim artists. The influence of Islamic art upon non-Muslim cultures is beyond the scope of this study, but the examples cited are worthy of inclusion due to their geometric character as well as their historical circumstance. This opening chapter tracks the history of Islamic geometric patterns through to achieving full maturity and for the most part leaves the later, more derivative manifestation of this tradition for future consideration.

Chapter 2 explores the varied discrete features that characterize this multifaceted discipline. Previous works have concentrated primarily on the variety of star types and regular polygons within specific patterns when categorizing geometric designs, and more recent studies have classified patterns according to their crystallographic plane symmetry group. Yet there are many more distinguishing characteristics that help to broaden an overall understanding and appreciation of this artistic tradition. Emphasis is given to the variety of repetitive stratagems employed by Muslim geometric artists. Of course the simple square, equilateral triangle, and regular hexagon were commonly employed as repeat units, and many very complex patterns employ these well understood structures. From as early as the eleventh century patterns with alternative repetitive structures entered the artist's repertoire. These included rectangular, rhombic, and irregular hexagonal repeats, as well as designs with rotational symmetry. The repetitive stratagems included in this chapter also include *oscillating square* designs and *rotating kite* designs. These are essentially orthogonal, but by incorporating alternating rhombi and squares in the former, and alternating kites and squares in the latter, it is possible to produce designs with seemingly incompatible regions of n-fold local symmetry into an otherwise fourfold structure. Considering the early period of origin, and the fact that there was no historical precedent that these Muslim geometric artists could have borrowed from, their familiarity with these diverse repetitive structures is surprisingly sophisticated and predates analogous repetitive structures from other cultures by many centuries. The intrinsic relationship between the n-fold symmetry of a given pattern and the proportions of its repeat unit are examined in detail. As patterns became increasingly complex, with multiple regions of local symmetry incorporated into a pattern matrix that was based on neither a square nor a triangular repeat, these regions of local symmetry were critical proportional determinants of the overall repetitive structure: be it rectangular, rhombic, or irregular hexagonal. This chapter also differentiates between patterns according to their numeric qualities and postulates an abbreviated descriptive nomenclature that is based upon rather basic geometric and numeric analysis. In discussing Islamic geometric patterns it is sometimes difficult to express concisely and with precision the qualities of a given pattern. This is especially true of the more complex designs. The approach to identifying the salient features of a given design that is advocated in this chapter is intended to promote both cogency in dialogue and clarity in understanding. Once again, this section moves from the simple to the complex, beginning with examples that employ single star forms or regular polygons into simple orthogonal or isometric repetitive structures, and ending with complex designs with multiple star forms within a single pattern matrix that repeat upon the less common grids mentioned above. In classifying these more complex structures, I have identified the conventions for including added regions of local n-fold symmetry of the primary stars. These are placed at key locations within the repetitive structure, such as the vertices of the repetitive grid, the vertices of the dual grid, upon the midpoints of the repetitive edges, and occasionally upon lines of radius within the field of the pattern matrix. Another category of design that is examined imposes a geometric motif into a repetitive grid that is typically incompatible with the symmetry of the motif: for example, the placement of octagons into an isometric structure. Muslim geometric artists produced many fine *imposed symmetry* patterns, and this variety of design has been largely ignored in previous studies. Classification according to the crystallographic plane symmetry group is critical to understanding the geometric underpinnings of this tradition. There are just 17 ways in which translation, rotation, reflection, and glide reflection dictate the repetitive covering of the two-dimensional plane. Other than those that have either rotational symmetry or cover a non-Euclidean surface, all historical Islamic geometric patterns adhere to one or another of these 17 plane symmetry groups. This section focuses primarily on the geometric characteristics of the plane symmetry groups as manifest within the Islamic geometric idiom, and only slight reference is made to the relative occurrence of different symmetry groups within this overall tradition. This work does not include a statistical analysis whereby individual designs from individual Muslim cultures provide the data points for determining shifting preferences for specific symmetry groups throughout the history of this artistic tradition. Such anthropological and ethnographical studies have been highly revealing of the artistry of other cultures, such as the textile arts of the Americas,[1] but as pertains to Muslim geometric art, this remains the work of future scholarship.

Perhaps the least apparent and least understood classification within the tradition of Islamic geometric pattern concerns design methodology. Considerable attention is given to this subject throughout this book. Historical examples of Islamic geometric art are reflective of multiple disparate forces all of which combine to create a given piece, be it an architectural panel or Quranic illumination. This influential confluence includes the aesthetic predilections of the specific cultural milieu, the personal preferences and aesthetic judgments of the artist balanced with the wishes of the patron or client, the technical and material constraints of the medium, and the stylistic impact of design methodology. Less complex patterns from this design discipline can often be created with more than a single methodology. However, as patterns increase in complexity, the formative method employed in their creation has a pronounced stylistic influence upon their aesthetic nature. The question of methodology is somewhat thorny in that there is surprisingly little historical evidence, and there are several competing theories as to the specific methodology that is most relevant to this tradition. I am a proponent of the polygonal technique as the design methodology that is most responsible for the tremendous diversity and complexity that are hallmarks of this form of art. I have come to this conclusion based upon several criteria: the preponderance of historical evidence from several sources, including Quranic manuscript illuminations, design scrolls (most notably the Topkapi scroll), and multiple architectural representations; the fact that the polygonal technique readily allows for the production of the four pattern families that are fundamental to this tradition; and the fact that only the polygonal technique is suitable for producing the plethora of especially complex patterns found in this tradition. The section of Chap. 2 that pertains to differentiation between design methodologies begins with the polygonal technique and discusses the historical evidence in considerable detail. This section also examines the key features that result from the employment of the polygonal technique, including the distinctive features of the four pattern families, and how they are the product of different types of pattern line application to the underlying generative tessellations; the systematic use of the polygonal technique in developing the five historical design systems—each with its own distinctive aesthetic merit; and the nonsystematic use of the polygonal technique wherein the connective polygons in each underlying generative tessellation are distinct unto themselves and will not recombine into other tessellations. This section on the polygonal technique also focuses on *additive patterns*, a method of increasing the complexity of otherwise less complex patterns that was especially popular, but by no means exclusive to the Ilkhanid dynasty of Persia.

---

[1] Washburn and Crowe 1988 and 2004.

The design methodology that has been promoted most widely to date as being primarily responsible for Islamic geometric patterns is referred to herein as *point-joining*. Multiple books by numerous authors have advocated for this method of creating patterns, but unless I am very much mistaken, none have given it a name *per se*. This methodology involves setting up a repetitive cell (such as a square), with a matrix of internal lines and circles that provide coordinate points that can be joined with lines that produce a pattern: hence the descriptive name *point-joining*. This technique has the benefit of producing patterns precisely and quickly at whatever scale is required. However, there are several significant problems with this methodological approach. From a practical perspective, point-joining has three main problems: (1) it does not lend itself to creating original designs, but is primarily useful in recreating existing patterns in a step-by-step fashion; (2) it is impractical for recreating complex patterns with multiple regions of local symmetry; and (3) the step-by-step construction sequence of each specific pattern must be individually memorized or kept in documentary form. From the historical perspective, there is scant evidence from surviving documents that the geometric artists of Muslim cultures used this methodology. By contrast, the polygonal technique is immanently suited to creating original designs, is the only methodology that allows for the creation of the most highly complex patterns found in this tradition, and its inherent flexibility allows the artist to create an unlimited number of designs without having to memorize construction sequences. What is more, there is significant evidence for its historical use. By contrast, the only known source of evidence for what I am herein referring to as point-joining is the anonymous Persian language treatise *On Similar and Complementary Interlocking Figures* in the Bibliothèque Nationale de France in Paris. The section of Chap. 2 that is concerned with point-joining addresses my speculations on the historical relevance of this document in considerable detail. This is an important treatise for many reasons, not the least of which is that it is the only known historical document that provides step-by-step written instructions for creating a number of Islamic geometric patterns. This feature has led some to conclude that the methodology demonstrated in this anonymous treatise is reflective of that used by Muslim geometric artists of the past. My hypotheses are somewhat less generous to the methodological significance of this document. The third methodological approach discussed in Chap. 2 makes use of geometric grids from which pattern lines are extracted. This methodology employs either the orthogonal or isometric grid, and many of the early patterns of low or moderate complexity can be constructed with this technique. It is worth noting that these same designs can often be alternatively created from either the polygonal technique or point-joining, and it is often impossible to know for certain which methodology the artists of a given example used. The use of the orthogonal grid will frequently produce patterns that require geometric adjustment to the proportions of the design. For example, when locating the points of an eight-pointed star upon the vertices of the orthogonal grid, the eight points will not be equal. In such incidences, it is necessary to engage in corrective measures to bring the design into compliance with the aesthetics of this tradition. In the hands of skilled practitioners, the orthogonal grid can be used to create geometric patterns of considerable sophistication. This variety of design and methodological approach has been thoroughly explored by Jean-Marc Castéra in his excellent book *Arabesque*,[2] and is therefore only touched upon in this work. Another methodological approach that is included in this section involves offsetting of the lines that make up a radii matrix so that two parallel lines are produced. These are then extended and trimmed with other offset and extended lines to create the finished pattern. I refer to this method as *extended parallel radii*. There are very few designs from the historical record that fall into this category, but their distinguishing aesthetic

---

[2] Castéra 1999.

qualities merit their inclusion in this work. The final methodology covered in this chapter is *compass-work*. Many of the earliest Islamic geometric patterns were created from a matrix of circles set upon a repetitive grid, and trimmed to create the final design. This technique produces designs with a distinctive character that often parallels the earlier Hellenistic geometric ornament from which it derived. Compass-work was soon surpassed by other more flexible design methodologies, but occasional examples are found throughout the historical record. This chapter on differentiation within this geometric idiom concludes with a description of the various conventions governing line treatment within this tradition.

Chapter 3 provides a thorough analysis of the methodological practices associated with the polygonal technique. The illustrations of historical examples that accompany this chapter do not attempt to reproduce the colors and secondary background motifs such as floral designs or calligraphic inserts. Rather, by showcasing just the geometric designs themselves these illustrations are intended to more effectively demonstrate the geometric quality of these patterns. The illustrations in Chapters 2 and 3 were produced by me using a combination of software programs, including AutoCAD®, Rhino3D®, and Adobe Illustrator®. The illustrations in Chapter 3 include multiple examples of underlying generative tessellations that are associated with the geometric designs themselves. In some cases, I have demonstrated how a single pattern can be produced from more than one underlying tessellation, and that these alternative tessellations frequently have a dual relationship. Here again, it is not always possible to know which underlying polygonal tessellation was used to create such examples. This chapter also examines various historical conventions for arbitrarily introducing design modifications that significantly alter the visual quality of the finished pattern. The subdivisions within this chapter follow an analytical approach rather than the chronological approach of Chap. 1. Within this chapter as a whole, and within each subdivision of this chapter, the cited examples move from more simple structures to the more complex; and as this is also a general trend within the historical development of this tradition, there is a degree of similitude between the sequential order of the examples from both these chapters. The opening of Chap. 3 is focused upon the five historical design systems. This is followed by the nonsystematic use of the polygonal technique. The penultimate section of this chapter is concerned with dual-level designs and the chapter concludes with the application of the polygonal technique to non-Euclidean surfaces such as domes, half domes, and spheres. I have provided each of these sections with ample illustrations to demonstrate the high degree of diversity, subtlety, and geometric sophistication that characterizes this tradition. I have occasionally included patterns of my own creation as a means of demonstrating the further potential of specific underlying polygonal tessellations, or specific methodological practices. In some cases, this opens new avenues for further design innovation, for example the creation of multilevel recursive designs with true quasicrystallinity. While this chapter is not intended as an instruction manual, it is nonetheless hoped that it will help provide a methodological understanding that can be applied by dedicated artists, designers, and architects who have an interest in augmenting their work with this exceptionally beautiful form of geometric art.

Chapter 4 has been contributed by Craig Kaplan of the Cheriton School of Computer Science at the University of Waterloo in Ontario. This chapter is concerned with the use of computer algorithms to generate Islamic geometric patterns. There is considerable interest in creating geometric designs in this fashion, and Craig Kaplan is the leading authority in this application of computer technology. The obvious advantage of using computer algorithms is the rapidity of results, and the concomitant ability to explore and compare many design options with great fluidity. The algorithmic procedure prescribed in this chapter employs the same variety of underlying polygonal tessellations as were used historically and is therefore a contemporary expression of an ancient methodological practice. What is more, as amply demonstrated, the synergistic dynamic between the inherent flexibility of the polygonal technique and the ability to explore new geometric territories through the power of the computer opens new vistas of creative innovation.

This book is the culmination of many years working with Islamic geometric patterns. As a child I was always attracted to geometric design and long before seeing my first Islamic geometric pattern in a book, I spent long hours drawing radial geometric patterns with stars as primary motifs. As a teenager I began to work seriously with geometric art. My father was a professor of organic chemistry, and happening upon one of his plastic templates used for drawing molecular diagrams essentially determined the trajectory of my life's work. This template had perforations for each of the polygons from the triangle to the dodecagon, and with this I was able to explore the tessellating properties of numerous polygons in diverse configurations. I soon discovered that I could place lines at strategic locations of these tessellations, such as the vertices and midpoints of the polygonal edges, and that discarding the initial tessellation would thereby create attractive geometric designs. In time, as a young teenager, I saw my first Islamic geometric patterns in a book on Islamic architecture and was amazed at the similarity, and in some cases exactitude, of my creations with the patterns of long past Muslim artists. Thus began my lifelong fascination with Islamic geometric design specifically, and Muslim cultures, art and history more generally. I've had the good fortune to live and travel widely amongst Muslim cultures and have developed great respect and appreciation for the people I have encountered along the way and have the privilege to call my friends. I received my master's degree in this field in 1983 from the Royal College of Art in London, after which I began my professional career as an ornamental design consultant specializing in architectural projects for Muslim clients. I have had the honor of working on many significant buildings, including the expansion of the Prophet's Mosque in Medina and the expansion of the Grand Mosque in Mecca. I also designed the ornament for the new minbar of the Kaaba in Mecca. For many years and on four continents, I have lectured and taught design seminars on the subject of Islamic geometric design. I have also contributed several publications in this field. However, I chose design consulting over academia as a career and have remained an unaffiliated scholar of this discipline. Throughout my career I have continued my research and exploration of Islamic geometric patterns, always striving to better understand the methodological working practices that sustained this tradition for such a long period, and to apply these techniques to an expanding repertoire of original designs. I have been particularly concerned with the gathering of evidence for the historical use of the polygonal technique. In 1981, while attending the Royal College of Art, I received a letter from my friend Carl Ernst, now the Kenan Distinguished Professor of Religious Studies at the University of North Carolina at Chapel Hill, recommending that I seek out the work of Ernest Hanbury Hankin. These publications were still relatively obscure at this time, and reading Hankin's work provided my first confirmation that the method of using underlying polygonal structures to extract geometric patterns was historical. Soon after this I had occasion to study an illuminated frontispiece from a fourteenth century Mamluk Quran at the British Library. The scribed lines of the underlying polygonal tessellation that produced the fivefold geometric pattern were faintly detectable beneath the painting when viewed from an oblique angle. By far the most significant source of confirmation was my good fortune in seeing and photographing the Topkapi scroll while it was on temporary display at the Topkapi Museum in 1986 during a business trip to Istanbul. Unfurled before me was a large catalogue of excellent geometric designs, many of which included their underlying generative tessellations. From the perspective of historical design methodology, the significance of this document cannot be overstated. During the years of ongoing research I also found multiple architectural sources of evidence, including a niche from the *iwan* of the Sultan al-Nasir Hasan funerary complex in Cairo (1356-63), several Mughal jali screens, a Seljuk fritware tile in the collection of the Los Angeles County Museum of Art, and a Karamanid walnut door in the collection of the Museum of Turkish and Islamic Art in Istanbul. For many years I kept my discoveries concerning the polygonal technique strictly confidential, but in the early 1990s I resolved to make my findings available to the public via lectures and design seminars. Eventually I turned my sights toward publication. The first draft of this book was completed and copyrighted in

2000. This manuscript was accepted for publication by Springer in 2007. The comments from one of Springer's anonymous reviewers were particularly persuasive and led to my completely rewriting the book into its current form over these past 10 years.

It is my pleasure to register my sincere appreciation of the many people and organizations that have been supportive of my research over the years. I will begin by thanking most sincerely Ann and George Hogle for paying my tuition fees to the Royal College of Art during my master's degree. Such artistic patronage and selfless generosity is rare, and I remain forever grateful. I thank the technical staff, lecturers, senior lecturers, and professors at the Royal College of Art. My work in the Department of Painting and Printmaking and the Department of Ceramic and Glass, and as a postgraduate Research Fellow prepared me for my career as a design consultant specializing in Islamic architectural ornament. I thank the organizers of the annual Bridges Conference who accepted my first paper on the subject of self-similarity within Islamic geometric art, and honored me as their plenary keynote speaker at their 2003 conference in Granada. In particular, I thank the late Professor Reza Sarhangi of the Department of Mathematics at Towson University in Maryland who was the heart and soul of the Bridges Organization, and remains an inspiration to everyone fortunate to have known him. He was a skilled designer of Islamic geometric patterns in his own right and was a supportive friend throughout the many years of knowing him. I extend my sincere thanks to my aforementioned friend Professor Carl Ernst for translating selections of relevant text from the *Fi tadakhul al-ashkal al-mutashabiha aw al-mutawafiqa* (On Similar and Complementary Interlocking Figures), an anonymous Persian language treatise at the Bibliothèque Nationale de France. I am grateful to the Aga Khan Visual Library for providing me with access to their extensive collection of photographs of Islamic architecture—both before and after this was made available as an online resource. I am particularly grateful to their former senior administrator Jeffery Spurr for having been so helpful during my initial research while visiting their substantial photographic archive in Cambridge, Massachusetts; and more recently to Michael Toler, the Archnet Content Manager, for his valuable assistance in obtaining photographic permissions. I am profoundly grateful to Professor Jan Hogendijk of the Mathematics Department at the University of Utrecht. His recommendation to Springer of my initial manuscript led directly to the publication of this book, and his publications and specialized knowledge of the history of Islamic mathematics have been a very helpful resource during the preparation of this manuscript. I am equally grateful to Professor Gülru Necipoğlu, the Director of the Aga Khan Program of Islamic Architecture at Harvard. Her seminal book on the Topkapi scroll is the single most important work on the subject of Islamic geometric ornament and has been of paramount importance to my own research—as doubtless that of many others. Her interest in my methodological approach during our first meetings at Harvard in the late 1990s was more helpful to me than she probably realizes. I am particularly grateful for her critique of my interpretation of the methodological significance of the anonymous Persian language treatise *On Similar and Complementary Interlocking Figures*. I thank Hüseyin Sen for his continued interest and support of my work as a lecturer and teacher of this geometric discipline. His dedication to the history of science in the Islamic world is detailed and inspiring. My participation in the Islamic geometric design program that he helped organize at the Lorentz Center at the University of Leiden was an exceptional experience, and his more recent leadership in organizing the *International Workshop on Geometric Patterns in Islamic Art* in association with the Istanbul Design Center has been deeply rewarding. I also thank the staff and administrators of the Istanbul Design Center for their overwhelming hospitality and professionalism. I am very grateful to Carol Bier for her patience and helpful insight during our many conversations relating to the content of this book. I would like to express my thanks and utmost respect to Jean-Marc Castéra for his second-to-none dedication to this discipline. Although we have different methodological approaches, our differences only bring us closer—would that this were the way of the world. I also extend my appreciation to other specialists of Islamic geometric design who have made valuable contribution to this field of study, including: Emil Makovicky, Anthony Lee, Peter Cromwell, Mirek Majewski, Rima Ajlouni, and Eric Broug. Their dedication is inspiring. I also give my special thanks to my close friends John Bussanich and Michael Baron for their ongoing interest and encouragement in the protracted progress of this book.

I would be remiss in failing to acknowledge my appreciation and gratitude to multiple authors of previous published works on the subject of Islamic geometric design. Of particular significance to my work in this field have been the publications of previously mentioned specialists, including Ernest Hanbury Hankin, Gülru Necipoğlu, Jean-Marc Castéra, Craig Kaplan, Anthony Lee, and Peter Cromwell. I highly recommend the work of each of these authors to those interested in expanding their understanding of this fascinating art form. I am also greatly indebted to Gerd Schneider for his exhaustive survey of Seljuk geometric patterns in Anatolia. I have relied heavily upon his catalogue of designs in my own formulations concerning the methodological analysis and historical development of the geometric ornament produced during the Seljuk Sultanate of Rum.

I would also like to acknowledge my students. Over the years I have had the privilege of instructing a number of very talented individuals, and the work of these students has frequently been inspiring and illuminating. I am always fascinated by the high level of idiosyncratic style that is possible within a discipline that can mistakenly be thought of as rigid and devoid of personal expression. In particular I want to extend my appreciation to Marc Palletier and Amina Buhler Allen. Their dedication to this discipline includes the education of young children in the use of polygonal systems to create Islamic geometric designs. Marc Pelletier has become a highly skilled designer of Islamic geometric patterns who has contributed several significant innovations to the systematic use of the polygonal technique. I am immensely grateful to him for putting me in touch with Sir Roger Penrose.

I am very grateful to my editors at Springer. They have been a pleasure to work with and their patience has been greatly appreciated. I could not be in better hands. I am solely responsible for the content of Chaps. 1, 2, and 3. This work covers a tremendous amount of territory and despite my best efforts, there are bound to be some mistakes. I take full responsibility for any and all such flaws, hope that these are of an insignificant nature, and thank my readers for their indulgence. I also hope that I have succeeded in not coming across as a methodological absolutist in my advocacy of the polygonal technique. It is always important to keep in mind the flexible nature of this design tradition, and the fact that many of the historical examples from this tradition can be created from more than a single technique. Mindful of this, I apologize if I occasionally overemphasize my commitment to the primacy of the polygonal technique with statements such as "created with the *system of regular polygons*;" especially if this appears to suggest that no other design methodology may have alternatively been used. This is especially relevant in examining the less complex patterns that can often be created with diverse approaches. In recognition of the relevance of other design methodologies, and acknowledging the valuable contributions of specialists who advocate for these, I welcome alternative views and further contributions to this subject.

As stated, there is a significant amount of repetition within this manuscript, with historical examples of given designs provided multiple times. Within the opening section of this book, this repetition has resulted in large part due to the need for historical context in emphasizing the aesthetic continuity between Muslim cultures. Similarly, within the section devoted to design methodology, the specifically historical examples that are included are additionally identified with their historical locations. Again, this is for the purpose of identifying the cultural context of these many patterns: in some cases specific to a single Muslim culture, and in others, used ubiquitously throughout the Islamic world. I therefore thank the readers of this book for patiently accepting the degree of repetition as a necessary vehicle for providing historical context and continuity.

I want to thank the many photographers who have agreed to have their work included in this book. In particular I am grateful to Thalia Kennedy for her photographs from Afghanistan and Central Asia, to Jean-Marc Castéra for his photographs from Iran and for providing a photograph of himself drawing a design freehand, to Tom Goris for his photographs of the northeast dome chamber of the Friday mosque at Isfahan, to Bernard O'Kane for allowing me to use his photographs from Afghanistan and Central Asia, and to Daniel Waugh for providing photographs from Iran and China. I am especially grateful to David Wade for making his substantial archive of photographs available to the public via his website, Pattern in Islamic Art, and for allowing me to use so many of his excellent photographs within this book. I am also indebted to Marcus Baron for providing the excellent renderings for the seven spherical designs in Chapter 3. I extend my sincere thanks to Craig Kaplan for agreeing to provide the concluding chapter of this book concerning the use of computer algorithms to generate Islamic geometric patterns. In addition to his work as a computer scientist, he is a highly skilled designer and geometrician and has been a valued friend and associate for many years. I am very pleased to include his most worthy contribution to this publication.

I am immensely honored and very grateful to Sir Roger Penrose for agreeing to write the foreword to this book. Sir Roger is the distinguished Emeritus Rouse Ball Professor of Mathematics of the Mathematical Institute at the University of Oxford. His reputation as one of the world's leading mathematical physicists and cosmological theorists precedes him, but some might be less aware of his pioneering work in geometry. In the early 1950s his forays into visual geometric illusions were an influence upon M. C. Escher. Of particular relevance among those interested in Islamic geometric design is Sir Roger's discovery of Penrose tilings in 1974. These are comprised of two prototiles that have edge conditions that force aperiodicity—known appropriately as Penrose matching rules. Penrose tilings have fivefold symmetrical characteristics, and in addition to aperiodicity they will recursively inflate and deflate to produce infinitely scalable self-similar structures. Penrose tilings anticipated the discovery of fivefold quasicrystalline compounds in nature, and more recently, and to considerable fanfare, have been attributed to the structure of several examples of Persian geometric design. My own analysis of these cited Persian examples reveals that each is in fact governed by translation symmetry and therefore does not meet the aperiodic criteria of quasicrystallinity. Yet as a design professional, it could be regarded as quixotic to challenge the scientists who first promulgated this theory. It is therefore the greatest satisfaction to have the support of Sir Roger Penrose who is without question the foremost authority on this subject. His interest and support of my work and propositions pertaining to Islamic geometric design means more than I can adequately express.

I am also without words to properly articulate my heartfelt appreciation to my family who have patiently endured my 18 years of work on this book. I thank my parents and siblings for their longstanding interest, as well as my wife's remarkable family for their sustained encouragement over the years. More than anyone else, I extend my love and appreciation to my ever patient and supportive wife, Shireen, without whom this book could not have been written. Likewise my heartfelt love and thanks to our daughter, Mehera, son-in-law, James, and to our beloved grandchildren for being constant sources of inspiration in every aspect of my life.

Finally, I thank my readers. I hope the historical, geometric, and methodological content of this work contributes to the greater understanding and appreciation of this remarkable and uniquely beautiful artistic tradition. For contemporary artists and designers of Islamic geometric patterns there is still a vast expanse of innovative potential waiting for exploration. This includes new and original systematic and nonsystematic patterns that build upon the repertoire and aesthetics of traditional work, designs with diminishing scale, parquet deformations as per the work of Craig Kaplan, multiple-level designs with self-similarity, aperiodic designs with true quasicrystallinity, and non-Euclidean designs with spherical, hyperbolic, and irregular topographical geometry. This world needs more beauty, and I hope that this book will help to inform and inspire those with a sincere desire to embark upon such explorations. And in doing so, it is my further hope that this book may contribute to the rekindling of this remarkable artistic discipline so that it can become an evermore-active vehicle of contemporary artistic expression.

Santa Fe, NM   Jay Bonner
June 2016

Contents

Photographic Credits

Note: All line drawings for the figures in Chapters 2 and 3 are by Jay Bonner. Those in Chapter 4 are by Craig Kaplan.

Bachmann, Walter (Courtesy of the Aga Khan Documentation Center at MIT): 86

Betant, Jacques (Aga Khan Trust for Culture-Aga Khan Award for Architecture): 69

Blair, Sheila and Jonathan Bloom: 38, 75

British Library Board, London, UK: 48, 65

Самый древний (Public domain, via Wikimedia Commons): 35

Castéra, Jean-Marc: 80, Fig. 79

Chester Beatty Library, Dublin, Ireland (© The Trustees of the Chester Beatty): 6

Creswell, K. A. C. (Ashmolean Museum; Creswell Archive, University of Oxford): 28, 34

Dumont, Jean-Guillaume: 67

Emami, Farshid (Courtesy of the Aga Khan Documentation Center at MIT): 23

Entwistle, Damian: 4

Ganji, Muhammad Reza Domiri: 91

Goncharov, Igor: 15

Goris, Tom: 18, 19, 21, 25, 26, 27, 30

Haddow, Scott: 58

Kennedy, Thalia: 13, 16, 20, 31, 32, 76, 78, 83, 84, 85, 94

Lewis, David: 54

Los Angeles County Museum of Art (Public Domain: www.lacma.org): 104

Majewski, Mirek: 44

Martin, Pete: 92

Mawer, Caroline: 39, 40

Metropolitan Museum of Art (Open Access for Scholarly Content [OASC], www.metmuseum.org): 3, 11, 47, 52

Mortel Richard: 24

O'Kane, Bernard: 1, 12, 14, 33

Osseman, Dick: 105

Rabbat, Nasser (Courtesy of the Aga Khan Documentation Center at MIT): 36, 37

Roudneshin, Reza: 17

Schastok, Horst P. (Courtesy of Special Collections, Fine Arts Library, Harvard University): 10

Sébah and Joaillier (Courtesy of Special Collections, Fine Arts Library, Harvard University): 2

Sönmez, Serap Ekizler: 42, 81

Tabba, Yasser (Courtesy of the Yasser Tabba Archive, Aga Khan Documentation Center at MIT): 41

Wade, David: 5, 7, 8, 9, 29, 43, 45, 49, 50, 51, 53, 55, 56, 57, 59, 60, 62, 63, 64, 66, 68, 71, 72, 73, 74, 77, 79, 89, 90, 95, 97, 98, 99, 100, 101, 102, 103

Waugh, Daniel C.: 22, 70, 82, 87, 96

Williams, John A. and Caroline: 46, 61

Yazar, Hatice (Courtesy of the Aga Khan Documentation Center at MIT): 88

The Historical Antecedents, Initial Development, Maturity, and Dissemination of Islamic Geometric Patterns

1.1   Geometry in Islamic Art

Since the earliest period of Islamic history the ornamental traditions of Muslim cultures have found expression in a highly diverse range of styles and media. Throughout this broad sweep of ornamental diversity and historical longevity there remained an essential Islamic quality that differentiates this tradition from all others. One of the primary characteristics responsible for such cohesion is the pervasive triadic nature of Islamic ornament. From its onset, this ornamental tradition employed three principal design idioms: calligraphy, geometry, and stylized floral.[1] It can be argued that figurative art depicting both human and animal forms is also characteristic of Islamic art. This additional feature of Islamic art requires brief mention, if only to legitimately dismiss it for the purposes of this discussion. During the Umayyad period figurative motifs were widely used in both architecture and the applied arts, and virtually all subsequent Muslim cultures used figurative depictions to a greater or lesser extent. Such work has always been anathema to Islamic religious sentiments and frequently to Muslim cultural sensibilities.[2] Even among the Umayyads, who inherited the figural traditions of the late antique period, the use of figurative depictions was invariably secular and often associated with courtly life. The eighth-century Umayyad palaces of Qusayr 'Amra and Khirbat al-Mafjar are replete with figurative decoration, the former carried out in fresco and the latter in mosaic and carved stucco. Such notable examples notwithstanding, the surviving religious architecture of the Umayyads is evidence of the interdiction in the use of human and animal depiction within mosques. It is significant that the Umayyad architectural motifs in the mosaics of the Great Mosque of Damascus and the Dome of the Rock were entirely devoid of human and animal figures; and the one area of carved stone ornament that is entirely without animal representation at the eighth century Umayyad palace of Qasr al-Mshatta is a wall directly adjacent to the mosque. The figurative restraint in the ornament of Mosques was adhered to strictly throughout succeeding Muslim cultures, and the continued use of human and animal figures was generally limited to the decoration of utility objects such as ceramic pottery, textiles, metal vessels, furniture, wood and ivory boxes, and the occasional architectural expression in murals and carved relief in such secular locations as palaces, private homes, and bath houses. However, with the exception of the miniature traditions,[3] this form of decoration was certainly never a primary feature in Islamic art or architectural ornament. For all of their beauty and refinement, the figurative aesthetics of the Persian, Mughal, and Turkish miniature traditions were for the most part insular, and did not significantly overlap with other artistic traditions within these Muslim cultures. Notable exceptions include the "miniature" style of the enameled *minai'i* ware of late twelfth- and thirteenth-century Kashan[4]; the so-called Kubachi painted ceramic vessels created in northeastern Persia during the Safavid period[5]; and the many Persian painted tile panels produced during the Qarjar period. Perhaps the greatest indication of the lesser role that figurative imagery played throughout the history of Islamic art and architecture is the fact that the non-miniature figurative art of Muslim cultures was not subject to the concerted effort

---

[1] Hillenbrand (1994a), 8.

[2] Allen (1988), 17–37.

[3] The depiction of the human figure in the fine art traditions of Persian, Turkish, and Mughal miniature painting is an exception to the conventions of human representation occasionally found in the ornamental and applied arts of various Islamic cultures. The reconciliation of these miniature traditions with the Islamic religion and Islamic cultures is a fascinating study, but outside the scope of this work.

[4] –Watson (1973–75), 1–19.
  –Watson (1985).
  –Hillenbrand (1994b), 134–41.

[5] Golombek et al. Chap. 4: "The Kubachi Problem and the Isfahan Workshop."

J. Bonner, *Islamic Geometric Patterns*, DOI 10.1007/978-1-4419-0217-7_1

toward continued refinement and stylistic development that is a hallmark of the calligraphic, geometric, and floral traditions. As such, with the exception of the miniature traditions, figurative art can be regarded as tangential rather than integral to Islamic art, and to have been occasionally employed rather than part of an ongoing developmental evolution.

It may seem remarkable that such an apparently limited palette of calligraphy, geometric patterns, and floral design should have provided the basis for such a rich and varied artistic tradition. Yet each of these separate disciplines benefits from unlimited developmental opportunities, and when used together provide an inexhaustible supply of aesthetic variation. The continued adherence to the triadic quality of Islamic ornament provided a governing mechanism whereby the aesthetic expressions of multiple Muslim cultures, spanning great divisions of distance and time, were able to be both culturally distinct yet identifiably Islamic. Similarly, this also served as a form of regulator, or cohesive principle, through which Muslim artists could appropriately assimilate specific ornamental conventions from non-Muslim cultures. This assimilative process contributed greatly to the tremendous stylistic diversity found in Islamic art and architectural ornament.

The historical development of all three of the primary ornamental idioms is characterized by an evolving refinement and increased complexity. This process was aided by any number of influences, not the least of which include contacts with other mature artistic traditions; concomitant improvements in fabricating technologies (e.g., a brocade loom allows patterns to be woven that would otherwise not be possible); vainglorious patronal expectations that commissioned works should exceed that of their predecessors or neighbors; and the natural tendency for an artist to strive for creative excellence by challenging personal limitations and pushing the boundaries of an artistic tradition. Such criteria are common to all cultures, but the refinement and growth in complexity within the ornamental traditions of Muslim cultures were also greatly aided by the ongoing fascination with and influence of geometry.

That geometry should be at the root of the geometric idiom goes without saying. Yet the role of geometry in the aesthetic development of both Islamic calligraphy and floral design was also of paramount importance. The tradition of Islamic calligraphy is, first and foremost, a book art. Within Muslim cultures, calligraphy is regarded as the highest art form, and the copying of the Quran is as much a spiritual discipline as it is an artistic activity. The creative heights to which Muslim calligraphers refined this tradition were directly driven by their need to adequately express their deep reverence for the Quran.[6] In copying the Quran,

calligraphers were motivated by the need for legibility and beauty. This twofold concern led to the continual refinement and eventual preference of the more easily read cursive scripts over the older and overtly angular *Kufi* scripts. The ability for geometry to positively influence the legibility and beauty of an artistic tradition characterized by non-repetitive rhythmic undulation might appear unlikely. However, the use of geometry as an underlying governing principle for the more legible cursive scripts was not just appropriate, but crucial to the beauty and longevity of this tradition. In the first half of the tenth century the renowned calligrapher Abu 'Ali Muhammad ibn Muqlah (d. 940) developed a system of calligraphic proportion that was applied to the development of the six principal cursive scripts.[7] His system was complex and highly rational, and so successful in creating balanced writing that the rules he established have been universally employed through successive centuries by Muslim calligraphers. The rules he established for cursive calligraphy relied upon the application of carefully contrived mathematical proportions. His standard unit was the rhombic dot, produced by moving the pen nib diagonally the same distance as the nib is wide. The rhombic dots were, in turn, used to determine the height of the *alif*, and the height of the *alif* was then used to determine the diameter of a circle, all of which were used to determine the precise proportions of each letter.[8] Ibn Muqla was *vizier* to three successive Abbasid Caliphs during a time when interest in mathematics and geometry was acute. The training he received in the geometric sciences is evident in the refinement he brought to bear upon the tradition of calligraphy, and is fully consistent with the intense interest in geometry during this period.[9]

The geometric angularity of *Kufi* is in marked contrast with the flowing movement of the cursive scripts, and together these stylistic trends create a dynamic complementarity that was used to great aesthetic effect. This is especially evident within the realm of Islamic architectural ornament. In contrast to the writing of the Quran, conventions for the use of text on buildings were less rigid, and the traditions of architectural calligraphy allowed greater latitude for ornamental stylization. Being less bound by governing rules, *Kufi* scripts were particularly suited to ornamental elaboration. From as early as the eighth century, the letters of the *Kufi* script were embellished with floral extensions that encroach upon the background space

---

[6] Schimmel (1990), 77–114.

[7] Each of the six principal cursive scripts (*al-Aqlam as-Sitta*) is associated with a student of Yaqut al-Musta'simi. These are *Naskh* ('Abdallah as-Sayrafi), *Muhaqqaq* ('Abdullah Arghun), *Thuluth* (Ahmad Tayyib Shah), *Tauqi* (Mubarakshah Qutb), *Rihani* (Mubarakshah Suyufi), and *Riqa* (Ahmad as-Suhrawardi). See Schimmel (1990), 22.

[8] Tabbaa (2001), 34–35.

[9] Tabbaa (2001), 34–44.

between the letters. Similarly, plaited *Kufi* intertwines the vertical letters into elaborate knots. In time, the traditions of foliated and plaited *Kufi* became highly elaborate, and a prominent feature of architectural ornament and decorative arts throughout the Islamic world.

The geometric quality of *Kufi* received its most extreme expression in the development of the principally epigraphic style of *Shatranji Kufi* (chessboard *Kufi*). This calligraphic style forces each letter of the alphabet to conform to the orthogonal grid; and the resulting geometric nature of this style endows it with a quality which appears, at first glance, more akin to geometric key patterns than written words. As an ornamental device, this expressly geometric calligraphic style is highly effective and can be found in buildings from al-Andalus to India. The orthogonal nature of *Shatranji Kufi* was ideally suited to the technical constraints that governed early Islamic brick ornament.[10] Among the earliest examples of this calligraphic style are the raised-brick ornament of the Ghaznavids and Seljuks in the regions of Khurasan and Persia. Two fine examples of this art are included in the ornament of the minaret of Mas'ud III in Ghazni, Afghanistan (1099-1115) [Photograph 1], and the interior façade of the Friday Mosque at Golpayegan, Iran (1105-18). Each of these examples place the bricks that define the calligraphy at 45° from the direction of the orthogonal script, and the direction of the background bricks perpendicular to those of the calligraphy. This creates the herringbone brick aesthetic that remained popular for many centuries, and received particular attention during the Timurid period, as well as subsequent Qajar and Uzbek periods. *Shatranji Kufi* is notoriously difficult to read, and its deliberate obscuration requires the viewer to stop and consider the text in an attempt to unveil its meaning. It is an interesting fact that the development of the virtually illegible *Shatranji Kufi* took place during the same period as the refinements to the legibility of the cursive scripts. Just as religious sentiments were an impetus for calligraphers to better reflect the gravity of the Quran by refining their art to be ever more beautiful and legible, it is possible that the developers of the willfully illegible *Shatranji Kufi* script may also have been motivated by religious convictions in creating an epigraphic corollary of the *Hadith* (saying of the Prophet) wherein Allah replies to the prophet David "I was a hidden treasure, and I longed to be known."[11]

The role of geometry within the traditions of Islamic floral ornament is primarily structural: providing symmetrical order upon which the stylized tendrils, flowers, and foliation rest. Most obvious are the innumerable examples of floral design

Photograph 1 *Shatranji Kufi* at the Minaret of Mas'ud III in Ghazni, Afghanistan (© Bernard O'Kane)

with reflective symmetry. Floral designs with bilateral symmetry are commonly used as infill motifs within the individual cells of a geometric pattern. The use of floral patterns as fillers in an otherwise geometric pattern was certainly part of the pre-Islamic, Late Antique ornamental vocabulary that assisted in the formation of Islamic art as a distinct tradition. However, with the Muslim development of increasingly sophisticated geometric patterns comprised of far more complex and diverse polygonal elements and multiple regions of diverse local symmetry, over time, the floral fillers followed this growth in complexity by becoming considerably more symmetrically varied than their antecedents. Both as polygonal fillers and as stand-alone motifs, floral designs with multiple lines of reflected symmetry were widely employed within Islamic architecture, manuscript illumination, and applied arts. Within architecture, floral designs with reflected symmetry were frequently used for dome ornamentation. In such examples the number of radial lines of symmetry will invariably be divisible by the number of side walls of the chamber that the dome is covering: e.g., if the plan of the chamber is a square, the reflected symmetry will be a multiple of 4.

The use of rotational symmetry was also common; and floral designs with twofold, fourfold, fivefold, sixfold, and eightfold rotational symmetry frequent this tradition. Such

---

[10] Schimmel (1990), 11.

[11] –Furuzanfar, Badi' al-Zaman (1956), no. 70.
  –Ernst (1997), 52.

designs were typically used as roundel motifs, on tiles, or as fillers within the background elements of geometric patterns. As with higher order reflective symmetries, rotational floral designs were also used for dome ornamentation—typically with 8-, 12-, or 16-fold symmetry, albeit less frequently.

Geometric patterns were occasionally provided with an additive floral device that meanders throughout the geometric design rather than being contained as fillers within the individual polygonal background elements. The movement and symmetrical order of this variety of floral design are always in strict conformity to the symmetry of the governing geometric pattern. This hierarchic relationship is visually emphasized by the fact that the floral element invariably flows beneath the geometric pattern. Several notable examples of this type of ornamental device include the wooden *mihrab* from the mausoleum of Sayyidah Nafisah, Cairo (1138-46); the central arch of the stucco *mihrab* at the Reza'iyeh mosque in Orumiyeh, Iran (1277); the carved stucco ornament above a niche in the Pir-i Baqran mausoleum in Linjan, Iran (1299-1311); the carved marble entry facade of the Hatuniye *madrasa* in Karaman, Turkey (1382); and the dome of the mausoleum of Sultan Qaytbay, Cairo (1472-74) [Photograph 2].

Since its onset, Islamic architectural ornament frequently made use of floral scrollwork border designs. This form of floral design was widely used in all media throughout the long history of this ornamental tradition, receiving distinctive interpretations throughout the breadth of Muslim cultures. Such designs employ a single repetitive unit to populate a linear spatial expansion, and without exception adhere to the symmetrical constraints of the *seven frieze groups* that are comprised of different combinations of translation symmetry, reflection symmetry, rotation symmetry, and glide reflection.[12] All linear repeat patterns, be it floral, geometric, figurative, etc., must conform to one or another of these seven frieze groups. There is no indication that Muslim artists, or indeed Muslim mathematicians, had specific knowledge of the seven frieze groups, but the inherent genius for empirical geometric exposition nonetheless led Muslim artists to create border designs that repeat according to the symmetry of each of the seven frieze groups.

Within the science of two-dimensional tiling, just as translation symmetry, reflection symmetry, rotation symmetry, and glide reflection provide the constraints for the seven frieze groups, so do they also provide for the symmetrical conditions for the *17 plane symmetry groups* (also referred to as the *wallpaper groups* or *plane crystallographic groups*).

Photograph 2 A dome with a geometric and floral design at the mausoleum of Sultan Qaytbay, Cairo (Sébah and Joaillier photograph, courtesy of Special Collections, Fine Arts Library, Harvard University)

All two-dimensional repetitive space filling follows the symmetrical order of one or another of the 17 plane symmetry groups. These were first enumerated by crystallographers and mathematicians in the late nineteenth and early twentieth centuries.[13] There is no evidence that Muslim artists or mathematicians were knowledgeable of this branch of crystallography. However, the art history of all pattern-making cultures is evidence of the fact that an artist does not have to understand the science of two-dimensional space filling in order to make efficacious use of its principles. Several studies have sought to identify examples of all 17 plane symmetry groups within the Islamic ornamental tradition,[14] or even within the single architectural complex of the Alhambra in Spain.[15] Within the Islamic floral idiom, foliage net designs

---

[12] –Weyl (1952).
   –Hargittai (1986).
   –Washburn and Crowe (1988).
   –Farmer (1996).

[13] –Fedorov (1891), 345–291.
   –Pólya (1924), 278–282.

[14] –Lalvani (1982).
   –Lalvani (1989).
   –Abas and Salman (1995).

[15] –Müller (1944).
   –Grünbaum, Grünbaum and Shepherd (1986).
   –Pérez-Gómez (1987), 133–137.

are always predicated upon one or another of these 17 plane symmetry groups, as are most of the many repetitive floral scrollwork designs.

The Muslim use of overtly geometric ornament dates back to the earliest period of military expansion. The rapid acquisition of territories previously governed by the Byzantines, Copts, and Sassanians availed the Muslim conquerors to a wide range of artistic and ornamental influences. These included several mature geometric design conventions that were readily appropriated into the ambitious architectural projects of the Muslim conquerors, and that were to prove highly influential to subsequent Muslim dynasties. In this way, the ornamental art of early Islamic cultures can be considered as inheritors of the geometric traditions of their conquered subjects, as well as progenitors of the extraordinary advances in the geometric arts that followed. Among the geometric design conventions that the Muslim conquerors inherited were stellar mosaics, compass-work compositions, braided borders, key patterns, and polygonal tessellations. Each of these continued to be used to a greater or lesser extent throughout the history of this ornamental tradition, and part of the genius of Muslim artistry was the ability to assimilate foreign artistic conventions by reworking them to fit within its own distinctive aesthetic. Under the patronage of the Umayyads, the inherited design conventions employed in the creation of stellar mosaics, compass-work mosaics, and polygonal tessellations were particularly relevant to the development of the preeminent form of overtly geometric Islamic ornament: the star patterns that characterize this ornamental tradition.

1.2   The Rise to Maturity

The history of Islamic geometric design can be regarded as a sequential evolution from simplicity to complexity. From its onset in the ninth and tenth centuries, this new form of ornament was characterized by an overall geometric matrix with primary stars or regular polygons located upon the vertices of a repetitive grid. The geometric designs from this early period have either threefold or fourfold symmetry: the former characterized by hexagons or six-pointed stars located on the vertices of either a triangular or a hexagonal repeat unit, and the latter generally characterized by eight-pointed stars, octagons, or squares placed on the vertices of a square repeat unit. Geometrically simple patterns of these varieties are found in several of the early monuments in the central and western regions of Abbasid influence, including the Great Mosque of Shibam Aqyan near Kawkaban in Yemen (pre-872); the mosque of ibn Tulun (876-79) in Fustat, Egypt (now part of greater Cairo); and the Baghdadi *minbar* (c. 856) at the Great Mosque of Kairouan in Tunisia.

Several techniques for creating geometric patterns appear to have been used historically, and many of the less complex geometric designs can be created from more than a single methodological approach.[16] It is therefore not always possible to know for certain which generative technique was used for a given historical example. Of particular interest, and the primary focus of this study, is the *polygonal technique*. Almost all of the early geometric patterns can be easily produced with this design methodology. However, placing stars or polygons in simple point-to-point configurations will also create many of the earliest patterns known to this tradition. Additionally, simple tracings upon the isometric grid will easily create many of the early threefold designs, as is similarly possible with the orthogonal grid for some of the least complex fourfold designs. The fact that the polygonal technique is a more demanding design methodology requiring two distinct steps would appear to argue for the greater relevance of less complicated and more immediate methodologies. However, the strength of the polygonal technique is in its inherent flexibility, providing the high level of design diversity and range of complexity that characterize this tradition. By the close of the tenth century geometric patterns were being created that were significantly more difficult to produce using other methodologies. With the growth in complexity, the polygonal technique became an increasingly important force behind the evolving sophistication of Islamic geometric star patterns that took place between the onset of this tradition during the ninth and tenth centuries and its full maturity during the thirteenth century.

The distinctive feature of this methodology is the employment of a polygonal tessellation that acts as a substructure from which the geometric pattern is derived. This process involves the placement of the pattern lines upon specified points along the edges of each polygon within an underlying generative tessellation. By the twelfth century, four distinct families of geometric pattern had evolved. Three of these are determined by placing crossing pattern lines that intersect on, or occasionally near, the midpoints of the underlying polygonal edges. The specific angle of these crossing pattern lines, referred to herein as the *angular opening*, determines the overall character of the design. For purposes of descriptive clarity these three families are referred to as *acute*, *median*, and *obtuse*. The fourth historically common pattern family places the pattern lines upon two points of each underlying polygonal edge, and is hence referred to as the *two-point* family. These two contact points are frequently determined by dividing the polygonal edge into either thirds or quarters. With rare exception, the

---

[16] A comparison and descriptive analysis of differing generative methodologies is covered in Chap. 2.

underlying polygonal tessellation is dispensed with after the pattern is created, leaving behind only the derived pattern with no easily discernable indication for how the pattern was constructed. Any one of the four pattern families can be used when extracting a geometric pattern from an underlying polygonal tessellation. The fact that each underlying formative tessellation can generate patterns from each family significantly augments the generative design potential of this methodology.

During the period of rising maturity, Muslim artists discovered several polygonal *systems* for creating geometric patterns. It is impossible to know for certain exactly when and where these systems were discovered, and without definitive historical evidence, it could be argued that use of these systems by Muslim artists remains conjecture. However, the systematic mode of the polygonal technique is the only practical explanation for the fact that such large numbers of patterns were created that strictly adhere to a common set of visual features that are associated with specific pattern families within one or another of these design systems. For example, the large number of fivefold *acute* patterns with ten-pointed stars that are ubiquitous to this ornamental tradition share very specific design characteristics within their pattern matrix, and these similarities are difficult, if not impossible, to explain with anything other than the existence of the *fivefold system* of pattern generation. Each of the historical design systems relies upon a limited set of polygonal modules that combine together in an edge-to-edge configuration to make the underlying generative tessellation. As described above, pattern lines in either the *acute*, *median*, *obtuse*, or *two-point* families are applied to the edges of these polygonal modules. The strength of designing patterns with polygonal systems is the ease of exploring new assemblages and resulting patterns. If one considers that the modules that make up each system can be combined in an infinite number of ways, and that each of the four pattern families can be applied to each tessellation, there are an unlimited number of geometric patterns available to each system.

The surviving architectural record indicates that the earliest methodological system to have been developed relies upon regular polygons to create the underlying generative tessellations. This is referred to herein as the *system of regular polygons*. The construction of geometric patterns from underlying tessellations made up of regular polygons appears to have begun in the ninth century and continued throughout the length and breadth of this ornamental tradition. From as early as the eighth century, Muslim artists employed tessellations made from the regular polygons as ornamental motifs in their own right. Noteworthy among the early examples of this form of geometric ornament are the Yu'firid ceiling panels of the Great Mosque of Shibam Aqyan near Kawkaban in Yemen (pre-871-72). Considering

the interest in polygonal ornament generally, it is entirely reasonable to allow for the inventive leap from using such tessellations as ornamental motifs to employing them as a substratum from which pattern lines can be extracted. As said, the precise date for the methodological discovery of using underlying tessellations to create geometric patterns is uncertain. This is due to the aforementioned fact that the simplicity of ninth- and tenth-century examples allow for their creation with either the polygonal technique, the iterative placements of simple star forms, or simply the tracing of lines from the isometric grid. What is certain is that almost all of the ninth- and tenth-century prototypical geometric patterns *can* be easily created using the polygonal technique. It is significant that when considered from the perspective of the polygonal technique, the underlying generative tessellations for virtually all of these early examples are comprised of regular polygons. As the use of this regular polygonal methodology became more sophisticated, the resulting geometric patterns became more diverse and more complex; and the prevalence of such patterns became sufficiently common to warrant their own descriptive classification: the *system of regular polygons*. The growth in complexity of geometric patterns made from the *system of regular polygons* is directly associated with the expansion of knowledge of the tessellating potential of the regular polygons.

Only five of the regular polygons will combine to uniformly fill the two-dimensional plane in an edge-to-edge configuration: the triangle, square, hexagon, octagon, and dodecagon [Figs. 89–91]. *Regular*, *semi-regular*, *two-uniform*, and *three-uniform* tessellations were all used historically to generate geometric patterns [Figs. 95–115]. Depending on the arrangement of the polygonal modules, patterns with either threefold or fourfold symmetry were constructed. Similarly, the variety of repeat units found within this system includes the equilateral triangle, square, and regular hexagon, as well as rectangles, rhombi, and non-regular hexagons. As this ornamental tradition matured the *system of regular polygons* occasionally included additional non-regular polygons. These non-regular modules are created as interstitial spaces in a tessellation of otherwise regular polygons. These interstice modules create distinctive pattern characteristics that augment the beauty of patterns made from this system [Figs. 116–122]. The historical record demonstrates a high level of symmetrical and repetitive variety within this design system, resulting in the surprising degree of design diversity that is emblematic of this systematic methodology.

The regular polygons that tessellate together include the octagon. However, unlike the other regular polygons from this system, the octagon only tessellates in one combination: the semi-regular 4.8^2 tessellation of squares and octagons [Fig. 89]. The octagon and square are also components of the fourfold system A [Fig. 130]; and patterns made from this semi-regular tessellation of octagons and squares are justifiably associated with either of these two systems. However, patterns such as the ubiquitous classic star-and-cross design [Fig. 124b] that are easily created from the 4.8^2 tessellation precede the earliest known patterns associated with the fourfold system A by some 200 years. Furthermore, the early examples of patterns created from the octagon and square share the same approximate time and place with other designs that are created from the system of regular polygons. Patterns created from the underlying 4.8^2 tessellation are therefore more appropriately considered as part of the group of designs that originate from the system of regular polygons. However, due to that fact that the octagon has only a single tessellation within the theoretically infinite number of possible combinations of the other regular polygons within this system, for the purposes of this discussion, patterns derived from this tessellation of octagons and squares are regarded as a separate generative category. It is worth noting that the design diversity produced from this one tessellation is truly remarkable, and its historical use very likely exceeds that of any other single underlying tessellation from this design tradition [Figs. 123–129].

Two of the polygonal systems used regularly throughout Muslim cultures have fourfold symmetry and employ the octagon as the primary polygonal module. These are referred to herein as the fourfold system A and the fourfold system B. Most of the patterns that these two systems create repeat upon the orthogonal grid, although patterns with 45° and 135° angled rhombic repeat units were occasionally employed, as were patterns with rectangular repeat units. Patterns with radial symmetry are also possible with these systems, although infrequently used. The fourfold system A has three modules that are regular polygons: a large octagon, a small octagon, and a square. Other than these, all of the polygons within this system are irregular [Fig. 130]. The geometric construction for each shape within this system is easily derived from the large octagon [Fig. 131]. The fourfold system A is comprised of a relatively large number of polygonal modules, resulting in a particularly high level of diversity in the potential underlying generative tessellations [Figs. 136–168]. The fourfold system B has fewer polygonal modules: allowing for less tessellating variation than that of the fourfold system A. This system is nevertheless responsible for a wide variety of distinctive and beautiful designs from the historical record [Figs. 173–186]. The octagon is the only regular polygon within the set of generative modules of the fourfold system B [Fig. 169]. The polygonal modules of this system are easily constructed from this primary polygon, or through identifying interstice regions through tessellating with the octagon and irregular pentagon [Fig. 170]. The large number of historical patterns that are associated with both these fourfold systems has by no means exhausted the generative potential for making new and original designs.

Ghaznavid and Qarakhanid artists produced the earliest patterns associated with the fourfold system A during the first quarter of the eleventh century. Seljuk and Ghurid artists adopted this methodological system within half a century, and the diversity of patterns created by these eastern cultures is remarkable. The rapid westward spread of Seljuk influence introduced this system to the Artuqids, Zangids, and the Seljuk Sultanate of Rum, by which time it had become part of the standard geometric design repertoire of these regions. This system was not adopted by the Fatimids of Egypt, and even their Ayyubid and Mamluk successors made significantly less use of this variety of pattern than their contemporaries to the north and east. By the fourteenth century, the fourfold system A was also an integral feature of the western Islamic ornamental tradition, and the number of examples found in Nasrid, Marinid, and Mudéjar monuments is remarkable. Artists working in the Maghreb developed this system to further levels of refinement and complexity through the incorporation of 16-pointed stars. Even more remarkable was the innovative use of this system to create the astonishing dual-level designs that are the earliest expressions of complex self-similar art ever produced.[17]

The architectural record indicates that development of the fourfold system B took place during the first half of the twelfth century. These earliest examples are Qarakhanid, Seljuk, and Ghurid, but there were far fewer patterns produced from this system during this early period in the eastern regions than those of the fourfold system A. The predominance of early patterns created from the fourfold system B is found in the western regions of Seljuk influence, and these were produced under the patronage of the Ildegizids, Zangids, Ayyubids, and the Seljuk Sultanate of Rum. The Mamluks were far more disposed to this system than to the fourfold system A. Following the innovations in the western regions under Seljuk influence, the fourfold system B was readily incorporated into the ornament of the eastern regions following the Mongol destruction, and fine examples were produced under Ilkhanid, Kartid, Muzaffarid, Timurid, and Mughal patronage. And in the western regions of the Maghreb, as with the fourfold system A, the Nasrids and Marinids also used this system widely in their architectural ornament. The rapid adoption of the fourfold system B into the body of geometric expression among diverse Muslim cultures suggests a transcultural mechanism wherein artistic innovations were willfully shared between artists under the patronage of both friendly and rival dynasties. At the very least, it must be concluded that the currency of artistic knowledge was highly valued and facilitated the movement of specialists from region to region.

The differences in appearance between the fourfold system A and the fourfold system B are readily apparent to a trained eye. Both incorporate eight-pointed stars and

---

[17] Bonner (2003).

octagons as standard features, and the vast majority of patterns from both systems repeat upon the orthogonal grid. However, the very different characteristics of the respective underlying polygonal modules from each system result in geometric designs with concomitant differentiation. In particular, the pentagonal and hexagonal modules from the *fourfold system B* create distinctive pattern qualities that are entirely dissimilar to the geometric characteristics associated with the *fourfold system A*. By the twelfth century, artists working with the *fourfold system B* discovered that the application of *acute* pattern lines to the elongated hexagonal modules could be varied to allow for the creation of octagons within the pattern matrix [Fig. 172].

Almost all of the innumerable patterns with fivefold symmetry and ten-pointed stars that are found throughout the Islamic world have their origin in the *fivefold system*. The repeat units of patterns generated from this system are predominantly either rhombic or rectangular. There are two rhombi associated with fivefold symmetry that function as repeat units for patterns made from this system [Fig. 5]: the wide rhombus with 72° and 108° included angles, and the thin rhombus with 36° and 144° included angles. The wide rhombus was used more extensively as a repeat [Figs. 232–240], but many patterns were also created that repeat with the thin rhombus [Figs. 241–244]. The proportions of the rectangular repeat units used with this system varied considerably [Figs. 245–256]. Less common are patterns with irregular hexagonal repeat units [Figs. 257–259], and those with radial symmetry [Fig. 260]. Occasionally, greater complexity was achieved through using several repetitive components within a single design, any one of which is able to create patterns on its own [Figs. 261–266]. In this study, these are referred to as *hybrid* patterns, and the earliest known example was produced by Seljuk artists for one of the recessed arches in the northeast dome chamber of the Friday Mosque at Isfahan (1088-89) [Fig. 261] [Photograph 25]. Most of the subsequent fivefold hybrid examples were produced under the patronage of the Seljuk Sultanate of Rum.

The *fivefold system* has a greater number of components than either of the two fourfold systems [Figs. 187–188]. The limited set of polygonal modules that comprise the *fivefold system* includes two that are regular: the decagon and pentagon. The polygonal modules of this system can be easily produced from the decagon [Fig. 189], or through interstice regions when tessellating with other modules [Figs. 190–191a]. Some modules can also be created from overlapping the pentagon or decagon, and a further set of components is created from the union of two conjoined decagons [Fig. 191b]. There are two edge lengths among the polygonal modules within this system: the shorter being the length of the edges of the regular decagon and pentagon, and the longer being equal to the distance from the center of the

decagon to one of its vertices. The ratio of these two edge lengths is the golden section (1:1.618033987...); and indeed, the proportional relationships inherent within fivefold geometric patterns are imbued with this geometric ratio [Fig. 195].

A subcategory of fivefold patterns forgoes the decagon within the underlying tessellation, thereby eliminating the characteristic ten-pointed stars from the overall pattern matrix. Such designs are referred to as *field patterns*, as the absence of the ten-pointed stars produces more uniform density within the pattern matrix [Figs. 207–220]. This variety of design was especially popular in the architecture of the Seljuk Sultanate of Rum. Such field patterns are both aesthetically distinct from standard fivefold patterns, and pleasing to the eye. The repeat units of fivefold field patterns are predominantly either rectangles or irregular hexagons.

The versatility and visual appeal of patterns made from the *fivefold system* led to its rapid spread throughout Muslim cultures; and outstanding examples are to be found in diverse ornamental media throughout the length and breadth of this ornamental tradition. The earliest extant fivefold geometric designs were produced by Seljuk artists during the close of the eleventh century. Within a decade, the architectural ornament of the Ghaznavids also incorporated patterns with fivefold symmetry. By the middle of the twelfth century Ghurid artists also made use of patterns created from the *fivefold system*, followed by the Qarakhanids some 30 years later. And as with the *fourfold systems A* and *B*, the *fivefold system* spread westward from Khurasan and Persia into regions under Seljuk influence, subsequently becoming an ubiquitous feature of the ornamental arts of Muslim cultures generally.

Among the most fascinating systematic geometric patterns to have been created by Muslim artists are a relatively small number of designs with sevenfold symmetry [Figs. 279–282 and 286–294]. However, the small number of surviving historical examples of such patterns begs the question as to the extent to which geometric artists were aware of the systematic repetitive potential of the underlying polygonal components that made up the generative tessellations. This variety of patterns is very beautiful, and had the systematic potential for these components been known by those artists working with geometric patterns generally; one would assume that, as with fivefold patterns, there would be far more examples found throughout the Islamic world. This paucity of examples appears to indicate the rarity of knowledge of this system among geometric artists. However tenuous our understanding of past sevenfold methodological knowledge is, it is nonetheless a fact that the relatively few sevenfold patterns in the historical record would have been relatively easy to create from a limited set of repetitive polygonal modules that include associated pattern lines in each of the four standard pattern families. The earliest known sevenfold geometric patterns include a Seljuk field pattern from the northeastern domed chamber (1088-89) of the Friday Mosque at Isfahan [Fig. 279], and two Ghaznavid examples from the minaret of Mas'ud III in Ghazna, Afghanistan (1099-1115) [Figs. 280 and 281]. Interestingly, these same monuments also include the earliest known examples of two-dimensional fivefold patterns. Each of these early Seljuk and Ghaznavid sevenfold patterns repeat upon irregular hexagonal grids. The underlying generative tessellation of the earlier Seljuk example employs two varieties of hexagon to create the sevenfold field pattern, and the hexagonal repeat unit is a product of the specific arrangement of underlying hexagons. The hexagonal repeat units of the two Ghaznavid examples have touching edge-to-edge heptagons placed at each vertex of the repeat unit. The interstice of these six edge-to-edge regular heptagons is comprised of two irregular pentagons that likewise touch edge to edge. The first of these Ghaznavid sevenfold geometric designs incorporates a set of primary pattern lines placed upon the vertices of the underlying heptagons, thereby producing a set of seven-pointed stars whose points touch those of adjacent seven-pointed stars [Fig. 280b]. The second set of pattern lines are placed upon two points of each heptagonal edge [Fig. 280c]. This is a remarkably complex design, especially considering its very early date. The second sevenfold design from Ghazna is no less impressive. The primary pattern lines of this design are located upon the midpoints of the heptagons in the same underlying tessellation [Fig. 281b], while the secondary pattern lines are an arbitrary addition that makes this design considerably more complex [Fig. 281c]. Approximately a hundred years later, artists in Anatolia created several sevenfold geometric patterns using the same underlying generative tessellation of heptagons and irregular pentagons. These three Anatolian examples differ in that they are less complex, and fully systematic—in that all of the pattern lines are the direct product of the underlying polygonal tessellation [Fig. 282]. In time, this *sevenfold system* developed in its increased use of a larger number of polygonal components with a resulting increase in complexity. A noteworthy feature that distinguishes these later examples from the earlier designs is the use of underlying tetradecagons (14-sided regular polygons) that produce 14-pointed stars. There is a marked increase in the number of polygonal modules associated with the *sevenfold system* over the other historical systems [Fig. 271], and as a general rule, the greater the number of sides to the primary polygon, the greater the number of modules within a given system. An added feature of the growth in complexity of this system was the use of additional repeat units beyond the initial elongated hexagon described above. These included patterns with rectangular repeats, and patterns based upon one or another of the three rhombi associated with 14-fold symmetry

[Fig. 10]. As with the other systems, the primary star forms (in this case 14-pointed stars) were typically placed upon the vertices of each repeat unit. These more complex sevenfold geometric patterns originated among the Mamluks in Egypt and the Levant during the fourteenth and fifteenth centuries, and to a lesser extent were also employed by a select number of artists working for the Ottomans and Timurids.

Perhaps the most remarkable historical use of generative polygonal design systems was in their application to multiple level designs. During the fourteenth and fifteenth centuries, the innovative dual-level use of the *system of regular polygons*, both fourfold systems and the *fivefold system*, brought about the last great creative leap in the historical development of Islamic geometric star patterns [Figs. 442–477]. Through careful manipulation of these polygonal systems, Muslim artists produced several varieties of geometric design that are consistent with the modern geometric criteria for self-similarity whereby an entity or a structure is recursively present within an analogous scaled-down substructure that, in turn, provides for the possibility of infinite further recursive iterations. While this recursive process is mathematically infinite, be it cosmological, geographical, biological, or anthropogenic, the manifestation of self-similar recursion is constrained by the medium in which it occurs. The historical examples of self-similar star patterns never exceed a single recursion, and are characterized by two levels of design: the visual character and methodological origins of each being either identical or very similar to the other. Can an object be self-similar if it has only a single recursion? The answer is yes, provided the relationship between both levels satisfies the criteria for self-similarity, and the recursion has the theoretical capacity for infinite reiteration. The recursive scaling ratio is always a product of the geometric schema, and the placement of the secondary pattern is determined through the application of scaled-down underlying polygonal modules from the same system that were used to create the primary design. These scaled-down elements typically place the primary polygonal modules, such as an octagon or a decagon, upon the crossing lines of the primary pattern. Most of the examples of Islamic self-similar ornament were fabricated in cut-tile mosaic, and a few examples were produced in wood. The fact that the Muslim artists responsible for these masterpieces of geometric art limited themselves to just two levels of self-similarity is more to do with the material constraints of their chosen medium than with any lack of geometric ingenuity.

Islamic self-similar design developed along two distinct historical paths. The earliest occurrence of such patterns was during the fourteenth century in the western regions of Morocco and al-Andalus under the patronage of the Marinid and Nasrid dynasties. A century later, highly refined self-similar patterns were introduced to the architectural ornament of Transoxiana, Khurasan, and Persia under rival

Timurid, Qara Qoyunlu, and Aq Qoyunlu patronage. It is unknown whether these two design traditions developed in isolation, or the preceding design methodologies from the Maghreb directly influenced the development of this design convention in the eastern regions. While the methodology in the creation of self-similar designs from both regions is essentially the same, their respective stylistic character is very different. As mentioned, this methodology is reliant upon the recursive tessellating properties inherent to these design systems. When considered from the perspective of Islamic art history, the self-similar designs created in these western and eastern regions represent the pinnacle of systematic geometric design, and, as said, the last great innovation in the illustrious tradition of Islamic geometric star patterns. As pertains to the history of mathematics, these fourteenth- and fifteenth-century designs are no less significant in that they appear to be the earliest anthropogenic examples of sophisticated self-similar geometry.

In addition to geometric star patterns being produced via a systematic design methodology, Muslim artists expanded the polygonal technique to include *nonsystematic* designs. These are generated from underlying tessellations that include polygons that are irregular and specific to the tessellation [Figs. 309–441]. In contrast to the various generative systems, many of the polygonal components of such tessellations will not reassemble into other tessellations, and as such, patterns made from this variety of underlying tessellation are therefore nonsystematic. One of the virtues of a systematic design methodology is the ease of creating new patterns through new assemblages of the polygonal modules. One has only to produce a new tessellation from a predetermined set of compatible decorated polygonal modules. The creation of nonsystematic geometric patterns is entirely different. Muslim artists developed a precise design methodology that produced a wide range of underlying tessellations with polygonal components that are specific to the construction. As with the systematic approach, each nonsystematic tessellation will produce geometric designs in each of the four pattern families. Although only conjecture, similarities between nonsystematic designs and those created from the *fivefold system* suggest the possibility that the mature expression of nonsystematic patterns was directly influenced by the aesthetics and working practices found within the *fivefold system*. Fundamental to the creation of nonsystematic underlying tessellations is the use of radii matrices as an initial foundation for the construction sequence. Evidence that radii matrices were an integral feature of the nonsystematic use of the polygonal technique is found in many of the geometric star patterns illustrated in the Topkapi Scroll. This is a unique and immensely important document in many respects, including the insight it provides into the methodology employed for constructing complex geometric star patterns. The maker of the Topkapi

Scroll used a steel graver to scribe non-inked "dead drawing" reference lines into the paper, and included among these barely visible scribed lines are radii matrices.[18] These articulate the regions of primary and secondary local symmetry, and relate directly to the construction of the underlying polygonal tessellations, most frequently illustrated in finely dotted lines of red ink, upon which the typically black pattern lines are positioned.

The tradition of nonsystematic geometric star patterns is immensely diverse and covers a wide range of symmetries and variety of repeat units. Most commonly, nonsystematic geometric patterns will repeat on either the isometric or the orthogonal grids. The least complex examples of this type of geometric design employ a single variety of star that is located upon each vertex of the repetitive grid [Figs. 309–345]. The number of points for these stars is governed by the number of angles at each vertex as a multiplier, with *n*-points being the product. In this way, patterns that repeat upon the isometric grid will typically have 6, 12, 18, 24 (etc.) pointed stars at each vertex [for example: Fig. 320], while the vertices of patterns that repeat on the orthogonal grid will typically have 8, 16, 24 (etc.) pointed stars [for example: Fig. 337]. The regular hexagonal grid was also employed, and such patterns will commonly have 6, 9, 12, 15, 18 (etc.) pointed stars at the vertices of this repetitive grid [for example: Fig. 313]. The isometric and orthogonal grids also provide for patterns with greater complexity that have additional varieties of local symmetry beyond those located at the vertices of the repeat unit [Figs. 346–411]. These are generally referred to as *compound patterns*, and the least complex will place additional stars at the vertices of the dual of the isometric or orthogonal grid—which is to say at the centers of each repeat unit. The dual of the isometric grid is the regular hexagonal grid, and examples of compound local symmetry for such patterns can include star combinations of 6 and 9 points, 12 and 9 points, 12 and 15 points, etc. [for example: Fig. 346]. The dual of the orthogonal grid is of course another orthogonal grid, and compound patterns of this variety will typically include star combinations of 8 and 12, 8 and 16, 12 and 16, etc. [for example: Fig. 379]. Still further complexity was achieved through additional centers of local symmetry being incorporated into the isometric or orthogonal repeat units. The locations for these additional regions of local symmetry are typically at the center points of each edge of the repeat unit [for example: Fig. 402], or within the field of the repeat unit [for example: Fig. 400]. It is worth mentioning that the center point of the repeat unit's edge is also the intersection of the grid and its dual [Fig. 1]. These additional locations provide the designer with greater latitude in determining the variety of local

---

[18] Necipoğlu (1995), 239–283.

symmetry and resulting star forms. When these additional star forms are located at the midpoint of the edge of the repeat unit, they tend to have an even number of points, while the use of additional local symmetries within the field of the repeat unit is less rigid.

Some of the most remarkable nonsystematic geometric patterns are characterized by their incorporation of two seemingly incompatible varieties of primary local symmetry. As mentioned above, compound patterns that repeat upon either the isometric or the orthogonal grids will most commonly place regions of local symmetry at the vertices of both the repetitive grid and its dual. The relationship between the grid and its dual provides for star forms at these locations that are compatible and predictable. By contrast, this more elusive variety of compound pattern brings together two varieties of n-fold local symmetry that would not ordinarily work with one another to fill the two-dimensional plane: for example 9- and 11-pointed stars [Fig. 431]. This variety of compound pattern typically employs either a rectangular grid [Figs. 412–428] or an elongated hexagonal repetitive grid [Figs. 429–439]. As with the more complex compound patterns that adhere to the isometric, regular hexagonal, and orthogonal grids, patterns that repeat with rectangular and elongated hexagons will occasionally incorporate additional centers of local symmetry upon the edges or within the field of the repeat unit [for example: Fig. 427].

The beauty of nonsystematic compound star patterns is, in large part, the direct consequence of their geometric sophistication. Indeed, this highly refined utilization of the polygonal technique is responsible for the creation of the most geometrically complex Islamic star patterns throughout the length and breadth of the Islamic world. Many patterns created from one or another of the historical design systems can also be produced via alternative methodologies, for example, point joining or through the use of grid-based constructions. However, these alternative methodologies become less and less relevant as complexity increases, and the only practicable method for constructing the considerably more complex nonsystematic designs with multiple regions of local symmetry is via the polygonal technique. Other historically demonstrable design methodologies do not have the flexibility to work seamlessly with the diverse complexities associated with multiple regions of local symmetry.

The continued development of the polygonal technique allowed Muslim artists to raise the geometric arts to an unsurpassed level. The versatility of this methodology facilitated the remarkable diversity that characterizes this tradition, including the discovery of new and ingenious repetitive formulae for covering the two-dimensional plane; the establishment of the four principal pattern families; the discovery of several tessellating *systems* that

employ a limited set of decorated polygons that iteratively combine in an infinite number of ways; the development of nonsystematic compound patterns wherein centers of differing local symmetry allow for the placement of different star types within a single pattern; and the discovery of the recursive application of the polygonal systems to create two-level geometric patterns that conform to the modern criteria of self-similarity.[19] Each of these is a separate and significant aspect of this overall tradition, and each is unlikely to have developed and flourished without the inherent flexibility of the polygonal technique.

1.3 Umayyads (642-750)

In 635, just 3 years after the death of the Prophet Muhammad, Muslim forces of the Rashidun Caliphate conquered the Byzantine vassal state of Syria. Within two years the Sassanian Empire of Persia fell, followed by Byzantine-held Egypt in 642. The succeeding Umayyad Caliphate continued this rapid expansion: taking control of a contiguous region from Spain and North Africa to the Indus River. The vast territorial expanse of this empire created the need for a more central administrative capital. This brought about the move of their capital from Medina to Damascus. The conquering of Byzantine Syria, Persia, and Egypt brought the Muslim conquerors into contact with several cultures with highly developed architectural and ornamental traditions. By contrast, the artistic heritage of the conquering Arabs was far less sophisticated. The Umayyad rise to power and wealth facilitated an ambitious emphasis upon the construction of large monumental buildings. They were prolific builders, and were quick to employ the superior skills of their non-Muslim subjects. When the Great Mosque of Kufa was rebuilt in 670, a Persian architect was employed who had worked for the Sassanid kings; and Persian masons were used in rebuilding the Kaaba in 684. Builders and craftsmen from Egypt, Greece, and Syria were employed in rebuilding the Masjid al-Nabawi (Prophet's Mosque) in Medina during the period of 707-709; and Coptic Christians from Egypt were likewise used in building both the al-Aqsa mosque in Jerusalem, and the Dar al-Imara palace from 709 to 715. Very little remains of the original al-Aqsa mosque, and among the most important existing examples of early Umayyad ornament, are the Qubbat al-Sakhra (Dome of the Rock) in Jerusalem (685-92), the Great Mosque of Damascus (706-15), the excavated palace of Khirbat al-Mafjar near Jericho in the Jordan Valley (739-44), and the archeological site of Qasr al-Mshatta in Jordan (744-50).

---

[19] Bonner (2003).

Photograph 3 A fifth-century Coptic textile with eight-pointed stars (The Metropolitan Museum of Art: Gift of George F. Baker, 1890: www. metmuseum.org)

The precise origins of Islamic geometric star patterns are impossible to establish categorically. There are too many ornamental influences, and too few remaining buildings or objects of art from the early formative period to know definitively when or precisely how this intrinsically Islamic ornamental convention began. The use of stars as a decorative motif was practiced by the pre-Islamic cultures of Byzantium, Coptic Egypt, and Sassanid Persia, and included their use as either singular motifs within a decorative schema or constellations wherein multiple stars provide the primary character of the design. Within the pre-Islamic Coptic textile tradition the eight-pointed star was frequently used as an independent element, often filled with an elaborate profusion of embroidered interweaving knot-work [Photograph 3]. A pre-fifth-century Hellenic mosaic pavement from the Sardinian town of Nora may be relevant to the later development of Islamic geometric star patterns. This design incorporates multiple eight-pointed stars composed of two interweaving squares that are placed upon a square grid in such a manner that two adjacent points from each star touch the equivalent points from each orthogonally placed neighboring star. The interstices of this stellar formation are regular octagons and rhombi. A conceptually similar design from a Roman settlement in El Djem, Tunisia (third century), uses 12-pointed stars in a similar arrangement [Photograph 4].[20] These also repeat upon the square grid and are orientated so that two adjacent points touch the equivalent two points of their orthogonal neighbor. This arrangement of 12-pointed stars results in the background shapes being rhombi and a cross-like element that is further filled with regular hexagons and central 4-pointed star. Prior to the advent of Islam, Byzantine artists continued working with the long-established conventions of Hellenic mosaics, including the geometric idiom that forms part of this overall tradition.[21] Exposure to the architectural ornament and mosaic pavements from the historic centers of Byzantine culture in the Middle East, such as Jerusalem, Antioch, Madaba, Tel Mar Elias, al-Maghtas, and Tell Hesban, would have familiarized the early Arab conquerors with the Hellenic practice of creating designs from an assemblage of stars. Moreover, the Umayyads had ready access to the aesthetic

---

[20] From the collection of the El Djem Archaeological Museum, El Djem, Tunisia.

[21] Kitzinger, Ernst. (1965), 341–352.

Photograph 4 A third-century Roman mosaic with 12-pointed stars from El Djem, Tunisia (© Damian Entwistle)

conventions of Byzantine artisans living in the newly conquered territories who were now their subjects. From a design standpoint, the primary difference between the Hellenic examples of patterns with multiple stars and those subsequently developed by Muslim artists is in the cohesiveness of the overall design. In the earlier Hellenic work, the stars are independent elements scattered across the plane in a repetitive staccato fashion, relating to one another through geometric proximity and similitude. By contrast, within the Islamic star pattern aesthetic, the lines of each star proceed outward to join with the similarly extended lines from adjacent stars to produce an interconnected network wherein each star is an integral part of a unified whole. Secondary to the geometric pattern itself is the treatment of the lines. More often than not, the geometric matrix was given an interweaving treatment wherein the pattern lines are widened to a desired thickness, often informed by material constraints, and made to flow over and under one another. Interweaving lines were a common feature of the pre-Islamic decoration of the Byzantines, Copts, and Sassanids, and were similarly employed in the braided borders, compass-

work motifs, and key patterns of the Umayyads. Over the centuries, the geometric star pattern aesthetic was broadened by the introduction of further forms of line treatment [Figs. 85–88], but the primacy of interweaving widened lines continued throughout the long history of this tradition and helped to provide aesthetic continuity within the ornamental arts of Muslim cultures for centuries to come.

The Umayyad innovation of applying Byzantine compass-work mosaic conventions to their pierced window grilles was another progenitor of the tradition of Islamic star patterns. The methodology used for constructing these Umayyad window grilles was described and aptly named by K. A. C. Creswell as *compass work*.[22] The Hellenic art of the Byzantines included a distinctive geometric device constructed from overlapping circles. This form of ornament was employed widely in the embellished mosaic pavements throughout the Hellenic world. The diverse range of ornamental motifs in the fourth-century mosaic paving at Mount

---

[22] Creswell (1969), 75–79.

Photograph 5 A fourfold compass-work design with eight-pointed stars from a pierced stone window at the Great Mosque of Damascus (© David Wade)

window grilles is fundamentally the same as seen in Hellenistic mosaics, the aesthetic effect is distinct and original. The interweaving line work of the mosaic pavements is heavily elaborated with secondary elements such as interior braided bands. By contrast, the stucco window grilles rely on a more austere geometric exposition that is highly effective and beautiful. It is possible that the inspiration for applying the compass-work design methodology of mosaic paving to Umayyad window grilles derived in part from Sassanid sources. Excavations at the Sassanid fortified township of Qasr-i Abu Nasr near Shiraz revealed a stucco window grille dated from the sixth or seventh century that, while very simple in its pierced honeycomb design, is identical in architectonic concept to later Umayyad window grilles. It is also significant that Sassanid artists were masters of carved stucco ornament. The mixed cultural milieu of the Umayyads wherein artists from Byzantium were working alongside those from Persia may have led to the amalgamation of these two separate ornamental traditions. Whoever the originators were, this innovation was undoubtedly driven as much by technical and functional constraints as by aesthetic consideration. Being a pierced window grille, the ratio of foreground to background had to be carefully determined so that adequate light would filter into the building, yet with interweaving line work thick enough, and so designed, as to provide adequate structural integrity. Adding to the more austere aesthetic is the fact that the resulting line work is too thin for much in the way of secondary elaboration: the added surface decoration frequently limited to a simple carved groove that creates the over/under interweave, and narrower incised grooves that run parallel to the central line work. The austere geometric aesthetic born from such constraints provided a very successful and, one can argue, much-needed counterpoint to the highly ornate Umayyad floral conventions. Furthermore, despite methodological differences, the overtly geometric aesthetic of these window grilles undoubtedly influenced the cultural predilections that eventually led to the development of Islamic geometric star patterns. However, the rudimentary geometry and simple techniques of construction for these early compass-work geometric patterns are in marked contrast to the geometric complexity of the tradition that was to follow.

The Umayyad compass-work window grille aesthetic was also appreciated in their western territories. Begun in the eighth century, the Great Mosque of Córdoba is one of the masterpieces of Umayyad architecture. The many pierced marble window screens that adorn this mosque include a compass-work example just north of the Puerta de San

Nebo, a Christian site in the mountains of Jordan, includes this variety of geometric motif. The Umayyads were quick to employ compass-work designs in their tesserae mosaic architectural decoration. Of particular note is the mosaic pavement in the Umayyad palace of Khirbat al-Mufjar near Jericho (724-43). This is of outstanding quality in both design and execution, and is the largest mosaic floor to have survived from antiquity. Among the multiple ornamental panels that make up this pavement are several that employ interweaving circular elements. The earliest extant Umayyad window screens with this type of geometric design were executed in stucco and several are found in the Great Mosque of Damascus (c. 715) [Photograph 5], and at Khirbat al-Mafjar.[23] While the general geometric schema of these

---

[23] Several stucco window grilles were found in the ruins of the Umayyad palace of Qasr al-Hayr al-Gharbi (724–727): now in the National

Museum of Damascus. The design of each of these is comprised of a central palm motif flanked by floral scrollwork rather than the overlapping circles under discussion.

Esteban[24] (855-6) that is very similar in conceptual design to the earlier Umayyad compass-work window grilles from their Syrian homelands.

The abundant application of a diverse range of geometric motifs in Umayyad ornament provides clear evidence of the Islamic fascination for applied geometry dating back to the earliest period of Islamic architectural accomplishment. Despite their appreciation for geometric ornament, none of the Umayyad geometric designs exhibited the distinctive qualities found in the mature tradition of Islamic geometric star patterns. Without a working knowledge of the precise methodology that allows for the creation of Islamic geometric star patterns this stylistic disconnect was inescapable. However, the Umayyad geometric convention of employing polygonal tessellations as ornament appears to have been critical to the later development of Islamic geometric star patterns. Their familiarity with polygonal tessellations is significant in that such knowledge was essential to the eventual development of the *polygonal technique* of geometric pattern generation wherein a polygonal tessellation is used as scaffolding upon which pattern lines are located, and, like scaffolding, discarded once the pattern is completed.

This form of geometric ornament utilizes an edge-to-edge configuration of one or more regular or irregular polygons to create a tessellating field pattern, typically with secondary floral designs contained within each polygonal cell. An early Umayyad example of this type of inherited ornament is found in the portal of the palace of Qasr al-Hayr al-Gharbi near Palmyra, Syria[25] (724-27). The carved stucco ornament in this highly elaborate entry portal includes two panels that employ regular hexagons and rhombi with 60° and 120° included angles that are the equivalent of the $3^2 6^2$-3.6.3.6 *two-uniform* tessellation of triangles and hexagons [Fig. 90]. As with other polygonal designs composed of regular hexagons, triangles, and double-triangle rhombi, this design can be easily constructed using the isometric grid. The fact of this form of isometric design being an established motif among the pre-Islamic peoples of the eastern Mediterranean is confirmed in the surviving ornamental ceilings from the second-century Roman ruins of Baalbek in Lebanon. The Umayyad mosaic pavement of the Khirbat al-Mafjar includes multiple panels with polygonal tessellations, including designs made up of elongated hexagons and squares, as well as panels with regular octagons and squares. Each of these tessellations received continued use by subsequent Muslim cultures. The Umayyads also combined simple polygonal tessellations with compass-work patterns. While both of these ornamental

themes were derived from pre-Islamic sources, their combined use was an original development. Umayyad examples of this form of geometric ornament are found in two of the pierced stucco window grilles at the Great Mosque of Damascus [Fig. 82c] and one of the window grilles from Khirbat al-Mafjar [Fig. 82d]. These two examples employ the 3.6.3.6 tessellation of triangles and hexagons as the polygonal component of the composition. The use of polygonal tessellations as an ornamental device continued among artists in later Muslim cultures, but the great innovation in the ornamental use of polygonal tessellations was the discovery of the *polygonal technique* wherein these tessellations could be used as generative structures from which geometric patterns were extracted.

1.4 Abbasids (750-1258)

The forces of Abu'l-Abbas as-Saffah (721-754) defeated the Umayyads in 750. This marked the beginning of the Abbasid dynasty: one of the longest lasting and most influential dynasties in Islamic history. The Abbasids were descended from the Prophet Muhammad through the Prophet's uncle, Abbas ibn Abd al-Muttalib (566-653); and this kinship to the Prophet allowed them to assert greater religious authority and right to the caliphate over that of their Umayyad predecessors. Over the centuries, the respect bestowed upon the Abbasid Caliphate was to have a profound and continuing influence on Islamic politics and culture, including the arts. With control over Egypt and North Africa to the west, and Persia, Khurasan, and much of Transoxiana to the east, the Abbasids chose to move their capital eastward to a place more central to their empire. In 762 Al-Mansur founded his capital of Baghdad. He brought hundreds of builders, engineers, and craftsmen to Baghdad from areas throughout his empire. It can be assumed that this influx of artists and architectural specialists into a single location would have contributed greatly in creating the atmosphere of ornamental innovation that took place at this time. With Baghdad as the center of the Abbasid Empire, Persian influence became a major aspect of Abbasid culture. Persian customs were adopted as part of royal protocol; Persians were placed in important positions of power and influence within the government and military; and Persian artistic and architectural traditions were enthusiastically embraced by the otherwise Arab culture of the Abbasids.

The earliest Islamic geometric star patterns date to the ninth century at a time when Baghdad was the preeminent center of Arab culture. The rise of the Abbasid Caliphate heralded a period of great sophistication and refinement, creating a legacy for which subsequent Islamic cultures, and indeed the entire world, must be forever indebted. Baghdad became the foremost center for Islamic religious

---

[24] Originally known as Bab al-Wuzara.

[25] The portal of the palace of Qasr al-Hayr al-Gharbi is now in the National Museum of Dasmascus.

studies and scholarly learning, attracting the most learned scholars and theologians from far and wide. It was during this early Abbasid period that the four primary orthodox Sunni religious doctrines were developed: *Hanafi, Maliki, Shafi'i*, and *Hanbali*. Great emphasis was given to the translation of earlier Greek texts; and these works laid the groundwork for the following 800 years of Muslim achievements in the sciences. Great advances in the fields of philosophy, chemistry, medicine, zoology, botany, mathematics, geometry, astronomy, geography, linguistics, and history augmented the course of human knowledge. Many of these scientific works were introduced to Europe in the fourteenth and fifteenth centuries, and provided a significant influence upon the Italian Renaissance. Indeed, it was largely through Arabic translations that Europeans regained their knowledge of Greek science and philosophy. The Abbasid cultural milieu provided the background for such important philosophical thinkers as al-Kindi (d. c. 874), al-Farabi (d. 951), al-Haytham (d. 1021), and ibn Sina (d. 1037). Similarly, the cultural richness of this period engendered the blossoming of Islamic mysticism with such luminaries as Rabia of Basra (d. 801), Bayazid Bastami (d. 874), al-Junayd Baghdadi (d. 910), and al-Hallaj (d. 922), to name but a few. This was also an environment in which poetry thrived. In fact, the lines of demarcation separating poetry, mysticism, philosophy, and science were not so clearly delineated as experienced in the present era.

The Abbasids were equally committed to the further development of the arts and architecture: calligraphy and Quranic illumination were developed into a discipline of great beauty and originality; new architectural forms were assimilated from a variety of pre-Islamic sources,[26] bringing ever-greater diversity to the Islamic architectural tradition; and aesthetic innovation within the ornamental arts benefited greatly from patronal attention. Architectural ornament was a primary beneficiary of this commitment to innovation: both in terms of an increased availability to wider range of materials and fabricating technologies, and in the ever-expanding diversity of decorative motifs and themes. This included the development of the incipient tradition of Islamic geometric star patterns during the ninth century. Over the course of some 300 years, this design tradition developed to its full maturity, characterized by exceptional versatility, great beauty, unparalleled geometric ingenuity, and pan-Islamic appreciation.

It is generally agreed that the sophisticated culture of Baghdad was central to the initial development of Islamic geometric star patterns. Even with the early rise in prominence of other early centers of Muslim culture such as Córdoba, Cairo, Shiraz, Nishapur, Bukhara, and Merv the preeminence of Baghdad as the seat of Abbasid religious authority and cultural influence remained undisputed. While the surviving brickwork, woodwork, and stucco ornament from such widespread locations as Kairouan, Cairo, Balkh, Na'in, Tim, Qala-i-Bust, Uzgen, Damghan, and Kharraqan provide some of the best evidence of the early development of Islamic geometric star patterns, the broad distribution of so many stylistically similar examples during the same approximate period argues for the centrality of Baghdad as the principle place of origin and dissemination for this discipline.[27] Furthermore, knowledge of the importance of Baghdad in the historical development of other allied and highly influential artistic traditions is well known. Notable examples include the calligraphic innovations of Abu 'Ali Muhammad ibn Muqlah (886-940), the inventor of the geometric system of calligraphic proportion that was critical to lifting this tradition to the level of fine art[28], the development of the highly distinctive beveled style of floral ornament (Samarra style C) that appears to have originated in nearby Samarra and was used widely throughout the vast regions of Abbasid influence[29], and the technically sophisticated lusterware ceramics that also developed in and around Samarra.[30] The case can similarly be made for Baghdad as an important center in the ongoing development of ornamental brickwork. Relatively little architecture survived the succession of Mongol invasions during the thirteenth and fourteenth centuries and Baghdad did not escape this destruction. While most extant pre-Mongol ornamental brickwork architecture is found in the regions of Persia, Khurasan, and Transoxiana, the fact that older, albeit less complex, examples of ornamental brickwork façades are found in locations near Baghdad supports the theory that this tradition grew out of the cultural vitality of Baghdad, and was disseminated from there to regions under Abbasid influence.

Another case for Baghdad as the principal place of origin in the development of Islamic geometric star patterns is the central importance of Baghdad in the study of mathematics and geometry during this period. These disciplines were provided a practical emphasis in such areas as geography, land surveying, navigation, taxation, commerce, and the arts. Abu al-Wafa al-Buzjani (940-998) was a leading mathematician of his time. As a young man he moved from Khurasan to Baghdad where he lived the remainder of his life. He is best known for his work with plane and spherical trigonometry. More prosaically, al-Buzjani was also concerned with

---

[26] For a detailed analysis of pre-Islamic influences upon the development of early Islamic architecture, see Hillenbrand (1994a).

[27] Necipoğlu (1995), 99–100.

[28] Schimmel (1990), 18–19.

[29] Creswell (1969), 75–79.

[30] Caiger-Smith (1985), 21–31.

the practical application of mathematics and geometry,[31] and is associated with the work *About that which the Artist needs to Know of Geometric Constructions*.[32] This work details practical solutions to geometric problems posed by members of the professional classes, including people working in the arts. Perhaps most significantly, the Abbasid caliph al-Mu'tadid (r. 892-902) founded a royal atelier within his palace dedicated to the furtherance of theoretical and practical sciences and their application to diverse artistic disciplines.[33] It was during this approximate period that geometric star patterns began to emerge as a distinct ornamental aesthetic. It is reasonable to speculate that this interaction between mathematicians and artists may have played an influential role in the development of the methodologies required in the construction of complex geometric star patterns. Certainly the place and time are significant.

The fact that most extant examples of early Islamic geometric star patterns are architectural should not lead one to conclude that this ornamental discipline developed solely as part of the architectural traditions of Islam. The book arts, and specifically the concerted attention paid to Quranic illumination, appear to have also played a significant role in the progressive development of Islamic geometric patterns. It is regrettable that so few Qurans have survived from the early formative period of this ornamental tradition, and knowledge of the degree of interplay between geometric artists working on Qurans and those working on buildings is limited to conjecture. However, the cultural centrality and royal patronage of this tradition, coupled with the few evidentiary examples that have survived, do indeed indicate the likelihood that these artists were involved in the development of geometric patterns. During the ninth and tenth centuries, the art of Quranic ornamentation evolved from simple border devices, emphasizing *surah* headings and *ayah* markers, to fully illuminated pages. The work of the great calligrapher ibn Muqla (d. 940) is an example of the successful application of geometric principles to the arts of the Baghdadi cultural milieu. He is known to have studied geometry, and his prescribed use of geometric proportion to perfect the cursive scripts profoundly influenced the trajectory of Islamic calligraphy. Preeminent among artists, calligraphers would have participated in the exploratory exchanges between scientists and other artists, and it is certainly possible that calligraphers were involved in discussions that may have assisted in the development of geometric star patterns. Significantly, the first such pattern

Photograph 6 A frontispiece comprised of two varieties of octagon from a Quran produced in Baghdad by Ibn al Bawwab (© The Trustees of the Chester Beatty Library, Dublin: CBL Is 1431, ff. 7b-8a)

known to have been created by a specific individual is from a matching set of illuminated frontispieces from the celebrated Baghdad Quran produced by 'Ali ibn Hilal, better known as ibn al-Bawwab (d. 1022). Like ibn Muqla, he is regarded as one of the great masters of Arabic calligraphy. His Baghdad Quran was produced in 1001 and includes several illuminated pages that are believed to be his own creations. Most of these illuminations are compass-work creations, but the matching frontispieces employ a beautifully executed geometric pattern comprised of a network of interweaving lines that create a series of large and small octagons[34] [Fig. 127c] [Photograph 6]. The pattern that ibn al-Bawwab incorporated into his surviving Quran is an alternative treatment of the well-known, and less complex, design comprised of octagons touching corner to corner. Creswell has written of the use of this less complex octagonal pattern in pre-Islamic architecture,[35] and cites the example of a ceiling coffer from the Great Temple of Palmyra

---

[31] Necipoğlu (1995), 123.

[32] *Kitab fima yahtaju ilayhi al-sani min a'mal al-handasa*, MS Persan 169, sec. 23, folios. 141b–179b, Bibliotheque Nationale, Paris.

[33] –Özdural (1995), 54–71.
    –Necipoğlu (1995), 123.

---

[34] Chester Beatty Library Ms. 1431, fol. 7b–8a.

[35] Creswell (1969), 77.

(c. 36). This simple octagonal design from Palmyra is easily constructed by iteratively applying octagons in a corner-to-corner orientation upon an orthogonal grid. This same basic octagonal pattern was widely used by many generations of Muslim artists, and in keeping with the rise to dominance of the polygonal technique within this pattern tradition, it can be conveniently produced through the application of pattern lines onto an underlying 4.8^2 tessellation of squares and octagons [Fig. 124c]. However, prior to the earliest known use of this very simple octagonal design by Muslim artists, ibn al-Bawwab had incorporated his more complex version into his celebrated Quran. By contrast, the added complexity of the pattern produced by ibn al-Bawwab is not so easily produced via simple iteration, and is more readily created from the underlying 4.8^2 tessellation. This more complex pattern can be extracted from the tessellation in either of the two pattern line arrangements [Figs. 127c and 128d]. His use of this design strongly suggests that ibn al-Bawwab was knowledgeable of the advances in geometric design methodology generally, and quite possibly the polygonal technique specifically, that were taking place during this period.

As with calligraphy and illumination, bookbinding also received decorative emphasis during the Abbasid period. Paper technology was introduced from China, allowing for books to be lighter weight than either parchment of papyrus. This new material was less susceptible to the adverse effects of humidity, providing greater technical efficiency and allowing for lighter bindings. Abbasid artists developed bindings that were made from leather-covered pasteboard. From as early as the ninth century, the leather coverings were decorated with blind tooling: a process whereby the leather binding was dampened and stamped with metal tools and dies, and burnished to completion. A surviving Aghlabid example of an early interweaving geometric pattern being used to decorate such a binding is from a ninth-century Quran in the library of the Great Mosque of Kairouan. The design is interesting in that it is an embossed leather representation of an ancient cane weave that is still used to this day in the furniture industry, wherein it is known as the standard cane weave. The geometric structure of this design is produced from a four-directional weave made up of parallel interweaving double lines in the vertical and horizontal directions, and over-under single lines in the diagonal directions. The interweaving lines create regular octagons that are located upon the vertices of the repetitive orthogonal grid. An unusual feature of this design is the nonuniform structure of the interweave, wherein the individual lines will skip over-over-under-under, rather than over-under-over-under: the standard of this tradition. In this respect, the bookbinding design is faithful to the cane weave. The interweaving aesthetic of this design is nevertheless similar to that of later Islamic geometric star patterns, and indeed, the geometric structure of this cane weave design relates directly to the classic star-and-cross design [Fig. 124b]: the

difference being that the horizontal, vertical, and diagonal lines from the Aghlabid book binding are continuous, and are widened to their maximum extent. This is similar to a design from a wooden ceiling at the Alhambra, but without the small arbitrarily included eight-pointed stars [Fig. 126b].

The aesthetic similarity between the pattern on the Aghlabid bookbinding and later Islamic geometric ornament is indicative of an emerging aesthetic orientation that took form under the auspices of Abbasid patronage during the ninth and tenth centuries. Among the earliest extant examples of Islamic geometric star patterns are the multiple pierced wood panels from the *minbar* (c. 856) at the Great Mosque of Kairouan. This *minbar* was manufactured in Baghdad and exported to North Africa. The sides of this *minbar* are a veritable cornucopia of early geometric design, and provide the best surviving evidence for the emerging geometric aesthetic of early Abbasid Baghdad. Among the diverse multitude of designs are key patterns, compass-work patterns, polygonal tessellations, and several panels with very basic prototypical star patterns. Each of these is very simple compared with the characteristic complexity that eventually became a hallmark of this tradition. The star patterns from the Kairouan *minbar* all have eight-pointed stars as their central feature, and repeat upon the square grid. One of these is a particularly early occurrence of the classic star-and-cross design that went on to become the most ubiquitous geometric star pattern throughout the Islamic world [Fig. 124b]. As discussed, using the polygonal technique, this classic orthogonal design can be easily created from the 4.8^2 tessellation of squares and octagons.

Being that the region of greater Baghdad during the ninth and tenth centuries was central to the development of several significant artistic traditions, that the arts were informed by mathematics and geometry under royal patronage, and that significant artistic trends and objects were exported from this region to diverse regions of Abbasid influence, it appears likely that the cultural milieu of Baghdad provided the impetus for the development of Islamic geometric star patterns, and that knowledge of this incipient tradition was dispersed widely from this region throughout the Islamic world. However, the few remaining Abbasid buildings from the region of Baghdad that date from the early formative period are devoid of geometric star patterns, and the Kairouan *minbar* notwithstanding, it is impossible to know categorically the extent of methodological knowledge of the polygonal technique enjoyed by the artists working in Baghdad during this period.

1.5   Tulunids (868-905)

In 868 Ahmad ibn Tulun, originally from Bukhara, was sent with an army from Iraq to Egypt to be deputy to the viceroy of Egypt. Within a short time he became the Abbasid governor of Egypt and Syria, founding the Tulunid dynasty. While only ruling until 905, the Tulunids had a significant influence on subsequent Egyptian ornament. The mosque of ibn Tulun (876-79) is located in Fustat, Egypt (now part of greater Cairo). The geometric patterns at the ibn Tulun mosque are collectively the most significant ninth-century examples from the western regions of Abbasid influence. Of particular note are the soffits of the multiple arches that surround the large courtyard. These are decorated with a wide variety of patterns, including compass-work, polygonal tessellations, and early examples of geometric star patterns. Like the Great Mosque of Kairouan, the ornament in the ibn Tulun mosque was transitional: incorporating elements from the earlier Umayyad period with more contemporary ornamental devices such as the Samarra floral styles and simple geometric star patterns. Several of these are noteworthy in that they are among the earliest extant Islamic geometric patterns to have threefold symmetry, although this cannot be regarded as innovative in that such patterns are known from the pre-Islamic architecture of the Byzantines and Sassanians. One of these soffit designs is comprised of interweaving rhombi and hexagons and can be regarded as pure polygonal ornament [Photograph 7]. One of the distinctive characteristics of this pattern is the six-pointed star motif that is placed upon the vertices of the isometric grid. This pattern of hexagons and six-pointed stars (without the rhombic emphasis) was used with great frequency throughout the Islamic world in succeeding centuries; so much so, in fact, that it can be regarded as the classic threefold pattern. This classic design is easily constructed using the polygonal technique from the hexagonal grid as the underlying generating tessellation [Fig. 95b]. As said, many of the simple threefold geometric patterns that are characterized by 60° and 120° angles can be constructed from either the *system of regular polygons*, by the simple assembly of design elements, or by simply tracing over the isometric grid. However, as this tradition developed, the increase in complexity required a constructive methodology that surpassed the limitations of simple grid tracing or assembly, but were amply met with the polygonal technique. One of the more intriguing soffit designs at the ibn Tulun mosque uses a 3.6.3.6 tessellation of triangles and hexagons as polygonal ornament, with interweaving circles of greater line thickness located at the centers of each hexagon [Photograph 8]. As each circle approaches the center of a hexagon, its curvature is tweaked toward this center, creating a distinctive flower with six petals. This beautiful design has an innovative playfulness that qualifies it as one of the outstanding ninth-century examples from this burgeoning tradition. The conceptual similarity of this design to two of the window grilles from the Great Mosque of Damascus and the one from Khirbat al-Mufjar is striking: all include the 3.6.3.6 polygonal tessellation with added circular elements

Photograph 7 A Tulunid threefold pattern with six-pointed stars from a carved stucco arch soffit at the Ibn Tulun Mosque, Cairo (© David Wade)

positioned at key points within the geometry of the tessellation. The straight-line component of this composition is equally attributable to either of the two categories: the *system of regular polygons* [Fig. 95d], or polygons as pattern. Several additional curvilinear patterns were used on the soffits of the ibn Tulun mosque, including one with threefold symmetry that was also constructed from the 3.6.3.6 tessellation: although this example does not include the tessellation as part of the finished design. The swing points for the compass were conveniently left as a subtle design feature, providing evidence of the 3.6.3.6 tessellation having been used to create the design. The center point of the semicircular pattern line is located upon the center point of each polygonal edge of the underlying 3.6.3.6 tessellation, and the radius of each curved pattern line is equal to half the polygonal edge length.

Several different *fourfold* designs that are easily constructed with the 4.8^2 underlying tessellation of squares and octagons are found at the ibn Tulun mosque, including

Photograph 8 A Tulunid threefold pattern with six-pointed stars from a carved stucco arch soffit at the Ibn Tulun Mosque, Cairo (© David Wade)

Photograph 9 A Tulunid fourfold pattern with eight-pointed stars that can easily be created from the 4.8^2 tessellation of squares and octagons from a carved stucco arch soffit at the Ibn Tulun Mosque, Cairo (© David Wade)

the classic star-and-cross pattern from one of the carved stucco arch soffits [Fig. 124b]. This pattern was contemporaneously used by Yu'firid artists in the ceiling ornament of the Great Mosque of Shibam Aqyan near Kawkaban in Yemen (pre-871-72). Another of the arch soffits at the ibn Tulun mosque is decorated with a distinctive interweaving design of eight-pointed stars and two sizes of square in the background [Fig. 129a] [Photograph 9]. This design is easily created by drawing the eight-pointed stars from vertex to vertex within the octagons of the underlying tessellation, resulting in a pattern composed of two sizes if interweaving squares that surround the eight-pointed stars. As noted, the methodology of the polygonal technique typically places the pattern lines upon the midpoints, or upon two points, of each edge of the generative polygons, and the use of polygonal vertices is relatively rare, and when found is almost always associated with the early formative period. The third design from the ibn Tulun arch soffits that is easily created from the

underlying 4.8^2 tessellation of squares and octagons is a compass-work design comprised of interweaving curvilinear pattern lines. This design replaces the 90° angles in the eight-pointed stars of the standard star-and-cross pattern with a network of *s*-curves that weave together to create eight-lobed rosettes within each underlying octagon, and full circles within each underlying square and at the center of each eight-lobbed rosette.

1.6   Umayyads of al-Andalus (756-929)

Muslim conquerors first landed in Spain in 711, and by 714 had wrested control of the greater portion of the Iberian Peninsula from the Christian Visigoths. Until the arrival of Abd er-Rahman I in 755, Islamic Spain was ruled by an assortment of governors under the authority of the Umayyad Caliphate in Damascus. Following the Abbasid overthrow of the Umayyads in Syria, the heirs to the Umayyad Caliphate were rounded up and executed. The only survivor was Abd er-Rahman I, the grandson of the last Umayyad Caliph. Along with many of his Syrian supporters, he successfully escaped to the Iberian Peninsula where he was accepted as the Amir in 756. This continuation of the Umayyad dynasty in al-Andalus was the beginning of one of the great epochs of Islamic civilization. Centered in Córdoba, this dynasty was to rule over most of the Spanish peninsula for over 250 years. In 929 the eighth Umayyad Amir, Abd er-Raham III, declared himself Caliph, directly challenging the authority of the Sunni Abbasids in Baghdad, and the Shi'a Fatimids in North Africa and Egypt.

The Great Mosque of Córdoba was founded by Abd er-Rahman I between 784 and 786, and expanded by the Umayyad Caliph al-Hisham II and his minister al-Mansur between 987 and 990.[36] This expansion included the introduction of a number of marble window grilles. In addition to the purely functional benefits of these grilles, their incorporation into this mosque may have served as a homage to the cultural greatness of their Umayyad ancestors from Syria, and more specifically to the window grilles of the Great Mosque of Damascus. These Iberian window grilles are designed in a variety of geometric styles, and included patterns easily created with the polygonal technique. One of these is a threefold pattern that can be produced from either the 6^3 hexagonal grid [Fig. 96e] or the 3.6.3.6 underlying tessellation [Fig. 99e], although the precise proportions of this example relate to the former generative schema. The earliest known example of this pattern is from one of the original window grilles at the al-Azhar mosque in Cairo (970-76). Later historical examples for the use of this pattern include one of the raised brick panels on the exterior of the Seljuk eastern tomb tower of Kharraqan (1067) [Photograph 17]; a pierced wood screen on the Seljuk *minbar* at the Friday Mosque of Abyaneh (1073); and a Fatimid pierced marble grille from the al-Aqmar mosque in Cairo (1125) that is stylistically identical to the earlier example from Córdoba. Two other surviving window grilles from Córdoba are in the collection of the Museo Arqueológico de Córdoba. These are thought to be from the same period as the examples from the Great Mosque of Córdoba, and possibly from the same workshop.[37] Like the above-cited window grille from the Great Mosque of Córdoba, one of these has the distinctive

quality of being made up of interweaving superimposed hexagons that are placed upon the isometric grid, yet their design characteristics are noticeably distinct from one another. And like the example in the Great Mosque of Córdoba, the isometric window grille from the Museo Arqueológico is easily created from the 6^3 grid of regular hexagons [Fig. 96d], and is conceptually the same, but with different proportions, to a pattern created from the 3.6.3.6 arrangement of triangles and hexagons [Fig. 99c]. The simple and easily discerned arrangement of interweaving hexagons is responsible for the compelling beauty of this pierced marble window grille, and it is not surprising that it was also used frequently throughout the history of Islamic ornament, including the Ghurid portal at the Friday Mosque at Herat (1200). The second marble window grille from this workshop in Córdoba places eight-pointed stars upon the orthogonal grid, and, like the previously referenced example from the ibn Tulun mosque, locates the pattern line upon the vertices of the underlying octagons within the 4.8^2 tessellation of squares and octagons. This design is methodologically identical to a later Ghurid raised brick panel on the exterior of the western mausoleum at Chisht, Afghanistan (1167) [Fig. 129c]. However, the design from Córdoba is differentiated by the additive inclusion of two varieties of semicircular scallops incorporated into the otherwise uninterrupted straight lines of the pattern. These scallops serve several functions: by touching their neighbors, they provide added structural integrity to the pierced marble; they open up an otherwise dense area of the design where three interweaving lines would otherwise touch at a single point (as per the example in Chisht); and their curvilinear quality adds visual dynamism to the overall design. Additional examples of this design, sans scallops, are found on the Ghurid minaret of Jam in central Afghanistan (1174-75 or 1194-95), and the *minbar* of the al-Aqsa mosque in Jerusalem (1187). Another example of a design from the Great Mosque of Córdoba that can be created from the underlying 4.8^2 tessellation is from the celebrated tessera mosaic *mihrab* (971) that was ordered by al-Hashim II [Fig. 126c]. This is a variation of the star-and cross design with curvilinear four-pointed stars within the underlying squares, and a second eight-pointed star within the primary eight-pointed star.

1.7   Abbasids in the Eastern Provinces

During the period when the tradition of geometric star patterns was advancing to full maturity, Persia, Khurasan, Sindh, and Transoxiana were beset with political turmoil. Out of this turmoil rose and fell a series of great empires. The Buyids wrested control over Persia and Iraq from the Abbasids, placing them in direct confrontation with the

---

[36] King Ferdinand III of Castile converted the Great Mosque of Córdoba into a cathedral in 1236.

[37] It is speculated that these two window screens may have been made for a country residence outside Córdoba: possibly that of al-Mansur. See Dodds [ed.] (1992), 252.

Samanids in Transoxiana and Khurasan. The Samanids suffered defeats at the hands of the Qarakhanids and Ghaznavids, the latter of whom were eventually defeated in turn by the Ghurids and Seljuks. The Seljuk overthrow of the Buyid dynasty, and the liberation of the caliphate in Baghdad, brought on the Sunni Revival. Following the defeat of the Buyids, the Seljuk sphere of influence spread across Persia into the Caucasus and Anatolia, al-Jazirah, Mesopotamia, much of Syria, and the Levant. The Qara Khitai defeated the Seljuks in the northern regions of Transoxiana, only to be overrun by the Khwarizmshahs who went on to defeat the last of the Great Seljuks and Ghurids, consolidating control over an empire that stretched across Persia and Khurasan, and northward across the vast regions of Transoxiana. Soon after their multiple victories, the Khwarizmshahs fell to the Mongol onslaught in the thirteenth century. Yet throughout this history of military conquest and political upheaval great cities thrived, intercontinental trade continued, great wealth was amassed, and the arts and sciences flourished. It was in this tumultuous yet culturally refined environment that the development of Islamic geometric star patterns matured beyond the simplistic modalities that were characteristic of the patterns used in the *minbar* of the Great Mosque of Kairouan and the arch soffits at the ibn Tulun mosque. Over time, the advances made in the eastern regions were disseminated throughout the Islamic world, where they were, in turn, readily incorporated into the palette of ornamental themes and applied to a broadening range of materials and techniques.

Following the defeat of the Umayyads, the Abbasids soon came under increased pressure to more effectively govern the vast regions of their empire by founding the more centrally located capital of Baghdad. As in North Africa and Egypt, in Persia, Khurasan, and Transoxiana, governorships were granted, leading to several powerful semiautonomous vassal states. During the ninth century, Abbasid suzerainty over its eastern provinces began to break apart. This challenge to the Abbasid authority in Baghdad did not stem solely from the desire for independence, but, in many cases, was driven by fundamental religious differences. The Mu'tazilite reform doctrine of the created Quran caused deep schisms within Abbasid Sunni orthodoxy; and further pressure resulted in the growing Shia movement that regarded the descendants of the Prophet Muhammad through the line of 'Ali ibn Abi Talib, the Prophet's son-in-law, as the only legitimate heirs to the caliphate. The erosion of the Abbasid dynasty in the ninth and tenth centuries led to the rise in importance of various regional centers; and this was to have a profound effect on the history and development of Islamic architecture and ornament.

The uncertainty as to exactly when, where, and under what circumstances the polygonal technique for creating geometric patterns originally developed is compounded by

the fact that the simplicity of the earliest geometric star patterns allows for their creation by other generative techniques beyond just the polygonal technique. The degree of overlap between competing methodologies, and the point at which the polygonal technique assumed its role as the preeminent design methodology, is, therefore, impossible to definitively determine. Regardless of whether this seminal design methodology first originated and possibly matured in and around Baghdad,[38] the architectural record clearly indicates that the earliest extant mature expression of this ornamental tradition is found in the eastern regions of the Islamic world. If the maturity of Islamic geometric patterns corresponds with the surviving architectural record, and indeed occurred in the eastern provinces, this shift in creative vitality would have paralleled the waning influence of the Baghdadi caliphate in the face of the de facto independence of those outlying regions that had previously come under direct Abbasid control.

Like the ibn Tulun mosque, the surviving architecture in the eastern regions of Abbasid suzerainty provides some of the best indications for the early use of the polygonal technique as a generative methodology during the initial developmental period of geometric star patterns. The ruins of the No Gumbad mosque in Balkh, Afghanistan (800-50), are extensively ornamented with carved stucco geometric and floral designs. Among the many ornamental motifs is an example of the classic star-and-cross design with eight-pointed stars at each vertex of the orthogonal grid [Fig. 124b] [Photograph 10]. It is significant that the use of this design at the No Gumbad mosque is contemporaneous with its use on the wooden *minbar* at the Great Mosque of Kairouan. Clearly, ninth-century Abbasid ornamental conventions disseminated quickly throughout their vast territories; helping to create an ornamental style that, while engendering distinct regional variations, nonetheless exhibited remarkable aesthetic cohesion. The No Gumbad mosque was built during the same approximate period as the floral examples found in excavations of the Bab al-'Amma (836-7) and the Bulawara Palace (849-59) in Samarra, Iraq. Both the floral infill of the geometric designs at the mosque of ibn Tulun in Egypt and the carved stucco floral infill designs in the No Gumbad mosque have much in common with the Samarra style B floral designs; providing added evidence of the rapid dissemination of newly developed ornamental innovations throughout Abbasid territories. The ninth-century incorporation of the Samarra floral conventions in regions as far flung as Cairo and Balkh, as

---

[38] For a detailed exposition on the importance of Baghdad in the development of Islamic science and mathematics, and the influence of these developments upon the origins of the geometric design idiom, see Necipoğlu (1995), 131–166.

Photograph 10 An early Abbasid example of the classic star-and-cross pattern at the No Gunbad in Balkh, Afghanistan (Horst P. Schastok photograph, courtesy of Fine Arts Library, Harvard University)

well as the utilization of the star-and-cross pattern during the same period in both the east and west, strongly supports the argument for the centrality of Baghdad in this process of dissemination. However, the surviving architectural record strongly indicates that the early developmental innovations and maturation of Islamic geometric star patterns took place primarily in the eastern regions of Khurasan and eastern Persia and spread westward during the period of Seljuk expansion.

Advances in the tradition of Islamic geometric patterns must be regarded in the context of the brickwork ornament of the eastern dynasties. The Persian term for this brickwork ornament is *banna'i*, or *work of brick builders*.[39] The Persian

term *hazarbaf* for woven rush matting is occasionally applied to brickwork when the design resembles this variety of interweaving woven structure. The earliest examples of Islamic ornamental brickwork are found in present-day Iraq, and make use of very simple geometric motifs that rely upon the rectilinearity of the brick module. Among the earliest surviving examples are the Baghdad Gate in Raqqa (772) and the Court of Honor at the desert palace of Ukhaidir (c. 764-778), some 120 miles south of Baghdad. Both these examples employ simple brickwork designs such as chevrons and swastikas inside a series of horizontally aligned blind arches. Of particular interest is the minaret of Mujda (mid-eighth century), situated between the two Abbasid palaces of Ukhaidir and Atshan. While little of this minaret still stands, and while the geometric brickwork is very basic, it is remarkable for its conceptual similarity to the beautiful ornamental brick minarets produced by the Ghaznavids and Seljuks some 300 years later.[40] In the eastern regions, the rise in sophistication of ornamental brickwork began with the Samanids and Buyids, and can be seen in such buildings as Samanid mausoleum in Bukhara (c. 914-43) and the Jurjir mosque in Isfahan (976-85). The rival Qarakhanids and Ghaznavids built upon the brick-building heritage of their predecessors; and the Ghaznavids in particularly were especially innovative in their use of this medium. Artists working for both of these dynasties pioneered the application of geometric star patterns to brickwork ornament. The three adjoining mausolea in Uzgen, constructed between 1012 and 1186, exhibit some of the finest Qarakhanid geometric ornament, while the brick minaret Mas'ud III in Ghazna, Afghanistan (1099-1115), has some of the most sophisticated geometric patterns of its period. Their Seljuk and Ghurid successors further expanded upon this decorative device, creating works of exceptional beauty and originality in such buildings as the northeast dome chamber of the Friday Mosque at Isfahan (1088-89) and the minaret of Jam in central Afghanistan (1174-75 or 1194-95). As this tradition matured the variety of patterns employed became increasingly diverse and complex. What began as simple key patterns and interlocking devices that firmly adhered to the 90° orthogonal angularity of the brick module transformed into an ornamental medium with tremendous design flexibility. The repertoire of the brick artist was expanded to include cast ceramic inserts, often with either a glazed or an unglazed decorative relief, as well as specially cut or specially molded bricks that allowed them to break free of the orthogonal rigidity that otherwise constrained this medium. In this way, the rise in technical

---

[39] The Persian terms *hazarbaf* and *parceh* are also used for brickwork ornament. It is interesting that these terms are also associated with the woven rush matting and textile industries.
    –Wolff (1966), 118.
    –Creswell (1969), 186.

[40] Hillenbrand (1994a), 144.

mastery of ornamental brickwork in Khurasan and Persia during the tenth, eleventh, and twelfth centuries provided an ideal vehicle for the growth in complexity of geometric patterns with angles other than 90°. This was equally the case for the brickwork application of cursive calligraphic scripts, increasingly elaborate forms of knotted and floriated *Kufi*, and the floral idiom. Added to this integral evolution of ornamental motifs and materials was the revival of glazed ceramic faience, and the continuation of carved stucco, carved stone, carved wood, and, less commonly, fresco painting. All of these architectural media were exceptionally well suited to the burgeoning tradition of geometric star patterns.

1.8   Samanids (819-999)

The Samanid Empire was founded with the appointment of four brothers to rule over the regions of Samarkand, Ferghana, Herat, and Shash (Tashkent) by the Abbasid Governor of Khurasan in 819. These brothers were granted their positions of leadership as an award for their military support in putting down a revolt against the caliph al-Ma'mun. The Samanid Empire reached its political and cultural height during the reign of Isma'il Samani (892-907). During this period the Samanids controlled a vast region that included most of modern-day Afghanistan, the eastern half of Iran, parts of Pakistan, and much of Tajikistan, Kyrgyzstan, Uzbekistan, Kazakhstan, and Turkmenistan. The Samanids originated from the region of Balkh, and were strict adherents of Sunni Islam. At the height of power they ceased paying tribute to the Abbasids, but continued to recognize the religious authority of the caliph in Baghdad. Like the Saffarids whom they vanquished, the Samanids revived Persian language and culture. Their first capital was Bukhara, and their principal cities were Samarkand, Herat, and Nishapur. Bukhara in particular became a great cultural center of learning and the arts, rivaling Baghdad, Cairo, and Córdoba. Such luminaries as 'Ali Sina Balkhi (Avicenna), Muhammad al-Bukhari, Rudaki, and Ferdowsi received patronage from the Samanid court. Clearly this was a highly sophisticated culture where religion, sciences, and arts flourished. Indeed, it was during the Samanid period that Nishapur became one of the great Islamic centers of ceramic art. The few Samanid buildings that have survived to this day give clear indication that this was also an architecturally innovative period, and written accounts from this period make repeated reference to the architectural wonders and great ornamental beauty of these early Islamic buildings.

Excavations of a private residence at Sabz Pushan outside Nishapur revealed a number of finely carved stucco panels dating from 960 to 985 during the period of Samanid rule over Nishapur.[41] These include a threefold geometric panel that is easily constructed from the 6^3 grid of underlying regular hexagons[42] [Fig. 96c] [Photograph 11]. This same pattern was used a century later by Seljuk artists at the eastern tomb tower at Kharraqan (1067-68). It is interesting to note that, like the two isometric window grille designs from Córdoba, this contemporaneous design is also made up of superimposed hexagons. The floral infill designs from Sabz Pushan are derivative of the Samarra style C—the beveled style. The carved stucco geometric ornament from Sabz Pushan also includes an interweaving classic star-and-cross linear border design [Fig. 124b]. The *pishtaq* of the mausoleum of Arab Ata (977-78) at Tim, Uzbekistan, 85 km southwest of Samarkand, employs several geometric designs in incised stucco, including a typical key pattern, an example of polygons as pattern, and two threefold geometric patterns easily constructed from the *system of regular polygons*. This Samanid building is also noteworthy for having the earliest extant example of a trilobed squinch. This particularly attractive solution to the structural challenge of placing a circular dome upon a square chamber became a regular feature of Seljuk brick architecture, achieving its apogee in the northeast and southwest domes of the Great Mosque of Isfahan (1072-92). This architectonic device is thought to be an important influence upon the development of *muqarnas* vaulting. The ornament of the mausoleum of Arab Ata is considerably more sophisticated than that of its better known predecessor, the Samanid mausoleum in Bukhara[43] (c. 914-43). This earlier example of Samanid funerary architecture is remarkable for its simple beauty wherein the entire surface of both the interior and exterior walls is replete with decorative brickwork. While the individual designs are, in and of themselves, very simple, the overall effect is of a wholly ornamented building. Such abundant use of ornamental texture was soon to become a predominant characteristic of Islamic architecture in the eastern regions. The mausoleum of Arab Ata, by contrast, limits its ornament to the front façade of the *pishtaq*, and employs geometric designs that are noticeably more complex. Of especial interest is the geometric pattern created from the *system of regular polygons* set within the arched tympanum above the entry *pishtaq* [Photograph 12]. This very successful design is conveniently created from the basic 6^3 underlying generative tessellation [Fig. 96g]. This same design was used by the Seljuks on both the eastern tomb

---

[41] Blair (1991), 55.

[42] This panel from Sabz Pushan, Nishapur, is now in the permanent collection of the Metropolitan Museum of Art, New York: Accession Number 40.170.442.

[43] The Samanid Mausoleum is also known as the Tomb of Ismail the Samanid.

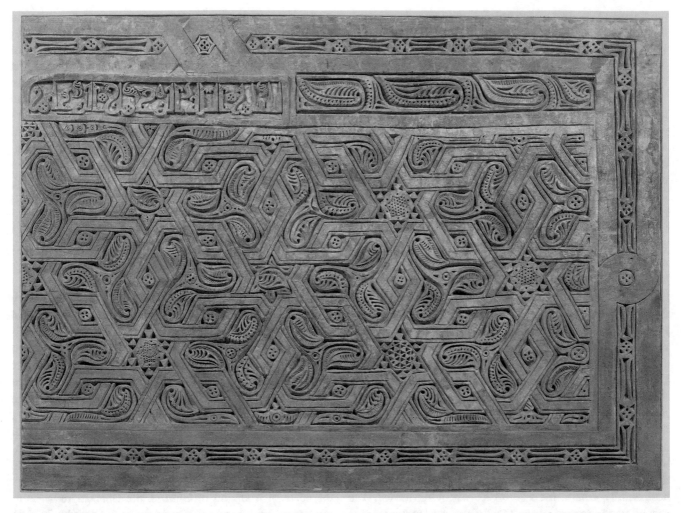

Photograph 11 A Samanid period carved stucco panel with six-pointed stars that is easily created from the *system of regular polygons* that was found at the Sabz Pushan excavation near Nishapur, Iran (The Metropolitan Museum of Art: Rogers Fund, 1940: www.metmuseum.org)

tower at Kharraqan, Iran (1067-68), and the minaret in Daulatabad outside Balkh, Afghanistan (1108-09), and by the Seljuk Sultanate of Rum at the Izzeddin Kaykavus hospital and mausoleum in Sivas (1217). The other isometric design used on the front façade of the mausoleum of Arab Ata can likewise be created from the 6^3 underlying tessellation [Fig. 95d], but is, in and of itself, a widened line version of simple 3.6.3.6 polygonal tessellation [Fig. 89]. This is one of the most widely used threefold patterns, with one of the earliest examples found at the ibn Tulun mosque in Cairo (876-79).

1.9   Buyids (945-1055)

While the Saminids were flourishing in the regions of Khurasan and Transoxiana, the rise of the Shia Buyids had tremendous impact upon the political and military authority of the Abbasid Caliphate. The Buyids were a Persian tribe originally from the mountainous region of Daylam, south of

Photograph 12 A Samanid threefold pattern with six-pointed stars easily created from the *system of regular polygons* in the tympanum over the *pishtaq* of the mausoleum of Arab Ata in Tim, Uzbekistan (© Bernard O'Kane)

the Caspian Sea. Around the year 932 they set out to conquer large areas of central Persia, and by 945 had conquered Baghdad, most of Iraq, Oman, and parts of Syria. Under the Buyids, the temporal power of the Abbasid Caliphate was reduced to a position of political subjugation.

Like the Saminids to the northeast, the Buyids were greatly influenced by earlier Persian culture. Under Buyid patronage, their capital cities of Isfahan and Shiraz became important centers of Islamic culture, with a great emphasis on the arts and architecture. Adud ad-Dawla, who reigned between 936 and 983, was a great patron of the arts and learning, and was reputed to have been an avid calligrapher. He was a prolific builder, and is reported to have ordered the construction of 3000 mosques in his lifetime, although this must certainly be an exaggeration.[44] Among the many architectural achievements of Adud ad-Dawla was his palace in Shiraz. This no longer exists, but was said to have 360 rooms, one for each day of the year, and each decorated in a differing style. For all the monumental architecture built by the Buyids, regrettably little has survived to the present.

The Friday Mosque at Na'in, Iran (960), was built by the Buyid Dynasty some 25 years after their seizing control of Baghdad. This mosque includes two carved stucco geometric star patterns that can be produced from the 4.8^2 tessellation of squares and octagons. One is a bold linear band treatment of the classic star-and-cross design [Fig. 124b] with floral background infill that is similar in concept to the Samarra style A. This linear border runs vertically and horizontally around the *mihrab* as a framing device. The second wraps around one of the circular supporting piers in the prayer hall and is interesting in that the interweaving pattern lines are curvilinear [Fig. 127b]. The fact that the two examples from the Friday Mosque in Na'in share the same generative polygonal tessellation would not appear to be coincidental, and are certainly not the only architectural examples where two or more patterns are placed in close proximity that share the same underlying generative tessellation. The curvilinear treatment of the design on the pier has a softening effect on the rigid angularity that is otherwise a standard feature of this tradition. This is an early curvilinear design easily produced from the polygonal technique, and while known in the work of succeeding Muslim cultures, the allure of such designs is augmented by virtue of their rarity.

The degree of Buyid involvement in the development of Islamic geometric patterns is difficult to establish due to the paucity of ornamental examples that have survived from this dynasty. The most significant example of Buyid involvement in the maturation of this geometric tradition is the above-cited illuminated frontispiece from the celebrated Quran created by ibn al-Bawwab [Figs. 127c and 128d] [Photograph 6]. This design of interweaving octagons set upon the orthogonal grid is significant in several respects: as an indication of the relevance of the book arts in the early development of Islamic geometric patterns; as an indication of the ongoing importance of Baghdad to the development of this tradition; and as an early example of a geometric pattern that likely employed the polygonal technique in its creation. While the Quran of ibn al-Bawwab was certainly a product of the Abbasid cultural continuity associated with Baghdad, the Buyid patronage of ibn al-Bawwab is nonetheless significant.

1.10　Ghaznavids (963-1187)

The Ghaznavid dynasty was founded by Turkic military commanders of the Saminids who, in 977, took control of the Samanid territories in Afghanistan, setting up their capital in Ghazna. While politically autonomous, as staunch Sunnis, they were closely allied with the Abbasids. Under the command of Mahmud of Ghazna, who ruled between 988 and 1030, the Ghaznavids were victorious over both the Buyids in central Persia, and the Saminids in Khurasan. At the height of their power, they governed over an immense empire encompassing much of Azerbaijan, Persia, Transoxiana, and Khurasan, as well as large portions of the Indus Valley and northern India. In 1040, just 10 years after the death of Mahmud of Ghazna, the Ghaznavids were to loose much of their western territories to the Seljuks. In 1161 they lost Ghazna to the Ghurids, a rival central Afghan dynasty. Following this defeat, the Ghaznavids moved their capital to Lahore, and held control of their Indian provinces until their final overthrow in 1182, again at the hands of the Ghurids.

The Ghaznavids were ambitious patrons of science and the arts. Abū Rayḥān al-Bīrūnī, one of the great Muslim polymath scientists, rose to prominence within the Ghaznavid cultural milieu, and it was the commission by Mahmud of Ghazni that prompted Firdausi to write his epic poem *Šāh-nāma*. The Ghaznavid Empire was immensely wealthy, both by virtue of their precious metal resources and the plunder they amassed in the conquering of northern India. This wealth was poured into architecture and the arts. Ghaznavid metalwork included highly refined work in silver and gold, as well as utilitarian objects in bronze. While little has survived the passage of time, their metalwork was likely to have also included architectural components such as lamps, locks, hinges, door-pulls, and knockers. The acclaimed Persian poet and scholar Nasir-i Khusraw wrote that large silver door-pulls produced in the workshops of Ghazna were sent to Mecca for the door to the

---

[44] Hillenbrand (1994a), 373.

Photograph 13 A Ghaznavid threefold pattern with six-pointed stars created from the *system of regular polygons* originally located at the South Palace at Lashgari Bazar near Bust, Afghanistan (© Thalia Kennedy)

Kaaba.[45] Nishapur continued to thrive as a center for ceramic production under Ghaznavid rule. Ceramic tiles with molded relief decoration and vivid turquoise glaze have been found in excavations at Ghaznavid sites. However, the relatively few surviving examples of Ghaznavid architecture are devoid of ceramic faience decoration. Nonetheless, the remaining examples of Ghaznavid architecture show a remarkable degree of ornamental sophistication, and with their Samanid antecedents it is not surprising that the Ghaznavid architectural aesthetic would be largely characterized by ornamental brickwork. The quality of Ghaznavid design and fabrication surpassed that of their Samanid predecessors and set the standard for the outstanding Ghurid and Seljuk architectural brickwork that followed. Ghaznavid artists also employed carved stucco and marble to great effect, as well as painted fresco—although very little has survived. Each of these media was used in giving expression to the remarkable innovations

in the art of geometric star patterns that transpired during this period. The remaining examples of Ghaznavid architecture include the Lashkar-i Bazar near Bust, Afghanistan (early eleventh century); the minaret and palace of Mas'ud III in Ghazna, Afghanistan (1099-1115); the ruins of the Ribat-i Mahi Caravanserai near Mashhad, Iran (1019-20); and the Arslan Jadhib tomb and minaret in Sangbast (997-1028).

Ghaznavid artists in Khurasan played a significant role in the development of geometric star patterns. It was under the auspices of this empire that the polygonal technique was expanded to include a greater range of geometric design, opening the door to the maturity and diversification of this tradition. These experimental innovations led to the creation of geometric patterns that expanded the stylistic boundaries and geometric underpinnings of this burgeoning tradition. An excellent Ghaznavid example of a geometric pattern derived from the *system of regular polygons* is a carved stone panel from the audience hall in the South Palace at Lashgari Bazar (completed in 1036), near Bust [Photograph 13]. This has a number of interesting and unusual

---

[45] Ward (1993), 57.

characteristics [Fig. 102a1]. The underlying polygonal tessellation is comprised of triangles and hexagons in a 3.6.3.6 configuration. What sets this pattern apart is the unusual manner in which the pattern lines relate to the underlying tessellation. The application of crossing pattern lines to the edges of each underlying polygon within a given tessellation typically employs the same angular treatment throughout. In this way, the pattern lines will relate equally to each underlying polygon in an identical manner. In the design from Lashgari Bazar, two different pattern line arrangements have been applied to the midpoints of the edges of alternating underlying hexagons: one set that cross with 60° angular openings, and another with 90° angular openings. This alternating innovation adds a further level of design diversity to what is already a highly versatile methodology. This alternating methodology never became widely practiced, and those patterns that employ this design technique are almost exclusively created from the *system of regular polygons*. This variant practice was mostly employed during

the formative years in the eastern regions, and by the time this design tradition reached its full maturity in the thirteenth and fourteenth centuries, such patterns were seldom used. The carved stone panel that employs this design dates from before the Ghurid destruction of Lashkar-i Bazar in 1151. This very distinctive Ghaznavid geometric pattern was also used on the door of the Zangid *minbar* at the al-Aqsa mosque in Jerusalem (1168-74) [Fig. 102a3]. The very unusual derivational methodology of this particular pattern suggests the possibility of the Zangid example being produced by an artist familiar with the panel at Lashkar-i Bazar: perhaps having fled the political turbulence in Khurasan during the period of Ghurid conquest, or conceivably on the pilgrimage route to Mecca via Jerusalem.

A carved stucco panel from the intrados of the arched portal at the Ribat-i Mahi Caravanserai near Mashhad, Iran (1019-20), employs a fourfold *acute* pattern with octagons at the vertices of the square repeat unit [Fig. 138b] [Photograph 14]. This is one of the earliest designs associated with

Photograph 14 A Ghaznavid carved stucco panel with octagons created from the *fourfold system A* in the arched portal at the Ribat-i Mahi Caravanserai near Mashhad, Iran (© Bernard O'Kane)

the *FourFold System A*, and uses only the large hexagon and square elements from the multiple components of this system. The lack of large octagons within the underlying modules qualifies this as a field pattern. *Acute* patterns within this system are characterized by 45° crossing pattern lines set on the midpoint of each edge of the underlying polygons. This example from the Ribat-i Mahi is unusual in that the underlying hexagonal modules employ two pattern line conditions at their edges. In addition to the 45° crossing pattern lines placed at the midpoints of the edges that are contiguous with other underlying hexagons, the hexagonal edges that are contiguous with the underlying squares have arbitrarily placed pattern lines with 90° angles. These 90° angled lines create a distinctive diagonally orientated square within the underlying square module, while the 45° crossing pattern lines create an octagon at each vertex where four elongated hexagons meet.

The significance of the Ghaznavid minaret of Mas'ud III in Ghazna, Afghanistan (1099-1115), looms large in the history of Islamic geometric star patterns. Like its nearby neighbor, the minaret of Bahram Shah (1117-58), this minaret is not associated with an adjacent mosque, and is possibly a victory tower commemorating successful military campaigns in the Indus Valley and northern India, and possibly inspired by their exposure to Hindu commemorative towers.[46] All that remains of the minaret of Mas'ud III is the magnificent stelliform shaft of the lower half, the upper cylindrical shaft having been destroyed by an earthquake in 1902. The lower shaft is an eight-pointed star in plan, and each pair of vertical flanges is divided into a series of ornamental panels of elaborate raised brick ornament; each divided in half at the 135° included angle of the eight-pointed star. The diverse ornamental treatment of these multiple raised brick panels includes herringbone *shatranji Kufi* calligraphy, knotted *Kufi*, and several linear bands of the classic fourfold star-and-cross design [Fig. 124b]. The horizontal grouping of eight geometric star patterns around the base of the shaft, as well as an array of eight similar patterns that circle the midsection of the shaft are remarkable for their level of complexity at this early date. Each of these 16 patterns has either fivefold or sevenfold symmetry, and includes patterns with 7- and 10-pointed stars [Fig. 206], 10- and 20-pointed stars, and 5- and 7-pointed stars [Figs. 280 and 281]. The fivefold patterns repeat on a rhombic grid, the 72° and 108° angles of which correspond to the symmetry of the decagon [Fig. 5a]. Throughout the subsequent history of this tradition, this was the most frequently used repetitive foundation for fivefold patterns, and these two-dimensional examples from Ghazni were among the earliest occurrences of patterns with fivefold symmetry, the only known earlier

examples being from the northeastern domed chamber of the Friday Mosque at Isfahan (1088-89). It is also significant that these Ghaznavid designs are, collectively, of considerably greater complexity than most all other Islamic geometric patterns from this same period, with the only contemporaneous examples of equal geometric complexity being the Seljuk work at the Friday Mosque in Isfahan, and the Friday Mosque at Barsain near Isfahan (1105). It would appear that the artist who designed the raised brick panels of the minaret of Mas'ud III was a pioneer of outstanding ability. The fivefold patterns of this minaret employ the polygonal technique in their construction [Fig. 206]. However, they differ from the contemporaneous Seljuk fivefold patterns in Isfahan, as well as subsequent fivefold geometric patterns generally, in that they do not employ a systematic methodology: relying upon a less rigid approach to the application of the pattern lines to the edges of the generative polygonal tessellation of decagons. These decagons are placed in a vertex-to-vertex arrangement that repeats upon a rhombic grid. This is in marked contrast to edge-to-edge polygons that eventually became standard practice for underlying generative tessellations. What is more, these decagons were kept as part of the completed design, thereby providing telltale evidence for the polygonal schema of this pattern. The seven-pointed stars within the pattern matrix are non-regular, but nonetheless add appreciably to the beauty of the design.

As with fivefold geometric patterns, the earliest extant sevenfold pattern is found within the northeastern domed chamber of the Friday Mosque at Isfahan. This single example is a field design of relative simplicity [Fig. 279]. By contrast, each of the two patterns with sevenfold symmetry from the minaret of Mas'ud III is considerably more complex. Both of these Ghaznavid examples utilize an elongated hexagon as the repeat unit [Figs. 280 and 281]. This has four 2/7 and two 3/7 included angles. In a manner that is similar to their neighboring fivefold designs, the application of the pattern lines to their respective generative polygonal tessellations was nonsystematic, involving a higher level of arbitrarily determined design components than frequently found in this tradition. The first of the sevenfold designs [Fig. 280] places a set of pattern lines that connect every other heptagonal corner, creating a matrix of seven-pointed stars that touch point to point. Into this matrix is added a secondary set of arbitrary pattern lines that complete the design. The first set of pattern lines in the second sevenfold pattern [Fig. 281] are placed upon the midpoints of the heptagonal edges and extend into the interstice region of twin pentagons. On its own, this initial set of pattern lines is a very acceptable *median* pattern that follows the midpoint conventions of this design tradition, and qualifies as being systematic. The artist responsible for this masterpiece of geometric design added a secondary set of pattern lines to

---

[46] Hoag (1977), 189.

the initial design, thereby making it far more complex, but also creating a design that more affectively balanced with the aesthetics of the neighboring geometric panels. The early occurrence of designs with fivefold and sevenfold symmetries, coupled with the comparatively greater complexity of the patterns themselves to other geometric star patterns of similar date, gives this building great significance to the historical development of Islamic ornamental art. Adding to this importance is the fact that, along with the single example from Isfahan, these sevenfold geometric designs predate the next earliest extant examples by approximately a hundred years.

1.11   Qarakhanids (840-1212)

The Qarakhanids began as a confederation of Turkic tribes who rose to power in Central Asia during the ninth century. At the close of the tenth century, Qarakhanid and Ghaznavid forces defeated the Samanids: with the Ghaznavids taking control of the Samanid territories in Khurasan and the Qarakhanids taking control of Transoxiana. The boundary between these two Turkic rivals was the Amu Darya (Oxus River). Their capitals included Kashgar in western China, Balsagun, and Uzgen in Kyrgyzstan. The Qarakhanids concentrated their power in Central Asia, and were rivals with the Seljuks as well as the Ghaznavids. In 1140 they became subjects of the Kara-Khitan Dynasty from northern China, and were finally defeated by the Khwarizmshahid Dynasty in 1212.

Qarakhanid architectural ornament generally followed the monochrome brickwork and stucco practices prevalent in eastern regions of the Islamic world during the eleventh and twelfth centuries. Along with the geometric ornament of the Ghaznavids, the Qarakhanids were among the first Muslim cultures to expand the repertoire of geometric design to include patterns of greater complexity and diversity. Despite the very few remaining Qarakhanid buildings, the range of extant geometric patterns provides strong evidence for the important role they played in the development of the mature style of Islamic geometric patterns. The architectural record indicates that the Qarakhanids were particularly fond of geometric patterns made from both the *system of regular polygons* and the *fourfold system A*. Examples of Qarakhanid patterns made from the *system of regular polygons* include a very simple design constructed from the underlying 6^3 tessellation of regular hexagons located within the corners of the quarter dome of the southern portal at the Maghak-i Attari mosque in Bukhara, Uzbekistan (1178-79). This is a *two-point* pattern that uses the 6^3 hexagonal grid both as part of the completed pattern and as the formative schema [Fig. 96f]. The anonymous southern tomb in the complex of three adjoining Qarakhanid mausolea in Uzgen (1186) has

two patterns with threefold symmetry that are constructed from the 3.4.6.4 underlying tessellation of triangles, squares, and hexagons [Fig. 89]. The first of these is located on the wide soffit of the entry arch and includes an overt expression of the underlying tessellation within the pattern itself. This pattern employs the square module of the generative tessellation as a primary feature of the completed design, thereby indicating the underlying triangles and hexagons as implied background elements [Fig. 104d]. The second pattern from the southern tomb at Uzgen to use the underlying 3.4.6.4 tessellation is located beneath the arch soffit on the sidewall of the arched portal [Fig. 105b]. As an added design feature, this example arbitrarily places six-pointed stars at the vertices of the isometric grid. This additive variation is similar in concept to a Ghurid example of the same design at the minaret of Jam, dating from just 20 years earlier [Fig. 105c]. The 3.4.6.4 pattern from Jam differs in that it places additive hexagons into these same positions. The northern tomb of the Jalal al-Din Hussein (1152-53) at this complex of three mausolea at Uzgen features a particularly delicate interpretation of the classic fourfold star-and-cross design on the intrados of the main arch of the portal. The eight-pointed stars touch point to point rather than their interweaving with one another. The visual impact of this less typical arrangement is augmented by a secondary interweaving motif of finer line thickness that results in an overall design that is unique, delicate, and extremely effective.

The earliest of the three adjoining mausolea at Uzgen, Kyrgyzstan, is the middle tomb of Nasr ibn Ali (1012-13). The entry portal of this tomb is framed with a *median* pattern in raised brick that was constructed from the *fourfold system A* [Fig. 159] [Photograph 15]. Along with the above-cited Ghaznavid design from the portal of the Ribat-i Mahi, this is one of the earliest examples of an Islamic geometric pattern that can be created from this generative system, and multiple later examples are found within the historical record. It is important to once again emphasize that when regarding more basic designs produced during the early formative period of this design tradition, it is impossible to know for certain which methodological practice was used in a given circumstance. This design from the tomb of Nasr ibn Ali could have been produced just as readily from either the grid method or the point joining technique as from the *fourfold system A* [Fig. 72]. An identical Seljuk use of this Qarakhanid design is found at the Sultan Sanjar mausoleum in Merv, Turkmenistan (1157). The points of the eight-pointed stars in the later example from Merv are irregular (as per Fig. 75a), and indicate that this example of the pattern may have been produced using the orthogonal graph paper technique. At the base of the sidewalls in the entry portal of the anonymous southern tomb at Uzgen is a small square carved stone panel with a pattern constructed from the

Photograph 15 A Qarakhanid raised brick pattern with eight-pointed stars created from the *fourfold system A* in the entry façade of the tomb of Nasr ibn Ali in Uzgen, Kyrgystan (© Igor Goncharov)

Photograph 16 A Qarakhanid pattern with eight-pointed stars created from the *fourfold system A* in the entry façade of the entry of the Maghak-i Attari mosque in Bukhara, Uzbekistan (© Thalia Kennedy)

fourfold system A [Fig. 160]. The underlying tessellation for this geometric pattern includes modules that are atypical to this system, but add a very acceptable dynamic to the completed design. This design dates from the early period of the *fourfold system A* when experimentation with both polygonal components and application of their associated pattern lines was prevalent. The ornamental banding in the nearby minaret of Uzgen (twelfth century) includes a geometric pattern created simply from the orthogonal grid that has design characteristics similar to the *fourfold system A* [Fig. 74]. The entry *pishtaq* of the Maghak-i Attari mosque in Bukhara has several raised brick geometric panels with patterns constructed from the *fourfold system A*. The two most basic of these are located closest to the ground. The absence of eight-pointed stars in both of these qualify them as field patterns, and the underlying generative tessellation is made up of just large hexagons and squares. One of these is a *median* pattern with 90° crossing pattern lines [Fig. 138c], and the other is a design comprised of superimposed dodecagons, each of which is centered upon the vertex of the four large hexagons [Fig. 138f]. These dodecagons relate to the underlying tessellation through 90° crossing pattern

lines located upon the midpoints of each hexagonal edge, and 120° crossing pattern lines are located at the midpoints of the square edges. The highest panel on this façade from Bukhara is a *median* pattern created from the *fourfold system A* with considerably greater complexity than its neighbors [Fig. 155]. The relationship between the pattern lines and the underlying generative tessellation is less consistent than normally found within this tradition: with some midpoints of the underlying polygonal edges having crossing pattern lines; others having lines that meet, but do not cross these midpoints; and still others having no pattern lines at all. The completed geometric design is as much a product of subjective artistic license as systematic methodology. Specifically, the completed design is the result of a subtractive process whereby pattern lines are strategically removed to create a new pattern matrix with background regions that would not have otherwise been there. While orthogonal, the repetitive schema of this design is comprised of the 4.8^2 semi-regular grid, upon which octagons are located at each vertex. The middle panel on the façade of the Maghak-i Attari mosque also employs a design created from the *fourfold system A* [Fig. 151] [Photograph 16]. This is an elegant *acute* pattern that is created from the square and triangle modules from the *fourfold system A*, and with added large octagons with sides equal to the longer edges of the triangle. The octagon with this edge size is atypical to patterns made from this system. The end result is an exceptional pattern comprised of two sizes of interweaving octagons, and arbitrarily added octagons set within the large octagons (not shown in Fig. 151).

The design used on each of the circular columns that flank the entry portal of the anonymous southern tomb at Uzgen is a *two-point* pattern generated from the *fourfold system B* [Fig. 176c]. The perpendicular parallel pattern lines at the center of the square repeat unit are an additive device that was particularly popular in the eastern regions: for example, the central region of the repeat unit for the Ghurid raised brick design from the Friday Mosque at Herat (1200) [Fig. 174b]. The underlying generative tessellation for this design employs octagons, small pentagons, and small hexagons from this system.

The back wall of the south entry portal at the Maghak-i Attari mosque in Bukhara has two adjacent carved stone relief panels with identical geometric patterns created from the *fivefold system*. This is an *obtuse* design with rectangular repeat units [Fig. 245a]. Along with the more complex contemporaneous Seljuk example from the Seh Gunbad in Orumiyeh, Iran (1180), these are among the earliest extant examples of purely systematic fivefold patterns that repeat upon a rectangular grid. With the adoption of the *fivefold system* by subsequent Muslim cultures, this Qarakhanid design became the most widely used fivefold pattern that repeats upon a rectangular grid. Several especially fine examples include an Ilkhanid arch soffit at the Friday Mosque at Ashtarjan, Iran (1315-16); a Timurid cut-tile panel from an entry portal at the Shah-i Zinda funerary complex in Samarkand, Uzbekistan (1386); and a Timurid running mosaic wainscoting panel at the Abdullah Ansari complex in Gazargah, Afghanistan (1425-27).

1.12 Great Seljuks (1038-1194)

The art and architecture of the Ghaznavids profoundly influenced their Seljuk and Ghurid successors. These rival dynasties vied for power within the tight confines of greater Khurasan. Each adhered to Sunni Islam, and each had a strong affinity with Persian customs and culture. The Seljuks rose to power as military commanders of the Qarakhanids who fought against the Ghaznavids. As an independent force, they conquered Merv and Nishapur in 1028-1029, followed by Ghazna in 1037. In 1038, Tughril adopted the title of Sultan of Nishapur: officially founding this immensely influential dynasty. In 1040, they defeated the Ghaznavids at the Battle of Dandanaqan, taking control of the Ghaznavid's western territories. Upon securing the

greater portion of Khurasan, Seljuk forces expanded their conquest further westward against the Buyids. Allied with the Abbasid Caliph, Tughril defeated the Buyid forces in Baghdad. Within 20 years of his declaring himself Sultan, Tughril had wrested control over a broad swath of land that extended from the Levant and most of Anatolia in the west, all of Persia, large tracks of Transoxiana in the north, to western Khurasan in the east.

The eleventh-century advances in geometric design methodology made in Khurasan and Transoxiana spread westward during the twelfth century. During the first half of the eleventh century the Ghaznavid Empire gained control over eastern Persia. This was rapidly eclipsed by the military successes of the Seljuks, whose rule and hegemony profoundly influenced the architectural ornament throughout their vast territorial holdings for over a century. It was during this period of Seljuk cultural dominance that complex geometric design became a dominant feature of the architectural ornament in their western territories. Furthermore, the twelfth-century westward spread of evermore complex geometric patterns is evidence that the polygonal technique—the only viable method of creating particularly complex patterns—was wholeheartedly embraced throughout the regions of Seljuk influence, beyond to Egypt, and across North Africa to Morocco and al-Andalus—effectively establishing a pan-Islamic geometric aesthetic. Throughout this westward expansion, the use of the polygonal technique continued to be employed as a primary methodology for creating geometric patterns.

Surviving examples of early Seljuk architectural ornament include numerous fine geometric patterns created easily with the polygonal technique. Of particular note are the two tomb towers of Kharraqan in Qazvin Province, Iran: the eastern tower (1067) and the western tower (1093-94). These two towers are decorated with a variety of geometric patterns executed in raised brick, including key patterns, polygonal tessellations as pattern, and an assortment of geometric star patterns. Several of these geometric patterns were used earlier at the mausoleum of Arab Ata [Fig. 96g] [Photograph 12], the Great Mosque of Córdoba [Fig. 96e], and Sabz Pushan near Nishapur [Fig. 96c] [Photograph 11], while other patterns from Kharraqan appear at their earliest known date. One of the most interesting geometric patterns from Kharraqan appears on the eastern tower, and is very likely the earliest surviving example of an Islamic geometric pattern with 12-pointed stars [Photograph 17]. The stars are located on each vertex of the isometric grid, and the underlying polygonal tessellation that produces this pattern is made up of triangles and dodecagons in a 3.12^2 configuration [Fig. 108a]. This exact same design was used as part of the interior ornament for the Friday Mosque of Golpayegan, Iran (1105-18); the Sayyid Ruqayya Mashhad in Cairo (1133); the Great Mosque at Kayseri, Turkey (1205); one

Photograph 17 A Seljuk example of a threefold pattern with 12-pointed stars created from the *system of regular polygons* at the eastern tomb tower at Kharraqan, Iran (© Reza Roudneshin)

of the Mamluk window grilles from the restoration of the Ibn Tulun mosque in Cairo (1296); and the interior of the Mamluk door (1303) of the Vizier al-Salih Tala'i mosque in Cairo. Indeed, over time, this design came to enjoy great popularity throughout the Islamic world. A particularly beautiful curvilinear example was used as an illumination in the celebrated 30-volume Quran of Uljaytu,[47] written and illuminated by 'Abd Allah ibn Muhammad al-Hamadani in 1313 [Fig. 108c]. Another early Seljuk pattern with 12-pointed stars that is easily created from the 3.12^2 tessellation is from the southern *iwan* of the Friday Mosque at Forumad in northwestern Iran (twelfth century) [Fig. 108d]. This example was also popularly used in later periods, including a frontispiece from a Baghdadi Quran illuminated by Muhammad ibn Aybak ibn 'Abdullah (1303-07), and a Mamluk stone mosaic panel from the Amir Aq Sunqar

funerary complex in Cairo (1346-47) [Photograph 45]. One of the very successful isometric patterns from the east tower at Kharraqan is easily made from the simple hexagonal grid with an additive six-pointed star motif at the centers of the underlying hexagon [Fig. 96h]. This was a very popular design that was used in many succeeding locations: including the wooden *minbar* of al-Aqsa mosque in Jerusalem (1168); the *mihrab* of the Lower Maqam Ibrahim at the citadel of Aleppo (1168); the entry portal of the Izzeddin Kaykavus in Sivas, Turkey (1217-18); and an archivolt at the Zahiriyya *madrasa* in Aleppo (1217). The wooden *minbar* at the Friday Mosque at Abyaneh, Iran (1073), includes another Seljuk pattern created from the 3.6.3.6 underlying tessellation that places six-pointed stars at the vertices of the isometric grid [Fig. 99a]. As with several other designs made from this system, this example is comprised of superimposed hexagons. The earliest extant example of this design is one of the window grilles at the Great Mosque of Córdoba (987-99), and over time, this came to enjoy wide popularity throughout the Islamic world. A raised brick border that surrounds the *mihrab* of the Friday Mosque of Golpayegan (1105-18) can also be derived from the 3.6.3.6 underlying tessellation of triangles and hexagons [Fig. 99b]. This *median* pattern places 90° crossing pattern lines at the midpoints of each edge of the underlying polygons. The resulting design is characterized by superimposed dodecagons that repeat upon the isometric grid. The earliest known use of this pattern is from a Fatimid window grille at the al-Azhar mosque in Cairo (970-72), and over time, it was widely used by succeeding Muslim cultures. A conceptually similar design with superimposed dodecagons can be created from the simple 6^3 tessellation of hexagons [Fig. 97c]. While the placement of the dodecagons within the pattern matrix is identical, their size relative to the isometric repeat is slightly larger. This produces differently proportioned concave octagons and ditrigonal shield-shaped background modules. This subtle variation was also widely used by diverse Muslim cultures, and two fine Seljuk examples include the surrounding border of the *pishtaq* of the Seh Gunbad in Orumiyeh, Iran (1180), and a carved stucco panel from the Friday Mosque at Forumad in Iran (twelfth century). Other notable examples of these closely related designs are found at the Sirçali *madrasa* in Konya, Turkey (1242-45); the Shah Rukn-i-'Alam tomb in Multan, Pakistan (1320-24);[Photograph 69]; and the fourteenth-century ceramic tile work added to the main *iwan* of the tomb of Abu Sa'id Abul Khayr in Mayhaneh, Turkmenistan.[48] A very beautiful Seljuk example of an additive variation of this pattern was used in the celebrated tympanum over the door in the main portal

---

[47] This Ilkhanid Quran is in the National Library in Cairo: 72, pt. 19.

[48] The Tomb of Abu Sa'id Abul Khayr in Mayhaneh, Turkmenistan, is known locally as the Tomb of Meana Baba.

Photograph 18   A Seljuk example of a threefold pattern with six-pointed stars from the northeast dome chamber in the Friday Mosque at Isfahan
(© Tom Goris)

of the Gunbad-i Surkh in Maragha, Iran (1147-48). This example is credited with being the earliest extant Islamic ornamental panel to incorporate glazed faience ceramics[49]: a precursor to the tradition of cut-tile mosaics wherein the whole ornamental surface is covered with specially cut ceramic pieces that fit together to make the design. Prior to this example, faience was used as an ornamental accent, frequently to emphasize the calligraphic component of the ornament. There are two separate additive motifs in this panel: the first being a series of superimposed nonagons in turquoise faience, and the second being a series of parallel pattern lines that emphasizes the hexagonal repetitive grid. A very similar additive design was used on the Kaykavus hospital in Sivas (1217-18), as well as the tomb of Sahib Ata in Konya (1283-93). The difference between these two later examples from the Seljuk Sultanate of Rum in Anatolia is that their secondary additive elements are parallel pattern lines that emphasize the isometric repeat rather than the hexagonal dual.[50]

A very simple, but highly effective, Seljuk geometric pattern from one of the multiple blind arches in the upper portion of the northeast dome chamber in the Friday Mosque at Isfahan (1088-89) is created from the 6^3 hexagonal grid [Photograph 18]. This example places 45° crossing pattern lines at the center points of each underlying hexagonal edge, and is hence categorized as a variation of the standard 30° angular openings of an *acute* design from this system [Fig. 95a]. These crossing pattern lines create six-pointed stars at the centers of each hexagonal repeat unit. It is surprising that this dynamic design was not used nearly as often as those composed of either 60° or 90° crossing pattern lines that were created from this same underlying tessellation. Other Seljuk examples of this pattern are found in one of the small blind arches in the upper *muqarnas* squinches of the Friday Mosque in Barsian, near Isfahan (1105), and in the carved stucco on the intrados of an arch at the Friday Mosque at Sin in Iran (1134). Contemporaneous with the example from Sin is an example from the niche of the Fatimid portable wooden *mihrab* of the Sayyid Ruqayya Mashhad in Cairo[51]

[49] Wilber (1939), 35.

[50] Schneider (1980), pattern no. 225.

[51] Currently in the collection of the Islamic Museum in Cairo.

(1133), and a later example from Cairo is from a Mamluk carved stone relief at the Imam al-Shafi'i mausoleum (1211).

The thin border that surrounds the entry door of the Gunbad-i 'Alaviyan in Hamadan, Iran (late twelfth century), employs a rather clever design with six-pointed stars and nonagons. This is relatively easy to construct from the 3.6.3.6 underlying tessellation of triangles and hexagons [Fig. 100c] [Photograph 22]. Later examples of this design were produced by artists working in Anatolia, and include a pattern in the Great Mosque of Divrigi (1228-29), the Muzaffar Barucirdi *madrasa* in Sivas (1271-72), and the central panels on the interior of the Mamluk door (c. 1303) at the Vizier al-Salih Tala'i mosque.

Like their Ghaznavid predecessors, Seljuk artists occasionally employed the atypical variation to this design methodology whereby two varieties of pattern-line configuration are applied to adjacent underlying polygonal cells of the same type. As mentioned, the earliest known example of such a design is the above-cited Ghaznavid pattern from the Audience Hall in the South Palace Lashkar-i Bazar (before 1036) [Fig. 102a1] [Photograph 13]. This uses the 3.6.3.6 underlying tessellation as its generative schema. A panel above one of the exterior blind arches of the west tower at Kharraqan (1093) appears to be the earliest Seljuk pattern to similarly employ differentiated treatments to adjacent polygonal cells from the underlying generative tessellation: in this case the simple 6^3 hexagonal grid [Fig. 98c]. The primary underlying hexagons have six-pointed stars with 60° crossing pattern lines located at the center points of the polygonal edges: the standard pattern line application of the *median* family. Each of these primary hexagonal cells is surrounded by six secondary underlying hexagons that place pattern lines that connect each vertex through the center of the underlying hexagon, as well as extend the 60° crossing pattern lines from the primary underlying hexagons. Somewhat surprisingly, this fine pattern is not known elsewhere within the historical record. A Seljuk example of a 3.6.3.6 design with alternating pattern application to the underlying hexagons is immediately adjacent to the door within the portal of the Seh Gunbad tomb tower in Orumiyeh, Iran (1180) [Fig. 101a]. This utilizes 90° crossing pattern lines located at the center points of the active underlying hexagons. The constructive methodology of this design is unusual. The pattern is created by extending the 90° crossing patterns that are placed upon the primary underlying hexagons into the pattern matrix where they are met by the extended 60° crossing pattern lines that originate from designated active underlying triangles. Locations of later examples of this Seljuk design include the Sirçali *madrasa* in Konya, Turkey (1242), and a variation from the Çifte Minare *madrasa* in Sivas, Turkey (1271) [Photograph 41].

Another type of atypical pattern line application simply widens the lines of the underlying tessellation itself, rather than following the standard convention of the pattern lines being located upon the midpoints of the generative polygonal edges. A Seljuk isometric design of this type was used on the façade of the west tower at Kharraqan [Fig. 110]. This design is created from the $3^2.4.3.4$-$3.4.6.4$ *two-uniform* tessellation of regular triangles, squares, and hexagons [Fig. 90]. All of the polygonal edges in this tessellation are widened to create the interweaving design except the coincident edges of the twin triangles. In this respect, the twin triangles are treated as a single rhombus. This Seljuk pattern from Kharraqan appears to be the earliest example of the use of a *two-uniform* tessellation in Islamic art. Another design created from a *two-uniform* underlying polygonal tessellation is one of the multiple patterns from the anonymous Persian language treatise *On Similar and Complementary Interlocking Figures* in the Bibliothèque Nationale de France in Paris.[52] It is speculated that this was produced circa 1300 and was influenced by earlier Seljuk and possibly Khwarizmshahid artistic practices and sources.[53] One of the designs included in this treatise is a *two-point* pattern created from the $3^3.4^2$-$3^2.4.3.4$ underlying tessellation of triangles and squares [Fig. 112d]. This is a rather remarkable orthogonal design that has oscillating squares and rotating kite motifs within each repetitive square component. While unknown to the architectural record, the aesthetic style of this design is very closely related to examples from the Khwarizmshahid [Fig. 112b] [Photograph 38] and Ilkhanid periods [Fig. 111].

Among the many Seljuk geometric patterns in the northeast dome chamber of the Friday Mosque at Isfahan (1088-89) is the earliest example of a design that employs the distinctive ditrigonal shield module in its underlying generative tessellation [Fig. 118a] [Photograph 19]. This underlying tessellation is, in and off itself, identical to the classic *median* pattern created from the 6^3 tessellations of hexagons [Fig. 95c]. This pattern is one of several patterns that decorate the series of small blind arches in the upper portion of the square base of the northeast dome chamber. This same underlying generative tessellation was used in several other locations to produce patterns that are very similar to the example from Isfahan. These include two examples from the Seljuk Sultanate of Rum: one from the *mihrab* of the Karatay *madrasa* in Antalya (1250) [Fig. 118d], and the other from the Ahi Serafettin mosque in Ankara (1289-90) [Fig. 118c]. A later Ottoman example of inferior quality was used in the *mihrab* of the Yesil mosque in Bursa, Turkey (1419-21). A Mamluk variation that can be created from this underlying tessellation was used in a window grille at the

---

[52] MS Persan 169, fol. 188b.

[53] –Özdulral (1996).

  –Necipoğlu [ed.] (Forthcoming).

Photograph 19 A Seljuk example of a threefold pattern with six-pointed stars and octagons from the northeast dome chamber in the Friday Mosque at Isfahan (© Tom Goris)

Tabarsiyya *madrasa* (1309) at the al-Azhar mosque in Cairo [Fig. 118b].

Surviving examples of Seljuk patterns that can be derived from the 4.8^2 underlying tessellation are relatively uncommon. An example of the classic star-and-cross design was used in the wooden ceiling of the Friday Mosque of Abyaneh (1073) [Fig. 124b]. A raised brick border from the Gunbad-i 'Alayvian in Hamadan (late twelfth century) employs a pattern that can be constructed in several ways, including from the 4.8^2 tessellation [Fig. 125c]; and a Seljuk carved stucco panel at the Tehran Museum can also be created from this tessellation [Fig. 127e]. It is important to stress that these latter two examples have alternative methods of construction that may well have been employed at the time of their creation.

Along with their Qarakhanid and Ghaznavid counterparts, Seljuk artists were among the first to explore the design potential of the *fourfold system A*. The diverse range of patterns used in the decoration of the two tomb towers at Kharraqan includes a very simple border design on the earlier eastern tomb tower (1067-68) that is conveniently constructed from this generative system and was used

widely by succeeding Muslim cultures [Fig. 138c]. The underlying polygonal tessellation that creates this pattern is comprised of just the elongated hexagonal and square polygonal modules. On its own, the basic tessellation of squares and elongated hexagons (but with different proportions) had been used as early as 300 years previous by Umayyad artists at Khirbat al-Mufjar, as well as by Abbasid artists at Samarra some 200 years previous. A contemporaneous Abbasid example of the ornamental use of this polygonal tessellation, with approximately the same hexagonal proportions as used in the *fourfold system A*, was used at the No Gunbad mosque in Balkh. The example from Kharraqan uses this tessellation to generate a *median* geometric design that places crossing pattern lines with 90° angular openings at the midpoints of each polygonal edge. The absence of large octagons within the underlying generative tessellation means that this pattern from Kharraqan does not have eight-pointed stars, and therefore qualifies as a field pattern. In addition to the polygonal technique, this well-known design can also be produced using either the orthogonal grid method or the point joining method [Fig. 77]. Field patterns associated with the *fourfold system A* were pioneered in the eastern regions during the early developmental period of this artistic tradition. The *acute* pattern created from this same underlying tessellation of large hexagons and squares was used by two western subordinate dynasties that were part of the sphere of Seljuk influence during the twelfth century [Fig. 138a]: the Artuqid *mihrab* of the Maqam Ibrahim at Salihin in Aleppo (1112) and the Tepsi minaret in Erzurum (1224-32) produced by the Saltukids. A *median* field pattern created from a tessellation of just large hexagons was used as a border that surrounds the entry portal of the Khwaja Atabek mausoleum in Kerman (1100-1150) [Fig. 137d].

A more complex Seljuk example of the early use of the *fourfold system A* is a raised brick pattern surrounding the midportion of the shaft of the minaret of the Friday Mosque at Damghan, Iran (1080) [Fig. 145]. It is interesting to note that this *median* design can be created from two separate sets of underlying tessellations from this same system: the first comprised of large octagons, large hexagons, and pentagons; and the second comprised of small octagons, small hexagons, pentagons, squares, and interstice rhombi. The reason for this unusual reciprocal feature is that the underlying polygonal edges can bisect the 90° crossing pattern lines

in two perpendicular directions. These two underlying generative tessellations are essentially duals of one another. This same design enjoyed great popularity among Seljuk artists, and the many examples include: the Friday Mosque at Golpayegan, Iran (1105-18), that is particularly interesting for its being interwoven into the ascending letters of a band of *Kufi* script; the minaret of Daulatabad outside Balkh, Afghanistan (1108-09) [Photograph 20]; the minaret of the Friday Mosque at Saveh, Iran (1110); the Friday Mosque at Sangan-e Pa'in (second half of the twelfth century); and the Friday Mosque at Gonabad, Iran (1212). The Ghurids used this design during the same early period in several panels from the minaret at Jam, Afghanistan (1174-75 or 1194-95). The *mihrab* of the Malik mosque in Kerman (eleventh–twelfth century) is decorated with a more complex *median* pattern created from the *fourfold system A* [Fig. 153]. This introduces the triangular module that is 1/8 of an octagon into an underlying tessellation of large octagons and pentagons. A strong visual feature of this design is the set of large orthogonally placed octagons that are orientated vertex to vertex. This orientation relates to the classic *obtuse* pattern of octagons and four-pointed stars that is derived

Photograph 20 A Seljuk example of a pattern with eight-pointed stars created from the *fourfold system A* on the minaret of Daulatabad outside Balkh, Afghanistan (© Thalia Kennedy)

from the 4.8^2 underlying tessellation of octagons and squares. A second *fourfold system A* pattern from the Friday Mosque at Gonabad is an *acute* pattern that is unusual in that it uses an eight-pointed star as a primary component of the underlying generative tessellation that is created from the arrangement of square modules [Fig. 147a]. The use of this eight-pointed star creates a design that is atypical to this generative system. It is interesting that essentially this same design can be produced from an altogether separate underlying tessellation of different components from this same system [Fig. 146]. Ordinarily, when a given design can be produced from two different underlying tessellations, they are duals of one another. In this case, the two generative tessellations are not duals. This alternative tessellation employs octagons, pentagons, and small hexagons that combine together to create a large dodecagonal interstice region at the center of each repeat. As with other systematic designs created during this formative period, these two examples from Gonabad exemplify the ongoing experimentation that led toward the full maturity of this design tradition.

Despite the Seljuks being highly influential innovators of the geometric idiom, examples of their extant architectural ornament do not include a representative quantity of patterns created from the *fourfold system B*. This is surprising in that their allied Zangid, Ildegizid, and Sultanate of Rum neighbors to the west made wide use of this variety of geometric design. One notable exception to the rarity of Seljuk designs created from this system is an example of the classic *acute* pattern found within the *mihrab* arch spandrels at the Friday Mosque at Sin, Iran (1134) [Fig. 173a]. This is not only the earliest known use of this highly popular Islamic geometric design, but also the earliest known example of a pattern constructed from the *fourfold system B*. Were it not for the Mongol destruction, it is possible that a far greater number of Seljuk *fourfold system B* designs may have survived to the present, and our knowledge of the origins and dissemination of this important variety of geometric design would be more complete.

The architectural record indicates that the Seljuks were also the first to develop geometric patterns created from the *fivefold system*. This methodological system for creating Islamic geometric patterns is of particular significance to the history of Islamic art and architecture. Over time, this form of design spread throughout the Islamic world, receiving ongoing innovative attention and lasting popularity. The earliest fivefold designs date from the close of the eleventh century, and within a hundred years this variety of systematic design was making full use of rhombic, rectangular, and hexagonal repeat units, as well as fully mature patterns in each of the four pattern families: *acute*, *median*, *obtuse*, and *two-point*. The earliest Islamic geometric patterns created from the *fivefold system* are three examples from the northeast dome chamber of the Friday Mosque at Isfahan (1088-

89). One of these three is the classic *obtuse* pattern that repeats upon a rhombic grid with 72° and 108° included angles [Fig. 229a] [Photograph 21]. This early example includes the additive star rosette infill of the ten-pointed stars that, in time, became a common feature of *obtuse* patterns [Fig. 221]. An interweaving version of this same design (without the additive infill) was used very soon after at the Friday Mosque at Golpayegan, Iran (1105-1118) [Fig. 229b], and indeed this design was used with great frequency throughout Muslim cultures. Other early Seljuk patterns created from the *fivefold system* that employ this same rhombic repeat unit include a *two-point* pattern in the magnificent entry tympanum at the Gunbad-i 'Alaviyan in Hamadan, Iran (late twelfth century) [Fig. 231d] [Photograph 22], and a classic *acute* pattern from the Friday Mosque at Gonabad (1212) [Fig. 226c] [Photograph 23]. Indeed, each of these three early Seljuk examples employs the same underlying generative tessellation. A late Abbasid *median* pattern at the mausoleum of 'Umar al-Suhrawardi in Baghdad (early thirteenth century) also uses the same underlying generative tessellation, but uses only selected midpoints of the underlying tessellation for locating the pattern lines [Fig. 228d]. This caliphal building dates to when Baghdad was no longer ruled by the Seljuks but was still under the aesthetic influence of Seljuk culture. A particularly complex Seljuk design that employs the rhombic repeat with 72° and 108° included angles wraps nine of the ten sides of the Gunbad-i Qabud in Maragha, Iran (1196-97) [Photograph 24].[54] The rhombic repeat unit of this design holds an unusually large number of polygonal modules that comprise the underlying generative tessellation [Figs. 239 and 240]. The continuous flow of this pattern across the nine sides of this decagonal tomb tower includes coverage of the ten engaged columns at each corner of the tomb tower, and is only discontinued on the side of the tower with the entry portal. This remarkable geometric design has been the subject of considerable interest in recent years, with arguments and counterarguments as to whether it is an example of quasicrystalline geometric design.[55] While this design has clear Penrose tiling characteristics, it nevertheless repeats in nine linear units, each of which is a unit cell, thereby disqualifying it from being an aperiodic quasicrystalline structure. Although the linear repeats of this design appear as rectangular, corresponding to the rectangular façades of the building, when considered more broadly it becomes clear that the actual repeat units are the fivefold

---

[54] Bier (2012).

[55] –Makovicky (1992), 67–86 and (2007).
      –Lu and Steinhardt (2007b), 1106–1110.
      –Cromwell (2009), 36–56.
      –Cromwell (2015), 1–15.

Photograph 21 A Seljuk example of the classic *obtuse* pattern with ten-pointed stars created from the *fivefold system* in the northeast dome chamber of the Friday Mosque at Isfahan (© Tom Goris)

rhombus with 72° and 108° included angles. The apparent complexity and randomness of the underlying tessellation bely what is actually a well-ordered geometric schema comprised of rings of ten edge-to-edge decagons placed upon each vertex of the rhombic grid. This arrangement of decagons is then treated to a secondary application of polygonal modules from the *fivefold system*: with most of the decagons being filled, and some remaining unfilled. To further complicate this secondary application, the infill of the secondary polygons only has reflection symmetry along the vertical line of axis within the rhombic repeat units. This is highly unusual, and creates a geometric pattern that is certainly eccentric, but not aperiodic. Still further complexity is achieved by the arbitrary infill of the ten-pointed stars within the remaining unfilled decagons with an additive infill motif that was popularly used among Seljuk artists in Persia and Anatolia [Fig. 224a]. This effectively disguises the ten-pointed stars, and transforms the overall design into a field pattern. This is the earliest extant example of this well-used transformative variation to the ten-pointed star. The pattern from the Gunbad-i Qabud incorporates yet a further degree of complexity through the introduction of a

secondary design element that is arbitrarily added into the pattern matrix [Fig. 67]. This is the most elaborate example of a Seljuk additive pattern, and the dual-level quality of this design can be regarded as an aesthetic precursor of the recursive geometric patterns that were developed in the same region some 250 years later.

Among the earliest examples of patterns created from the *fivefold system* that repeat upon the more acute rhombic grid of 36° and 144° angles [Fig. 5b] is a remarkable *two-point* design from the late Abbasid main entry portal of the Mustansiriyah *madrasa* in Baghdad (1227-34) [Fig. 243b]. This was built after the collapse of the Seljuk Empire, but during the period when the Seljuk artistic heritage was still influential on the ornamental arts of Baghdad. As with other late Abbasid geometric designs that survived the Mongol destruction of Baghdad in 1258, this *two-point* pattern is highly innovative. This is one of the earliest designs to employ truncated decagons within its underlying generative tessellation. What is more, the angles of the applied pattern lines to each of the two points of the underlying polygonal edges have 54° angles of declination rather than the 72° or 36° that are standard among *two-point* patterns created from the *fivefold system*.

Photograph 22 A Seljuk example of the classic *two-point* pattern with ten-pointed stars created from the *fivefold system* in the arched tympanum of the Gunbad-i 'Alaviyan in Hamadan, Iran (© Daniel C. Waugh)

Designs created from the *fivefold system* that employ rectangular repeat units also appear to have been a Seljuk innovation. The *median* field pattern from the Khwaja Atabek mausoleum in Kerman (1100-1150) [Fig. 211] is interesting not just for its early date, but also for its unusual geometry. The inventive arrangement of the polygons that comprise the underlying tessellation employs just one module from the *fivefold system*: the 1/10 decagonal triangle. By placing two of these triangles edge to edge along their long edges, and applying the 72° crossing pattern lines to the short edges (as per convention), the pattern lines allow for the creation of a distinctive trefoil device within the two adjacent triangles [Fig. 188]. Another Seljuk example with rectangular repeat units is an *obtuse* border design that frames the *pishtaq* of the Seh Gunbad in Orumiyeh, Iran (1180). This places ten-pointed stars at the vertices of each rectangular repeat unit. This design is similar to a later Seljuk Sultanate of Rum example from the Sirçali *madrasa* in Konya (1242-45) [Fig. 247].

In addition to the above-cited *obtuse* pattern from one of the small blind arches in the northeast dome chamber of the Friday Mosque at Isfahan, there is also a very interesting *acute* pattern created from the *fivefold system* in another of the set of arches that surround the dome [Fig. 261b] [Photograph 25]. This design is remarkable in that it is the earliest example of a hybrid design known to this tradition. Of the three varieties of repetitive cell that comprise this design, the most visually apparent is the large central pentagon, the base of which rests upon the horizontal spring line of the arch. Attached to the four exposed edges of this pentagon are rhombi with 72° and 108° included angles. It is noteworthy that the pattern contained within these rhombic regions is the classic *acute* design, and the occurrence of this rhombic motif is the earliest known representation of this classic *acute* pattern, albeit not as a continuous surface coverage in its own right. The pattern within the large central pentagon is noteworthy on two counts: it is the earliest example of a fivefold design with rotation symmetry, and it is the earliest fivefold pattern to employ the motif of a central pentagon surrounded by 5 nine-sided flattened five-pointed star motifs that are derived from the five underlying irregular pentagons [Fig. 261a]. The use of two or more otherwise

Photograph 23 Seljuk unglazed brick and terra-cotta geometric ornament from the exterior façade of the Friday Mosque at Gonabad, Iran (photograph by Farshid Emami; courtesy of the Aga Khan Documentation Center at MIT)

independent repetitive cells to create a hybrid geometric design with greater complexity was used in several Anatolian locations during the Seljuk Sultanate of Rum [Figs. 262–265]. Marinid artists in Morocco employed this practice at a later date, and a small number of examples were also produced by both Mamluk artists in Egypt and Mughal artists in India. However, this design from the northeast dome chamber of the Friday Mosque in Isfahan appears to be the first historical hybrid design that employs more than a single repetitive cell within a single construction. The sophistication of this hybrid design presupposes an earlier origin of the *fivefold system* than the 1088-89 date of the northeast dome chamber in Isfahan; and one can assume that prior to the inventive discovery that otherwise distinct repeat units were able to work together to produce a more complex geometric design, artists would have already been familiar with the independent application of these individual repeat units for standard surface coverage. It is important to note that in analyzing this hybrid design a certain amount of conjecture has been used to fill the two-dimensional plane beyond the obvious central pentagon and adjacent rhombic cells. The artist who created this remarkable design may well

have used a different combination of repetitive cells in the peripheral regions that extend beyond the central pentagon and contiguous rhombi. Indeed, the artist did not have to work with a continuous two-dimensional coverage at all, and may have just worked with the three rhombic cells on each side of the central pentagonal cell. This would have been enough to complete the design. It is interesting to consider that one way or another, an artist clever enough to have developed the use of the pentagon and rhombi with 72° and 108° included angles used in this design would have likely also discovered the need of the further rhombus with 36° and 144° included angles for full two-dimensional coverage. These more acute rhombi are implicit within this construction (for example, the upper point of the ten-pointed star at the apex of the arch), and were the artist who devised this design aware of the more acute rhombus, this individual may have been the first to discover the contiguous tiling potential of these two "Penrose rhombi." Although these rhombi have the ability for non-periodic application, or even aperiodic tiling with Penrose's matching rules [Fig. 480], the historical examples of Islamic geometric designs are invariably periodic with translation symmetry,

Photograph 24   A Seljuk dual-level design from the façade of the Gunbad-i Qabud in Maragha, Iran, wherein the primary pattern is created from the *fivefold system* (© Richard Mortel)

and there is no evidence that Muslim artists were aware of the non-periodic potential of the design methodology they employed. The application of repetitive pentagonal and rhombic cells in the design from the northeast dome chamber has bilateral symmetry that reflects upon the vertical line that bisects the arch. The applied repetitive cells do not fall into a recognizable periodic structure. This is due to the fact that there is too little information within the arch to determine whether there is a larger meta repeat that is unseen. As said, the artist may have just added rhombic cells to the central pentagonal cell until the arched region was covered. If the latter, which would seem likely, the three varieties of repetitive cell within this example could be extended outward from the line of symmetry to produce either a periodic repeat with translation symmetry or a non-periodic structure without translation symmetry. However, considering the very limited cellular exposition in this design, the question of whether this design is one or the other is essentially moot.

Still greater evidence for the significance of the northeast dome in the Friday Mosque at Isfahan to the history of Islamic geometric art is once again found in the multiple blind arches that surround the cupola. Like the fivefold examples cited above, one of these arches contains the earliest example of a sevenfold pattern known to this ornamental tradition [Fig. 279] [Photograph 26]. Unlike the nonsystematic Ghaznavid sevenfold designs from the minaret of Mas'ud III in Ghazni (1099-1115), this Seljuk pattern is systematic, which is to say that the underlying generative polygonal modules that make up the particular tessellation are part of a limited set of sevenfold elements that tessellate in innumerable ways [Fig. 271]. This underlying generative tessellation is made up of just two of these polygonal modules, both being irregular hexagons of differing proportions. At this early stage of development, it is impossible to know to what extent the artist was aware of these two generative hexagons as systematic modules with greater

Photograph 25   A Seljuk hybrid *acute* design with ten-pointed stars created from the *fivefold system* in the northeast dome chamber of the Friday Mosque at Isfahan (© Tom Goris)

tessellating potential. However, in light of the fact that the two fivefold examples from this same chamber are most distinctly systematic, it can be assumed that the same artist would have also known the systematic potential of the sevenfold polygonal modules. The example from the northeast dome chamber is an *acute* pattern with crossing pattern lines of 51.42857...°. This angular opening is determined from the inherent geometry of the heptagon. The absence of regular polygons, such as heptagons or tetradecagons (14 sides), within the underlying tessellation, and the concomitant absence of star forms with matching radial symmetry within the generated pattern, places this example into the *field pattern* category. The only other historical example of this sevenfold pattern is, significantly, an illustration from the anonymous Persian treatise *On Similar and Complementary Interlocking Figures* at the Bibliothèque Nationale de France in Paris.[56] This illustrated example, and its accompanying step-by-step instructions, is all the more

interesting in that the underlying generative polygonal tessellation is visually represented and textually described. Such depiction of the generative schema is extremely unusual, and this illustration is hence one of the rare examples of a primary source for the historicity of the polygonal technique.

In addition to creating designs from underlying polygonal tessellations that were systematic, Seljuk artists also derived patterns from tessellations that were nonsystematic. Among the earliest Seljuk examples of nonsystematic pattern making is also from one of the blind arches that surround the northeast dome chamber of the Friday Mosque at Isfahan (1088-89) [Photograph 27]. The underlying generative tessellation for this design places edge-to-edge regular pentagons upon each triangular edge of the isometric grid [Fig. 309a]. This creates two interstice elements that are specific to this pentagonal arrangement: a six-pointed star at the vertices of the isometric grid, and an irregular ditrigon at the centers of each triangular repeat unit. The *acute* design that was extracted from this underlying tessellation is very successful, with incorporated regular heptagons within the

---

[56] MS Persan 169, fol. 192a.

Photograph 26   A Seljuk *acute* pattern created with underlying modules included in the *sevenfold system* from the northeast dome chamber of the Friday Mosque at Isfahan (© Tom Goris)

pattern matrix. Of particular interest is the pattern treatment within the central ditrigonal element. This motif—with slightly different proportions—became a relatively common feature within the *system of regular polygons* [Figs. 117–120]. Zangid artists used a variant of this nonsystematic design at the Nur al-Din Bimaristan in Damascus (1154) [Fig. 309c], but what is especially interesting is that the anonymous Persian treatise *On Similar and Complementary Interlocking Figures* also contains an illustration of a close variation of the nonsystematic design from the northeast dome chamber[57] [Fig. 309b]. The occurrence of these two examples in both locations provides further evidence of a likely direct association between this manuscript and the architectural ornament of this portion of the Friday Mosque in Isfahan.

Among the most common nonsystematic patterns are those with just 12-pointed stars as the higher order star form. As with designs with 12-pointed stars created from the *system of regular polygons*, this variety of nonsystematic

pattern will invariably place the 12-pointed stars upon the vertices of the repetitive grid. Being that 12 is divisible by 6, 4, and 3, nonsystematic designs with 12-pointed stars can repeat on the regular hexagonal grid, the orthogonal grid, and the isometric grid. Seljuk locations of fourfold nonsystematic patterns with 12-pointed stars include the arched tympanum over an entry gate near the northeastern *iwan* at the Friday Mosque at Isfahan[58] (after 1121-22) [Fig. 335a]; the minaret of the Great Mosque of Siirt, Turkey (1129) [Fig. 335b]; and a pair of bronze doors from the Seljuk *atabeg* of Cizre, Turkey (thirteenth century) [Fig. 337]. An Artuqid example, with strong Seljuk influences, is found in the carved stucco back wall of the *mihrab* niche of the Great Mosque of Silvan, Turkey (1152-57) [Fig. 336a]. Two Seljuk-influenced examples of nonsystematic threefold patterns with just 12-pointed stars located at each vertex of the isometric grid are found on the Mengujekid *minbar* at the Great Mosque of Divrigi in Turkey (1228-29). The *acute* design within the triangular side panel employs an

---

[57] MS Persan 169, fol. 193a.

[58] Ettinghausen, Grabar and Jenkins-Madina (2001), 141, pl. 215.

Photograph 27 A Seljuk nonsystematic threefold *acute* pattern with six-pointed stars and heptagons in the northeast dome chamber of the Friday Mosque at Isfahan (© Tom Goris)

underlying tessellation that separates the underlying dodecagons with edge-to-edge pentagons [Fig. 300a *acute*]. Immediately adjacent to this is a vertical panel with an *acute* design created from a modified version of this same underlying tessellation wherein the pentagons are truncated into trapezoids [Fig. 320]. The use of two designs with such closely related methodological origin would appear to have been a willful decision on the part of the artist. Other examples of Seljuk-influenced threefold patterns that place 12-pointed stars at the vertices of the isometric grid include the carved stucco ornament of the Abbasid Palace of the Qal'a in Baghdad (c. 1220) [Fig. 300b *acute*] [Photograph 28], and a carved stucco wall panel at the Mustansiriyah in Baghdad (1227-34) [Fig. 300a *acute*]. These two late Abbasid buildings were constructed only decades after the overthrow of Seljuk dominion over Baghdad, and have strong stylistic affiliations with Seljuk ornament. The Friday Mosque at Barsian (1105) is remarkable in a number of respects. The dome and supporting *muqarnas* squinches in this Seljuk building are remarkably similar to that of the southwest dome chamber of the Friday Mosque in Isfahan

(1086). These two buildings were constructed within 20 years of one another and are in relatively close proximity. Like the Seljuk ornament in Isfahan, the Friday Mosque in Barsian also contains several interesting nonsystematic geometric designs. Two of the arched *muqarnas* faces in the *mihrab* of this mosque are decorated with nonsystematic orthogonal patterns with eight-pointed stars. One of these places eight-pointed stars on the vertices of the square repeat, and octagons at the center of the repeat [Fig. 331a]. This is a standard feature of patterns created from the *fourfold system B*. However, the three varieties of irregular pentagon and the irregular triangles are nonsystematic, and the *acute* pattern that the underlying polygonal tessellation produces is unusual. Another one of the *muqarnas* arch faces in the *mihrab* of the Friday Mosque at Barsian is decorated with a nonsystematic orthogonal design comprised of five-, six-, seven-, and eight-pointed stars [Fig. 332a]. Of these, only the six- and eight-pointed stars have regular rotational symmetry. This same design was used in several other locations that were strongly influenced by the Seljuks, including the Danishmend portal of the Great Mosque of

Photograph 29   A Seljuk nonsystematic fourfold *acute* pattern with 8- and 12-pointed stars from the Friday mosque at Isfahan (© David Wade)

Photograph 28   A late Abbasid nonsystematic threefold *acute* pattern with 12-pointed stars from the Abbasid Palace of the Qal'a in Baghdad (photograph by K. A. C. Creswell; © Ashmolean Museum, University of Oxford)

Niksar, Turkey[59] (1145); the Ildegizid façade of the Mu'mine Khatun mausoleum in Nakhichevan, Azerbaijan (1186); and the Qara Qoyunlu portal of the Great Mosque in Van, Turkey (1389-1400). These later examples only differ in their widened interweaving line treatment [Fig. 332b, c]. The same underlying generative tessellation that produced both of these examples was used by Mamluk artists some 200 years later to produce a more complex design at the Amir Sarghitmish *madrasa* in Cairo (1356) [Fig. 332e]. In addition to the examples from Barsian, Nakhichevan, and Niksar, Seljuk artists produced several additional patterns with five-, six-, seven-, and eight-pointed stars; and this seems to have been a somewhat popular geometric theme. The design in the *mihrab* arch spandrel at the Gar mosque (1121-22) in the outskirts of Isfahan employs such a design, although the amount of geometric information contained within each triangular panel is insufficient to definitively determine either the repeat pattern or the underlying

polygonal structure. Nonetheless, the five-, six-, seven-, and eight-pointed star structure is apparent within the limited context. Similarly, among the ornamented arched *muqarnas* faces in the exterior façade of the Gunbad-i Qabud in Maragha (1196-97) is the repetitive use of a design with this same combination of star forms. And once again, such a design was used at the Izzeddin Kaykavus hospital in Sivas, Turkey (1217-18).[60]

The southern interior corner of the southeastern *iwan* of the Friday Mosque at Isfahan includes a small blind arch decorated with a Seljuk example of a carved stucco nonsystematic compound pattern comprised of 8- and 12-pointed stars [Photograph 29]. This mosque went through multiple restorations and additions by subsequent dynasties, and the dating of specific unattributed features is frequently problematic.[61] That said, this example is stylistically similar to the Seljuk geometric ornament within the nearby northeast dome chamber. This design can be constructed from either of the two underlying polygonal tessellations: one with edge-to-edge dodecagons and octagons with concave hexagonal interstice regions [Fig. 379d], and the other with dodecagons and octagons separated by a matrix of irregular pentagons and barrel hexagons [Fig. 379f]. A later Seljuk example of this same pattern was used in an exterior border that runs vertically along the sides of the north *iwan* of the Friday Mosque at Gonabad (1212) [Fig. 379e] [Photograph 23]. Indeed, multiple examples of this same *acute* design

---

[59] Schneider (1980), pattern no. 352.

[60] Schneider (1980), pattern no. 351.

[61] Ettinghausen, Grabar and Jenkins-Madina (2001), 140–143.

were used widely by succeeding Muslim cultures [Photograph 46]. A slightly later example of this same *acute* pattern was used in the carved stucco ornament at the Abbasid Palace of the Qal'a in Baghdad (c. 1220). A remarkable example of an orthogonal pattern with compound local symmetries is found on one of the small arched surfaces of the *muqarnas* hood in the *mihrab* of the Friday Mosque at Barsian (1105). This unglazed ceramic mosaic ornament is an *acute* design that combines 12- and 16-pointed stars [Fig. 392b]. This compound orthogonal design has considerably greater complexity than other contemporaneous orthogonal designs.

The Seljuks excelled in creating complex nonsystematic geometric patterns from a relatively early date. At its most sophisticated, this variety of geometric pattern will frequently include seemingly irreconcilable combinations of star forms, and will frequently require repeat units other than the standard triangle, square, or regular hexagons. The tradition of especially complex patterns with compound local symmetries reached full maturity under the auspices of the Seljuk Sultanate of Rum in Anatolia and the Mamluks in Egypt during the thirteenth and fourteenth centuries; but the antecedents and earliest examples of this variety of design were established during the twelfth century by the Great Seljuks and their *atabeg* subordinates. In seeking an understanding of the historical development of particularly complex geometric patterns with multiple regions of differentiated local symmetry, it is important to take into account the tremendous loss of early monuments in Transoxiana, Khurasan, Persia, and Iraq through natural disasters, neglect, and especially the Mongol destruction during the thirteenth century. As per the previous cited example, the Friday Mosque at Barsian (1105) is of particular significance to the early history of this variety of geometric pattern. The *mihrab* of this mosque is framed by a very interesting nonsystematic design comprised of seven- and nine-pointed stars that repeats upon an elongated hexagonal grid [Fig. 429]. This is an early example of a geometric design that fills the two-dimensional plane by virtue of a geometric ploy whereby the numeric quality of the alternating star forms is one numerical step above and below the number of stellate points of a more common and convenient design with singular repeating stellations, such as 6-, 8-, 10-, or 12-pointed stars. For example, the fact that six-pointed stars will conveniently repeat is an indication that a compound pattern can be created that employs both five- and seven-pointed stars. The pattern from the *mihrab* of the Friday Mosque at Barsian applies this *principle of adjacent numbers* to the repetitive convenience of the octagon, indicating the potential for a successful, if considerably less geometrically convenient, repetitive pattern with seven- and nine-pointed stars. It is impossible to know whether Muslim artists of the past were aware of this as a design principle per

se, or whether their creation of such patterns comprised of 5- and 7-, 7- and 9-, 9- and 11-, or 11- and 13-pointed stars was purely serendipitous. In addition to the orthogonal design with 12- and 16-pointed stars cited above, other examples of particularly complex geometric patterns are included among the arches of the *muqarnas* hood in the *mihrab* from the Friday Mosque at Barsian. These include a pattern comprised of 13-pointed stars, and another comprised of 11- and 12-pointed stars. The limited amount of geometric information contained in each of these two examples is insufficient to conclusively determine either the repetitive structure or the complete underlying generative tessellation, and it is possible that the artist distributed 11-, 12-, and 13-pointed stars into these two small arched regions without their being part of of an actual repetitive structure.

Other Seljuk examples of nonsystematic compound patterns include a design with five-, six-, and seven-pointed stars in the arch spandrels at the top of each exterior wall of the decagonal façade on the Gunbad-i Qubad in Maragha, Iran (1196-97), and an adjacent pattern with eight- and nine-pointed stars that frames the *muqarnas* arch at the top of each exterior wall of the façade of the same building. It is interesting to note what would appear to be the deliberate decision by the artist to juxtapose the pattern with five-, six-, and seven-pointed stars with a pattern comprised of sequenced eight- and nine-pointed stars. As with the exceptionally large repeat unit of the fivefold *obtuse* design that surrounds this tomb tower, the use of two adjacent complex designs that have continuous sequenced numeric qualities is very unusual, and emphasizes the unique character of this building.

Among the greatest achievements of Seljuk geometric artists is the pioneering application of geometric patterns onto the surfaces of domes. Subsequent Muslim dynasties followed in this design convention, and exceptional examples with greater complexity, were produced by the Zangids and Ayyubids in Syria, the Nasrid and Christian *Mudéjar* artists in Spain, the Mamluks in Egypt, the Muzaffarids and Timurids in Persia and Central Asia, and the Mughals in India. By comparison, the early work of the Seljuks appears simplistic. Indeed, the elaborate ribbed vault of the Sultan Sanjar mausoleum in Merv, Turkmenistan (1157), for all its boldness and beauty, does not exhibit particular geometric complexity. This design employs an eight-pointed star at the apex, and the design unfolds upon an eightfold radial division of the domical surface. The earlier Seljuk dome of the Friday Mosque at Golpayegan (1105-18) similarly places an eight-pointed star at the apex, and employs eightfold radial segmentation of the surface. Surrounding the raised brick central eight-pointed star are 8 seven-pointed stars, followed downward by 8 five-pointed stars, and culminating at the periphery with a ring of 8 eight-pointed stars divided in half [Fig. 491]. This stellar matrix,

Photograph 30 A Seljuk domical geometric design governed by dodecahedral symmetry in the northeast dome chamber of the Friday Mosque at Isfahan (© Tom Goris)

while still rather simple when compared to the non-Euclidean work of subsequent generations of Muslim artists, has all the visual characteristics of a pattern that was produced using the methodology of the polygonal technique. The dome at Golpayegan is the earliest extant example in Islamic architecture of the application of a geometric design to the surface of a dome using radial gore segments as the repetitive device. However, Seljuk artists were also pioneers of the other principal method of applying geometric patterns onto domical surfaces: the use of polyhedral geometry as the repetitive strategy for controlled domical surface coverage. The earliest use of polyhedra for creating a non-Euclidean geometric design is from the northeast dome of the Friday Mosque at Isfahan (1088-89) [Photograph 30].[62] This dome is remarkable on several counts, not the least of which is the fact that the magnificent dome is decorated with a *two-point* pattern derived from the underlying geometry of the dodeca-hedron [Fig. 496]. The dodecahedron is comprised of 12 pentagonal faces, and the application of *two-point* pattern lines onto each underlying domical pentagon creates the distinctive and unusual fivefold symmetry of this dome. The pentagonal faces of the dodecahedron are spherically projected onto the curved surface of the dome, and it is important to point out that the use of the dodecahedron would ordinarily produce a hemispherical dome. However, the curvature of the northeast dome rises to an apex. While the applied *two-point* pattern is unquestionably derived from the dodecahedron, this otherwise spherical surface has been modified to emphasize the characteristic ascendancy of the traditional Persian pointed dome. The use of underlying generative pentagons projected to a domical surface aligns this example with patterns produced from the *fivefold system*: the difference being that the two-dimensional plane requires at least one other module from the *fivefold system* to accompany the generative pentagons, while the dodeca-hedron is a spherical tessellation of regular pentagons alone. Along with the two previously discussed fivefold patterns within the small blind arches immediately beneath this dome, the ornament of the northeast dome chamber in the Friday Mosque at Isfahan has the distinction of having the earliest extant examples of Islamic geometric designs with fivefold symmetry: predating both the Ghaznavid nonsys-tematic fivefold patterns on the minaret of Mas'ud III

---

[62] Bonner (2016).

[Fig. 206] and the Seljuk example from the interior wall of the Friday Mosque at Golpayegan [Fig. 229b] by 10–20 years. It has been postulated that the great Persian mathematician and poet, 'Umar Khayyam, may have designed the northeast dome of the Friday Mosque at Isfahan.[63] Certainly he was living in Isfahan at the time of this dome's construction, and enjoyed the scientific patronage of Taj al-Mulk who commissioned the dome. As a prominent mathematician of his time, 'Umar Khayyam would have been very familiar with polyhedral geometry and spherical projection: a requisite of the designer of this important monument.[64] If true, and especially in light of the relationship between the *two-point* geometric pattern on the dome and those employed within the eight recessed arches of the domed chamber, 'Umar Khayyam may have been highly significant not just as a mathematician and poet, but also to the historical development of the polygonal technique: the design methodology most responsible for the mature style of Islamic geometric design. Such a confluence of mathematics, poetry, and geometric art is a delight to the imagination.

The fivefold domical geometric design in the northeast dome chamber of the Friday Mosque at Isfahan, together with the series of geometric patterns placed within the eight recessed arches, represents a remarkable advance in the historical development of Islamic geometric design. The many "first occurrences" present in this chamber opened the door to the fully mature geometric design practices that soon followed. As such, the importance of these patterns to the history of Islamic geometric art is paramount, and firmly establishes Seljuk artists as fundamental innovators in the furtherance of this tradition. The design innovations that were first introduced during the construction of this building include the first use of underlying ditrigonal modules within the *system of regular polygons* [Fig. 118a] [Photograph 19]; the earliest occurrence of the classic fivefold *obtuse* design [Fig. 229a] [Photograph 21]; the earliest fivefold *acute* design, in this case a hybrid design with multiple repetitive cells [Fig. 261] [Photograph 25]; the earliest fivefold *two-point* design, in this case on the dome [Photograph 30] [Fig. 496]; the first pattern with sevenfold symmetry created from the *sevenfold system* [Fig. 279] [Photograph 26]; the earliest example of a nonsystematic design [Fig. 309a] [Photograph 27]; and the first occurrence of a domical geometric pattern that uses a polyhedron as its repetitive schema

[Fig. 496] [Photograph 30]. What is more, the 3 fivefold designs and the 1 sevenfold pattern are the earliest sophisticated examples of these two types of symmetry known to have been produced by humankind the world over. It is doubtful that any other single room, or even individual building within the totality of Islamic architecture, had such a profound significance to the historical development of Islamic geometric art.[65]

1.13   Ghurids (1148-1215)

Following their defeat by the Seljuks at the Battle of Dandanaqan (1040) and the loss of their vast western territories, the Ghaznavids were forced to negotiate a peace treaty with the Seljuks that brought relative stability to Khurasan for approximately a hundred years. In 1150 Ghazna fell to the Ghurid forces of Ala'uddin Hussain. Within two decades the Ghaznavids were driven from their homelands in Khurasan to their eastern territories in Sindh, and in 1187 the Ghurids further defeated the Ghaznavids in Lahore, bringing an end to the Ghaznavid Empire. The Ghurids are thought to have been Tadjiks of eastern Iranian origin that migrated to their homeland of Ghur, in central Afghanistan, at an undetermined time. Ghur is mountainous and provided an ideal defensive location against the largely unsuccessful attempts to conquer this region by the Ghaznavids and Seljuks. Following the final defeat of the Ghaznavids at Lahore, the Ghurids expanded their empire to include most of modern-day Afghanistan and Pakistan, as well as much of northern India. Their first capital was Firuzkuh (present-day Jam) but as they spread eastward they also established capitals in Ghazna, Lahore, and eventually Delhi. The Ghurids were avowed Sunnis and recognized the religious authority of the Abbasid Caliph in Baghdad. As with the Ghaznavids, Persian cultural affinities flourished under Ghurid rule, and great emphasis was placed upon poetry, literature, and arts. Ghurid control over their Afghan territories came to an end in 1215 following their defeat by the Khwarizmshahs, but control over their eastern territories in the Indian subcontinent was maintained through the assumption to power of the Mamluk Sultanate of Delhi.

Like the Seljuks, the Ghurids fully embraced the dynamic architectural practices of the Ghaznavids. Like other Muslim dynasties, the Ghurids approached the design of their architectural monuments, in part, as a way of commemorating their ascendancy as a sovereign force, as well as glorifying Islam within their territories with large non-Muslim populations, such as the Indus Valley and northern India. Significant Ghurid architectural monuments in Khurasan

---

[63] –Grabar (1990), 85, note 5.
  –Özdural (1998), 699–715.
  –Hogendijk (2012), 37–43.

[64] The works of the mathematician and astronomer, Abu al-Wafa Buzjani (940–998), would have been familiar to 'Umar Khayyam, and of especial relevance to this discussion would have been his work on right-angled spherical triangles and spherical trigonometry.

[65] Bonner (2016).

Photograph 31   A Ghurid *two-point* pattern originally located in Lashkar-i Bazar, Afghanistan, that is easily created from the *system of regular polygons* (© Thalia Kennedy)

include portions of Lashkar-i Bazar, the arch at Bust, the two mausolea at Chisht, the Shah-i Mashhad in Gargistan, the Friday Mosque at Herat, and the minaret of Jam. This latter building is located deep in the Ghur Mountains of central Afghanistan at the confluence of the Hari and Jam rivers. It stands 65 m in height and is the second tallest historical minaret in the Islamic world.

The surviving Ghurid monuments in Khurasan are relatively few, but the sophistication of the ornamental design and the quality of execution are equal to the finest work of the Ghaznavids. As with the Seljuks, the Ghurids added to the established ornamental practices of the Ghaznavids by introducing turquoise glazed faience into their exterior façades: enlivening key ornamental components, such as calligraphy, with vivid color in an otherwise monochrome aesthetic. This innovative approach to architectural ornament was no less focused upon the further development of geometric design, and the use of the *system of regular polygons* continued as a primary methodology employed in

the creation of geometric patterns. The carved stucco ornament from a Ghurid *mihrab* at Lashkar-i Bazar[66] (after 1149) employs a threefold geometric design that is easily constructed from an underlying 3.6.3.6 tessellation of triangles and hexagons [Fig. 99d] [Photograph 31]. This is a *two-point* pattern that locates the pattern lines upon two points on each underlying polygonal edge rather than the more common single midpoint location. As demonstrated in this example, the occurrence of multiple closed-loop elements within the pattern matrix is a typical and distinctive feature of the *two-point* family of geometric patterns. This same pattern was used in several locations historically, including an earlier panel in a wooden *maqsura* from the mausoleum of the Seljuk *atabeg* Sultan Duqaq in Damascus (1095–1104). The Ghurid arch soffit of an *iwan* at the Friday

---

[66] In the collection of the National Museum of Afghanistan in Kabul, Afghanistan.

Mosque in Herat, Afghanistan (1200), employs a raised brick *two-point* pattern of superimposed hexagons that is easily constructed from the 6^3 tessellation of regular hexagons [Fig. 96d]. The earliest known use of this design was by the Umayyads of Spain in a tenth-century marble window grille. Like much of the initial Ghurid ornament of this monument, this pattern was covered with later Timurid cut-tile mosaic, but eventually revealed through degradation of the Timurid work. The arch spandrel immediately above this *iwan* employs another threefold geometric pattern that can likewise be constructed from an underlying 6^3 tessellation of regular hexagons [Fig. 98a]. As with several earlier Ghaznavid and Seljuk examples, this pattern applies two varieties of pattern line into adjacent underlying hexagons. The primary underlying hexagons dictate the design by centrally placing six-pointed stars with 60° crossing pattern lines upon the midpoints of each primary hexagon's edges, while four of the edges of each secondary underlying hexagon also place 60° crossing pattern lines upon the midpoints, and a pattern line is drawn between two opposite vertices. This pattern closely resembles a later Ottoman design from the Great Mosque of Bursa (1396-1400) that also is created from alternating active and passive underlying hexagons [Fig. 98b]. The greater regularity in the size of the background elements of the Ottoman design produces a more successful design. The remote ruins of the Shah-i Mashhad in Gargistan, Afghanistan (1176), contain a profuse assortment of ornamental motifs, including knotted *Kufi* calligraphy, highly elaborate braided borders, simple floral designs, and an assortment of geometric patterns. These include several threefold patterns that can be constructed with the *system of regular polygons*. The ornament on this remote Ghurid *madrasa* is mostly of exquisitely executed raised brickwork and molded terra-cotta tiles and inserts. A threefold pattern on one of the remaining *iwans* can be easily produced from the 3.4.6.4 arrangement of triangles, squares, and hexagons as the underlying generative tessellation [Fig. 104c]. This design is reminiscent of the window grille from Córdoba in that both are made up solely of superimposed hexagons in rotation around a hexagonal nodal center. However, in the case of this pattern from Shah-i Mashhad, each of the superimposed hexagons is elongated rather than regular. The façade of the western mausoleum at Chisht, Afghanistan (1167), employs a threefold pattern that can also be made from the 3.4.6.4 underlying tessellation [Fig. 105a]. This raised brick pattern was used inside one of the large exterior blind arches at the side of the main portal. The parallel pattern lines that characterize this design emphasize both the hexagonal grid and its isometric dual. This pattern is simply derived by initially placing a hexagon within each underlying triangle. An identical Seljuk example of this same design was used in the Friday Mosque at Gonabad, Iran, some 50 years later (1212). The

minaret of Jam (1174-75 or 1194-95) includes a variation of the 3.4.6.4 design from Chisht. The difference between these two Ghurid examples is that the pattern from Jam has an added hexagonal element at each of the vertices of the triangular repeat [Fig. 105c]. The addition of this hexagon is independent of the underlying tessellation and was a purely arbitrary decision on the part of the artist. This minaret also employs a threefold pattern with 12-pointed stars within one of the panels at the base of the structure that can be easily constructed with the 4.6.12 underlying tessellation of squares, hexagons, and dodecagons [Fig. 109f]. This example is largely characterized by large dodecagons within the pattern matrix. Another Ghurid decagonal pattern that was used more or less contemporaneously at both the minaret of Jam and the Shah-i Mashhad is a fourfold design that repeats on the orthogonal grid, with the underlying dodecagons placed at the vertices of the square repeat units [Fig. 120]. An unusual feature of the underlying tessellation that produces this design is the further infill of each dodecagon with four ditrigons and four triangles. This infill allows for the rather ingenious transformation of the finished pattern from what would have been 12-pointed stars at the vertices of the square repeat unit to regular octagons. This same distinctive design was used on the *minbar* of the Alaeddin mosque in Konya (1219-21) some 50 years later.

One of the Ghurid carved stucco panels from Lashkar-i Bazar[67] (after 1149) is comprised of a network of superimposed four-pointed stars. This is easily created from the 4.8^2 underlying tessellation of squares and octagons [Fig. 128c]. Artists used this rather simple orthogonal design approximately 100 years later during the Seljuk Sultanate of Rum at the Karatay *madrasa* (1251-55). A raised brick design with eight-pointed stars from the façade of the western mausoleum at Chisht can be produced with several methodologies, including the polygonal technique, whereby the design is easily generated from the underlying 4.8^2 tessellation of squares and octagons [Fig. 129c]. The Ghurids used this same design on the minaret of Jam (1174-75 or 1194-95), and the wide-ranging popularity of this design is attested to by its use in one of the marble window grilles at the Great Mosque of Córdoba (980-90), and the *minbar* of the al-Aqsa mosque in Jerusalem (1168). The minaret of Jam also employs an example of the classic star-and-cross design that is likewise easily created from the 4.8^2 tessellation [Fig. 124b].

Multiple Ghurid examples of raised brick patterns that are easily created from the *fourfold system A* include a simple *median* border device on the façade of the western mausoleum at Chisht (1167), as well as at the Shah-i Mashhad in

---

[67] In the collection of the National Museum of Afghanistan in Kabul, Afghanistan. See Crane and Trousdale (1972), 215–226.

Photograph 32 A Ghurid *two-point* pattern at the Friday Mosque at Herat, Afghanistan, created from the *fourfold system B* (© Thalia Kennedy)

Gargistan, Afghanistan (1176) [Fig. 138c]. This is predated by an identical Seljuk example at the eastern tomb tower in Kharraqan by 100 years (1067). The minaret of Jam in central Afghanistan (1174-75 or 1194-95) has several panels with a *median* pattern that enjoyed ongoing use by many Muslim cultures [Fig. 145]. This design was also used previously by Seljuk artists at both the minaret of the Friday Mosque at Damghan, Iran (1080), and the minaret at Daulalabad, near Balkh, Afghanistan (1108-09) [Photograph 20]. This popular design is closely related to another Ghurid raised brick example from the eastern tomb tower at Chisht, Afghanistan (1197) [Fig. 143a]. This design from Chisht was used 18 years earlier by Saltukid artists at the Great Mosque of Erzurum, Turkey (1179).

An outstanding Ghurid example of a pattern created from the *fourfold system B* is found in the Friday Mosque at Herat, Afghanistan (1200) [Fig. 174b] [Photograph 32]. The specially cut raised brickwork is augmented with circular turquoise glazed plugs set within the backgrounds of the eight-pointed stars and pentagons. This is a *two-point* pattern that includes an arbitrary treatment of the central region of the repeat unit. This design can also be created from the simple

4.8² underlying tessellation of squares and octagons [Fig. 128h]. However, the use of the underlying tessellation from the *fourfold system B* more specifically relates to the geometric composition of the design. The introduction of the arbitrary square element within the pattern matrix was a popular device in *two-point* patterns created from this system. This is associated with the cluster of five pentagons within the underlying generative tessellation. Examples of other patterns that exhibit this distinctive feature include the Qarakhanid southern anonymous mausoleum in Uzgen, Kyrgyzstan (1186) [Fig. 176c]; the Mamluk painted ceiling of the Sultan al-Mu'ayyad Shaykh complex in Cairo (1415-22) [Fig. 176b]; the Sidi Madyan mosque in Cairo (1465) [Fig. 176a]; Bimarhane hospital in Amasya, Turkey (1308-09) [Fig. 174c]; and the Aqbughawiyya *madrasa* in the al-Azhar mosque in Cairo (1340) [Fig. 174a].

The soffit of the Ghurid Arch at Bust, Afghanistan (1149), is beautifully decorated with a pattern that is easily made from the *fivefold system* [Fig. 226c] [Photograph 33]. This is a masterpiece of monochrome architectural ornament: both for its early innovative use of fivefold geometric design and for the precision of the specially cut raised brickwork and the refinement of the vegetal insert plugs that rest below the surface of the geometric pattern. The repeat unit for this geometric design is a rhombus with 72° and 108° angles. This remarkable design was produced at a time when patterns created from the *fivefold system* were just beginning to enter the lexicon of Islamic ornamental motifs, and is the earliest known example of the classic *acute* design that, over time, became ubiquitous to this tradition. The underlying tessellation for this *acute* pattern is comprised of just three polygonal modules: the decagon, pentagon, and barrel hexagon. This same underlying generative tessellation was responsible for the Seljuk *obtuse* pattern used some 60 years earlier in one of the blind arches in the northeast dome chamber of the Friday Mosque at Isfahan (1088-89) [Fig. 229b] [Photograph 21], and the Seljuk *two-point* pattern in the tympanum over the entry of the Gunbad-i 'Alaviyan in Hamadan, Iran (late twelfth century) [Fig. 231d] [Photograph 22]. When considering the history of this classic *acute* pattern, it is highly significant that the rhombic elements with 72° and 108° included angles that are included in the repetitive make up the fivefold hybrid design in the northeast dome chamber in Isfahan are ornamented with the same *acute* pattern, albeit in association with the pattern lines placed within the other repetitive hybrid components [Fig. 261].

1.14   Ildegizids (1136-1225)

The Ildegizids of Azerbaijan came to power as Seljuk *atabegs* in 1136. They gained independence from the Seljuks in 1194, and at the height of their power the Ildegizids

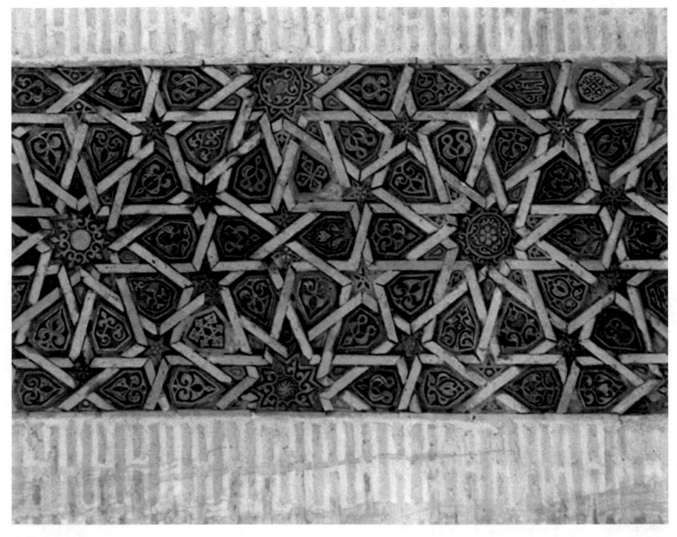

Photograph 33 A Ghurid example of the classic *acute* pattern created with the *fivefold system* from the soffit of the Ghurid Arch at Bust, Afghanistan (© Bernard O'Kane)

controlled the region stretching from Isfahan in the southeast to the borders of the Kingdom of Georgia to the northwest. The Khwarizmshahs overthrew them in 1225. As with the Seljuks, they were of Turkic origin with strong affiliations for Persian culture and language, and like their Seljuk suzerains, the Ildegizids were patrons of the geometric arts. Indeed, several outstanding examples of Islamic geometric art, created from the polygonal technique, were produced by this culture.

The use of the *system of regular polygons* figured into their geometric aesthetic. The façade of the Mu'mine Khatun mausoleum in Nakhichevan, Azerbaijan (1186), includes two patterns made from this system. One of these is the frequently used pattern of superimposed interweaving hexagons easily derived from the 3.6.3.6 underlying tessellation, the earliest known example of which is one of the marble window grilles at the Great Mosque Córdoba [Fig. 96e]. This is located in the upper spandrels of the

three-tiered *muqarnas* arch at the top of the Mu'mine Khatun mausoleum. The second example is also associated with the 3.6.3.6 tessellation, but with different pattern lines applied to alternating underlying hexagonal and triangular modules [Fig. 101d]. This is located above the portal, and the approximate proportions of this design can also be produced from the isometric grid [Fig. 73c]. Interestingly, this design from the Mu'mine Khatun mausoleum is very similar to the pattern over the portal at the nearby mausoleum of Yusuf ibn Kathir in Nakhichevan (1161-62) [Fig. 101c]. This variation was also used on a carved stone lintel above the Zangid portal of the Bimaristan Arghun at the citadel of Aleppo[68] [Photograph 36], and a Seljuk Sultanate of Rum stone relief at the Hatun Han near Pazar, Turkey (1238-39). The

---

[68] Terry Allen identifies the origin of the portal as predating the rest of the Bimaristan Arghun. See Allen (1999).

mausoleum of Yusuf ibn Kathir is a small Ildegizid tomb
tower with an octagonal plan. The design in one of the
exterior façades exhibits the generative 3.6.3.6 tessellation
along with included octagons placed upon the vertices of the
hexagons and triangles [Fig. 51c]. The harmonious placement of octagons within a threefold design is an example of
an *imposed symmetry* design, and works by virtue of the two
perpendicular lines of reflective symmetry at each 3.6.3.6
vertex. This design can also be created from an underlying
3.4.6.4 generative tessellation [Fig. 107a], in which case the
3.6.3.6 motif is an arbitrary inclusion. It is interesting to
compare this with the geometric structure of the pattern in
the adjacent façade. In contrast to the 3.6.3.6 design that
places arbitrary fourfold elements (octagons) into the otherwise threefold structure of the design, the pattern inside the
adjacent panel arbitrarily places a threefold motif of
six-pointed stars and a surrounding hexagon into a fourfold
orthogonal pattern matrix [Fig. 52b]. This is accomplished
by placing the threefold elements upon the midpoints of each
edge of the square repeat unit. This juxtaposition of imposed
symmetry designs with converse symmetrical characteristics
and arbitrary inclusions indicates an admirable and adroit
playfulness on the part of the artist.

Ildegizid artists did not make frequent use of the *fourfold
system A*. One fine example is a *median* field design created
from an underlying tessellation of large and small hexagons,
pentagons, and interstice rhombi [Fig. 141]. As with three of
the above-cited Ildegizid examples, this is also from one of
the façades of the mausoleum of Yusuf ibn Kathir. This
design was subsequently used during the Seljuk Sultanate
of Rum at the Haunt Hatun in Kayseri (1238), and the Haci
Kiliç mosque and *madrasa* also in Kayseri (1249).[69]

The Mu'mine Khatun mausoleum makes use of several
designs created from the *fourfold system B*. The more basic
of these is the classic *acute* pattern that was used with great
frequency throughout the Islamic world, and had already
been featured some 50 years previous at the Friday Mosque
at Sin (1134) [Fig. 173a]. The most remarkable of the *fourfold system B* designs at the Mu'mine Khatun mausoleum is
an *acute* pattern located in one of the long vertical panels
decorating the exterior façade of the tomb [Fig. 182]. This is
a hybrid design characterized by the employment of two
separate repetitive cells within the overall schema: the
square and rhomb. There is a long history of Islamic geometric patterns that fill the two-dimensional plane with more
than a single variety of repetitive cell: the earliest extant
example being the above-cited fivefold hybrid design from
the northeast dome chamber in Isfahan [Photograph
25]. There are two ways of tessellating the plane using just
squares and rhombi. One of these places the rhombi in a

Photograph 34 A late Abbasid hybrid *acute* pattern from the
Abbasid Palace of the Qal'a in Baghdad that employs both square and
rhombic repetitive elements and is created with the *fourfold system A*
(photograph by K. A. C. Creswell; © Ashmolean Museum, University
of Oxford)

pinwheel-like rotation around each square, and the other
places the rhombi and squares in a tessellation of alternating
linear bands. The repetitive structure of the example from the
Mu'mine Khatun mausoleum is the latter type, with rhombi
that have 45° and 135° included angles. Any variety of rhomb
will tessellate with a square in such a linear arrangement: the
requirement being that the length between the two obtuse
angles of the rhomb be equal to the edge length of the square,
thus defining the coincident edges of the square and triangle
produced from the half rhombi. In order for the pattern to flow
seamlessly across these two repeat units it is necessary for the
underlying polygonal tessellation to share the same distribution of polygonal modules along the coincident edges of each
repeat unit. This hybrid design from the Mu'mine Khatun
mausoleum was also used at the Abbasid Palace of the
Qal'a in Baghdad (c. 1230) [Photograph 34]. It is worth noting
that each of these repetitive cells will work independently to
fill the two-dimensional plane. In the case of the example
from the Mu'mine Khatun mausoleum, the square repeat unit
is the classic *acute* pattern that, on its own, is found on the

---

[69] Schneider (1980), pattern no. 281.

same building high above the entry door [Fig. 173a]. The rhombic repetitive element at the Mu'mine Khatun mausoleum was likewise used independently within this tradition: for example, at the Izzeddin Kaykavus hospital and mausoleum in Sivas, Turkey (1217) [Fig. 181]. However, this example from the Mu'mine Khatun mausoleum is the first to use this *fourfold system B* repetitive rhombic element, and the first design from the *fourfold system B* to employ a hybrid approach to filling the two-dimensional plane with two distinct repetitive cells.

Among the diversity of geometric patterns that adorn the Mu'mine Khatun is an outstanding panel created easily from the *fivefold system*. This example employs a rhombic repeat unit with 72° and 108° included angles, and, like other designs produced with this system, places the primary underlying decagons upon the vertices of the rhombic grid [Fig. 232g]. This is an *acute* pattern derived from an underlying tessellation of decagons, elongated hexagons, trapezoids, and large concave hexagons with edges that are equal to the long edge of the trapezoid [Fig. 232h]. It is worth noting that this design can also be created from an underlying tessellation of just decagons and concave hexagons [Fig. 232f], as well as an arbitrary modification of the classic *median* pattern [Fig. 227e]. Generative ambiguity of this nature is not uncommon with patterns created from the *fivefold system*; and while this often makes it impossible to know categorically which of two, or even several, underlying polygonal tessellations the original designer employed, it by no means undermines the legitimacy of this methodological practice. Rather, this exemplifies the inherent flexibility of the *fivefold system*. The rhombic underlying tessellation of decagons, barrel hexagons, trapezoids, and large concave hexagons that produces this design is a modification of a rhombic underlying tessellation of decagons, barrel hexagons, and six contiguous pentagons that surround a small rhombi [Fig. 223]. This configuration of six pentagons surrounding a small rhombus was used frequently by geometric artists working with the *fivefold system*. However, it is only well suited to the production of *obtuse* and *two-point* patterns, rather than *acute* and *median* patterns. For acceptable patterns in these latter two families the underlying pentagons are truncated into trapezoids that lend themselves to the pattern characteristics of the *acute* and *median* families. There are two conventions for this truncation [Figs. 198 and 199]: one that truncates just four of the pentagons (leaving a large rhombic interstice region), and the other that truncates all six pentagons (leaving a large concave hexagonal interstice region). With *median* patterns, it is also possible to adjust the pattern lines themselves rather than the underlying tessellation [Fig. 199]. The design from the Mu'mine Khatun is the earliest known example of a fivefold *acute* pattern that utilizes this very effective

adjustment to the six clustered pentagons. This design was widely used, and later examples include a Zangid entry door at the Awn al-Din Mashhad in Mosul, Iraq (1248); part of the Mamluk exterior carved stucco ornament on the drum of the dome at the Hasan Sadaqah mausoleum in Cairo (1315-21); a Mamluk wooden door on the *minbar* of the Sultan Qaytbay funerary complex in Cairo (1472-74); a Mamluk cupboard door at the Qadi Abu Bakr Muzhir complex in Cairo (1479-80); and a Mamluk wooden panel along the stair railing of the *minbar* at the Amir Azbek al-Yusufi complex in Cairo (1494-95).

The range of remarkable geometric patterns employed in the exterior ornament of the Mu'mine Khatun mausoleum includes several designs that are created from nonsystematic underlying polygonal tessellations. The most basic of these is the well-used *acute* pattern of 12-pointed stars placed upon the vertices of the orthogonal grid with octagons at the centers of each square repeat [Fig. 336a]. The many other locations of this pattern include the Seljuk minaret of the Great Mosque of Siirt, Turkey (1129); the Artuqid *mihrab* niche at the Great Mosque of Silvan, Turkey (1152-57); and the Seljuk Sultanate of Rum cenotaph at the Izzeddin Kaykavus hospital and mausoleum in Sivas, Turkey (1217). A far less common nonsystematic design from the Mu'mine Khatun mausoleum is a fourfold *acute* pattern that places eight-pointed stars at the vertices of the square repeat unit, six-pointed stars at the midpoints of the repeat unit, and an irregular octagon at the center. The underlying tessellation for this pattern is made up of just three elements: the regular octagon, regular hexagon, and irregular pentagon [Fig. 178c]. This same underlying tessellation was used to create a very similar ceramic panel at the Altinbugha mosque in Aleppo (1318), as well as the design on the exterior of the Mamluk door (1303) at the Vizier al-Salih Tala'i mosque. Another fourfold nonsystematic *acute* pattern from the Mu'mine Khatun mausoleum uses underlying octagons at the vertices of the orthogonal grid that are separated along the edge of the square repeat unit by two regular hexagons rather than just the one from the previous example [Fig. 332a, c]. It must be assumed that the use of these two patterns on the Mu'mine Khatun mausoleum—one with a single underlying hexagon and the other with two underlying hexagons separating the underlying octagons—was a willful and subtle artistic act on the part of the artist. The polygonal matrix within the central region of the repeat employs irregular heptagons, irregular pentagons, and a square at the center of the repeat unit. This same underlying tessellation was first used by Seljik artists at the Friday Mosque at Barsian (1105), and at a later date by Mamluk artists to create a significantly more complex design for a window grille at the Amir Sarghitmish *madrasa* in Cairo (1356) [Fig. 332e]. Of particular interest in the history of this ornamental tradition is a compound pattern from the

Photograph 35 An Ildegizid nonsystematic *acute* pattern from the Mu'mine Khatun mausoleum in Nakhichevan, Azerbaijan, with 11- and 13-pointed stars (photograph by Самый древний (Own work) [Public domain], via Wikimedia Commons)

Mu'mine Khutun mausoleum with 11- and 13-pointed stars that is among the most complex nonsystematic geometric designs produced in the long history of this tradition [Fig. 434b] [Photograph 35]. The eccentricity of a design comprised of 11- and 13-pointed stars might appear to challenge the limits of two-dimensional space filling. However, as with the Seljuk pattern with 7- and 9-pointed stars from the *mihrab* of the Friday Mosque at Barsian [Fig. 429], the *principle of adjacent numbers* indicates that the practicality of making designs with 12-pointed stars allows for the likelihood of a successful pattern being created with 11- and 13-pointed stars. And indeed, this pattern from the Mu'mine Khatun is exceptionally successful. This pattern has the unusual characteristic of repeating with either an elongated hexagonal grid with the 13-pointed stars at each vertex or an alternative elongated hexagonal grid with 11-pointed stars at each vertex [Fig. 434a]. What is more, these two repetitive grids are perpendicularly orientated duals of one another. This pattern is beautifully balanced, pleasing to the eye, and

a masterpiece of geometric art. While other highly complex geometric patterns with perpendicularly arranged hexagonal dual repeat units were produced subsequently, this particular example is unique to the Mu'mine Khatun mausoleum, and does not appear to have been used by succeeding artists.

1.15 Artuqids (1102-1409)

The Artuqids began as military commanders of the Seljuks in Damascus, and rose to power as the Seljuk governors of Jerusalem. Their rule over eastern Anatolia, northern Syria, and al-Jazirah (northern Iraq) vacillated between independence and as vassals to their Seljuk, Zangid, Ayyubid rivals, and eventually the more powerful dynasties of the Sultanate of Rum, Ilkhanids, and Timurids. Their principle capital was Diyarbakir in southeastern Anatolia, which benefited from considerable architectural and artistic patronage. A surviving working drawing for the construction of the bronze doors for the Diyarbakir Palace were produced by a known individual and are adorned with a geometric pattern. This design, along with the accompanying instructions for casting, is the work of Ismail ibn al-Razzaz al-Jazari, and is part of his celebrated *Book on the Knowledge of Ingenious Mechanical Devices*[70] (1206). This geometric pattern is easily created from the 3.4.6.4 underlying tessellation of triangles, squares, and hexagons [Fig. 105e]. It is interesting to note the contrast between the proportional imprecision within his illustration—especially noticeable among the eight-pointed stars—and the geometric accuracy that characterizes the innumerable historic examples of Islamic geometric art. As made clear by the text that accompanies this illustration, this is a working drawing intended to be merely indicative of the final palace door. The text that accompanies this drawing explains, *"In the drawing I have not aimed for completeness. My purpose was to present a general arrangement so that it can be understood in the whole and in detail."*[71] Very few Artuqid buildings have survived intact to the present day, and our knowledge of their use of geometric design is slight at best. A carved stone relief panel at the Great Mosque of Dunaysir in Kiziltepe, Turkey (1204), employs a pattern that directly relates to the 4.8^2 tessellation of underlying octagons and squares [Fig. 129b]. This is a subtractive variation of the well-known design that locates the pattern lines upon the vertices of the generative tessellation [Fig. 129c]. The Artuqid niche of the *mihrab* at the Maqam Ibrahim at Salihin in Aleppo[72] (1112) is decorated with an *acute* field pattern constructed from just the large hexagonal and square modules of the *fourfold system A* [Fig. 138a]. This simple design places octagons on the vertices of the orthogonal grid, and was occasionally used by subsequent dynasties. The Artuqid use of patterns derived from the *fivefold system* includes a classic *acute* pattern that repeats on the rhombic grid of 72° and 108° angles from the surrounding border of the *mihrab* at the Great Mosque of Dunaysir in Kiziltepe, Turkey (1204) [Fig. 226c]. The wall of the *mihrab* niche at the Great Mosque of Silvan in southeastern Turkey (1152-57) employs a nonsystematic fourfold *median* pattern that places 12-pointed stars on the vertices of the orthogonal grid and octagons at the centers of each square repeat unit [Fig. 336a]. The underlying generative tessellation for this pattern is comprised of dodecagons and two varieties of irregular pentagons. This *median* example from the Great Mosque of Silvan is the earliest known pattern made from this tessellation.

1.16 Zangids (1127-1250)

During the twelfth century the Zangids became one of the primary powers in the region of al-Jazirah and Syria. Their founder, Imad al-Din Zengi, was Kurdish, and their rise to power began as the Seljuk *atabegs* of Mosul. Their greatest leader was Nur al-Din whose successes against the crusader kingdoms substantially increased their territorial dominion and aided their standing among Sunni Muslims throughout this region. Nur al-Din's forces were successful in overthrowing the Fatimids in Egypt in 1169, although he died before consolidating Egypt into the Zangid Empire. His death facilitated the rise of Ṣalāḥ ad-Dīn Yūsuf ibn Ayyūb and the founding of the Ayyubid successors to Zangid and Fatimid rule. Following the loss of Syria to the Ayyubids in 1183, the Zangids held on to their northern Iraqi territories until their demise in the mid-thirteenth century. The Zangids were great patrons of architecture, and many outstanding Zangid architectural monuments have survived into the modern era. Zangid architecture is best represented in Aleppo and Damascus in Syria, and to a lesser extent Mosul in Iraq. The refinement of the geometric ornament, including their magnificent use of *muqarnas*, attests to the level of interest that Zangid patrons had for this idiom. As demonstrated by the many examples of geometric pattern, their Seljuk origins allowed for the direct assimilation of the precise knowledge of geometric design methodology.

Like other Muslim cultures of the same period, the Zangids made frequent use of patterns constructed from the *system of regular polygons*. A pattern inside the niche of the wooden *mihrab* at the Lower Maqam Ibrahim at the citadel of Aleppo (1168) is easily created from an underlying 6^3 tessellation of regular hexagons [Fig. 96h]. The first known use of this design was approximately a hundred years earlier on the Seljuk eastern tomb tower at Kharraqan, and subsequent generations of Muslim artists made regular use of this geometric pattern. A Zangid use of this same design is found on the outer panels of the *minbar* doors at the al-Aqsa mosque in Jerusalem[73] (1168-74). The pattern which adorns the inside surfaces of these same *minbar* doors is created from the underlying 3.6.3.6 tessellation [Fig. 102a3]. This pattern employs two varieties of applied crossing pattern lines to the midpoints of the underlying polygonal edges. As mentioned above, the first known use of this design is from the Ghaznavid South Palace at Lashkar-i Bazar in Afghanistan (before 1036) [Fig. 102a1] [Photograph 13]. The Zangid portal of the otherwise Mamluk Bimaristan

---

[70] Istanbul, Topkapi Sarayi Müzesi Kütüphanesi, MS A. 3472, fols. 165v–166r.

[71] Necipoğlu (1995), 150–152.

[72] Allen (1999), Chap. 2.

[73] The al-Aqsa Mosque in Jerusalem is primarily a Fatimid building. However, the *minbar* was commissioned by the Zangid ruler Nur al-Din in 1168, placing it within the sphere of Seljuk influence. See Tabbaa (2001), 86–88.

Photograph 36 An incised stone panel in the Zangid portal of the Bimaristan Arghun in Aleppo with a threefold *median* pattern easily created from the *system of regular polygons* (photo by Nasser Rabbat, courtesy of the Aga Khan Documentation Center at MIT)

Arghun in Aleppo[74] contains a large bold rectangular panel above the entry door that is decorated with a design also created from the underlying 3.6.3.6 tessellation of triangles and hexagons [Fig. 101c] [Photograph 36]. Like the pattern from the interior surface of the doors of the al-Aqsa *minbar*, the construction of this design uses two types of crossing pattern line applied to the alternating underlying hexagons and triangles. This pattern is incised into the alternating colors of *ablaq* stonework in a manner wherein the zigzag divisions between the alternating dark and light stones are determined by the lines of the geometric pattern. This is a highly sophisticated, and virtually unique, ornamental device that indicates the origin of this portal to the high period of Zangid architectural ornament during the second half of the twelfth century. An incised stone panel placed over the door in the portal at the Adilliyya *madrasa* in Damascus (1172-74[75]) is a very fine example of Zangid geometric ornament. This is an isometric design that is

made from the 3.12^2 underlying tessellation of dodecagons and triangles. This pattern is created by placing 60° crossing pattern lines upon the centers of each coinciding polygonal edge [Fig. 108a]. The earliest known use of this design is from the raised brick ornament of the eastern tomb tower at Kharraqan (1067) [Photograph 17], and over time it became widely circulated throughout Muslim cultures. The use of incised lines within stonework ornament was characteristic of Zangid and later Ayyubid architectural ornament. This masonry technique requires considerably less time than carving stone in high relief. The loss of clarity associated with the highlights and shadows of high-relief carved stone was compensated by the introduction of pigments into the incised lines. This has its own bold aesthetic appeal that, to the modern eye, appears overtly graphical. This technique was used primarily on exterior façades, and over time had the disadvantage of loosing its boldness in color contrast if the paint was not refreshed periodically. Without the paint, the incised lines loose their boldness and are experienced more as a subtle presence, with just a vestige of its former ornamental impact. An outstanding example of a Zangid orthogonal pattern created from the *system of regular*

---

[74] Allen (2003).

[75] Different dates for this portal have been posited. See Allen (1999).

polygons was used on the side panel of a wooden *minbar* commissioned by Nur al-Din in 1186, and likely made in Aleppo[76] [Fig. 113c]. This pattern is created from the $3.4.3.12\text{-}3.12^2$ *two-uniform* tessellation with 120° crossing pattern lines placed upon the midpoints of the underlying polygonal edges. This Zangid pattern may have served as an inspiration to later Mamluk artists whose widespread use of this design includes at least two examples of *minbar* side panels: at the Sultan al-Nasir Muhammad ibn Qala'un at the citadel of Cairo (1295-1303), and the Amir Altinbugha al-Maridani mosque in Cairo (1337-39).

The railing of the celebrated al-Aqsa *minbar* (1187) contains an outstanding Zangid geometric pattern that is generated from the 4.8^2 underlying tessellation of squares and octagons. This is an early example of a subtractive variation of the classic star-and-cross pattern [Fig. 126f] that was used in a number of locations historically, including an almost contemporaneous design from the façade of the southern portal at the Qarakhanid Maghak-i Attari mosque in Bukhara (1178-79), and one of the panels on the interior of the Mamluk door (1303) of the Vizier al-Salih Tala'i mosque in Cairo. One of the side panels of this same Zangid *minbar* at the al-Aqsa mosque in Jerusalem uses a *two-point* pattern from the same 4.8^2 underlying tessellation [Fig. 128g]. This *two-point* pattern has essentially the same geometry as a design that employs the vertices of this same generative tessellation [Fig. 129c]. A Ghurid example of this alternative derivation can be found at the roughly contemporaneous western mausoleum at Chisht, Afghanistan (1167).

There are few Zangid examples of geometric patterns created from the *fourfold system A*. One notable example is a design used for several window grilles at the Nur al-Din Bimaristan in Damascus (1154) [Fig. 161] [Photograph 37]. This design is created from an underlying tessellation of octagons, truncated octagons, and eight-pointed stars. The use of the truncated octagons as underlying generative modules is an unusual feature more typically associated with later *fourfold system A* patterns from the Maghreb, and indicates a significant innovative imagination on the part of the artist who created this design. The widened interweaving lines are unusual in that some of the pattern lines are widened through their being offset on both sides equally, and others are widened through being offset in one direction only. This produces an irregularity to the otherwise predictable design, and is responsible for the distinctive visual character of these windows.

Zangid artists were among the earliest to make use of the *fourfold system B*. The first known use of this

Photograph 37 A Zangid window grille at the Nur al-Din Bimaristan in Damascus with a fourfold *median* pattern created with the *fourfold system A* (Photo by Nasser Rabbat, courtesy of the Aga Khan Documentation Center at MIT)

methodological system is from the above-mentioned *mihrab* in the Friday Mosque at Sin (1134). By the second half of the twelfth century the *fourfold system B* had been readily adopted by artists working in the western regions of Seljuk influence. The earliest Zangid examples include the same *acute* design that had been used at Sin, and was to become the most widely used pattern created from this design system [Fig. 173a]. This is located on the base of the minaret of the Great Mosque of Nur al-Din at Mosul (1170-72). Its utilization throughout Muslim cultures and artistic media, coupled with its easily recognizable distribution of eight-pointed stars and octagons, qualifies this as the classic *fourfold system B* pattern. Zangid artists were also among the first to use the innovative additive device whereby octagons can be incorporated into the pattern matrix through adjusting the *acute* pattern lines within the underlying large hexagonal modules [Fig. 172b]. Over time, among succeeding Muslim cultures, this became a standard feature of *acute* patterns

---

[76] This *minbar* is in the collection of the Hama National Museum in Syria.

created from this system. Both the framing border of the wooden *mihrab* in the Lower Maqam Ibrahim in the citadel of Aleppo (1168) and the large triangular side panel of the wooden *minbar* at the al-Aqsa mosque in Jerusalem (1187) employ a particularly beautiful pattern with this octagonal characteristic [Fig. 177b]. The Zangid ruler Nur al-Din commissioned both of these superlative examples of Muslim woodworking.

The al-Aqsa *minbar* also employs one of the earliest Zangid examples of a pattern constructed from the *fivefold system*. This is a classic *acute* pattern that repeats upon a rhombic grid of 72° and 108° angles [Fig. 226c]. As discussed above, this design can be traced back to one of the repetitive cells in the hybrid design at the northeast dome chamber of the Friday Mosque at Isfahan (1088-89) [Fig. 261] [Photograph 25]; and other early examples include the Ghurid Arch at Bust, Afghanistan (1149) [Photograph 33], and the Seljuk façade of the Friday Mosque at Gonabad, Iran (1212) [Photograph 33]. This predates the panel from the al-Aqsa *minbar* by only 38 years. The interior of the main entry door of the Imam Awn al-Din Mashhad in Mosul (1248) is decorated in raised copper with a very-well-conceived *acute* pattern made from the *fivefold system* [Fig. 232g]. The earliest example of this popular design is from the Ildegizid tomb of Mu'mine Khatun in Nakhichvan, Azerbaijan (1186). As discussed previously, the underlying tessellation of this pattern employs a modification to the cluster of six pentagons at the center of each rhombic repeat unit. This configuration of six pentagons makes very satisfactory *obtuse* and *two-point* patterns, but is not suited to creating acceptable *acute* and *median* designs. In order to overcome this design limitation, the cluster of six pentagons is modified in either of the two ways [Figs. 198 and 199]. As with the Azerbaijani design, the fivefold pattern from the Awn al-Din Mashhad truncates each of the six clustered pentagons into six adjacent trapezoids, creating the distinctive concave hexagonal feature at the center of each rhombic repeat unit.

The Zangids made occasional use of geometric patterns that were created from nonsystematic underlying polygonal tessellations. The exterior of the bronze door at the Nur al-Din Bimaristan in Damascus (1154) is decorated with a threefold pattern comprised of five- and six-pointed stars [Fig. 309c]. A very similar Seljuk design was used earlier in the northeast dome chamber of the Friday Mosque at Isfahan (1088-89) [Photograph 27]. As explained, the underlying tessellation that produces this pattern places pairs of coinciding pentagons upon each edge of the triangular repeat unit in such manner that they create irregular hexagonal ditrigons within the centers of the triangular repeat units. This pentagonal configuration also produces a six-pointed star at each vertex of the isometric grid. The pattern lines are placed upon the midpoints of each underlying pentagonal

edge, and the 36° angular opening of the crossing pattern lines creates a very acceptable *acute* pattern. An incised pattern on the archivolt surrounding the *muqarnas* hood of the Zangid portal at the Bimaristan Arghun in Aleppo makes use of a nonsystematic border pattern of alternating nine- and six-pointed stars. This is a poorly conceived design that has the further problem of being in very poor repair: the linear repetitive structure being obscured beyond interpretation. A far more successful, if common, Zangid nonsystematic geometric pattern was used on the double-entry doors of this same portal. These are decorated with interweaving bronze straps that create an *acute* pattern with 12-pointed stars set on an orthogonal grid [Fig. 337]. The underlying polygonal tessellation for this design is a modification of the tessellation used to generate the *median* pattern in the contemporaneous Artuqid *mihrab* niche of the Great Mosque of Silvan, Turkey (1152-57) [Fig. 336a]. Just as the cluster of six regular pentagons within the *fivefold system* can be modified through truncating the pentagons into trapezoids, so too is this possible with the cluster of four irregular pentagons with coinciding edges that are found in this Artuqid example. These trapezoids are responsible for generating the four dart motifs that are a distinctive feature of this design. Other locations that employed this design include the Ildegizid exterior ornament of the Mu'mine Khatun mausoleum in Nakhichevan, Azerbaijan (1186); the Zangid entry doors and incised stone ornament at the Zahiriyya *madrasa* in Aleppo; the Mamluk mausoleum of Sultan al-Zahir Baybars in Damascus (1277-81); a pair of wooden doors from the *atabeg* of Cizre in southeastern Turkey (thirteenth century); the *minbar* of the Mamluk funerary complex of Sultan al-Zahir Barquq in Cairo (1384-86); and the *minbar* doors at Amir Taghribardi funerary complex in Cairo (1440). Another example of a nonsystematic Zangid pattern with 12-pointed stars is from a *mihrab* archivolt at the Upper Maqam Ibrahim at the citadel of Aleppo (c.1214). The original mosque at this site was rebuilt by Nur al-Din following a fire in 1212.[77] An interesting feature of the underlying tessellation for this pattern is the truncation of three irregular pentagons into three trapezoids that surround a central triangle [Fig. 320b]. The geometric pattern employed over this *mihrab*, when considered in the abstract, is both beautiful and conceptually satisfying. However, the execution of this pattern on the *mihrab* archivolt is imprecise and clumsy. The incised carving is poorly executed and the geometric distortion that is required when applying a linear motif onto a curve is, in this example, inelegant, and stands in marked contrast to the vastly superior execution in the Zangid archivolt pattern at the Zahiriyya *madrasa* produced just 3 years later. Far more

---

[77] Allen (1999), Chap. 8.

pleasing subsequent examples of this lovely threefold pattern include a thirteenth-century Christian *khachkar* stele from Dsegh, Armenia,[78] and two vertical panels from a fourteenth-century wooden door from Fez, Morocco.[79]

The most remarkable Zangid geometric pattern created from a nonsystematic underlying tessellation was reported by Ernst Herzfeld to have been used on a pair of doors at the Lower Maqam Ibrahim at the Aleppo citadel[80] (1168). This *acute* design is comprised of an ingenious combination of 12-, 11-, and 10-pointed stars arranged in linear vertical bands. The repeat unit is a long rectangle with 12-pointed stars at the vertices, 10-pointed stars at the midpoints of the long sides of the rectangular repeat, and two 11-pointed stars centered between the 10- and 12-pointed stars within the rectangular repeat [Fig. 427]. This design is equal to the most complex designs from the Anatolian Sultanate of Rum and Mamluks of Egypt. A pattern on the portal of the kiosk at Erkilet near Kayseri, Turkey (1241), employs the same exact sequence of linear star forms [Fig. 425]. Despite accolades to the contrary,[81] the pattern that was reported to have been in the doors of the Lower Maqam Ibrahim contains areas where the pattern matrix is problematic: specifically, the crowding of the pattern lines at the centers of the triangular regions defined by the 11- and 12-pointed stars. The overall balance of the conceptually similar Sultanate of Rum design from Erkilet is far superior to the example recorded by Herzfeld at the Lower Maqam Ibrahim.

During the twelfth century, the Zangids experimented with the Great Seljuk innovations in applying geometric patterns onto domical surfaces. The Zangid artists working in this specialized field of ornament were woodworkers, and were arguably the most skilled craftsmen working within the Zangid artistic milieu. This form of design and fabricating technology is highly specialized and required a practical knowledge of spherical geometry. The exceptional woodwork from this period includes a quarter dome hood in the *mihrab* niche of the Lower Maqam Ibrahim in the citadel of Aleppo (1168): signed by Ma'ali ibn Salam: clearly a master of this art. This quarter dome is ornamented with a beautiful geometric pattern derived from a spherical projection of underlying polyhedral geometry. As with the Seljuk

ornament of the northeast dome of the Friday Mosque at Isfahan, the use of polyhedral geometry as an organizing principle for domical ornament is unusual, but very effective. The great majority of domes with geometric decoration are based upon segmented radial gores, and only a relatively small number of domes utilize the geometry of Platonic or Archimedean polyhedra. The quarter dome hood of the *mihrab* niche in the Lower Maqam Ibrahim employs a pattern with five- and six-pointed stars that is based upon a spherical projection of the truncated icosahedron as the formative underlying structure [Fig. 498]. The truncated icosahedron is comprised of 12 pentagons and 20 hexagons in a 5.6^2 configuration at each vertex. As a spherical projection, this polyhedron is often associated with the soccer ball. The geometric pattern in this quarter dome places 60° crossing pattern lines at the midpoints of the coinciding pentagonal and hexagonal edges. This creates a *median* pattern that flows across the spherical surface in a very cohesive and pleasing fashion, placing five-pointed stars within the projected pentagons, and six-pointed stars inside each projected hexagon.

1.17   Fatimids (909-1171)

The Fatimid dynasty was founded by Kutama Berbers from the northeastern coastal region of Algeria in the early tenth century. They rose to prominence rapidly, and by the second half of the tenth century they conquered Egypt, founding Cairo as their capital in 969. By the close of the tenth century they controlled an empire stretching from the Maghreb in the west, across all of North Africa and Egypt, Sicily, and portions of southern Italy, Malta, the northern Red Sea coastal region of the Hijaz, and the Levant. The Fatimid Caliphate was founded in 909 and rivaled the Abbasid Caliphate until the Zangids defeated the substantially weakened Fatimids in 1169. The Fatimids were Shi'a Muslims, and were ruled by members of the Ismaili sect. In contradistinction to rival empires, their governance recognized merit over heredity in rewarding advancements. With aptitude as the primary qualifier, Sunni Muslims, Christians, and Jews were entrusted with high levels of responsibility and authority. Under Fatimid patronage, Cairo became one of the great cities of the world. The great wealth amassed through sea fairing trade in the Mediterranean and Indian Ocean, as well as with China enabling extravagant building projects and widespread support for the arts. With their Mediterranean origins and influential communities of Byzantine and Coptic Christians, the aesthetic sensibilities of the Fatimids were distinct from the prevalent artistic trends occurring in the eastern regions of the Islamic world. Within the minor arts, representational motifs with animal and human figures were used more widely than similar work from eastern Muslim

[78] Azarian (1973), pl. 58.

[79] Dar al-Athar al-Islamiyya Kuwait National Museum.

[80] This pair of doors is no longer present at the Lower Maqam Ibrahim in Aleppo. Ernst Herzfeld published a drawing and description of this pattern, and this is the only record of its existence. See Herzfeld (1954–56), Fig. 56.

[81] "It is the most complicated design ever produced by that branch of art. The almost unsolvable problem of a design based on horizontal groups of 11-pointed stars is solved by alternative intercalation of a parallel group of 12-pointed and one of 10-pointed stars between them": Herzfeld (1943), 65.

cultures, often exhibiting a level of facility and playfulness that contradicts the commonly held view that the depiction of human and animal forms is anathema to all Muslim cultures. While still placing great emphasis upon calligraphy, Fatimid artists generally favored the floral idiom over the geometric in the ornamentation of their architectural monuments. The extent to which Fatimid floral preferences originated from an antipathy toward the cultural mores of their neighboring Sunni rivals or simply a genuine partiality toward this more naturalistic idiom is uncertain.[82] The ornament of the earlier Fatimid period embraced the Samarra floral styles exemplified at the ibn Tulun mosque, as well as the aesthetics of their North African ancestral homelands.[83] The Fatimid eschewal of geometric designs and *muqarnas* vaulting began to change during the twelfth century. The gradual incorporation of the advances to the geometric arts that were carried out under the auspices of the Ghaznavids, Ghurids, Qarakhanids, and Seljuks appears to have been initially advanced by non-imperial patronage,[84] and only later fully adopted into the fabric of Egyptian aesthetics. What is certain from the historic record is that the Fatimids incorporated geometric patterns and *muqarnas* vaulting with less frequency than the contemporaneous Sunni cultures to the north and east. Despite this floral predilection, the twelfth-century geometric ornamental advances nonetheless made their way into the fabric of Fatimid culture—however tenuously—and these examples include a number of patterns of great beauty.

The majority of Fatimid geometric designs were produced from the *system of regular polygons*. One of the earliest is a pattern made up of superimposed interweaving dodecagons from a window grille in the chamber of the Dome of al-Hafiz at the al-Azhar mosque in Cairo (970-72) that is easily created from the 6^3 hexagonal grid [Fig. 97c]. This same design was used some 10–20 years later in one of the window grilles at the Great Mosque of Córdoba, and indeed was widely used throughout Muslim cultures. A window grille from the al-Aqmar mosque in Cairo (1125) uses a design that can also be derived from the 6^3 hexagonal grid. This is comprised of superimposed hexagons [Fig. 96e], and also shares provenance with one of the window grilles in the Great Mosque of Córdoba (980-90). The carved stucco *mihrab* of the al-Amri mosque in Qus, Egypt (1156), appears to employ a 3.4.6.4 design that extends the coincident triangular and square edges of the generative tessellation to create a very successful design with 12-pointed stars within each hexagon [Fig. 106b]. This derivation is unusual in that the 12-pointed stars are not created from an underlying dodecagon. This design is

very similar to an example from the Seljuk Sultanate of Rum that employs a 4.6.12 underlying generative tessellation [Fig. 109a], and indeed the Fatimid design can also be made from this tessellation. The rear portion of the portable Fatimid *mihrab* at the Sayyid Ruqayya Mashhad in Cairo (1133) includes a threefold pattern with 12-pointed stars that is generated from the 3.12^2 tessellation of triangles and dodecagons [Fig. 108a]. This is the same pattern that was used for the first time on the eastern tomb tower at Kharraqan some 66 years earlier [Photograph 17]. In this Fatimid example, an arbitrary 6-pointed star motif has been added into the center of each 12-pointed star. This is a sixfold example of a form of additive pattern modification that was more commonly applied to *obtuse* patterns created from the *fivefold system* [Fig. 224b]. Later examples of this design—without the modification—include a Zangid incised stone panel at the Adilliyya *madrasa* in Damascus (c. 1172), and the top and bottom panels from the Mamluk double doors of the Vizier al-Salih Tala'i mosque in Cairo (1303). The same portable wood *mihrab* of the Sayyid Ruqayya Mashhad in Cairo employs a lovely geometric design that dominates the front surface framing the niche. This design can be created from either of the two underlying polygonal tessellations: the 3.6.3.6 semi-regular tessellation with alternating active and passive hexagonal and triangular cells [Fig. 101b], and a *three-uniform* tessellation of triangles, squares, and hexagons in a $3^4.6\text{-}3^3.4^2\text{-}3^2.4.3.4$ arrangement, wherein the 60° crossing pattern lines are applied to the midpoints of just selected polygonal edges [Fig. 114c]. This is an unusual and highly inventive use of the *system of regular polygons*. A Fatimid pattern from the triangular side panel of the wooden *minbar* (1091-92) of the Haram al-Ibrahimi in Hebron, Palestine, can be created from at least three different underlying tessellations: the simple 6^3 tessellation of regular hexagons [Fig. 98d]; the 3.4.6.4 tessellation of triangles, squares, and hexagons [Fig. 106c]; and the $3^4.6\text{-}3^3.4^2\text{-}3^2.4.3.4$ *three-uniform* tessellation of triangles, squares, and hexagons [Fig. 114a]. While it is impossible to know which method the artist used in any given example, the high degree of available methods to create just a single design speaks to the flexibility of this design methodology. Another historical occurrence of this design is from the *minbar* of the al-Amri mosque in Qus, Egypt (1156). The niche of the above-cited portable wooden *mihrab* at the Sayyid Ruqayya Mashhad in Cairo also employs a very simple *acute* design created from the 6^3 hexagonal grid as an underlying generative structure [Fig. 95a]. This pattern places 30° crossing pattern lines at the midpoints of each hexagonal edge, resulting in a design comprised of six-pointed stars and distinctive ditrigonal shield shapes. This very becoming, if rather simple, design never generated the level of pan-Islamic interest as its close relatives that can be created from the same regular

---

[82] Tabbaa (2001), 80–84.

[83] Ettinghausen, Grabar and Jenkins-Madina (2001), 195.

[84] Bloom (1988), 27–28.

hexagonal grid with 60°, 90°, and 120° crossing pattern lines. Earlier Seljuk examples of this simple design are found at the northeast domed chamber of the Friday Mosque at Isfahan (1089-90) [Photograph 18], and the Friday Mosque at Sin, Iran (1133).

Fatimid geometric patterns created from the underlying 4.8^2 tessellation of squares and octagons include a fine example of the classic star-and-cross design from the stucco *mihrab* of the Umm Kulthum and al-Qasim Abu Tayyib mausoleum in Cairo (1122) [Fig. 124b]. This bears an unmistakable resemblance to the aesthetic treatment of the much earlier example of this pattern on the arch soffits at the No Gunbad mosque in Balkh, Afghanistan (800-50) [Photograph 10]. The multiple small circles applied linearly within the interweaving straps of the design, as well as the circularity of the floral motifs within each eight-pointed star, are so similar that it is possible that the artist who designed this Fatimid *mihrab* may have been familiar with the earlier Afghan example. A more complex geometric design created from the same underlying 4.8^2 tessellation is found in the Fatimid *mihrab* from the mausoleum of Sayyidah Nafisah in Cairo (1138-46) [Fig. 127d]. This same design was used by Mengujekid artists at the Great Mosque of Divrigi, Turkey (1228-29), and later still by Mamluk artists on the minaret of the Sultan Qaytbay funerary complex in Cairo (1472-74).

Fatimid artists appear to have limited their use of the polygonal technique to the *system of regular polygons*. They did not make use of either the *fourfold system A* or *B* for creating geometric designs, and their use of the *fivefold system* was rare at best.[85] Similarly, the Fatimids did not incorporate nonsystematic patterns within their architectural ornament. These omissions are surprising considering the widespread adoption of all of these design methodologies by their Muslim neighbors to the north and east during the eleventh and twelfth centuries. This may have been a deliberate rejection based upon the association of such designs with their Seljuk and Abbasid Sunni rivals, or it may simply have been due to the absence of knowledge of these more advanced methodologies within the community of artists working under Fatimid patronage. Whichever the case, the fact of the virtually exclusive use of the *system of regular polygons* to create their geometric ornament can be regarded, at least in part, as more a willful continuation of the earlier methodological practices and geometric aesthetic

of their Tulunid predecessors, and less an influence by the art of their Sunni rivals.

1.18 Ayyubids (1171-1260)

In 1169, the Zangid ruler Nur ad-Din Zangi sent his general Asad ad-Din Shirkuh on a campaign to overthrow the Fatimids in Egypt. Shirkuh died soon after his successful defeat of the Fatimids, and was succeeded by his nephew Ṣalāḥ ad-Dīn Yūsuf ibn Ayyūb. While establishing relative autonomy in Egypt, Ṣalāḥ ad-Dīn remained faithful to Nur ad-Din. Upon the death of the Fatimid Caliph, at Nur ad-Din's request, Ṣalāḥ ad-Dīn reestablished the authority of the Abbasid Caliphate in Baghdad, returning Egyptian rule to Sunni Islam. Ṣalāḥ ad-Dīn's military successes against the Crusader Kingdoms won him the lasting respect of Muslims throughout this region. Following the death of Nur ad-Din in 1174, Ṣalāḥ ad-Dīn's rise to power as the first Ayyubid Sultan was more the result of the high esteem to which he was held throughout the Zangid territories than through military conquest. Like his Zangid predecessors, Ṣalāḥ ad-Dīn was Kurdish, and his superior military tactics and honorable conduct of warfare won him the lasting respect of his Christian Crusader adversaries, many of whom honored him as a paragon of knightly virtue. At the height of their power, the Ayyubids controlled a region stretching from Tripoli in the west, across the North African coastal zone, all of greater Egypt and Nubia, large portions of the Arabian Peninsula including Yemen, the Levant, Syria, al-Jazirah, and much of southeastern Turkey.

Considering the close historical connection between the Zangid and Ayyubid dynasties, it is not surprising that Ayyubid architecture and ornament was, in essence, a furtherance of the Zangid aesthetic practices and preferences.[86] The architectural attention paid to Aleppo and Damascus by the Zangids continued under the Ayyubids, and with Egypt now integrated into the sphere of Sunni influence, the new construction commissioned by Ayyubid patrons spread this distinctive style to Cairo. It was during this period that several ornamental devices became prevalent in the architecture of Cairo, including *muqarnas* vaulting and *ablaq* masonry: the bold use of alternating light and dark stone that originated in Syria in the early twelfth century became an important ornamental feature of Ayyubid, Mamluk, the Sultanate of Rum, and Ottoman architecture. And under Ayyubid patronage in Cairo, the fledgling attention paid to

---

[85] One possible example of the Fatimid use of the *fivefold system* is a window grille in the northeast wall at the al-Hakim Mosque in Cairo. This is an *acute* dart motif generated from just the barrel hexagon, and one of the simplest fivefold field patterns. However, it is very likely that this window grille dates to the post-earthquake restoration by Amir Baybars al-Jashankir in 1303, or the restorations by Sultan al-Nasir Hasan in 1360.

[86] For detailed accounts of Zangid and Ayyubid architecture and architectural ornament in Aleppo and Damascus, see
    –Allen (1999).
    –Tabbaa (2001).

geometric design during the Fatimid period received far greater prominence and influence.

The Ayyubids continued and refined the established conventions of geometric pattern making, and in this way, the *system of regular polygons*, both of the fourfold systems, as well as the *fivefold system* were all used in their architectural ornament. As well as being innovators, many examples of patterns that were used by previous Muslim dynasties were likewise incorporated into the Ayyubid ornamental milieu. These include a star-and-cross pattern on a barrel vault in the Burg al-Zafar in Cairo (1176-79) [Fig. 124b]. An Ayyubid example of the well-used pattern made from the 6^3 underlying tessellation of hexagons that was first used at the eastern tomb tower in Kharraqan, as well as previously by both Fatimid and Zangid artists, was used on an arched portal of the Zahiriyya *madrasa* in Aleppo (1217) [Fig. 96h]. The Ayyubid use of this design is noteworthy for the highly unusual manner in which the pattern continues across the 90° change in angle between the archivolt and intrados of the arch. As per the convention established by their Zangid predecessors, this pattern is expressed in an incised line technique, and the absence of contrasting pigment within the incised lines makes this pattern relatively difficult to discern from a distance. The successful application of what would otherwise be a linear band of geometric pattern onto the curve of the archivolt requires considerable skill. This example from the Zahiriyya *madrasa* accomplishes the requisite distortion so successfully that the finished design appears completely natural, as if the pattern should always appear in this fashion. What makes this curvilinear distortion all the more remarkable is the fact that the matching pattern on the surface of the intrados is purely linear in its layout. While the geometric pattern itself is not particularly complex, and was certainly well known by the time this example was produced, the artist responsible for this archway was clearly endowed with considerable geometric skill and ingenuity.

The teakwood cenotaph at the Imam al-Shafi'i mausoleum in Cairo (1211) is decorated with two designs with 12-pointed stars made from the *system of regular polygons*. One of these is the same threefold design that was first used at the eastern tomb tower at Kharraqan (1067) [Fig. 108a] [Photograph 17]. This is a *median* pattern that is derived from the 3.12^2 semi-regular underlying tessellation of triangles and dodecagons. This Ayyubid example from the Imam al-Shafi'i mausoleum uses the same 6-pointed star additive motif within the center of the 12-pointed stars as the Fatimid example from the Sayyid Ruqayya Mashhad in Cairo (1133). Despite the 78 years separating their production, the close physical proximity of these two Cairene examples may explain their similarity. It is the juxtaposed presence of the second pattern from the cenotaph at the Imam al-Shafi'i mausoleum in Cairo that makes these two

patterns exceptional. The second pattern has fourfold symmetry, and is created from a *two-uniform* underlying tessellation in a $3.4.3.12$-3.12^2 configuration [Fig. 113a]. The edges of both the square repeat unit of the fourfold pattern and the triangle repeat of the threefold pattern have the same arrangement of edge-to-edge dodecagons, and hence the pattern lines that are generated from these tessellating dodecagons are likewise identical upon their respective repetitive edges. While it is certainly possible that the artist responsible for the cenotaph at the Imam al-Shafi'i mausoleum merely replicated these two patterns from two earlier local buildings, the remarkable concordance in the edge configuration of the respective repeat units indicates the artist's knowledge of the special geometric relationship between these two patterns, and that their selection was not coincidental. Had the artist wanted, these two repeat units could have been used together to create a single hybrid composition, and indeed several historical examples of hybrid designs made up of both square and triangular repeat units are known, and invariably, the edge configurations are, per force, identical [Fig. 23].

As stated, Ayyubid artists made use of the 4.8^2 underlying tessellation of squares and octagons to create a particularly bold example of the classic star-and-cross *median* design that covers the surface of a barrel vault at the Burg al-Zafar in Cairo (1176-93) [Fig. 124b]. A later example of their use of this well-known pattern was used in the inlaid stone ornaments of the Sharafiyya *madrasa* in Aleppo (1242). The Firdaws *madrasa* in Aleppo (1235-36) employs an unusual variant of the classic *median* design created from this tessellation that uses 60° crossing pattern lines at the midpoints of each underlying polygonal edge [Fig. 126a]. This 60° angular opening is more commonly associated with isometric patterns that have triangles and hexagons within their underlying polygonal matrix, and the use of this angular opening within this orthogonal design produces a pleasing alternative to the standard star-and-cross design. An Ayyubid example of an *acute* pattern created from the 4.8^2 underlying tessellation is found at the Sahiba *madrasa* in Damascus (1233-45). This differs from the standard *acute* design [Fig. 124a] through the incorporation of small eight-pointed stars within the underlying square modules [Fig. 125b].

An interesting orthogonal design with 12-pointed stars was used in the Ayyubid wooden *mihrab* (1245-46) from the Halawiyya mosque and *madrasa* in Aleppo. The underlying tessellation for this pattern places a dodecagon upon each of the four vertices of the square repeat unit. These are edge to edge with an octagon located at the center of the repeat. This arrangement of dodecagons and octagons produces concave hexagonal interstice regions [Fig. 333a]. This design is unusual in that the octagon does not play a direct formative role in deriving the pattern. Rather, the pattern lines within these polygons and the concave interstice regions are continuations of the 60° crossing pattern lines created from the dodecagons. This same underlying tessellation of edge-to-edge dodecagons and octagons will produce many very acceptable geometric designs [Figs. 379–382].

A small stone lintel over a door at the Sahiba *madrasa* in Damascus is decorated with a simple pattern derived from the *fourfold system A*. This is an *acute* field pattern that makes use of only the square and large hexagon in its underlying generative tessellation [Fig. 138a]. As with other Zangid and Ayyubid examples, the ornamental carving in this panel is incised into the stone, requiring far less time and cost than carved high relief. This same field pattern was used to decorate the *mihrab* niche of the Sharafiyya *madrasa* in Aleppo (1242). Both of these Ayyubid examples may have been inspired by the identical *acute* field pattern that was used over a century earlier in the Artuqid *mihrab* niche at the Maqam Ibrahim at Salihin in Aleppo (1112), and subsequent examples include a Seljuk Sultanate of Rum exterior faience border design at the Sirçali *madrasa* in Konya (1242).

The Ayyubids used the *fourfold system B* more widely than the *fourfold system A*. The Farafra *khanqah* in Aleppo (1237-38) employs the well-known *acute* pattern created from the *fourfold system B* that makes use of just the underlying octagons and small pentagons from this system [Fig. 173a]. This classic example from the Farafra *khanqah* is from a wooden soffit over one of the door openings. The vertical flanking panels on the *mihrab* at the Zahiriyya *madrasa* in Aleppo (1242) are decorated with the same design, as is the niche of the wooden *mihrab* (1245-46) of the Halawiyya mosque and *madrasa* in Aleppo. These three identical examples were produced in Aleppo within 10 years of one another, and it is certainly possible that a single artist was responsible for each. Over time, the popularity of this very-well-balanced pattern spread widely throughout the Islamic world. A particularly fine example of an Ayyubid *obtuse* design made from the *fourfold system B* is a very bold *ablaq* geometric pattern used at the top of the portal façade at the Palace of Malik al-Zahir at the citadel of Aleppo (before 1193[87]) [Fig. 175d]. This same pattern is found at the Taybarsiyya *madrasa* (1309) in the al-Azhar mosque in Cairo, and in the Great Mosque at Bursa in Turkey (1396-1400). The polygonal modules that make up the underlying tessellation that creates this pattern are the octagon, small hexagon, and pentagon. It is perhaps significant that this underlying tessellation was used only 7 years previously to produce a Qarakhanid *two-point* pattern at the southern

anonymous mausoleum at Uzgen (1186) [Fig. 176c]. An Ayyubid example of another design created from a similar underlying tessellation created from the *fourfold system B* is from the pierced marble balustrades on the minaret of the Aqsab mosque in Damascus (1234) [Fig. 177a]. This is an *acute* pattern created from an underlying tessellation that replaces the small hexagons with the large hexagons from this system. The proportions of the pentagons in this example are specific to the arrangement of octagons and hexagons, and are unique to this single tessellation. As such, this new pentagonal element is not typical to the set of standard polygonal modules that comprise the *fourfold system B*. The most complex Ayyubid design created with this system is from the niche wall of the *mihrab* at the Imam al-Shafi'i mausoleum in Cairo (1211). This is a *two-point* pattern that is as outstanding for its beauty and ingenuity, and likely influenced the complex *two-point* aesthetic of the succeeding Mamluks. This pattern places 16-pointed stars at the vertices of the orthogonal grid, a cluster of four pentagons at the center of each square repeat unit, and an 8-pointed star within each quadrant of the repeat unit [Fig. 185b]. The eight-pointed stars are located at the vertices of the 4.8^2 tessellation of octagons and squares, and indeed this is a governing structural feature of this design. This same underlying tessellation was used by Nasrid artists to create an equally fine *acute* pattern that was used in a cut-tile mosaic panel in the Sala de las Aleyas at the Alhambra[88] (fourteenth century).

The portal of the Malik al-Zahir in Aleppo contains a fourfold pattern created out of the arbitrary placement of six-pointed stars upon the midpoints of each edge of the square repeat units [Fig. 52a]. This is an example of a class of geometric design that imposes radial symmetry—in this case sixfold—into a repetitive structure that is generally incompatible—in this case orthogonal. The extension of the lines of the six-pointed stars creates the overall pattern matrix, and includes the four-pointed star at the center of each repeat unit, the square at the vertices of the repeat unit, and the four irregular octagons that surround the four-pointed stars. A remarkable and highly distinctive feature of this *imposed symmetry* pattern is the continuous application of the incised pattern across the dark and light colored alternating *ablaq* masonry voussoirs that surround the doorway. It is interesting to note the geometric similarity between this imposed symmetry design and the fourfold pattern from the mausoleum of Yusuf ibn Kathir in Nakhichevan, Azerbaijan (1161-62) [Fig. 52b]. The earlier

---

[87] For details on the dating of this portal, see Allen (1999), Chap. 5.

[88] This mosaic panel was moved to the Christian chapel of the Mexuar by Morisco artists during the sixteenth century, and now resides at the Museo Nacional de Arte Hispanomusulmán in Granada. See Dodds [ed.] (1992), 374–375.

Ildegizid design also arbitrarily places six-pointed stars in the same location of the square repeat. However, the smaller size of these stars provides for the inclusion of bounding hexagons at these same locations, which in turn creates an irregular eight-pointed star at the center of each repeat.

The use of geometric patterns in Ayyubid architecture predominantly relied upon designs created from the *system of regular polygons* and the *fourfold system B*. The extent to which this was due to aesthetic preferences or some other more prosaic reason remains unclear. It may simply be that the Ayyubid architectural designers who were most successful in receiving patronage had a comparatively limited knowledge of the broad range of available design methodologies and consequent pattern types. Whatever the reason, there are relatively few Ayyubid examples of patterns produced from the *fivefold system*, or more complex patterns created from nonsystematic underlying polygonal tessellations. One Ayyubid example of a fivefold *acute* pattern is an incised stone surround of a domical hood from a courtyard portal at the Palace of Malik al-Zahir at the citadel of Aleppo. This design has many of the characteristics of the classic fivefold *acute* pattern that repeats upon a rhombic grid [Fig. 226c]. Regrettably, very little of the original geometric panel has survived, and it is impossible to determine the full systematic character of the design, or even whether the repeat unit is rhombic or rectangular.

Among the few examples of nonsystematic Ayyubid geometric pattern is a carved stone design found in the city walls of the Bab Antakeya in Aleppo (1245-47). This is a very common *median* pattern with 8- and 12-pointed stars that was used subsequently by other Muslim cultures [Fig. 380b], but deviates from the standard design by introducing curvilinear lines within the central region of each 12-pointed star. This design was subsequently used by Ilkhanid and Timurid artists and a particularly fine later example is a Muzaffarid cut-tile mosaic panel in the lower section of the southern *iwan* at the Friday Mosque of Yezd (c. 1365).

The Ayyubids inherited the traditions of domical geometric ornament from their Zangid predecessors. One of the earliest Ayyubid geometric domes is located within the hood of the *mihrab* (before 1205) at the al-Sharafiyah *madrasa* in Aleppo. As with both the Seljuk northeast dome of the Friday Mosque at Isfahan, and the quarter dome in the *mihrab* of the Lower Maqam Ibrahim, this example from the al-Sharafiyah *madrasa* uses polyhedral geometry as the fundamental structure upon which the geometric pattern rests [Fig. 500]. Specifically, the domical portion of the *mihrab* niche is based upon a spherical projection of the octahedron. This Platonic solid is comprised of eight triangular faces, with four triangles at each of the six vertices. This quarter dome employs just two of the triangular projections of the octahedron, equaling just 1/4 of the spherical surface that results after both horizontal and

vertical divisions. An inscription on this *mihrab* credits it to 'Abd al-Salâh Abû Bakr,[89] and the concept for his design is both simple and elegant. Being that the 1/4 sphere is comprised of two projected equilateral triangles, the artist employed a two-dimensional design that uses a triangle as its repetitive unit, and applied this to the three-dimensional triangular surface of the octahedron. The two-dimensional progenitor is a well-known nonsystematic pattern that places a dodecagon on each vertex of the isometric grid, and a ring of 12 pentagons around each dodecagon, three of which are clustered at the centers of the triangular repeats [Fig. 300a *acute*]. Locations of earlier examples of this two-dimensional design include the *mihrab* of the Great Mosque at Niksar, Turkey (1145). The quarter dome in the *mihrab* of the al-Sharafiyah *madrasa* replaces the 12-pointed stars that rest upon the vertices of the two-dimensional originator with 8-pointed stars. This is due to there being just four triangles at each vertex, and each triangular corner having just two points for the star. In two dimensions, these two points are arrayed around each vertex 6 times, making 12 points, whereas on the octahedron they are only repeated 4 times, resulting in 8 points. It is interesting to note that when the same triangular repeat is applied to the 20 triangular faces that make up the icosahedron; each of the 12 vertices becomes the host for 10-pointed stars.

The geometric design in the highly refined quarter dome of the Zangid wooden *mihrab* at the Lower Maqam Ibrahim in the citadel of Aleppo was undoubtedly the inspiration for the quarter dome of the magnificent Ayyubid wooden *mihrab* (1245-46) of the Halawiyya mosque and *madrasa* in Aleppo. This later *mihrab* is the work of Abu al-Husayn bin Muhammad al-Harrani 'Abd Allah bin Ahmed al-Najjar: a master of his art. Like the previous example from the al-Sharafiyah madrasa, the geometric design of this dome is derived from the octahedron, and likewise places eight-pointed stars at the vertices of the four repetitive triangular faces. The central region of each spherically projected triangular face is populated with nine-pointed stars, and the eight- and nine-pointed stars are separated by a network of five-pointed stars and darts [Fig. 501]. The earlier Seljuk and Zangid polyhedral designs were derived from the actual faces of their respective polyhedra—in effect, using the polyhedral faces as the underlying generative tessellation. By contrast, both of these Ayyubid domical hoods employ their projected repetitive faces as a substructure for their applied underlying generative tessellation. This is a spherical analogue to the very common application of underlying generative tessellations to triangular repetitive cells on the two-dimensional plane. Indeed, when used two-dimensionally, the pattern contained within each

---

[89] Allen (1999), Chap. 8.

triangular repeat from the domical hood in the mihrab of the Halawiyya mosque and madrasa produces a very successful design with 9- and 12-pointed stars that was used by Mengujekid artists at the Great Mosque and hospital of Divrigi (1228-29) less than twenty years previous [Fig. 346a]. As with the example from the al-Sharafiyah madrasa the 12-pointed stars of the two-dimensional analogue are replaced with eight-pointed stars by virtue of there being four triangles at each vertex rather than six.

An Ayyubid quarter dome hood that rests upon two tiers of *muqarnas* in the stone portal of the Farafra *khanqah* in Aleppo (1238) is decorated with an incised geometric design that is based upon the cubeoctahedron. The vertices of the cubeoctahedron have bilateral symmetry, and are composed of two opposing squares that alternate with two opposing triangles. This pattern places a six-pointed star at each vertex, and orientates the star to coincide with the bilateral symmetry of the vertex. The placement of the six-pointed stars onto the vertices creates a subtle incompatibility wherein the symmetry of the sixfold division does not precisely reconcile with the rigid symmetry of the cubeoctahedron. The pattern lines that stem from, and interact with, these six-pointed stars are consequently ill suited to comfortably fill the regions defined by the projected squares and triangles of the cubeoctahedron. This creates noticeable irregularities that result in a design that, however ambitious, is considerably less elegant than that of the quarter dome of the wooden *mihrab* (1245-46) at the Halawiyya mosque and *madrasa*.

1.19 Khwarizmshahs (1077-1231)

The Khwarizmshah lineage is traced back to pre-Islamic times. Their homeland is the vast Khwarizm oasis formed by the Amu Darya delta immediately south of the Aral Sea in present-day Turkmenistan and Uzbekistan. Throughout the reign of the Samanids, Qarakhanids, Ghaznavids, Seljuks, and Qara Khitai they maintained their rule over Khwarizm through military strength and by accepting the suzerainty of sequential empires. As vassals to the Seljuks and Qara Khitai they became immensely powerful. During the last decade of the twelfth century the Khwarizmshahs defeated the Qara Khitai forces in Transoxiania, and within decades they came to rule over an immense territory that stretched from Azerbaijan and portions of Iraq in the west, across all of Persia, through all of Afghanistan in the east, and Transoxiana in the north. Within just 5 years of their victory over the Ghurids in Khurasan, in 1220 the Mongol onslaught brought the Khwarizmshah Empire to a crushing defeat.

Few extant examples of Khwarizmshahid architecture remain, but those that have survived clearly indicate the influence of Seljuk aesthetics. Khwarizmshahid architectural

decoration continued the Seljuk practice of raised brick and faience ceramics. Indeed, both the mausoleum of Sultan Tekesh in Konye-Urgench, Turkmenistan (c. 1200), and the Zuzan *madrasa* in northeastern Iran (1219) are notable for their exuberant use of these ornamental media. The Zuzan *madrasa* is one of the earliest extant Muslim buildings to have expanded the use of ceramic faience ornament beyond a single color. The color palette of this building includes turquoise, dark blue, and white combined with unglazed terra-cotta. This building is also remarkable for the monumental scale of the *iwan*: prefiguring the architectural predilections of later Muslim cultures of this region. The exterior façade of the Zuzan *madrasa* has two interesting patterns made from the *system of regular polygons* and executed in raised brick. One of these is a *two-point* pattern with threefold symmetry that is created from the 4.6.12 semi-regular underlying tessellation of squares, hexagons, and dodecagons. This pattern is a remarkable concatenation of interweaving dodecagons, octagons, and arbitrary six-pointed stars [Fig. 109e]. The second is also a *two-point* pattern comprised from a *two-uniform* underlying tessellation made up of hexagons, squares, and triangles in an orthogonal $3^2.4.3.4$-$3.4.6.4$ configuration [Fig. 90]. This design from the Zuzan *madrasa* repeats upon a root-4 (double-square) rectangular grid [Fig. 112b] [Photograph 38]. Were it not for the 90° rotation of the underlying hexagonal modules, and the alternating orientation of the six-pointed stars that results from this rotation, this noteworthy pattern would repeat on a standard square grid. A very similar design, albeit with a significant difference in the line treatment, was used by post-Ilkhanid artists in the mausoleum of Muhammad Basharo in the remote village of Mazar-i Sharif in Tajikistan[90] (1342-43) [Fig. 111a]. The similarity in aesthetic character between these two *two-point* designs results from their identical pattern line application to the square modules of the underlying tessellation. Both utilize *two-uniform* tessellations with $3^2.4.3.4$-$3.4.6.4$ vertices, the difference being that the *two-uniform* tessellation from the Zuzan *madrasa* is orthogonal, while that of the mausoleum of Muhammad Basharo is isometric.

The ornament of the Zuzan *madrasa* includes a round faïence panel with a highly original geometric design that places octagons upon the vertices of the $3^2.4.3.4$ tessellation of triangles and squares [Fig. 103] [Photograph 39]. The two edge-to-edge triangles produce an underlying rhombic component, and their rotational orientation around each square element produces the repetitive structure common to all

---

[90] Not to be confused with the city of Mazar-i Sharif in Afghanistan. The village of Mazar-i Sharif, where the Mausoleum of Muhammad Basharo is sited, is approximately 25 km east of Penjikent, Tajikistan, and located in the Zarafshan River valley.

Photograph 39 A Khwarizmshahid oscillating square pattern created from the *system of regular polygons* located at the Zuzan *madrasa* in Iran (© Caroline Mawer)

Photograph 38 A Khwarizmshahid *two-point* pattern created from the *system of regular polygons* located at the Zuzan *madrasa* in Iran (© Sheila Blair and Jonathan Bloom)

oscillating square designs [Figs. 23–26]. Patterns based upon the $3^2.4.3.4$ tessellation are unusual, and more readily place 12-pointed stars at the vertices; such as an example from the Topkapi Scroll[91] [Fig. 23d]. The use of octagons at the vertices of this tessellation from the Zuzan *madrasa* does not readily conform to the angles of the underlying vertex configuration, and the success of their use is reliant upon the bilateral symmetry at each vertex.

Among the multiple roundel motifs that make up the highly elaborate interior frieze at the Zuzan *madrasa* are two identical *obtuse* patterns, with radial symmetry, created from the *fivefold system* [Fig. 260b] [Photograph 40]. As is often the case with fivefold *obtuse* patterns, this design can be created from two distinct underlying tessellations: one that places pattern lines with 108° angular openings into underlying network of a central decagon surrounding by pentagons, barrel hexagons, and thin rhombi, and the other with 72° angular openings applied to an underlying tessellation comprised of a central decagon surrounded by long hexagons and concave hexagons from the *fivefold system*. *Obtuse* patterns from this system invariably have 108° angular openings, while those with 72° are typically associated with the *median* family. However, in tessellations comprised of just decagons, long hexagons, and concave hexagons, and devoid of pentagons and barrel hexagons in particular, the 72° crossing pattern lines replicate the distinctive character of the *obtuse* family. For this reason, regardless of the specific underlying generative tessellation that the artist employed to create this example, it is rightfully identified as an *obtuse* design.

[91] Necipoğlu (1995), diagram no. 35.

Photograph 40 A Khwarizmshahid *obtuse* pattern created from the *fivefold system* located at the Zuzan *madrasa* in Iran (© Caroline Mawer)

1.20 Seljuk Sultanate of Rum (1077-1307)

In 1071 the Seljuks defeated the Byzantine forces at the battle of Manzikert, near Lake Van in eastern Anatolia. Within a decade, the Seljuk military commander Suleyman bin Kutalmish (d. 1086) declared himself Sultan over the conquered Byzantine territories in Anatolia, and established his capital in Isnik. By the middle of the twelfth century, from their capital in Konya, the Seljuk Sultanate of Rum controlled all of central Anatolia and portions of both the Mediterranean and Black Sea coastlines. In the thirteenth century, at the height of their power, the Seljuk Sultanate of Rum averted the loss of their empire by becoming vassals to the conquering Mongols. As such, they prevented the high level of destruction that befell so many of the great cities and centers of culture across Iraq, Persia, Khurasan, and Transoxiana. In marked contrast to the Mongol destruction in the eastern regions, the Seljuk architectural heritage in Anatolia remained relatively intact, and many of the most significant extant examples of Seljuk geometric ornament are from this region. Anatolia fell from Seljuk rule through internal strife and the rising power of rival dynasties: the last Seljuk Sultan of Rum being murdered in 1307.

The architecture of the Sultanate of Rum is predominantly stone. This was the material of choice for the Christian population of Anatolia and Armenia. The Seljuk conquerors of these regions readily adopted the highly evolved stone masonry practices of their new subjects. Over time, the ornamental use of stone in the architecture of the Sultanate of Rum developed a level of sophistication that is unsurpassed in the historical use of this material. Like the brick architecture of the eastern regions, the carved stone ornamental façades built by Seljuk artists in Anatolia were primarily monochrome with high relief; creating an aesthetic effect of stunning boldness that was largely reliant upon texture, light, and shadow. Carved stone provided an ideal medium for monumental epigraphy, floral design, and complex geometric patterns. The Seljuk architecture of Anatolia continued the practice of augmenting important parts of a building with colorful faience mosaics. The faience *mihrab* and dome of the Alaeddin mosque in Konya (c. 1219-21) is lavishly decorated with floral and geometric designs, as well as both cursive and knotted Kufi script. The density of the faience, nearing total coverage, is a forerunner of the *muarak* cut-tile mosaic aesthetic that came to prominence under the auspices of the Muzaffarids and Kartids approximately a century later in Persia and Khurasan respectively, from whom it was readily adopted by the Timurids. Other superb examples of faience mosaic from Seljuk Anatolia are found at the Izzeddin Kaykavus hospital and mausoleum in Sivas (1217) and the Sirçali *madrasa* in Konya (1242). Rather than being used as an accent for especially important parts of a building, such as a *mihrab*, the faience mosaics in both of these latter examples is applied throughout the main façade of the building. These are examples of transformative architectural ornament, creating a polychromatic aesthetic from what had been primarily monochromatic.

The architectural ornament of the Seljuk Sultanate of Rum exhibits multiple examples of geometric patterns that were used earlier in the eastern regions, including patterns made from the *system of regular polygons*. Several of these are sufficiently unusual as to strongly suggest direct inspiration rather than their independent re-creation, thereby substantiating the westward transmission of methodological knowledge. A possible example of such influence is an ornamental panel from the Izzeddin Kaykavus hospital and mausoleum in Sivas, Turkey (1217) [Fig. 106a], that closely resembles one of the patterns created from the 3.4.6.4 underlying tessellation at the Qarakhanid anonymous tomb at Uzgen, Kyrgyzstan (1186) [Fig. 104d]. The example from Uzgen was produced some 30 years earlier. Another unusual pattern made from the *system of regular polygons* that was used in both Khurasan and Anatolia is derived from the unusual pattern line extraction process wherein alternating underlying hexagons from the 3.6.3.6 tessellation are treated differently. The narrow border pattern that surrounds the door at the Seh Gunbad tomb tower in Orumiyeh, Iran (1180) [Fig. 101a] is also found at the Great Mosque of Niksar, Turkey (1145), and an outstanding variation is

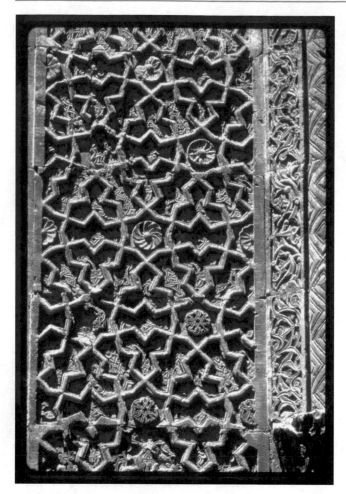

Photograph 41 A Seljuk Sultanate of Rum threefold pattern that is easily created from the *system of regular polygons* located at the Çifte Minare *madrasa* in Sivas, Turkey (courtesy of the Yasser Tabba Archive, Aga Khan Documentation Center at MIT)

from the portal of the Çifte Minare *madrasa* in Sivas, Turkey (1271) [Photograph 41]. The only differences between these examples are slight variations in the angles of the pattern lines, and the absence of the small triangle centered within the trefoil elements. This difference is purely stylistic, and results from an arbitrary aesthetic determination on the part of the artist rather than directly determined by the design methodology. Further evidence of the westward migration of artistic practices comes from an inscription on a faience panel at the Sirçali *madrasa* in Konya, Turkey (1242), wherein it is stated "made by Muhammad, son of Muhammad, son of Othman, architect of Tus." This artist traveled from Khurasan to Anatolia during the period of instability following the Mongol invasion.[92] During this tumultuous period, the exodus of those artists who were privy to the methodology used to create complex geometric

designs shifted the nexus of innovation firmly westward. As discussed above, during the twelfth century knowledge of the polygonal technique for designing geometric patterns had spread from Khurasan and Transoxiana westward into Ildegizid Azerbaijan, Seljuk and Artuqid Anatolia, Seljuk Iraq and al-Jazirah, as well as Egypt, Syria, and North Africa under successive Fatimid, Zangid, and Ayyubid rule. During the period of destruction, and the destabilized aftermath wrought by the Mongols throughout Transoxiana, Khurasan, Persia, and Iraq during the greater part of the thirteenth century, the western regions that either accepted the suzerainty of their Mongol overlords or successfully repelled the Mongols militarily became the primary benefactors of this ongoing design tradition. The Seljuk Sultanate of Rum and their rivals, the Mamluk Sultanate in Egypt were of particular importance in this ongoing westward growth in interest in the further development of Islamic geometric art.

Along with more complex varieties of geometric pattern, the *system of regular polygons* received considerable innovative attention from artists working in Anatolia during the Sultanate of Rum. The considerable quantities of monuments that have survived from Seljuk Anatolia provide a remarkable diversity of Islamic geometric patterns. Within just the *system of regular polygon*s there are dozens of beautiful patterns originating from this cultural milieu. A distinctive design that can be conveniently created from the underlying 6^3 hexagonal grid was used at the Great Mosque at Divrigi (1228-29) [Fig. 96i]. This design was used with some frequency by artists in Anatolia, including in the carved stone façade of the Hatuniye *madrasa* in Karaman (1382). A more complex design also created from the simple 6^3 grid is found in the south portal of the Great Mosque at Bayburt[93] (1220-35) [Fig. 97b]. This pattern is characterized by superimposed interweaving nonagons, set in rotation around the vertices of the visually exposed hexagonal grid, producing six-pointed stars at each vertex of the isometric dual grid. The same design was used as a border design at the Çifte Minare *madrasa* in Erzurum (late thirteenth century), and in the faience mosaic *mihrab* of the Ahi Serafettin mosque in Ankara (1289-90). A faience mosaic panel from the Ali Tusin tomb tower in Tokat, Turkey (1233-34), employs an isometric design that is easily created from the underlying 3.6.3.6 tessellation of hexagons and triangles[94] [Fig. 99f]. This is a very simple and well-balanced design comprised of meandering lines that do not lead back to themselves to make closed circuits. It is surprising that Muslim geometric artists did not use this very becoming pattern more frequently. The *mihrab* of the Great Mosque of Siirt (1129) uses an isometric design produced from the

[92] Wilber (1939), 40.

[93] Schneider (1980), pattern no. 198.

[94] Schneider (1980), pattern no. 250.

3.6.3.6 tessellation that is made up of two interlocking elements: the six-pointed star and a motif with threefold rotation symmetry [Fig. 100a]. This distinctive pattern is part of a group of similar interlocking isometric designs found within the historical record. A hallmark of these designs, including this example, is their ability to easily be created from the isometric grid [Fig. 73a]. As is often the case, it is impossible to know for certain which methodology the artist working in Siirt used to create this fine pattern. A much later example of this interlocking design is found at the tomb of I'timad ad-Dawla in Agra, India (c. 1628-30). Another example of an Anatolian pattern made from the 3.6.3.6 underlying semi-regular tessellation was employed on the northern portal at the Great Mosque at Divrigi[95] (1228-29) [Fig. 100c]. This is a Mengujekid monument that is noteworthy for the unique and highly elaborate baroque quality of the floral ornament. Yet the many examples of geometric design within this building conform to the aesthetic practices of their neighboring artists working under the patronage of the Seljuk Sultanate of Rum. A later Seljuk example of this pattern was used at the Muzaffer Barucirdi *madrasa* in Sivas (1271-72). A design that uses alternating active and passive hexagons and triangles from the 3.6.3.6 underlying tessellation was widely used by artists working in the Seljuk Sultanate of Rum [Fig. 101c]. Examples include the portal of the Great Mosque of Niksar (1145); the portal of the Alaeddin mosque in Konya (1219-21); and the Huand Hatun complex in Kayseri (1237). This design was also used at the Zangid portal of the Bimaristan Arghun at the citadel of Aleppo (twelfth century) [Photograph 36].

A well conceived 3.4.6.4 pattern from the north portal of the Sungur Bey mosque in Nigde (1335) places nonagons upon the vertices of the hexagonal grid[96] [Fig. 107c]. These nonagons are located on the centers of the triangles of the underlying 3.4.6.4 tessellation, with superimposed hexagons added into the pattern matrix. It is interesting to note that the perpendicular arrangement of the *two-point* pattern lines that cross the shared edges of the triangular and square modules produces a similar aesthetic to the design created from the same 3.4.6.4 underlying tessellation at the Chaghatayid mausoleum of Tughluq Temür in Almaliq, western China (1363) [Fig. 105d] [Photograph 70]. A border pattern in the east *iwan* of the Çifte Minare *madrasa* in Erzurum (late thirteenth century) employs a very successful isometric pattern made from the 3.4.6.4 underlying tessellation of triangles, squares and hexagons.[97] This design is conceptually similar to the above referenced pattern at the Great Mosque at Bayburt that places six nonagons in rotation

around each hexagonal repeat [Fig. 97b]: the difference being that the design from the Çifte Minare *madrasa* places six *octagons* around each hexagon. This is achieved by locating the center of each octagon upon the center of each underlying square of the 3.4.6.4 tessellation [Fig. 107a]. The application of octagons into an isometric repetitive structure qualifies this as an *imposed symmetry* design wherein a star or polygon with an *n*-fold rotational symmetry—in this case eightfold—is placed within a seemingly incompatible repetitive structure—in this case isometric. A 3.4.6.4 design from the Cincikh mosque in Aksaray (1220-30) [Fig. 107b] is conceptually similar to the above-cited example from the Çifte Minare *madrasa*: a similar distribution of superimposed octagons has hexagons added into the pattern matrix.[98] The simple addition of the hexagons augments the complexity considerably, and radically changes the overall appearance of the design. Another innovative Anatolian geometric pattern created from the 3.4.6.4 underlying tessellation is from the portal of the Gök *madrasa* and mosque in Amasya, Turkey (1266-67).[99] The angular openings of the applied pattern lines of this design are determined by the placement of regular octagons within the square modules of the underlying tessellation [Fig. 104b]. This lovely design was used nearly 240 years later in the stone mosaic ornament in the *mihrab* niche of the Sultan Qansuh al-Ghuri complex in Cairo (1503-05).

Patterns with 12-pointed stars created from underlying dodecagons were also widely used by the Seljuk Sultanate of Rum. Two of the earliest Anatolian examples of the well-known threefold pattern created from the underlying semi-regular 3.12^2 tessellation are found at the Great Mosque of Kayseri (1205) and the *mihrab* of the Great Mosque of Akşehir near Konya (1213) [Fig. 108a]. These were preceded by the example of this pattern at the eastern tomb tower at Kharraqan by approximately 150 years [Photograph 17]. Several nearly identical examples of an isometric pattern with 12-pointed stars that are easily created from the 3.4.6.4 underlying tessellation are found in the city of Ahlat on Lake Van in eastern Turkey. These include examples from the Usta Sagirt tomb[100] (1273) [Fig. 105g], the Huseyin Timur tomb[101] (1279) [Fig. 105i], and a number of the highly ornamented gravestones for which this town is renowned. A variation of this pattern from the Seljuk Sultanate of Rum was widely used in the Maghreb [Fig. 105h].

---

[95] Schneider (1980), pattern no. 206.

[96] Schneider (1980), pattern no. 226.

[97] Schneider (1980), pattern no. 227.

[98] Gerd Schneider illustrates the close relationship between these examples in his patterns 227 and 228. However, in keeping with the totality of his illustrations, he does not provide generative methodologies or underlying formative structures. Schneider (1980).

[99] Schneider (1980), pattern no. 215.

[100] Schneider (1980), pattern no. 397.

[101] Schneider (1980), pattern no. 403.

This places an arbitrary eight-pointed star within the underlying square elements. Among the many examples of this variation are a *zillij* mosaic panel at the Alcazar in Seville (fourteenth century), and the carved stucco ornament from the Córdoba Synagogue (1315). Each of these *two-point* patterns is constructed in an identical manner, and only differs in the arbitrary stylistic treatment of the pattern elements within the underlying square and triangular modules. The construction of this particular group of 3.4.6.4 patterns is not limited to the polygonal technique, and can also be produced using the methodology of *extended parallel radii*. Artists working under patronage of the Seljuk Sultanate of Rum were pioneers of this alternative technique for generating geometric patterns; and though this methodology was never as widely utilized as the polygonal technique, it is certainly possible that the multiple examples of these designs from Ahlat were generated using this alternative technique [Figs. 80 and 81]. Patterns created from the 4.6.12 tessellation are less common, and two fine examples include a pattern from the drum of the dome of the Izzeddin Keykavus hospital and mausoleum in Sivas[102] (1217-18) [Fig. 109d] and an Ottoman carved stone relief panel from the Hasbey Darül Huffazi *madrasa* in Konya[103] (1421) [Fig. 109a]. A very fine orthogonal pattern created from the 3.4.3.12-3.12^2 *two-uniform* tessellation was used at the Hasbey Darül Huffazi Han near Kayseri (1235-41) [Fig. 113c]. Other examples of this pattern from the Seljuk Sultanate of Rum are found at the Sultan Han in Kayseri (1232-36), and the tomb of Sultan Mesud in Amasya (fourteenth century). This pattern was used at an earlier date on the side panels for a Zangid *minbar* commissioned by Nur al-Din in 1186, as well as many subsequent examples produced by Mamluk artists.

The high degree of artistic innovation during the Sultanate of Rum is reflected in a group of patterns created from the *system of regular polygons* that employ non-regular modules within their otherwise fully regular underlying tessellations. An orthogonal design from the *mihrab* of the Alaeddin mosque in Konya (1219-21) is the earliest of many Sultanate of Rum examples of an unusual fourfold pattern made up of dodecagonal underlying polygonal elements that are filled in with four ditrigonal shield shapes and four triangles[104] [Fig. 120]. Without the ditrigons and triangular inclusions, this would be the 3.4.3.12-3.12^2 *two-uniform* generative tessellation that was commonly used for creating Islamic geometric patterns [Fig. 113]. This same design was used some 40 years earlier at the Shah-i Mashhad and the minaret of Jam in Afghanistan: providing further evidence for the westward dissemination of specific designs into the eastern areas of Seljuk influence. A fine example of a threefold pattern that incorporates this same ditrigonal module within the underlying generative tessellation is located in the *mihrab* of the Yelmaniya mosque in Cemiskezck, Turkey[105] (1274) [Fig. 117]. The underlying generative tessellation for this pattern is comprised of dodecagons, squares, and triangles, with irregular ditrigons at the center of each triangular repeat unit. This same design was subsequently used in a number of historical locations, including the *mihrab* of the Aqburghawiyya *madrasa* (1340) at the al-Azhar mosque in Cairo and the Ottoman ornament of the Dome of the Rock in Jerusalem. Another threefold pattern that uses this ditrigonal module in its underlying generative tessellation is from the portal of the Huand Hatun complex in Kayseri (1237), and was also used in the *mihrab* of the Ahi Serafettin mosque in Ankara (1289-90) [Fig. 118c]. This design can also be constructed using the 6^3 tessellation of hexagons [Fig. 97a]. The ditrigons in this generative tessellation are identically proportioned to those from the Yelmaniya mosque in Cemiskezck. However, in this design the angles of the applied crossing pattern lines are determined by the placement of octagons upon the underlying polygonal vertices with 90° angles. The underlying tessellation that produces this design is actually a well-known *median* pattern in its own right: created from the simple hexagonal grid [Fig. 95c]. This use of an existing geometric pattern as an underlying generative structure is unusual, but not unique. An Ottoman design from the *mihrab* of the Yesil mosque in Bursa (1419-21) is identical to the example from the Huand Hatun complex except that six-pointed stars arbitrarily replace the hexagonal elements within the pattern matrix. A very similar pattern created from the same underlying tessellation to the example from the Huand Hatun complex is from one of the Mamluk window grilles at the al-Anzar mosque in Cairo [Fig. 118b]. A further example from the Karatay *madrasa* in Antalya, Turkey (1250) employs a variation within the underlying six-pointed star module that is derived from the inclusion of pentagons and a central hexagon into the underlying tessellation [Fig. 118d]. Another pattern from the Seljuk Sultanate of Rum that adds non-regular modules into the underlying tessellation of regular polygons is from the tomb of Seyit Mahmut Hayrani in Aksehir near Konya[106] (1275) [Fig. 122]. This elegant orthogonal pattern is created from an arrangement of underlying squares and triangles such that the interstice regions provide for the creation of four coinciding irregular pentagons. The cluster of four pentagons has shared characteristics with some *fourfold system B* designs, except

---

[102] Schneider (1980), pattern no. 256.

[103] Schneider (1980), pattern no. 413.

[104] Schneider (1980), pattern no. 298.

[105] Schneider (1980), pattern no. 441.

[106] Schneider (1980), pattern no. 325.

that the proportions of the pentagons are slightly different, and do not tessellate systematically.

As with other Muslim cultures, patterns that relate to the 4.8^2 underlying tessellation were commonly used by the Seljuk Sultanate of Rum. An otherwise standard *acute* pattern [Fig. 124a] from the minaret of Hotem Dede in Malatya (twelfth century) radically alters the appearance of this standard design through a subtractive variation [Fig. 125a], while the additive variation of this same standard design from the Sirçali *madrasa* in Konya (1242-45) likewise alters the visual quality of the original design significantly [Fig. 125d]. Several unique, if very simple, *two-point* designs were also likely produced from this tessellation, including a pattern comprised of superimposed four-pointed stars from the Karatay *madrasa* in Konya (1251-55) [Fig. 128c]; a pattern made up of two sizes of octagons from the Divrigi hospital (1228-29) [Fig. 128e]; and a pattern composed of superimposed concave hexagons from the Great Mosque of Divrigi (1228-29) [Fig. 128f]. Another example from the Great Mosque of Divrigi is an *obtuse* pattern that employs arbitrarily placed eight-pointed stars into the underlying octagonal regions [Fig. 127d]. Fatimid artists used this pattern over half a century earlier in the wooden *mihrab* of the mausoleum of Sayyidah Nafisah in Cairo (1138-46), and a later Mamluk example was used on the upper shaft of a minaret at the Sultan Qaytbay funerary complex in Cairo (1472-74).

The architectural ornament of the Seljuk Sultanate of Rum made wide use of the *fourfold system A*, including examples that were already known in the eastern regions, and many that the historical record suggests were original to artists working in Anatolia. The least complex designs created from this system are field patterns. These eschew the use of the underlying large octagons, and hence have no eight-pointed stars within their pattern matrix. As discussed, the earliest examples of field patterns created from the *four-fold system A* were produced by Ghaznavid, Qarakhanid, Ghurid, and Seljuk artists working in the eastern regions, and artists working under the patronage of the Seljuk Sultanate of Rum readily adopted this category of geometric design. Examples of such field patterns include multiple examples of a very basic pattern comprised of interlocking concave octagons, the earliest example of which is found in the *mihrab* of the Great Mosque of Niksar (1145) [Fig. 136b]. This pattern is generated from a tessellation of just the small hexagons from the *fourfold system A*. At least two field patterns were used during the Seljuk Sultanate of Rum that are easily produced from an underlying tessellation of large hexagons and squares. Each was used in multiple locations over a wide span of time. The first of these is an *acute* pattern that was used at the Tepsi minaret in Erzurum (1124-32). This was built by the Saltukids within decades of its first use by the Artuqids at the Maqam Ibrahim at Salihin

in Aleppo (1112) [Fig. 138a]. The Saltukids were an Anatolian *beylik* that were allied with the Great Seljuks in Persia, and who were overthrown by the rise of the Seljuk Sultanate of Rum. The second pattern created from this underlying tessellation is a well-known *median* design that was used throughout the Islamic world. Among its earliest locations in Anatolia is in the portal of the Alay Han near Aksaray (1155-92) [Fig. 138c]. A variation of this *median* pattern was used in the *mihrab* of the Great Mosque of Erzurum (1179) [Fig. 139]. A *median* field pattern that can be created from an underlying tessellation of large hexagons and pentagons was also used widely by artists in Anatolia [Fig. 140]. The earliest known example is from the Sultan Han in Kayseri (1232-36). Generally, the greater the number of underlying polygonal modules used to create a design the more complex the resulting pattern. A very successful *median* field pattern from the Huand Hatun complex in Kayseri (1237) makes use of four different underlying modules, and is one of the more complex field patterns created from the *fourfold system A* [Fig. 141].[107] This design was used soon after at the Haci Kiliç mosque and *madrasa* also in Kayseri (1249), but its earliest use appears to have been by Ildegizid artists at the mausoleum of Yusuf ibn Kathir in Nakhichevan (1161-62).

Original Seljuk Sultanate of Rum patterns created from the *fourfold system A* include a large number of *median* designs with 90° crossing pattern lines placed at the midpoints of each underlying polygonal edge. This provides these patterns with similar features that are easily recognized as a family. Examples of this family include a design from the Great Mosque at Erzurum (1179) that was used subsequently in many locations in Anatolia[108] [Fig. 143a]; a border design on the exterior façade of the Alay Han, 35 km northeast of Aksaray, Turkey[109] (1155-92) [Fig. 148]; a border design placed around the drum of the dome at the Gök *madrasa* in Amasya (1266-67) [Fig. 143b]; and a wooden screen railing in the Esrefoglu Süleyman Bey in Beysehir, Turkey (1296-97) [Fig. 142]. An original *acute* carved stone border from the portal façade of the hospital at the Çifte *madrasa* in Kayseri[110] (1205) is unusual in that it uses two types of crossing pattern line. This design can be created from either of two underlying tessellations [Figs. 146 and 147a]. Interestingly, these are not duals of one another. A variation of this design was used at the Friday Mosque in Gonabad, Iran (1212), some 7 years after the example from Kayseri. The distinctive quality of this design, and their close dates of production, implies a direct causal relationship between these two examples of this pattern. A fine example

---

[107] Schneider (1980), pattern no. 281.

[108] Schneider (1980), pattern no. 302.

[109] Schneider (1980), pattern no. 303.

[110] Schneider (1980), pattern no. 306.

of a unique Sultanate of Rum *two-point* pattern was used on the *iwan* of the Sirçali *madrasa* in Konya[111] (1242) [Fig. 152]. This example is unusual in that the edge length of the underlying octagons matches the longer rather than the smaller edge of the triangular module from this system, resulting in a proportionally larger underlying octagonal module and consequent eight-pointed star motif. This same underlying tessellation was used 63 years earlier by Qarakhanid artists to create a pattern comprised of two sizes of superimposed octagons at the Maghak-i Attari mosque in Bukhara, Uzbekistan (1178-79) [Fig. 151].

All of these examples repeat upon the standard orthogonal grid with eight-pointed stars located at the vertices of the repeat. By contrast, a remarkable orthogonal *median* pattern created from the *fourfold system A* located in the *mihrab* in the mosque of the Huand Hatun complex in Kayseri[112] (1237) places the primary eight-pointed star motifs at the vertices of a square and rhombus symmetrical structure rather than at the vertices of the more broad scaled square repeat [Fig. 156]. The 64.4712...° and 115.5288...° included angles of the rhombi are eccentric, and do not readily conform to eightfold geometry, yet in this instance they combine with the square repetitive cells to produce a design that is as visually successful as it is unusual. These rhombi are placed in rotation around each square in the same manner as the $3^2.4.3.4$ semi-regular tessellation of regular triangles and squares [Fig. 89]. The application of the underlying polygonal modules from the *fourfold system A* to this repetitive structure is unconventional in two respects: (1) the eight-pointed stars within the pattern matrix are generated with two alternating underlying polygonal arrangements, and (2) the layout of the underlying tessellation does not have coinciding edges that symmetrically align with the square and rhombic coinciding edges. In contradistinction to other patterns that have square and rhombic hybrid structures—such as the aforementioned design from the Mu'mine Khatun mausoleum in Nakhichevan, Azerbaijanin [Fig. 182]—the layout of the underlying polygonal modules of this design from the Huand Hatun mosque prevents either the squares or rhombic elements from functioning as repeat units on their own. Another unusual feature of this design is the discrepancy between the *plane symmetry group* of the underlying polygonal tessellation and that of the pattern itself. Ordinarily, both the underlying tessellation and its generated design will share an identical plane symmetry group (unless and until a design's crossing pattern lines acquire chirality through their being provided with an interweaving treatment). The underlying tessellation that creates this design from the Huand Hatun falls into the

cmm plane symmetry group, while the pattern itself is in the *p4g* group. These unique geometric characteristics qualify this example as perhaps the most symmetrically complex pattern created from the *fourfold system A* throughout the length and breadth of the Islamic ornamental tradition. However, despite this extremely eccentric geometric character, it is nonetheless very balanced and pleasing to the eye.

Artists working under the auspices of the Sultanate of Rum experimented with geometric patterns that employ an additive swastika device within the square components of particular designs. Most of these are based upon patterns that were created from the *fourfold system A*, although this same additive device was also applied to other varieties of design: the operative qualifier being the presence of squares within the pattern matrix. Most of the Anatolian designs with this variety of additive treatment are rather simplistic,[113] but a particularly sophisticated example was used in the faïence ceramic ornament of the Karatay *madrasa* in Konya[114] (1251-55) [Fig. 150a]. This variety of additive motif became especially popular under Timurid patronage.

In keeping with Zangid and Ayyubid practices to the south, and unlike the aesthetic predilections of the Great Seljuks to the east, the *fourfold system B* was more widely used than the *fourfold system A* under the patronage of the Sultanate of Rum. The classic *acute* pattern from this system [Fig. 173a] was used ubiquitously[115], with the earliest known Anatolian example located on the *minbar* of the Great Mosque of Aksaray (1150-53). This was produced within two decades of its first apparent use in the arch spandrels of the *mihrab* at the Friday Mosque at Sin, Iran (1134). This design uses only the underlying octagons and pentagons from this set of modules. An *acute* design that is similar in appearance, but slightly more complex, was used at the Sultan Han in Aksaray[116] (1229) [Fig. 177a]. The underlying tessellation that creates this pattern incorporates the same octagons and pentagons, but with elongated hexagons separating the octagons. Tessellations with this configuration of polygons, albeit with variable proportions to the hexagonal module, were used to create a wide variety of patterns in each of the four pattern families [Figs. 175–178]. The Zangid artists who produced the *minbar* (1187) for the al-Aqsa mosque used this same underlying tessellation for the notable *acute* pattern that adorn the triangular side panels [Fig. 177b]. The difference between the visual characteristics of the earlier Zangid *acute* pattern and that

---

[111] Schneider (1980), pattern no. 236.

[112] Schneider (1980), pattern no. 330.

[113] Schneider (1980), pattern numbers 91–98.

[114] Schneider (1980), pattern no. 91.

[115] Schneider (1980), pattern no. 321. Schneider identifies no less than 38 examples of this pattern scattered throughout the many monuments built by the Sultanate of Rum: pp. 183–4.

[116] Schneider (1980), pattern no. 217.

of the Sultan Han results from the treatment of the applied pattern lines to the underlying elongated hexagonal modules [Fig. 172], and the treatment of the pattern lines within the cluster of four pentagons at the center of the square repeat unit. Another popular Anatolian Seljuk *acute* pattern created by the *fourfold system B* employs rhombic repeat units with 45° and 135° included angles, and eight-pointed stars at the rhombic vertices [Fig. 181]. The earliest known use in Anatolia is in the portal of the Izzeddin Keykavus hospital and mausoleum in Sivas[117] (1217-18). However, this repetitive element was used in conjunction with a square repetitive element to create the distinctive hybrid design some 30 years earlier at the Ildegizid tomb of Mu'mine Khatun in Nakhichevan, Azerbaijan (1186) [Fig. 182]. This rhombic design was widely used by artists during the Sultanate of Rum. An aesthetically similar *acute* design, also from the Izzeddin Keykavus hospital and mausoleum in Sivas, incorporates far more geometric information within each square repetitive unit, and like its rhombic neighbor, employ the same pattern line variation within the underlying elongated hexagons that create the distinctive octagons within the pattern matrix[118] [Fig. 179a]. This orthogonal design places eight-pointed stars upon the vertices of the 4.8^2 semi-regular grid, and as with the rhombic design from this same location, many subsequent examples of this design were used by Anatolian artists during this period. Another Anatolian design created from the same underlying tessellation as the orthogonal design from the Izzeddin Keykavus hospital and mausoleum in Sivas is an *obtuse* pattern from the Hudavent tomb in Nidge (1312) [Fig. 179b]. A *two-point* pattern from a portal at the Bimarhane hospital in Amasya (1308-09) [Fig. 174c] is nearly identical to the notable Ghurid *fourfold system B* raised brick panel in the Friday Mosque at Herat [Photograph 32] from just over a hundred years earlier [Fig. 174b]. The unusual quality of this design argues against independent development, and for the possibility that knowledge of this design was imported from Khurasan to Anatolia.

The quantity of geometric patterns created from the *fivefold system* by artists working in Anatolia under the Sultanate of Rum is vast, and far exceeds the confines of this study. As one would expect, the classic *acute* pattern from this system [Fig. 226c] was used multiple times in the architecture of the Sultanate of Rum[119], with the earliest known Anatolian example located in the *minbar* (1155) of the Alaeddin mosque in Konya. The underlying polygonal

tessellation that produces the classic *acute* pattern is also responsible for equally classic patterns in each of the other pattern families [Figs. 85–88]. The earliest Anatolian *obtuse* design created from this tessellation is from the Great Mosque of Siirt (1129): just decades later than the earliest known occurrence at the Friday Mosque at Isfahan in Iran [Fig. 229b] [Photograph 21]. And the earliest Anatolian use of the *two-point* pattern created from this tessellation is from the Huand Hatun complex in Kayseri (1237): only decades later than its earliest known use at the Gunbad-i 'Alaviyan in Hamadan, Iran (late twelfth century) [Fig. 231d] [Photograph 22]. As with other varieties of geometric pattern, the spirit of experimentation among artists in the Sultanate of Rum was widely applied to the *fivefold system*, with the result of there being a greater concentration of diverse fivefold patterns in Anatolia than found in any other extant Islamic architectural tradition. This fivefold diversity included designs with very broad repetitive structures, multiple examples of field patterns, many repetitive strategies, and various additive treatments to the ten-pointed stars that are inherent to this system.

Artists working under the patronage of the Seljuk Sultanate of Rum produced many original patterns from the *fivefold system* that repeat upon the rhombic grid of 72° and 108° angles. Many of these were used multiple times throughout Anatolia; some were adopted by succeeding Muslim cultures; and others are only known to exist in a single location. Examples of this variety of original fivefold pattern include: the Huand Hatun complex in Kayseri[120] (1237) [Fig. 235d]; the Agzikara Han near Aksaray (1231-40) [Fig. 233b]; and a pattern from the portal of the Gök *madrasa* in Tokat[121] (1275-80) [Fig. 234b]. The design from the Gök *madrasa* is unusual in that the underlying polygons applied to the central region of the rhombic repeat create an interstice region that is filled by simply extending the pattern lines from the adjacent underlying polygons into the open region. More complex fivefold patterns that repeat upon the rhombic grid of 72° and 108° included angles include an outstanding *median* pattern from the Sultan Han in Kayseri[122] (1232-36) [Fig. 237] [Photograph 42]. The ten-pointed stars located at the vertices of the repetitive grid are not usual to the *median* family. Typically, these will have 72° crossing pattern lines placed at the midpoints of each decagonal edge. However, rather than decagons, the underlying tessellation of this example from Kayseri has large ten-pointed star interstice regions at each vertex of the repetitive grid. The 72° crossing pattern lines that are placed upon

[117] Schneider (1980), pattern no. 322.

[118] Schneider (1980), pattern no. 320.

[119] Schneider (1980), pattern no. 219. Schneider identifies 45 examples of the fivefold classic *acute* pattern in the many Anatolian Seljuk buildings he studied.

[120] Schneider (1980), pattern no. 279.

[121] Schneider (1980), pattern no. 377.

[122] Schneider (1980), pattern no. 392.

Photograph 42 A Seljuk Sultanate of Rum *median* pattern created from the *fivefold system* located at the Sultan Han in Kayseri, Turkey (© Serap Ekizler Sönmez)

the edges of this interstice region are extended to create the ten-pointed stars that are atypical to this pattern family.

Seljuk Sultanate of Rum patterns that repeat upon the more acute fivefold rhombus with 36° and 144° included angles include a very pleasing *acute* pattern from the Huand Hatun complex (1237) that uses irregular pentagons within the underlying generative tessellation that are not a part of the standard modules used in this design system [Fig. 242]. Later examples of this design include a panel from a Mamluk *minbar* in the collection of the Victoria and Albert Museum[123] (c. 1300), and a Kartid pair of wooden doors of the mausoleum of Shaykh Ahmed-i Jam at Torbat-i Jam in northeastern Iran (1442-45). A very successful *obtuse* pattern that repeats upon this more acute rhombus is from the

portal of the Muzaffar Barucirdi *madrasa* in Sivas[124] (1271-72) [Fig. 241b]. This same design was employed by Mamluk artists at the *mihrab* of the Amir Altinbugha al-Maridani mosque in Cairo (1337-39).

Artists working in the Seljuk Sultanate of Rum frequently used rectangular repeat units of various proportions when making patterns from the *fivefold system*. One of the most basic fivefold rectangular designs, with the least amount of geometric information contained within the repeat unit, is found at the Sultan Han in Aksaray (1229) [Fig. 245a]. The underlying polygonal tessellation for this *obtuse* design places decagons at the vertices of the rectangular repeat units, with two edge-to-edge pentagons separating the decagons along the short edge of the repeat, and barrel hexagons in the long dimension [Fig. 203]. This arrangement creates the cluster of six pentagons at the center of the repeat unit: a configuration that produces very acceptable *obtuse* and *two-point* patterns, but requires adjustment for acceptable *acute* and *median* designs [Figs. 197 and 198]. It should be noted that this design can also be created with equal ease by using an underlying tessellation of contiguous decagons in the short dimension of the repeat, and the concave hexagon separating the decagons in the long dimension. The earliest known use of this very popular rectilinear pattern was at the Maghak-i Attari mosque in Bukhara, Uzbekistan (1179-79), and the variety of later locations include the Amir Zadeh mausoleum in the Shah-i Zinda funerary complex in Samarkand, Uzbekistan (1386); the Abdullah Ansari complex in Gazargah near Herat, Afghanistan (1425-27); the Gur-i Amir complex in Samarkand (1403-04); and the tomb of Akbar in Sikandra, India (1613). Rectangular fivefold designs that are original to the Sultanate of Rum include an *obtuse* design from the *iwan* of the Sirçali *madrasa* in Konya[125] (1242-45) [Fig. 247]; an *obtuse* design from the *iwan* of the Yusuf ben Yakub *madrasa* in Cay[126] (1278) [Fig. 249]; and a *median* pattern from the *mihrab* of the Külük mosque in Kayseri[127] (1280-90) [Fig. 251]. A design from a door panel of the Hekim Bey mosque in Konya[128] (1270-80) has the unusual feature of transitioning from the *acute* family at the ends of the very elongated rectangular repeat unit to the *median* family throughout the rest of the rectangular repeat [Fig. 269]. This is achieved through the use of two scales of underlying polygonal modules. Patterns with variable scaled underlying polygonal modules are extremely rare, and

---

[123] A nineteenth century reproduction of the original panel is in the collection of the Victoria and Albert Museum, London; museum number 887–1184.

[124] Schneider (1980), pattern no. 374.

[125] Schneider (1980), pattern no. 376.

[126] Schneider (1980), pattern no. 380.

[127] Schneider (1980), pattern no. 370.

[128] Schneider (1980), pattern no. 388. This door panel currently resides in the Ince Minare *madrasa* History Museum in Konya, Turkey.

appear to be exclusive to Turkey. A later example is from an Ottoman wooden door at the Sultan Bayezid II Kulliyesi in Istanbul (1501-06) [Fig. 270].

Field patterns made from the *fivefold system* have a distinct quality that sets them apart from non-fivefold varieties of field pattern, and indeed, other fivefold patterns with their characteristic ten-pointed stars. These will often employ rectangular or hexagonal repeat units with a minimum of geometric information. Examples of the former include a *median* pattern on an arch at the Kayseri hospital[129] (1205-06) [Fig. 209]; a *two-point* design from the *iwan* of the Great Mosque at Malatya[130] (1237-38) [Fig. 207]; and an *obtuse* pattern that is used as a linear band at the Haci Kiliç *madrasa* in Kayseri (1275) [Fig. 208]. Rectangular field patterns with greater complexity include a *median* pattern from the west portal of the Huand Hatun complex in Kayseri[131] (1237) [Fig. 210]. Field patterns with diversely proportioned small hexagonal repeat units were also well known to this tradition, and examples include the Sitte Melik tomb in Divrigi[132] (1196) [Fig. 213]; a *median* design from a courtyard portal at the Sultan Han in Kayseri[133] (1232-36) [Fig. 216]; the Huand Hatun complex in Kayseri[134] [Fig. 218]; the Great Mosque in Malatya[135] (1237-38) [Fig. 220]; the Hekim Bey mosque in Konya[136] (1270-80) [Fig. 219]; and the Çifte Minare *madrasa* in Erzurum[137] (late thirteenth century) [Fig. 215]. Two very nice linear border designs with hexagonal repeat units were created from the same underlying tessellation: one of these is a *median* pattern from the Alaeddin mosque in Konya [Fig. 214a], and the other is an *obtuse* pattern from the Çifte Minare *madrasa* in Erzurum [Fig. 214c]. Yet another *fivefold system* field pattern with a hexagonal repeat from the Çifte Minare *madrasa* in Erzurum has characteristics that are equally *obtuse* (pentagons), and *median* (kite shapes) [Fig. 215]. An interesting field pattern with minimal geometric information within each repetitive element is found at the Sahib Ata mosque in Konya[138] (1258) [Fig. 211]. Like the above-cited hexagonally repeating designs from the Huand Hatun and Sultan Han in Kayseri, the underlying tessellation of this *median* design employs a distinctive kite-shaped module that is 2/5 of the decagon [Fig. 188]. The difference with this

design from Konya is that the pattern is created exclusively from the kite-shaped underlying polygonal module. This is achieved by setting the kite-shaped polygons into alternating linear bands with coincident long edges. The earliest example and likely progenitor of this design was produced by Seljuk artists at the Khwaja Atabek mausoleum in Kerman (1100-1150). This group of field patterns demonstrates the diversity of methods the *fivefold system* offers for filling the two-dimensional plane with a single atypical repeat unit.

Artists working in the Seljuk Sultanate of Rum occasionally introduced either of two additive motifs into the ten-pointed stars of fivefold *median* Patterns [Fig. 224]. This type of pattern variation was developed by the Seljuks in Persia and the earliest known use is found at the Gunbad-i Qabud in Maragha (1196-97) [Fig. 240] [Photograph 24]. This additive technique has the affect of transforming the overall design into a field pattern. A very successful Anatolian Seljuk example of this additive technique is from the Huand Hatun complex in Kayseri (1237) [Fig. 257c]. The modification to the standard *median* ten-pointed stars in this pattern replaces them with a five-pointed star motif [Fig. 224b]. This results in a highly cohesive design that is unsurpassed in beauty by the many outstanding field patterns produced in Anatolia during this period. It is interesting that this simulated field pattern from Kayseri is one of the only Anatolian Seljuk fivefold designs with underlying decagons in the generative tessellation that repeats upon a hexagonal grid. Another remarkable example of a fivefold *median* pattern that arbitrarily fills the ten-pointed stars in a similar fashion is from the Karatay *madrasa* in Konya (1251-55) [Fig. 238]. This example repeats upon the fivefold rhombus with 72° and 108° included angles, and the number of underlying polygonal modules within this repeat is significantly greater than usual. This results in a rather complex design whose initial complexity is augmented through the additive treatment of the decagonal regions. A fundamental feature of the governing structure used in the creation of this design is the placement of a ring of ten edge-to-edge decagons at each vertex of the rhombic repeat [Fig. 238a]. These decagons are either filled with further underlying polygonal modules [Fig. 238b], or with the arbitrary modification to the generated ten-pointed stars that introduces a pentagon at the center of each decagon [Figs. 224a and 238c]. An interesting feature of this pattern is the ten-pointed star rosette within each ring of ten decagons. These introduce the characteristics of the *acute* family into what is otherwise a *median* pattern, and the overall affect is highly successful. It is interesting to note that the initial layout of the ring of ten decagons placed at each vertex of the rhombic repeat was also used some 50 years previously on the celebrated decagonal tomb tower of the Gunbad-i Qabud in Maragha (1196-97) [Figs. 239 and 240] [Photograph 24]. While the initial

---

[129] Schneider (1980), pattern no. 363.

[130] Schneider (1980), pattern no. 362.

[131] Schneider (1980), pattern no. 367.

[132] Schneider (1980), pattern no. 367.

[133] Schneider (1980), pattern no. 369.

[134] Schneider (1980), pattern no. 360.

[135] Schneider (1980), pattern no. 365.

[136] Schneider (1980), pattern no. 368.

[137] Schneider (1980), pattern no. 204.

[138] Schneider (1980), pattern no. 361.

decagonal layout is identical, both of these examples incorporate very different secondary polygonal infill into their decagons and interstice regions, as well as very different locations for the arbitrary modifications to their ten-pointed stars. However, the conceptual similarity suggests a direct influence of the earlier upon the latter.[139]

Artists working in the Sultanate of Rum augmented the complexity of patterns created from the *fivefold system* by combining otherwise stand-alone repeat units into single hybrid constructions. In its broad context, the overall repeat for each of these examples is a rectangle; but these broad rectangular repeats are the direct product of, and best understood as, a tessellating conglomerate of smaller repetitive units. For such designs to be successful, the pattern-lines located upon the *n*-length and *x*-length edges of each independent repetitive cell must precisely match: which is to say, the underlying polygonal modules that are placed upon each repetitive edge of equal length must have the same coinciding edge configuration. In this respect, these fivefold hybrid patterns employ the same principle that was used in the above-cited *fourfold system B* Ildegizid hybrid example from the Mu'mine Khatun tomb tower in Nakhichevan, Azerbaijan (1186) [Fig. 182]. In keeping with the rich diversity of innovative fivefold patterns at the Huand Hatun complex in Kayseri, this monument also includes a fivefold hybrid design in one of its portals[140] [Fig. 262d]. This exceptional example is a *median* pattern, and employs two repetitive units: a rhombus with 72° and 108° included angles, and an elongated hexagon. It is worth noting that each of these will tessellate independently. Indeed, the rhombic component is the classic *median* pattern used with great frequency throughout the Islamic world, including at the Huand Hatun complex [Figs. 87 and 227a]. As with other fivefold *median* patterns produced during the Seljuk Sultanate of Rum, this design arbitrarily modifies the standard ten-pointed stars with a central pentagon surrounded by five rhombi and distinctive trefoil elements [Fig. 224a]. This effectively transforms the original design into a field pattern. A hybrid design from the Izzeddin Kaykavus hospital and mausoleum in Sivas[141] (1217) [Fig. 263c] employs four repetitive elements: a small rhombus of 72° and 108° included angles, a larger rhombus of the same proportion, a more acute rhombus with 36° and 144° included angles, and a triangle that is half the acute rhombus, which is to say a 1/10 segment of the decagon [Fig. 263a]. Each of these three rhombi, with their associated pattern lines, was used on its own for surface

coverage within this Anatolian design tradition. The pattern within the small rhombus is the above-mentioned classic *median* design; the larger rhombic repeat element was used on its own at the Huand Hatun complex in Kayseri (1237) [Fig. 235d]; and the acute rhombus was used on its own at the Muzaffar Barucirdi *madrasa* in Sivas (1271-72) [Fig. 241b]. This example from the Izzeddin Kaykavus hospital and mausoleum is historically significant in that it is the earliest example of a hybrid design that overtly employs more than two repetitive cells within its overall structure. The most complex fivefold hybrid designs are two examples from the Karatay Han (1235-41), 50 km east of Kayseri [Figs. 264c[142] and 265c[143]]. Both of these are *acute* patterns and share several of the same repetitive units, and their similarity clearly indicates that the same artist produced both. The first of these employs four repetitive elements: the rhombus with 72° and 108° included angles; the rhombus with 36° and 144° included angles; a triangle with the proportions of 1/5 of a pentagon, which is half the more obtuse rhombus; and an elongated hexagon with the same proportion as the barrel shape from the polygonal modules of the *fivefold system* [Fig. 264a]. The pattern within the more obtuse rhombus in this set of repetitive elements is the classic fivefold design that was used ubiquitously during the Seljuk Sultanate of Rum, and harkens back to the outstanding design used on the Ghurid soffit of the Arch at Bust, Afghanistan (1149) [Figs. 85 and 226c], and earlier still to one of the repetitive cells within the hybrid design in the northeast dome chamber at the Friday Mosque at Isfahan (1088-89) [Fig. 261] [Photograph 25]. These same four repetitive units, with the same pattern line application, were also used in the second hybrid design from the Karatay Han in Karadayi, but with the further addition of a rectangular and elongated hexagonal element [Fig. 265a]. Until the development of fivefold dual-level patterns in fourteenth-century Spain and fifteenth-century Persia, these hybrid patterns from the Sultanate of Rum represent the most sophisticated examples of Islamic geometric design created from the *fivefold system*.

Artists working in the Sultanate of Rum either appropriated or rediscovered the simple, but elegant, method of creating an underlying tessellation from six regular heptagons placed together in an edge-to-edge arrangement with bilateral symmetry. By drawing lines that connect the centers of the heptagons, an elongated hexagonal repeat unit is established. The interstice of these six heptagons is comprised of two irregular pentagons that meet edge to edge in the center of the heptagon cluster. As detailed above, this same arrangement of heptagons was first used in the Ghaznavid minaret of Mas'ud III in Ghazni,

---

[139] The author is indebted to both Emil Makovicky and Jean-Marc Castéra for independently discovering the geometric similarity between these two fivefold patterns. See Castéra (2016).

[140] Schneider (1980), pattern no. 366.

[141] Schneider (1980), pattern no. 382.

[142] Schneider (1980), pattern no. 386.

[143] Schneider (1980), pattern no. 387.

Afghanistan (1099-1115), approximately 100 years earlier. The two Ghaznavid patterns derived from this tessellation are considerably more complex [Figs. 280 and 281]. These two examples also include a large number of pattern lines that are arbitrarily placed within the pattern matrix rather than the strict product of a systematic schema. As such, while they are sevenfold patterns, they do not fall into the category of having been created from the *sevenfold system*. By contrast, the three Anatolian Seljuk designs created from this same underlying tessellation are among the earliest examples of patterns created from underlying polygonal modules that eventually became recognized as components of the *sevenfold system*. The rarity and simplicity of sevenfold patterns created by artists working in the Sultanate of Rum suggests that these artists were not fully aware of the systematic potential of these underlying modules. This is in clear distinction from the artists working under the Mamluks some 150 years later when this system came to full maturity. The three Anatolian examples are an *acute* pattern from the *mihrab* of the Great Mosque of Dunaysir in Kiziltepe[144] (1204) that was also used at the Alaeddin mosque in Nidge (1223) [Fig. 282a]; an *obtuse* design from the Eğirdir Han[145] (1229-36) [282c]; and a *two-point* pattern from the façade of the Great Mosque of Malatya[146] (1237-38) [Fig. 282d]. These were produced within some 30 years of one another, and may well have been the product of the same artist, lineage, or atelier. Although it does not appear to have been used historically, the *median* pattern created from this same underlying tessellation is equally acceptable [Fig. 282b]. However, this *median* design was the foundational basis for one of the highly complex sevenfold designs from the minaret of Mas'ud III [Fig. 281b].

The architectural ornament of the Seljuk Sultanate of Rum includes a large number of very fine geometric patterns that are nonsystematic. These range from more basic designs that were first used by earlier Muslim cultures to highly innovative original constructions that are among of the most complex nonsystematic geometric patterns from the totality of this artistic tradition. Unlike their neighbors to the east, the continuance of the Seljuk Sultanate of Rum in the face of the Mongol onslaught of the thirteenth century insured that there was no consequent interruption in the developmental continuity of the geometric arts in Anatolia. On the contrary, the Sultanate of Rum and the Mamluks directly benefited from the exodus of skilled artists and craftspersons fleeing the Mongol destruction. Along with their Mamluk contemporaries, artists working under the patronage of the Seljuk Sultanate of Rum were responsible

for bringing the nonsystematic use of the polygonal technique to its full geometric sophistication; and the nonsystematic geometric ornament of subsequent Muslim cultures, for all its originality and aesthetic distinction, never surpassed the innovative developments of these two important dynasties.

Among the nonsystematic designs that repeat upon the hexagonal grid are a number of interesting patterns with nine-pointed stars. Six nonagons will cluster when placed edge to edge in sixfold radial symmetry. The central region of an underlying tessellation constructed from this configuration is an interstice six-pointed star, and the pattern lines that extend into this interstice region likewise form a six-pointed star [Fig. 310]. This simple tessellation was used to create a very successful *acute* pattern located in the *Turkish triangle* pendentives in the dome of the *mihrab* at the Alaeddin mosque in Konya[147] (completed in 1219-21). The placement of nonagons on the vertices of the hexagonal grid also allows for their being separated by a ring of nine pentagons. As with the previous example, this arrangement creates an underlying six-pointed star interstice region at the center of each hexagonal repeat unit. An *acute* design created from this closely related underlying tessellation was also included in the Turkish triangle pendentives at the Alaeddin mosque.[148] [Fig. 312c]. This same underlying tessellation was used to create an equally successful *median* pattern that was used in several locations, including the Alay Han near Aksaray (1155-92) [Photograph 43]; the Huand Hatun in Kayseri; and the Agzikara Han[149] [Fig. 312b]. These differ from the example in the *Turkish triangles* in the treatment of the pattern lines within the central region, as well as slight variations in the angular opening of the crossing pattern lines. An *acute* design from the Izzeddin Kaykavus hospital and mausoleum in Sivas (1217) employs an underlying tessellation that is essentially the same, with nonagons at the vertices of the hexagonal grid that are separated by mirrored pentagons. However, the central region of this example places six contiguous barrel hexagons around a regular hexagon at the center of each repetitive hexagonal cell [Fig. 313c]. The ornament of the Great Mosque of Malatya includes an unusual design that employs underlying nonagons placed on the vertices of the isometric grid, with equilateral triangles separating each nonagon. Despite this placement, due to the nonagon's odd number

---

[144] Schneider (1980), pattern no. 216.

[145] Schneider (1980), pattern no. 205.

[146] Schneider (1980), pattern no. 209.

[147] Schneider (1980), pattern no. 359.

[148] Schneider (1980), pattern no. 218 (pl. 19 and 34).

[149] Schneider (1980), pattern no. 218 (pl. 34). Schneider compares the similarity between the nonagonal pattern from the Alaeddin Mosque in Konya with those from the Huand Hatun and Agzikara Han in this figure. However, he does not identify the reason for their similarity: that being the single underlying polygonal tessellation that is responsible for both these *acute* patterns.

Photograph 43 Seljuk Sultanate of Rum geometric panels in carved stone relief at the Alay Han near Aksaray, Turkey (© David Wade)

of sides, this design repeats upon a rhombic grid with 60° and 120° angles [Fig. 311]. This example from the Great Mosque of Malatya is a *median* pattern with an unusual threefold rotational devise generated from the ditrigonal hexagons that are edge to edge with the three nonagons and their triangles.[150] Another nonagonal design from the Seljuk Sultanate of Rum is from the Gök *madrasa* and mosque in Amasya, Turkey (1266-67). This separates each underlying nonagon with a barrel hexagon, and places a ring of 12 pentagons that surround a central irregular dodecagonal interstice region. In an innovative tour de force, the *acute* pattern lines in the dodecagonal interstice region place regular octagons into the pattern matrix [Fig. 315]. One of the most interesting patterns created from an underlying tessellation that utilizes nonagons at the vertices of the regular hexagonal grid is a second such example from the Great Mosque of Malatya.[151] This rather exceptional *median* pattern employs both nine- and seven-

pointed stars in a fashion that is reminiscent of the aesthetic quality of *median* patterns created from the *fourfold system A* [Fig. 318]. The use of these two star forms is an example of the *principle of adjacent numbers* wherein the ease of generating repetitive patterns with the eight-pointed star indicates that successful patterns can also be made with nine- and seven-pointed stars. As with other nonagonal designs, the nine-pointed stars are located at the vertices of the hexagonal grid, while a ring of 6 seven-pointed stars rests within the field of the hexagonal repeat unit, and a six-pointed star is located at the center of each repeat unit. The underlying polygonal matrix that connects the nonagons and heptagons is comprised of irregular pentagons and hexagons that cleverly imitate those of the *fourfold system A*.

Like other preceding and neighboring Muslim cultures, the Sultanate of Rum also employed nonsystematic patterns that place 12-pointed stars on the vertices of the isometric grid. The underlying tessellation in one of the most basic patterns of this type places a ring of pentagons around each dodecagon, with three pentagons meeting at the center of each triangular repeat. One of the earliest examples of an *acute* pattern made from this underlying tessellation is from

---

[150] Schneider (1980), pattern no. 211.

[151] Schneider (1980), pattern no. 356.

Photograph 44 A threefold Seljuk Sultanate of Rum nonsystematic *two-point* pattern with 7- and 24-pointed stars from the Esrefoglu Süleyman Bey mosque in Beysehir, Turkey (© Mirek Majewski)

the *mihrab* of the Great Mosque of Niksar in north central Turkey[152] (1145) [Fig. 300a *acute*]. This was produced under the patronage of the Danishmend Dynasty: early rivals of the Seljuks in Anatolia. Later Anatolian examples of this design include the cenotaph of the Izzeddin Kaykavus mausoleum in Sivas (1217-18), and the side panel of the *minbar* at the Great Mosque of Divrigi (1228-29). A variation of this *acute* pattern is created from truncating the three coinciding pentagons in the underlying tessellation such that they become trapezoids with coincident edges with the central equilateral triangle[153] [Fig. 320]. Anatolian examples with this design variation include the vertical side panel of the Mengujekid *minbar* in the Great Mosque of Divrigi (1228-29), and in a portal niche at Çifte Minare *madrasa* in Erzurum (late thirteenth century). Another underlying tessellation with dodecagons placed upon the vertices of the isometric grid is from the kiosk of the Keybudadiya at Kayseri (1224-26). The

acute pattern generated from this underlying tessellation was widely used throughout Muslim cultures[154] [Fig. 321b *acute*]. The underlying tessellation of this example also places a cluster of three coincident pentagons at the center of each triangular repeat unit, and introduces an elongated hexagon that separates the dodecagons. Another early use of this *acute* pattern that dates to the same approximate period is a carved stucco panel at the Abbasid Palace of the Qal'a in Baghdad (c. 1220) [Photograph 28]. Artists working for the Sultanate of Rum also created patterns that place 24-pointed stars onto the vertices of the isometric grid. The portal of the Nalinci Baba tomb and *madrasa* in Konya (1255-65) is decorated with a very beautiful *two-point* pattern that incorporates seven-pointed stars into the pattern matrix that surround 24-pointed stars [Fig. 327]. This exceptional design was also used in the *mihrab* niche at the Esrefoglu Süleyman Bey mosque in Beysehir, Turkey (1296-97) [Photograph 44].

---

[152] Schneider (1980), pattern no. 402.

[153] Schneider (1980), pattern no. 398.

[154] Schneider (1980), pattern no. 401.

Isometric patterns from the Seljuk Sultanate of Rum frequently employed more than a single region of local symmetry. Compound patterns with 12-pointed stars at the vertices of the isometric grid and 9-pointed stars within the centers of each triangular repeat were especially popular. The most commonly used underlying tessellation with this form of compound symmetry separates the dodecagons from the nonagons with a ring of irregular pentagons, and places an elongated hexagon between the nonagons. Patterns from the Seljuk Sultanate of Rum that are created from this underlying tessellation include an *obtuse* design from the *mihrab* of the Great Mosque of Aksehir[155] (1213) [Fig. 347a] and an *acute* design from a faience ceramic panel on the façade of the Cincikli mosque in Aksaray[156] (1220-30) [Fig. 346a]. The design from Aksaray is the earliest known use of this *acute* pattern, and over time this was used throughout the Islamic world. A significantly more complex *median* pattern with the same combination and location of 12- and 9-pointed stars was used in a portal at the Susuz Han in the village of Susuzköy[157] (1246) [Fig. 354d].

Artists working under the Seljuk Sultanate of Rum created a number of isometric designs with significantly greater complexity in their diversity of local symmetries. In addition to the vertices and centers of the isometric grid, such designs will place additional regions of local symmetry upon the midpoints of the triangular repeat, and occasionally into the field of the repeat. Multiple examples of a particularly ambitious *acute* design with 9-, 10-, 11- and 12-pointed stars placed onto these locations include a stone relief panel from the Egridir Han[158] (c. 1229-36); the courtyard portal at the Seri Han near Avanos (1230-35); and a framing border in the entry to the mosque at the Karatay Han (1235-41). This design places the 12-pointed stars at the vertices of the isometric grid, 9-pointed stars at the center of each triangular repeat, 10-pointed stars upon the midpoints of each triangular edge, and 11-pointed stars within the triangular field[159] [Fig. 367]. The technically demanding construction of this complex design, coupled with the closeness in age and proximity of these three examples argues for each to have been the product of a single workshop. Another example with a similar geometric arrangement of star forms is from a gravestone in Ahlat[160] (thirteenth–fifteenth centuries). This is an *acute* pattern that also places 12-pointed stars at the vertices of the isometric grid, but nonagons rather than

9-pointed stars at the centers of each triangular repeat, and 8-pointed stars rather than 10-pointed stars at the midpoints of each repetitive edge, and serendipitous heptagons into the field of the design [Fig. 361]. This is the only known historical example of this exceptionally well-balanced design. An outstanding isometric pattern with multiple centers of local symmetry was used in the *mihrab* niche of the Great Mosque of Ermenek (1302). This is an *acute* design that places 24-pointed stars upon the vertices of the grid, 12-pointed stars at the centers of each triangular repeat unit, and 8-pointed stars upon the midpoints of each edge of the repeat unit[161] [Fig. 365].

Artist working in the Seljuk Sultanate of Rum were equally innovative in their focus upon nonsystematic patterns based upon the orthogonal grid. As with nonsystematic isometric patterns, these most commonly place 12-pointed stars upon the vertices of each repeat unit—in this case squares. The underlying polygonal tessellation for the most basic of such patterns places a ring of pentagons around each dodecagon, with four of these pentagons meeting at the centers of the square repeat. An early Anatolian example is found on the minaret of the Great Mosque of Siirt[162] (1129) [Fig. 335b], dating from less than 10 years after the earliest known use of this design in the northeastern *iwan* of the Friday Mosque at Isfahan [Fig. 335a]. Multiple instances of this ever-popular design were used subsequently by artist working for the Seljuk Sultanate of Rum. A variation of this pattern truncates the cluster of four underlying pentagons at the centers of the square repeat units [Fig. 337]. An example of an *acute* pattern created from this variation is found on the pair of bronze doors from the Anatolian Seljuk *atabeg* of Cizre, Turkey (thirteenth century), and the earliest known use is from a Zangid bronze door at the portal of the Bimaristan Arghun in Aleppo (twelfth century).

Orthogonal patterns with multiple centers of local symmetry were widely employed by the Sultanate of Rum. The most common nonsystematic pattern of this type places 12-pointed stars on the vertices of square repeat units, and 8-pointed stars at the center. The earliest known example of this variety of pattern was created by Mengujekid artists in the portal of the Kale mosque in Divrigi[163] (1180-81) [Fig. 379b]. Multiple later examples of orthogonal designs with 8- and 12-pointed starts were used both in Anatolia and throughout Muslim cultures. A very pleasing *acute* pattern with 16-pointed stars on the vertices of the square repeat unit and 8-pointed stars at the center of each repeat was used in

---

[155] Schneider (1980), pattern no. 412.

[156] Schneider (1980), pattern no. 358.

[157] Schneider (1980), pattern no. 414.

[158] Now spolia in the city walls of Egridir.

[159] Schneider (1980), pattern no. 418.

[160] Schneider (1980), pattern no. 407.

[161] Schneider (1980), pattern no. 435.

[162] Schneider (1980), pattern no. 408.

[163] Schneider (1980), pattern no. 406.

the *mihrab* of the Keykavus hospital in Sivas[164] (1217-18) [Fig. 389a]. This same design was used by 'Abd Allah ibn Muhammad al-Hamadani in the illumination of the 30 volume Quran of Uljaytu (1313), and by *Mudéjar* artist in a window grille at the ibn Shushen Synagogue of Toledo (1180), referred to today as the Santa Maria la Blanca.

As with isometric designs, more complex patterns made from nonsystematic orthogonal underlying polygonal tessellations will frequently incorporate additional areas of local symmetry at the midpoints of the square repeat units, and within the field of the repeat. Geometric artists from the Seljuk Sultanate of Rum were particularly resourceful in producing designs of this type. An outstanding case in point is a pattern from the Kayseri hospital (1205-06) that places 12-pointed stars on the vertices of the square repeats, octagons at the center of the repeat, 10-pointed stars at the midpoint of each edge of the repeat, and 9-pointed stars within the field[165] [Fig. 400]. This same design was used in several later Anatolian locations, including the Agzikara Han near Aksaray (1231-40), and the Ince Minareli *madrasa* in Konya[166] (1264-65). Several examples of orthogonal compound patterns were created that have 16-pointed stars at the vertices of the square repeat, 8-pointed stars at the centers of the repeat, 12-pointed stars at the midpoints on each edge of the repeat, and 10-pointed stars within the field of the repeat. Notable among these is from the *iwan* of the Kemaliya *madrasa* in Konya[167] (1249) [Fig. 404].

Artists working for the Sultanate of Rum also created nonsystematic patterns with compound symmetry that employed repetitive schema other than the isometric and orthogonal grids. This variety of pattern is especially complex, and is generally comprised of three types: those that have rectangular repeat units, those with elongated hexagonal repeat units, and those that are characterized by linear bands of primary star forms. Technically, this last category repeats with an especially broad rectangle, but the visual quality is sufficiently distinct from other rectangular designs as to warrant its own separate consideration. An early Anatolian example of a nonsystematic rectangular design with 10- and 12-pointed stars was used on the wooden *minbar* of the Great Mosque at Aksaray[168] (1150-53) [Fig. 414]. The dual of a rectangle is an identical rectangle; and the repeat unit for this design can be regarded equally as having either the 10- or 12-pointed stars placed upon the vertices, with the other star form located at the center of the repeat unit. This

example is an *acute* pattern, and the underlying generative tessellation makes equally successful designs with each of the other three pattern families, although none are known within the historical record [Fig. 415]. This design with 10- and 12-pointed stars is the only known architectural example, although it is interesting that the same design is illustrated in the anonymous Persian treatise *On Similar and Complementary Interlocking Figures*,[169] as well as in the Topkapi Scroll.[170] A very pleasing Mengujekid *acute* pattern that borders the interior of a window at the Great Mosque of Divrigi (1228-29) places 12-pointed stars on the vertices of the rectangular repeat units, 8-pointed stars at the midpoints of the long edges of each repeat, and two 9-pointed stars within the field of each repeat[171] [Fig. 421]. Another design from this general region that repeats upon a rectangular grid is a highly complex *acute* pattern with 10- and 11-pointed stars that was used on a stone *khatchkar* in Noravank, Armenia, created by Momik, a monk and artist who worked between the years 1282 and 1321 [Fig. 423]. This is not strictly speaking the product of the Sultanate of Rum. However, this tradition of Armenian Christian commemorative stone crosses was greatly influenced by the carved stone ornament of the Anatolian Seljuks. Their incorporation of Islamic geometric and floral design motifs is in aesthetic conformity with the contemporaneous work of their Anatolian neighbors. Among the many geometric patterns that were used on Armenian *khachkars* are several with complex geometry. This example by Momik is particularly complex, and one of the earliest signed examples of such a pattern. It is also one of the most sophisticated examples of the non-Muslim adoption of Islamic geometric art, even if rather disproportionate in the relative sizes of the five-pointed stars and the shape of the 11-fold rosettes. An example of an Anatolian compound pattern that employs an elongated hexagon as the repeat unit is found in the courtyard portal of the Karatay Han[172] (1235-41). This pattern employs 9-pointed stars at the vertices of the hexagonal repeat unit, with 12-pointed stars at the midpoints of the two opposite parallel edges of the repeat, an 8-pointed star at the center of the repeat, and four 10-pointed stars within the repetitive field [Fig. 439]. An outstanding design that is characterized by linear bands of star-forms arranged in an alternating sequence of 12-, 11-, 10-, 11-, and 12-pointed stars was used on the portal of the Kiosk at Erkilet near Kayseri[173] (1241) [Fig. 425]. This highly complex pattern was used a second time in Kayseri: in the *mihrab* of the Çifte

---

[164] Schneider (1980), pattern no. 423.

[165] Schneider (1980), pattern no. 429.

[166] This stone panel currently resides in the Museum of Wooden Artifacts and Stone Carving in Konya: collection number 157092.

[167] Schneider (1980), pattern no. 427.

[168] Schneider (1980), pattern no. 416.

[169] Bibliothèque Nationale de France, Paris, MS Persan 169, fol. 195b.

[170] Necipoğlu (1995), diagram no. 44.

[171] Schneider (1980), pattern no. 421.

[172] Schneider (1980), pattern no. 420.

[173] Schneider (1980), pattern no. 417.

Kümbet (1247): the work almost certainly of the same artist. This is an identical numeric sequence to the earlier, and inferior, Zangid design that was reported by Ernst Herzfeld to have been used on a pair of doors from the Lower Maqam Ibrahim in the citadel of Aleppo[174] (1168) [Fig. 427].

Artists working in the Seljuk Sultanate of Rum also applied their knowledge of geometric design to the decoration of domes and semi-spheres. During the same general period that Ayyubid artists were working in this same specialized discipline, artists in Anatolia created several fine examples that utilized both radial and polyhedral geometry. The renowned faience mosaic dome of the Karatay *madrasa* in Konya (1251-52) is an overt homage to the number 24: with a complex geometric matrix of multiple 24-pointed stars applied within the 24 gore segments that provide the repetitive schema for this domical ornament. There are a number of examples of polyhedral ornament from this dynasty that apply geometric designs onto the surfaces of domical hemispheres that protrude from the ornamental design of their otherwise two-dimensional backgrounds. Most of these historical examples are carved stone and are based upon the geometry of the dodecahedron and include: a pattern that places five-pointed stars associated with the *acute* family into each projected pentagonal face at the Huand Hatun complex in Kayseri (1237); a second example from the Huand Hatun complex in Kayseri that places five-pointed stars from the *median* family onto each pentagonal face [Fig. 497]; and a *two-point* pattern from the Sahib Ata mosque in Konya (1258) that is identical in geometric concept to the ornament in the northeast dome of the Friday Mosque at Isfahan[175] [Fig. 496] [Photograph 30]. A more complex example of one of these projecting hemispherical stone ornaments is constructed from the spherical projection of an underlying truncated cube in a portal at the Susuz Han in the village of Susuzköy (1246) [Fig. 499].

1.21 Mamluks of Egypt (1250-1517)

The Mamluk dynasty of Egypt was founded by former Turkic slaves who gained positions of military and political power during Ayyubid rule. Their loyalty to the Ayyubid Sultans, and their military prowess, made the Mamluk martial guard a crucial aspect of the Ayyubid governance. Many Mamluk members of the military were awarded freedom from slavery, and appointed to high-ranking positions within government. With the collapse of the Ayyubids, these highly placed political and military professionals assumed

governance. The Mamluk Empire lasted for over 250 years. At its peak, this great empire included all of Egypt, part of Libya to the west, Nubia to the south, the Hijaz to the east, and Palestine, Syria, and part of southern Anatolia to the north. Evidence of their military strength was the defeat of the invading Mongol forces of Hulagu Khan at the battle of Ain Jalut, near Nazareth, in 1260, bringing an end to the Mongol's westward expansion in the Levant.

The Mamluk tactics at the battle of Ain Jalut were devised by the military commander Baybars al-Bunduqdari, who also led the vanguard of the Mamluk forces. Following this victory, he succeeded to the position of Sultan. Baybars proved to be as adept in diplomacy as he was in battle. When the Mongols conquered Baghdad in 1258, they executed the Abbasid Caliph al-Musta'sim, along with most of his family. In 1261, Baybars offered a surviving descendant of al-Musta'sim refuge in Cairo. This invitation led to the reestablishment of the Abbasid Caliphate in the new location of Cairo. Baybars extended his dominion to include the Hijaz region of the Arabian Peninsula. As the protector of the holy cities of Mecca and Medina, victor over the invading Mongols, and benefactor to the transplanted caliphate, Baybars became one of the most greatly respected Muslim leaders of his time. Equally important in spreading his reputation among Sunni Muslims were his many victories over the Christian crusader kingdoms. During Mamluk reign, Cairo maintained its exalted reputation and position of importance throughout the Islamic world.

The Mamluk dynasty was responsible for some of the most beautiful art and architecture of the Islamic world. Their artists worked in all media, and at a level of skill that was unsurpassed. Mamluk patronage gave particular attention to the book arts, and the Quranic calligraphy, illumination, and bookbinding of this period represent one of the high points of this most important Islamic art. Great emphasis was also given to calligraphic epigraphy, and, as with other Muslim cultures, such inscriptions were often elaborated with highly refined floral backgrounds. A supremely beautiful example of this form of ornament is found in the Sultan al-Nasir Hasan funerary complex in Cairo (1356-63), where a continuous running band of calligraphy and floral ornament surrounds the interior in an embrace of Quranic revelation. This is one of the most beautiful examples of *Kufi* script with floral ornament from anywhere in the Islamic world.

The Mamluk metal work of Cairo and Damascus rivaled the best of Mosul, Tabriz, Shiraz, or Herat. Under the Mamluks, the Mosul style of inlaying bronze vessels with silver and gold achieved further refinements. To this end, many of the finest metalworkers from Mosul are known to have relocated to Damascus and Cairo during the Mamluk period. All manner of vessels were produced under Mamluk

---

[174] Herzfeld (1954–6), Fig. 56.

[175] Schneider (1980), pattern numbers 437, 438, and 439.

patronage, including vases, basins, lamps, candle holders, incense burners, pen and ink holders, ewers, as well as weapons and scientific instruments. Mamluk metalwork was held in the very highest regard throughout the Islamic world, as well as in Europe: as exemplified by the use of an especially fine Mamluk basin, called the Baptistere de St Louis, as a baptismal bowl for the kings of France. As with metal work, Mamluk knotted carpets were the equal of the finest carpets from al-Andalus and Persia.

Mamluk architecture is one of the great Islamic classical styles, and the exceptional beauty of historical Cairo is primarily due to its Mamluk heritage. The Mamluk architectural style was a direct beneficiary of Ayyubid and Zangid architectural traditions, with stone remaining the primary material for both construction and ornamentation. The earlier conventions of *ablaq* polychrome stone ornament was fully embraced by Mamluk artists, and the exuberant *ablaq* vegetal designs that were created during this period represent one of the pinnacles of floral ornamental expression throughout the Islamic world. The Mamluks also expanded upon the Ayyubid and Zangid practice of applying geometric patterns to the carved stone the quarter domes of entry portals and *mihrab* niches to include the application of geometric patterns onto the entire exterior surfaces of domes. The Mamluks rose to power during the period of upheaval in Transoxiana, Khurasan, and Persia brought on by the invading Mongols. Like the Seljuks of Rum, and as stated, the Mamluks benefited from the influx of artists fleeing the Mongol onslaught. Several eastern architectural features were introduced into Egypt during this period. The grand entry portal of the Sultan al-Nasir Hasan funerary complex in Cairo (1356-63) has several characteristics that are more common to Persia: including its monumental size and height, and the use of recessed spiral columns at its corners. Originally this great entry portal had twin minarets on each side: another distinctive Persian feature. These were discarded following the collapse of one of the minarets soon after completion, killing many orphaned children in an adjacent school. Rather than rebuilding the fallen minaret it was considered more prudent to remove the remaining minaret. Eastern influences on Mamluk Quranic illumination include the occasional incorporation of distinctive Mongol floral devices such as stylized lotus and peony flowers. It is an interesting fact that these Mongol influenced floral motifs rarely found expression in Mamluk architectural ornament.

As with Quranic illuminations, Mamluk architecture made full use of the fully mature tradition of Islamic geometric design. The exterior of Mamluk monumental architecture was frequently ornamented with very bold geometric patterns. These geometric patterns were simple, and their very large scale gives emphasis to the monumentality of the building itself. This was a Fatimid ornamental devise that the Mamluks further refined, and provides an architectural

façade with an ornamental boldness that can be appreciated from a considerable distance. Within the interior of Mamluk buildings, geometric patterns were also used widely. The Egyptian tradition of pierced geometric window grilles was continued, but the geometric patterns used by the Mamluks were more complex than those used in earlier times. Complex geometric patterns were also regularly used in the inlaid marble ornament of *mihrab* and fountains. Among the most noteworthy incorporation of geometric patterns are panels from the exceptionally beautiful wooden *minbars* for which the Mamluks are renowned. As with other aspects of their architectural ornament, this focus upon wooden *minbars* was inherited from their Zangid and Ayyubid predecessors. Rather like the contemporaneous carved stone ornament of the Seljuk Sultanate of Rum, this rich tradition is characterized by the use of a wide variety of very complex geometric patterns. These *minbars* are masterpieces of design and craftsmanship, and rank among the finest examples of Islamic art,[176] and the application of geometric patterns within Mamluk *minbars* represent one of the most sophisticated expressions of the geometric idiom from the whole of the Islamic world.

Mamluk geometric artists built upon the practices inherited from their predecessors, and applied the polygonal technique to new heights of sophistication and complexity. Their work with two-dimensional systematic pattern making continued with the widespread use of the *system of regular polygons*, both fourfold systems and the *fivefold system*, as well as the use of diverse nonsystematic designs already known to this ornamental tradition. What is more, Mamluk artists were responsible for bringing the *sevenfold system* of pattern generation to full maturity; and the relatively small number of patterns that were created from this system are remarkable for their beauty and ingenuity. To a very limited extant, this system was adopted by Ottoman and Timurid arts. Like their contemporaries in the Seljuk Sultanate of Rum, Mamluk geometric artists also produced many outstanding examples of highly complex non-systematic designs with multiple centers of differentiated symmetry. These compound patterns represent the full maturity of this nonsystematic ornamental tradition. The innovation of Mamluk artists is also exemplified in the many stone domes and quarter domes ornamented with geometric designs. Their work with highly complex nonsystematic patterns, applied geometric patterns onto domical surfaces, together with their development of the *sevenfold system*, is evidence of the important contributions by Mamluk artists to the diversity, maturity and richness of Islamic geometric ornament.

---

[176] Atil (1982), 195–196.

The Mamluk use of patterns created from the *system of regular polygons* was widespread and diverse. Many of these designs were already well known throughout the Islamic world. In addition to the beauty that less complex, easily comprehended, and immediately recognizable designs contribute to an overall ornamental schema, such geometric patterns can be regarded as a unifying device that helped establish an aesthetic continuity and cultural affiliation among preceding Muslim cultures. For example, numerous previously established patterns that are easily created from the 6^3 tessellation of regular hexagons were used as architectural ornament during the Mamluk period, including the relatively uncommon *acute* design with 30° crossing pattern lines used in the exterior carved stone ornament at the Imam al-Shafi'i mausoleum in Cairo (1211) [Fig. 95a], and the simple *median* design with 60° angular openings used in the stone window grilles at the Sultan Qala'un funerary complex in Cairo (1284-85) [Fig. 95b] [Photograph 55]. This latter example is a classic threefold *median* pattern that was used universally by Muslim cultures. The design of the window grilles immediately adjacent to this example from the Sultan Qala'un funerary complex is far less common [Fig. 99f] [Photograph 55]. This adjacent *two-point* pattern is directly associated with the 3.6.3.6 underlying tessellation of triangles and hexagons, and its use at the Sultan Qula'un funerary complex was some 50 years after its use by artist during the Seljuk Sultanate of Rum at the Ali Tusin tomb tower in Tokat, Turkey (1233-34). Complementing the wide use of well-known designs made from the *system of regular polygons*, Mamluk artists also used this system to create new and original geometric patterns. A very successful *two-point* pattern that appears to be derived from the 3.6.3.6 underlying tessellation was employed in the mosaic spandrel above the *mihrab* niche of the Aqbughawiyya *madrasa* (1340) at the al-Azhar mosque in Cairo [Fig. 100d]. This unusual design utilizes the hexagons within the generative tessellation as part of the completed pattern. A more complex pattern produced from this same 3.6.3.6 underlying tessellation was used as a border design that surrounds a door at the *manzil* (house) of Zaynab Khatun in Cairo (1468) [Fig. 102b]. This rather unusual pattern is comprised of superimposed dodecagons and ditrigonal shield shapes: the latter being generated by applying the 90° crossing pattern lines at the midpoints of alternating underlying hexagons, and allowing these crossing pattern lines to extend into the adjacent triangles and hexagons until they meet with other extended pattern lines, and the former being the product of simply applying dodecagons so that they cross the underlying triangles in an aesthetically acceptable fashion. A considerably more complex Mamluk design created from the 3.6.3.6 underlying tessellation is from the central panels of the double doors at

the Vizier al-Salih-i Tala'i mosque in Cairo [Fig. 100c]. These Mamluk doors were added to this Fatimid mosque during its restoration following an earthquake in 1303. This design is unusual in that it incorporates nonagons centered upon each underlying triangular module. The Mengujekids of Anatolia used this same pattern many decades earlier at the Great Mosque and hospital of Divrigi in Turkey (1228-29), as did Seljuk artists in a narrow border at the Gunbad-i 'Alaviyan in Hamadan, Iran (late twelfth century) [Photograph 22]. Particularly successful examples of Mamluk 3.4.6.4 designs include a *median* pattern from the *mihrab* spandrel of the Aydumur al-Bahlawan funerary complex in Cairo (1364) [Fig. 104a], and a stone mosaic *obtuse* pattern from the *mihrab* niche at the Sultan Qansuh al-Ghuri complex in Cairo (1503-05) [Fig. 104b]. The earliest known use of this latter pattern is from the Gök *madrasa* and mosque in Amasya, Turkey (1266-67). A strong characteristic of this pattern is the application of octagons within the square modules of the underlying tessellation. Many examples of patterns created from the 3.12^2 underlying tessellation were employed by Mamluk artists, including a *median* pattern from the *mihrab* arch spandrel of the Amir Salar and Amir Sanjar al-Jawli funerary complex in Cairo (1303-04) [Fig. 108a]. The quarter dome hood of the *mihrab* niche also employs this design. However, the artist naively forced this two-dimensional design onto the spherical surface, thereby causing significant distortion. This forced fit is surprising in that this pattern could have uniformly fit the domical surface had the artist employed either an octahedral or icosahedral layout of the multiple triangular repetitive units. Two exquisite stone mosaic panels at the Amir Aq Sunqar funerary complex in Cairo (1346-47) were created from this same underlying tessellation. One of these is a *two-point* pattern [Fig. 108f], and the other is an *obtuse* pattern [Fig. 108d] [Photograph 45]. The occurrence of these two mosaic panels with their shared generative origin would appear to be a deliberate, if subtle, feature of the ornamental schema, and provides peripheral evidence for the use of the polygonal technique within this tradition. This *two-point* pattern was also used by Mamluk artists at the Amir Aq Sunqar funerary complex in Cairo (1346-47), as well as during the Ilkhanid period on an illuminated frontispiece of a Baghdadi Quran illuminated by Muhammad ibn Aybak ibn 'Abdullah (1306-07). The same *obtuse* pattern that was used at the Amir Aq Sunqar funerary complex was later used on a pair of wooden cupboard doors at the Sultan Qansuh al-Ghuri complex in Cairo (1503-05). Perhaps the most renowned Mamluk geometric pattern easily created from the 3.12^2 tessellation is a frontispiece from the 30-volume Quran written and illuminated by 'Abd Allah ibn Muhammad al-Hamadani in 1313. The visual appeal of this outstanding illumination is augmented by the curvilinear

Photograph 45 A threefold Mamluk *obtuse* pattern with 12-pointed stars that can be created from the *system of regular polygons* located at the Amir Aq Sunqar funerary complex in Cairo (© David Wade)

treatment of the pattern lines[177] [Fig. 108c]. An *obtuse* pattern derived from the 4.6.12 underlying tessellation was particularly popular among Mamluk artists. This pattern places octagons within the square modules of the generative tessellation [Fig. 109b]. The many Mamluk buildings that employed this pattern include: one of the exterior carved stucco roundels at the base of the dome at the Amir Sanqur al-Sa'di funerary complex in Cairo (1315); the entry door of the Amir Ulmas al-Nasiri mosque and mausoleum in Cairo (1329-30); the entry door of the Sultan Qansuh al-Ghuri *madrasa* (1501-03); and the entry door of the Sultan Qansuh al-Ghuri Sabil-Kuttab in Cairo (1503-04). The shared patron and time period of the latter two examples indicates the likelihood of their being produced by the same artist.

In addition to the use of regular and semi-regular tessellations, Mamluk artists also made frequent use of *two-uniform* and *three-uniform* tessellations when using the *system of regular polygons*. The 3.4.3.12-3.12^2 tessellation was especially relevant to the Mamluk geometric idiom. A particular *obtuse* pattern created from this tessellation was used with great frequency by Mamluk artists [Fig. 113c].

The 120° crossing pattern lines are easily determined by the application of regular hexagons placed within each underlying triangles, and by applying lines that skip one polygonal edge within the dodecagon—as per a 12-*s*2 pattern line application. Mamluk examples of this pattern include the side panel of the *minbar* at the mosque of Sultan al-Nasir Muhammad ibn Qala'un at the citadel of Cairo (1295-1303); a window grille at the Amir Sanqur al-Sa'di funerary complex in Cairo (1315); the side panels of the *minbar* at the Amir Altinbugha al-Maridani mosque in Cairo (1337-39); the arch spandrel over the *mihrab* at the Araq al-Silahdar mausoleum in Damascus (1349-50); a frontispiece in an illuminated Quran[178] written by Ya'qub ibn Khalil al-Hanafi in 1356; and a stone mosaic floor at Fort Qaytbey in Alexandria (c. 1480). A very beautiful *two-point* pattern [Fig. 113e] made from this same 3.4.3.12-3.12^2 *two-uniform* tessellation was used on the side panels of the *minbar* at the Amir Azbak al-Yusufi complex in Cairo (1494-95) [Photograph 46], as well as in the *minbar* railing at the mosque of Amir Qijmas al-Ishaqi in Cairo (1479-81). And an eccentric *median* pattern created from the 3.4.3.12-3.12^2 was used in

---

[177] Cairo, National Library, 72, pt. 19.

[178] Cairo, National Library, 8, ff. IV-2r.

Photograph 46   Mamluk *minbar* at the Amir Azbak al-Yusufi complex in Cairo (© John A. and Caroline Williams)

the stone *minbar* of the Zawiya wa-Sabil Faraj ibn Barquq in Cairo (1400-11) [Fig. 113f]. This design employs two distinct pattern line treatments within the alternating dodecagons: a feature quite common in the Maghreb, but very unusual in Mamluk ornament. A Mamluk pattern that can be created from a *three-uniform* tessellation was used in the window grilles of the main façade at the Sultan Qala'un funerary complex in Cairo (1284-85). The application of 60° crossing pattern lines into the $3^4.6\text{-}3^3.4^2\text{-}3^2.4.3.4$ tessellation of triangles, squares, and hexagons produces this outstanding *median* design [Fig. 114b]. It is worth noting that this pattern can also be created from the 4.6.12 tessellation of squares, hexagons, and dodecagons [Fig. 109c]. When using this underlying tessellation to generate the design, the central six-pointed stars inside each of the underlying dodecagons are an arbitrary modification of what would otherwise be 12-pointed stars. The fact that this pattern can be created from more than just one underlying tessellation demonstrates the inherent methodological flexibility of the polygonal technique. These two derivations have slightly different proportions within the extracted pattern lines, but those created from the *three-uniform* tessellation precisely match the proportions and pattern density of the window

grille in the Sultan Qala'un funerary complex. This design shares characteristics with the pattern from the earlier Zangid portal of the Bimaristan Arghun at the citadel of Aleppo (twelfth century) [Fig. 101c] [Photograph 36].

Mamluk artist occasionally employed the previously mentioned ditrigonal module that is part of the *system of regular polygons*. This hexagonal module has three 90° included angles that alternate with three 150° included angles. A three-fold *median* pattern that incorporates this module into its underlying generative tessellation was used in the *mihrab* of the Amir Altinbugha al-Maridani mosque in Cairo (1337-39), as well as in one of the small blind arches within the Mamluk *mihrab* niche of the Aqbughawiyya *madrasa* (1340) at the al-Azhar mosque in Cairo [Fig. 117]. The closeness in time and location suggests that these two designs may have been the work of the same artist or atelier. The underlying tessellation is comprised of dodecagons located at the vertices of the isometric grid, separated by a vertex-to-vertex square surrounded by four coincident triangles. The ditrigon is located at the center of each triangular repeat unit, and can be regarded as the interstice of the regular polygonal modules. A fourfold pattern from the side panels of the *minbar* (c.1300) at the Vizier al-Salih Tala'i mosque in Cairo uses four radially arrayed underlying ditrigonal modules within alternating dodecagons of the otherwise $3.4.3.12\text{-}3.12^2$ generative tessellation [Fig. 119]. The pattern lines that are generated from the cluster of four ditrigons create an octagon at the center of the repeat unit. This same pattern was used as a border in the mosaic *mihrab* of the Mamluk Tabarsiyya *madrasa* (1309) at the al-Azhar mosque in Cairo. This same location has a second pattern that also employs the ditrigon within its underlying generative tessellation. A window grille from this *madrasa* employs a design that is created from an underlying tessellation that places six ditrigons around and interstice six-pointed star [Fig. 118b]. This underlying tessellation is, itself, the classic *median* pattern created from the 6^3 tessellations of hexagons [Fig. 95c]. The earliest known pattern created from this underlying tessellation of ditrigons and six-pointed stars was produced by Seljuk artists working on the northeast dome chamber of the Friday Mosque at Isfahan (1088-89) [Fig. 118a] [Photograph 19].

One of the most elegant examples of the standard *acute* pattern created from the 4.8^2 tessellation of squares and octagons is from the stucco window grille on the façade of the entry court at the Sultan Qala'un funerary complex in Cairo (1284-85) [Fig. 124a]. Mamluk artist used the well-known subtractive version of the standard *median* pattern created from this tessellation on the door of the Vizier al-Salih Tala'i mosque in Cairo (1303) [Fig. 126f], and an example of an exceptional variation to the standard *obtuse* pattern created from this tessellation surrounds the upper shaft of a minaret at the Sultan Qaytbay funerary complex in Cairo (1472-74) [Fig. 127d].

The Mamluks were less disposed toward patterns created from the *fourfold system A*. Of the relatively few patterns from this system, most were recreations of existing patterns that had been used by prior Muslim cultures. Such examples include a *median* pattern surrounding the circular shaft in the upper portion of the minaret at the Amir Taghribardi funerary complex in Cairo (1440) that was used by Qarakhanid artists nearly 300 years previously at the Maghak-i Attari mosque in Bukhara (1178-79) [Fig. 151] [Photograph 16]. An example of a Mamluk *median* field pattern created from this system frames an entrance to the Khan al-Sabun in Aleppo (1492) [Fig. 138c]. This is the ubiquitous design first found at the Seljuk east tomb tower at Kharraqan (1067-68). A *median* pattern on the minaret of the Attar mosque in Tripoli, Lebanon (1350), appears to be an original construction, although it is not overly complex and may well have been used previously. While this design repeats upon an orthogonal grid, the center points of the eight-pointed stars are placed upon the vertices of the 4.8^2 tessellation of squares and octagons [Fig. 154]. Patterns that use this repetitive schema are most frequently produced from the *fourfold system B* [Figs. 179 and 180], and one of the relatively few additional occurrences of the Mamluk design in the Attar mosque is at the Mughal tomb of I'timad al-Daula in Agra (1622-28) [Photograph 73].

The Mamluk use of the *fourfold system B* was far more pervasive than that of the *fourfold system A*. It would appear significant that the ornament of the Fatimids, Zangids, and Ayyubids all shared in the relative absence of geometric patterns constructed from the *fourfold system A*. Reasons for this bias are lost to history, but one can surmise that the small body of artists working for successive dynasties in this geographical region, and who were the inheritors of the polygonal technique as a principle design methodology, were substantially less familiar with this particular system than their eastern counterparts.

Among the many Mamluk patterns created from the *fourfold system B* are multiple examples of the classic *acute* design derived from the underlying tessellation of just octagons and irregular pentagons [Fig. 173a]. An early Mamluk example of this *acute* design was used in the pierced stone window grilles of the Sultan Qala'un funerary complex in Cairo (1284-85) [Photograph 55]. Later Mamluk examples of this well-known pattern include the lower mosaic panels of the *mihrab* niche of the Mamluk Taybarsiyya *madrasa* (1309) at the al-Azhar mosque in Cairo; the minaret of the Aydumur al-Bahlawan funerary complex in Cairo (1364); and a curvilinear variation from a carved stone relief panel at the entry of the Qadi Nur al-Din mosque in Cairo (1466). A *two-point* pattern made from the same underlying tessellation [Fig. 173d] was used in the *mihrab* niche of the Mamluk Aqbughawiyya *madrasa* (1340) in the al-Azhar in Cairo, as well as the entry portal

of the Ashrafiyya *madrasa* in Jerusalem (1482). A variation to this *two-point* pattern was used in the magnificent painted ceiling of the Sultan al-Mu'ayyad Shaykh complex in Cairo (1415-22) [Fig. 174a]. A stylistically similar *two-point* pattern created from an underlying tessellation of octagons, pentagons, and elongated hexagons was used in the entry portal of the Sidi Madyan mosque in Cairo (1465) [Fig. 176a]; and the same underlying tessellation was used to derive a very pleasing *obtuse* pattern in two adjacent upper panels of the Mamluk mosaic *mihrab* at the Taybarsiyya *madrasa* (1309) in the al-Azhar mosque in Cairo [Fig. 175d]. The earliest known use of the *obtuse* design from the Taybarsiyya *madrasa* is from the Ayyubid portal façade at the Palace of Malik al-Zahir at the citadel of Aleppo (before 1193). Like their counterparts in other Muslim cultures, in using the *fourfold system B* Mamluk artists employed the attractive variation to the *acute* pattern line application into the underlying elongated hexagonal module that provides for the creation of regular octagons within the pattern matrix [Fig. 172b]. One of the most outstanding Mamluk examples of this type of *fourfold system B* design is from the wooden *minbar* (1296) at the mosque of ibn Tulun in Cairo[179] [Fig. 179a]. This *minbar* was part of the restoration of the mosque by Sultan Lajan (r. 1296-1299) stemming from his gratitude at having successfully escaped his enemies by hiding in the derelict mosque. This pattern places eight-pointed stars upon the vertices of the 4.8^2 tessellation. This same *acute* pattern was used during the Mamluk period in several instances; including a very beautiful illuminated frontispiece of a Mamluk Quran[180] (before 1369), and an exterior stone panel at the Cathedral of St. James in Jerusalem that was likely produced by local Armenian stone carvers during the Mamluk period. A carved stone lintel above a recessed bay at the Sultan Qaytbey Sabil in Jerusalem (1482) is interesting in that the lines of the *obtuse* pattern are irregularly placed within portions of the underlying polygonal tessellation [Fig. 184]. This irregularity results from the application of 90° crossing patterns lines at select locations within the otherwise pattern matrix of 118° crossing pattern lines. This variation in pattern angles creates an unusual dynamic that is very successful, and has analogous aesthetic characteristics with *obtuse* patterns created from the *fivefold system*. All of these cited examples employ the orthogonal grid in their repetition. However, Mamluk artists working with this system occasionally created patterns that repeat on a rhombic grid. A panel above the *mihrab* at the al-Mar'a mosque in Cairo (1468-69) is a

---

[179] The panel from this *minbar* is in the collection of the Victoria and Albert Museum, London: museum number 1051–1869.

[180] Mamluk *mashaf*: Quranic manuscript No. 16; Islamic Museum, al-Aqsa Mosque, al-Haram al-Sharif, Jerusalem.

Photograph 47 Mamluk inlaid stone panel with a *two-point* pattern created from the *fivefold system* (The Metropolitan Museum of Art: Gift of the Hagop Kevorkian Fund, 1970: www.metmuseum.org)

case in point [Fig. 183]. This rather clever *two-point* pattern utilizes the octagon, pentagon, hexagon, and rhombus as underlying modules in the pattern construction.

The Mamluk use of geometric patterns created from the *fivefold system* was pervasive and incorporated the full range of diversity in repetitive schema. As with other Muslim cultures, the most commonly used fivefold repeat unit was the more obtuse rhombus with 72° and 108° angles. Among these are multiple examples of patterns created from the most commonly employed underlying tessellation of decagons, pentagons, and barrel-shaped elongated hexagons [Fig. 226a]. The Mamluk use of the classic *acute* pattern that is created from this underlying tessellation was less frequent than that of other Muslim cultures. Two examples of this design are found on Mamluk doors: one at the al-Azhar mosque, and another that is currently in the courtyard of the French Embassy in Giza [Fig. 226c]. A Mamluk carved stone relief panel on the main façade of the Sultan Qansuh al-Ghuri complex in Cairo (1503-05) makes use of a widened line version of the *median* design created from this underlying tessellation [Fig. 87]. Multiple examples of the *obtuse* design produced by this tessellation were used during the Mamluk period, and examples include a window grille within the dome of the Sultan Qala'un funerary complex in Cairo (1284-85), and several blind arches surrounding the drum of the dome at the Amir Sanqur al-Sa'di funerary complex in Cairo (1315) [Fig. 229b]. Similarly, the Mamluks were particularly disposed toward the *two-point* pattern made from this same tessellation, and examples include a panel in the entry portal of the Qadi Abu Bakr

Muzhir complex in Cairo (1479-80); an inlaid stone panel from the Sultan Qaytbay Sabil-Kuttab in Cairo (1479); and a contemporaneous Mamluk polychrome stone mosaic panel in the collection of the Metropolitan Museum of Art, New York City[181] [Fig. 231d] [Photograph 47]. Mamluk geometric artist frequently used a pattern that was first used on the Ildegizid mausoleum of Mu'mine Khatun in Nakhichevan, Azerbaijan (1186), although they were more likely influenced by less distant Ayyubid or Zangid examples such as that found at the Imam Awn al-Din Meshhad in Mosul (1248) [Fig. 232g]. This pattern also repeats with the 72° and 108° rhombus, and can be produced from either of two separate underlying tessellations: from the tessellation of just decagons and concave hexagons [Fig. 232f], and from a tessellation of decagons, barrel hexagons, and trapezoids that surround a large concave hexagon [Fig. 232h]. As mentioned previously, this design can also be produced in yet a third manner: through manipulating the *median* pattern lines from the standard design created from the most basic rhombic underlying tessellation of decagons, pentagons, and barrel hexagons [Fig. 227e]. Among the many Mamluk locations of this design are the Sultan Qala'un funerary complex in Cairo (1284-85); Amir Sanqur al-Sa'di funerary complex in Cairo (1315); the Hasan Sadaqah mausoleum in Cairo (1315-21); the Sultan Qaytbay funerary complex in Cairo (1472-74);

---

[181] Metropolitan Museum of Art: gift of the Hagop Kevorkian Fund; 1970.327.8.

Photograph 48 Mamluk illuminated frontispiece from a Quran commissioned by Sultan Faraj ibn Barquq with a design created from the *fivefold system* (British Library Board: BL Or. MS 848, ff. 1v-2)

the Qadi Abu Bakr Muzhir complex in Cairo (1479-80); and the Amir Azbak al-Yusufi complex in Cairo (1492-95) [Photograph 46]. An underlying tessellation of decagons, pentagons, barrel hexagons, and small rhombi was used to create a fine *obtuse* pattern that was used in the *mihrab* niche of the Sultan Qala'un funerary complex in Cairo (1284-85), as well as an illuminated frontispiece in the 30-volume Mamluk Quran commissioned by Sultan Faraj ibn Barquq (1399-1411) [Fig. 233b] [Photograph 48]. As with other examples, this design can alternatively be created from the dual of this tessellation: in which case the generative tessellation is comprised of decagons, elongated hexagons, and concave hexagons. Either of these same dualing tessellations will produce a very satisfactory *two-point* pattern that was used in the *mihrab* at the Sultan Qansuh al-Ghuri complex in Cairo (1503-05) [Fig. 233e]. Another very successful pattern that repeats on this rhombus was used on a pair of matched frontispieces from a Quran (1313) originally owned by Sultan Nasir al-Din Muhammad, and illuminated by Aydoğdu

bin Abdullah al-Badri and Ali bin Muhammad al-Rassam[182] [Fig. 235c]. The pattern from this Quran can also be made from two distinct underlying polygonal tessellations made from the components of the *fivefold system*. Both derivations are equally valid, and the original artist is as likely to have used one as the other. This same design was used during the Seljuk Sultanate of Rum at the Huand Hatun complex in Kayseri (1237) [Fig. 235d]. A highly complex fivefold pattern that uses the same obtuse rhombic repeat with 72° and 108° included angles is found on one of the metal doors of the *madrasa* of Qadi Abu Bakr ibn Muzhir[183] (1479-80) [Fig. 267]. This pattern is distinctive for its use of 20-pointed stars: each placed upon a vertex of the rhombic repeat. Each of the 20-pointed stars is surrounded by ten 10-pointed stars. Interestingly, these 10-pointed stars are located upon the vertices of a secondary tessellation of decagons and concave hexagons. These two distinct repetitive cells identify this as the only known Mamluk example of a hybrid design.

Mamluk artists also produced a variety of geometric patterns that employ the more acute fivefold rhombus comprised of 36° and 144° angles. The historical occurrence of fivefold patterns that repeat with this rhombus are significantly less common, and the Mamluk examples are a testament to the exploratory approach to geometric design among these artists. A fine example of a *two-point* pattern that repeats on this more acute rhombus was used on the bi-fold doors of a *minbar* that was commissioned by Sultan Qaytbey[184] (r. 1468-96) [Fig. 244]. Another example of this variety of fivefold repeat was used in at least two Mamluk locations: a *minbar* door panel of uncertain provenance in the Victoria and Albert Museum in London[185] and the *minbar* railing at the *khanqah* and mosque of Sultan al-Ashraf Barsbay funerary complex in Cairo (1432-33) [Fig. 242b]. The earliest known examples of this pattern are two contemporaneous locations: a pair of Kartid wooden doors at the Turbat-i Shaykh Ahmad-i Jam in Torbat-i Jam in northwestern Iran (1236); and in the carved stone portal of the Huand Hatun *madrasa* in Kayseri in central Anatolia (1237). This *acute* pattern is unusual in that the underlying generative tessellation includes irregular pentagons and associated rhombi that give the pattern lines associated with these modules qualities that are characteristic of the

---

[182] This Quran is in the collection of the Museum of Turkish and Islamic Arts; Sultanahmet, Istanbul, Turkey: Museum Inventory Number 450.

[183] Mols (2006), cat. no. 46/1, pl. 191–194.

[184] This *minbar* is in the collection of the Victoria and Albert Museum in South Kensington, London: 1050: 1 to 2–1869.

[185] This fivefold pattern is from a nineteenth-century copy of the original Mamluk *minbar* door panel, and is part of the collection of the Victoria and Albert Museum in South Kensington, London: 887–1884.

median pattern family. The application of this pattern to *minbar* doors is particularly appropriate in that the typical proportions of a door necessitate long and narrow panels, and the more acute proportions of this rhombus fit nicely within these design constraints. An *obtuse* pattern from a Mamluk mosaic panel that repeats with this same acute repeat unit was used in the *mihrab* niche of the Amir Altinbugha al-Maridani mosque in Cairo (1337-39) [Fig. 241]. Once again, this pattern can be made from either of two dualing tessellations with equal facility. Two matching stone relief panels above the door at the southeast entrance of the Amir Qijmas al-Ishaqi mosque in Cairo (1479-81) employ an interesting *two-point* pattern that repeats upon the more acute rhombi [Fig. 243d]. The floating rhombic elements that separate the ten-pointed stars give this design a somewhat non-cohesive quality. This example is not known to have been used elsewhere.

Like the artists in preceding dynasties, Mamluk artists also applied the *fivefold system* to designs that repeat upon a rectangular grid. The wall of the *mihrab* niche at the Qadi Abu Bakr Muzhir complex in Cairo (1479-80) includes a relatively simple *two-point* pattern executed in polychrome stone mosaic [Fig. 245c]. A considerably more complex *two-point* pattern, with much broader rectangular repeats, was use to decorate a side panel of the *minbar* commissioned by Sultan Qaytbay[186] (r. 1468-96) [Fig. 248]. This same design was also used on the *minbar* of the Amir Qijmas al-Ishaqi mosque in Cairo (1479-81), as well as the *minbar* at the Amir Azbak al-Yusufi complex in Cairo (1494-95) [Photograph 46]. Both these *minbars* utilize this design on the side panels adjacent to the platform, and the fact that Sultan Qaytbay and Amir Qijmas were contemporaries indicates that the same geometric artists likely worked on both. This rectangular *two-point* pattern has the interesting feature of a ten-pointed star being placed at the center of the repeat unit with radii that are not aligned with the radii of the ten-pointed stars at the corners of each repeat unit. This skewed orientation provides an unusual and dynamic quality. Other Mamluk *two-point* patterns with non-aligned radii between the ten-pointed stars at the vertices of those at the centers of the rectangular grid include a second design from the Amir Qijmas al-Ishaqi mosque (1479-81) [Fig. 252] and a carved stone lintel at the Sultan Qaytbay Sabil in Jerusalem (1482) [Fig. 250d]. The underlying generative tessellation that produces this latter design was also used by the Mamluks at two other locations: an *obtuse* design on the stone *minbar* of the Sultan Barquq mausoleum in Cairo (1384-86) [Fig. 250f] [Photograph 57], and an *acute* pattern in a bronze window grille at the al-Azhar mosque in Cairo

[Fig. 250b]. The artist who created this *acute* design recognized the inherent problem when applying 36° crossing pattern lines to the long hexagon within this system [Fig. 187]. Rather than adjusting the underling tessellation itself, this artist arbitrarily changed the angles of the lines within this module. At first glance, this appears to be an acceptable solution. However, upon closer inspection, the break in the angles of the pattern lines is awkward and poorly resolved, and does not follow the well-established conventions for fivefold patterns. A more acceptable Mamluk *acute* pattern that repeats upon a rectangular grid was used on the *minbar* door (thirteenth century) of the otherwise Fatimid mosque of Vizier al-Salih Tala'i [Fig. 255]. Indeed, this is an exceptional example of a complex fivefold *acute* pattern, and was used many hundreds of years later by Mughal artists at the tomb of Akbar in Sikandra, India (1612).

The Mamluks rarely used hexagonal repeat units with the *fivefold system*. Of the few examples are two identical raised stone panels in the entry portal of the Ashrafiyya *madrasa* in Jerusalem (1482) [Fig. 258]. The *two-point* design of these two panels follows the occasional Mamluk convention of representing only a minimum portion of the design. In the case of this design from Jerusalem, the limited view of the pattern obscures the clarity of the total repetitive unit. If the featured image is reflected and repeated with translation symmetry, as per convention, the repeat unit for the total design is hexagonal. The interstice region at the center of the repeat unit creates pattern elements that are atypical of the *fivefold system*, as are the very close parallel lines placed within the underlying rhombic modules. The rectangular cropping of this example cleverly divides the underlying rhombi in half, thereby eliminating the problem of the two overly close parallel lines [Fig. 258b].

Mamluk artists were less disposed toward fivefold field patterns than their contemporaries from Anatolia. Among the relatively rare Mamluk examples of this variety of five-fold design is a simple, but pleasing, carved stone relief above the door at the entry portal of the Amir Ghanim al-Bahlawan funerary complex in Cairo (1478) [Fig. 212]. This is a *median* pattern that repeats upon a hexagonal grid with *two-point* characteristics that result from the pattern line application to the long edges of the underlying triangles. This form of *median* pattern line treatment can be traced back to the Khwaja Atabek mausoleum in Kerman (1100-1150) [Fig. 211], and was a particularly popular ornamental devise among artists working in the Seljuk Sultanate of Rum.

Mamluk geometric patterns created from the *system of regular polygons*, both fourfold systems and the *fivefold system*, for the most part continued the working practices and aesthetic predilections of their Zangid and Ayyubid predecessors. Yet Mamluk artists were among the most innovative in the long history of this design tradition. This

---

[186] This *minbar* is in the collection of the Victoria and Albert Museum in South Kensington, London: 1050: 1 to 2–1869.

innovation is particularly evident in three areas of geometric design: the application of geometric patterns to the surfaces of domes; the further development of complex non-systematic geometric patterns with multiple regions of differentiated local symmetry; and the bringing to maturity the class of geometric patterns that are created from the *sevenfold system* of pattern generation. Earlier examples of sevenfold patterns were used in Seljuk Persia and Turkey, but these are so few in number that it is impossible to determine the extent to which the responsible artists knew that the underlying polygonal elements with their associated pattern lines formed part of a comprehensive system, much like the *fivefold system*. As cited above, these earlier Seljuk examples employed only four polygonal modules: two varieties of irregular hexagon in the Persian example, and the heptagon and irregular pentagon in the three examples from the Sultanate of Rum. This paucity of underlying modules argues against the artist's knowledge of this as a distinct *system* per se, as does the fact that there are so few examples of sevenfold designs from this earlier period. Had there been knowledge of the *sevenfold system* at this earlier time, one would assume that there would be many more examples of such patterns in the historical record. Rather, it appears that these Seljuk artists, in their quest to apply the polygonal technique to the creation of sevenfold patterns, happened upon several of the underlying polygonal tessellations that were, in time, discovered to be part of a comprehensive *sevenfold system*.

The Mamluk development of sevenfold patterns began in the early fourteenth century, and from their earliest examples, the underlying generative polygons included a greater diversity than previous sevenfold designs. In particular, the underlying generative tessellations of these Mamluk examples included the 14-sided tetradecagon at the vertices of the repetitive grid. This module is responsible for the 14-pointed stars that characterize patterns created from this system in its fully mature expression. The earliest Mamluk example of a sevenfold *median* pattern is from a carved stone lintel on the south elevation of the Qawtawiyya *madrasa* in Tripoli, Lebanon[187] (1316-26) [Fig. 286a]. Just as there are two rhombi associated with fivefold symmetry [Fig. 5], there are three rhombi associated with sevenfold symmetry [Fig. 10]. However, only the two more obtuse sevenfold rhombi were used historically as repeat units. The example from Tripoli repeats on the more acute of these two rhombi: comprised of 2/14 and 5/14 included angles. A variation of the same underlying generative

tessellation is associated with a *median* pattern on the carved stone exterior façade of the Amir Qijmas al-Ishaqi mosque in Cairo (1479-81) [Fig. 286b]. In keeping with a common Mamluk decorative convention pertaining to geometric designs, only a portion of the overall design is shown. This same design, in its full reveal, was used at a somewhat earlier date by Timurid artists to create a carved stucco wall panel at the Amir Burunduq mausoleum at the Shah-i Zinda complex in Samarkand (1390-1420). A side panel from the *minbar* at the Sultan Barsbay complex at the northern cemetery in Cairo (1432) employs a fine sevenfold *obtuse* pattern that repeats upon this same rhombus[188] [Fig. 287b]. This was copied for the entry door of the Hanging Church in Cairo (al-Mu'allaqa), a Coptic church dedicated to St. Mary. This door is stylistically Mamluk in both the sophistication of the design and woodwork. A subtractive variation of this pattern was used on an earlier Ottoman door panel at the Bayezid Pasa mosque in Amasya, Turkey (1414-19) [Fig. 287c].[189] An example of a sevenfold *obtuse* pattern that repeats upon the more obtuse rhombus comprised of 3/14 and 4/14 included angles was used on the double doors of the *minbar* at the Haram al-Ibrahimi in Hebron, Palestine [Fig. 290]. While this *minbar* is Fatimid (1092), and brought to Hebron by the great Ayyubid leader Ṣalāḥ ad-Dīn, some components are clearly later Mamluk additions. In particular, the style of the patterns used in the *minbar* doors and back panel of the platform are of Mamluk origin. A similar sevenfold pattern was used during the Ottoman period on the incised marble ceiling of the small rectangular water feature within the courtyard of the Suleymaniya mosque in Istanbul (1550-58) [Fig. 289] [Photograph 81]. This later Ottoman example can be created from a very similar underlying polygonal tessellation wherein the concave decagonal interstice regions remains free of additional polygonal modules, and the kite motifs within this region have a distinctive *two-point* quality. The same more obtuse rhombic repeat unit was used for two outstanding Mamluk sevenfold designs that are created from the same underlying polygonal tessellation: an *obtuse* pattern from the *minbar* door of the 'Abd al-Ghani al-Fakri mosque in Cairo (1418) [Fig. 292a], and a *two-point* pattern from the large congregational Quran Stand at the Sultan Qahsuh al-Ghuri

---

[187] This design was illustrated in the Monument Survey of Tripoli, Lebanon by Hala Bou Habib, Karl Sharro, and Hind Abu Ibrahim for the American University of Beirut, Department of Architecture, 1991 and 1992.

[188] Bourgoin (1879), pl. 166. As with all Bourgoin's illustrations, this pattern is not shown with its formative structure.

[189] This pattern is also identical to a pattern in raised brick in one of the ground-level blind arches in the courtyard of the Mustansiriyah in Baghdad (1227–34). This building stems from the late Abbasid period just decades prior to the Mongol conquest. However, the incorporation of this sevenfold design appears to date from the nineteenth-century Ottoman restoration of this building. The earlier Ottoman provenance of this sevenfold pattern appears to have been the source of influence for the example at the Mustansiriya in Baghdad.

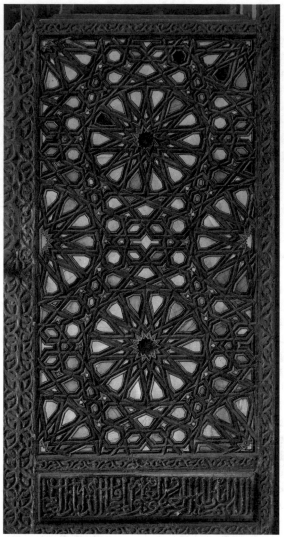

Photograph 49 A Mamluk entry door at the Sultan Qansuh al-Ghuri complex in Cairo with a pattern created from the *sevenfold system* (© David Wade)

Photograph 50 A side panel from the Mamluk *minbar* at the Sultan al-Mu'ayyad Shaykh complex in Cairo with a pattern created from the *sevenfold system* (© David Wade)

complex in Cairo[190] (1503-05) [Fig. 292b]. The underlying generative tessellation for the obtuse example from the 'Abd al-Ghani al-Fakri mosque was first attributed to this design by Ernest Hanbury Hankin.[191] One of the entry doors at the Sultan Qansuh al-Ghuri complex in Cairo, Egypt (1503-05) [Photograph 49] is also decorated with a fine sevenfold pattern that can be produced from either of two underlying tessellations [Fig. 288]. As with designs produced from the *fivefold system*, patterns made from the *sevenfold system* can

also repeat upon a rectangular grid, although only one such pattern is known from the historical record: a fine Mamluk example from one of the side panels of the *minbar* at the Sultan al-Mu'ayyad Shaykh complex in Cairo[192] (1415-22) [Fig. 294] [Photograph 50]. This design places 14-pointed stars at the vertices of the rectangular grid, the proportions of which are nearly a square. A 14-pointed star is also placed at the center of each rectangular repeat unit. The two underlying mirrored contiguous triangles that separate the underlying tetradecagon at the center of the repeat from those at the corners create a skewed orientation between the 14-pointed stars. This snub-like quality contributes to the powerful dynamic of this exceptional geometric pattern.

---

[190] Bourgoin (1879), pl. 168.

[191] In *The Drawing of Geometric Patterns in Saracenic Art* Hankin illustrates this underlying tessellation along with its associated *obtuse* pattern lines, but does not attribute the historical location of this design. As with his other published pattern analyses, he does not represent the polygonal elements used in creating his design examples as being part of a systematic methodology for pattern generation. In analyzing this design, it is likely that Hankin worked from the pattern collection of Joules Bourgoin (1879), plate 167. Hankin (1925a).

[192] Bourgoin (1879), pl. 169.

The significance of the *sevenfold system* is belied by the paucity of examples from the historical record. The development of this system represents a landmark achievement both for the beauty of the resulting designs and their geometric ingenuity. The small number of known examples warrants the inclusion of a further design created from this system that was recorded by Jules Bourgoin in his nineteenth-century collection of Islamic geometric patterns: *Les Eléments de l'art arabe: le trait des enterlacs.*[193] Unfortunately, Bourgoin did not provide the provenance of the designs he recorded. However, he worked principally in Egypt, and one must assume that the stylistic similarities to known Mamluk designs created from this system indicate a Mamluk provenance. The unidentified sevenfold pattern in Bourgoin's collection repeats upon the more obtuse rhombus, and is a very successful and well-balanced design. This is an *acute* pattern that uses an underlying tessellation that is closely related to those employed in the examples from the Suleymaniya mosque in Istanbul [Fig. 289] [Photograph 81], and the Haram al-Ibrahimi in Hebron [Fig. 290]. The difference between the underlying tessellations of these three designs is in the infill treatment of the identical concave decagonal regions.

Like their contemporaries in the Seljuk Sultanate of Rum, Mamluk artists applied the polygonal technique to a vast number of highly diverse nonsystematic geometric patterns. As inheritors of the artistic traditions of their Zangid and Ayyubid predecessors, Mamluk artists made frequent use of geometric designs that were already well known within the Islamic world. Yet the Mamluks contributed greatly to the full maturity of this design idiom by also developing original compound patterns with multiple centers of local symmetry, often with diverse repetitive schema. As in the Seljuk Sultanate of Rum, the creative attention of Mamluk artists was responsible for one of the last great innovative periods for nonsystematic geometric pattern making. Both of these cultures survived the onslaught of the Mongols and were able to continue the process of design innovation uninterrupted by political and cultural chaos. What is more, and as stated previously, the arts of both cultures benefited from the influx of artists fleeing the destruction in the east while seeking patrons in more stable Muslim lands.

Two identical adjacent Mamluk panels with a particularly successful nonsystematic design comprised of nine-pointed stars placed at the vertices of the hexagonal grid were used above the door in the entry portal at the Ashrafiyya *madrasa* in Jerusalem (1482). This is a *two-point* pattern created from an underlying tessellation of nonagons surrounded by a ring of nine irregular pentagons. This arrangement of pentagons creates an interstice six-pointed star at the center of each

Photograph 51 A side panel from the Mamluk *minbar* at the Sultan Qaytbay funerary complex in Cairo with a nonsystematic threefold *acute* pattern with six- and nine-pointed stars (© David Wade)

hexagonal repeat unit [Fig. 313b]. This *two-point* pattern appears to be unique to this location. However, an *acute* pattern made from this same underlying tessellation was used on the side panel of the *minbar* at the Sultan Qaytbay funerary complex in Cairo (1472-74) [Fig. 312a] [Photograph 51]. The close proximity in time and location between these two Mamluk examples raises the possibility of their being created by the same artist. This same underlying generative tessellation was used by earlier artists from the Seljuk Sultanate of Rum to create a *median* pattern that was first used at the Alay Han near Aksaray (1155-92) [Fig. 312b] [Photograph 43], as well as by later Shaybanid artists in the creation of an *obtuse* pattern that was used in several locations, including the Kukeltash *madrasa* in Bukhara, (1568-69) [Photograph 83]; the Nadir Diwan Begi *madrasa* and *khanqah* in Bukhara; and the Tilla Kari *madrasa* in Samarkand (1646-60) [Fig. 313a]. Separated over distance and time, this underlying tessellation was used to create designs in each of the four pattern families.

---

[193] Bourgoin (1879), pl. 165.

Photograph 52 A door panel from the *minbar* of the Amir Qawsun mosque in Cairo with a nonsystematic threefold *acute* pattern with 12-pointed stars (The Metropolitan Museum of Art: Edward C. Moore Collection, Bequest of Edward C. Moore, 1891: www.metmuseum.org)

As with other Muslim cultures, the isometric grid was widely used by Mamluk artists for the repetition of nonsystematic geometric patterns; especially for designs with 12-pointed stars placed at the vertices of triangular repeat units. An underlying polygonal tessellation with dodecagons surrounded by a ring of 12 irregular pentagons was as widely employed by Mamluk artists as by the artists of other Muslim cultures. A fine *acute* example was used to decorate a pair of bifold doors thought to be from the *minbar* of the Amir Qawsun mosque in Cairo[194] (1329-1330) [Fig. 300a *acute*] [Photograph 52]. Another underlying generative tessellation that places dodecagons at the vertices of the isometric grid connects each dodecagon with elongated hexagons, and places three edge-to-edge irregular pentagons at the center of each triangular repeat unit [Fig. 321a]. This underlying tessellation was used by Mamluk artists to create a variety of designs, including an *acute* pattern from a door at the Al-Azhar mosque in Cairo [Fig. 300b *acute*]; an *obtuse* design from the stone window grilles of the Sultan Qala'un funerary complex in Cairo (1284-85) [Fig. 321f]; a *two-point* pattern that was used on both the *minbar* rail and an interior wooden door at the Sultan al-Mu'ayyad Shaykh complex in Cairo (1415-22); the portal of the Ribat Khawand Zaynab in Cairo (1456); as well as a carved stone lintel at the Ashrafiyya *madrasa* in Jerusalem (1482) [Fig. 300b 2-point]. A variation of this tessellation, with pentagons that are truncated into trapezoids also makes fine designs

[Fig. 321i]. The *acute* pattern that is created from this modified tessellation [Fig. 321j] can also be produced from the 3.12^2 tessellation from the *system of regular polygons* [Fig. 108d]. This design was used by Mamluk artists in a mosaic panel from the Amir Aq Sunqar funerary complex in Cairo (1346-47), as well as in the doors of a cupboard at the Sultan Qansuh al-Ghuri complex in Cairo (1503-05).

The Mamluk use of nonsystematic patterns that repeat upon the isometric grid frequently included more complex geometric representations with multiple regions of diverse local symmetry. By far the most common Mamluk example of such a design was the already well established *acute* pattern that places 12-pointed stars at the vertices of the isometric grid, and 9-pointed stars at the centers of each triangular repeat unit [Fig. 346a]. Two Mamluk examples of this *acute* pattern include: a particularly fine example from a bronze door at the Sultan al-Zahir Baybars *madrasa* in Cairo[195] (1262-63), and a very becoming curvilinear example from one of the bronze doors of the Sultan al-Nasir Hasan funerary complex in Cairo (1356-63). The *two-point* pattern created from the same underlying tessellation was also used by Mamluk artists, and examples include one of the side panels of the *minbar* at the Qadi Abu Bakr Muzhir complex in Cairo (1479-80), as well as a side panel from the *minbar* at the Sultan Qansuh al-Ghuri complex (1503-05) [Fig. 347b] [Photograph 53]. The geometric logic of these patterns calls for the vertices of the isometric

[194] This pair of *minbar* doors is in the collection of the Metropolitan Museum of Art in New York City: accession number 91.1.2064.

[195] This Mamluk bi-fold door presently serves as the entry door of the French Embassy in Egypt.

Photograph 53 A side panel from the Mamluk *minbar* at the Sultan Qansuh al-Ghuri complex in Cairo with a nonsystematic threefold *two-point* pattern with 9- and 12-pointed stars (© David Wade)

Photograph 54 A Mamluk bronze entry door at the Sultan al-Zahir Barquq *madrasa* and *khanqah* in Cairo with a nonsystematic threefold *acute* pattern with 12- and 18-pointed stars (© David Lewis)

grid to be populated by *n*-pointed stars whose points are multiples of 6, and the *n*-pointed stars at the vertices of the hexagonal dual of the isometric grid (i.e., the centers of each triangular repeat) to be multiples of 3. Mamluk artists created a number of more complex isometric compound patterns that follow this design stratagem. An example with 12-pointed stars at the vertices of the isometric grid (2 stellar points × 6), and 15-pointed stars at the vertices of the dual hexagonal grid (5 stellar points × 3) was used on a carved stone lintel at the Qartawiyya *madrasa* in Tripoli, Lebanon (1316-26) [Fig. 355d]. This is a *median* pattern that employs a variation to the 15-pointed stars that are typical of Mamluk *median* designs.[196] An *acute* design from the

minbar railing in the Amir Qijmas al-Ishaqi mosque in Cairo (1479-81) places 18-pointed stars at the vertices of the isometric grid and 9-pointed stars at the vertices of the hexagonal dual grid [Fig. 357]. This also introduces octagons at the midpoint intersections of these dual grids that function similarly to the roughly contemporaneous isometric design from the Sultan Qaytbay funerary complex in Cairo (1472-74) [Fig. 109b]. Another outstanding Mamluk *acute* design with 18-pointed stars located at the vertices of the isometric grid was used in the bronze entry door of the Sultan al-Zahir Barquq *madrasa* and *khanqah* in Cairo (1384-1386) [Photograph 54]. As distinct from the previous

---

[196] The author has extrapolated the reconstruction of this pattern from the Qartawiyya Madrasa in Tripoli, Lebanon, from an indistinct photograph taken by Hana Alamuddin [Aga Khan Visual Archive, Massachusetts Institute of Technology; catalogue number IHT0078]. This is the only image of this compound isometric pattern that the author has been able to find. The analysis represented in Fig. 355d is

based upon the inherent logic of the 15- and 12-fold regions of local symmetry as exemplified in the indistinct proportions indicated within this photograph. A closer examination of this example may reveal slightly different angles in the crossing pattern lines, and pattern line relationships.

example, this places 12-pointed stars on the vertices of the dual-hexagonal grid [Fig. 359].

The Mamluks also showed great versatility in creating nonsystematic geometric patterns that repeat upon the orthogonal grid. At the most basic level, these will place primary star forms solely upon the vertices of this grid, and these stars will have *n*-fold rotational symmetry that is a multiple of 4. Most Islamic geometric patterns with 8-pointed stars are created from either the *fourfold systems A* or the *fourfold system B*, and the primary stars of most nonsystematic designs that repeat on the square grid are 12-pointed. An example of a Mamluk nonsystematic design with primary eight-pointed stars located at the vertices of the orthogonal grid is from a window grille at the Amir al-Sayfi Sarghitmish *madrasa* in Cairo (1356). The primary underlying octagons are separated along the edges of the repeat unit by two regular hexagons with coincident edges bisecting the midpoint of each edge of the square repeat unit. The underlying polygonal infill of the repeat unit is comprised of four irregular pentagons, and four irregular heptagons that are clustered around a single square at the center of the repeat unit. This creates a very-well-balanced pattern composed of five-, six-, seven-, and eight-pointed stars [Fig. 332e]. The earliest known pattern created from this same underlying tessellation is from the Seljuk ornament in the *mihrab* of the Friday Mosque at Barsian (1105) [Fig. 332a]. The most common Mamluk nonsystematic design with only a single primary region of local symmetry places 12-pointed stars at the vertices of the orthogonal grid. Each underlying generative dodecagon is surrounded by a ring of 12 edge-to-edge pentagons. This tessellation creates an *acute* pattern that was well known throughout the Islamic world [Fig. 335b], and three representative Mamluk examples include one of the carved stone lintels at the Khatuniyya *madrasa* in Tripoli, Lebanon (1373-74); the *minbar* door at the Amir Taghribardi *madrasa* complex in Cairo (1440); and a side panel from the *minbar* at the Sultan Barsbay complex at the northern cemetery in Cairo (1432). This latter example is noteworthy for the curvilinear treatment of the pattern lines. A *two-point* pattern produced from the same underlying tessellation was used in the *minbar* of the Amir Qijmas al-Ishaqi mosque in Cairo (1479-81) [Fig. 336d]. An *acute* pattern created from a simple variation to this underlying tessellation also enjoyed popularity among Mamluk artists. This variation calls for the truncation of the four clustered pentagons at the center of the square repeat: transforming the four adjacent five-pointed stars into four dart motifs [Fig. 337]. The many Mamluk examples of this design include a bronze entry door, and incised stone border around one of the interior doorways at the Zahiriyya *madrasa* and mausoleum of Sultan al-Zahir Baybars in Damascus (1279); and a fourteenth- or fifteenth-century Egyptian *minbar* panel of very high quality construction.[197]

Another underlying tessellation with dodecagons at the vertices of the orthogonal grid separates these dodecagons with squares, thereby creating large interstice regions in the centers of the square repeats. These are divided into a cluster of four irregular pentagons [Fig. 334]. This tessellation was used to create an *acute* pattern in one of the window grilles at the Sultan Qala'un funerary complex in Cairo (1284-85) [Photograph 55]. As with other patterns that incorporate the cluster of four underlying pentagons, this configuration generates an octagon within the pattern matrix that is a primary feature of this design. It is interesting to note that the artist working on adjacent windows of the Sultan Qala'un funerary complex juxtaposed this design with another pattern that also places an octagon within the center of the square repeat that is created from four coincident pentagons. The pattern in this adjacent window is the classic *fourfold system B acute* pattern [Fig. 173a] [Photograph 55]; and whereas the octagons are located at the same respective positions within these two window grilles, the vertices of the former have 12-pointed stars, while those of the latter have eight-pointed stars. Two very becoming examples of Mamluk nonsystematic orthogonal designs with singular primary star forms were employed as the frontispieces in a Quran (1369) commissioned by Sultan Sha'ban for the *madrasa* founded by his mother in Cairo.[198] [Fig. 344d]. The first of these is a *median* pattern created from an underlying tessellation that separates the 16-gons with barrel hexagons, and an atypical infill comprised of further barrel hexagons, pentagons, quadrilateral kites, and a central square. The treatment of the 16-pointed stars in this design was modified in a manner common among Mamluk artists working with *median* patterns [Fig. 344c]. The second such design from this Quran also places 16-gons at the vertices of the orthogonal grid. Each 16-gon is surrounded by a ring of 16 edge-to-edge pentagons, which in turn are surrounded by eight barrel hexagons. At the center of each repeat unit is a cluster of four contiguous pentagons. The arrangement of these underlying polygons produces four irregular octagons within each square repeat. The large degree of distortion in these octagons would ordinarily produce unsatisfactory pattern conditions. However, the artist who created this pattern devised a very clever, and visually acceptable solution that is unique to this ornamental tradition [Fig. 345].

As with isometric patterns, the Mamluks made wide use of nonsystematic orthogonal patterns with differentiated regions of local symmetry. The most basic of these are derived from an underlying tessellation that places octagons at the vertices of the square repeat, separated by regular

---

[197] In the collection of the Royal Museum of Scotland: museum inventory number A.1884.2.1.
[198] Cairo National Library; 7, ff. IV-2r.

Photograph 55 Mamluk pierced window grilles from the Sultan Qala'un funerary complex in Cairo (© David Wade)

hexagons on each midpoint of the repetitive edge [Fig. 178]. Once again, there is a cluster of four pentagons at the center of each repeat. This tessellation is similar in concept to those from the *fourfold system B* [Figs. 175–177] except that the elongated hexagons are replace with regular hexagons. The proportions of the pentagons are, by necessity, changed to suit these new circumstances, and as a consequence become nonsystematic. An outstanding *acute* pattern created from this underlying tessellation was used on the exterior of the Mamluk door (1303) at the Vizier al-Salih Tala'i mosque in Cairo [Fig. 178a]. Depending on the angular openings of the crossing pattern lines, this underlying tessellation will create designs with quite different visual qualities, as seen in another Mamluk example from a glazed ceramic panel at the Altinbugha mosque in Aleppo (1318) [Fig. 178b]. The earliest known pattern created from this tessellation is from the façade of the Ildegizid mausoleum of Mu'mine Khatun in Nakhichevan, Azerbaijan (1186) [Fig. 178c]. The example from the Vizier al-Salih Tala'i mosque employs more acute crossing pattern lines that are 32.2042°: an angle that is determined by drawing lines that connect two adjacent vertices of the underlying hexagon with the midpoint of the opposite edge of the same hexagon. As applied to the cluster of four pentagons, this produces the

distinctive flattened octagon at the center of each square repeat unit. The application of the crossing patterns lines to the examples from the Mu'mine Khatun and Altinbugha mosque predominantly employ 60° angular openings, which qualifies them as *median* patterns. As with multiple other examples, the crossing pattern lines at the cluster of four pentagons of the design from the Altinbugha mosque replaces the 60° angular openings with 45° angular openings, thus creating regular octagons at these central locations.

Perhaps more than in any other Muslim culture, Mamluk artists applied the greatest design diversity to orthogonal patterns with multiple centers of local symmetry. Of particular significance was their use of designs with 8- and 12-pointed stars in each of the four pattern families that are derived from either of two underlying tessellations: that of dodecagons placed at the vertices of the square grid, with edge-to-edge octagons located at the centers of each square repeat, and concave hexagonal interstice regions, and that of proportionally smaller dodecagons and octagons placed at the same locations, but separated by pentagons and barrel hexagons. These two underlying tessellations have a dual relationship with one another, and the wide diversity of resulting patterns can be created from either with equal facility [Figs. 379–382]. The Mamluk use of patterns that can be created from either of these underlying tessellations exceeded that of other Muslim cultures, with designs in each of the four pattern families represented within their ornamental canon. Mamluk examples of *acute* patterns generated from either of these underlying tessellations include a curvilinear *acute* pattern from a bronze entry door at the Sultan al-Nasir Hasan funerary complex in Cairo (1356-63), and a square panel in the railing of the *minbar* at the Amir Azbek al-Yufusi complex in Cairo (1494-95) [Fig. 379e] [Photograph 46]. The triangular side panels from the *minbar* of the Sultan Mu'ayyad mosque in Cairo (1415-21) employ a typical Mamluk variation to the standard *median* pattern that changes the character of the 12-pointed stars by replacing the crossing pattern lines that are ordinarily located at the midpoints of the dodecagonal edges with an arbitrary star rosette surrounded by darts. In accordance with Mamluk geometric aesthetics, this also introduces heptagonal elements into the pattern matrix [Fig. 380e]. Examples of Mamluk *obtuse* patterns created from either of these two underlying tessellations include the blind arches surrounding the exterior drum of the dome at the Hasan Sadaqah mausoleum in Cairo (1315-21), and a window grill in the drum of the dome at the contemporaneous Amir Sanqur al-Sa'di funerary complex in Cairo (1315) [Fig. 381b]. Their proximity in location and date, and similarity in media and design expression, suggests the likelihood of their being produced by the same artist or atelier. A *two-point* pattern that can be derived from either of these tessellations was used in the side panels of the *minbar* at the Princess Asal Bay mosque in Fayyum, Egypt (1498) [Fig. 382b]. A considerably more complex Mamluk orthogonal *two-point* pattern with 8- and 12-pointed stars is derived from an underlying tessellation that is also comprised of dodecagons, octagons, elongated hexagons, and pentagons; although in this case the pentagons are clustered in an edge-to-edge arrangement around a thin rhombus. This configuration of pentagons is a corollary with the *fivefold system*, and similarly only works (without adjustment) with *obtuse* and *two-point* patterns [Fig. 197]. This was used in the triangular side panel of the *minbar* at the Sultan Qaytbay funerary complex in Cairo (1472-74), as well as in a carved stone panel of Mamluk origin at the al-Azhar mosque in Cairo [Fig. 383].

Mamluk artists also made common use of several compound patterns that place 16-pointed stars at the vertices of the orthogonal grid. The least complex of these incorporate eight-pointed stars at the centers of the square repeat unit. An *acute* pattern on the small side door into the interior of the *minbar* at the Sultan Mu'ayyad mosque in Cairo is just such a design [Fig. 388] [Photograph 60]. As with the *median* design on the triangular side panels of the same *minbar* with 8- and 12-pointed stars [Fig. 380e], this design can be generated from two distinct underlying tessellations.

The first of these uses 16-gons placed at the vertices of the repetitive grid and coinciding octagons located at the center of each square repeat. The alternative underlying tessellation is comprised of 16-gons and octagons that are separated by a network of pentagons. This underlying tessellation can be modified so that the eight pentagons that surround each octagon are truncated into trapezoids, creating a square with triangles placed on each edge that is located at the center of the square repeat unit. This variation replaces the eight-pointed stars with octagons. A Mamluk pattern that employs this modified underlying tessellation was used in the ornament of the Sultan al-Nasir Hasan funerary complex in Cairo [Fig. 391]. Several varieties of orthogonal compound design exhibiting 12- and 16-pointed stars were also produced by Mamluk artists. One of the most interesting is created from an underlying tessellation comprised of dodecagons and 16-gons separated by rings of pentagons that was used on the original bronze entry doors of the Sultan al-Nasir Hasan funerary complex in Cairo (1356-63) [Fig. 392d].[199] This example employs a variation of the otherwise standard *median* pattern that has the added feature of placing heptagons within the pattern matrix. The 12- and 16-pointed star rosettes of this design are modified in the common Mamluk fashion whereby the ring of five-pointed stars is transformed into darts that make up an alternative ten-pointed star [Fig. 223]. Truncating the six pentagons that surround the small rhombi can modify the underlying tessellation that otherwise creates this example from the Sultan al-Nasir Hasan funerary complex; thereby producing an altogether new pattern [Fig. 393]. This alteration of the underlying generative tessellation follows the convention established within the *fivefold system* by changing the pattern lines to conform with the cluster of six truncated pentagons. [Fig. 198]. Mamluk designs that employ this modified underlying tessellation include a window grill at the mosque of Altinbugha al-Maridani in Cairo (1337-39) [Photograph 56]; a curvilinear variation in one of the bronze entry doors of the Sultan al-Nasir Hasan funerary complex in Cairo (1356-63); and the triangular side panel of a wooden *minbar* (1468-96) commissioned by Sultan Qaytbay and currently on display at the Victoria and Albert Museum in London. It is interesting to note that the earliest use of this design appears to be Seljuk: in one of the large *muqarnas* panels of the *mihrab* hood in the Friday Mosque at Barsian, Iran (1105). Further complexity is provided to patterns with 12- and 16-pointed stars by the incorporation of two mirrored 7-pointed stars into the pattern matrix. These are located at the midpoints of each edge of the square repeat unit. The underlying tessellation that creates this design

---

[199] Moved in 1416–17 to the Sultan Mu'ayyad Mosque in Cairo where it functions as the main entry door to this day. See Mols (2006), 214.

Photograph 56 A Mamluk window grille at the mosque of Altinbugha al-Maridani in Cairo with an *acute* pattern comprised of 12- and 18-pointed stars (© David Wade)

separates the dodecagons and 16-gons with barrel hexagons, with a cluster of ten pentagons that surround two edge-to-edge irregular heptagons [Fig. 395]. This underlying tessellation was used by Mamluk artists to create several *two-point* designs, including the triangular side panel of the stone *minbar* of Sultan al-Zahir Barquq complex in Cairo (1384-86) [Photograph 57]; the triangular side panel of the wooden *minbar* at the Amir Qijmas al-Ishaqi mosque in Cairo (1479-81); and one of the side panels of the wooden *minbar* at the Sultan Qansuh al-Ghuri complex in Cairo (1503-05). This underlying tessellation was also used to create the carved stone *median* pattern at the base of the minaret at the Mughulbay Taz mosque in Cairo (1466) [Fig. 396b].

The Mamluks did not produce as many geometric patterns with more than two primary regions of differentiated symmetry as their contemporaries in the Seljuk Sultanate of Rum. A stunning exception is an *acute* pattern from one of the side panels of the *minbar* at the Sultan Qaytbay funerary complex at the northern cemetery in Cairo (1472-74). This exceptionally well-balanced design places 16-pointed stars upon the vertices of the orthogonal grid, 12-pointed stars at the center of each repeat unit, and 10-pointed stars on the midpoints of each edge of the repeat unit [Fig. 402].

Examples of nonsystematic compound patterns with repetitive grids that are neither isometric nor orthogonal are less common among the Mamluks than their contemporaries in the Seljuk Sultanate of Rum. Among the few examples are two designs from the Sultan al-Nasir Hasan funerary complex in Cairo (1356-63). The first of these is a very nice *acute* field pattern from an incised stone border that places octagons at the vertices of a rectangular repeat unit, as well as at the center of the repeat [Fig. 412] [Photograph 58]. This border pattern is located in the entry portal of this mosque. The second example from the Sultan al-Nasir Hasan funerary complex is immediately adjacent to the first, adorning the back wall of a niche on the sidewalls of the main entry portal, and is fashioned in inlaid polychrome stone [Fig. 413] [Photograph 58]. This design is noteworthy in that it is one of the few historical examples of a geometric pattern that expressly shows the underlying generative tessellation as part of the ornament. As such, this panel provides important historical evidence for the use of the polygonal technique of geometric pattern construction. Mamluk examples of more complex nonsystematic compound patterns that do not repeat with either the isometric or orthogonal grids are unusual. A rhombic *acute* example is found on the wooden entry door at the Zaynab Khatun Manzil (house) in Cairo (1468). This places 24-pointed stars on the vertices of the rhombus, with 12-pointed stars at the midpoints of each edge of the rhombic repeat, and 8-pointed stars within the field. The proportions of this rhombic repeat are governed by $2 \times 3/24$ and $2 \times 9/24$ included angles. While the complexity of this Mamluk pattern is the equal of earlier examples created under the auspices of Seljuk patronage, this design from the Zaynab Khatun Manzil entry door is poorly proportioned and ill conceived: in large part due to the position of the underlying octagon in relation to the underlying 24-gons and dodecagons. Their respective radii are not congruent, or near enough to appear as such. A far more successful Mamluk nonsystematic compound *acute* pattern was used on the entry door of the 'Abd al-Ghani al-Fakhri mosque in Cairo (1418). This pattern repeats upon a rectangular grid, with 10-pointed stars located at the vertices of the rectangle, 10-pointed stars at the midpoints of the long edge of the repeat, and two 11-pointed stars within the field of the repeat [Fig. 417]. This is perhaps the most successful Mamluk geometric pattern with complex local symmetries, and is superior in overall balance to the only other known geometric pattern comprised of 10- and 11-pointed stars: the *acute* pattern produced by the Armenian Christian monk Momik (between 1282-1321) for one of his stone *khachkar* crosses in Noravank [Fig. 423].

During the later Mamluk period, many stone domes with highly ornate exterior surfaces were constructed in Cairo. Noteworthy among these are a small number of domes that

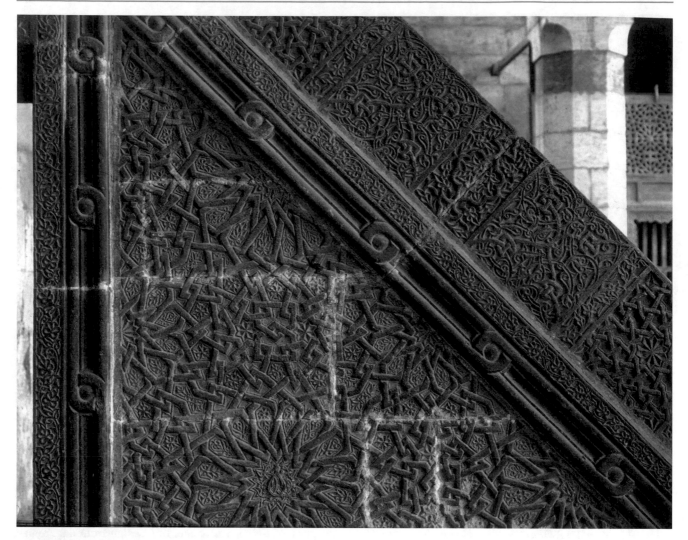

Photograph 57 A side panel of the Mamluk *minbar* at the Sultan al-Zahir Barquq complex in Cairo with a *two-point* pattern comprised of 7-, 12-, and 18 pointed stars (© David Wade)

are ornamented with geometric designs in high relief. This was a new and distinctly Mamluk development. Other subsequent Muslim cultures, such as the Safavids, also decorated the exterior of their domes with geometric designs. However, what makes the Mamluk domes so distinctive is their monochrome aesthetic: emphasizing the design with high relief and the consequent play of light and shadow. These Mamluk geometric domes invariably have radial symmetry wherein the gore segments are provided with an underlying generative tessellation upon which the geometric design is constructed. This is a highly specialized application of the polygonal technique requiring an additional facility with three-dimensional geometry and especially close collaboration with the architectural designer.

Mamluk examples of domical geometric ornament include three domes at the Sultan al-Ashraf Barsbay funerary complex at the northern cemetery in Cairo (1432-33).

The geometric design on the dome over the Sultan Barsbay mausoleum is a dense interweave that is reminiscent of a basket weave [Photograph 61]. This pattern places a ring of half eight-pointed stars around the periphery of the dome, ascending to second ring of eight-pointed stars, followed by sequential rings of seven-, six-, and five-pointed stars [Fig. 493a]. The initial double course of eight-pointed stars is the well-known classic *acute* pattern created from the tessellation of octagons and squares [Fig. 124a]. The transmission through the eight-, seven-, six-, and five-pointed stars results from the diminishing width of the gore segment, and the narrow proportions of this gore segment result from the 20-fold radial segmentation of the dome's surface. In this way, the geometric design culminates at the apex in a 20-pointed star. Immediately adjacent to this dome is the much smaller dome of the Amir Gani Bak al-Ashrafi mausoleum. This hosts a curvilinear design comprised of six half 12-pointed stars around the periphery, and a ring of six

Photograph 58 A Mamluk niche in the entry portal of the Sultan al-Nasir Hasan funerary complex in Cairo with a nonsystematic pattern that includes its underlying generative tessellation, as well as a nonsystematic border pattern that surrounds the niche (© Scott Haddow)

10-pointed stars in the mid-section of the pattern matrix. These primary regions of local symmetry are connected with 6-pointed stars and octagonal regions, culminating in a 12-pointed star at the apex. The third geometric dome at this complex covers an anonymous tomb of a Barsbay family member. The geometric design on this dome is the most successful in its overall balance, and is comprised of a ring of six half 12-pointed stars surrounding the periphery of the dome, followed by a ring of six 8-pointed stars, and a second ring of six 12-pointed stars, culminating in a 12-pointed star at the apex [Fig. 493b]. Perhaps the most notable Mamluk geometric dome is at the Sultan Qaytbay funerary complex in the northern cemetery in Cairo (1472-1474) [Photograph 2]. The powerful visual appeal of this dome is the result of the augmentation of the geometric design with a meandering floral devise that fills the background. The geometric design itself is less complex than the three examples from the Sultan Barsbay funerary complex, but the addition of the floral element provides a richer overall ornamental affect. This geometric pattern places a ring of eight half 10-pointed stars at the base of the dome, followed by a ring of eight

9-pointed stars, a further ring of sixteen 5-pointed stars, and culminating in a 16-pointed star at the apex of the dome [Fig. 493c]. The exterior treatment of this dome is an outstanding example of Mamluk monochrome geometric and floral ornament.

Mamluk artists also continued the Zangid and Ayyubid tradition of decorating the interior quarter domes above entry portals and *mihrab* niches with geometric patterns. Several of the earlier Mamluk examples were clumsy in their use of geometry: for example the *mihrab* hood at the Haram al-Ibrahimi in Hebron, Palestine (fourteenth century), is decorated with a geometric design that forces hexagons onto an orthogonal repeat. The further incompatibility of the orthogonal grid with the surface of a sphere results in a pattern that is poorly conceived and rife with geometric inconsistencies. The forced application of otherwise two-dimensional patterns with triangular repeat units onto quarter dome surfaces was also occasionally practiced, and examples include the *mihrab* hood at the Amir Salar and Amir Sanjar al-Jawli funerary complex in Cairo (1303-04) and the *mihrab* hood at the Faraj ibn Barquq Zawiya and Sabil in Cairo (1408). However, none of these examples comes close to equaling the sophistication in design and construction of the earlier geometric quarter domes of the Zangids and Ayyubids.

Two relatively simple Mamluk geometric semidomes are found in Cairo that were constructed within a decade of one another: a *two-point* pattern in the *mihrab* niche at the al-Mar'a mosque in Cairo (1468-69) and a curvilinear *acute* pattern comprised of six-pointed stars in the hood of the entry portal at the Timraz al-Ahmadi mosque in Cairo (1472). Each of these exhibits a bold simplicity that is sufficiently distinct from other examples to suggest the possibility of their being designed by the same individual. In contrast to the above examples, the interior geometric ornament of several Mamluk niches rivals the complexity and sophistication of the finest Zangid and Ayyubid work. The stone mosaic *mihrab* hood from the Amir Qijmas al-Ishaqi mosque in Cairo (1479-81) places half 12-pointed stars at the base of the quarter dome, 10-pointed stars at the middle of the radial gore, and an 8-pointed star at the apex [Fig. 493d]. The stone quarter dome in the entry portal of the Amir Ahmad al-Mihmandar funerary complex in Cairo (1324-25) is one of the most spectacular examples of Egyptian three-dimensional geometric art [Photograph 59]. The archivolt that surrounds the quarter dome is decorated with a *median* pattern comprised of 10- and 11-pointed stars that are centered on the edge of the arched opening to the hood. The pattern turns the corner of the archivolt to continue directly onto the surface of the quarter dome and the pattern on the domical surface has nine- and ten-pointed stars. This is an immensely complex geometric schema that is unique to this location, but was likely inspired by the archivolt of the

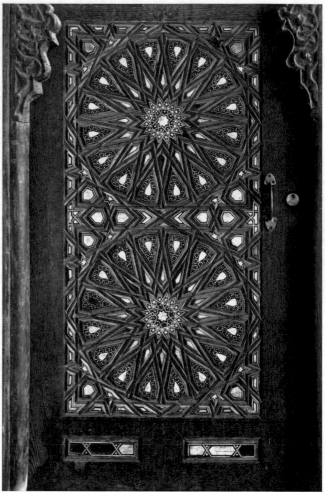

Photograph 59 A Mamluk quarter dome in the entry of the Amir Ahmad al-Mihmandar funerary complex in Cairo with a surrounding nonsystematic *median* pattern comprised of 10- and 11-pointed stars, and a contiguous pattern on the domical surface with 9- and 10-pointed stars (© David Wade)

Photograph 60 A Mamluk side door to the *minbar* of the Sultan Mu'ayyad mosque in Cairo with a nonsystematic *median* pattern comprised of 8- and 16-pointed stars (© David Wade)

Ayyubid entry portal at the Zahiriyya *madrasa* in Aleppo (1217). While this earlier example also turns the corner at the arch opening, the pattern only continues onto the ascending plane of the intrados of the interior side of the arch rather than onto an actual domical surface.

It is noteworthy that almost all of the many Mamluk examples of geometric domical ornament, be they the exceptional monochrome exterior domes or the quarter domes in *mihrab* and entry portals, are based upon decorated gore segments and consequent radial symmetry. Mamluk artists were either ignorant of the design potential of the Platonic and Archimedean polyhedra or they preferred the aesthetic qualities of decorated gore segments. For the most part, these artists did not follow the earlier polyhedral precedents of their Zangid and Ayyubid predecessors. A very beautiful Mamluk exception to this deficiency is from the domical hood of the entry portal at an anonymous mausoleum in the Nouri district of Tripoli, Lebanon. This spherical design is

based on the geometry of the cubeoctahedron. Unlike its Levantine wood predecessors, this example of polyhedral-geometric design was produced in polychromatic inlayed stone as per the gore segmented *mihrab* hood at the Amir Qijmas al-Ishaqi mosque. This example from Tripoli is a *two-point* pattern that places ten-pointed stars at the vertices of the cubeoctahedron, and eight-pointed stars at the centers of each projected square face of the polyhedra [Fig. 502]. Each polyhedral vertex is made up of two squares and two triangles in a 3.4.3.4 configuration, and the ten-pointed stars are suited to this location by virtue of three points being allocated to each square corner, and two points to each triangular corner. The underlying generative tessellation that is applied to both varieties of polyhedral face is comprised of decagons and octagons that are separated by a connective polygonal network of pentagons and barrel hexagons. This is one of the most complex and beautiful polyhedral domes known to the historical record.

Photograph 61 The dome over the Mamluk mausoleum of Sultan al-Ashraf Barsbay with a geometric pattern comprised of ascending eight-, seven-, six-, and five-pointed stars (© John A. and Caroline Williams)

1.22 Design Developments in the Western Regions

The history of the art and architectural ornament of al-Andalus is a complex interplay of cross-cultural influences. From as early as the eighth century, following the Muslim conquest of the Iberian Peninsula, the aesthetic predilections of the Umayyads helped to create an ornamental style that was distinct from other Muslim cultures. The successive Almoravid and Almohad Berber invasions from North Africa introduced a more austere approach to architectural ornament that was in part informed by their Abbasid affiliations. Throughout this history was the ongoing interaction with indigenous Christian artists, and the *Mudéjar* ornamental style that grew out of this cultural overlap continued long after the forced departure of Muslims and Jews from Spain. From the thirteenth to fifteenth centuries the

inherently Muslim ornamental style of al-Andalus was equally embraced by Christian and Jewish patronage. The community of artists and master craftsmen engaged in this work included both Muslims and Christians, and the ecumenism between these three faiths allowed for artists to work for patrons who adhered to a different faith.

The rise of the Almoravids in the mid-eleventh century brought about cultural and artistic change in al-Andalus. This Berber tribe had come to dominate much of North Africa, and in 1090 they defeated the Umayyad Caliphate in Spain. The Almoravids were Sunnis, and accepted the authority of the Abbasid Caliphate in Baghdad. The coming to power of a dynasty with strong religious and political affiliations with the Abbasids introduced North Africa and Spain to a wealth of new ornamental design traditions. This was particularly the case with an increased emphasis on geometric ornament. Prior to the Almoravids, the use of geometric patterns in Spain primarily followed the earlier example set by the Umayyads in Damascus: the application of interlaced geometric patterns into window grilles. A number of very beautifully carved marble window grilles dating from the late tenth century demonstrate the continuity of this use of geometric patterns. However, these later examples differ in that they are not the products of compass work, but are significantly more sophisticated in the methods used in construction. The mature nature of these window grille designs raises an interesting question: If artists working during the late tenth century Umayyad Caliphate of al-Andalus were able to create complex and beautiful geometric patterns, why was it that these patterns were only applied to window grilles, and not more generally to areas of architectural mass? Nor was there any appreciable utilization of geometric patterns to the minor arts during the Umayyad period. It would appear that this conscious limitation to the application of geometric patterns was, on the one hand, born out of a wish to remain true to their earlier Syrian roots, and on the other hand, a possible desire to assert their cultural distinction from the might of the Abbasid empire to the east. By contrast, the Almoravids appear to have been both familiar with, and sympathetic to the ornamental developments that were taking place in Baghdad and the Islamic east, for it was during this period that *muqarnas* vaulting was first introduced to the Maghreb and al-Andalus; and it was also during this period of Sunni ascendancy that the tradition of complex geometric pattern making finally became a central aspect of the ornamental arts of the western regions.[200]

The Tiafa kingdoms in al-Andalus continued the aesthetic approach to architectural ornament of their Umayyad

---

[200] Necipoğlu (1995), 101.

predecessors. Many fine examples of geometric design were incorporated into the fabric of their building, and patterns created from the *system of regular polygons* were especially common. A fine, if geometrically simple, example is from the entry door of the Aljafería Palace in Zaragoza that is created from the 6^3 hexagonal grid, and comprised of six-pointed stars and hexagons [Fig. 95b]. The exterior façade of the nearby Cathedral of San Salvador includes a *Mudéjar* brickwork pattern created from the 3.6.3.6 underlying tessellation of triangles and hexagons. This design is comprised of superimposed hexagons [Fig. 99e]. This pattern adheres to the specific proportions that result from the 3.6.3.6 generative schema rather than the alternative 6^3 derivation that produces an almost identical design [Fig. 96e]. The architectural ornament of the succeeding Almoravids continued the stylistic developments of their Umayyad and Tiafa predecessors: the abundant use of the floral idiom with rudimentary geometric patterns created mostly from the *system of regular polygons*. The Almoravid recognition of Abbasid religious authority and cultural affinity may have facilitated the introduction of more overtly geometric forms of ornament, such as *muqarnas*. Unfortunately, most Almoravid architecture was destroyed by their Almohad conquerors. Of that which remains, the carved stucco ornament in the Great Mosque of Tlemcen (1136) stands out as especially beautiful. The dome in front of the *mihrab* is made up of 12 rib-arches that form a 12-pointed star at the center of the dome. Within the center of the 12-pointed star is a fine example of early Maghrebi *muqarnas* vaulting. To add to the stunning visual effect of this dome, the floral ornament inside the 12 arches is pierced so that sunlight can filter through to the interior space. This is one of the most outstanding domes produced in the western Islamic regions.

The Almohads were also a Berber tribe who originated from the Atlas Mountains of North Africa. Like the Almoravids, they were Sunni Muslims. They nevertheless had strong religious differences with Almoravids, as well as with the Abbasid Caliphate in Baghdad. Led by their religious zealotry, they won victories against the Almoravids in both North Africa and Spain during the first quarter of the twelfth century. The Almohads ruled over the Maghreb from their capitol in Marrakesh, and in al-Andalus from their capital in Seville. The rise of the Almohads had a strong influence on the art and architecture of al-Andalus and the Maghreb. The Almohads regarded the opulence of Almoravid ornament as decadent, and an indication of their moral and religious inferiority. Their puritanical zeal led to the destruction of almost all Almoravid architecture. However, their dislike for architectural splendor notwithstanding, they constructed many beautiful examples of Islamic architecture, albeit in a more austere and restrained style. While the Almohads employed both geometric patterns and

muqarnas vaulting, the level of sophistication was still far short of the contemporaneous work of the Zangids and Ayyubids to the east.

The Nasrids rose to power following the defeat of the Almohads in al-Andalus by the Christian forces from the northern regions of the Spanish peninsula. The Nasrids founded their capitol in Granada in 1238. The Nasrid dynasty lasted nearly 250 years, until it was finally defeated by the Christian *reconquista* in 1492. It was at this time that Muslim rule over the last stronghold of Iberian territory ended, drawing to a close almost 800 years of Muslim dominance in al-Andalus. The western style of Islamic ornament reached its full maturity during the Nasrid period. Perhaps as a reaction to the austerity of the Almohad ornamental style, the ornament of the Nasrids was especially ornate, and replete with highly refined detail. The Nasrids rejoiced in creating an architectural feast for the eyes. Virtually no space was left unadorned, and beautiful calligraphic inscriptions were accented through the use of complex geometric patterns, *muqarnas* vaulting, and beautifully stylized floral designs. The greatest Nasrid architectural monument to survive to the present day is the Alhambra in Granada. This is both a fortress and a palace, and was the seat of Nasrid power. It stands out as one of the most remarkable and beautiful architectural complexes in the world. A wide variety of ornamental materials were used throughout this complex, including carved and painted stucco, ceramic tile and cut-tile mosaic, carved and painted wood, pierced stone, carved stone, and cast bronze. At the hands of less skilled artists, such an exuberant use of so wide a range of materials might be expected to create a cacophony of ornamental overload. It is a remarkable attribute of the Nasrid ornamental tradition that these diverse materials and abundant design elements were harmoniously brought together to create an architectural environment that is as tranquil as it is luxuriant.

The Nasrids of al-Andalus were closely allied with the Marinid dynasty in North Africa. The Marinids were a Berber tribe from the Sahara desert who entered Morocco in 1216. By 1250 they had taken control of Fez, where they founded their capital. The exceptional beauty of this city is a legacy of Marinid rule. In 1269 they brought an end to the Almohad dynasty by conquering their capital of Marrakesh. During the Marinid period, the city of Fez became a great center of Islamic culture. By the middle of the fourteenth century, the architecture of the Marinids had reached a level of beauty and refinement that was equal to that of the Nasrids in al-Andalus. Of particular note are the al-'Attarin *madrasa* in Fez (1323), and the Bu 'Inaniyya *madrasa* in Fez (1350-55). The architectural ornament in both these buildings is fully mature, and along with that of the Alhambra, represents the fulfillment of the western style of Islamic design. The Marinids of the Maghreb and the Nasrids of al-Andalus were close allies and trading partners. The close political and cultural relations between these neighboring dynasties included the ability for skilled artists to receive patronage on both sides of the Straits of Gibraltar, and resulted in the aesthetic synthesis of style and quality in the ornament in these two regions during this period.

The Nasrids and Marinids were prodigious patrons of the geometric arts. With the exception of the *sevenfold system*, Nasrid and Marinid artists created many fine examples of geometric patterns that are associated with each of the polygonal systems. What is more, in addition to Muslim patronage, Christian and Jewish patrons also commissioned geometric designs of the highest quality. Of the countless number of patterns created from the *system of regular polygons* one example was particularly popular: a design created from the 3.4.6.4 underlying tessellation that is characterized by 6- and 12-pointed stars, with arbitrary 8-pointed stars introduced into the underlying square modules [Fig. 105h]. This pattern was employed widely in the western regions during this period, and examples include a cut-tile mosaic panels at the Alcazar in Seville (fourteenth century), and a carved stucco panel in the Córdoba Synagogue (1315). Several examples of essentially the same design produced under the patronage of the Sultanate of Rum are found in Ahlat, eastern Turkey, from approximately 40 years earlier [Fig. 105gi]. A later Alawid design from the Moulay Ismail Palace in Meknès, Morocco (seventeenth century) employs a becoming additive variations within the 12-pointed stars of an *obtuse* pattern created from the 3.12^2 generative tessellation [Fig. 108e]. A fine Nasrid *obtuse* pattern created from the $3.4.3.12$-3.12^2 *two-uniform* tessellation was used in the *zillij* cut-tile mosaic of the Alhambra [Fig. 113d]. This design is identical to contemporaneous Mamluk examples from Egypt [Fig. 113c] except for the arbitrary incorporation of an eight-pointed star within the underlying square modules.

Nasrid and Marinid examples of patterns created from the 4.8^2 underlying tessellation of squares and octagons are plentiful. Indeed, each of the four pattern families, along with countless stylistic variations, is represented in the derivation of patterns from this single tessellation. One of the more basic *zillij* mosaic panels at the Alhambra includes the standard *acute* pattern created from this tessellation [Fig. 124a]; and one of the wooden ceilings from the Alhambra is ornamented with an elaborated version of the classic star-and-cross *median* pattern [Fig. 126b]. Examples of the standard *obtuse* pattern include a wooden screen from the Sultan's Palace in Tangier [Fig. 124c]; and an exceptional Nasrid example of the standard *two-point* design created from the 4.8^2 underlying tessellation was used on a silk brocade weaving dating from the fourteenth century[201]

[Fig. 124d]. The Bu 'Inaniyya *madrasa* in Fez (1350-55) includes several patterns that can be easily produced from the 4.8^2 tessellation. These are provided with greater complexity through arbitrary additions to the pattern matrix [Fig. 126e, g, h], while a more simplified variation of the above-referenced *two-point* design used on the Nasrid silk brocade enjoyed great popularity among artists of both cultures [Fig. 128a]. Many of the patterns associated with the 4.8^2 underlying tessellation can be created with alternative methodologies, and it is often impossible to know for certain how a given example was produced. A pattern from the Alhambra with *two-point* characteristics may well have been created using the orthogonal graph paper technique instead [Fig. 128b], and a very simple design with octagonal characteristics from a tile mosaic panel in Fez may, or may not, have employed this tessellation in its creation [Fig. 127f]. A design that places the pattern lines upon the vertices of the underlying 4.8^2 tessellation was used at both the Alhambra and the Alcazar in Seville (1364-66) [Fig. 129d]. This design is identical to a much earlier Ghurid pattern [Fig. 129c] except that it has arbitrary small eight-pointed stars placed at the underlying square modules.

A cut-tile mosaic panel from the Alhambra utilizes an unusual underlying generative tessellation comprised of dodecagons, equilateral triangles and rhombic interstice modules [Fig. 339]. This places the dodecagons onto the vertices of the orthogonal grid in a vertex-to-vertex rather than the more typical edge-to-edge orientation. Each square repeat unit has four equilateral triangles that are coincident with the dodecagons and meet at the center of the repeat. This underlying generative tessellation is atypical of the *system of regular polygons*, and has characteristics that are akin to nonsystematic design methodology. The pattern generated from this tessellation is in the *acute* family. The designer of this cut-tile mosaic panel incorporated multiple variations to the treatment of the pattern lines within the dodecagons [Figs. 340 and 341]. These variations are independent of the underlying tessellation, and are arrived at via arbitrary design considerations that adhere to the geometric aesthetics of the Nasrids and Marinids. While such variations to the primary star form were a common feature of geometric patterns in the Maghreb, this particular example is worthy of note for its use of multiple variations within a single design. As such, this mosaic panel from the Alhambra exemplifies the diversity and playfulness of the western geometric design tradition.

Patterns associated with the *fourfold system A* were especially popular among Nasrid and Marinid artists. Indeed, their use of this system exceeded other Muslim cultures in the creation of especially complex and sophisticated patterns. This increased complexity is due to two primary factors: very broad repeat units comprised of large numbers of underlying polygonal modules, and an innovative approach to varying the applied pattern lines through

---

[201] In the collection of the Metropolitan Museum of Art, New York, Fletcher Fund, (1929), 29.22.

arbitrary design decisions. These western practices created patterns with significantly greater visual diversity and design variation. This is in marked distinction to the more uniform aesthetics of eastern Muslim cultures wherein the repetitive structure was more readily evident. Multiple examples of this more complex use of the *fourfold system A* were seen at the Alhambra in Granada. A fine example of the incorporation of arbitrary variations in the application of the pattern lines to diverse modules in the underlying tessellation is demonstrated in one of the wooden ceilings at the Palace of the Myrtles (1370) at the Alhambra. This is an *acute* design that places 45° crossing pattern lines placed at the midpoints of each short edge of the underlying polygonal modules, and 90° crossing pattern lines at the midpoints of each longer polygonal edge [Fig. 149b]. The standard *acute* pattern created from this underlying tessellation is certainly satisfactory [Fig. 149a], but the visual character of this design from the Palace of the Myrtles is pleasingly altered by adjusting the application of the pattern lines within the underlying square modules.

Artists in the Maghreb built upon the standard set of underlying polygonal modules that comprise the *fourfold system A*, thereby increasing the design potential of this system. These additional design modules are easily derived through either truncating the large and small octagons, or by identifying interstice regions when tessellating with the standard modules [Fig. 158]. An outstanding wooden door from the Alhambra demonstrates the efficacy of these additional polygonal modules as used by Nasrid and Marinid master artists [Fig. 162] [Photograph 62]. This example also demonstrates a less dogmatic approach to the application of the pattern lines to the underlying tessellations by artists of this region. This is similar to the pattern lines associated with the underlying square regions of the above referenced design from the Palace of Myrtles [Fig. 149b], although in the example from the wooden door from the Alhambra, the arbitrary pattern line adjustments apply to the truncated octagons, trapezoids, and triangles. In each case, the angles and placement of the applied *obtuse* patterns lines are determined more by their associated neighbors than by the pedantic iterative placement of the same applied pattern lines onto a given polygonal module in all circumstances and locations. This is a more creative process wherein the artist is actively making arbitrary decisions that nonetheless conform to the geometry of the underlying tessellation. The design of the door at the Alhambra utilizes two distinct rectangular repetitive elements; and either of these will work on their own as perfectly acceptable patterns. Many of the *zillij* mosaic panels at the Alhambra have increased complexity resulting from the incorporation of added polygonal modules and similar variations within the applied pattern lines. An especially beautiful example from the Hall of Ambassadors places pattern variations into selected octagons within the

Photograph 62 Nasrid wood joinery from a door at the Alhambra with an *obtuse* pattern created from the *fourfold system A* (© David Wade)

underlying tessellation [Fig. 163]. This panel also arbitrarily rotates the central eight-pointed star by 22.5°, causing the set of parallel lines within this star to be out of sync with the rest of the overall design. This dynamic central feature is illustrative of the flexible approach to geometric design as practiced in the Maghreb.

Nasrid and Marinid artists developed a variation to patterns created from both the fourfold systems that incorporate 16-pointed stars into the pattern matrix. This was achieved through the discovery that the modules of these systems can be circularly arranged to have 16-fold symmetry, allowing for further infill of the 16-pointed star motif. This polygonal arrangement is very elegant, and it is surprising that artists from other Muslim cultures did not discover this inherent, if obscure, 16-fold capacity within the fourfold systems. As pertains to the *fourfold system A*, this variety of pattern invariably employs the expanded set of generative polygonal modules. A *Mudéjar* stucco window grille from the Sinagoga del Tránsito in Toledo (1360) has a very fine example of such a pattern among its many window grilles [Fig. 165]. A similar design was used in the upper portion of three identical adjacent window grilles at the Alhambra [Fig. 166]. The lower portion of these windows adds a rectangular repetitive element beneath the upper square that is created from the expanded modules of the *fourfold system A*. Both the square and rectangular repetitive elements from this window can be used independently as repeat units. Two additional examples of this variety of design were found within the *zillij* mosaic wainscotings at the Alhambra, and include an example with moderate complexity (as measured by the amount of geometric information within the square repeat unit) [Fig. 167], and a design composed of considerably more underlying polygonal components within the square repeat unit, with consequent increased overall complexity [Fig. 168].

Patterns created from the *fourfold system B* were also well known in the Maghreb, including multiple examples of designs that had been used previously by Muslim cultures in the east. Yet even the use of established designs was often imbued with innovative flourish. A case in point is a wooden ceiling at the Alhambra that incorporates curvilinear elements into an otherwise well-known *acute* pattern that had been widely used by earlier Muslim cultures [Fig. 177d]. Moreover, and in keeping with Maghrebi practices, this design is provided greater complexity through the application of two distinct varieties of pattern line application to the underlying polygonal edges to create the overall design. A Marinid *acute* design from the portal of the Sidi Abu Madyan mosque in Tlemcen, Algeria (1346), is produced from the same underlying tessellation, except with the small hexagons rather than the large hexagons. This design arbitrarily extends the lines within the pentagons to create an eight-pointed star at the center of each repeat unit surrounded by four dart motifs [Fig. 175b].

Perhaps the most distinctive augmentation of the *fourfold system B* by Nasrid and Marinid artists are a category of *acute* designs that add 16-pointed stars into the pattern matrix. These are identical in methodological concept to designs with 16-fold regions of local symmetry that are created with from the *fourfold system A* [Figs. 165–168]. In fact, both rely on a circular arrangement of alternating octagons (or elements that fill the octagon) and hexagons within their set to create the regions with 16-fold symmetry. There are more modules within the expanded *fourfold system A*, and consequently far greater tessellating potential for creating patterns with 16-pointed stars. By contrast, the more limited set of modules within the *fourfold system B* provides less design potential, and explains why there are fewer designs created from this system than from the *fourfold system A*. Nonetheless, the designs with 16-pointed stars produced from the *fourfold system B* are exceptionally elegant, albeit generally less complex. A *zillij* panel from the

Photograph 63 A Nasrid *zillij* mosaic panel from the Alhambra with an *acute* pattern made up of 8- and 16-pointed stars created from the *fourfold system B* (© David Wade)

Alhambra is a representative case in point [Fig. 185a] [Photograph 63]. This *acute* design can be differentiated from similar examples created from the *fourfold system A* by virtue of the pentagons along the edges of the repeat unit. The proportions of these are specific to the *fourfold system B* [Fig. 170]. As with other examples, the underlying tessellation places octagons upon the vertices of the 4.8^2 grid of squares and octagons. This same design, with or without variations, was used in many locations in Spain and Morocco, including a *zillij* panel from the Sa'dian tombs in Marrakesh, Morocco (sixteenth century), and an Alawid stucco ceiling at the Moulay Ishmail mausoleum in Meknès (seventeenth century). A very becoming *acute* pattern created from the *fourfold system B* was used in a wooden ceiling at the Bu 'Inaniyya madrasa in Fez (1350-55) [Fig. 186] [Photograph 64]. This juxtaposes the 16-pointed stars in a very similar fashion as the above-cited *zillij* panel from the Alhambra, except that it cleverly repeats upon a rhombic grid with 16-pointed stars at the vertices, and equally upon

Photograph 64 A Marinid wooden ceiling at the Bu 'Inaniyya madrasa in Fez, Morocco, comprised of an *acute* pattern with 8- and 16-pointed stars that is created from the *fourfold system B* (© David Wade)

the hexagonal dual grid with the 8-pointed stars located at the vertices.

The Nasrid and Marinid use of the *fivefold system* was less pervasive than the two fourfold systems. Nonetheless, many examples were used in the architectural ornament of these allied cultures. However, Maghrebi artists working with this system did not apply the same level of innovation and variation that was a hallmark of their work with the *system of regular polygons* and both fourfold systems. Specifically, the many western examples of patterns created from the *fivefold system* do not employ analogous variations to the primary ten-pointed stars. What is more, the variety of fivefold design was invariably limited to the *acute* family with 36° crossing pattern lines. In these regards, their use of this system was generally less innovative than the fivefold work of the eastern regions. A typical example of a Nasrid fivefold design is from one of the stucco window grilles at the Alhambra. Like so many of the fivefold examples from this region, this is a classic *acute* pattern [Fig. 226c], but has the added feature of the design being modified at the periphery to create a framing parallel line motif, referred to as the *river*, that is typical of the geometric ornament of the

Maghreb.[202] This feature strays noticeably from the pattern lines of the actual *acute* design at the lower portions of the panel where the lines of the framing motif do not conform with the five directions of the pattern lines in the design itself, as well as in the upper arched portion of the window grill were the parallel framing lines within the river are somewhat at odds with the pattern itself. Despite the small sacrifices to the integrity of the *acute* pattern lines that comprise the standard design, this framing device is visually attractive. Two carved stucco arched panels in the courtyard façade of the Bu 'Inaniyya *madrasa* in Fez (1350-55) are decorated with a fivefold pattern that is an exception to the relative lack of Maghrebi innovation with this system. This design places 20-pointed stars at the vertices of the rectangular repeat, as well as at the center of the repeat unit [Fig. 268]. The incorporation of 20-pointed stars into a system that is ordinarily limited to 10-pointed stars is analogous to the Maghrebi introduction of 16-pointed stars into both the fourfold systems. This is a rare design phenomenon,

---

[202] Castéra (1996).

and one of the only other instances of such a pattern is a Mamluk example from the Qadi Abu Bakr ibn Muzhir in Cairo (1479-80) [Fig. 267]. Indeed, this Marinid fivefold pattern is a masterpiece of geometric art. While the overall repeat is a rectangle, the internal repetitive structure is comprised of rhombi and decagons. The pattern within the rhombic elements is the classic *acute* pattern constructed from underlying decagons, pentagons, and barrel hexagons [Fig. 226b]. The use of these two repetitive elements within a broader repeat unit conforms with the relatively small group of fivefold hybrid patterns created within this tradition [Figs. 261–268].

Another notable exception to the relative lack of fivefold innovation in the western region is found in several *zillij* dual-level panels at both the al-'Attarin *madrasa* (1323) [Fig. 476] and Bu 'Inaniyya *madrasa* (1350-55) in Fez [Fig. 474]. While still of the *acute* family, these reach high degrees of added complexity through very broad repeat units comprised of large numbers of underlying polygonal modules. This variety of pattern is characterized by two levels of design that are differentiated through the application of color. Most Maghrebi dual-level designs were created from the *fourfold system A*, but a very few were created from the *fivefold system*. While the visual character is distinct, the methodology of the dual-level tradition in the Maghreb is essentially the same as that of the eastern regions that followed some hundred years hence. The history of this variety of geometric design, in both the western and eastern regions, is examined later in this chapter.

The Marinid and Nasrid use of nonsystematic patterns was almost exclusively limited to designs that repeat upon either the orthogonal or isometric grids. As such, previously originated compound patterns with 8- and 12-pointed stars, and 9- and 12-pointed stars, were commonly used by the local Muslim, Christian and Jewish communities alike. Maghrebi examples of orthogonal patterns with 8- and 12-pointed stars are mostly of the *acute* family [Fig. 379h] and include a series of stucco window grilles from the Alhambra; a *Mudéjar* painted wood panel from the Casa de Pilatos in Seville (sixteenth century); and a series of stucco window grilles at the Sinagoga del Tránsito in Toledo (1360). Similarly, the most common nonsystematic isometric examples are *acute* patterns with 9- and 12-pointed stars, including several cut-tile mosaic panels at the Alhambra, and an illuminated frontispiece from a Moroccan Quran written for the Sharifi Sultan 'Abd Allah ibn Muhammad[203] (1568) [Fig. 346a] [Photograph 65]. The most outstanding Maghrebi example of a nonsystematic geometric design is from two identical adjacent window grilles also found at the Sinagoga del Tránsito. This design is comprised of 8-, 14-,

Photograph 65 A Sharifi illuminated frontispiece from a Quran produced for Sultan 'Abd Allah ibn Muhammad in Morocco comprised of a nonsystematic *acute* pattern with 9- and 12-pointed stars (British Library Board: BL Or. MS 1405, ff. 370v-371r)

and 18-pointed stars that repeat upon a rectangular grid [Fig. 419]. Although it may not appear so at first glance, this is a hybrid design that utilizes two distinct rectangular repeat units, either of which will cover the plane on its own. The complexity of this remarkable design rivals that of the most complex compound patterns from the Seljuk Sultanate of Rum.

In al-Andalus the tradition of using geometric patterns as ornament for domes was roughly contemporaneous with the Mamluk practice: the principle difference being the partiality toward wood rather than either carved stone or stone mosaic, and the application to interior surfaces exclusively. The Islamic geometric pattern in the wooden *artesonado* cupola over the Capilla de Santiago at the Convent of Las Huelgas near Burgos, Spain (late thirteenth century), is an early Andalusian example of a form of ceiling vault that, while not domical, is closely related in aesthetic character and fabricating technology to the true wooden domes that soon followed. This variety of pseudo-dome is comprised of a series of flat trapezoidal panels that connect along their nonparallel edges, and is generally surmounted by a flat

---

[203] London; British Library, Or. 1405, ff. 370v–371r.

Photograph 66   The Nasrid octagonal cupola of the *madrasa* of Yusuf I at the Alhambra with an *acute* pattern produced from the *fivefold system* (© David Wade)

ceiling panel at the apex to enclose the cupola. Multiple examples of this form of truncated pyramidical ceiling are found at the Alhambra, including the square-based cupola from the Hall of the Ambassadors (c. 1354-91) with its highly ornate matrix of 16- and 8-pointed stars; the octagonal cupola of the *madrasa* of Yusuf I (c. 1333-54) with 12- and 8-pointed stars; a small stucco cupola with an *acute* design produced from the *fivefold system* [Photograph 66]; and the 16-sided cupola of the Torre de las Damas at the Palacio del Partal (c. 1302-09) upon which the 16 trapezoids are decorated with 8-pointed stars joined together in a appealing geometric matrix that continues onto the 16-sided flat panel at the apex with a bold 16-pointed star. The multitude of sides in this cupola gives it the feel of a true dome with both vertical and horizontal curvilinearity. The most visually arresting domical geometric ornament in al-Andalus is in the Hall of Ambassadors in the Alcázar of

Seville (1364) [Photograph 67]. This palace was built by Pedro I (Pedro the Cruel) on the site of an earlier Almohad palace, and the geometric dome in the Hall of the Ambassadors is the work of Diego Roiz. This dome is a significant example of the Islamic geometric idiom, even if created by a Christian artist. However, the geometric pattern appears more complex than it actually is, and for all its beauty, it is not without problems. The basic iterative unit is the same rhombic repeat as found in the classic fivefold *acute* pattern [Fig. 226c]. This rhomb has been slightly distorted to fit the curvature of the dome, and is repetitively placed upon the surface of each 1/12 gore segment of the dome. There is a basic problem with this approach: the rhombic repeat units do not fit accurately into the precise curvature of the 1/12 segment, requiring the geometric pattern lines to be inelegantly truncated where they cross the edges of each gore segment. The design only repeats radially

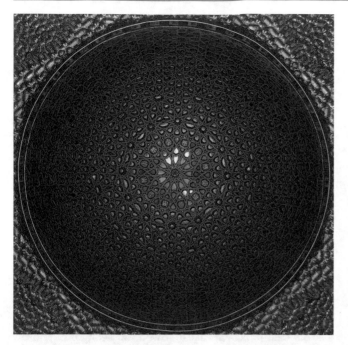

Photograph 67   The *Mudéjar* dome over the Hall of Ambassadors in the Alcázar of Seville with an *acute* pattern produced from the *fivefold system* (© Jean-Guillaume Dumont)

by virtue of each truncated line meeting an identical reflected line along the joint of each gore segment; thereby establishing reflected symmetry and a semblance of order. The one area that falls far short of appearing purposeful is in the midsection of the overall geometric matrix wherein the centers of the 12 horizontally arrayed 10-pointed stars fall just inside the edges of each segment such that, when reflected, an elongated 11-pointed star is thereby created. The frustration caused by this ring of 12 irregular stars is ameliorated by the beauty of the 12-pointed star at the apex of the dome, and the overall beauty of the complete dome. A *Mudéjar* wooden dome at the Casa de Pilatos in Seville (sixteenth century) has a similar ceiling that was doubtless inspired by the example from the Hall of Ambassadors. It is interesting that this later and smaller example suffers from the exact same problem, complete with the central ring of 12 distorted 11-pointed stars. The two Nasrid wooden domes in the projecting porticos of the Court of Lions (c. 1354-91) at the Alhambra are of comparable beauty to the geometric dome in the Hall of Ambassadors in Seville, but far exceed this dome in geometric ingenuity. The geometry of one of these hemispherical domes is particularly interesting in that its repetitive structure is polyhedral rather than based on radial gore segments. The artist who devised this dome worked with a polyhedral subdivision of the surface of the sphere into 6 squares, 40 isosceles triangles, and 8 interstice triangles distributed in rhombic pairs around the hemispherical base whose proportions are determined by the distribution of the squares and isosceles triangles. As detailed by

Emil Makovicky,[204] this polyhedral face configuration has two vertex conditions, and is similar to the octacapped truncated octahedron[205] except that it utilizes two types of triangle (one of which is equilateral) rather than a single non-equilateral triangle. The genius of the geometric pattern in this dome is the placement of regular 11-pointed stars at the vertices of the square and triangular spherical faces. When used as a repeat unit on the two-dimensional plane, the angles of both the triangle and square allow for the placement of 12-pointed stars at their corners: each 90° corner of the square having 3 of the 12 points, and each 60° corner of the triangle having just 2 of the 12 points. Similarly, a rhombus with two 30° acute angles and two 150° obtuse angles will receive a single point of a 12-pointed star at the acute angles, and five points at the obtuse angles. In short, in two dimensions, each of these polygonal faces is compatible with 12-fold symmetry at the vertices. As applied to the sphere, these three projected polygons tessellate together in a manner that creates 11-fold symmetry at both vertex conditions: square with four isosceles triangles $[3 + (2 \times 4) = 11]$, and five isosceles triangles with the acute angle of the rhomb $[1 + (2 \times 5) = 11]$. This is an ingenious solution to the covering of a dome with a geometric design. A similar approach was employed in the geometric ornament of the long wooden barrel vault in the Sala de la Barca (c. 1354-91) that is adjacent to the Hall of the Ambassadors at the Alhambra. The geometric schema of this vault is unique in Islamic architectural history. This barrel vault is capped on each end with half of a hemispherical dome, and the geometric pattern that runs the length of the barrel vault is adapted to seamlessly continue onto the surface of the domical regions at each end of the vault. If one were to remove the barrel vault and bring the two half domes together, a single hemispherical dome would be created with pattern qualities that closely resemble those of the dome in the Court of Lions. In fact, the repetitive square motif in the Sala de la Barca is identical to the central square motif in the dome at the Court of Lions. That said, the polyhedral structure of these two domical designs from the Alhambra are geometrically distinct from one another: the example from the Court of Lions being a distorted octacapped truncated octahedron, and the quarter domes at the Sala de la Barca being based upon the geometry of the rhombicuboctahedron. While the spherical projection of the square repeat units in the dome at the Court of Lions places the 11-pointed stars at the vertices, the example from the Sala de la Barca places the 12-pointed stars at the center of each repetitive square. The placement of the primary stars at the centers of each square repeat unit dispenses with the

---

[204] Makovicky (2000), 37–41.

[205] O'Keeffe and Hyde (1996).

need for the vertices to have 11-fold symmetry, and provides for a simple continuation of the design into the projected equilateral triangles. The similarity in polyhedral concept, and the proximate time and place of these two examples of Nasrid domical ornament is strong evidence of their having been designed and produced by the same individuals.

Another example of Nasrid polyhedral geometric ornament is from the Jineta sword of Muhammad XII, the last Nasrid ruler also known as Boabdil.[206] The hilt of this sword is elaborately ornamented with carved ivory and cloisonné enamel, much of which includes the common star-and-cross pattern that is easily created from the 4.8^2 tessellation of squares and octagons [Fig. 124b]. The spherical pommel at the end of the hilt is ornamented with a three-dimensional corollary of the star-and-cross pattern that is created from a spherical projection of the truncated cube. Each vertex of this polyhedron is comprised of an equalateral triangle and two octagons in a 3.8^2 configuration that is analogous to the two-dimensional 4.8^2 tessellation of squares and octagons. Whereas the underlying square of this two-dimensional tessellation is responsible for the fourfold cross element within the standard star-and-cross design, the underlying triangular component of the spherical projection produces the threefold analogue found in this sword pommel. This is the only known historical example of a spherical geometric design used in immediate proximity to its two-dimensional analogue.

A distinctive feature of the later style of geometric ornament in the western regions was the development and utilization of geometric patterns with star forms placed at key nodal points that have an unusually large number of points.[207] Most of the designs of this variety have fourfold symmetry and repeat upon the square grid. As such, the stellar rosettes will commonly have 16, 20, or 24-points; and stars with higher numbers, such as 40- and 64-points, were also used. Less common are patterns of this type that employ the hexagon as a repeat unit; with large numbered star forms comprised of multiple of 3 (e.g., 18-, 21-, and 24-pointed stars). This variety of geometric design developed after the expulsion of Muslims from Spain following the final *reconquista*, and is mostly associated with the work of Moroccan artists during and after the Alawid dynasty. A typical example of this regional style is a wooden ceiling at the Bahia Palace in Marrakech (nineteenth century) that employs an orthogonal design with 8- and 16-pointed stars, and a large 32-pointed star at the center of the design. Generally, the tradition of Islamic geometric patterns is

remarkably cohesive throughout the Muslim cultures of the Turks, Persians and Arabs.[208] The aesthetic differentiation in the use of geometric patterns between Muslim cultures is primarily determined by four interdependent factors: (1) available or preferred media; (2) specific fabrication technologies; (3) cultural predilections favoring different varieties of geometric pattern; and (4) different conventions for expressing and embellishing the design (e.g., thin or thick line weights, interweaving vs. interlocking characteristics, arbitrary additive elements). The primarily Alawid geometric style that uses higher numbered star forms is one of the few design traditions that is uniquely distinctive in geometric structure and resulting aesthetic character, and is exclusive to a single region and period of time.

1.23 Further Design Developments in the East After the Mongol Destruction

The societal upheaval that followed the Mongol overthrow of the Khwarizmshahs and the Abbasid Caliphate in the first half of the thirteenth century greatly impacted the arts and architecture throughout the regions of Transoxiana, Khurasan, Persia, and much of Iraq. Many great centers of Muslim culture, such as Merv, Samarkand, Bukhara, Herat, Balkh, Tus, and Nishapur experienced extensive destruction. In 1258, Baghdad was sacked by Hulagu Khan (d. 1265), the grandson of Chingiz Khan and founder of the Ilkhanid dynasty. The caliph al-Musta'sim was executed: ending the 500 years of religious authority and political hegemony of the Abbasid Caliphate of Baghdad. The westward expansion of Hulagu Khan's army was eventually checked in 1260 at the battle of Ain Jalut in eastern Galilee by the Mamluk forces of Sultan Baybar I (d. 1277). The number of buildings and works of art destroyed during this period of Mongol upheaval is incalculable, and a detailed and accurate knowledge of the early developmental history of Islamic art and architecture, including geometric patterns, is consequentially forever compromised. The Mongols generally

---

[206] In the collection of the Museo del Ejército, Madrid: no. 24.902.

[207] For more information on this distinctive regional style see:
  –Piccard (1983).
  –Castéra (1996).

[208] The cultural adoption of Islamic geometric patterns as a primary ornamental devise was primarily promulgated under the auspices of Turkic, Persian, and Arab patronage. The ornamental traditions of the Muslim populations in the more peripheral regions of sub-Saharan Africa, southeastern Europe, central Russia, the southern portion of the Indian subcontinent, southeast Asia, and China utilized this design aesthetic to a far less degree. When geometric patterns were employed, they were, more often than not, simplistic and derivative. All of these more peripheral cultures had their own distinctive and rich ornamental traditions that would have satisfied the aesthetic expectations of their artists and patrons. However, it is possible that the wider incorporation of more sophisticated Islamic geometric patterns would have likely appealed to these Muslim cultures had their artists been privy to the very specific design methodology required for their production.

spared those cities that did not resist them and accepted their authority; and even in cities such as Merv in Turkmenistan, where wholesale genocide was tragically employed as a military tactic, there is historical evidence of the lives of some artists having been spared.[209] Those artists who survived the Mongol destruction found themselves living in a world where the established system of patronal support was broken. Many artisans fled their homelands to settle in more stable regions,[210] and it can be assumed that these refugees would have included specialists in the geometric arts. The Seljuk Sultanate of Rum and the Mamluks of Egypt were among the direct beneficiaries of this artistic exodus: the former by virtue of their acceptance of Mongol suzerainty, and the latter by having conclusively repelled the Mongol advance on the Levant and Egypt at Ain Jalut.

As a direct result of the societal chaos that followed the Mongol invasion, the developmental momentum of Islamic geometric design that continued under Mamluks and Seljuk Sultanate of Rum patronage was arrested throughout Transoxiana, Khurasan, and Persia. The early Mongol rulers of the Ilkhanid Dynasty favored Christianity and Buddhism over Islam. It was not until Ghazan Khan succeeded the Ilkhanid throne in 1293 that Islam once again became the state religion of the vast region under Ilkhanid control, and the quintessentially Islamic artistic conventions that preceded the Mongol invasion began to reassert themselves into the new cultural paradigm. The reign of Ghazan Khan eventually reestablished many of the vanquished cities as important centers of Islamic culture, attracting tremendous wealth through public works and active trade, particularly with the Yuan Dynasty of China. Ghazan Khan was an avid and enthusiastic builder, purportedly ordering the building of a mosque and bathhouse in every town: with the proceeds from the bathhouses used to support the mosques. He moved the capitol of his empire to Tabriz, which became an influential center of Islamic arts and culture. His greatest architectural undertaking was the Ghazaniyya (1297-1305): his palace complex in Sham outside of Tabriz. Very little has survived to the present, but in its day it was vast on an unprecedented scale, with "monasteries, *madrasas*, a hospital, library, philosophical academy, administrative palace, observatory and palatial summer residences, as well as arcades and gardens of exceptional charm."[211] The

mausoleum of Ghazan Khan in the Ghazaniyya is reported to have had a richly ornamented dome some 45 m in height, which 14,000 people worked on over a period of 4 years to complete.[212] Ghazan Khan was succeeded by his younger brother, Sultan Uljaytu Khudabanda, who was also a dedicated patron of the arts. The mausoleum of Uljaytu in Sultaniya, Iran (1307-13), was similarly lavish. Much of this building is still standing, and is regarded as the most significant extant Ilkhanid building, and one of the most important examples of Islamic funerary architecture in Iran.

Despite the tremendous Ilkhanid emphasis on architectural projects, the level of geometric sophistication of their architectural ornament did not generally parallel the work of their Mamluk and Seljuk neighbors to the west. Without doubt, the geometric ornament at such buildings as the Sultaniya is of great beauty, and exhibits a distinctive Ilkhanid aesthetic. However, the ornamental originality is primarily in the use of materials and color rather than a pioneering approach to geometric design. As such, their use of geometric patterns was, for the most part, informed by the geometric ornament of their pre-Mongol predecessors, and included patterns made from the *system of regular polygons*, both fourfold systems, and the *fivefold system*. Nonsystematic patterns were also widely utilized, including examples with compound symmetry. Most of these were patterns that had already been used previously by Muslim artists, and tend to be less complex designs such as orthogonal patterns with 8- and 12-pointed stars, and isometric patterns with 9- and 12-pointed stars.

The later Ilkhanids and their Muzaffarid successors in central and southern Persia had an aesthetic predilection for additive geometric patterns. This type of pattern is created by applying additional pattern elements into an existing design, resulting in a heightened level of geometric complexity. This additive practice is relatively simple and does not require particular skill or specialized knowledge, and was used to a limited extent by earlier Muslim cultures: for example, the Seljuk arched panel over the entry door at the Gunbad-i Surkh in Maragha (1147-48). As typical of later Ilkhanid and Muzaffarid additive geometric designs, the additive elements of this Seljuk example are differentiated with color. Among the more significant Ilkhanid additive patterns are several examples from the mausoleum of Uljaytu at Sultaniya (1313-14). These include two designs created from the simple hexagonal grid: a *median* pattern with 90° crossing pattern lines that was also used by Ilkhanid artists at the Khanqah-i Shaykh 'Abd al-Samad in Natanz,

---

[209] The Persian historian Ata al-Mulk Jujayni wrote in his account of the Mongols, *Ta'rikh-i jahan-gusha* (History of the World Conqueror) that the order was given for the whole population of Merv, including women and children, to be put to death except for 400 artisans.

[210] An inscription on a panel of faience mosaic at the Sirçali Madrasa in Konya, Turkey (1242) states that the work was carried out by "Muhammad, son of Muhammad, son of Othman, architect of Tus." See Wilber (1939), 40.

[211] Pope (1965), 171.

[212] Wilber (1955), 124–126.

Iran (1304-25), into which octagons are added at the centers of each intersection of the primary pattern [Fig. 64], and a very basic *median* design with 60° crossing pattern lines that places additional 6-pointed stars at the same centers as the original 6-pointed stars, but rotated 30°, thereby creating 12-pointed stars in an isometric arrangement [Fig. 65]. One of the most outstanding additive designs at the Uljaytu mausoleum is a *median* pattern made from the *fourfold system A* [Fig. 66]. This design repeats on a rhombic grid, and the primary motif on its own was used subsequently by Timurid artists in the Bibi Khanum in Samarkand, Uzbekistan (1398-1404) [Fig. 157b]. An arch spandrel from the Uljaytu mausoleum contains an additive *two-point* design created from a nonsystematic underlying polygonal tessellation of octagons surrounded by coinciding triangles, pentagons and squares [Fig. 331b] [Photograph 68]. This is a rare example of an Ilkhanid design created from an underlying tessellation that appears to have no prior use. Furthermore, this is unusual in that additive patterns are almost always elaborations of patterns that were created from one or another of the generative systems, whereas this example is nonsystematic. The resulting design is arguably the most elaborate Ilkhanid additive pattern.

The use of geometric patterns by Muslim cultures in the regions affected by Mongol conquest, albeit largely derivative of earlier work, is nonetheless refined and beautiful. While lacking the creative vitality and methodological innovation of the contemporaneous work of Egyptian and Anatolian artists, the quality of execution was outstanding. This is especially the case with the increased application of geometric patterns to the burgeoning tradition of cut-tile mosaic that took place among the Muzaffarid, Kartid, Qara Qoyunlu, Aq Qoyunlu, and Timurid successors to the Ilkhanids. These cultures continued the prolific use of systematic geometric methodologies, and innumerable examples are found in diverse media. Designs created from the *system of regular polygons* were especially popular and many fine examples were employed throughout this vast region by succeeding dynasties. The original creation of most of these designs took place during the period of high innovative development in the twelfth and thirteenth centuries, and predates the period of the Mongol destruction. Notable examples of earlier designs created from the *system of regular polygons* that were incorporated into the post-Mongol work in the eastern regions include an Ilkhanid isometric design from the 30-volume Quran (1310)

Photograph 68 An Ilkhanid cut-tile mosaic and stucco arch spandrel from the mausoleum of Uljaytu at Sultaniya, Iran, with a nonsystematic additive *two-point* pattern (© David Wade)

commissioned by Sultan Uljaytu that has the precise proportions of the pattern derivation associated with the 3.6.3.6 underlying tessellation[213] [Fig. 99c], as distinct from the proportions created from the use of the 6^3 hexagonal grid [Fig. 96d]; and a Tughluqid raised relief ceramic panel from the tomb of Shah Rukn-i 'Alam in Multan, Pakistan (1320-24), that has the proportions of the 3.4.6.4 derivation of this otherwise similar design [Fig. 107d] [Photograph 69], as distinct from the proportions produced from other underlying polygonal tessellations [Figs. 97c and 99b]. A *two-point* pattern from a Chaghatayid ceramic relief panel at the mausoleum of Tughluq Temür in the ancient city of Almaliq (present-day Huocheng) in western China (1363). This is created from the 3.4.6.4 underlying tessellation [Fig. 105d] [Photograph 70]. This is very similar in structure to the somewhat more complex Qarakhanid *two-point* pattern at the southern anonymous tomb at Uzgen dating from approximately 350 years earlier [Fig. 105b], as well as a

Ghurid *two-point* pattern from the minaret of Jam (1174-75 or 1194-95) [Fig. 105c]. A Khoja Khanate *two-point* design from the Apak Khoja mausoleum in Kashi, China (c. seventeenth century), employs a wooden window grille created from the 3.6.3.6 tessellation that is identical to a carved stone relief pattern from an Armenian *khatchkar* dating to the fourteenth century [Fig. 100b]. The design of the window grille from Kashi is very similar to a fine pattern created from the 3^2.4.3.4-3.4.6.4 *two-uniform* underlying tessellation that was used in the entry portal of the post-Ilkhanid mausoleum of Muhammad Basharo in the village of Mazar-i Sharif, Tajikistan (1342-43) [Fig. 111a]. This example is a *two-point* pattern that is similar in concept to the above-referenced Chaghatayid design from the mausoleum of Tughluq Temür in western China (Fig. 105d). The mausoleum of Muhammad Basharo has an immediately adjacent second design that is a variation to the 3^2.4.3.4-3.4.6.4 *two-uniform* design [Fig. 111a], except that it replaces the central

Photograph 69 A Tughluqid ceramic panel from the tomb of Shah Rukn-i 'Alam in Multan, Pakistan, with a pattern comprised of superimposed dodecagons that is easily created from the *system of* *regular polygons* (© Aga Khan Trust for Culture-Aga Khan Award for Architecture/Jacques Betant [Photographer])

---

[213]Calligraphed by 'Ali ibn Muhammad al-Husayni in Mosul (1310). British Library, Or. 4945, ff. IV-2r.

Photograph 70 A Chaghatayid ceramic panel from the mausoleum of Tughluq Temür in Huocheng, western China, with a *two-point* pattern that is easily constructed with the *system of regular polygons* (© Daniel C. Waugh)

cluster of underlying triangles, squares, and central hexagon with a single dodecagon [Fig. 111b]. This modification adroitly provides for the attractive 12-pointed star located at the center of the ornamental panel.

Among the Ilkhanid examples of design created from the *system of regular polygons* are two noteworthy Quranic frontispieces that utilize the 3.12^2 polygonal tessellation. The first of these is a *two-point* pattern from a Quran illuminated in Baghdad by Muhammad ibn Aybak ibn 'Abdullah[214] (1306-07) [Fig. 108f]; and the second is a curvilinear design from the 30 volume Quran of Uljaytu,[215] illuminated by 'Abdallah ibn Muhammad al-Hamadani (1313) [Fig. 108c]. Other post-Mongol patterns that feature 12-pointed stars and are created from the *system of regular*

polygons include an Ilkhanid triangular pendentive for one of the vaults at the mausoleum of Uljaytu in Sultaniya that is decorated with the well-known isometric *median* pattern created from the 3.12^2 underlying tessellation [Fig. 108a]; an *obtuse* design in one of the ceiling vaults at this same mausoleum [Fig. 108d]; and a Qara Qoyunlu arched entry portal of the Great Mosque at Van in eastern Turkey (1389-1400) that is decorated with a very nice representation of the equally well-known orthogonal *median* pattern created from the $3.4.3.12$-3.12^2 underlying *two-uniform* tessellation [Fig. 113a]. A Timurid cut-tile mosaic design from the exterior of the Abu'l Qasim shrine in Herat, Afghanistan (1492) employs a *median* pattern created from the $3.4.3.12$-3.12^2 that is very similar to the far more widely used design from the Great Mosque of Van [Fig. 113b]: the difference being in the pattern line treatment within the underlying triangular and square elements. A rare example of a *three-uniform* pattern created from the *system of regular polygons* was used by Qara Qoyunlu artists at the Great Mosque of Van in eastern Turkey (1389-1400) [Fig. 115]. This is a particularly complex *median* pattern created from the 3^6-3^3 $.4^2$-$3^2.4.12$ tessellation.

There were many designs created in the eastern regions after the Mongol destruction that are easily created from the 4.8^2 underlying tessellation. A beautifully executed Chaghatayid ceramic border in the tomb of Tughluq Temür in Almaliq (Huocheng), China (1363), is an example of the classic star-and-cross pattern [Fig. 124b]; and a Timurid example of this same pattern from the Ghiyathiyya *madrasa* in Khargird, Iran (1438-44), is provided with greater complexity by emphasizing the generative tessellation equally with the final pattern [Fig. 126d]. This design also has a secondary eight-pointed star with 45° included angles arbitrarily added inside each eight-pointed star of the *median* pattern. An Ilkhanid illuminated frontispiece to a Quran (1304) employs a version of the standard *obtuse* design with pattern lines that extend into the underlying squares: creating two sizes of octagon within the pattern matrix [Fig. 127a]. Muzaffarid examples of the star-and-cross design include a very fine cut-tile mosaic border at the Friday Mosque at Yazd (1324) that is further developed with additive pattern lines that interweave with the standard design. The Tughluqid use of the star-and-cross pattern at the Adina mosque in Pandua, West Bengal, India (1375) serves as an overall textural background to the exterior façade. The combination of the small scale of design and low-level relief provides the carved stone with a subtle aesthetic unlike that of other Muslim cultures. Mughal artists also made wide use of patterns created from this underlying tessellation, including: a painted mural with the standard *acute* pattern from the tomb of Jahangir in Lahore (1637) [Fig. 124a]; and the simple but elegant red and white octagonal paving at the Taj Mahal in Agra (1632-53) [Fig. 127f].

---

[214] Chester Beatty Library Ms. 1614 (Arberry No. 92).
[215] Cairo National Library; 72, pt. 19.

Both the fourfold systems were less widely used by Ilkhanid and Muzaffarid artists, but regained popularity under the patronage of succeeding dynasties of the Kartids and the Timurids. The many exquisite designs in the ceiling vaults at the Ilkhanid mausoleum of Uljaytu in Sultaniya include the ever-popular *median* pattern created from the *fourfold system A* that was used extensively by Seljuk and Ghurid artists in Khurasan as early as the late eleventh century [Fig. 145]. Ilkhanid artists used this same design in two additional locations: at the Mashhad-i Bayizid Bastami in Bastam, Iran (1300-13), and in the carved stucco *mihrab* of the Imamzada Rabi'a Khatun shrine in Ashtarjan, Iran (1308). Later examples of this design include a very beautiful Mughal inlaid stone panel at the mausoleum of Akbar in Sikandra, India (1613), and a contemporaneous carved stone border that surrounds a window in the Bayt Ghazalah private residence in Aleppo created during the Ottoman period. A Muzaffarid *median* pattern from the Friday Mosque at Kerman (1349) is unusual in that it uses 60° crossing pattern lines [Fig. 144a]. This arrangement is more typical of designs with 6- and 12-pointed stars. An interesting *median* field pattern was created by Qara Qoyunlu artists for the *mihrab* arch spandrels at the Great Mosque at Van [Fig. 138d]. This shares the same variation to the standard pattern line application within the large hexagonal modules as the design in the wooden railing at the Esrefoglu Süleyman Bey in Beysehir, Turkey (1296-97) [Fig. 142]. In fact, these two designs are identical except that the example from the Esrefoglu Süleyman Bey has elegantly incorporated eight-pointed stars within the pattern matrix. Under the Timurids and Shaybanids, as well as the later Safavids and Qarjars, the *fourfold system A* was applied widely to the glazed *banna'i* brickwork façades that feature prominently in the architecture of these cultures. Part of the aesthetic of this brickwork tradition is the emphasis on designs that are comprised of just vertical, horizontal and 45° diagonal lines expressed via the orthogonal layout of the brick modules. This layout provides for smooth edges for the pattern lines that run vertically and horizontally, but stepped edges for those lines that run diagonally. This creates a very distinct visual quality that softens the rigidity of the geometric design. What is more, the limitation to just four directions of pattern line is ideally compatible with the constraints of *median* patterns created from the *fourfold system A*. Innumerable examples of this variety of ceramic ornament were used since the fourteenth century, and typical examples include a Timurid arched panel in the exterior façade of the Bibi Khanum mosque in Samarkand (1398-1405) [Fig. 157b] and part of the Shaybanid exterior façade of the Tilla Kari *madrasa* in Registan Square, Samarkand (1646-60) [Fig. 138c] [Photograph 71]. An especially common practice was the application of this form of *fourfold system A* design to arched tympanums in the back walls of

Photograph 71 Shaybanid polychrome brickwork from the Tilla Kari *madrasa* in Samarkand with a *median* pattern created from the *fourfold system A* (© David Wade)

entry *iwans*. Timurid examples of this form of architectural ornament are found at the Gawhar Shad mosque in Mashhad, Iran (1416-18); the Ulugh Beg *madrasa* in the Registan Square in Samarkand (1417-20); and the Khwaja Akhrar funerary complex in Samarkand (1490). Later Shaybanid examples are found at the Shir Dar *madrasa* in Registan Square, Samarkand (1619-36) [Photograph 72], and the Tilla Kari *madrasa* (1646-60) in the same square in Samarkand. Both of these introduce a central eight-pointed star within a *median* field pattern [Fig. 145]. Among the many Mughal examples of patterns created from the *fourfold system A* is a fine stone mosaic panel from the tomb of I'timad al-Daula in Agra (1622-28) [Fig. 154] [Photograph 73]. The underlying generative tessellation for this *median* design is easily created by placing coinciding octagons at the vertices of the 4.8^2 tessellation of octagons and squares, and infilling the central region with an underlying octagon surrounded by eight pentagons. A particularly eccentric pattern that can be created from an underlying tessellation of

Photograph 72 Shaybanid polychrome brickwork from the Shir Dar *madrasa* in Samarkand with a *median* pattern created from the *fourfold system A* (© David Wade)

eight squares in eightfold rotation around eight-pointed star interstice regions was used in the entry portal of the Task-Kala caravanserai in Konye-Urgench, Turkmenistan (fourteenth century) [Fig. 147b]. This building was built under either Chaghatayid or Sufid patronage, but is in the emerging Timurid style. The arrangement of squares in the underlying tessellation is identical to that of a pattern used more than a hundred years previously at both the Çifte *madrasa* in Kayseri, Turkey (1205), and the Friday Mosque in Gonabad, Iran (1212) [Fig. 147a]. The application of the pattern lines to the underlying tessellation in the example from Konye-Urgench is highly unusual, and not in keeping with the standard methodological practices associated with the *fourfold system A*. The pattern is initiated by first placing regular hexagons at key locations within the eightfold geometric structure, and extending the lines of these hexagons until they meet with other extended lines. While aesthetically pleasing, the resulting pattern is visually distinct from more overtly systematic orthogonal designs. An example of a Timurid design created from the *fourfold system A* that repeats on a rectangular grid is found in a border design at the Shah-i Zinda complex in Samarkand [Fig. 164] [Photograph 74]. This *median* pattern places eight-pointed stars at

the vertices of the rectangular grid, as well as at the vertices of the rectangular dual grid. In fact, the geometric information contained within the repeat unit and the dual repeat unit are identical.

Designs created from the *fourfold system A* were occasionally given an additive treatment that incorporates a swastika device within square components of the pattern matrix. This variety of additive variation was used to a limited extend by Seljuk artists working in the Sultanate of Rum, but greater variation and ingenuity was employed by post-Mongol artists, particularly during the Timurid period. A fine, if rather predicable, example encompasses the marble shaft of a column found at the Gawhar Shad *madrasa* and mausoleum in Herat (1417-38) [Fig. 150b]. The Topkapi Scroll illustrates several examples of this variety of additive feature. One is particularly interesting in that it repeats upon a rhombic grid[216] [Fig. 157a]. The designer of this pattern used the implicit squares contained within the elongated hexagonal modules in the pattern to incorporate the swastika motif.

---

[216] Necipoğlu (1995), diagram number 67.

Photograph 73 A Mughal inlaid stone panel at the tomb of I'timad al-Daula in Agra, India, with a *median* pattern created from the *fourfold system A* (© David Wade)

Examples of the Ilkhanid use of the *fourfold system B* include the classic *acute* pattern created from underlying octagons and pentagons located in another of the ceiling vaults at the mausoleum of Uljaytu in Sultaniya (1307-13) that was used some 175 years previously at the Friday Mosque at Sin, Iran (1134) [Fig. 173a], and an *acute* pattern created from the underlying tessellation of octagons, pentagons, and elongated hexagons in the arched tympanum over the entry door of the round tomb tower of Hulagu Khan's sister in Maragha, northeastern Iran (thirteenth century) [Fig. 177c]. A Muzaffarid design from the exterior façade of the Friday Mosque at Kerman (1349) has a *median* pattern created from an underlying tessellation that is essentially the same accept that it uses the small hexagons rather than the large hexagons from this system [Fig. 175c]. A painted fresco in the mausoleum of Shaykh Ahmed-i Jam at Torbat-i Jam in northeastern Iran (1442-45) is a fine Kartid design that utilizes the variation to the pattern lines within the hexagon that allow for regular octagons within the design [Fig. 179a] [Photograph 75]. Timurid examples created from this system are mostly derivative of earlier work, and include: several mosaic panels with the classic *acute*

pattern at the Abdulla Ansari complex in Gazargah near Herat, Afghanistan (1425-27) [Fig. 173a] [Photograph 76]; and a rhombic *acute* pattern produced in carved stucco from the mausoleum of Amir Burunduq in the Shah-i Zinda in Samarkand (1390) [Fig. 181]. This was used earlier by artists in the Seljuk Sultanate of Rum at the Izzeddin Keykavus hospital and mausoleum in Sivas (1217-18). The Mughals made occasional use of the *fourfold system B*, and most of these examples are the classic *acute* design produced from just the underlying tessellation of octagons and pentagons [Fig. 173a]. A distinctive example in high-relief carved stone is found in the Agra Fort (1565-73). However, the most remarkable Mughal use of this classic *acute* design is a marble *jali* screen from the tomb of Salim Chishti at Fatehpur Sikri (1605-07) [Photograph 77]. This is one of several Mughal *jali* screens that prominently portray the generative tessellation as part of the completed design. In addition to being stunningly beautiful, these are important examples of historical evidence for the use of the polygonal technique as a traditional design methodology.

The use of the *fivefold system* enjoyed continued popularity throughout the eastern regions during this period, and

Photograph 74 Timurid cut-tile mosaic ornament from the Shah-i Zinda complex in Samarkand with a *median* pattern created from the *fourfold system A* (© David Wade)

included examples of established designs in all four of the pattern families, as well as original patterns with diverse repeat units comprised of a relatively large number of underlying polygonal modules. Artists working during the period that followed the Mongol destruction were evidently very familiar with the subtleties of the *fivefold system*; as evidenced by the large number of original patterns created from this system, and the high level of innovation that was applied to these designs. As pertains to the tradition of Islamic geometric patterns, the mausoleum of Uljaytu in Sultaniya, Iran (1307-13), is the most important surviving building from the period of Ilkhanid reconstruction, and several very successful fivefold designs were produced for this monumental tomb. A particularly beautiful example is an unglazed raised brick geometric design with unglazed ceramic cast relief inserts in the background [Fig. 246]. The raised relief geometric design, and the cast inserts are elegantly separated by a thin outline of *lājvard* (dark blue) glazed ceramic. This Ilkhanid example is an *obtuse* pattern

that can be created with equal ease from either of two dualing underlying tessellations. The design methodology of an Ilkhanid faïence mosaic panel at the Gunbad-i Gaffariyya in Maragha (1328) is unusual in that the pattern lines are placed upon the vertices of the underlying tessellation rather than their standard placement upon the midpoints [Fig. 259]. The resulting design is equally unusual, and while the pentagrams are akin to those of the standard *acute* pattern, their vertex-to-vertex orientation within each underlying pentagon is virtually unique,[217] and does not conform to any of the four pattern families. Another atypical aspect of this design is the convergence of multiple non-coincident pattern lines upon a single point rather than the standard crossing of two pattern lines. Except for the pattern lines within the underlying decagonal modules, this feature disallows the pattern lines from interweaving with one another. A Muzaffarid *acute* design that surrounds the north portal of the courtyard of the Friday Mosque at Kerman (1349) has a particularly large rectangular repeat unit with considerably more underlying polygonal modules than was typically employed within this tradition [Fig. 254c]. This is a supremely successful fivefold Islamic geometric pattern. A Kartid cut-tile mosaic panel at the Shamsiya *madrasa* in Yazd (1329-30) employs a *median* pattern created from an unusual underlying tessellation of modules in the *fivefold system* [Fig. 236]. Rather than the standard underlying decagon, the decagons in this example are proportioned to the width of the short half of the underlying wide rhombus from the *fivefold system*. The matrix of the resulting *median* pattern has distinctive large decagons located at each vertex of the repeat unit, and the ten-pointed stars within these decagons are as becoming as they are atypical. This same distinctive *median* design was used by Timurid artists on an arch soffit at the Ulugh Beg *madrasa* in Samarkand (1417-20) approximately 100 years later. The unusual qualities of this design are unlikely to have been independently derived, and it would appear likely that this design variation was directly influenced by an Anatolian design from the Sultan Han in Kayseri (1232-36) that applies the same treatment to the ten-pointed stars [Fig. 237] [Photograph 42]. The identical underlying decagonal condition, with its distinctive ten-pointed star, was also used in a Timurid *median* border pattern in the cut-tile mosaic ornament at the Imam Reza shrine complex in Mashhad, Iran (fourteenth century) [Fig. 253]. This design is made more dynamic by alternating the unconventional

---

[217] Another example of this unusual arrangement of acute five-pointed stars is found in the pattern that fills the tympanum of the arched entry portal at the hospital of the Great Mosque of Darussifa in Divrigi, Turkey (1228–29): although this Mengujekid pattern is simplistic by comparison.

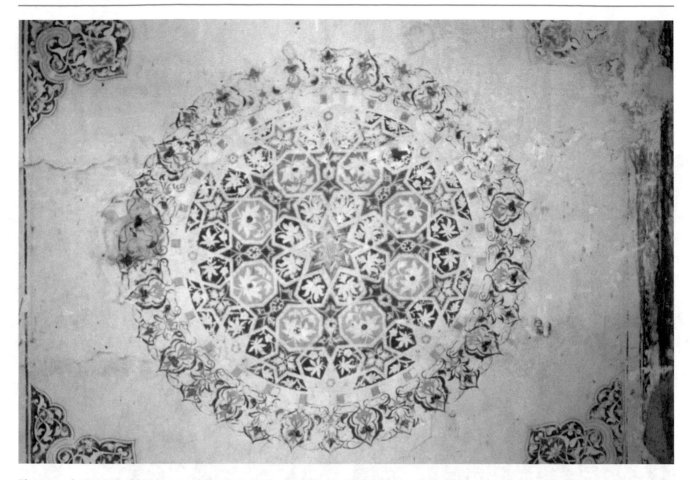

Photograph 75   A Kartid fresco roundel at the mausoleum of Shaykh Ahmed-i Jam at Torbat-i Jam in northeastern Iran that employs a *median* pattern created from the *fourfold system B* (© Sheila Blair and Jonathan Bloom)

ten-pointed stars with those created from the standard decagon from this system. The Anatolian example from the Sultan Han, together with the above cited Kartid example from the Shamsiya *madrasa* in Yazd and this design from Mashhad are the only known fivefold patterns that utilize this unusual decagonal feature within the underlying generative tessellation, and this rarity suggests they may share a common origin, perhaps through association with the same *tumar*. A comparatively simple Timurid *acute* pattern that repeats upon a rectangular grid was used in a cut-tile mosaic panel at the Shah-i Zinda complex in Samarkand (fourteenth century) [Fig. 254]. While far less complex, this is nonetheless an elegant design that, surprisingly, was not more widely used. A panel from a Shaybanid wooden door at the Kukeldash *madrasa* (1568-69) in the Lab-i Hauz complex in Bukhara employs a very successful *acute* pattern with a comparatively large number of polygonal modules used in its underlying generative tessellation [Fig. 256] [Photograph 78]. This design includes the distinctive partial ten-pointed star motifs, in this case 2/10 and 3/10 [Fig. 196], that are a frequent feature of fivefold *acute* patterns that originated during the later period in the eastern regions. This same

design was also used on a wooden door panel produced during the Janid Khanate at the Bala Hauz mosque in Bukhara (1712). A similarly proportioned rectangular repeat is found on a Mughal inlaid stone *acute* border design from the mausoleum of Akbar in Sikandra, India (c. 1612) [Fig. 255]. This design places ten-pointed stars at the vertices of the rectangular repeat, as well as at the center of the repeat unit. As is often the case with such patterns, at first glance, this arrangement of primary star forms gives the impression of repeating on a rhombic grid. However, the tenfold radii at these locations do not align with one another; and their skewed orientation causes the repetitive structure to be more accurately defined as rectangular. This design has the distinctive and unusual feature wherein the geometric information contained within the rectangular repeat unit is identical to that of the dual repeat. A very beautiful inlaid stone panel from the mausoleum of Humayun in Delhi (1562-72) is a relatively rare example of the use of *two-point* methodology among Mughal artists [Fig. 245b] [Photograph 79]. One of the pierced marble *jali* screens at the I'timad al-Daula in Agra (1622-28) makes use of a fascinating   rendering   of   the   classic   *acute*   pattern

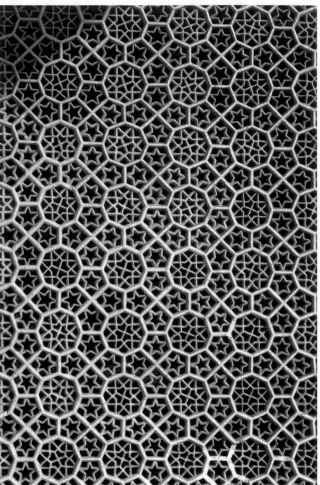

Photograph 76 Timurid cut-tile mosaic ornament from the Abdulla Ansari complex in Gazargah, Afghanistan, with an *acute* pattern created from the *fourfold system B* (© Thalia Kennedy)

Photograph 77 A Mughal pierced marble *jali* screen at the tomb of Salim Chishti at Fatehpur Sikri, India, that employs an *acute* pattern created from the *fourfold system B* along with its underlying generative tessellation (© David Wade)

[Fig. 226c] that is one of the most widely used patterns created from the *fivefold system*. This example is unusual in that it prominently incorporates the generative tessellation with the standard *acute* pattern into the finished screen. This Mughal exposure of the underlying generative tessellation is also found in the above-cited *fourfold system B* design from the mausoleum of Salam Chishti in Fatehpur Sikri (1605-07) [Photograph 77], and like the earlier example from Fatehpur Sikri, this is an important piece of historical evidence for the use of the polygonal technique in generating Islamic geometric patterns.

In addition to patterns that repeat upon both rhombic and rectangular grids, artists working in the eastern regions following the Mongol destruction occasionally applied the *fivefold system* to patterns with radial symmetry. Among the most common are the secondary patterns of dual-level designs that are incorporated into the primary pattern elements with five or tenfold rotational symmetry; for example pentagons, decagons, five-pointed stars, and ten-pointed

stars. Many examples of this variety of fivefold radial design were used in the dual-level designs produced by Timurid and Qara Qoyunlu artists [e.g. Fig. 453]. The Topkapi Scroll is replete with examples of this form of radial design application [Fig. 22]. Another type of radial design places two-dimensional patterns created from the *fivefold system* onto domical surfaces. This method of domical geometric ornament makes use of eight 1/10 segments of a tenfold pattern for application to the eight gore segments of a dome. The resultant distortion is minimal and undetectable to the eye. Ernest Hanbury Hankin first identified this form of domical ornament when writing about the Samosa Mahal at Fatehpur Sikri, India[218] (sixteenth century) [Fig. 21]. The Safavid exterior ceramic ornament of the large dome at the Mashhad-i Fatima in Qum (c. 1519) employs the same

---

[218] Hankin (1925a), Figs. 45–50.

Photograph 78 Shaybanid wood joinery from a door panel at the Kukeldash *madrasa* in Bukhara, Uzbekistan, that employs an *acute* pattern created from the *fivefold system* (© Thalia Kennedy)

decorative methodology in its lower portion, but breaks from the *fivefold system* in the upper quarter as it approaches the apex. The Topkapi Scroll appears to have another example of this form of fivefold domical ornament, although only the 1/10 decagonal triangle is represented. Without any associated text, it is impossible to know for certain whether this was intended for use on a dome[219] [Fig. 260e].

The method of employing more than one repetitive cell, each with its own geometric pattern, into a single larger hybrid fivefold construction was a practice first developed by Seljuk artists in Persia, and later in Anatolia [Figs. 261–265]. Still later, Mamluk and Marinid artists engaged in this practice to a lesser extent, but with exceptional results [Figs. 267 and 268]. Mughal artists also experimented with fivefold hybrid designs, although such work is comparatively rare. A fine example was used in the stone mosaic façade of the I'timad al-Daula in Agra (1622-28) [Fig. 266]. This exceptional *acute* design has the unusual characteristic of having regions within the design that have rotational point symmetry. The hybrid repetitive cells that comprise this design are of three types: a rectangle that includes the point symmetry, a rhombus with 72° and 108° angles, and a half rhombus.

Ilkhanid artists devised an additive treatment to fivefold patterns that was popularly adopted by artists in several subsequent eastern dynasties. This variety of additive pattern places arbitrary pattern lines into the standard design in such

---

[219] Necipoğlu (1995), diagram number 90a.

Photograph 79 A Mughal inlaid stone panel at the mausoleum of Humayun in Delhi with a *two-point* pattern created from the *fivefold system* (© David Wade)

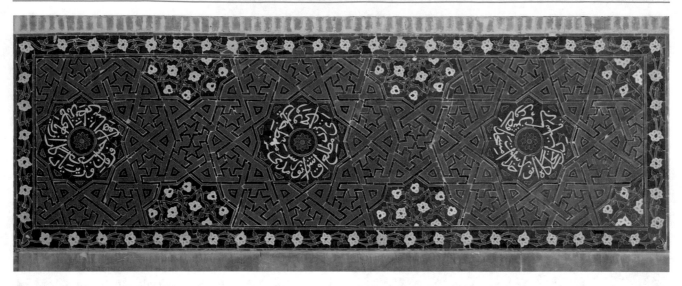

Photograph 80 Muzaffarid cut-tile mosaic ornament at the Friday Mosque at Yazd, Iran, that employs an additive variation of an *obtuse* pattern created from the *fivefold system* (© Jean-Marc Castéra)

manner as to fill the background regions with a meandering mazelike device. An early example of this variety of additive design that employs the classic fivefold *acute* pattern as its starting motif was used in a mosaic panel at the mausoleum of Uljaytu in Sultaniya [Fig. 226d]. This fivefold additive design is similar in concept to the fourfold patterns with additive swastikas that were also popular among the later eastern cultures in Persia, Khurasan, and Transoxiana [Figs. 150a, b and 157a]. A Muzaffarid cut-tile mosaic panel from the Friday Mosque at Yazd (1324) [Photograph 80] is an outstanding example of an additive *obtuse* design with the swastika aesthetic [Fig. 230]. This same design was used many centuries later by Safavid artists at the Shah mosque in Isfahan (1611-38), and an example is included in the repertoire of designs illustrated in the Topkapi scroll.[220] And just as Seljuk artists in Anatolia were among the first to develop this variety of fourfold additive design, so also were they early developers of the use of swastika additive elements within the *fivefold system*.[221]

An artist working during the Ottoman period employed the *fivefold system* to create a rather remarkable design wherein the individual polygonal modules that comprise the underlying generative tessellation transition between two distinct scales. This unusual design technique was used in a door panel from the Sultan Bayezid II Kulliyesi in Istanbul (1501-06) [Fig. 270]. The use of differently scaled polygonal modules within a single generative tessellation is conceptually the same as the earlier Seljuk example from the Hekim Bey mosque in Konya (1270-80) [Fig. 269],

although the scaling factor of the Ottoman example is considerably larger. While the earlier Seljuk design transitions between *acute* pattern lines in the smaller modules and *median* pattern lines within the larger polygonal modules, the Ottoman example employs *acute* pattern lines within the smaller underlying polygons and *two-point* pattern lines within the larger underlying polygons. Due to the use of the 72° angular openings in the *two-point* pattern line application, the Ottoman example includes five-pointed stars typically associated with the *median* family. In this way, this design contains pattern characteristics of the *acute*, *two-point*, and *median* families within a single construction.

By far the most sophisticated designs to be created from the *fivefold system* in the post Mongol eastern regions are a series of highly complex dual-level patterns produced under Qara Qoyunlu and Timurid patronage. These designs also employ two scales of generative polygons; although rather than the modules transitioning between scales within a single tessellation, these dual-level designs apply a smaller secondary tessellation to an already created primary pattern—thereby creating the secondary pattern within the overall design. The history of this class of Islamic geometric design is examined later in this chapter, and the methodology is detailed in Chap. 3.

Examples of geometric patterns created from the *sevenfold system* that originate from the eastern regions following the Mongol destruction are quite rare. Among the relatively few is a carved stucco relief panel from the Timurid mausoleum of Amir Burunduq at the Shah-i Zinda complex in Samarkand (1390-1420) [Fig. 286b, c]. This *median* pattern can be produced from either of two distinct underlying tessellations. Mamluk artists utilized a very similar generative schema in at least two locations: a 14-*s2 obtuse* design

---

[220] Necipoğlu (1995), diagram no. 8.
[221] Schneider (1980), pattern no. 73.

from the Qawtawiyya *madrasa* in Tripoli, Lebanon (1316-26) [Fig. 286a], and a design from the Amir Qijmas al-Ishaqi mosque in Cairo (1479-81) that is identical to the earlier Timurid example. A fine example of 14-*s*3 pattern created from this system was used in several of the deeply recessed blind arches in the courtyard of the Timurid shrine complex of Imam Reza in Mashhad, Iran (1405-18) [Fig. 293c]. The pattern line application to the pentagons and barrel hexagons is analogous to the *median* design within the *fivefold system*. Ottoman artists also produced fine patterns from the *sevenfold system*. These include a door panel from the Bayezid Pasa mosque in Amasya, Turkey (1414-19) [Fig. 287c]. This example uses a subtractive variation that produces a distinctive trefoil motif. This design, without the subtractive treatment, was also used by Mamluk artists in one of the side panels of the *minbar* at the Sultan Barsbay complex at the northern cemetery in Cairo (1432), as well as in the entry door of the Hanging Church (al-Mu'allaqa) in Cairo [Fig. 287b]. Another fine Ottoman example is a 14-*s*6 *acute* design from the incised marble ceiling in the small rectangular water feature within the courtyard of the

Suleymaniya mosque in Istanbul (1550-55) [Fig. 289] [Photograph 81].

The significant innovations in creating evermore complex and varied nonsystematic geometric patterns among artists working under Mamluk and Sultanate of Rum patronage was not, for the most part, equaled by artists in the eastern regions following the Mongol destruction. Even with the reestablishment of societal stability during the fourteenth century, the post-Ilkhanid cultures of Transoxiana, Khurasan, Persia, and Iraq, never placed the degree of emphasis upon highly complex geometric design as practiced contemporaneously by their fellow artists in Egypt and Anatolia. It can be assumed that in the wake of the loss of methodological knowledge following the Mongol destruction, the necessary skills for creating highly complex nonsystematic designs were slow to return to these eastern regions. The Ilkhanids and their successors relied heavily upon systematic design methodologies, and the post-Mongol eastern examples of nonsystematic patterns are mostly recreations of existing designs rather than expressive of an innovative spirit. This general de-emphasis toward complex nonsystematic geometric design continued into the

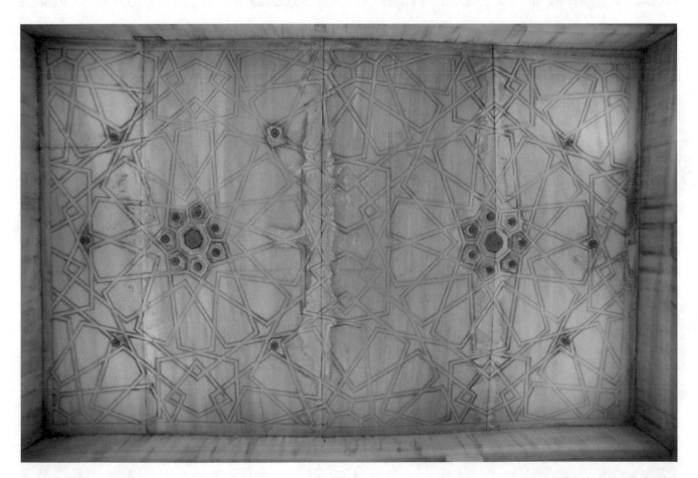

Photograph 81 An Ottoman incised stone ceiling in a water feature of the courtyard at the Suleymaniya mosque in Istanbul that employs an *acute* pattern created from the *sevenfold system* (© Serap Ekizler Sönmez)

gunpowder dynasties of the Ottomans, Safavids, and Mughals, well after cultural stability and trade throughout these regions had allowed for the aesthetic predilections and artistic practices of neighboring cultures to be more widely known. It is reasonable to speculate that the relative paucity of especially complex nonsystematic patterns in the eastern regions following the Mongol destruction was as much to do with an aesthetic preference for more easily ascertained geometric constructions as with a loss in methodological knowledge. Indeed, these two conditions would appear to be intimately entwined.

Among the previously originated post-Mongol nonsystematic patterns with isometric symmetry are many with a single primary star form, and many with multiple primary star forms. Examples of the former include a *median* pattern with 12-pointed stars in one of the ornate Ilkhanid vaults at the mausoleum of Uljaytu in Sultaniya, Iran (1307-13) [Photograph 82], created from an underlying tessellation of just dodecagons and pentagons [Fig. 300a *acute*]. One of the most remarkable post-Mongol examples of this same design is from a pierced *jali* screen in one of the marble brackets at the tomb of Salim Chishti at Fatehpur Sikri (1605-07). This example is significant in that it overtly includes the generative tessellation within the finished screen. As such, this is an important source of historical evidence for the nonsystematic use of the polygonal technique as the preeminent geometric design methodology. Another isometric example with 12-pointed stars in one of the ceilings at the mausoleum of Uljaytu can be created from either of two underlying tessellations. The first separates the underlying dodecagons with barrel hexagons and places three contiguous pentagons at the center of each triangular cell that are truncated into trapezoids [Fig. 321j]. This design can also be created from the 3.12^2 tessellation of triangles and dodecagons [Fig. 108d]. A Qara Qoyunlu *obtuse* design created from the same nonsystematic underlying tessellation was used in a cut-tile mosaic wainscoting within the *iwan* of the Imamzada Darb-i Imam in Isfahan (1453). The standard design was modified such that the 12-pointed stars within each underlying dodecagon become 6-pointed stars, providing the visual quality of a field pattern [Fig. 321f]. An almost identical example of this modified design is found at the Sultan Qala'un funerary complex in Cairo (1284-85). Multiple examples of a design the employs nine-pointed stars at the vertices of the hexagonal grid, and six-pointed stars at the center of each repeat unit were used in the eastern

Photograph 82   An Ilkhanid vault in the mausoleum of Uljaytu in Sultaniya, Iran, with multiple geometric designs (© Daniel C. Waugh)

Photograph 83 Shaybanid cut-tile mosaic ornament from the Kukeltash *madrasa* in Bukhara, Uzbekistan, with a nonsystematic *obtuse* pattern comprised of six- and nine-pointed stars (© Thalia Kennedy)

Photograph 84 A Timurid cut-tile mosaic panel at the Abdulla Ansari complex in Gazargah, Afghanistan, that employs a nonsystematic threefold *median* pattern with 6- and 18-pointed stars (© Thalia Kennedy)

regions during this later period. A nonagonal design that places nine-pointed stars upon the vertices of the hexagonal grid was used by Shaybanid artists at both the Kukeltash *madrasa* in Bukhara (1568-69) [Photograph 83], and the Tilla Kari *madrasa* in Samarkand (1646-60) [Fig. 313a]. The underlying generative tessellation for this *obtuse* pattern is the same by Mamluk artists for creating a *two-point* pattern at the Ashrafiyya *madrasa* in Jerusalem (1482) [Fig. 313b], as well as an *acute* design produced during the Seljuk Sultanate of Rum and located at the Izzeddin Kaykavus hospital and mausoleum in Sivas (1217) [Fig. 313c]. One of the most remarkable isometric designs with only a single primary star form is from a Timurid cut-tile mosaic panel at the Abdulla Ansari complex in Gazargah near Herat, Afghanistan (1425-27) [Photograph 84]. This pattern places 18-pointed stars upon the vertices of each triangular repetitive cell, and is an original design not known to have been used elsewhere [Fig. 322]. The distinctive visual character of this design appears more complex than the rather simple underlying tessellation might suggest.

Inherited nonsystematic isometric designs with more than one region of local symmetry include multiple examples of the most commonly used design comprised of 9- and 12-pointed stars. One of the ceiling vaults at the mausoleum of Uljaytu includes an *acute* pattern of this variety [Fig. 346a]. The *median* pattern created from this underlying tessellation was also used by Mughal artists in an inlaid stone panel from the Friday Mosque at Fatehpur Sikri (1566) [Fig. 346b], and the *obtuse* pattern created from this underlying tessellation was used by Qara Qoyunlu artists at the Great Mosque at Van in eastern Turkey (1389-1400), by Mughal artists at the tomb of I'timad al-Daula in Agra (1622-28), and by Timurid artists at the Abdulla Ansari complex in Gazargah near Herat, Afghanistan (1425-27) [Fig. 347a] [Photograph 85]. A particularly beautiful Qara Qoyunlu *median* design from the stucco ornament of the Great Mosque of Van in eastern Turkey (1389-1400) places 12-pointed stars at the vertices of the isometric grid, and 10-pointed stars upon the midpoints of each triangular edge

Photograph 85 A Timurid cut-tile mosaic border at the Abdulla Ansari complex in Gazargah, Afghanistan, that employs a nonsystematic threefold *obtuse* pattern with 9- and 12-pointed stars (© Thalia Kennedy)

[Fig. 363] [Photograph 86]. This example is unique to this location. For the points of the 10- and 12-pointed stars to meet, they must be distorted slightly, causing these primary stars to be non-regular. This is an atypical feature of this design tradition, and could be considered a flaw were it not for the strong visual appeal of this design.

Among the many examples of orthogonal nonsystematic designs with a single primary region of local symmetry used by later Muslim cultures in the eastern regions is an outstanding illumination from the celebrated Ilkhanid 30-volume Quran written and illuminated by 'Abd Allah ibn Muhammad al-Hamadani in 1313.[222] This *acute* design places 12-pointed stars upon the vertices of the orthogonal grid and octagons at the centers of each square repeat unit, and is created from an underlying tessellation of dodecagons and two varieties of non-regular pentagons [Fig. 335d]. Like the majority of nonsystematic designs of this post-Mongol period, this design had been used earlier: two examples being from the Great Mosque of Siirt in Turkey (1129),

and the Mu'mine Khatun in Nakhichevan, Azerbaijan (1186). The same underlying tessellation that created this illuminated example was used to produce a *median* pattern at the Friday Mosque in Kerman, Iran, during the Qarjar period [Fig. 336a]. One of the earliest examples of this popular design is from the Artuqid *mihrab* in the Great Mosque of Silvan in Turkey (1152-57). Another especially beautiful Ilkhanid orthogonal design incorporates twelve 7-pointed stars that surround the 12-pointed stars placed upon each vertex of the square grid [Fig. 342]. This is a *median* pattern that was used in one of the ceiling vaults at the mausoleum of Uljaytu in Sultaniya (1313-14) [Photograph 87] and does not appear to have been used elsewhere. The underlying tessellation is comprised of 12 non-regular edge-to-edge heptagons that surround each vertex of the orthogonal grid. These successive rings of heptagons create a cluster of four heptagons surrounding a square at the centers of each repeat unit. The application of the pattern lines to this underlying star produces a 12-pointed star that shares an aesthetic treatment with many patterns in the eastern regions.

Preexisting nonsystematic patterns with multiple regions of local symmetry were also frequently employed in the

---

[222] This Quran is often given the appellation of the Uljaytu Quran. National Museum, Cairo; 72, part. 22.

Photograph 86 Qara Qoyunlu carved stucco ornament at the Great Mosque at Van in eastern Turkey that employs a nonsystematic three-fold *median* pattern comprised of 10- and 12-pointed stars (photo by Walter Bachmann, courtesy of the Aga Khan Documentation Center at MIT)

eastern regions following the Mongol invasion. Among the most common are designs with 8- and 12-pointed stars, and noteworthy examples include two Ilkhanid *obtuse* patterns [Fig. 381b]: one from a ceiling vault at the mausoleum of Uljaytu at Sultaniya (1307-13), and the other from a cut-tile mosaic border in the entry portal of the Gunbad-i Gaffariyya in Maragha (1328). An Ilkhanid variation of this *obtuse* design, also at the mausoleum of Uljaytu, modifies the 12-pointed stars so that they become 6-pointed [Fig. 381e]. This sixfold modification is analogous to the more common convention established within the *fivefold system* [Fig. 224a]. The standard *obtuse* design was also used in the Qarjar compound entry portal of the Aramgah-i Ni'mat Allah Vali shrine in Mahan, Iran, and by Mughal artists at the tomb of Akbar in Sikandra, India (1612). *Acute* examples [Fig. 379] created from this same underlying tessellation include a Muzaffarid cut-tile mosaic panel from the Friday Mosque at Yazd (1324), and a Kartid painted ceiling at the Shamsiya *madrasa* in Yazd (1329-30) that employs an atypical curvilinear treatment within the 12-pointed stars. The Friday Mosque in Yazd also includes a *median* pattern created from this underlying tessellation [Fig. 380b].

Despite the preponderance of nonsystematic designs with earlier origins, the later orthogonal patterns from the eastern regions that have more than one variety of local symmetry include a number of very beautiful examples that appear to be original rather than recreations of earlier work. The In'juid tympanum in the east portal of the Friday Mosque at Shiraz (1351) is decorated with an unusual *median* design comprised of 8- and 12-pointed stars that is created from an underlying tessellation of dodecagons separated by elongated hexagons along the edges of the square repeat unit, and a central array of eight rhombi that collectively create an 8-pointed star at the center of the underlying tessellation [Fig. 384a]. The neighboring Muzaffarids used a variation of this same unusual pattern some 15 years later in a cut-tile mosaic border at the Friday Mosque at Yazd (1365), and there is also a representation of this pattern included in the Topkapi scroll[223] [Fig. 384b]. Considering the relative proximity in time and place, it is likely that there was a direct causal influence of the earlier upon the latter. Another fine example with 8- and 12-pointed stars is a *median* pattern from the Ulugh Beg *madrasa* in Samarkand (1417-20) [Fig. 386] [Photograph 88]. This Timurid cut-tile mosaic panel has several similarities with the previous example from Shiraz: specifically the underlying polygonal origin of the eight-pointed stars within the central regions. However, the underlying tessellation for this design contains many more polygonal elements, and the resulting pattern has significantly greater geometric information within the square repeat unit. This lovely orthogonal pattern shares distinctive characteristics with a fivefold design that was also used at the Ulugh Beg *madrasa* [Fig. 236]. Despite the differences in their respective symmetry, both of these *median* patterns employ regions in their underlying tessellations comprised of four coincident rhombi; both have principle regions of local symmetry that are separated in the same fashion by two underlying pentagons that are rotated so that their vertices are orientated toward the centers of the neighboring primary polygons rather than their edges; and the applied pattern lines within these underlying star forms in both designs have the same atypical visual character. There can be no doubt that the artist responsible for these two exceptional designs employed them within the same building with full knowledge of the geometric concordance between these otherwise disparate varieties of geometric design. A Timurid cut-tile mosaic panel from the main entry *iwan* at the Ulugh Beg *madrasa* in Samarkand (1417-20) uses an orthogonal *acute* design that places 16-pointed stars at the vertices of the square grid and 8-pointed stars at the centers of each repeat unit [Photograph 89]. The underlying tessellation for this design places a ring of pentagons around both the octagon

---

[223] Necipoğlu (1995), diagram no. 72d.

Photograph 87 An Ilkhanid vault in the mausoleum of Uljaytu in Sultaniya, Iran, with a fourfold nonsystematic *median* pattern comprised of 7- and 12-pointed stars (© Daniel C. Waugh)

and 16-gon [Fig. 389b]. A more complex orthogonal design with 16-pointed stars placed at the vertices of the square grid is illustrated in the Topkapi Scroll.[224] This remarkable *median* pattern incorporates four 13-pointed stars within the field of each square repeat unit [Fig. 398]. The matrix of edge-to-edge polygons that connect the tridecagons (13-gons) and hexadecagons (16-gons) is comprised of barrel hexagons and pentagons.

At least three examples of nonsystematic hybrid designs were produced in the eastern regions following the Mongol invasion. A Jalayirid border design that surrounds the arch and arch spandrel at the Mirjaniyya *madrasa* in Baghdad (1357) is cleverly comprised of both square and triangular repeat units. The geometric patterns within each of these repeat units were well known at the time that this was constructed, but their combined use within a single design was unusual. The triangular repetitive element contains an *acute* pattern created from the underlying tessellation of dodecagons surrounded by a ring of 12 pentagons, three of which are clustered at the center of the repeat [Fig. 300a

acute], and the square elements contain an *acute* pattern with the same edge configuration of dodecagons and pentagons within its underlying tessellation, and a cluster of four coincident pentagons at the center [Fig. 335b]. This Jalayirid hybrid design was likely inspired by the late Abbasid hybrid pattern in the carved stucco ornament in the entry portal of the Mustansiriyah *madrasa* in Baghdad (1227-34). This earlier Baghdadi example is conceptually similar in its combined use of regular triangular and square repetitive elements to populate the border that surrounds the arched entry. However, the design of the earlier Abbasid example uses different patterns within the two repetitive elements, both of which can be used independently to cover the plane through translation symmetry. The triangular elements conform with the well known *acute* design that is created from and underlying tessellation of dodecagons separated by barrel hexagons, with three pentagons clustered at the center of the triangle [Fig. 321b], while the pattern within the square element is derived from an underlying tessellation that shares the same edge configuration, but includes an octagon at the center of the square that is surrounded by eight pentagons [Fig. 379f]. It is interesting to note that the *acute* hybrid design from the earlier Abbasid example is

Photograph 88 A Timurid cut-tile mosaic border at the Ulugh Beg *madrasa* in Samarkand that employs a nonsystematic *median* pattern made up of 8- and 12-pointed stars (photo by Hatice Yazar, courtesy of the Aga Khan Documentation Center at MIT)

essentially identical to one of the other post-Mongol hybrid designs from the eastern regions: an example from the Topkapi Scroll[225] [Fig. 23d]. The only difference between these historical examples is in the angle of the applied pattern lines within the pentagonal elements of the underlying tessellation: the version from the Topkapi Scroll having angles that are more readily associated with the *median* pattern family. The prominent arc that runs through the illustrated example in the Topkapi Scroll suggests the intended use within an arched tympanum. The layout of the triangles and squares follows the $3^2.4.3.4$ semi-regular tessellation wherein mirrored triangles are placed in rotation around each square [Fig. 89]. The third hybrid design from the post-Mongol period in the eastern regions is also found in the Topkapi Scroll.[226] This uses the same arrangement of

triangles and squares, but with a much simpler application of pattern lines into these repetitive elements [Fig. 23g]. The pattern contained within the triangle is identical to the most basic isometric *median* design governed by 90° crossing pattern lines [Fig. 95c], while the pattern within the square elements is identical to the classic star-and-cross *median* pattern that is ubiquitous to this tradition. These two elements work together by virtue of their both placing 90° crossing pattern lines at the midpoints of each repetitive module, and when placed together in this fashion, non-regular seven-pointed stars are produced at each vertex of the repetitive grid. The earliest known use of this simple hybrid design is from the Malik mosque in Kerman, Iran (eleventh century).

Although less common than isometric and orthogonal nonsystematic patterns, artists working in the post-Mongol eastern regions produced a variety of noteworthy nonsystematic designs with less typical repeat units. While fewer in number than found in the work of their Seljuk neighbors in Anatolia, the level of beauty and sophistication occasionally rivaled those from the Sultanate of Rum. Unlike the especially complex nonsystematic designs created by Seljuk artist in Anatolia, patterns with greater complexity from the eastern regions rarely have more than two primary varieties of local symmetry. As in earlier Muslim cultures, this variety of design utilizes a diverse range of repetitive schema that includes rectangles, rhombi, and radial symmetries. The use of non-regular hexagons as repeat units does not appear to have been practiced in the post-Mongol eastern regions. The Mughal inlaid stone ornament in the Friday Mosque at Fatehpur Sikri (1566) includes a very beautiful border design comprised of 14-pointed stars that repeats on a rhombic grid.[227] Ordinarily, patterns with these features are created from the *sevenfold system*. However, this design separates the underlying tetradecagons located on each vertex with a square. This arrangement dictates the proportions of the underlying elongated hexagonal and pentagonal elements that complete the generative tessellation; and while these elements work well together to create this lovely *median* pattern, they do not reassemble into addition tessellations, and are not, therefore, part of a systematic methodology.

Some of the most complex post-Mongol eastern nonsystematic geometric patterns are found in the Topkapi scroll. This anonymous scroll, or *tumar*, dated to the fifteenth or sixteenth century, is thought to have originated in central or western Iran, and reflects the ongoing influence of Timurid aesthetics within this region.[228] It is of added

---

[225] Necipoğlu (1995), diagram no. 35.

[226] Necipoğlu (1995), diagram no. 81a.

[227] Hankin (1925a), Fig. 34, pl. VII.

[228] Necipoğlu (1995), 37–38.

Photograph 89 A Timurid cut-tile mosaic arch spandrel at the Ulugh Beg *madrasa* in Samarkand that employs a nonsystematic *acute* pattern made up of 8- and 16-pointed stars (© David Wade)

significance in that the artist or artists who produced this scroll frequently illustrated the underlying generative tessellation in addition to the geometric patterns themselves. In many cases the underlying tessellations are overtly illustrated as dotted red lines, and in other cases more subtly indicated with non-inked "dead" lines scribed with a steel point. An *acute* design that repeats upon a rectangular grid places 12-pointed stars at the vertices of the grid and 10-pointed stars at the center of each rectangular repeat.[229] Conversely, the dual of this repetitive grid places the 10-pointed stars at each rectangular vertex, and the 12-pointed stars at the centers [Fig. 414]. This *acute* pattern also appears in the anonymous Persian treatise *On Similar and Complementary Interlocking Figures*,[230] and the earliest known architectural example is the product of Anatolian Seljuk artists working at the Great Mosque at Aksaray (1150-53). One of the Topkapi Scroll designs indicated with bold arcs for use within an arched tympanum employs 8-, 10-, and 12-pointed stars.[231] However, this *median*

pattern is poorly conceived, with strained symmetrical relationships between the primary underlying polygons. This creates multiple distortions throughout the underlying polygonal network and, consequently, the resulting geometric pattern. A far more successful complex design—in fact, one of the most remarkable patterns with just two regions of local symmetry in the history of this tradition—has the distinction of being the only known historical example of a design comprised of 9- and 11-pointed stars[232] [Fig. 431]. The repeat for this *acute* pattern is an elongated hexagon that places the 11-pointed stars on each vertex of the repetitive grid. Remarkably, the dual of this grid is also an elongated hexagon, but of differing proportions and orientated perpendicularly. This dual grid has the nine-pointed stars located upon its vertices. Either of these hexagons can equally be regarded as the repeat unit. It is interesting to note that this pattern shares a remarkable correspondence with two other examples from the historical record: the Seljuk border design from the *mihrab* of the Friday Mosque at Barsian, Iran (c. 1100) that employs 7- and 9-pointed stars [Fig. 429d]; and one of the patterns on

[229] Necipoğlu (1995), diagram no. 44.

[230] Bibliothèque Nationale de France, Paris, MS Persan 169, fol. 195a.

[231] Necipoğlu (1995), diagram no. 39.

[232] Necipoğlu (1995), diagram no. 42.

the exterior of the Mu'mine Khatun mausoleum in Nakhichevan, Azerbaijan (1186), is made up of 11- and 13-pointed stars [Fig. 434] [Photograph 35]. Each of these repeats with dual-elongated hexagons with one primary star form placed on the vertices of one hexagonal grid, and the other star form placed upon the vertices of the perpendicular dual-hexagonal grid. What is more, each of these three designs exhibit the *principle of adjacent numbers* wherein the convenience of 8-pointed stars anticipates the example with 7- and 9-pointed stars; the ease of making patterns with 10-pointed stars paves the way for the example with 9- and 11-pointed stars; and the flexibility of designing with 12-pointed stars allows for the example with 11- and 13-pointed stars.

A number of very fine nonsystematic geometric designs with radial symmetry were produced during this period in the eastern regions. Of particular note are a series of designs that fill the flat horizontal star shaped soffits within the outstanding Safavid *muqarnas* vault in the southeast *iwan* of the Friday Mosque at Isfahan. These soffit elements include four-, five-, seven-, eight-, and ten-pointed stars: each decorated with radial geometric patterns that are appropriate to the symmetry of the given star. The geometric

design inside the bounding 7-pointed star soffit is an *obtuse* design that places a 14-pointed star at the center of the design, with seven 11-pointed stars placed at the acute included angles of the bounding 7-pointed star. There are partial nine-pointed stars at each of the seven reflex angles of the bounding seven-pointed star. As with the *fivefold system*, this design can be created from either of two underlying tessellations [Fig. 440]. Another soffit in this *muqarnas* ceiling is a bounding ten-pointed star containing a *median* pattern that places a ten-pointed star at the center, surrounded by a ring of 10 seven-pointed stars, with partial ten-pointed stars at vertices of the obtuse angles of the star panel, and partial seven-pointed stars at the reflex angles of the ten-pointed star panel [Fig. 441] [Photograph 90]. The use of 90° crossing pattern lines in association with the radial configuration of seven-, nine-, and ten-pointed stars provides this design with the visual character of a *median* pattern created from the *fourfold system A* [Figs. 154 and 159].

Like their Mamluk contemporaries, post-Mongol artists in Persia, Khurasan, and Transoxiana produced many outstanding examples of domical geometric ornament. And like many of the examples produced by the Mamluks, these utilize radial gore segments as the repetitive units upon

Photograph 90   Detail of a ten-pointed star soffit from the Safavid *muqarnas* in the southeast *iwan* of the Friday Mosque at Isfahan that employs a nonsystematic radial *median* pattern with seven- and ten-pointed stars (© David Wade)

Photograph 91 Muzaffarid dome at the Friday Mosque at Yazd, Iran, with a geometric design comprised of ascending 6-, 7-, 6-, 5-, and 4-pointed stars, and culminating in a 16-pointed star at the apex (© Muhammad Reza Domiri Ganji)

which the nonsystematic underlying polygonal tessellations are applied. Among the earlier examples produced after the Mongol devastation is a shallow dome from the Ilkhanid tomb of Uljaytu in Sultaniya, Iran (1313-14), that is an 8-pointed star in plan with a 16-pointed star at the apex surrounded by sixteen 7-pointed stars in the field. This is a very shallow dome and the geometric design merely projects the otherwise two-dimensional pattern onto the slight curvature of the vault. Muzaffarid geometric domes were produced in cut-tile mosaic (*muarak*), and excellent examples include: the main interior dome at the Friday Mosque at Yazd (1324) [Photograph 91] with sixteen half 6-pointed stars at the base, ascending to a ring of sixteen 7-pointed stars, followed by sixteen 6-, sixteen 5-, and finally sixteen 4-pointed stars, with a 16-pointed star at the apex [Fig. 495a]; and a niche hood from this same building in Yazd that transitions the classic fivefold *acute* pattern on the walls of the niche onto a domical surface with a ring of 9-pointed stars, followed by two rings of 7-pointed stars, with an 8-pointed star at the apex. A magnificent cut-tile mosaic geometric dome with gore segmentation was

produced by artists during the short lived Sufid Dynasty at the mausoleum of Turabek-Khanym in Konye-Urgench, Turkmenistan (1370). This places twelve half 10-pointed stars around the periphery, ascending to another ring of twelve 10-pointed stars, followed by a ring of twelve 9-pointed stars, and surmounted by a 24-pointed star at the apex [Fig. 495b] [Photograph 92]. A relatively simple Muzaffarid geometric design in a quarter dome hood of a niche at the Friday Mosque at Kerman, Iran (1349), places half 8-pointed stars at the base, 5-pointed stars in the field, and a partial 12-pointed star at the apex, and a second quarter dome example at the same building in Kerman places 6-pointed stars at the base and an 8-pointed star at the apex. A rare Qara Qoyunlu geometric quarter dome is found in the hood of an arched niche at the Muzaffariyya mosque in Tabriz (1465). This example utilizes ten- and nine-pointed stars in fine quality cut-tile mosaic. A relatively simple, but powerful Timurid example is found on the interior of the dome at one of the anonymous mausolea at the Shah-i Zinda funerary complex in Samarkand (1385). This has an eight-pointed star at the apex whose lines descend

Photograph 92 Sufid cut-tile mosaic dome at the mausoleum of Turabek-Khanym in Konye-Urgench, Turkmenistan, with a geometric design comprised of ascending 10-, 10-, and 9-pointed stars, with a 24-pointed star at the apex (© Pete Martin)

into a simple geometric matrix. Later Safavid examples include the late-sixteenth-century decoration on the interior of a dome at the Friday Mosque at Saveh, Iran. This is comprised of a ring of 8 ten-pointed stars at the base, ascending to a ring of nine-pointed stars, followed by more ten-pointed stars, then seven-pointed stars, with an eight-pointed star at the apex. The exterior of this dome is also ornamented with a geometric design: with a ring of half 12-pointed stars at the base, ascending to a ring of 8-pointed stars, 11-pointed stars, 9-pointed star, and a 12-pointed stars at the apex [Fig. 495c]. The renowned geometric design of the exterior dome at the Aramgah-i Ni'mat Allah Vali in Mahan, Iran (1601) [Photograph 93] places half 8-pointed stars at the base, ascending to a ring of 10-pointed stars, followed by 9-pointed stars, 11-pointed stars, 12-pointed stars, and 9-, 7-, and 5-pointed stars; with a 12-pointed star at the apex [Fig. 495d]. The significant distortion in the n-fold symmetry of the stars in this design is only a minimal distraction from its great beauty. The Ottoman aesthetic did not generally include the application of geometric patterns onto the surfaces of domes. A rare exception is the exterior dome of the Haydar Khanah in Baghdad (1819-27) that is simply made up of several bands of

six-pointed stars. Stylistically, this has more in common with Safavid than Ottoman aesthetics.

The Mughals in the Indian subcontinent also used radial gore segments for decorating a number of their geometric domes. In his praiseworthy early twentieth century article *The Drawing of Geometric Patterns in Saracenic Art*,[233] E. H. Hankin describes the interior geometric decoration of several Mughal domes from Fatehpur Sikri in India (1570-80). Hankin's work is of primary historical interest to the study of Islamic geometric star patterns in that it represents the first European discovery of the polygonal technique as a generative methodology. Hankin concludes his paper with an analysis of several designs that were applied to domes at Fatehpur Sikri,[234] and demonstrates the ingenious traditional technique used by Mughal artists for applying the patterns to domical surfaces. Each of these makes use of the *fivefold system*, and utilizes just eight segments of a tenfold radial geometric design [Fig. 21]. Applying a two-dimensional 1/10 radial segment to a three-dimensional 1/8 domical

[233] Hankin (1925a).

[234] Hankin (1925a), pl. XIII, Figs 45–50.

Photograph 93 The Safavid dome at the Aramgah-i Ni'mat Allah Vali in Mahan, Iran, with a geometric design comprised of ascending 8-, 10-, 9-, 11-, 12-, 9-, 7-, and 5-pointed stars, with a 12-pointed star at the apex (© Aga Khan Trust for Culture-Aga Khan Award for Architecture/Khosrow Bozorgi [Photographer])

gore segment is an effective means of introducing the beauty of fivefold geometric patterns onto domical surfaces with minimal distortion. Examples of this form of domical design methodology are exclusive to Mughal India, with one notable exception being the large Safavid polychromatic dome at the Mashhad-i Fatima in Qum, Iran (seventeenth century). The exterior cut-tile mosaic decoration of this dome uses the latter truncated technique described by Hankin; although this dome breaks with the regularity of the ten-pointed stars in the uppermost portion of the dome.

1.24   Dual-Level Designs

The last great innovations in the tradition of Islamic geometric design were advances in the development of dual-level designs during the fourteenth and fifteenth centuries. This form of design further elaborates a primary geometric pattern with the inclusion of a smaller scaled secondary pattern of the same or similar variety. The Muslim proclivity for dual-level ornament precedes the development of mature dual-level geometric design by many hundreds of years. In addition to geometric design, the dual-level aesthetic found expression in both the floral and calligraphic idioms. Within the floral tradition, dual-level designs reached a high level of maturity during the fourteenth century under Timurid patronage wherein the primary and secondary motifs are differentiated by both scale and contrasting depths of relief. Timurid dual-level floral designs are typically monochromatic by virtue of their being carved from a single material such as wood or marble. Exceptional marble examples were occasionally used on sarcophagi during the fifteenth century, including that of Ghiyathuddin Mansur at the *madrasa* of Sultan Husain Mirza Baiqara in Herat (1492-93), and several from the Abdullah Ansari funerary complex in Gazargah near Herat, Afghanistan (1425-27) [Photograph 94]. Safavid dual-level floral designs are also of particular significance, especially the style that places a secondary floral scrollwork motif at the center of the background of the primary scrollwork design. This form of Safavid floral ornament was commonly carried out in cut-tile mosaic, such as that of the exterior dome of the Mardar-i Shah *madrasa* in Isfahan (1706-14) [Photograph 95]. Calligraphic dual-level examples are primarily architectural, where greater design flexibility and stylistic variation was accepted over the more constrained requirements of Quranic calligraphy. Dual-level calligraphy often places smaller scale *Kufi* script in the upper area of a calligraphic composition so that it runs through the ascending letters, such as the *alif*,[235] of a cursive primary text such as *Thuluth*. A fine Ilkhanid example of this variety of dual-level ornament is found in the carved stucco calligraphic band at the Friday Mosque at Varamin in Iran (1322). Muslim artists began experimenting with dual-level geometric designs as early as the ninth century, and two early examples include a window grille in an arch soffit [Photograph 8] at the mosque of ibn Tulun in Cairo (876-79). Within the geometric idiom, prior to the mature dual-level styles, many of the dual-level designs achieve their secondary component via additive processes. The most sophisticated example among these earlier patterns is the aforementioned fivefold field pattern that surrounds the exterior of the Gunbad-i Qabud tomb tower in Maragha, Iran (1196-97) [Fig. 67] [Photograph 24]. The aesthetics of this dual-level additive pattern anticipates the fully mature style developed approximately 250 years later in the same general region. Other early dual-level geometric patterns that can be

---

[235] The *alif* is the first letter of the Arabic alphabet. It is an ascender that is made from a single vertical stroke.

Photograph 94 A Timurid dual-level floral design from a sarcophagus at the Abdullah Ansari funerary complex in Gazargah, Afghanistan (© Thalia Kennedy)

regarded as formative to this tradition include: the ornamental exterior of the minaret of the Yakutiye *madrasa* in Erzurum, Turkey (1310); and an exterior panel from the Ilkhanid minaret of the Qabr Dhu'l Kifl shrine near Hillah, Iraq (1316), wherein a simple threefold geometric design is placed within the triangulated *Kufi* script. Each of these examples is visible from far and near: "allowing for the dynamics of scale to provide travelers with a progressive appreciation of the primary design from a relatively great distance, and the secondary elements upon closer proximity."[236] While calligraphic and floral expressions of dual-level ornament are exceptionally beautiful, they did not significantly enhance the aesthetic importance of these two ornamental modalities within the overall history of Islamic art and architecture. By contrast, the dual-level innovations that were applied to the geometric arts eventually led to an altogether new form of geometric design that is a significant historical addition to the breadth of this ornamental tradition.

The mature style of Islamic dual-level geometric design developed along two distinct historical paths. The earliest occurrence of such patterns was during the fourteenth century in the western regions of Morocco and al-Andalus under patronage from the Marinid and Nasrid dynasties. A century later, fully mature dual-level geometric designs were introduced to the architectural ornament of Transoxiana, Khurasan, and Persia under rival Timurid, Qara Qoyunlu, and Aq Qoyunlu patronage. It is unknown whether these two design traditions developed independently of one another or whether the preceding design methodologies from the Maghreb had a causative influence upon the development of these design conventions in the eastern regions. While the systematic methodology in the creation of dual-level designs from both regions is essentially the same, their respective aesthetic characteristics are very different. When considered from the perspective of Islamic art history, the tradition of

---

[236] Bonner (2003), 3.

Photograph 95 A Safavid dual-level floral design from a dome at the Mardar-i Shah *madrasa* in Isfahan (© David Wade)

dual-level design represents the pinnacle of systematic geometric pattern making, and was the last great innovation in the illustrious tradition of Islamic geometric star patterns. As pertains to the history of mathematics, many of the fourteenth- and fifteenth-century Islamic dual-level designs are consistent with the modern geometric criteria for self-similarity: the property of an object or overall structure to have an identical or analogous scaled-down substructure that, in the abstract, is or can be recursively scaled down ad infinitum. These dual-level designs are especially significant in that they appear to be the earliest anthropogenic examples of sophisticated self-similar geometry.[237]

Dual-level patterns invariably employ one of the established generative systems for creating both the primary and secondary designs. As such, historical dual-level patterns always have threefold, fourfold or fivefold symmetry. The artists working with this methodology never applied the *sevenfold system* in creating dual-level patterns, although this generative system is also well suited for such use.[238] The

self-similarity within the fully mature tradition of dual-level patterns is of growing interest to contemporary artists, art historians, and mathematicians alike. This remarkable artistic tradition is the direct result of the recursive manipulation of the generative polygonal modules that comprise these modular systems whereby proportionally scaled-down polygonal modules are applied into the structure of the primary design. While self-similar recursive processes are theoretically infinite—be they cosmological, geographical, biological, or anthropogenic—the practical manifestation of self-similar recursion within the arts is constrained by the medium in which it occurs. The historical examples of this variety of Islamic geometric design never exceed a single recursion; with both design levels employing constituent modules from the same set of pre-decorated underlying polygons. In this way, the scaled-down recursive use of the same set of generative polygonal modules, with the same family of pre-applied pattern lines, is responsible for the self-similarity. Can an object be self-similar if it has only a single recursion? The answer is yes, provided that the relationship between both levels satisfies the criteria for

---

[237] Bonner (2003).

[238] One has to assume the likelihood that the artists who developed the systematic dual-level methodology were unfamiliar with the *sevenfold system* of pattern generation.

–Bonner and Pelletier (2012), 141–148.
–Pelletier and Bonner (2012), 149–156.

self-similarity, and the recursion has the theoretical capacity for infinite scaled-down iteration. The recursive character of Islamic self-similar geometric designs can be identified as substitution tilings that are based upon *n*-inflation symmetry being applied within the primary underlying polygonal tessellating modules. Among the historical examples of Islamic dual-level geometric design, this inflationary process invariably takes place within a repetitive unit cell, and the resulting self-similarity is, therefore, not quasiperiodic,[239] nor is it the product of Penrose matching rules. Rather, the historical examples of self-similarity within this design tradition are comprised of "motifs of different scales [that] resemble each other in style or composition but are not replicas."[240] It is important to note that despite the high level of sophistication, there is no evidence to suggest that the artists responsible for this design tradition had any prescient knowledge or concept of self-similar geometry per se, just as there is no evidence that they were familiar with modern concepts of aperiodicity or quasicrystallinity. That said, the generative and recursive capabilities of the various polygonal systems have tremendous potential for contemporary designers who are interested in producing true aperiodic, quasicrystalline and self-similar designs with multiple levels of recursion.

Not all mature dual-level geometric designs satisfy the criteria for self-similarity. Many examples will use a different family of pattern in the primary and secondary levels. Strictly speaking, this difference in pattern families precludes such examples from qualifying as self-similar. However, the iterative use of the same polygonal tessellating modules at both levels allows for the design methodology to be regarded as self-similar, if not the design itself. Most of the examples of mature dual-level geometric ornament in both the east and the west are architectural, and were

fabricated in cut-tile mosaic. A number of examples in the east were also produced in wood and on paper. The examples from the Topkapi Scroll are particularly significant in that they reveal the systematic polygonal methodology behind their construction. In all cases, the fact that the Muslim artists responsible for these masterpieces of geometric art limited themselves to just two levels of design is more to do with the material constraints of their chosen medium than any lack of geometric ingenuity.

Muslim artists developed four distinct varieties of self-similar design. For purposes of clarification, these are being identified as *types A, B, C*, and *D*.[241] Each of the first three is from the eastern regions, and the fourth is from the Maghreb. *Type A* designs are characterized by the primary design expressed as a bold single line of contrasting color, with the reduced scale secondary pattern filling the *entire background* of the primary design. This variety of dual-level design typically locates scaled-down stars upon the vertices of the primary design. The earliest example of a *type A* design is from one of the Ilkhanid ornamental vaults at the mausoleum of Uljaytu in Sultaniya (1307-13) [Photograph 96]. Indeed, this is one of the earliest examples of a true dual-level geometric design from the eastern regions, and represents the transition toward the fully mature style. The secondary level is a *median* pattern with 10- and 12-pointed stars upon the isometric grid, and the primary design is created by emphasizing through relief selected lines of this grid such that the classic threefold *median* pattern with 60° crossing pattern lines is produced [Fig. 95b]. The use of the 10-pointed stars at given vertices creates problems in the pattern alignment between both levels of the design, and could have been avoided through the use of a more compatible isometric design with, for example, just 12-pointed stars. Outstanding examples of fully mature *type A* designs are found in a wide variety of architectural locations, and significant examples include an Qara Qoyunlu cut-tile mosaic arched panel at the Imamzada Darb-i Imam in Isfahan[242] (1453-54) [Photograph 97] wherein both the primary and

---

[239] It has been suggested that the Persian artists responsible for a dual-level pattern within an arch spandrel at the Imamzada Darb-i Imam in Isfahan applied quasiperiodic substitution rules while designing this example of dual-level geometric design; and that these artists may have had specific knowledge of the science of quasiperiodicity some 500 years before the discoveries of Sir Roger Penrose in the 1970s. However, the fact that the recursive use of the *fivefold system* of pattern generation *can* be used to create true quasiperiodic designs does not mean that the dual-level use of this system at the Imamzada Darb-i Imam *is* actually quasiperiodic. A rudimentary examination of the cited example reveals that both levels of the overall design repeat within the same rhombic unit cell: the very definition of periodic tiling. The claim to have found quasicrystallinity in the design from the Imamzada Darb-i Imam is based upon overlooking the unit cell in favor of arbitrarily isolating and analyzing limited portions of the overall structure. See Lu and Steinhardt (2007a). See also:
  –Makovicky and Hach-Ali (1996), 1–26.
  –Saltzman (2008), 153–168.
  –Cromwell (2009), 36–56 and (2015).
[240] Cromwell (2009), 47.

---

[241] In an earlier publication the author identified just three varieties of Islamic geometric self-similar design, but has since identified a fourth historical variety as a hybrid of his original *type A* and *type B*. As such, in this work the hybrid form is designated as *type C*, and the former *type C* is now renamed as *type D*. See Bonner (2003).

[242] The Imamzada Darb-i Imam employs a second example of this particular fivefold *type A* dual-level design in a pair of arch spandrels. This is a vastly inferior representation of this fine design, with multiple mistakes in the application of the secondary design. It is also poorly constructed with grossly disproportional polygonal figures, such as the pentagons, in the primary design. Its poor construction and myriad mistakes in the layout of the secondary elements lead one to assume that this was produced by a separate set of artists possibly working at a later date.

Photograph 96 An Ilkhanid vault in the mausoleum of Uljaytu in Sultaniya, Iran, with a central *type A* dual-level geometric design (© Daniel C. Waugh)

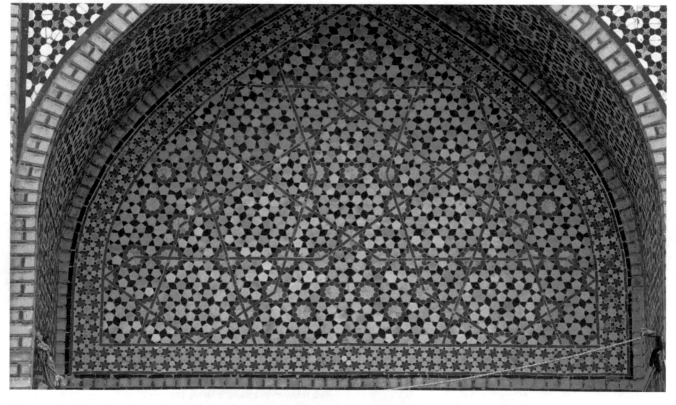

Photograph 97 A Qara Qoyunlu cut-tile mosaic arch at the Imamzada Darb-i Imam in Isfahan that employs a *type A* self-similar dual-level design that is constructed from the *fivefold system* (© David Wade)

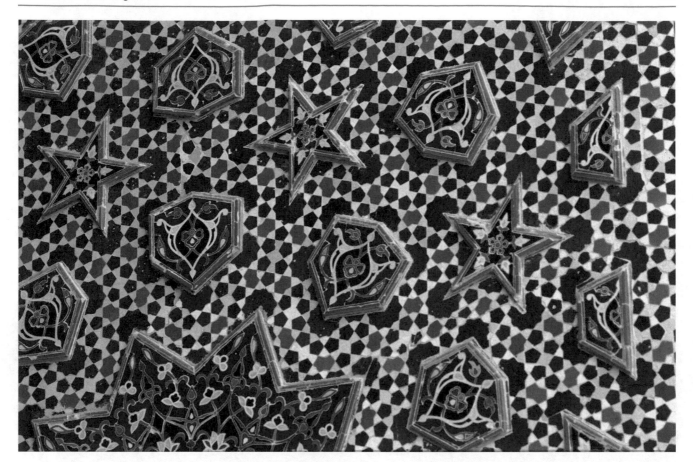

Photograph 98 A Qara Qoyunlu cut-tile mosaic panel at the Imamzada Darb-i Imam in Isfahan that employs a *type B* self-similar dual-level design that is constructed from the *fivefold system* (© David Wade)

secondary levels are fivefold *obtuse* patterns [Fig. 451]; a Safavid cut-tile mosaic panel from the Madar-i Shah *madrasa* in Isfahan (1706-14) wherein the primary level is the classic fivefold *acute* pattern, and the secondary level is an *obtuse* pattern [Fig. 453]; and an Aq Qoyunlu arch spandrel (c. 1475) at the Friday Mosque at Isfahan. *Type B* designs are characterized by widened primary pattern lines, with an analogous scaled-down secondary pattern placed within the widened primary design. The specific proportion of the widened line is geometrically determined to allow for the application of the secondary polygonal modules, with the primary polygons generally placed at the vertices of the widened lines. The polygonal background regions of *type B* designs are typically filled with either floral or calligraphic motifs. Exceptional architectural examples of this variety of dual-level design include a Timurid cut-tile mosaic border in the southern *iwan* of the courtyard at the Gawhar Shad mosque in Mashhad (1416-18) that is constructed from the *fourfold system A* [Fig. 460]; a Qara Qoyunlu cut-tile mosaic panel from the Imamzada Darb-i Imam in Isfahan that is created from the *fivefold system* wherein the widened primary design is an *acute* pattern, with the secondary infill design from the *obtuse* family [Fig. 463] [Photograph 98];

and an Aq Qoyunlu cut-tile mosaic panel created from the *fourfold system A* at the Friday Mosque of Isfahan wherein the widened primary design of octagons, concave octagons, and four-pointed stars is filled with a secondary design of eight-pointed stars[243] [Fig. 462] [Photograph 99]. A simplified form of *Type B* dual-level design was used frequently during the fifteenth and sixteenth centuries. These utilize either the isometric or the orthogonal grid as the basis for the primary design, and the widened line effect is achieved by isolating selected cells within the grid, and placing predesigned geometric patterns with either triangular or square repeat units into the selected cells. Despite the simplicity of this technique, these dual-level designs can be very beautiful, and especially fine examples include: a threefold Timurid cut-tile mosaic panel from the Friday Mosque at Varzaneh in Iran (1442-44) that places 12-pointed stars upon the vertices of the isometric grid of the primary design [Fig. 457]; a threefold Janid cut-tile mosaic border from the

---

[243] This may have been produced during the sixteenth century during Safavid rule.
    –Necipoğlu (1995), 37.

Photograph 99 An Aq Qoyunlu cut-tile mosaic panel from the Friday Mosque at Isfahan that employs a *type B* dual-level design that is constructed from the *fourfold system A* (© David Wade)

Nadir Divan Beg in Bukhara (1622-23) [Photograph 100] that places secondary 6-pointed stars at each prescribed vertex of the primary isometric grid and 9-pointed stars within the secondary pattern matrix [Fig. 455]; and a four-fold Aq Qoyunlu[244] cut-tile mosaic panel in the southwest *iwan* of the Friday Mosque at Isfahan (c. 1475) that places 12-pointed stars on the vertices of the rotating kite primary grid. As with some hybrid designs (e.g. Fig. 23), this example has the further quality of combining triangular and square repetitive cells in creating the widened line effect of the rotating kite primary design. *Type C* dual-level designs are essentially a fusion of *types A* and *B* in which the primary design is widened and filled with a secondary design in exactly the same fashion as *type B*, but the secondary design continues to flow into the background regions of the widened primary design, thus filling the entirety of the overall design with secondary patterning much like *type A* designs. Differentiation between the two levels of design is achieved in two ways: through emphasizing the widened lines of the primary design, and through coloring the secondary pattern within

the widened lines differently from the secondary pattern inside the background regions. The comparatively few examples of *type C* designs from the historical record include: an outstanding fourfold Muzaffarid cut-tile mosaic panel (1470) over the eastern entry portal of the Friday Mosque at Yazd wherein both the widened primary design and secondary design are created from the *fourfold system A*[245]; a fivefold Safavid cut-tile mosaic arched panel from the Mardar-i Shah in Isfahan wherein the widened primary design is the classic fivefold *obtuse* pattern, and the secondary design is also an *obtuse* pattern [Fig. 468]; and a Shaybanid wooden ceiling at the Khwaja Zayn al-Din mosque and *khanqah* in Bukhara (c. 1500-50) wherein the primary design is a standard threefold *median* pattern with 60° crossing pattern lines that is created simply from the regular hexagonal grid [Fig. 95b], and the secondary design is a simple device that places six-pointed stars at each vertex of the widened primary design. The wooden ceiling's bold

---

[244] This may date from the Safavid period.

[245] Peter Cromwell's detailed methodological analysis of the *type C* dual-level design in the entry portal of the Friday Mosque at Yazd demonstrates the use of the *fourfold system A* in its creation. See Cromwell (2012a), 159–168.

Photograph 100 A Janid cut-tile mosaic border at the Nadir Divan Beg in Bukhara, Uzbekistan, that employs a *type B* dual-level design (© David Wade)

relief provides the differentiation between the secondary design in the widened lines and those of the background regions. This is a noteworthy feature of many dual-level designs with widened lines (*types A* and *B*), and was used very successfully at the mausoleum of Uljaytu in Sultaniya, as well as the dual-level designs at both the Friday Mosque at Isfahan and the nearby Imamzada Darb-i Imam. *Type D* designs were developed in Morocco and al-Andalus in the fourteenth century: preceding the mature dual-level traditions in the eastern regions by approximately a century.[246] The designs from the Maghreb are invariably expressed in *zillij*—the Moroccan tradition of cut-tile mosaic. While the basic methodology in creating the two

levels through the application of scaled-down polygonal modules to strategic locations of the primary design is the same, this western dual-level tradition differs in the manner of emphasizing the two levels. In the Maghreb, the primary design is expressed exclusively through the contrasting color of the background areas of the secondary design. The secondary pattern in *type D* designs is an interweaving widened line that is given its own distinct mosaic color, typically white, within the overall color scheme. The primary design is differentiated from the secondary design by providing the requisite secondary background elements their own color. Depending on the color of the mosaic pieces that emphasize the primary design, the dual-level quality can be either bold or subtle. The color distribution of the remaining secondary background elements is determined according to the aesthetic predilections of the artist. Almost all of the dual-level designs in the Maghreb are created from the *fourfold system A*, but the *fivefold system* was used in at least two locations. Especially fine examples include a fourfold *Mudéjar* wall panel in the Patio de las Doncellas at the Alcazar in Seville[247] (1364-66) [Fig. 470] [Photograph 101]; a fourfold Nasrid wall panel from the Alhambra in Granada [Fig. 472] [Photograph 102]; and a fivefold Marinid wall panel from the Bu 'Inaniyya *madrasa* in Fez (1350-55) [Fig. 474]. The primary design in this last example is actually an arrangement of decagons that touch corner to corner, separated by interstice regions in the shape of non-regular four-pointed stars. Although the primary design is not a traditional fivefold geometric pattern, the method of highlighting the decagonal design through the background coloring of the secondary design is the same as used in the *type D* fourfold designs of the Maghreb. The use of tessellating decagons as a primary design was also used by the Marinids in a more complex dual-level *zillij* panel from the al-'Attarin *madrasa* in Fez (1323) [Fig. 476] [Photograph 103]. The Marinids and Nasrids were closely allied and artists were known to have traveled across the Straights of Gibraltar to work in both al-Andalus and Morocco. This explains the remarkable unanimity in the architectural ornament of these two cultures generally, as well as the exactitude in stylistic interpretation of dual-level designs more specifically. While this form of architectural ornament survived among succeeding dynasties in Morocco, with few exceptions the dual-level design methodology of the Nasrids did not survive the final *reconquista*, and the post-1492 art of the *Mudéjar* Christians in al-Andalus never reached the level of geometric sophistication as experienced under the courtly patronage of the Nasrids.

---

[246] The large number of examples of *type D* dual-level patterns at the Alhambra has led Jean-Marc Castéra, a renowned specialist in Islamic geometric art, to refer to this variety of design as the *Alhambra Technique*. See Castéra (1996), 276–277.

[247] –Makovicky and Hach-Ali (1996), 1–26.
   –Bonner (2003), 10–11.

Photograph 101 A *Mudéjar zillij* mosaic panel at the Patio de las Doncellas at the Alcazar in Seville, Spain, that employs a *type D* self-similar dual-level design constructed from the *fourfold system A* (© David Wade)

The mature tradition of dual-level design developed in early fifteenth century Persia under the auspices of the Timurids, and, by mid-century, additional patronage of both the Qara Qoyunlu and Aq Qoyunlu. Stylistically, the ornament of this period falls within the prevailing Timurid aesthetic, and despite political tension between these powerful rival dynasties, at least one artist is known to have received commissions from all three: Sayyid Mahmud-i Naqash.[248] The vast majority of Islamic geometric patterns used as architectural ornament are unsigned. To a large extent this anonymity remains true of the dual-level designs in the eastern regions. However, the significance of Sayyid Mahmud-i Naqash is not just that he is one of the few individuals known by name to have worked within the geometric design tradition generally. His association with the

development of the mature style of dual-level designs in the eastern regions is of particular art historical relevance. Relatively few architectural monuments were decorated with this methodologically complex art form, and it appears likely that only a select corps of elite artists possessed the requisite skills to create these dual-level designs. The earliest example of a fully mature dual-level geometric design from the eastern regions appears to be the fourfold *type B* border design in the *iwan* at the Gawhar Shad mosque in Mashhad[249] (1416-18) [Fig. 460]. The earliest known piece to have been signed by Sayyid Mahmud-i Naqash is the threefold *type B* panel from the Friday Mosque at Varzaneh[250] (1442-44) [Fig. 457]. The rarity of these dual-level designs, together with the timeframe of these two Timurid monuments, suggests that Sayyid Mahmud-i Naqash was likely affiliated with the master artist responsible for the work in Mashhad: perhaps as an apprentice working in the atelier that produced the earlier work in Mashhad. His name also appears in the outstanding cut-tile mosaics (c. 1475) of the Friday Mosque at Isfahan. This work was created under the patronage of the Aq Qoyunlu ruler Uzun Hasan. A *type A* dual-level border design in the northwest *iwan* of the Friday Mosque at Isfahan is identical to the earlier unsigned Qara Qoyunlu fivefold design in the arch at the Imamzada Darb-i Imam (1453-54) [Photograph 97]. These buildings are only some 300 m apart and were built within 20 years of one another, and it appears likely that the multiple examples of dual-level design at both these monuments were created by Sayyid Mahmud-i Naqash, or at least artists working within the same atelier or guild. This is supported by the fact that the diversity of dual-level work in both these buildings is of the highest caliber of design and execution, and appears to be the work of a single individual or guild. What is more, the dual-level work of Sayyid Mahmud-i Naqash "deserves recognition not just as a great artist and designer, but also as a pioneer of self-similar geometry some 500 years ahead of his time."[251]

The methodological practices responsible for the remarkable rise in dual-level maturity and sophistication that occurred in the fifteenth century under the guidance of a relatively small number of artists working in Mashhad and Isfahan continued for some hundreds of years in the eastern regions. Yet this distinctive form of design was not widely distributed throughout the monuments of successive dynasties. Rather, additional locations of dual-level panels is limited to only a handful of buildings, including the Darb-i Kushk in Isfahan (1496-97); Khwaja Zayn al-Din mosque and *khanqah* in Bukhara (c. 1500-50); the Nadir Divan Beg in Bukhara (1622-23); and the Madar-i Shah in Isfahan

---

[248] Hutt and Harrow (1979), 61–65.

[249] O'Kane (1987), 70.

[250] Hutt and Harrow (1979), 61.

[251] Bonner (2003), 5.

Photograph 102 A Nasrid *zillij* mosaic panel at the Alhambra in Granada, Spain, that employs a *type D* self-similar dual-level design constructed from the *fourfold system A* (© David Wade)

(1706-14). This suggests that the requisite methodology for creating these dual-level designs was not widely known amongst artists of the period, but was preserved over time through a more restricted master-to-student inherited methodological lineage. It is likely that *tumar* design scrolls contributed to this transference of knowledge. The main purpose of these scrolls appears to have primarily been as an *aide memoire* for master artists. However, it is likely that in addition to serving as a reference manual, these scrolls may also have been used for teaching. As such, they would have been an important facet in the preservation, dissemination and transference of specific patterns, as well as design methodology more generally.[252] These scrolls were made by gluing new sheets of paper onto the end of an already existing scroll, effectively lengthening the scroll with added designs. This additive process allows for the strong possibility that these scrolls were added to over time by successive owners, rather than the product of only a single individual.

Very few *tumar* design scrolls, such as the Topkapi Scroll, are known publicly and available for study by contemporary historians. Of those that are known, and at the time of writing, only the Topkapi Scroll illustrates dual-level geometric designs. Outside the architectural record, the Topkapi Scroll is the largest and most important repository of dual-level design from the eastern regions. Because the dual-level designs in the Topkapi Scroll also show the underlying generative tessellations (as overt solid or dotted lines differentiated by color, or by more subtle non-inked "dead" lines scribed into the surface of the paper with a steel graver) this document is exceptionally important as the only historical evidence of the polygonal design methodology behind the creation of these designs. The Topkapi Scroll illustrates seven dual-level geometric designs. Five of these are *type A* designs, and two are *type B*. Several of these exhibit the qualities of self-similarity and are the equal in design ingenuity to the architectural examples produced by Sayyid Mahmud-i Naqash. The *type A* designs are all five-fold (nos. 28, 29, 31, 32, and 34)[253] and follow the same formula of utilizing *median* patterns at both the primary and secondary levels that was used by Sayyid Mahmud-i Naqash

---

[252] Necipoğlu (1995), Chap. 1.

[253] The author is using the diagram numbers for each separate design as attributed by Gülru Necipoğlu: See Necipoğlu (1995).

Photograph 103 A Marinid *zillij* mosaic panel at the al-'Attarin *madrasa* in Fez, Morocco, that employs a *type D* dual-level design constructed from the *fivefold system* (© David Wade)

at the Friday Mosque at Isfahan, and later at the Imamzada Darb-i Imam [Photograph 97]. This recursive similitude qualifies these five designs as self-similar. Only design no. 28 expressly represents—in dotted red lines—the underlying generative tessellation. The underlying generative tessellations for designs 29, 31, 32, and 34 are indicated by un-inked scribed lines that are only visible through close inspection. Design nos. 38 and 49 are *type B* designs: the former comprised of threefold symmetry, and the latter of fivefold symmetry. With the added element of color as part of the composition, design no. 38 is the most visually arresting. This colorization may indicate an intended application to cut-tile mosaic. This example is the more simple variety of *type B* design wherein the widened pattern line is produced from a tessellation of repetitive polygonal cells: in this case triangles, squares and hexagons [Fig. 458]. The secondary infill pattern is applied to just the triangles and squares,

with the hexagons being open background elements. The geometric designs that are applied to both these repeat units were well known throughout Muslim cultures. The nonsystematic threefold *median* pattern made from an underlying tessellation comprised of a ring of 12 pentagons surrounding the dodecagons [Fig. 300b *median*], and the nonsystematic fourfold *median* pattern created from an underlying tessellation of dodecagons and octagons separated by pentagons and hexagons [Fig. 379f]. The use of these two repeat units is also represented in the single-level hybrid *median* design no. 35 from the Topkapi Scroll [Fig. 23d–f]. The method of transitioning from the interior pattern of decorated triangles and squares with open hexagons to the surrounding rectangular border is both clever and beautiful [Fig. 459a]. Similar formulae for bordering *type B* and *type C* designs were commonly employed within the architectural record. The no. 49 *type B* fivefold design in the Topkapi Scroll is a magnificent example of Islamic dual-level design, and, indeed, the most complex *type B* design from the historical record [Fig. 465]. The surrounding rectangular border is resolved in the same fashion as the previous threefold *type B* design from this scroll, albeit with rectangular repeat units rather than squares and triangles [Fig. 466d].

1.25 The Adoption of Islamic Geometric Patterns by Non-Muslim Cultures

Throughout its long and illustrious history, the evolution of Islamic ornament into its many and varied branches has been greatly influenced by the artistic conventions of non-Islamic cultures. The genius of Muslim artists to assimilate and reorient foreign design elements into their own distinctive ornamental tradition can be traced back to the earliest Islamic period. For example, the application of the Hellenistic geometric compass-work technique to the pierced stone window grilles of the Great Mosque of Damascus was an early Islamic innovation of great visual impact, as well as of continuing influence to subsequent Islamic cultures. Similarly, the highly stylized carved stucco vegetal ornament of Samarra appears to have been influenced by earlier Hellenistic and Sassanian vegetal motifs[254]: and as with pierced window grilles, the stucco design innovations of Samarra were to have a lasting influence on Islamic ornament well into the fifteenth century. Many centuries later, the Mongols introduced Chinese and Indian design motifs into the ornamental vocabulary of Islamic artists and designers.

Just as Islamic cultures were able to assimilate many of the artistic and architectural conventions of non-Muslim

---

[254] Allen (1988), 1–15.

peoples with whom they had close contact; it was inevitable that the rich and varied beauty of Islamic ornament should, in turn, become an influence to the art and architecture of many non-Muslim cultures. Non-geometric examples of such influence include the Tulunid form of the Samarra beveled style being used in the architectural ornament of the Egyptian Coptic community. Most notably, this style was employed in the Coptic monastery of Dayr as-Suryani (914) in the Wadi Natrun.[255] A strong late Fatimid influence is evident in the *muqarnas* vault in the Cappella Palatina in Palermo, Sicily (c. 1140). This church was built for Roger II, the Norman king of Sicily, and is also remarkable for the use of *Kufi* calligraphy, and distinctively Fatimid stylization of the painted human figures that adorn the *muqarnas*. But for the use of human figures, this Sicilian *muqarnas* parallels contemporaneous examples from Egypt.

In Spain, where Muslim, Christian, and Jewish communities lived side by side for many centuries, the degree of cultural interaction was to have a profound effect upon both the Christian and Jewish artistic practices of the region. Islamic geometric and floral patterns were freely used by Christians and Jews alike. Except in the use of Hebrew rather than Arabic, the Jewish synagogues in al-Andalus were stylistically Moorish in every respect. Only a few have survived relatively untouched by later Christian acquisition. Both the Santa Maria la Blanca (1180) and Nuestra Señora del Transito (1360) in Toledo were originally synagogues. The Christian use of Hispano-Moresque ornamental devices was mostly the work of *Mudéjar* artists and craftsmen. These were Muslims who lived in areas under Christian control. *Mudéjar* art also refers to the continuation of Islamic ornament after the final surrender of the Nasrids in Granada in 1492, and the expulsion of virtually all remaining Muslims and Jews from Spain. At its best, this highly influenced form of Christian architecture is often indistinguishable from the work from contemporaneous Arab patrons. Many magnificent buildings were built by *Mudéjar* craftsmen; two very noteworthy examples being the Palace of the Alcazar in Seville (1364-66), and the Convento de la Concepcion Francisca in Toledo (1311). Of particular note is the wooden geometric dome in the Hall of the Ambassadors at the Palace of the Alcazar in Seville [Photograph 67]. This distinctly Islamic styled geometric dome is the work of Diego Roiz, a Christian *Mudéjar* artist who was clearly well versed in the Islamic geometric idiom. To a limited extent, the *Mudéjar* use of Islamic geometric patterns even made its ways to the New World. An outstanding example is from the entry door of the Cathedral of Santa Domingo in Cusco, Peru (c. 1560-1654). This employs the classic nonsystematic pattern that places 12-pointed stars

upon the vertices of the isometric grid, and 9-pointed stars at the center of each triangular repeat. The Christian tradition of early Spanish manuscript illumination was particularly influenced by Islamic work. Rather than being carried out by *Mudéjar* craftsmen, the work of the Leonese School of manuscript illumination was created by Christian monks who, through prolonged close cultural contacts, were greatly influenced by Islamic artists. This style of highly influenced Christian art is referred to as *Mozarab*; from the Arabic word *mustarib*, which translates as Arabized. The general layout, and especially the interweaving geometric border designs of the Moralia in Iob, written in the monastery of Valeranica in 945, is a good example of the Islamic influence upon Mozarab art.

Just as the Seljuks in Anatolia were influenced by the stone masonry traditions of Armenian Christians, the Islamic geometric and floral ornament of the Seljuks had a very distinctive reciprocal impact on the carved stone ornament of the Armenians. This is especially apparent in the remarkable tradition of *khachkar* stela. These are large rectangular stone obelisks, at least twice their height as width, that invariably employ a central cross in deep relief as a primary motif. The cross is often winged and resting upon a circular rosette, and framed in a border of geometric and floral designs. (The reverse sides are provided with inscriptions.) *Khachkars* were presented to the church by patrons and benefactors in commemoration of a person or event, and as a means of securing religious favor. In such circumstances, these monuments were often set into the walls of churches. *Khachkars* also served as grave markers, and were set upon tombs in churchyards or exposed to the elements in open fields. It is noteworthy that the large number of Ahlatshah Muslim tombstones in Ahlat, Turkey are of the same approximate size and shape as the nearby Armenian *khachkars*, and the ornamental treatment of these tombstones is remarkably similar—except for the absence of any figurative elements, and Christian symbols. This similarity suggests the possibility of a reciprocal influence between these cultures. The Ahlatshahs were an Anatolian Turkish *beylik* closely allied with the Great Seljuks of Persia who ruled the region northwest of Lake Van that bordered on Armenia during the twelfth century. The tradition of *khachkars* developed during the second half of the ninth century, at a time when the Armenians had won back their independence from the Abbasid Caliphate. However, during the period of Seljuk dominion over Anatolia, the Armenian tradition of ornamental stone carving took on distinctively Seljuk characteristics. The nature of the geometric and floral ornament that was used on the *khachkars* of this period is, in many respects, identical to that of their neighboring Seljuk rivals to the south. The interweaving knotted borders, simple geometric field patterns, and meandering floral designs found on these monuments could easily be mistaken for the work of Muslim

---

[255] Kuhnel (1962), 58.

artists. For example, the floral and geometric patterns on the outer frame of the thirteenth century Siroun *Khachkar* from Toumanian are classically Seljuk in stylization, as is the floral rosette beneath the cross, and the *bandi-rumi*, or Anatolian knot-work, in the crown. One of the few Armenian ornamental devices that is stylistically distinct from Seljuk work is the occasional transformation of the interweaving lines of the geometric pattern into the floral design. The *khachkar* tradition is regarded as having reached its stylistic perfection in the thirteenth and fourteenth centuries in the Sunik and Azizbekov regions of Armenia. The work of one man in particular is of especial importance. Momik (1282-1321) was a renowned scribe, painter, architect, and sculptor, and was responsible for some of the finest *khachkars* found within this tradition. The church in Noravank has an especially refined example of his work that prominently features a very complex nonsystematic geometric pattern made up of 8-, 10-, and 11-pointed stars [Fig. 423]. This pattern approaches the sophistication of the geometric work produced during the same period by neighboring artists working for the Seljuk Sultanate of Rum. The geometric sophistication of this pattern indicates that Momik likely received training from Muslim artists in the traditional methods of constructing complex geometric patterns, despite his being a Christian. It is worth noting however that this design is not without problems. Generally, the variable size and distribution of the background elements are unbalanced, causing regions of greater and lesser pattern density. Of particular concern are the distorted geometric rosettes that surround each 11-pointed star.

Throughout the Islamic world, many non-Muslim minority communities adopted the ornamental conventions of the Muslim culture they lived in. The Coptic churches of Cairo frequently employ Islamic geometric patterns in their architectural ornament. For example, the Hanging Church (al-Mu'allaqa) dedicated to St. Mary is replete with a diversity of geometric designs, as well as floral motifs and *muqarnas* that is stylistically identical to the contemporaneous work in the mosques of Cairo. The architectural ornament of the Armenian Christian community in the New Julfa district of Isfahan fully embraces the practices and aesthetics of the Safavids, including the wide use of geometric patterns. As in Spain, synagogues in cities such as Fez, Tunis, Cairo, Baghdad, Istanbul, Isfahan, Kabul, and Samarkand frequently employ Islamic geometric patterns in their ornamentation.

In India, the architectural style of the Mughals had a tremendous influence on Hindu and Sikh architecture. In particular, the Hindu Rajput princes in Rajasthan were greatly influenced by Mughal courtly life, and freely adopted Mughal customs and practices. The art and architecture of the Rajputs was virtually a complete abandonment of earlier Hindu forms in favor of the Mughal style. Except for the fact that they were built by Hindu rulers, cities such as Udaipur, Jodhpur, Jaipur, and Jaisalmer are essentially Mughal in conception. And much like the later ornament of the Mughals themselves, the eighteenth and nineteenth century Rajput architecture and ornament tended toward an overabundant decadence. The Sri Harmandir Sahib in Amritsar is the principal temple of the Sikh religion. This three-story building sits upon an island in the center of a reservoir, and the architectural style is strongly derivative of late Mughal work. The Hindu and Jain adoption of Mughal architectural standards included geometric design, most notably in their use of pierced stone *jali* screens.

A number of European artists have shown an interest in Islamic design. Both Leonardo di Vinci (1452-1519) and Albrecht Durer (1471-1528) produced remarkable meandering concatenations: pen and ink rosettes of complex interweaving lines that are highly reminiscent of Islamic floral designs.[256] Hans Holbein (1497-1543) incorporated his studies of arabesque ornament into his paintings. In the nineteenth and early twentieth centuries numerous studies and collections of Islamic geometric and floral ornament were assembled by European scholars for the purposes of inspiring the ornamental arts of Europe. These include: Jules Bourgoin, Prisse d'Avennes, Owen Jones, E. Hanbury Hankin, and Archibald H. Christy. These collections were an integral aspect of the nineteenth-century Orientalist movement. More recently, the twentieth-century Dutch artist M.C. Escher (1898-1972) was greatly influence by the geometric ornament he encountered while visiting the Alhambra in Spain. Truly, the geometric ornament of Islamic cultures served as a source of inspiration for countless generations of artists, designers, and craftsmen throughout the Islamic world and beyond.

1.26 The Decline of Islamic Geometric Patterns

The gradual decline in the use of Islamic geometric pattern began with the three modern era Muslim empires: the Ottoman Turks, Persian Safavids, and Indian Mughals. Each of these cultures continued to employ geometric designs in their art and architecture; yet the spirit of innovation was substantially lost. With several notable exceptions, such as the geometric domes of the Safavids and Mughals, geometric pattern construction became highly derivative of previous work. During the sixteenth and seventeenth centuries, floral ornamentation was progressively given far greater emphasis over geometric design in each of these three empires. The reason for this aesthetic shift is unclear. What

---

[256] Coomaraswamy (1944), 109–28.

is certain is that creative vitality within the geometric idiom was highly reliant upon the very specific methodological knowledge of the polygonal technique. As patrons increasingly favored floral ornament, those artists knowledgeable of the polygonal technique would have found less work, and fewer apprentices to carry this tradition forward. Over time, this break with the past regrettably led to the inability of Muslim artists to create new and original geometric patterns, eventually relegating geometric patterns to the mere making of copies.

The early Ottoman use of geometric patterns continued the tradition inherited from the Seljuks, although in a less grand scale and with less geometric sophistication. By the end of the fifteenth century the Grand Ottoman style, exemplified in the works of Sinan, utilized the ornamental quality of geometric patterns to a far lesser extent. The emphasis of the floral idiom by the Ottoman Turks seemed to know no bounds: with textiles, metalwork, leatherwork, bookbinding, stained glass, painting and illumination, carved ivory, jewelry, and inlaid woodwork all receiving the prodigious floral talents of the Ottoman artists. Nevertheless, many fine examples of geometric design were created during the sixteenth century, albeit mostly derivative of earlier work. Inlaid and joined wood were especially popular media for their applied geometric designs, as seen in many of the finest examples of Ottoman *minbars*, doors, and furniture of this period.

Being from the same region as the Seljuk Sultanate of Rum, it is surprising that Ottoman artists were not more inclined to adopt the sophisticated nonsystematic design methodologies used in Anatolia by their predecessors. Yet the Ottoman use of geometric patterns tended toward more conventional designs with less symmetrical ambiguity. This would appear to be due to an aesthetic predilection toward more easily ascertained geometric structures that are less visually demanding. And as stated, it is also possible that knowledge of the more complex methodologies was not transferred to subsequent generations of artists. Of the relatively few examples of more complex Ottoman geometric design, the side panels beneath the platform of the wooden *minbar* at the Great Mosque of Bursa (1396-1400) is particularly significant. The long vertical panel adjacent to the wall is made up of an *acute* pattern with five vertically arranged stars from top to bottom in the following order: a 10-pointed star at the top, followed by another 10-pointed star, followed by two 9-pointed stars, and an 11-pointed star at the base. This design does not have a repeat unit per se. Rather, the underlying generative tessellation is arranged top to bottom without a satisfactory resolution at the edges, thus causing the pattern to be cut off in a fashion that would not repeat nicely were the panel repeated horizontally or vertically. This is an atypical feature that suggests the artist was less skilled than previous artists who produced patterns

with multiple regions of differentiated symmetry. The triangular side panel of the *minbar* at the Great Mosque of Bursa employs an orthogonal *acute* pattern that places 12-pointed stars at each corner of the square repeat unit, an octagon at the center of the repeat unit, a 10-pointed stars at the midpoint of each edge of the repeat unit, and 9-pointed stars within the field of the repeat unit [Fig. 400]. This same orthogonal design was used at the Kayseri hospital in Kayseri, Turkey (1205-06), the Agzikara Han near Aksaray, Turkey (1231-40), and the Çifte Kumbet in Kaysari (1247). However, in the later Ottoman example of this design the artist choose to replace the original octagons at the center of each square repeat unit with eight-pointed stars. This design variation is not the product of an adjustment to the underlying generative tessellation, and the arbitrary replacement of the eight-pointed stars into this central location is consequentially forced and clumsy in appearance. These two examples from this minbar indicate that Ottoman artists, for all their brilliance in other artistic disciplines, were not equally skilled in the methodology of more complex geometric patterns with multiple regions of differentiated symmetry.

The earliest significant Mughal building is the tomb of Humayun in Delhi. This beautiful building was constructed in 1560, and combines Persian elements with distinctive Indian influences. This combination of influences created a bold new architectural style. In time, Mughal architecture developed into one of the most refined and beautiful traditions within the Islamic world. Mughal architecture reached its full maturity during the reign of Jalal al-Din Akbar (1556-1605). His fortress/palace complex at Fatehpur Sikri, built between 1570 and 1580, is one of the great Islamic monuments in all of India, and is of equal importance and beauty to the other two great Islamic fortress/palaces complexes which have survived to the present day: the Alhambra in Spain, and the Topkapi in Istanbul. The most celebrated Mughal building is the Taj Mahal in Agra. This building was built by Shah Jahan between 1632 and 1647, and is the tomb of his wife Mumtaz. Like that of the earlier Indian Sultanate period, the architectural decoration of the Mughals was mostly stone. The Mughals introduced several important ornamental elements from Persia; including more complex geometric patterns, star vaulting, and distinctive floral styles. One of the primary Mughal uses of geometric patterns was in their pierced stone *jali* screens. These were made from either marble or red sandstone. Pierced stone screens have always been a popular form of geometric ornament within Islamic cultures; and, as mentioned earlier, many examples exist from the earliest Islamic period. The Mughals refined this tradition to a remarkable degree. Of particular interest to the question of design methodology are the aforementioned *jali* screens that employ the underlying generative tessellation along with the geometric pattern that the tessellation creates. As discussed previously, the most innovative Mughal application of geometric patterns was to the interior surfaces of domes. Yet the variety of geometric patterns that were most commonly used by Mughal artists were not particularly complex, and with rare exceptions were already well known in Persia and Transoxiana.

The last great Persian architectural and ornamental traditions were those of the Safavid dynasty. This was an important time in the history of Persian culture. It was during this period that Shia Islam became the official religion of Persia. The Safavids are descendants of Shaikh Safi al-Din (d. 1334), the founder of the Safawiyya Sufi order in Azerbaijan. The Safavid dynasty was founded by Shah Ismail who lived between 1501 and 1524. He was a popular leader who united Persia under a single leadership. Shah Ismail claimed to be a direct descendent from Ali, the son-in-law of the Prophet Mohammad, providing Shah Ismail and all subsequent Safavid rulers a religious authority that was strongly embraced by his Persian subjects. The architectural ornament of the Safavids is primarily characterized by the abundant use of floral designs and calligraphy. The preferences for floral and calligraphic ornament notwithstanding, many exceptional examples of geometric domes and panels were produced by the Safavids.

Following the fall of the Safavids in the first half of the eighteenth century, Persia, Afghanistan, and Transoxiana came under the rule of several rival dynasties, including the Qarjars, Durannis, and Uzbeks. The architectural ornament of the Qarjars and the Shaybanid Uzbeks in particular was strongly influenced by the work of their Timurid predecessors, and the use of geometric patterns took on greater emphasis than during the Safavid period. However, knowledge of the methodology for the polygonal technique appears to have been lost, as the geometric work of these cultures is, at best, derivative of earlier geometric design. Most of this ornament is undertaken in ceramic tile, and as with the late Ottoman use of ceramics, the color palette was radically altered by the introduction of European ceramic colors that were unknown in this region previously. Adding to this change was an emphasis on more naturalistic floral designs such as found at the Vakil mosque in Shiraz (1766). The architectural decoration of the Qarjars is, generally, of much poorer quality in design and technique than the work of earlier Persian traditions. The baroque-inspired floral designs, as well as the new color palette, give many Qarjar buildings an overworked decadent quality. The Uzbek architectural focus upon continuing the Timurid aesthetic was more successful than the parallel attempts by the Qarjars. This was dealt a crippling blow with their defeat and occupation by the Russians in the nineteenth century.

By the nineteenth century, the unfortunate decline of Islamic geometric patterns had become irreversible throughout the Islamic world. In addition to the huge areas under Turkish, Persian, and Indian influence, the Islamic regions of Central Asia and North Africa also lost the vitality of this tradition. The history of the decline of geometric patterns in Morocco is similar to that of Uzbekistan. Although the architectural decoration of Morocco has continued to use geometric designs up to the present day, Moroccan artists and designers also lost their skills in creating new and original geometric patterns from the polygonal technique. The architectural record indicates that the significant decline of this methodological tradition in Morocco was well advanced by the eighteenth century. As in other Muslim cultures, most Moroccan geometric art of the eighteenth and nineteenth centuries was derivative of earlier work, and appears to have been created from less innovative design methodologies such as the grid paper technique, and assemblages of *zillij* tesserae into different known arrangements. With the loss of knowledge for constructing original complex geometric patterns from the polygonal technique, artists throughout the Islamic world ended up with little alternative but to *copy* existing patterns from the past. Of course the copying of geometric patterns is a perfectly acceptable traditional practice, and can be traced back to the earliest Islamic period. Patterns found in the compasswork window grilles of the Great Mosque of Damascus were also used in the carved stucco arch soffits in the mosque of Ibn Tulin: and without doubt, specific geometric patterns were used repeatedly throughout Muslim cultures. However, a vital artistic tradition cannot be sustained and advanced by mere copying. Without the methodological knowledge required for the creation of new and original geometric patterns being handed-down to successive generations of artists, this tradition sadly slid from decline to inexorable demise. Yet through reawakening the traditional design methodologies that engendered the creative vitality that sustained generations of Muslim geometric artists, this remarkable artistic discipline can once again provide inspiration to new generations of artists and designers.

Differentiation: Geometric Diversity and Design Classification

2.1 Need for Classification

The wide-ranging diversity of Islamic geometric patterns is a testimony to the degree of understanding that early Muslim pattern artists had of geometry and symmetry. Their inspired use of geometry led to the development of multiple varieties of pattern, symmetrical stratagems, and generative methodologies; the likes of which no other ancient culture came close to equaling in ingenuity and beauty. The diversity and complexity of this design tradition make it difficult to categorize, and indeed, no systematized method of comprehensive classification has been established. At best, writers and scholars addressing this subject employ descriptive analysis; for example, "The design . . . is a fully developed star pattern based upon a triangular grid. Its primary unit is a six-pointed star inscribed within a hexagon, which is surrounded by six five-pointed stars whose external sides form a larger hexagon."[1] However, detailed descriptions rarely elucidate beyond the visually obvious star types and square or triangular repeat units. Other fundamental features are frequently unaddressed when examining a given geometric pattern, including the symmetrical schema for more complex designs, the crystallographic plane symmetry group, the generative methodology, the incorporation of culturally associated additive features and treatments, and identification of the specific pattern family. The absence of appreciation for these less obvious, but nonetheless significant design features obscures the extraordinary scope of this design tradition, and it is only through a more nuanced and differentiated approach to this study, with its myriad cultural and geometric attributes, that a thorough understanding and appreciation of Islamic geometric patterns can be achieved.

The benefits of a more comprehensive approach to the classification of Islamic geometric patterns are wide ranging,

and highly relevant to historians of Islamic art and architecture, as well as to contemporary artists, designers, and architects who use such designs in their work. In addition to the more general enhanced appreciation of the width and breadth of this ornamental tradition, the highly detailed classification of geometric patterns according to their overall symmetry, repetitive schema, numeric qualities, generative methodology, family type, and additive pattern variations and treatments has very specific relevance to each Muslim culture and dynasty. From the perspective of art and architectural history, the ascription of these differentiated qualities to the ornamental use of geometric designs allows for a far greater understanding of the artistic practices of a given Muslim culture, as well as an enhanced comparative appreciation for the subtle differences between the design conventions of neighboring and succeeding cultures. What is more, the categorization of the diverse geometric characteristics that comprise this design tradition provides the necessary methodological knowledge for those who wish to more fully explore the range of possibilities and unlimited potential for creating fresh original geometric designs that these historical methodologies still offer. It is only through such knowledge that this once great ornamental tradition can be rekindled into a contemporary artistic movement endowed with creative vitality.

Despite the expressed rationale for a more detailed categorization of Islamic geometric patterns, there is no evidence to suggest that Muslim designers of the past were particularly concerned with a need to systematically organize their geometric patterns into differentiated categories. The design scrolls that have survived to the present day are a random collection of diverse ornamental motifs that include *Kufi* calligraphy, *muqarnas*, star net vaulting, domical gore segments with geometric designs, and a wide variety of two-dimensional geometric patterns. The fact that these pattern scrolls have no logical sequence in the placement of their many individual designs obviously does not imply that Muslim designers had no appreciation for geometric

---

[1] This quotation references a geometric pattern used on a door at the Bimaristan al-Nuri in Damascus (1154). Tabbaa (2001), 88.

© Jay Bonner 2017
J. Bonner, *Islamic Geometric Patterns*, DOI 10.1007/978-1-4419-0217-7_2

differentiation within this ornamental tradition. On the contrary, the full range of sophistication in this Islamic design tradition, in and of itself, provides clear evidence that Muslim artists had a highly sophisticated knowledge of geometric diversity, but did not require this knowledge to be outwardly systematized. The history of collecting and classifying Islamic geometric patterns is closely associated with nineteenth-century orientalism: frequently with the objective of making illustrated representations of specific patterns available to Western artists working with this novel aesthetic.[2] The publication of *The Grammar of Ornament* by Owen Jones in 1856 included numerous geometric designs from Muslim sources.[3] The organizing principle behind this work was loosely ethnographical rather than geometric; with chapters dedicated to *Arabian*, *Turkish*, *Moresque*, *Persian*, and *Indian* ornament, and the examples of geometric design within these sections are arbitrarily sequenced alongside their floral and calligraphic neighbors. The earliest work to organize geometric patterns into geometric categories was published in 1879 by the ornamental theoretician and architect Jules Bourgoin.[4] The 190 geometric designs that comprise this collection are divided into eight numeric and geometric categories: hexagonal patterns; octagonal designs; dodecagonal designs; patterns with two different star forms; designs with squares and octagons; patterns with three and four different star forms; sevenfold patterns; and fivefold patterns with ten-pointed stars. While these categories may seem somewhat limited today, at the time this collection was a significant contribution to the spread of interest in this subject throughout Europe, and continues to be a standard reference book for Islamic geometric pattern to this day.[5] The history of the classification of Islamic geometric design is an interesting study in itself, and has mostly built upon the overtly obvious categories identified by Bourgoin. This organizational refinement began during the beginning of the last quarter of the twentieth century with the publication of several books on the

subject of Islamic geometric patterns.[6] For the most part, these more recent studies have included the ordering of patterns that repeat upon the isometric and orthogonal grids by complexity, as well as patterns that have fivefold symmetry. In the isometric examples the least complex designs are comprised of triangles, hexagons, and six-pointed stars. These are followed by patterns that place increasingly large star forms upon the vertices of the repetitive grid (and/or its hexagonal dual) whose local symmetry is always a multiple of 3: e.g., 9-, 12-, and 15-pointed stars. In some studies, recognition is also given to patterns with greater complexity that exhibit more than a single region of local symmetry, for example, the well-known designs with 9- and 12-pointed stars. Orthogonal patterns are similarly organized by repeat unit and increasing complexity: the least complex being comprised of squares, octagons, and eight-pointed stars, followed by more complex designs with star forms that are multiples of 4. The more thorough studies include designs with more than one region of local symmetry, such as 8- and 12-, 8- and 16-, as well as 8- and 24-pointed stars.[7] The most comprehensive twentieth-century catalogue of Islamic geometric design was published by Gerd Schneider in 1980.[8] This study focuses exclusively on the geometric ornament of the Seljuk Sultanate of Rum, under whose patronage this ornamental tradition produced many of the most sophisticated and complex geometric designs. Schneider illustrates 440 patterns that are not differentiated according to their repetitive structure, but placed within a broad set of visually explicit categories that include square *Kufi* calligraphy; orthogonal brick designs; domical brick designs; three-, four-, five-, and six-fold swastika designs; border designs; additive designs; superimposed polygonal designs; star patterns with extended points; patterns made up of a single repetitive device; patterns with 6-pointed stars; patterns with 6- and 12-pointed stars; patterns with hexagonal centers; fourfold patterns with square centers; 8-pointed star patterns with octagons; complicated 8-pointed star patterns; 9-pointed star patterns; pentagonal designs with 5- and 10-pointed stars; 10-pointed star patterns; 12-pointed star patterns; patterns with 8- and 12-pointed stars; patterns with 9- and 12-pointed stars; 12-pointed star patterns with additional star forms; 12- and 14-pointed star patterns; 16-pointed star patterns with additional star forms; 15- and 18-pointed star patterns with additional star forms; 24-pointed star patterns; and geometric patterns on domes and hemispheres. Many of

---

[2] Necipoğlu (1995), Chapter 4. *Ornamentalism and Orientalism: the Nineteenth and Early Twentieth Century European Literature*, 61–87.

[3] Jones (1856).

[4] Bourgoin (1879).

[5] The ongoing availability of Bourgoin's work is due to its being kept in print as part of the Dover Pictorial Archive Series (printed without original text). In creating the illustrations for his book, Bourgoin does not appear to have used a traditional methodology for recreating the patterns in his collection. As a consequence, the proportions within many of his illustrations—especially those with greater complexity—are inaccurately represented, and have clearly discernable distortion. Being that this has been an artist's reference for over 150 years, the direct copying of such problematic designs has occasionally promulgated these errors by their application within the applied and architectural arts.

---

[6] –Critchlow (1976).
  –El-Said and Parman (1976).
  –Wade (1976).

[7] Wade (1976), 63–79.

[8] Schneider (1980).

the designs illustrated in this valuable study are not represented elsewhere. Equally impressive is the work of Jean-Marc Castéra, dating from 1999, which focuses upon the geometric design diversity found in the Moroccan ornamental tradition.[9] This work includes orthogonal designs with increasingly large numbers of primary star forms (up to 96-pointed stars), the hexagonal family, the pentagonal family, and patterns with two varieties of star. This is also one of the earliest publications to categorize a subset of dual-level geometric designs, herein referred to as Type D, as having self-similar properties: a type of design created from what Castéra calls the "Alhambra method."[10] The differentiation between systematic and nonsystematic generative methodologies was first introduced by the author in 2003,[11] with the identification for the first time of four historical systems used for creating geometric patterns: one that produces patterns that can be either threefold or fourfold; two that produce patterns that are fourfold; and one that produces patterns that are fivefold. Further, this work also identified three geometrically and aesthetically distinct varieties of dual-level design with self-similar characteristics that were reliant upon these systems for their creation. This was expanded upon in 2012 to include the historical use of a system that generates sevenfold geometric patterns.[12] The application of Islamic geometric patterns to the parameters of the 17 plane symmetry groups is a particularly interesting development in the efforts toward methodical categorization. Beginning in 1944 a number of mathematicians and crystallographers have published works devoted to this topic.[13] Of especial note is the work of Syed Jan Abas and Amer Shaker Salman, dating from 1995, that identifies some 248 Islamic geometric patterns with their respective crystallographic plane symmetry group.[14]

The abounding diversity of this design tradition necessitates categorization according to several criteria. The standard classification of Islamic geometric patterns has provided a useful means for descriptive dialogue, and is certainly relevant to art historians and contemporary artists alike. However, this does not provide any insight into the methods used in the creation of these designs. The categorization according to methodology and pattern family

that concludes this chapter is a subject that has been largely overlooked by previous studies, but is fundamental to the thorough understanding of this ornamental tradition.

2.2 Classification by Symmetry and Repetitive Stratagems

In examining this tradition, the most fundamental category to which all patterns must ascribe is *geometric symmetry*. Most Islamic geometric patterns exhibit threefold, fourfold, or fivefold symmetry, although other more obscure symmetrical systems were also developed within this tradition. Directly related to a pattern's symmetry is its *repeat unit*. Islamic geometric patterns are able to continuously fill the plane through the repetitive use of a single element. These repeat units will always contain the minimum portion of a pattern that is able to seamlessly fill the plane through repetitive edge-to-edge translation symmetry. In this way, the repeat unit is essential to a pattern's ability to successfully fill two-dimensional space. The laws that govern the science of repetitive two-dimensional space filling, or tiling, are universal, and apply no less to Islamic geometric design than to any other pattern-orientated ornamental tradition. All periodic covering of the two-dimensional plane must conform to the symmetrical determinants of one or another of the 17 plane symmetry groups. Yet these limits offer tremendous scope for symmetrical and aesthetic diversity. And no artistic tradition explored symmetrical potential with the degree of passion and ingenuity as those of successive Muslim cultures.

A regular polygon is defined as having equal included angles and common edge lengths. As illustrated in Fig. 1, only three of the regular polygons are able to infinitely cover a plane on their own: the triangle, square, and hexagon. Figure 1a illustrates the isometric grid made from equilateral triangles, along with its hexagonal dual (green); Fig. 1b shows the orthogonal grid made from squares, with the dual grid (green) being the same orthogonal grid; and Fig. 1c shows the hexagonal grid made from regular hexagons, along with the isometric dual grid (green). The majority of Islamic geometric patterns repeat upon either the isometric or the orthogonal grids. Each vertex of the isometric grid has sixfold symmetry comprised of six acute angles of 60° where the six edge-to-edge equilateral triangles meet. The dual of the isometric grid is the grid of regular hexagons. This grid has threefold symmetry that results from the three coincident hexagons with 120° included angles that meet at each vertex. Islamic geometric patterns that repeat upon the isometric grid will invariably exhibit sixfold symmetry at the vertices of this grid, and threefold symmetry at the centers of each triangular repeat unit—which is to say the vertices of the dual-hexagonal grid. As such, these patterns have regions of both sixfold and

[9] Castéra (1996).

[10] Castéra (1996), 276–277.

[11] Bonner (2003).

[12] –Bonner and Pelletier (2012).
   –Pelletier and Bonner (2012).

[13] –Müller (1944).
   –Weyl (1952).
   –Lalvani (1982).
   –Lalvani (1989).
   –Lovric (2003), 423–431.

[14] Abas and Salman (1995).

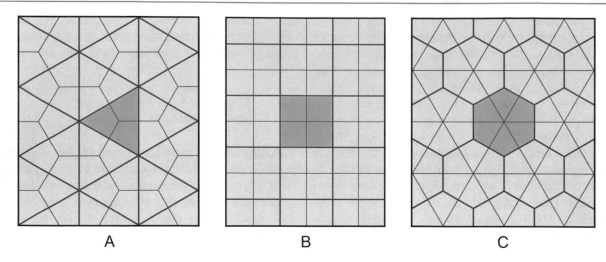

A   B   C

Fig. 1

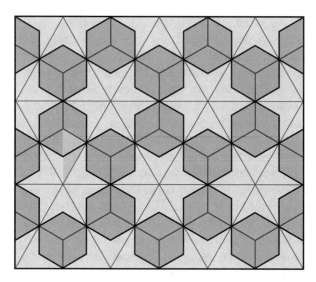

Fig. 2

threefold symmetry. However, for purposes of convenience, this category of geometric design is referred to as simply threefold. This is acceptable due to the fact that, three being a devisor of six, the sixfold vertices also have threefold symmetry. More complex threefold patterns will place higher ordered star forms at the vertices of either or both these grids, and the number of points of these stars will always be a multiple of the threefold or sixfold symmetry of the vertex. The vertices of the orthogonal grid have fourfold symmetry resulting from the four coincident squares, with 90° included angles, that meet at each vertex; and the dual of the orthogonal grid is an identical orthogonal grid whose vertices are located at the center of each square repeat unit of the original grid. Patterns that employ square repeat units are therefore referred to as fourfold. Similarly with threefold patterns, fourfold designs will often place star forms at the vertices of both the orthogonal grid and its dual that are multiples of 4, thus creating regions of higher order local symmetry.

It is important to differentiate between the repeat unit of a given pattern and its *fundamental domain*. The fundamental domain is the minimal essential repetitive component of a design. By the singular or combined functions of rotation, reflection, and glide reflection, the fundamental domain will populate the repeat unit. It is remarkable how little visual information is contained within the fundamental domains of many highly successful, albeit less complex geometric designs. Figure 2 illustrates the classic threefold *median* pattern comprised of 6-pointed stars located upon the vertices of the isometric grid. The relationship between the design and both the isometric grid (green) and its hexagonal dual (red) is clearly evident; and indeed, the triangles or hexagons are equally capable of being used as the underlying generative polygons responsible for this design. The fundamental domain for this pattern is a right scalene triangle (blue) with a single applied pattern line. This is reflected and then rotated to complete the repeat unit: ×3 for the

triangle, and ×6 for the hexagon. Figure 3 illustrates the classic fourfold star-and-cross *median* pattern that places eight-pointed stars at the vertices of the orthogonal grid (green) and fourfold crosses at the vertices of the dual of this grid (red). The fundamental domain is a right isosceles triangle (blue) with just two applied pattern lines. By reflecting the fundamental domain upon its hypotenuse, and rotating this four times at the vertex of the dual grid, the square repeat unit will be completed. Alternatively, rotating four times at the vertex of the repeat unit will fill a unit cell of the dual grid.

As said, the historical record is rife with Islamic geometric patterns based upon threefold and fourfold symmetry that respectively utilize the triangle, hexagon, and square as repeat units. Yet as early as the eleventh century Muslim artists began working with distinctive patterns characterized by fivefold and even sevenfold symmetry. This was made possible through the use of rhombic, rectangular, and elongated hexagonal repeat units and resulting repetitive grids with proportions that directly relate to fivefold and sevenfold symmetry. What is more, the proportions of these types of alternative repeat units could also conform to symmetries more commonly associated with the regular polygons. In this way, it was possible to create patterns with higher order star forms with points that are multiples of 3 and 4 (for example: 8, 12, 15, 16, 18, and 24) that were not confined to the isometric and orthogonal grids, yet often shared visual characteristics with their more conventional counterparts. These three less common repetitive stratagems are elongated corollaries of the three grids produced from the regular polygons: the isometric grid sharing properties with rhombic grids; the orthogonal grid with rectangular grids; and the regular hexagonal grid with elongated hexagonal grids. Changing the edge lengths and/or included angles of the polygonal components of these three regular grids such that the new angles and edge lengths correspond with the inherent proportions of specified polygons opened this tradition to the creation of designs with all manner of symmetries, including fourfold patterns with 8-pointed stars set upon a rhombic grid; fivefold patterns with

10-pointed stars set upon both rhombic and rectangular grids; sevenfold patterns with 14-pointed stars set upon both rhombic and rectangular grids; and patterns with 12-pointed stars set upon a rectangular grid. This more flexible approach to repeat units with specific inherent proportional properties also allowed for the creation of more complex designs with multiple centers of local symmetry that would ordinarily be incompatible. Such designs are invariably nonsystematic and include a pattern with 7- and 9-pointed stars set upon an elongated hexagonal grid; a pattern with 9- and 11-pointed stars set upon an elongated hexagonal grid; and a pattern with 11- and 13-pointed stars that is also set upon an elongated hexagonal grid.

The isometric grid is made up of three sets of parallel lines. By removing one of these sets a rhombic grid is produced. Each rhombus becomes a repeat unit with the proportion of two edge-to-edge equilateral triangles. The location and number of vertices remain identical to the original isometric grid. Figure 4 illustrates a very successful example of a class of pattern that uses this rhombic repetitive schema by placing nine-pointed stars at each isometric vertex. Whereas nines will work nicely at the vertices of the regular hexagonal grid, they do not conform to the vertex constraints of the regular triangular grid (because 9 is not evenly divisible by 6). The placement of nines upon the vertices of the rhombic grid elegantly overcomes this limitation. The fundamental domain for this design is an equilateral triangle (blue) that is reflected to create the rhombic repeat unit with translation symmetry. Figure 5 illustrates the proportional determinants for the two rhombi with fivefold symmetry that were used historically for patterns with 5- and 10-pointed stars. The opposing included angles of both these rhombi are multiples of 36°: a 1/10 division of the circle. Figure 5a illustrates the wide rhombus with two opposing acute angles with 2/10 included angles, and two opposing obtuse angles with 3/10 included angles. The acute included angles of the thin rhombus in Fig. 5b are a 1/10 segment, and the obtuse angles are 4/10 segments. Figure 6 illustrates the obtuse and acute fivefold grids that these two rhombi produce; and Fig. 7a shows how the wide rhombi relates to the pentagon, and Fig. 7b demonstrates how the thin rhombi relates to the decagon. Figure 8 illustrates two geometric patterns that repeat with these two rhombi.

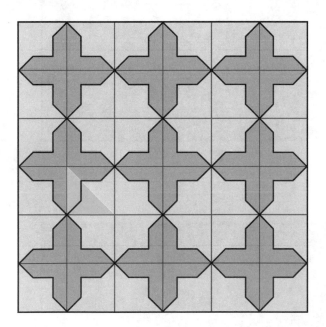

Fig. 3

Fig. 4

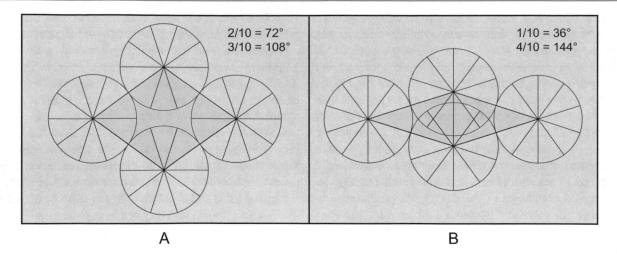

Fig. 5

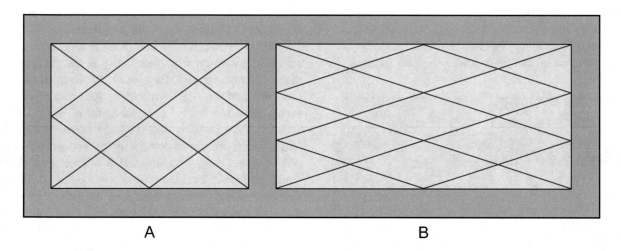

Fig. 6

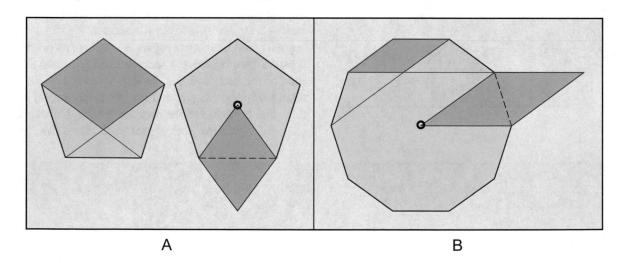

Fig. 7

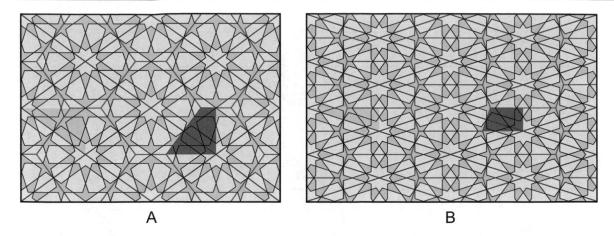

Fig. 8

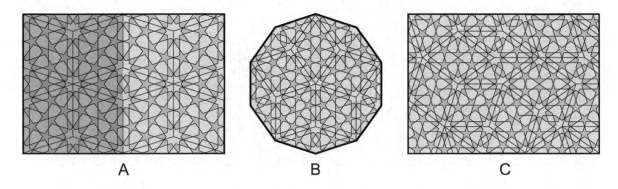

Fig. 9

Figure 8a is the classic fivefold *acute* pattern that repeats on the obtuse rhombic grid (red) and has a dual-hexagonal repetitive grid (green). The pattern in Fig. 8b (by author) repeats upon the acute rhombic grid (red) and also has a dual-hexagonal repetitive grid (green). The fundamental domain of both varieties of rhombic repeat unit is a 1/4 triangular segment (blue) that requires reflection 4×, while the fundamental domain of both types of the dual-hexagonal repeats is a 1/4 quadrilateral (orange) that also requires reflection 4×. Figure 9 demonstrates how these 2 fivefold rhombi can be used together to tessellate the plane in several fashions. The combined use of more than a single repetitive element qualifies each of the three examples in this figure as a hybrid design. Figure 9a is an example of a periodic tessellation with translation symmetry that is provided by a rectangular repeat unit (shaded) made up of four obtuse rhombi and two acute rhombi; Fig. 9b is an example of a radial tessellation; and Fig. 9c is an example of a non-periodic tessellation devoid of translation symmetry.[15]

The success of such hybrid designs is conditioned upon the applied pattern lines along the edge of each rhombus being congruent. Although no historical examples of fivefold hybrid designs that use just these two rhombi are known, several periodic fivefold hybrid designs were produced that employ multiple repetitive elements, including rhombi, pentagons, triangles, and non-regular hexagons, always with the requisite matching edge configurations within the applied pattern lines [Figs. 261–268].

The same rhombic repetitive logic applies to the generation of sevenfold geometric patterns. Figure 10 illustrates the proportional determinants for the three rhombi with seven-fold symmetry, with the obtuse rhombus in Fig. 10a being

---

[15] These 2 fivefold rhombi are the same as those identified by Sir Roger Penrose in his groundbreaking research into aperiodic tilings. However, the application of the geometric patterns to the two rhombi in Fig. 9

does not include Penrose's matching rules for forced aperiodicity and the design in Fig. 9c is therefore referred to herein as non-periodic rather than aperiodic. While never occurring within the historical record, it is certainly possible to populate these 2 fivefold rhombi with patterns that conform to the Penrose matching rules, thereby forcing the geometric design to be aperiodic [Figs. 480 and 482].

–Penrose (1974), 266–271.
–Gardener, Martin (January 1977), "Extraordinary nonperiodic tiling that enriches the theory of tiling," *Scientific America*, pp. 110–121.
–Penrose (1978), 16–22.

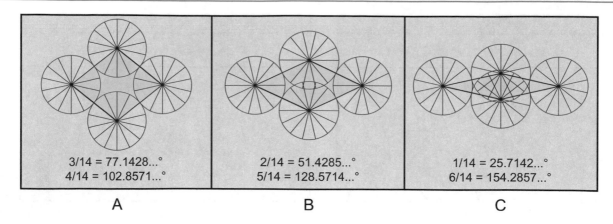

3/14 = 77.1428...°   2/14 = 51.4285...°   1/14 = 25.7142...°
4/14 = 102.8571...°   5/14 = 128.5714...°   6/14 = 154.2857...°

A   B   C

Fig. 10

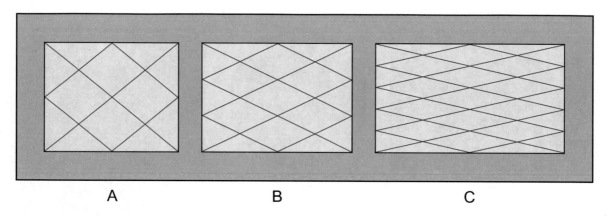

A   B   C

Fig. 11

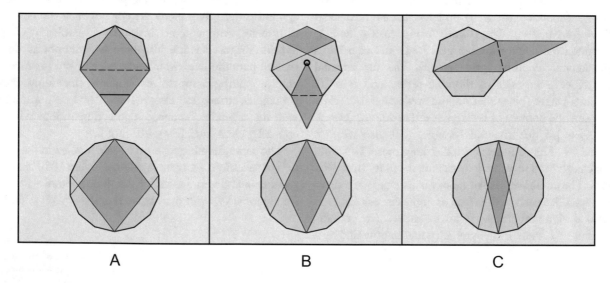

A   B   C

Fig. 12

comprised of 3/14 and 4/14 included angles; the median rhombus in Fig. 10b having 2/14 and 5/14 included angles; and the acute rhombus in Fig. 10c having 1/14 and 6/14 included angles. Figure 11 shows the three rhombic grids that these three rhombi produce. Figure 12 demonstrates several simple methods for creating the 3 sevenfold rhombi from the heptagon and tetradecagon. Figure 12a shows two ways of creating the obtuse rhombus; Fig. 12b shows two

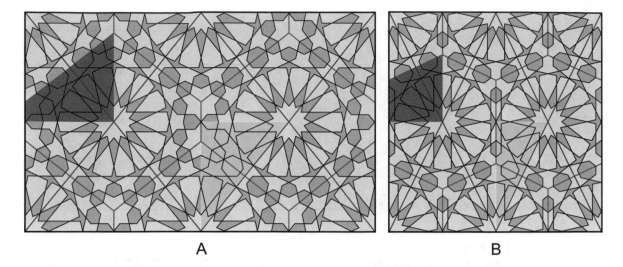

A   B

Fig. 13

ways of producing the median rhombus; and Fig. 12c shows two ways of producing the acute rhombus. Figure 13 illustrates two historical examples of patterns that employ the obtuse and median sevenfold rhombic repeat units. The design in Fig. 13a is from the 'Abd al-Ghani al-Fakhri mosque in Cairo (1418). This repeats with either the obtuse rhombic grid (red) or the dual-hexagonal grid (green). The fundamental domain of the rhombic repeat unit is a 1/4 segment right triangle (blue) that requires reflection 4× to fill the repeat. The fundamental domain of the dual-hexagonal repeat unit is a 1/4 segment quadrilateral (orange) that also requires reflection 4×. The design in Fig. 13b is from the Bayezid Pasha mosque in Amasya, Turkey (1414-19). This utilizes the median sevenfold rhombic grid (red), and also has a hexagonal dual grid (green). Like the previous example, the fundamental domain of the rhombic repeat is a 1/4 segment right triangle (blue) that requires reflection 4× to fill the repeat, and the fundamental domain of the dual-hexagonal repeat (orange) is a 1/4 segment quadrilateral that also requires reflection 4×. The acute sevenfold rhombus does not appear to have been used historically. As with the fivefold rhombi, the 3 sevenfold rhombi can be used with one another to create more complex periodic [Fig. 284], radial [Fig. 285], and non-periodic hybrid designs,[16] and indeed, these two historical examples have the requisite identical edge configuration to produce hybrid variations [Fig. 26d]. However, no examples of sevenfold hybrid designs are known from the historical record.

The use of rectangular repeat units for geometric designs with symmetries that do not readily conform with either the isometric or the orthogonal grids began in Khurasan during the late twelfth century. Figure 14 illustrates such an

Fig. 14

example from the Maghak-i Attari mosque in Bukhara, Uzbekistan (1178-79). This is one of the earliest examples of a pattern that repeats upon a rectangular grid, and is also one of the least complex rectangular fivefold patterns. This design places ten-pointed stars at the vertices of the rectangular grid (red), and the specific proportions of the rectangular repeat unit are determined by the arrangement of the underlying generative polygonal modules from the *fivefold system* that are responsible for this pattern [Figs. 203 and 245a]. As with the orthogonal grid, the dual of a rectangular grid is the same rectangular grid (green). Fundamental domains for designs that utilize rectangular repeat units are almost always a 1/4 rectangular segment (blue) that fills the repeat unit through reflection 4×.

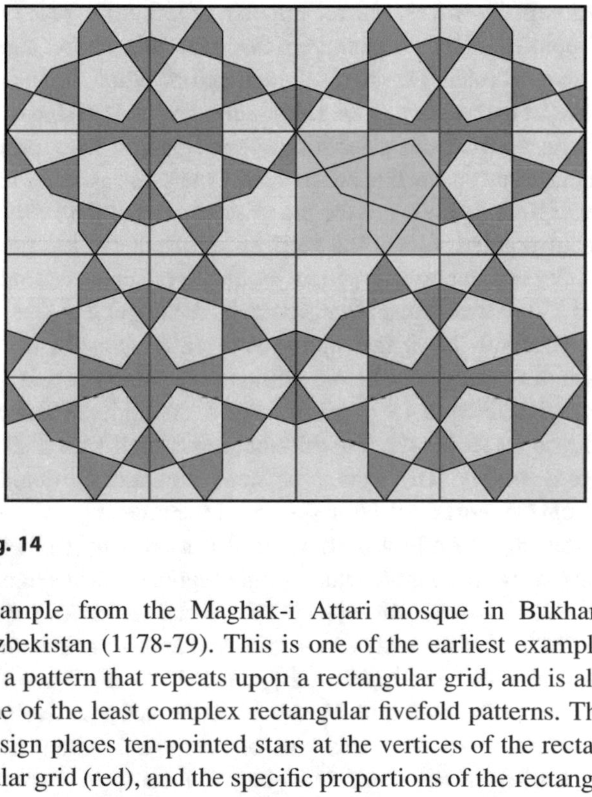

Figure 15 illustrates a design from the Sultan al-Mu'ayyad Shaikh complex in Cairo (1412-22) that is created from the *sevenfold system* and repeats upon both a

---

[16] Pelletier and Bonner (2012), 141–148.

Fig. 15

rectangular grid (red) and a hexagonal grid (green) [Fig. 294] [Photograph 50]. The hexagonal repeat unit is half the area of the rectangle, and is the true minimal repetitive cell. However, the rectangular repeat units place the 14-pointed stars at their vertices, and this is frequently more convenient for practical application. When considered from the perspective of the rectangular grid, this design has a second unusual characteristic: applied pattern lines on the rectangular repeats that are precisely the same as on its dual grid. (Note: were it not for the skewed orientation between the 14-pointed stars at the vertices and center of each rectangular repeat unit, this pattern would repeat upon a rhombic grid.) The fundamental domain of the rectangular repeat is a quadrilateral (blue) that must rotate 180° upon the center point of the long edge before filling the remaining repeat unit through reflection 4× to fill the repeat. As said, the hexagonal grid (green) is the true minimal repeat unit with translation symmetry. This shares the same fundamental domain, but only requires reflection 4× to fill the repeat.

Rectangular repeat units were also used with geometric designs that have more than a single region of local symmetry. Such patterns will typically place one variety of star at the vertices of the rectangular grid, and another star form at the center of each repeat unit; which is to say, upon the vertices of the dual grid. Figure 16 illustrates a particularly successful example of this type of compound pattern from the *minbar* of the Great Mosque of Aksaray in Turkey[17] (1150-53). This places 12-pointed stars upon the vertices of the primary grid (red), and 10-pointed stars on the vertices of the identically proportioned dual grid (green). The proportions of this rectangular repeat unit are the direct product of the correlation between the 12- and 10-fold local symmetries as they relate to the underlying polygonal tessellation that generates this design [Fig. 414]. The

Fig. 16

fundamental domain (blue) of this pattern is a 1/4 segment of the repeat unit that fills the unit by reflection 4×.

The use of non-regular hexagonal repeat units encompasses a wide variety of design types, including systematic and nonsystematic patterns, more simplistic field patterns, and very complex patterns with compound local symmetries and multiple star forms. The discovery that this repetitive

---

[17] Schneider (1980), pattern no. 416.

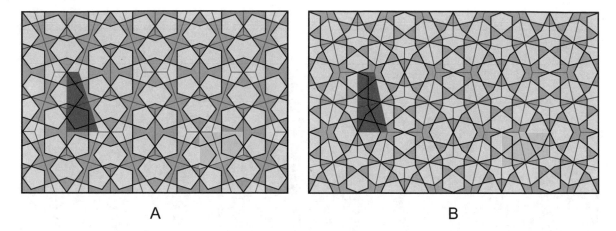

A   B

Fig. 17

stratagem was applicable to symmetries that do not conform to the convenient tessellating properties of the regular triangle, square, and regular hexagon can be traced back to the sevenfold design used by Seljuk artists in the northeast dome chamber of the Friday Mosque at Isfahan (1088-89) [Fig. 279] [Photograph 26], and the sevenfold designs created by Ghaznavid artists on the tower of Mas'ud III (1099-1115) [Figs. 280 and 281]. Figure 17 illustrates two rather simple, but nonetheless elegant, field designs that repeat upon non-regular hexagonal grids. Figure 17a is created from the *fivefold system*, and Fig. 17b from the *sevenfold system*. Similar to rhombic repeat units, each of the included angles of the hexagonal repeat units for both of these patterns (red) are multiples of a 10- and 14-fold division of a circle, respectively. The perpendicularly orientated dual of each of these hexagonal grids is also a hexagonal grid (green), and their included angles are likewise multiples of 10- and 14-fold divisions of a circle. Both the repetitive grid and its dual for each of these patterns have their own quadrilateral fundamental domain, and each fills the repeat unit through reflection 4×. Both of these patterns are from the Seljuk Sultanate of Rum: the fivefold pattern from the Sitte Melik tomb in Divrigi (1196) [Fig. 213], and the sevenfold design from the Great Mosque of Dunaysir in Kiziltepe (1204) [Fig. 282a]. Many of the more complex patterns that utilize a non-regular hexagonal repeat unit will have a combination of differing star forms that are seemingly irreconcilable in their geometric symmetry. Figure 18 is a remarkable design from the Mu'mine Khatun in Nakhichevan, Azerbaijan (1186). This design has two regions of local symmetry: 13-fold placed upon the vertices of the hexagonal primary grid (red), and 11-fold located at the vertices of the perpendicularly orientated hexagonal dual grid (green). The combination of equal numbers of 13- and 11-pointed stars requires a geometric dexterity that pushes the limits of two-dimensional space filling. The fundamental domain for each type of hexagonal repeat is a right-angled

quadrilateral that is a 1/4 segment of their respective repeat unit requiring reflection 4× to fill their respective repeat unit. Figure 19 illustrates the origin of the included angles of the 13-fold and 11-fold hexagonal repeat units from this design. Figure 19a illustrates a 1/13 division of a 13-fold tridecagon. Four of the included angles of the 13-fold hexagonal repeat unit are made up of three-and-a-half 1/13 segments, and two are made up of six 1/13 segments. Figure 19b illustrates a 1/11 division of a 11-fold hendecagon. Four of the included angles of the 11-fold hexagonal repeat unit are comprised of three 1/11 segments, while the remaining two included angles have five 1/11 segments.

The use of parallelograms as a repetitive device in the Islamic geometric tradition is extremely rare. One such example is from the Khwaja Atabek mausoleum in Kerman (1100-1150) [Fig. 211]. Figure 20 illustrates several repetitive features of this design. Figures 20a and b illustrate two distinct chevron repeat units, each comprised of mirrored parallelograms. The fundamental domains for these repeat units are rotated 180° and then reflected to fill the chevron repeat units. Figure 20c shows a 1/5 decagonal kite repetitive element that must be rotated 180° for edge-to-edge translation symmetry. The fundamental domain for this kite is a 1/10 triangular segment of a decagon that is reflected to fill the repetitive element.

Most motifs with a radial symmetry in Islamic ornament tend to be floral. However, Muslim designers also created many geometric radial patterns; mostly created from one or another of the polygonal systems (although the *sevenfold system* does not appear to have been used for such designs). Radial geometric designs that are applied to the gore segments of domes are a special category within this tradition. Whether on a two-dimensional plane or on the curved surface of a dome, a pattern with radial symmetry is significantly different from patterns that employ translation symmetry to cover the plane. Radial patterns work by dividing a circle into *n* number of equal divisions, and treating each

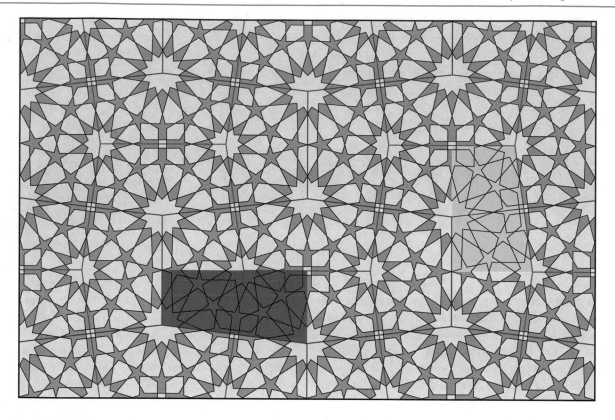

Fig. 18

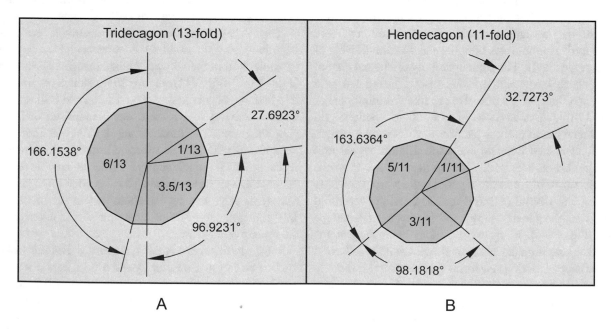

Fig. 19

segment as a distinct repetitive element that is copied and rotated *n* times around the center of the circle. In this way, the pattern is made to repeat through rotation along the radius of the circle. The Islamic conventions for applying radial geometric patterns onto domes most commonly employ 8-, 12-, 16- or 24-fold gore segments. Each of these relates comfortably to the square or octagonal base upon which domes most commonly rested. In his 1925 publication *The Drawing of Geometric Patterns in Saracenic Art*, E. H. Hankin demonstrates a Mughal

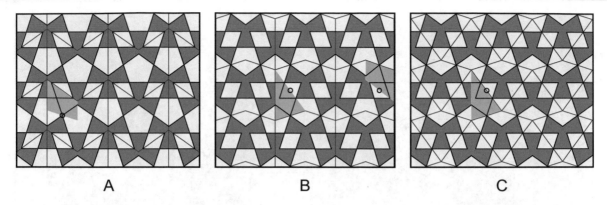

Fig. 20

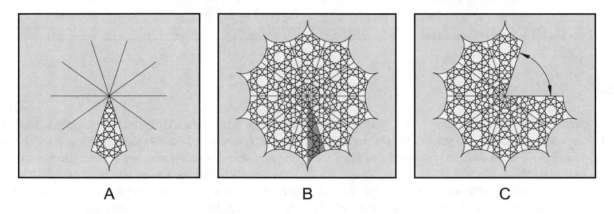

Fig. 21

technique for designing the gore segments of four domes in Fatehpur Sikri, India.[18] Each example that Hankin cites is created from the *fivefold system*, and characterized by the judicious use of ten-pointed stars. Figure 21 illustrates his analysis of the dome in the Samosa Mahal at Fatehpur Sikri, India (sixteenth century). This illustration shows how the geometric pattern is designed to fit the 1/10 division of a circle, and demonstrates how this segment can be arrayed around the central point ten times to create a very satisfactory radial pattern. Figure 21b shows how the fundamental domain of this radial design (blue) is half of the 1/10 segment divided through its central axis and reflected to fill the segment. As reported by Hankin, the Mughal technique for applying such patterns onto the three-dimensional interior surface of a dome called for the removal of two of the ten segments, and adjoining the remaining eight segments into a conical form that could then be applied to the curved surface of the dome with minimal distortion. This technique has the benefit of maintaining the integrity of this type of fivefold pattern even while being applied to a three-dimensional surface. Other than the minimal distortion, the only real

change to the pattern is that the central star will have eight points rather than the original ten. A feature of this methodology is the fact that the curvature of the dome is a direct result of the chosen geometric design. This is distinct from the more common approach to designing domical geometric patterns wherein the design is applied to a predetermined gore segment. Each segment of the design from the Samosa Mahal has acute projections at the periphery that, when applied to the dome, extend downward into eight arched pendentives. Historical examples of geometric designs with radial symmetry are far less common than two-dimensional patterns that repeat with translation symmetry. Some of the more interesting examples of two-dimensional radial design are from the flat stellate soffits that were incorporated into Persian *muqarnas* vaults during the Safavid period [Figs. 440 and 441], and from the secondary infill of dual-level designs [Fig. 447]. Figure 22 illustrates two radial designs with tenfold rotation symmetry from the Topkapi Scroll. Figure 22a is an *obtuse* pattern from the ten-pointed star component of a dual-level design from this collection of designs,[19] and Fig. 22b is the full dodecagonal infill of the

---

[18] Hankin (1925a), Figs. 45–50.

[19] Necipoğlu (1995), diagram no. 29.

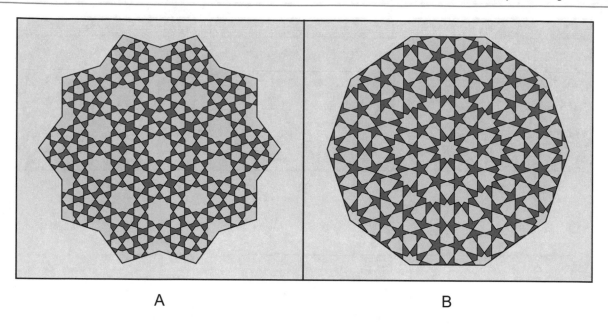

A   B

Fig. 22

1/10 segment of an *acute* pattern[20] that was possibly intended for use on a dome—as per the Mughal technique illustrated in Fig. 21. Both designs were produced from the *fivefold system* of pattern generation.

Another means of incorporating unusual symmetrical relationships in Islamic geometric patterns utilizes a particular tessellation of squares and rhombic repetitive units. This variety of tessellation is characterized by the square elements oscillating in orientation, and the rhombi being placed in an alternating perpendicular layout. Designs based upon this configuration of squares and rhombi are orthogonal, but eccentric; and are herein referred to as *oscillating square* patterns. The geometric structure of this variety of Islamic geometric pattern invariably adheres to the *p4g* plane symmetry group. This rather simple geometric repetitive device was occasionally used as ornament in and of itself, and an early example is found in the carved stucco panels of the Khirbat al-Mafjar outside Jericho (eighth century). However oscillating square tessellations can also provide the repetitive structure for more complex Islamic geometric patterns. In this class of design, the angular proportions of the rhombic elements will always inform the geometric characteristics of the completed pattern. Figure 23 illustrates the geometric principle behind two historical oscillating square patterns. Figure 23a illustrates the square-within-a-square motif. In this particular oscillating square tessellation, the rhombi are made up of two contiguous equilateral triangles, and the distribution of squares and rhombi is effectively the $3^2.4.3.4$ semi-regular tessellation [Fig. 89]. Oscillating square patterns

are characterized by multiple lines of symmetry, leading to a surprising number of equally valid repeat units with translation symmetry for a single design. Figure 23b places diagonal lines (green) within the square elements of this tessellation. This produces a grid comprised of concave octagonal shield shapes (orange) that tessellate through 90° rotation. The fundamental domain (blue) is rotated 4× to populate the minimal square region, and this is reflected 4× to complete a square repeat with translation symmetry. Each of these square repeat units has 16 fundamental domains. However, this is not the minimal repeat unit with translation symmetry. This grid also produces two types of hexagonal repeat unit (brown) with translation symmetry, each comprised of just eight fundamental domains. These are identical except for their respective 90° orientations, and the proportions of each are based upon 105° and 150° included angles. Figure 23c places an alternative set of lines (green) within the oscillating square elements that bisect the midpoints of each set of parallel edges. This creates the dual of the original square and rhombic grid (green) in Fig. 23a, and is similarly comprised of alternating concave octagonal shield repetitive units (orange), although with very different proportions as those from Fig. 23b. The fundamental domain (blue) for this dual grid is likewise a 1/4 segment of the minimal square region that is rotated 4× and reflected 4× to complete the square repeat unit with 16 fundamental domains. This dual grid also produces two types of hexagonal repeat units (brown) that are perpendicularly orientated and comprised of just eight fundamental domains. Like the regular hexagon, these non-regular hexagonal repeat units have 120° at each included angle: the difference being the two edge lengths rather than uniform

---

[20] Necipoğlu (1995), diagram no. 90a.

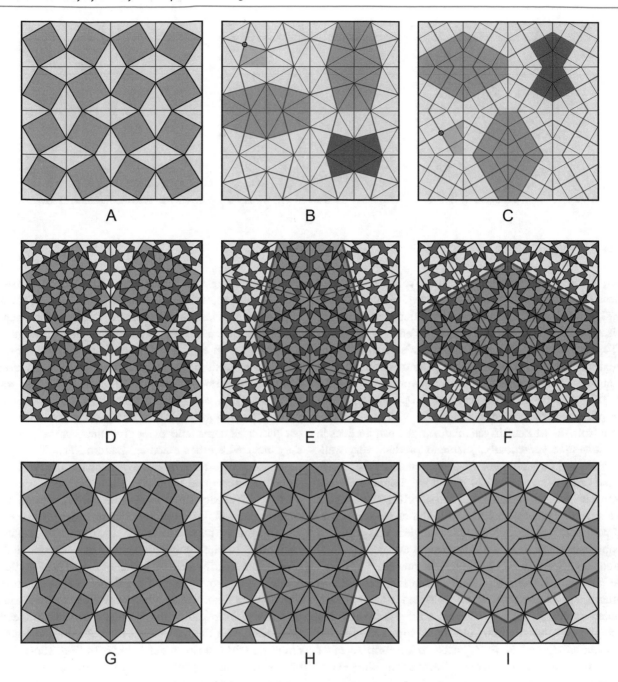

Fig. 23

edges. The two historical designs in this figure are imbued with each of these repetitive characteristics. The pattern in Fig. 23d through **f** is from the exterior stucco ornament of the Mustansiriyah *madrasa* in Baghdad (1227-34), as well as the Topkapi Scroll.[21] Figure 23d emphasizes the oscillating square and rhombic repetitive cells that govern the geometry of this pattern. The applied pattern lines within both the oscillating squares and rhombi are able to fill the plane

independently with very acceptable designs, and their combined use in these examples qualifies this example as a hybrid design. The pattern within just the rhombus is a variant of an isometric nonsystematic design with 12-pointed stars at the vertices of the isometric grid [Fig. 321b], while the design within each square cell is a very well-known nonsystematic design that places 12-pointed stars at the vertices of the orthogonal grid and 8-pointed stars at the center of each repeat unit [Fig. 379b]. Figure 23e demonstrates the placement of this pattern within the hexagonal repeat unit with 105° and 150° included angles, and Fig. 23f shows how the pattern fits

---

[21] Necipoğlu (1995), diagram no. 35.

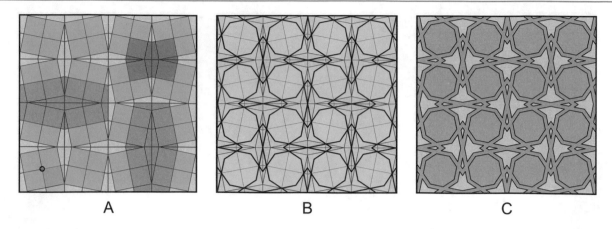

Fig. 24

within the repetitive hexagon with exclusively 120° included angles. The design with non-regular seven-pointed stars in Fig. 23g through **i** is found in several locations, including the Malik mosque in Kerman, Iran (eleventh century), as well as the Topkapi Scroll.[22] Figure 23g places this pattern into the oscillating squares and rhombi. The pattern is composed of 90° angular openings placed at the midpoints of the $3^2.4.3.4$ grid. This produces the non-regular seven-pointed stars that are a primary feature of this design. As with the previous pattern from the Mustansiriyah *madrasa*, the pattern lines in both the square and rhombic elements produce very-well-known designs on their own: the squares making the classic star-and-cross design [Fig. 124b], and the rhombi making a pattern with point-to-point six-pointed stars [Fig. 95c]. Figure 23h shows this pattern placed within the hexagonal repeat unit with 105° and 150° included angles, and Fig. 23i demonstrates the placement of this pattern into the repetitive hexagon comprised of just 120° included angles.

As mentioned, the proportions of the rhombic cells in oscillating square patterns are not restricted to 60° and 120° included angles. Muslim geometric artist discovered that the angles of the rhombic elements within oscillating square tessellations can be adjusted to conform to other polygonal symmetries, thereby introducing the visual characteristics inherent to these forms. Figure 24 demonstrates an oscillating square pattern from the Sultan Han in Aksaray, Turkey[23] (1229). The rhombic repetitive cells in Fig. 24a have 22.5° and 157.5° included angles. The fundamental domain (blue) is rotated 4× to fill the square cell that is reflected 4× to produce a square repeat unit with translation symmetry that is made up of 16 fundamental domains. As with the previous designs, the dual of this tessellation (green) provides for the two perpendicular

elongated hexagonal repeat units (brown) comprised of eight fundamental domains. This figure also illustrates the concave octagonal shield element (orange) that requires alternating 90° rotations to cover the plane. And as with the examples in Fig. 23, this oscillating square grid will also repeat with perpendicular hexagonal grids created from diagonal lines placed within each oscillating square (not shown). Figure 24b illustrates how the pattern can be derived from simply placing octagons within each of the oscillating squares and extending the pattern lines until they meet with other extended pattern lines. The specific proportions of the rhombi provide for the pattern lines to extend uninterrupted from octagon to octagon through the center point of each rhombus. Figure 24c is a representation of the historical design with widened pattern lines. Figure 25 illustrates two additional historical examples of oscillating square tessellations that use rhombi that are proportioned differently from that of the double-equilateral triangle. The proportions of the rhombic elements used in Fig. 25a are associated with sevenfold symmetry and can be derived from either the heptagon or the tetradecagon [Fig. 12c], with the acute angles being 1/14 divisions of a circle, and the obtuse angles being 6/14 [Fig. 10c]. The use of this rhombus elegantly provides for the incorporation of regular seven-pointed stars within a pattern matrix that is otherwise orthogonal in structure. This design was used in numerous locations, including the Mirjaniyya *madrasa* in Baghdad (1357), and the Amir Qijmas al-Ishaqi mosque in Cairo[24] (1479-81). Figure 25a illustrates the fundamental domain (blue) that is rotated 4× to populate the square that is then reflected 4× to produce a square repeat unit with translation symmetry. This repeat has 16 fundamental domains and, as with the previous examples, is not the

[22] Necipoğlu (1995), diagram no. 81a.

[23] Schnieder (1980), pattern no. 297.

[24] Bourgoin (1879), pl. 170.

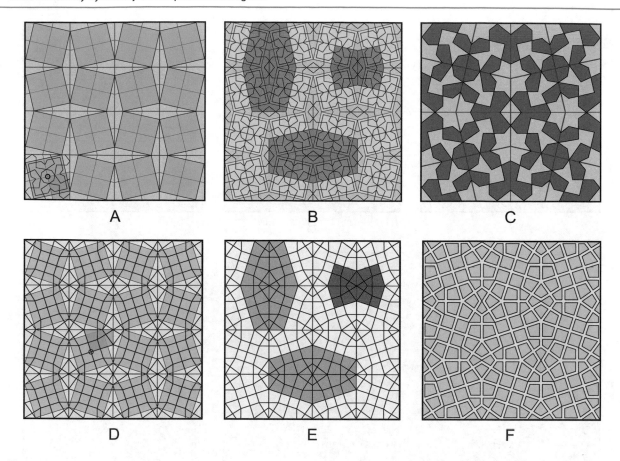

Fig. 25

minimal repeat unit. Figure 25b illustrates the two perpendicular hexagonal grids (green) that, together, are the dual of the square and rhombus tessellation. The two elongated perpendicular hexagons (brown) are the minimal repeat units with translation symmetry, comprised of eight fundamental domains. This figure also illustrates the repetitive shield element (orange) that requires alternating 90° rotations to cover the plane. This pattern will also repeat with the two perpendicularly oriented hexagonal grids created from the diagonal lines applied to each oscillating square (not shown). Figure 25c is a representation of the very successful historical design based upon this repetitive geometric schema. Figures 25d through f illustrate an interesting oscillating square raised brick design from the western tomb tower at Kharraqan in northwestern Iran (1093). This incorporates squares, near-regular pentagons, and near-regular triangles into the pattern matrix. Figure 25d illustrates the oscillating squares located within the orthogonal repetitive element. The fundamental domain (blue) requires rotation 4× to populate each square cell, and this, in turn, is reflected 4× to produce a square repeat unit with translation symmetry. The included angles of the rhombi are 36° and 144°: equaling 1/10 and 4/10 segments of the decagon. The applied pattern lines include lines from the dual grid that bisect the

midpoints of the oscillating squares, as well as an arbitrary network of pentagons, squares, and rhombi that complete the design. While visually becoming, the aesthetics of this design are atypical to this tradition. Figure 25e illustrates both orientations of the identical elongated hexagonal repeat unit. These are minimal repeat units with translation symmetry, and comprised of eight fundamental domains. This figure also illustrates the repetitive shield element (orange) that requires alternating 90° rotation to fill the plane. This is comprised of just four fundamental domains. The alternative hexagonal repeat unit produced from diagonal lines within each oscillating square also works as a repeat unit with translation symmetry (not shown). Figure 25f represents the widened line expression of this design as per the historical example.

Although not known to the historical record, a variation of oscillating square designs will make interesting patterns with unique repetitive structures. Figure 26a demonstrates the ability of two varieties of rhombus to tessellate together in a similar manner as the squares and rhombi of oscillating square configurations. In this variation, obtuse rhombi replace the square modules, and acute rhombi are placed on each of the edges of the obtuse rhombi such that they are in a rotational pinwheel arrangement. This can be thought of as a

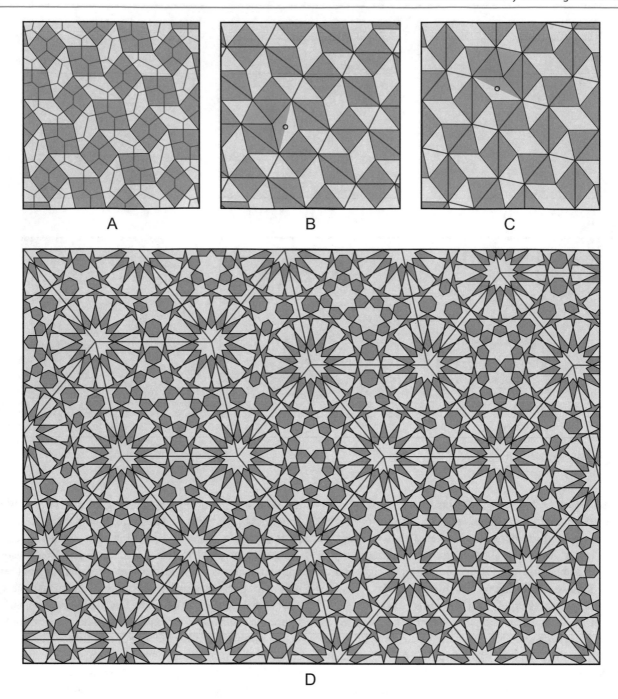

Fig. 26

skewed oscillating square tessellation, and the angular proportions of both the rhombi in this particular example are derived from sevenfold symmetry [Figs. 10a and b]. The dual of this grid (green) creates an interesting tessellation of irregular hexagons. Figure 26b illustrates how each hexagonal repeat unit (red) contains the area of two obtuse rhombi and two acute rhombi. Because of the skewed nature of the rhombic tessellation, the hexagonal repeat does not have reflection symmetry, and the fundamental domain (blue) requires rotation ×2 (point symmetry) to fill the repeat.

This changes the *p4g* plane symmetry group typical of oscillating square patterns to *p2*: one of the least common plane symmetry groups among Islamic geometric patterns. Figure 26c shows how the repetitive grid for this rhombic tessellation can also be oriented in a roughly perpendicular direction. This is analogous to the perpendicular hexagonal repeats of standard oscillating square designs. Figure 26d applies a sevenfold geometric pattern to each of the two varieties of rhombus in this unusual tessellation. The combined use of two types of rhombus qualifies this as a hybrid design (by author). Although examples of this variety of hybrid design are unknown within the historical record, this is a successful variation of traditional design methodology. The pattern placed within the obtuse rhombi is from the 'Abd al-Ghani al-Fakhri mosque in Cairo (1418) [Fig. 13a], and the pattern within the acute rhombi is from the Bayezid Pasa mosque in Amasya, Turkey (1414-19) [Fig. 13b].

Other historical examples of particularly fine oscillating square designs include a Khwarizmshahid example from the Zuzan *madrasa* in northeastern Iran (1219) [Fig. 103] [Photograph 39], and a remarkably complex Anatolian Seljuk example from the Huang Hatun complex in Kayseri (1237) created from the *fourfold system A*[25] [Fig. 156]. As with the above-mentioned oscillating square patterns from the Mustansiriyah *madrasa* in Baghdad and the Malik mosque in Kerman, the design from the Huang Hatun complex in Kayseri is also represented in the Topkapi Scroll.[26] Other oscillating square patterns in the Topkapi Scroll include a simple but affective design that incorporates floating squares and rhombi in a matrix of four-pointed stars with swastika centers[27], and a design that is suitable for polychrome *ablaq* inlaid stone that places swastikas inside the square elements and utilizes rhombi with 45° and 135° included angles.[28] While oscillating square patterns are essentially fourfold in that they repeat upon a square grid, their distinctive arrangement of local symmetries creates the topsy-turvy quality that is a hallmark of this unusual category of Islamic geometric design. It is worth noting that this same repetitive schema can be used to create substantially more complex designs than those found within the historical record. This category of design requires the star or regular polygon at the center of the square cell to have fourfold symmetry, and for a set of four additional centers of local symmetry to be located upon the edges of the square cell. These additional centers of local symmetry are required to have bilateral symmetry so that they mirror upon the square edges. This type of geometric construction is relatively unexplored, and lends itself to contemporary pattern making [Figs. 406–411].

Another historical method for introducing seemingly incompatible symmetries into an orthogonal repetitive structure employs the placement of four quadrilateral kites rotated around a central square. As with oscillating square designs, the *rotating kite* motif is mirrored into adjacent square cells, creating an overall reciprocating structure, and like oscillating square designs, this closely related variety of Islamic geometric pattern is invariably of the *p4g* plane symmetry group. This is a well-known ornamental

motif in its own right, but was occasionally used as a repetitive stratagem for more complex designs. Figure 27a demonstrates two simple methods of constructing a common form of the rotating kite motif: one from a single square, and the other from a 3 × 3 grid of nine squares. In addition to the two mirrored 90° included angles, these kites have acute angles of 53.1301...° and obtuse angles of 126.8698...°. Figure 27b shows the structural composition of the standard rotating kite design, and examples with this proportion and widened line thickness abound, including a Mughal high-relief red sandstone panel at the Agra Fort (1550). Among the earliest examples are several Ghurid raised brick panels from the exterior of both the western (1167) and eastern (late twelfth century) mausolea at Chisht in Afghanistan. These have wider pattern lines, but are otherwise identical. The fundamental domain (blue) is rotated 4× to create a square that is then reflected 4× to create a square repeat unit with translation symmetry. Figure 27c represents a Seljuk variation from the brickwork façade of the western tomb tower at Kharraqan, Iran (1093). This utilizes the same fundamental domain (not shown). As with oscillating square designs, the included angles of the kites can be adjusted for specific proportions and symmetries. The proportions of these two examples are the most commonly found within the historical record, and are characterized by the length of the square elements being equal to the short edges of each kite. Figure 28 illustrates two rotating kite patterns that utilize a kite and square tessellation with 60° and 120° included angles accompanying the two obligatory 90° angles. Rather than being used as the design itself, these two examples make use of this repetitive schema to construct a far more elaborate design. Just as the included angles of the rhombic elements in oscillating square patterns can be adjusted to conform to *n*-fold symmetry, the acute and obtuse angles of the kites in rotating kite designs can also be adjusted to accommodate local symmetries that are ordinarily incompatible with orthogonal patterns. While the 90° included angles of the kite are invariable, the *n*-fold symmetry of the kite's acute and obtuse included angles is required to be even numbered, thus imposing fourfold reflective symmetry on the pattern elements centered upon the vertices of each kite's acute and obtuse included angles. This includes *n*-fold symmetries that are not divisible by 4, such as 6 and 10. Figure 28a is from a stone *jali* screen at the Taj Mahal in India (1632-48). The 60° and 120° included angles of each kite provide the sixfold local symmetry at each vertex of the orthogonal grid. This allows for six-pointed stars to be located at each vertex of the kite's acute and obtuse angles, and the orientation of these stars is rotated by 90° from alternating vertices. The design in Fig. 28b is from the Topkapi Scroll.[29] This employs the

---

[25] Schneider (1980), pattern no. 330.

[26] Necipoğlu (1995), diagram no. 61.

[27] Necipoğlu (1995), diagram no. 41.

[28] Necipoğlu (1995), diagram no. 69b.

[29] Necipoğlu (1995), diagram no. 59.

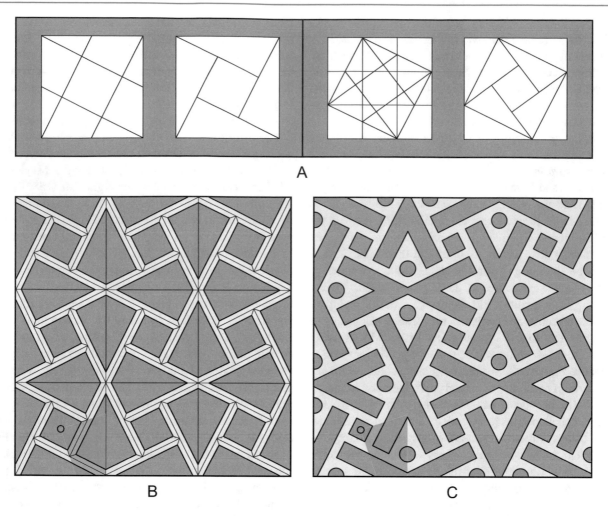

A

B

C

Fig. 27

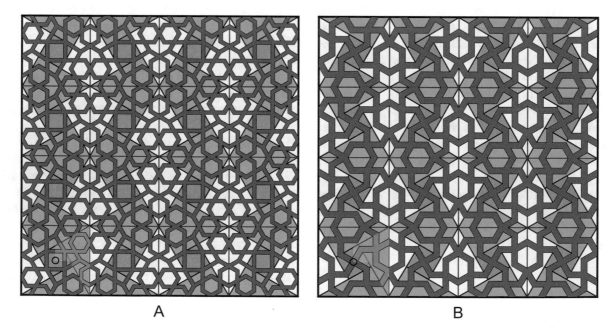

A

B

Fig. 28

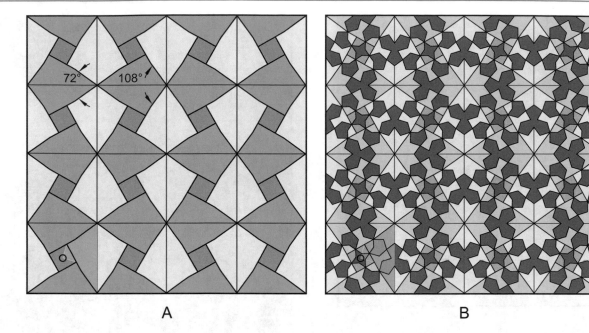

Fig. 29

same repetitive schema, and is also characterized by an alternating distribution of six-pointed stars. The fundamental domain for each of these designs (blue) is rotated 4× followed by reflection 4× to fill the square repeat with translation symmetry. Figure 29 illustrates a design based upon the repetitive structure of rotating kites that has ten-pointed stars placed at the vertices of the orthogonal grid, and four-pointed stars within the square elements of the tessellation. This unusual pattern is also from the Topkapi Scroll.[30] Figure 29a indicates the 72° and 108° included angles of the kites that correspond to tenfold symmetry. Each of the lines emanating from the square elements are slightly kinked so that they are not collinear, with 6.7783...° off 180°, providing each kite with six sides rather than four. While somewhat forced, this allows for the tenfold symmetry of the acute and obtuse angles to combine with a large fourfold center within each square cell, which in turn produces a four-pointed star that is balanced with the other elements within the pattern matrix. This example is testament to the flexible methodological practices employed by artists engaged in this tradition. Figure 29b shows the design from the Topkapi Scroll along with its repetitive schema. As with the six-pointed stars from the previous example, the ten-pointed stars are placed in 90° alternating orientation at the vertices of the orthogonal grid. As with other examples of this variety of pattern, the fundamental domain for this design (blue) rotates 4× followed by reflection 4× to fill the square repeat with translation symmetry.

2.3   Classification by Numeric Quality

Another means of classifying Islamic geometric patterns takes into account their prevalent numeric qualities. Because of the variables within this design tradition, this type of classification requires descriptive text rather than a singular nomenclature. When categorizing geometric patterns from this perspective, the numbers of points found in the characteristic star forms with n-fold rotation symmetry are particularly significant. The least complex and easiest to describe are those patterns with only a single variety of primary star; for example, the classic star-and-cross pattern that can be described as a *fourfold pattern, with point-to-point eight-pointed stars that repeat upon an orthogonal grid*: or more concisely, 8s on squares [Fig. 3]. As patterns become more complex, the identification of their numeric qualities becomes a useful tool for differentiating the particular attributes of a given design, as well as qualifying the scope and potential within this tradition overall. In addition to star forms, many patterns will incorporate regular polygons as key elements of the design that are located at the vertices of the primary grid or its dual. In abbreviating the numeric description of a given design that includes such polygons it is useful to follow the nomenclature of Gerd Schneider[31] by using Roman numerals for distinguishing these primary regular polygons. This is especially helpful in differentiating between regular polygonal features and stars with n-fold

---

[30] Necipoğlu (1995), diagram no. 72c.

[31] Schneider (1980).

Fig. 30

Fig. 31

Fig. 32

symmetry. Figure 30 illustrates a widely utilized fourfold design comprised of eight-pointed stars and regular octagons. Describing this pattern as having *fourfold symmetry, with eight-pointed stars placed at the vertices of the orthogonal grid and octagons upon the vertices of the dual grid*, or simply 8s on squares/VIIIs at center, does not uniquely apply to this pattern alone, but identifies it within a category into which only a select number of other patterns fall [Figs. 173a, 175a, 176b, 177a, 177c, etc.]. Figure 31 illustrates a fourfold design that repeats on a rhombic grid (red) with 45° and 135° angles. This was used in several locations historically, including: the Lower Maqam Ibrahim in the citadel of Aleppo, Syria; and the Izzeddin Kaykavus hospital and mausoleum in Sivas, Turkey (1217). This design places eight-pointed stars on the vertices of a rhombic grid, and octagons upon the vertices of the hexagonal dual grid (green): or 8s on rhombic vertices/VIIIs on hexagonal dual vertices [Fig. 181]. This example also illustrates how the duals of rhombic grids are always hexagonal grids.

Not all Islamic geometric patterns employ star motifs. Some are composed of a repetitive field of polygonal forms. Such *field patterns* are most commonly made up of either threefold or fourfold symmetry, although fivefold field patterns are also well known, and especially appealing. Figure 32 shows a well-known threefold pattern comprised of two sizes of hexagons, the larger placed at the vertices of the isometric grid and the smaller upon the vertices of the hexagonal dual grid: or VIs on triangle/smaller VIs at center [Fig. 96d]. Figure 33 illustrates a fourfold field pattern comprised of two sets of differently sized octagons, one set placed upon the vertices of the orthogonal grid, and the other on the vertices of the orthogonal dual grid: or simply, VIIIs

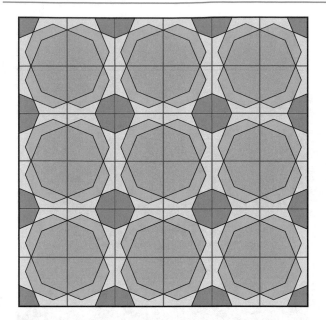

Fig. 33

Fig. 34

on square/smaller VIIIs at center. This was used in the celebrated Baghdad Quran (1001) produced by Ibn al-Bawwab [Photograph 6] [Figs. 127c and 128d]. Figure 34 illustrates a fivefold field pattern from the Great Mosque at Malatya (1237-38). This design can be described as a matrix of regular pentagons, concave octagonal shields, kites, and decagonal hourglass figures that will repeat upon several alternative grids with translation symmetry. These include a rectangular grid (red); the rectangular dual of this grid (green), both with eight fundamental domains; and a grid of hexagonal repeat units (brown) made up of just four fundamental domains (blue). The lack of higher order polygons or stars at the vertices of the different repetitive cells makes it more difficult to ascribe an abbreviated description with the tools discussed thus far. To populate the rectangular repeat units, the fundamental domains are rotated 2× and then reflected 4×, and to fill the hexagonal repeat unit the fundamental domain is simply reflected 4×. The plane symmetry group is *cmm*. The lack of stellar centers and the similarity in size of the polygonal elements give this example a pleasing homogeneous aesthetic that is a common quality of fivefold field patterns [Fig. 220].

As discussed previously, there is a direct corollary between the *n*-fold rotational symmetry at the vertices of a repetitive grid and the numeric quality of geometric star patterns. With threefold patterns, the vertices of the isometric grid support the application of stars with *n*-fold rotational symmetry that are multiples of 6. The vertices of the hexagonal dual grid similarly support stars whose points are multiples of 3. The simplest threefold star patterns employ six-pointed stars; and more complex designs will have higher numbered stars, such as 9, 12, and 15. The most

Fig. 35

Fig. 36

Fig. 37

common of the more complex threefold designs employ 12-pointed stars placed at the vertices of the isometric grid. Figure 35 is just such a design, and can be concisely described simply as 12s on triangle, for which it is one among many designs that fit this simple description [Figs. 300, 320, 321, etc.]. This example was used above the entrance to the tomb of Umar al-Suhrawardi in Baghdad (1234) [Fig. 300a, *two-point*]. Figure 36 shows a similar design [Fig. 321j], but with a curvilinear treatment. This is from a Turkish miniature (1558) painted during the reign of Süleyman the Magnificent.[32] This is a threefold curvilinear pattern, with 12-pointed stars at the vertices of the isometric grid: or simply curvilinear 12s on triangle. While less frequently used, patterns with nine-pointed stars at their repetitive vertices are particularly interesting. Figure 4 is an example of this type of pattern from the Great Mosque at Malatya (1237-38) comprised of threefold symmetry, with nine-pointed stars at the vertices of a rhombic grid: or simply 9s on rhombus [Fig. 311]. As mentioned, threefold patterns with *n*-pointed stars that are higher multiples of 6 and 3 were also widely used. Figure 37 shows an exquisite design with 24-pointed stars in the vertices of the isometric grid and 7-pointed stars within the field: or 24s on triangle/7s in field. This pattern was executed in the carved stone relief

of the portal at the Nalinci Baba tomb and *madrasa* in Konya, Turkey (1255-65), and in the cut-tile mosaic *mihrab* niche at the Esrefoglu Süleyman Bey mosque in Beysehir, Turkey (1296-97) [Fig. 327] [Photograph 44].

Numeric description becomes especially relevant when differentiating patterns with more than a single region of local symmetry. Figure 38 is a classic threefold compound pattern used throughout the Islamic world that uses 12-pointed stars set upon the vertices of the isometric grid, and 9-pointed stars at the vertices of the hexagonal dual grid: or 12s on triangle/9s at center [Fig. 346a]. Figure 39 illustrates a more complex threefold compound pattern from a carved stone lintel at the Qartawiyya *madrasa* in Tripoli, Lebanon (1316-26). This pattern has 12-pointed stars at the vertices of the isometric grid and 15-pointed stars at the vertices of the hexagonal dual grid: or simply, 12s on triangle/15s at center [Fig. 355d].

Similar to threefold designs, the application of stars to the vertices of the orthogonal grid, as well as to the center of each square repeat unit, will invariably exhibit *n*-fold

---

[32] Süleymanname: *Presentations of gifts to Süleyman the Magnificent on the occasion of the circumcision of his sons Bayezid and Cihangir in 1530* by Ali b. Amir beg Sirvani. Topkapi Museum, Istanbul TKS H. 1517. See: Rogers and Ward (1988), 45c (f. 360a).

Fig. 38

rotational symmetry that is a multiple of 4. In this way, patterns with 8-, 12-, and 16-pointed stars are very common, and higher order stars, such as those with 24 points, are not unusual. Figure 40 is an illustration of a fourfold pattern that was used throughout the Islamic world and is made up of 12-pointed stars placed at the corners of a square repeat unit with a 4-pointed star placed at the center of the repeat: or simply, 12s on square/4s at center [Fig. 113a]. Figure 41 illustrates an orthogonal design with 16-pointed stars at the vertices of the orthogonal grid with octagons at the vertices of the dual grid: or 16s on square/VIIIs at center. This fine design was used to illuminate a Mamluk Quran commissioned by Sultan Sha'ban in Cairo[33] (1369) [Fig. 344d].

Many of the more complex fourfold geometric patterns will incorporate higher order star forms at the vertices of both the orthogonal grid and its orthogonal dual: each constrained by the same multiple-of-four numeric mandate. Figure 42 shows a variant of a compound pattern with 12-pointed stars at the vertices, and 8-pointed stars at the center points: or just 12s on square/8s at center. This particular version of this well-known design is located at the Kale mosque in Divrigi (1180-81) [Fig. 379b]. Figure 43 is a fourfold compound pattern with 16-pointed stars at the vertices and 8-pointed stars in the centers: or 16s on square/8s at center. This example was used in the Quran of Uljaytu[34]

Fig. 39

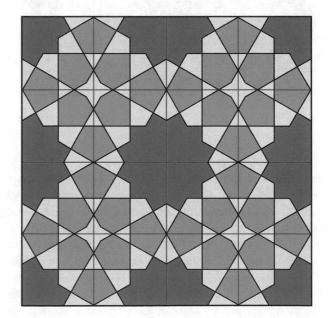

[33] Cairo, National Library, 7, ff. IV-2r.

[34] This Ilkhanid Quran is in the National Library in Cairo: 72, pt.19.   **Fig. 40**

Fig. 41

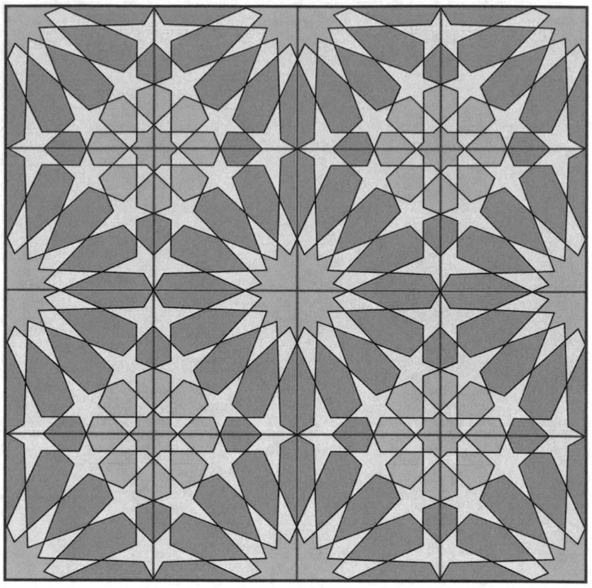

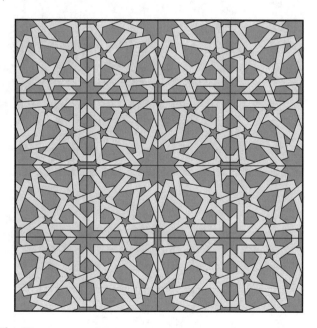

Fig. 42

Fig. 43

Fig. 44

(1313), written and illuminated by 'Abd Allah ibn Muhammad al-Hamadani [Fig. 389a]. Figure 44 shows a considerably more complex fourfold compound pattern with 16-pointed stars at the vertices of the orthogonal grid and 12-pointed stars at the vertices of the dual grid: or 16s on square/12s at center. This beautiful pattern was used on the minaret of the Mughulbay Taz mosque in Cairo (1466) [Fig. 396b]. Figure 45 shows a fourfold design with 20-pointed stars at the vertices of the orthogonal grid, and 8-pointed stars in the center of each square repeat unit: or 20s on square/8s at center. However, it is relevant to further note that the eight-pointed star at the center of the repeat unit

is an arbitrary feature. This exceptional Sa'did pattern is from the Badi' Palace in Marrakesh, Morocco (1578-1594). These examples are but a few of the vast number of fourfold designs with two primary regions of local symmetry employed frequently within the tradition of Islamic geometric patterns. It is worth noting that both of the previous examples employ seven-pointed stars within their pattern matrix, but these have not been included in categorizing according to their primary stars. This is due to the fact that in both cases the seven-pointed stars are not regular, are not placed upon nodal centers, and are, therefore, not primary star forms.

Fig. 45

Fig. 46

Historically, star patterns with fivefold symmetry are typically limited to a single variety of primary star: ten-pointed. Because these patterns are derived from the *fivefold system* they are limited by the polygonal modules that comprise this system, including the decagon as the generative module for the ten-pointed stars. Occasionally, 20-pointed stars were incorporated into patterns created from this system: creating fivefold designs with two varieties of primary star. Figure 46 is a very successful design that places 20-pointed stars at the vertices of a rectangular grid, 20-pointed stars at the vertices of the rectangular dual grid, and a network of 10-pointed stars upon the repetitive edges and within the field of the design: or more concisely, 20s on rectangle/20 at center/10s on edges/10s in field [Fig. 268]. As with many, but not all, patterns that repeat with a rectangular grid and have the same star at the center of the repeat as at the vertices, the pattern within the repeat is exactly the same as that of the dual. This feature can be described as *self-dualing*. This outstanding fivefold pattern was employed in the ornament of the Bu 'Inaniyya *madrasa* in Fez (1350-55). Figure 13a illustrates a Mamluk pattern from the 'Abd al-Ghani al-Fakhri in Cairo (1418) that is created from the *sevenfold system*, and characterized by two varieties of primary star: the 14-pointed stars located on the vertices of the rhombic grid, and the 7-pointed stars placed upon the vertices of the hexagonal dual grid. This can be abbreviated as 14s on rhombus/7s on hexagonal dual [Fig. 292a].

Among nonsystematic designs with two varieties of primary star are those that are neither threefold nor fourfold, and utilize other repetitive structures such as rectangular grids and irregular hexagonal grids. Figure 16 shows a design from the Great Mosque of Aksaray in Turkey (1150-53) that has 12-pointed stars at the vertices of its rectangular grid and 10-pointed stars at the vertices of the rectangular dual grid: or simply, 12s on rectangle/10s at center [Fig. 414]. Figure 47 shows one of the more geometrically complex nonsystematic designs from the Topkapi Scroll.[35] This repeats with equal efficiency upon either the irregular hexagonal grid with 11-fold proportional angles at the vertices (red), or the perpendicular irregular hexagonal dual grid with 9-fold proportional angles at the vertices (green). This allows for the placement of 11-pointed stars at the vertices of the former irregular hexagonal grid, and 9-pointed stars at the vertices of the latter irregular hexagonal dual grid: or 11s on hexagons/9s on dual hexagons [Fig. 431]. Figure 18 is a conceptually similar design from the Mu'mine Khatun in Nakhichevan, Azerbaijan (1186), with 13- and 11-pointed stars at the vertices of the dualing hexagonal grids: or 13s on hexagons/11s on dual hexagons [Fig. 434].

Triangles and squares as repeat units also support considerably more complex compound patterns with three or more regions of local symmetry. These will often have unusual, and seemingly irreconcilable, combinations of star forms. As with less complex compound patterns, these will place appropriately numbered *n*-pointed stars at the vertices of

---

[35] Necipoğlu (1995), diagram no. 42.

Fig. 47   **Fig. 48**

both the repetitive grid and its dual. However, these more complex compound patterns are fixed upon these locations by placing added primary stars with regular *n*-fold rotation symmetry upon the edges of the repeat unit and/or along the bisecting radii of the repeat unit. The number of points for the stars at these secondary locations is less constrained by predetermined local symmetries, often resulting in star forms with unexpected numeric qualities. Figure 48 shows a threefold design from the Karatay Han near Kayseri, Turkey (1235-41) that has 12-pointed stars at the vertices of the triangular repeat unit, 9-pointed stars at the center of the repeat, 10-pointed stars at the midpoint of each edge of the repeat, and 11-pointed stars upon the bisecting radii of each corner of the repeat unit. This can be described more concisely as 12s on triangle/9s at center/10s on edge/11s on bisecting radii (not shown) [Fig. 367]. The stars that are located at the midpoints of the repetitive edges of such patterns are required to be even numbered, such as the ten-pointed stars in this example, while those located upon

the bisecting radii can be either even or odd numbered. Figure 49 shows an orthogonal design from the Agzikarahan near Aksaray, Turkey (1231), that places 12-pointed stars at the vertices of the square repeat unit, an octagon at the center of the repeat, 10-pointed stars at the midpoints of the edges of the repeat, and 9-pointed stars along the bisecting diagonals of the repeat unit. This can be described more briefly as 12s on square/VIII at center/10s on edges/9s on diagonals [Fig. 400]. Figure 50 illustrates a design with three varieties of higher order star that repeats upon a rectangular grid of unusually long proportions. This was reportedly used at the Lower Maqam Ibrahim in the citadel of Aleppo[36] (1168). This highly complex nonsystematic pattern places 12-pointed stars at the vertices of the rectangular repeat unit

---

[36] The wooden panel described and drawn by Herzfeld is no longer present at the Lower Maqam Ibrahim, and its current location is unknown. Herzfeld (1954-56), Fig. 56.

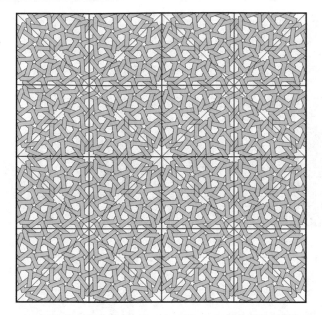

Fig. 49

Fig. 50

(red), 10-pointed stars at the midpoints of each long edge of the repeat, and two 11-pointed stars within the field of the pattern matrix, or 12s on rectangle/10s on long edges/11s in field [Fig. 427]. The plane symmetry group of this design is *pmm*, and the fundamental domain (blue) is a rectangle that is reflected 4× to fill the repeat unit.

Designs that use repetitive stratagems that allow *n*-pointed stars, with an otherwise incompatible number of points, to be placed at the vertices of the orthogonal grid can also be categorized according to their numeric qualities. As discussed earlier, oscillating square patterns and rotating kite designs will occasionally have local symmetries such as 6-, 7-, 8-, 10-, and 12-fold. The design in Fig. 23d through **f** is an oscillating square pattern with 12-pointed stars at vertices of the square and rhombus tessellation, and 8-pointed stars at the center of each square element. However, from the perspective of the overall orthogonal repeat, this design places 12-pointed stars upon each edge of the square repeat and 8-pointed stars at the centers: or 12s on square edges/8s at centers. Other historical oscillating square and rotating kite designs can be described in a similar fashion: the design in Fig. 23g through **i** can be described as irregular 7s on square edges/IVs on centers; Fig. 24 as VIIIs at centers; Fig. 25c as 7s on square edges/VIIIs at centers; Fig. 28a as alternating 6s on squares/IVs at center; Fig. 28b as alternating 6s on squares; and Fig. 29 as alternating 10s on square/4s at center.

Another category of design that elegantly utilizes local symmetries that are ordinarily incompatible with the repetitive structure is *imposed symmetry* designs. These do not have oscillating characteristics, but achieve their inclusion of otherwise atypical regular polygons or stars by (1) only using forms that have two perpendicular lines of reflected symmetry, and (2) placing the imposed stars or polygons upon the edges of the repeat unit rather than the vertices. Figure 51 illustrates three related imposed symmetry designs that introduce octagons into an isometric structure: each octagon being placed at the midpoint of each repetitive triangular edge. Figure 51a shows a design from the Çifte Minare *madrasa* in Erzurum, Turkey (late thirteenth century), and is comprised exclusively of superimposed octagons. The size and distribution of the octagons are determined by the constraints of the underlying 3.4.6.4 generative grid [Fig. 107a]. The included angles of the octagons produce the ditrigonal hexagons at the centers of each triangular repeat. Figure 51b shows a design from the Cincikh mosque in Aksaray, Turkey (1220-30). This maintains the identical octagonal structure as in Fig. 51a, but with the addition of hexagons into the superimposed polygonal design matrix [Fig. 107b]. Figure 51c shows a design from

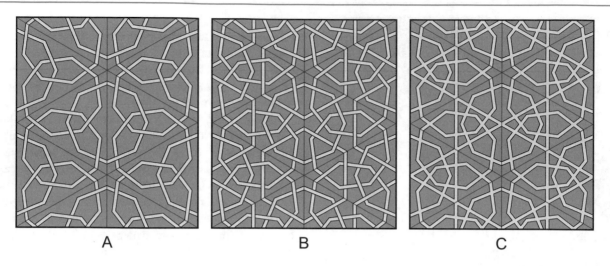

Fig. 51

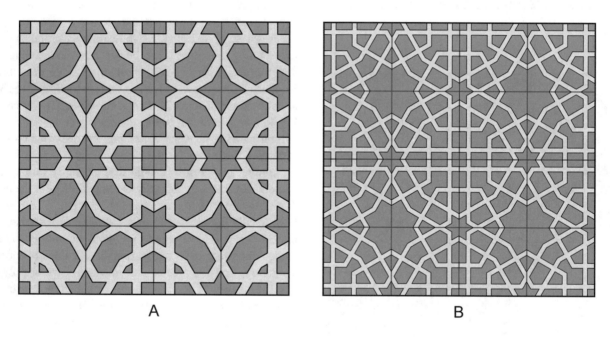

Fig. 52

the mausoleum of Yusuf ibn Kathir in Nakhichevan, Azerbaijan (1161-62). This also maintains the same octagonal structure, but includes the 3.6.3.6 tessellation of triangles and hexagons into the design matrix. In this case, the octagons are located at the vertices of the 3.6.3.6 tessellation. As such, this tessellation can be regarded as equally generative of the overall design as the 3.4.6.4 tessellation. Figure 52 represents two orthogonal imposed symmetry designs that are generated from the deployment of six-pointed stars upon the midpoints of each square repeat unit. Figure 52a shows a design from the original portal of the Palace of Malik al-Zahir at the citadel of Aleppo (before 1193). The parallel lines of the six-pointed stars extend

outward to create a four-pointed star at the center of the square repeat, an irregular octagon centered upon the corners of the repeat, and the small square at the corners of the repeat unit: IVs on square/6s on edge/4s at center. Figure 52b shows a design from the mausoleum of Yusuf ibn Kathir in Nakhichevan (1161-62) that is similarly produced from the extension of the parallel lines of the identically placed six-pointed stars. However, the smaller size of these stars provides for the inclusion of a hexagon that bounds each six-pointed star. The corners of this hexagon, together with the extended lines of the six-pointed stars, create an irregular eight-pointed star at the center of each repeat unit, or IVs on square/6s on edge/irregular 8s at center.

2.4 Classification by Plane Symmetry Group

In the late nineteenth century, scientists working in the field of crystallography determined that there are just 17 symmetrical conditions by which the plane can be periodically tiled. The two-dimensional periodic space filling characteristics of Islamic geometric patterns are, ipso facto, governed by the constraints of these 17 plane symmetry groups. As such, the inherent symmetry of all two-dimensional Islamic geometric patterns conforms to an imposition of a fundamental domain to the singular or combined isometric forces of translation, rotation, reflection, and glide reflection. This is not to suggest that artists knowingly applied these four isometric functions to pre-identified fundamental domains as part of their generative methodology. Rather, these are inherent geometric features that govern all periodic two-dimensional space filling and are, therefore, more relevant to the geometric analysis of these patterns than to questions of design methodology and historicity.

The crystallographic discoveries advanced by pioneering scientists including Yevgraf Fyodorov, Arthur Schönflies, William Barlow, and later George Pólya[37] were soon to find artistic expression. George Pólya is particularly relevant for his pronounced influence on Maurits Cornelis Escher.[38] Escher traveled twice to the Alhambra in Spain and was heavily influenced by the geometric designs there, recording many patterns in his workbooks. The same year of his first visit to the Alhambra (1924) he was sent a copy of Pólya's publication that included line drawings of repetitive patterns in each of the 17 plane symmetry groups, some of which were derived from Muslim architectural sources. Pólya and especially Escher appear to be the first individuals to examine Islamic geometric designs from the perspective of their crystallographic group. Later ethnomathematical studies of Islamic geometric design focused more specifically upon their crystallographic characteristics,[39] and historical examples of all 17 plane symmetry groups have been identified. The works of Syed Jan Abas and Amer Shaker

Salman,[40] as well as Emil Makovicky,[41] are particularly significant to this study.

Figure 53 shows a flowchart that identifies the four isometric conditions of rotation, reflection, glide reflection, and translation for each of the 17 plane symmetry groups,[42] and from which existing designs can be analyzed to readily identify their specific symmetry group. Figure 54 represents a geometric design from each of the plane symmetry groups with 120° rotational centers and/or 60° rotational centers. The *p3* symmetry group has three types of 120° rotation center (threefold), and is without reflections or glide reflections. Islamic geometric patterns that conform to this group are uncommon (the example shown is the author's creation). The *p31m* symmetry group has three types of 120° rotation center (threefold), and three directions of reflection. The lines of reflection form the isometric grid (red) and two of the points of rotation are located at the center of each triangular cell, while the third is located at each vertex of this grid. This structure also has three directions of glide reflection with lines that are parallel to and located in the middle of adjacent parallel lines of reflection. The design representing this symmetry group is a design that is easily created from the 6^3 underlying tessellation. The *p3m1* symmetry group has three types of 120° rotation center (threefold) and three directions of reflection that comprise the isometric grid. Each point of rotation is located at a vertex of this grid. This structure has three directions of glide reflection that are identical to the previous group. Mughal artists frequently used the design representing this symmetry group in the production of *jali* screens. The additive threefold lines at the center of each six-pointed star provide the stone screen with greater uniformity in the size of the openings, as well as greater structural integrity, an important consideration in this pierced stone medium. This additive device also changes the plane symmetry group of the well-known pattern of superimposed dodecagons from *p6m* to *p3m1*. The *p6* symmetry group has one variety of 60° rotation center (sixfold), and two types of 120° rotation center (threefold), and a single 180° rotation center. There are no reflections or glide reflections. The design that represents

[37] –Fedorov (1891), 345–291.
   –Schönflies (1891).
   –Barlow (1894), 1–63.
   –Pólya (1924), 278–298.

[38] Schattschneider (1990).

[39] –Müller (1944).
   –Bixler (1980).
   –Lalvani (1982).
   –Grünbaum, Grünbaum, and Sheppard (1986), 641–653.
   –Mamedov (1986), 511–529.
   –Pérez-Gómez (1987), 133–137.
   –Lalvani (1989).
   –Chorbachi (1989), 751–789.
   –Abas and Salman (1995).
   –Lovric (2003), 423–432.

[40] Abas and Salman (1995).

[41] –Makovicky and Makovicky (1977), 58–68.
   –Makovicky (1989), 955–999.
   –Makovicky (1994), 1–16.
   –Makovicky (1995), 1–6.
   –Makovicky (1997), 1–40.
   –Makovicky (1998), 107–127.
   –Makovicky (1999), 143–183.

[42] This flowchart replicates that of Donald Crowe, Department of Mathematics, University of Wisconsin-Madison, and is included in his book on symmetry in cultural artifacts: See Washburn and Crowe (1988).

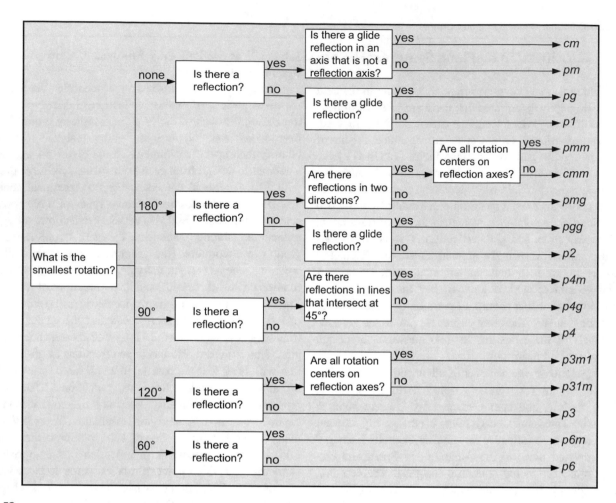

Fig. 53

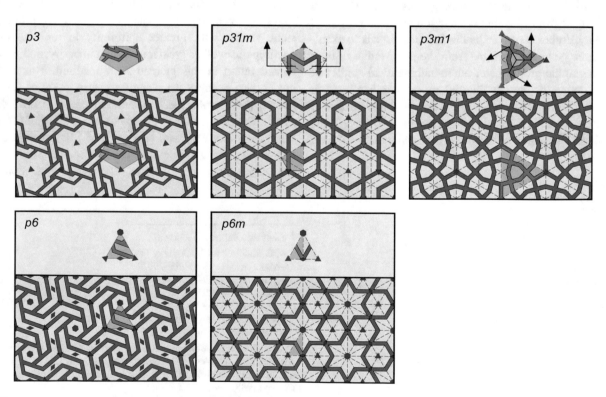

Fig. 54

Fig. 55

this symmetry was used widely throughout Muslim cultures, and one of the earliest examples is from the brickwork ornament of the western tomb tower at Kharraqan, Iran (1093). The *p6m* symmetry group has one variety of 60° rotation center (sixfold), two types of 120° rotation center (threefold), which are perpendicularly orientated; and three types of 180° rotation center. This group has six directions of reflection, and six directions of glide reflection (not shown) that are parallel with, and centered between, the lines of reflection. The example shown is one of the most common threefold geometric patterns. Figure 55 represents a geometric design from each of the plane symmetry groups with 180° rotational centers and/or 90° rotational centers. The *p2* symmetry group has four types of 180° rotation center, with no reflections or glide reflections. Islamic geometric patterns

structured on this symmetry group are unusual. Among the more interesting examples are a variety of square *Kufi* calligraphic designs, in this case with a simple *Allah* motif (the example shown is the author's creation). The *pgg* symmetry group has two types of 180° rotation center, with two glide reflections in perpendicular directions. There is no reflection symmetry. The example shown is a well-known key pattern with swastikas in glide reflection. The *pmg* symmetry group has two types of 180° rotation center, with parallel lines of reflection in just one direction. It also has glide reflections that are perpendicular to the lines of reflection, and the rotation centers are located on the lines of glide reflection. Islamic geometric designs with this symmetry group are ordinarily very simple. The example shown is from the Khwaja Atabek mausoleum in Kerman (1100-1150) and is one of the more complex historical designs with this crystallographic structure. The *pmm* symmetry group has four types of 180° rotation center, each located at a vertex of the perpendicular lines of reflection. There are no glide reflections. The example shown is ubiquitous throughout the Islamic world. The *cmm* symmetry group has four types of 180° rotation center: two located on the vertices of the perpendicular lines of reflection, and two that are not located on lines of reflection. The example shown is a very common fivefold *obtuse* pattern that repeats upon a rhombic grid. The *p4* symmetry group has two types of 90° rotation center, and two types of 180° rotation center (twofold). There are no lines of reflection or glide reflection. The example shown is well known from the historical record. The *p4g* symmetry group has two types of 90° rotation center (fourfold), and two types of 180° rotation center (twofold). The 90° rotation centers are not located on lines of reflection, while the 180° rotation centers are located on the vertices of the orthogonal lines of reflection. There are diagonally oriented glide reflections that run halfway between the vertices of the lines of reflection (not shown). This design was used in numerous locations historically.

The *p4m* symmetry group has two types of 90° rotation center (fourfold), two types of 180° rotation center (twofold), and four directions of reflection. All rotation centers are located at the vertices of the lines of reflection. This design is from a frontispiece from a Quran produced in 1001 by ibn al-Bawwab (d.1022). Figure 56 represents a geometric design from each of the plane symmetry groups with no rotation symmetry. The *p1* symmetry group relies solely upon translation symmetry, with no rotations, reflections, or glide reflections. Islamic geometric patterns based upon this group are very rare, and the example shown (by author) avoids reflection symmetry by the introduction of chirality with the interweaving lines. The *pg* symmetry group is defined by glide reflection only, with no rotation or reflection. This variety of pattern is also very rare in Islamic geometric design, and the illustrated example (by author) is a rather complex design that is otherwise indicative of the early brickwork ornament of Khurasan. The *pm* symmetry group only has parallel lines of reflection, with no rotation or glide reflection. This design places an additive pentagonal device within the otherwise central ten-pointed stars. Without this added fivefold device there would be the additional

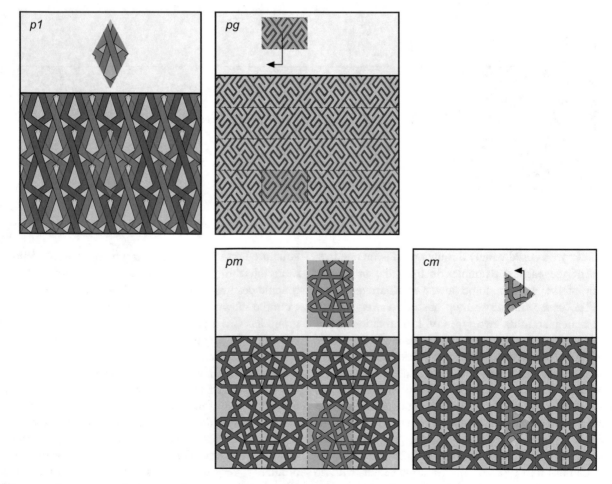

Fig. 56

lines of reflection and points of rotation of the *pmm* symmetry group. The *cm* symmetry group has parallel lines of reflection and parallel lines of glide reflection located halfway between the reflections. There are no points of rotation. Islamic ornament in this group is predominantly floral, such as certain ogee designs. The classic geometric patterns of Muslim cultures do not ordinarily conform to this symmetry group, although decent designs are possible (example shown by author).

It is beyond the scope of this work to quantify the distribution of historical geometric designs within a given Muslim culture, let alone the totality of Islamic art, according to their plane symmetry group. However, without question, certain isometric transformations occurred with greater frequency within this tradition, while others are less common or very rare. According to Abas and Salman, the *p6m* and *p4m* symmetries are the most widely distributed; the *cmm*, *pmm*, and *p6* are also significantly represented; the *p4*, *p31m*, *pm*, and *p3m1* are significantly fewer; and the *p4g*, *p3*, *cm*, *p2*, *pgg*, *pmg*, *p1*, and *pg* are very rare.[43] The question of why certain symmetry groups were favored over others appears to have more to do with methodological practices than aesthetic predilections. The vast majority of Islamic geometric patterns are readily created from the polygonal technique wherein a tessellation of diverse edge-to-edge polygons is used to extract the design. The symmetry group of an underlying generative tessellation directly determines the symmetry group of the extracted pattern. This is not to say that the two are always identical, especially when additive design features alter or cancel the rotation and reflection, or lines of reflection are annulled through the introduction of chirality with interweaving lines. Creating successful polygonal tessellations that are well suited to extracting patterns that conform to the aesthetic standards of this tradition typically involves the placement of higher order primary polygons at strategic locations of the repetitive grid. These invariably have *n*-fold rotation symmetry and their placement at the vertices, centers, and edges of the repeat unit insures compliance with those symmetry groups that are similarly structured, and is generally less suited to symmetry groups without rotation or reflection. Field patterns created from the polygonal technique eschew regions of local symmetry created from higher order polygons affiliated with strategic locations within the repeat. This lack of affiliation occasionally allows field patterns to be structured upon symmetry groups that are less common to this tradition. Islamic geometric designs that are not created from the polygonal technique will also occasionally employ these less commonly used symmetry groups. These can include key patterns, designs with swastika motifs, and square *Kufi* brickwork designs.

2.5 Classification by Design Methodology

2.5.1 The Polygonal Technique

The aesthetic character of a given geometric design is greatly determined by the method used in its creation. Generative methodology is therefore an important criterion for better understanding of Islamic geometric patterns. Some of the less complex geometric patterns are able to be produced from more than a single generative methodology, and it is not always possible to ascertain with certainty which was used in the creation of a particular historical example. As stated previously, surviving evidence indicates that the most widely used and, therefore, historically relevant design methodology was the *polygonal technique*, wherein strategic points of a polygonal tessellation, such as the midpoints of each polygonal edge, are used to locate pattern lines, after which the tessellation is discarded, leaving behind the completed design. Depending on the angles of the applied pattern lines, multiple designs can be created from a single underlying tessellation. Insofar as Islamic geometric patterns are concerned, no other design methodology provides the level of flexibility and consequent design diversity, and other approaches used over the centuries are of significantly less importance to this overall tradition.

It appears that the artists responsible for the development and furtherance of Islamic geometric patterns were discriminating in their need to balance generational transferral with protection of the highly specialized design practices required of this art form. There are no known historical sources that speak to the methodological secrecy employed by individuals, ateliers, and artists' guilds employed in the geometric arts. One must assume that the ongoing development of geometric design flourished under the same sort of protectionist control as other arts reliant upon patronage for their survival. This might explain the paucity of geometric artists' reference scrolls (*tumar*) and design manuals currently known to art historians. Of the few such documents, one is particularly significant in that it is very likely the earliest depiction of a geometric pattern accompanied by its underlying generative polygonal tessellation. Figure 57 illustrates a design created from one of the many figures contained in the anonymous Persian language treatise titled *On Similar and Complementary Interlocking Figures* in the

---

[43] The methodology behind the gathering of the data points for this statistical analysis of the distribution of the 17 symmetry groups within the tradition of Islamic geometric patterns is not provided in this study. See Abas and Salman (1995), 138.

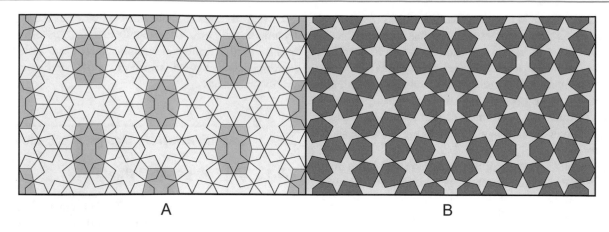

A   B

Fig. 57

Bibliothèque Nationale de France in Paris.[44] This pattern is all the more remarkable in that its only known architectural use is from one of the blind arches in the Seljuk northeast dome chamber of the Friday Mosque in Isfahan[45] (1088-89) [Photograph 26]. Figure 57a represents the manuscript's depiction of the polygonal tessellation comprised of two types of irregular hexagon, as well as the generated *acute* pattern whose intersecting lines rest upon the midpoints of each polygonal edge: the classic formulation of the polygonal technique. Figure 57b illustrates the pattern on its own. This heptagonal design is all the more interesting in that it is the earliest known example of a design created from the *sevenfold system* of pattern generation, and its use in Isfahan precedes later extant examples created from this system by 100 years. One of the remarkable features of *On Similar and Complementary Interlocking Figures* are the written instructions that accompany most of the illustrations, and the step-by-step instructions that accompany this figure are revealing in that they provide instructions for the creation of the polygonal tessellation, but not the pattern that this tessellation creates. The absence of secondary instructions for the application of the pattern lines onto the tessellation may indicate that this process was a given: sufficiently understood so as not to warrant further instruction. The many illustrations and instructions for geometric patterns in this anonymous manuscript are better known as one of the very few historical sources of evidence for what is herein referred to as the point-joining methodology. The historical relevance of this aspect of the manuscript is examined below; but the inclusion of this one representation of the polygonal technique is significant for four reasons: (1) it is one of the earliest known

examples of a pattern accompanied by its underlying generative tessellation; (2) it includes written instructions for creating the generative tessellation; (3) it is one of the earliest examples of a pattern created from the *sevenfold system* of pattern generation; and (4) it is one of the very few historical documents that overtly demonstrates the polygonal technique.

On its own, the example of the polygonal technique from the anonymous manuscript might be regarded as merely an interesting anomaly. However, in association with the many additional examples from diverse media, wide-ranging regions, and over prolonged periods of time, the validity of the polygonal technique as the preeminent historical methodology used in creating Islamic geometric patterns becomes unassailable. The earliest architectural examples include numerous patterns that maintain the generative tessellation as part of the completed design. Even during the eleventh and twelfth centuries when this ornamental tradition was in the process of rapid development, it was far more common for the generative tessellations to be discarded after completion of the design process. However, some early examples of patterns created from the *system of regular polygons* include the generative tessellation within the completed design. The least complex of these are based upon the 6^3 tessellation of regular hexagons, and include a Qarakhanid *two-point* pattern from the southern portal of the Maghak-i Attari mosque in Bukhara, Uzbekistan (1179-79) [Fig. 96f], and a Sultanate of Rum *two-point* pattern from the Great Mosque of Bayburt in northeastern Turkey (1220-35) [Fig. 97b]. A Mamluk *two-point* design from the *mihrab* of the Aqbughawiyya *madrasa* (1340) at the al-Azhar mosque in Cairo similarly expresses the 3.6.3.6 generative tessellation as part of the completed design [Fig. 100d]. The design of a brickwork panel in the portal of the anonymous southern tomb in the complex of three adjoining Qarakhanid mausolea in Uzgen (1186) includes the depiction of its 3.4.6.4 generative tessellation [Fig. 104d]. Several patterns that overtly express their 4.8^2 generative tessellation of squares and octagons are known to the historical record,

---

[44] MS Persan 169, fol. 192a.

[45] The author is indebted to Professor Jan Hogendijk at the University of Utrecht for pointing out the connection between the panel with sevenfold symmetry at the Friday Mosque at Isfahan and the design from folio 192r in the anonymous manuscript at the Bibliothèque Nationale de France in Paris.

including a Timurid variation of the classic star-and-cross pattern from the Ghiyathiyya *madrasa* in Khargird, Iran (1438-40) [Fig. 126d].

One of the most compelling examples of architectural evidence for the polygonal technique is from the main entry portal of the Sultan al-Nasir Hasan funerary complex in Cairo (1356-63). In one of the sidewalls of this Mamluk *iwan* is an arched niche with a *muqarnas* hood. This niche is decorated with an interesting nonsystematic pattern with six- and eight-pointed stars that repeats on a rectangular grid [Photograph 58]. The pattern in this niche is produced in white marble inlaid into a beige limestone background. The artist also inlaid a black stone representation of the generative tessellation of octagons, distorted hexagons, and rhombi [Fig. 413]. This is significantly different from the previously cited examples in that the geometric pattern is distinctly independent of the generative tessellation rather than being incorporated into the finished design. The presence of the tessellation is highly unusual in that it reveals the methodological key to this design specifically, and to almost all Islamic geometric patterns generally.

In addition to the panel from the Sultan al-Nasir Hasan funerary complex in Cairo, the most overt architectural examples of geometric designs accompanied by their generative tessellations come from several *jali* screens from Mughal India. A marble *jali* in the tomb of I'timad al-Daula in Agra (1622-28) expresses the generative tessellation in high relief as the primary visual motif, and the resulting geometric design as secondary elements. This example is the classic fivefold *acute* pattern created from the *fivefold system* [Fig. 226c]. Additional Mughal examples are located in the *jali* screens of the tomb of Salim Chishti at Fatehpur Sikri (1605-07), including a very-well-known *acute* pattern created from the *fourfold system B* [Fig. 173a] [Photograph 77]; a widely used nonsystematic *acute* design with 12-pointed stars on vertices of the isometric grid [Fig. 300a *acute*]; and an unusual example wherein the fivefold pattern generator is, itself, a field pattern created from the *fivefold system*. This field pattern is made up of just two design elements, pentagons and hourglass decagons. The simplicity of this design allows for it to be used as a generative tessellation for the secondary pattern.

Evidence of the polygonal technique is occasionally found in objects that employ comparatively complex polygonal tessellations without the presence of one of the geometric designs that can be generated from the tessellation. A particularly early example of such an item is a Persian fritware tile (c. 1250-1300) in the collection of the Los Angeles County Museum of Art[46] [Photograph 104]. The

date of origin suggests that this is either late Khwarizmshahid or early Ilkhanid. The molded relief decoration boldly depicts a polygonal tessellation comprised of dodecagons, decagons, and nonagons, with concave hexagonal secondary interstitial polygons that function within the tessellation much like the concave hexagons within the *fivefold system* [Fig. 232]. There are no known historical designs created from this nonsystematic polygonal tessellation. Rather than the midpoints of the polygonal edges being used to locate geometric pattern lines, these points determine the construction of a floral design. This is the only known example of an Islamic floral design being extracted from a complex polygonal substructure, and the fact that the polygonal midpoints are similarly used as location points is significant. Another significant example of a complex nonsystematic polygonal tessellation being used as ornament without the depiction of one of the geometric designs that the tessellation can create is from a Karamanid walnut door from the Imaret mosque in Karaman, Turkey[47] (1433) [Photograph 105]. The repetitive structure is orthogonal, and the local regions of symmetry are 8- and 12-fold, separated by irregular pentagons and barrel hexagons. This particular nonsystematic tessellation is one of the most commonly employed historically, and was used to produce innumerable geometric designs in all four of the principal pattern families throughout the Islamic world [Figs. 379–382].

By far the most convincing evidence of the polygonal technique as the primary historical method used by artists for creating complex Islamic geometric patterns is the Topkapi Scroll.[48] According to Gülru Necipoğlu, the prominent authority on the historical significance of this scroll:

The Topkapi Scroll was probably compiled in the late fifteenth or sixteenth century somewhere in western or central Iran, possibly in Tabriz, which served as a major cultural capital under the Ilkhanids, the Qaraqoyunlu, and the Aqqoyunlu Turkman dynasties, as well as the early Safavids. Its geometric designs in all likelihood were produced under Turkman patronage, but an early Safavid date is also a possibility as the international Timurid heritage would still have been very much alive.[49]

The Topkapi Scroll contains 157 different designs that represent the full range of geometric ornament in the regions directly influenced by Timurid aesthetics. These include *muqarnas* vaulting, star-net (*rasmi*) vaulting, geometric ornament for domes, *Kufi* script, square or chessboard (*shatranji*) *Kufi* motifs, and numerous examples of geometric patterns. Among the many geometric patterns is a wide range of diverse types, including three designs produced

---

[46] Los Angeles County Museum of Art, the Madina Collection of Islamic Art, gift of Camilla Chandler Frost (M.2002.1.285).

[47] In the collection of the Museum of Turkish and Islamic Arts, Istanbul, Turkey, accession no. 244.

[48] Topkapi Palace Museum Library MS H. 1956.

[49] Necipoğlu (1995), 37–38.

Photograph 104 Persian fritware relief tile with a polygonal tessellation comprised of quarter dodecagons and a half decagon as the primary ornament, and a floral motif with symmetry that is governed by the polygonal structure (The Los Angeles County Museum of Art: the Madina Collection of Islamic Art, gift of Camilla Chandler Frost (M.2002.1.285): www.lacma.org)

from the *fourfold system A* (nos. 1, 67, 61)[50]: one from the *fourfold system B* (no. 57); eight designs with rhombic repeat units made from the *fivefold system* (nos. 8, 52, 53, 54, 55, 62, 64, 73); six with rectangular repeats from the *fivefold system* (nos. 33, 48, 50, 56, 58, 60); five *Type A* dual-level designs created from the *fivefold system* (nos. 28, 29, 31, 32, 34); one *Type B* dual-level design produced from the *fivefold system* (no. 49); one Type B dual-level design that uses hybrid square and triangle repetitive elements with 8- and 12-pointed stars (no. 38); an additional hybrid design with square and triangle repeats with 8- and 12-pointed stars (no. 35); two nonsystematic designs with 12-pointed stars located at the vertices of the isometric grid, one with rotating square swastikas (nos. 63 and 70); two nonsystematic orthogonal compound patterns, one with 8- and 12-pointed stars (no. 72d), and the other with 13- and 16-pointed stars (no. 30); three nonsystematic designs that do not use either

---

[50] The indicated numbers in this paragraph follow the numbering protocol in the Topkapi Scroll—Geometry and Ornament in Islamic Architecture. See Necipoğlu (1995).

Photograph 105 A Karaminid walnut door from the Imaret mosque in Karaman, Turkey, that depicts a nonsystematic tessellation associated with the polygonal technique that includes partial octagons and dodecagons surrounded by pentagons and barrel hexagons (© Dick Osseman)

the isometric or the orthogonal grids, including one with 8-, 10-, and 12-pointed stars (no. 39), one with 10- and 12-pointed stars (no. 44), and one with 9- and 11-pointed stars (no. 42); one rotating kite design with 6-pointed stars (no. 59); two oscillating square patterns with swastikas (nos. 41 and 69b); three designs with forced 10-pointed stars in a square repeat unit (nos. 66, 68, 72c); and three designs for

application onto domical surfaces, including two created from the *fivefold system* (nos. 4, 90a), and one compound design with 8- and 10-pointed stars (no. 10b). The Topkapi Scroll is drawn primarily in black and red ink. These two colors are used to differentiate the features of a given illustration. This frequently involves the contrast between the pattern and its generative tessellation. Further differentiation is occasionally achieved through using dotted lines. What is more, many of the geometric patterns that do not overtly show the generative tessellation in ink reveal this important methodological feature in finely scribed layout lines made with a steel stylus and referred to as *dead drawing*. With the exception of the oscillating square, rotating kite, and forced patterns, virtually all of the geometric designs in the Topkapi Scroll visually represent their generative tessellation in either ink or inscribed lines. This is by far the largest known single repository of geometric designs represented in association with their underlying generative tessellations. The fact that this scroll was an artist's reference intended for practical application is an incontrovertible evidence for the polygonal technique being the preeminent methodology employed in the creation of Islamic geometric patterns during the time and place of the scroll's production, and by extrapolation, to this tradition more generally. The designs from the Topkapi Scroll range in complexity between the more basic systematic patterns and those that are highly complex with more than one region of local symmetry, as well as dual-level designs with self-similar characteristics. As demonstrated so aptly in this scroll, the polygonal technique is uniquely capable of creating these exceptionally complex designs.

Further scroll evidence for the historical use of the polygonal technique is found among the scroll fragments at the Institute of Oriental Studies in Tashkent. These range in date between the fifteenth to seventeenth centuries. Like the Topkapi scroll, these depict a combination of vaulting systems for three-dimensional application, and two-dimensional geometric patterns, and also include the use of colored inks and scribed lines. Of particular interest to the question of design methodology are a series of geometric patterns that include the underlying generative tessellation. These examples are estimated to date from the sixteenth or possibly seventeenth century, and employ only black ink with the underlying tessellation represented in the un-inked incised lines produced with a steel stylus.[51] The geometric patterns include *median* and *two-point* designs created from the *fivefold system*, as well as a nonsystematic *median* pattern with 9- and 12-pointed stars [Fig. 346b].

Much like the incised lines from the Topkapi and Tashkent scrolls, Quranic illuminators also used a steel stylus to lay out their designs prior to painting the final illumination. The relevance of Quranic illumination to the understanding of traditional design methodology has not received the research it deserves. The likely significance of

this artistic discipline in providing further corroboration of the prevalent use of the polygonal technique is found in an outstanding Mamluk illuminated frontispiece (c. 1399-1411) at the British Library [Photograph 48].[52] This is decorated with an *obtuse* pattern created from the *fivefold system* [Fig. 233b] that was used in many locations over the years, including the Izzeddin Kaykavus hospital and mausoleum in Sivas, Turkey (1217-18); the Agzikarahan in Turkey (1242-43); and the Sultan Qala'un funerary complex in Cairo (1284-85). Upon close inspection with oblique lighting, the fine incised lines beneath the paint that were used for laying out this illumination are faintly detectable with the naked eye. In this example, these painted-over incised lines reveal the underlying generative tessellation that was used to produce the pattern [Fig. 233c]. Unless this example is an anomaly, considering that this is just one of a very large number of illuminated pages with geometric ornament, it is entirely possible that a study of Quranic examples will reveal further evidence that illuminators used this design methodology when laying out their compositions.

The last piece of evidence for the historicity of the polygonal technique comes from the published observations of Ernest Hanbury Hankin, a bacteriologist working in India in the latter part of the nineteenth century. His observations of a deteriorating stucco ceiling in a bathhouse (*hammam*) at Fatehpur Sikri led to his discovery that Islamic geometric designs were constructed from underlying polygonal tessellations:

> During visits to Fathpur-Sikri many years ago, I spent much time in measuring the angles and making tracings of these designs but always failed to find any rational scheme by which they could be constructed. At last, by good fortune, I happened to enter a small Turkish bath attached to Jodh Bai's Palace. It had previously been inhabited by Indians, who had only just been evicted, and I was probably the first European to visit the place. In one of the rooms of the bath was a half dome decorated by a straight-line pattern. In addition to the pattern, some faint scratches were discovered on the plaster. Obtaining a table and chair and a piece of tracing paper I succeeded in making a copy. On closer examination these scratches were found to be parts of polygons, which, when completed, surrounded the star-shaped spaces of which the pattern was composed, and it turned out that these polygons were the actual construction lines on which the pattern was formed.[53]

Hankin first published his finding in a 1905 article in the Journal of the Society of Arts[54] entitled *On some Discoveries of the Methods of Design employed in Mahomedan Art*, and in greater detail in his 1925 article *The Drawing of Geometric Patterns in Saracenic Art* for the Memoirs of the Archeological Survey of India. In these

---

[51] There are relatively few sources of photographs of the geometric patterns from the Tashkent scrolls. See
  –Rempel' (1961).
  –Necipoğlu (1995), 12–13.

[52] British Library, London, BL Or. MS 848, ff. 1v-2.

[53] Hankin (1925a), 3–4, no. 15.

[54] Hankin (1905), 461.

works and others,[55] Hankin describes in considerable detail the design process for the polygonal technique, and analyzes a number of historical designs. He is the first European to have discovered this design methodology, yet the significance of his discoveries has had far less impact than deserved upon the prevailing views regarding traditional design methodologies that came after him. Since the turn of the millennium, recognition of the polygonal technique has gradually gathered momentum for its historicity and methodological flexibility.[56]

Regarded as a group, the above-cited methodological examples provide compelling evidence for the efficacy and historicity of the polygonal technique as a primary tool for generating the diverse range of designs that characterize this geometric art form. While the more basic designs can often be produced with alternative methods, the polygonal technique is the only methodology that will produce the more complex patterns within this tradition. This preponderance of evidence provides the certainty that the polygonal technique was the preeminent design methodology used historically. Without this evidence, the relevance of this method of creating geometric designs would be based solely upon common sense and experience.[57]

The earliest Islamic geometric patterns are easily created from regular, semi-regular, and occasionally *two-uniform* tessellations comprised of regular polygons. These polygons include the triangle, square, hexagon, octagon, and dodecagon. Being that the octagon is limited to only one semi-regular tessellation, and that other than the square, it will not tessellate with any of the other regular polygons, the 4.8^2 semi-regular tessellation is, for practical purposes, regarded within this study as its own group with its own distinctive aesthetic merits. This is why, for the purposes of this discussion, the octagon is not included within the modules of the *system of regular polygons* [Fig. 92]. The 4.8^2 tessellation is one of the most widely used, versatile, and prolific

underlying generative substrates used within this ornamental tradition [Figs. 124–129]. Figure 58 demonstrates just two of the many designs that can be produced from the orthogonal 4.8^2 semi-regular tessellation of octagons and squares. Figure 58a shows the classic *median* star-and-cross pattern used frequently throughout the Islamic world. Employing the polygonal technique to produce this pattern involves drawing lines through the edge at every second midpoint within the octagons, creating two superimposed squares with 90° crossing pattern lines at each midpoint. These lines are trimmed so that the interior octagons are converted to the characteristic eight-pointed stars. The design in Fig. 58b is also very well known within this tradition, and is produced in a similar fashion, except that the applied pattern lines transect the midpoints of every third octagonal edge: creating 45° crossing pattern lines at these midpoints, and four-pointed stars within each underlying square. The interior octagonal region is similarly trimmed to create the eight-pointed stars. The 45° crossing pattern lines in Fig. 58b qualify this as an *acute* pattern.

As mentioned, there are four standard techniques for extracting geometric designs from underlying polygonal tessellations. In addition to the *acute* and *median* families illustrated in Fig. 58, this tradition also includes *obtuse* and *two-point* patterns. Each of these four families has an identifiable visual quality that is independent of its symmetrical characteristics or repeat unit. These appear with such regularity, and are sufficiently distinct from one another that it is appropriate for each to be included as a distinct category of pattern created from the polygonal technique. Figure 59 highlights the distinctive features of each of these four pattern families. For the purposes of demonstration, the four examples shown are created from the *fivefold system*, but the four pattern families are equally relevant to all systematic and nonsystematic geometric designs created from the polygonal technique. What is more, the visual characteristics of the pattern elements created from the *fivefold system* have direct analogs to those created from other types of geometric design, and the examples provided in this illustration are therefore representative of this tradition generally. Figure 59a demonstrates the characteristics of the *acute* family, with stars, darts, kites, and bilateral shield-shaped hexagons. Quick visual references for identifying *acute* designs are the acute angles of the points that surround the primary star forms, as well as the acute angles of the five-pointed stars. Figure 59b demonstrates the more open character of the pattern elements of the *median* family, especially as pertains to the points of the primary stars and five-pointed stars. The angles of the stars, darts, overlapping darts, kites, and shields are recognizably less acute, and fall between the angles of the *acute* family and the *obtuse* family: hence the name *median*. Figure 59c demonstrates the characteristics of the *obtuse* family. The pattern elements in this type of

---

[55] In addition to the two articles mentioned above, E. Hanbury Hankin, M.A., Sc.D., also published occasional articles concerning Islamic geometric pattern derivation in the Mathematical Gazette. I am indebted to Dr. Carl Ernst for first bringing the work of Hankin to my attention in 1980. See

–Hankin (1925b), 371–373.
–Hankin (1934), 165–168.
–Hankin (1936), 318–319.

[56] –Bonner (2003).
–Kaplan (2005).
–Lu and Steinhardt (2007a).

[57] In 1987 I had the good fortune to see and photograph the Topkapi Scroll while it was on temporary display at the Topkapi Museum in Istanbul. Other than the publications of Ernest Hanbury Hankin, this was my first corroboration that the polygonal methodology I had developed independently, and had been employing as an artist for many years, was, in fact, historical.

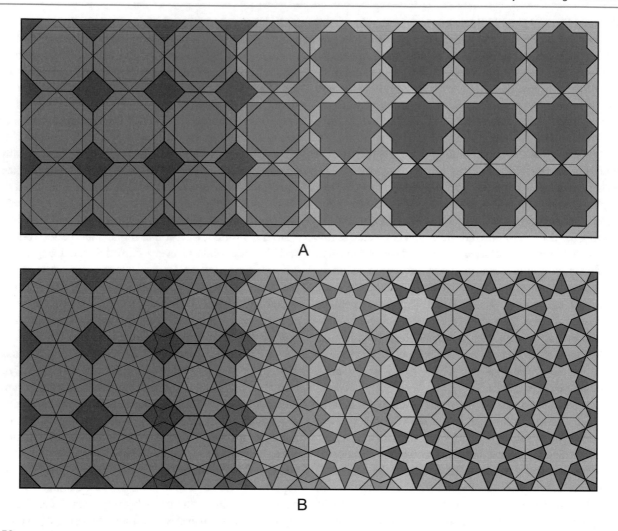

A

B

Fig. 58

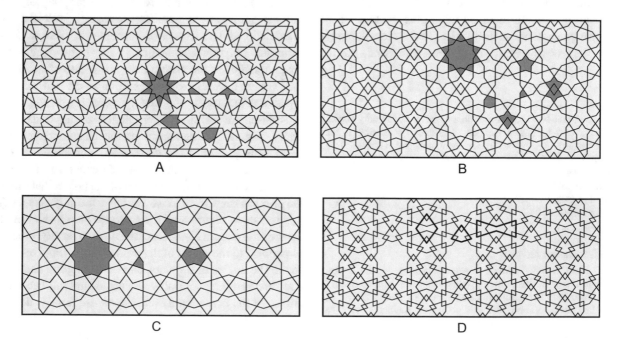

A

B

C

D

Fig. 59

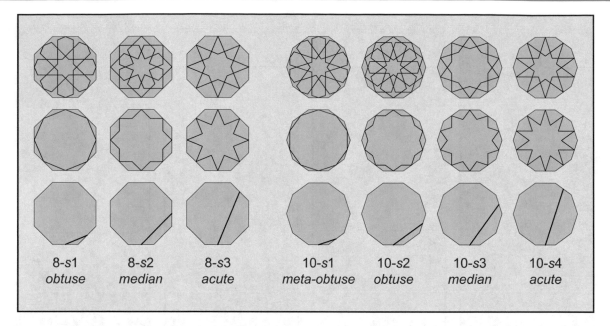

Fig. 60

pattern will typically have primary stars with points that are appreciably more obtuse, pentagons rather than five-pointed stars, kites, concave octagons, and distinctive hourglass polygons with ten sides. Figure 59d demonstrates the characteristics of the *two-point* family. This type of pattern is recognizable for its matrix of overlapping closed polygons, including kites, rhombi, and concave hexagons. Each of these families is subject to considerable stylistic variation and additive treatment, especially to the primary star forms.

In creating the primary stars, the application of pattern lines is most frequently determined by drawing lines that connect the midpoints of the primary underlying polygons. Figure 60 illustrates this process as applied to octagons and decagons as representative examples. The four pattern families are only descriptive of an aesthetic quality. For a more precise design classification it is often helpful to define the specific variety of star contained within the primary underlying polygons. The method illustrated roughly follows the nomenclature of Anthony J. Lee by identifying the number of sides of the primary polygon in relation to the number of sequential sides in midpoint-to-midpoint line application for creating a given star.[58] This way of identifying primary star forms is especially relevant to the regularity of systematic patterns. However, it is important to note that the application of pattern lines to primary underlying polygons does not always follow the convenient midpoint-to-midpoint method in this illustration. In some cases, especially with nonsystematic designs, the

supplemental angles of the pattern lines that are placed at the midpoints of each edge of the primary underlying polygons are not determined by a straight line that connects to another midpoint. Rather, the precise angle of the pattern lines that are applied to these points, that ultimately determines the visual character of the primary star, is arrived at through aesthetic evaluation on the part of the artist. This decision is greatly influenced by how the extended lines behave within the adjacent secondary underlying polygonal cells. When this aesthetic approach is used, the identifying nomenclature of Fig. 60 is not applicable.

Every underlying generative tessellation is capable of producing a pattern from each of the four families. However, this is a nuanced discipline and not all of the patterns so generated will be acceptable to the aesthetic standards of this tradition. Prior to the maturity of Islamic geometric patterns, the approach to applying pattern lines onto underlying tessellations was less codified and more experimental. During the twelfth and thirteenth centuries, as part of the overall maturing of this artistic tradition, these four pattern families were established as distinct methodological aspects of the polygonal technique, each producing designs that were recognizably distinct from one another, and each with its own aesthetic appeal. What is more, the aesthetic predilections of different Muslim cultures favored, to a lesser or greater extent, specific pattern families over others, as well as certain additive variations that were frequently applied to these designs. The *acute*, *median*, and *obtuse* families differ according to the angle of the crossing pattern lines that are located at, or near, the midpoints of each underlying polygonal edge, while the *two-point* family has applied pattern lines placed on two points of each edge. Figure 61 illustrates

---

[58] Lee (1995), 182–197.

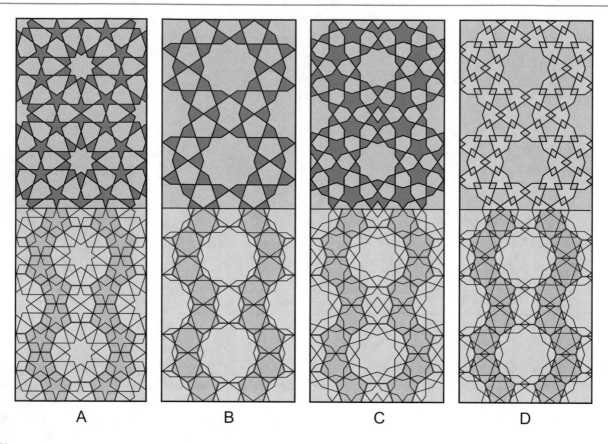

A   B   C   D

Fig. 61

these four pattern families as associated with an underlying tessellation of decagons, pentagons, and hexagons that repeat upon a rhombic grid. This is one of the most basic rhombic repeats produced from the *fivefold system*, and each of the four patterns created from this tessellation was used widely. Figure 61a shows the *acute* pattern created from this tessellation. The crossing pattern lines of *acute* patterns created from the *fivefold system* have a 36° angular opening at each midpoint of the polygonal edge. The bisector of the angular opening is perpendicular to the polygonal edge. These crossing pattern lines are easily determined by their transecting every second midpoint of the pentagons (5-*s*2), and every fourth midpoint of the decagons (10-*s*4). Figure 61b illustrates the application of the pattern lines of the *obtuse* family wherein the pattern lines transect the underlying pentagonal edges at adjacent midpoints (5-*s*1), and the decagons at every second midpoint (10-*s*2), creating crossing pattern lines at these midpoints with 108° angular openings. As stated, one of the visual characteristics of *obtuse* patterns is the occurrence of pentagons nested within the pentagons of the underlying tessellation—creating a more open aesthetic. Figure 61c shows the *median* pattern created from this tessellation. As the name implies, the angle of the crossing pattern lines is between the *acute* and *obtuse* angles. Within the *fivefold system* this is 72°. These lines are

conveniently determined by transecting every third midpoint of the decagon (10-*s*3). Figure 61d illustrates the *two-point* pattern created from this tessellation. This variety of design employs two points on each underlying polygonal edge rather than just one, and the resulting designs are almost always given a widened line or interweaving line treatment (rather than the colored tiling treatment in this illustration). The above-mentioned angular openings in the four pattern families mentioned above are standard to the *fivefold system*. In other systems, and indeed in nonsystematic designs, the angles of the crossing pattern lines employed in each pattern family will vary according to the inherent geometry of the system. For example: in the *system of regular polygons* the *acute*, *median*, and *obtuse* angular openings are typically 60°, 90°, 120° respectively, whereas those of both fourfold systems will have 45°, 90°, and 135°, respectively. In each case, the aesthetic character of each pattern family is essentially the same.

The extraordinary design diversity provided by the polygonal technique necessitates further subcategorization beyond the four standard pattern families. As mentioned previously, Islamic geometric patterns created from the polygonal technique fall into two distinct categories: systematic and nonsystematic. Differentiation between these two types of design is a primary form of classification and is fundamental to a thorough understanding of this tradition. Systematic designs employ a limited set of polygonal modules, with associated applied pattern lines, that are assembled into different tessellations to create myriad designs. There are five polygonal design systems that were used historically: I have named these the *system of regular polygons*, the *fourfold system A*, the *fourfold system B*, the *fivefold system*, and the *sevenfold system*. The numeric values in the names of the four-, five-, and sevenfold systems reference the smallest value of rotational symmetry within the primary star forms (other than 2). The patterns that Hankin analyzed in his publications include both systematic and nonsystematic examples. Hankin's groundbreaking work on the polygonal technique did not identify systematic characteristics or differentiate between these two categories of design. As such, he does not appear to have recognized the systematic nature of the underlying polygonal modules in many of his reconstructions. The use of methodological systems provided geometric artists with a fast and accurate means of producing new and original geometric designs with great ease. With few exceptions, each of the five pattern families has a specific set of pattern lines associated with each polygonal module. The simplicity of creating systematic geometric patterns explains the vast number of examples from all but the *sevenfold system*. The patterns produced from this latter system are very beautiful, and the paucity of examples found within the historical record is more likely due to a limited number of artists trained in this system rather than any aesthetic distaste for this variety of design. With the exception of the *sevenfold system*, these design systems were first discovered by the author as systems *per se* in the late 1970s and early 1980s while working on polygonal design methodologies. These findings were first recorded in an unpublished manuscript in 2000,[59] and later published in 2003 in the paper *Three Traditions of Self-Similarity in Fourteenth and Fifteenth Century Islamic Geometric Ornament*.[60] In addition to Hankin, several authors had previously identified some of the underlying polygonal modules that make up the *fivefold system*, but only in relation to individual tessellations rather than as components of a flexible modular system with associated pattern lines.[61] More recently, in 2007 a limited subset of the *fivefold system* received significant public acclaim as the methodological basis employed in the production of an allegedly quasicrystalline design at the Imamzada Darb-i Imam in Isfahan some 500 years before the discovery of fivefold aperiodic tilings by Sir Roger

Penrose.[62] The first published account of the *sevenfold system* as a historical design methodology was in 2012.[63]

Figure 62 illustrates a design, along with its generative tessellation, from each of these five polygonal systems. Figure 62a shows a design created from the *system of regular polygons* that is located at the Gök *madrasa* and mosque in Amasya, Turkey (1266-67); Figure 62b shows a design created by the *fourfold system A* that was used widely, with a particularly early example at the eastern tomb tower at Kharraqan, Iran (1067-68); Fig. 62c shows a design produced by the *fourfold system B* that was used ubiquitously by Muslim cultures; Fig. 62d shows a design created by the *fivefold system* from the Patio de las Doncellas at the Alcazar in Seville (1364); and Fig. 62e shows a design created by the *sevenfold system* that comes from the Sultan al-Mu'ayyad Shaykh complex in Cairo (1412-22). Except for the *fourfold system B*, which is more restricted due to the smaller number of modules, the polygonal modules in each of these systems can be assembled in an infinite number of tessellations, providing for an unlimited number of possible geometric patterns. And, as demonstrated, depending upon the angular opening of the crossing pattern lines located on the edges of the underlying polygons, no less than four distinct designs can be produced from any single tessellation, thus augmenting the already significant design potential within each of these systems.

Another variety of design classification is the differentiation within dual-level designs. These are associated most directly with the use of one or another of the design systems. As discussed in the previous chapter, these place scaled-down secondary modules from a given system into the pattern matrix of a design that was created from the same set of non-scaled-down modules. This variety of design is especially beautiful, and is the last of the great innovations associated with Islamic geometric patterns. Furthermore, many of these designs have geometric self-similarity whereby the qualities of the primary pattern are replicated within the scaled-down secondary pattern, and this recursive diminution can, in theory, be applied *ad infinitum*.

[59] Bonner (2000).

[60] Bonner (2003), 1–12.

[61] –Wade (1976) (after Hankin).
  –Pander (1982).
  –Makovicky (1992), 67–86.

[62] In 2007 Paul Steinhardt and Peter Lu published a paper citing their discovery of a set of "girih tiles" that share the inflation symmetry characteristics of the set of two prototiles with matching rules discovered by Sir Roger Penrose in the 1970s. The five "girih tiles" presented by the authors are, in fact, a subset of the ten polygonal modules with associated pattern lines detailed in my 2003 paper. Lu and Steinhardt's pattern lines for each "girih tile" are identical to the pattern lines of the *median* pattern family (with characteristic 72° crossing pattern lines located at the midpoints of each polygonal edge) that is described in detail in my 2003 paper. See
  –Bonner (2003).
  –Lu and Steinhardt (2007a).

[63] –Bonner and Pelletier (2012), 141–148.
  –Pelletier and Bonner (2012), 149–156.

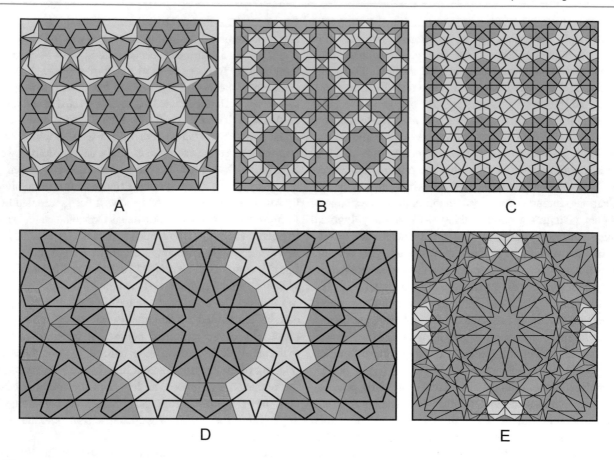

Fig. 62

This variety of Islamic geometric design has four distinct classifications: *type A, type B, type C, and type D* [Fig. 442].[64]

Nonsystematic designs created from the polygonal technique utilize underlying tessellations that include irregular polygons with proportions that are specific to the circumstances of the tessellation, and will not reassemble into other tessellations. This variety of geometric design is characterized by greater geometric complexity, often combining multiple centers of higher order local symmetry. As mentioned previously, the characteristic *n*-pointed stars are typically placed at the vertices of the repeat unit, and greater complexity is frequently achieved through placing further higher order star forms at the center of the repeat, at the midpoints of the edges of the repeat, and/or within the field of the repeat. These higher order stars are the product of their associated *n*-sided primary polygons within the underlying generative tessellation. A polygonal matrix comprised of smaller polygons, such as irregular pentagons and hexagons, separates the primary polygons from one another. Figure 63 illustrates three nonsystematic orthogonal tessellations in

sequential levels of complexity. Figure 63a shows dodecagons at the vertices of the square repeat unit with a connecting matrix of pentagons. Four of the pentagons (yellow) are clustered at the center of the repeat and have a different proportion than the two separating the dodecagons (dark blue). A feature of this underlying tessellation is the *ring of pentagons* that surrounds each dodecagon. This is a common motif in both systematic and nonsystematic tessellations, and reliably provides distinctive and desirable visual characteristics to the completed designs in each of the four pattern families. This tessellation was used to produce a number of very fine geometric designs [Figs. 335 and 336], including an *acute* pattern from the Great Mosque of Siirt in Turkey (1129); a *median* pattern from the Great Mosque of Silvan in Turkey (1152-57); and an *obtuse* pattern that was used frequently throughout the Islamic world. Figure 63b also employs dodecagons at the vertices of the square repeat, with added octagons at the centers of each repeat. This tessellation has a ring of pentagons around each octagon, as well as a ring of pentagons and barrel hexagons around each dodecagon. This ring of pentagons with included hexagons is also commonly encountered in both systematic and nonsystematic underlying tessellations. This tessellation was used to create very successful designs in all four pattern

---

[64] See footnote 241 from Chap. 1.

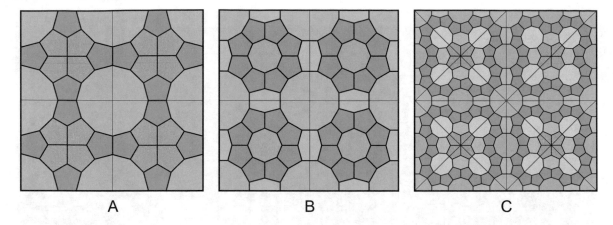

A   B   C

Fig. 63

families [Figs. 379–382]. The two regions of local symmetry allow for the 8- and 12-pointed stars that characterize these patterns. In addition to the dodecagons placed on the corners of the square repeat, Fig. 63c includes decagons placed at the midpoints of each repeat and enneagons placed upon the diagonal of the repeat unit. This is one of the most complex orthogonal generative tessellations employed within this overall tradition, and the *acute* pattern that this tessellation produces was used in several locations by Seljuk artists working during the Sultanate of Rum [Fig. 400]. The earliest example is from the Kayseri hospital (1205-06). This underlying tessellation also creates very attractive designs from the other three pattern families, although no historical examples are known. Patterns produced from this tessellation combine 9-, 10-, and 12-pointed stars.

A further subcategory of design created by the polygonal technique achieves greater complexity through added secondary pattern elements to an already existing design. Almost all such *additive patterns* are produced from one or another of the design systems: most frequently the *system of regular polygons* or the *fourfold system A*, although the *fivefold system* was occasionally used for additive modification. As with so many geometric design innovations, *additive patterns* were initially developed under the auspices of Seljuk influence, and early examples are found at the Gunbad-i Surkh in Maragha, Iran (1147-48), and the Gunbad-i Qabud in Maragha, Iran (1196-97). This additive practice was especially popular among artist working under the Ilkhanids in Persia. Figure 64 illustrates an Ilkhanid additive pattern from the portal of the Khanqah-i Shaykh 'Abd al-Samad in Natanz, Iran (1304-25), that is created from a very simple design from the *system of regular polygons*. The primary *median* pattern (blue) is generated from the underlying 6^3 tessellation of regular hexagons [Fig. 95c], and the additive component places octagons at each vertex of the primary design. The incorporation of octagons into a design with sixfold symmetry works by virtue of the 180° rotational symmetry at the vertices of the

Fig. 64

Fig. 65

primary design. Without the primary design, this arrangement of octagons placed upon the isometric grid is identical to the imposed symmetry design in Fig. 51a. The design in Fig. 65 is an Ilkhanid additive pattern from the interior of the Friday Mosque at Varamin, Iran (1322). The initial *acute*

Fig. 66

pattern is created from the same simple hexagonal grid as the previous example, the difference being in the 60° angular openings of the crossing pattern lines at the midpoints of each underlying hexagonal edge [Fig. 95b]. The addition of a second 6-pointed star in 30° rotation placed on the same center point as in the original design creates a more complex pattern of 12-pointed stars. Figure 66 shows an outstanding Ilkhanid additive pattern from the mausoleum of Uljaytu in Sultaniya, Iran (1305-1313). The original design (blue) is created from the *fourfold system A* [Fig. 157b], and the secondary additive elements provide this otherwise rather simple design with a feeling of far greater complexity. Figure 67 is a detail of an *obtuse* additive pattern created from the *fivefold system* that was used at the Gunbad-i Qabud in Maragha, Iran (1196-97) [Photograph 24]. This is one of the most ambitious examples of additive pattern making, and is characterized by its unusually large repeat unit [Figs. 239 and 240]. This design anticipates the dual-level aesthetic that developed in the same approximate region some 250 years later.

Additive patterns are similar to a less common class of patterns that are comprised of two superimposed designs that are otherwise distinct from one another. And like additive patterns, most of these are derived from the *system of regular polygons*. These *superimposed patterns* are most common to Anatolia during the Sultanate of Rum, and Gerd Schneider provides multiple examples in his book devoted to the

geometric ornament of this region.[65] Figure 68 illustrates a threefold superimposed pattern with one component being the classic threefold *acute* pattern with six-pointed stars [Fig. 95b], but with additive six-pointed star rosettes, created from the 6^3 hexagonal grid, and the second a well-known design of superimposed dodecagons created from the 3.6.3.6 underlying tessellation of triangles and hexagons [Fig. 99b]. This very fine superimposed pattern was used during the Sultanate of Rum at the Karatay Han (1235-41), 50 km east of Kayseri, Turkey,[66] as well as by the Timurids on a door from the mausoleum of Sayf al-Din Bakharzi in Bukhara[67] (fourteenth century).

To summarize, only the polygonal technique has been demonstrated to provide for so many fundamental features of this ornamental tradition, including a method for producing nonsystematic geometric patterns with greater complexity characterized by multiple centers of local symmetry; the occurrence of a multitude of designs with identical visual characteristics that result from the use of systematic polygonal methodologies; the extraordinary range of symmetrical and repetitive diversity that results from manipulations of this

---

[65] Schneider (1980), pl. 22–23.

[66] Schneider (1980), pattern no. 253.

[67] In the collection of the Victoria and Albert Museum, London, acc. No. 437–1902.

Fig. 67

Fig. 68

design methodology; the four pattern families that are ubiquitous to this tradition; and the means for producing highly complex dual-level designs with recursive characteristics.

2.5.2 The Point-Joining Technique

In addition to the polygonal technique, many of the less complex Islamic geometric patterns can also be produced from alternative design methodologies. This is especially true with the more basic patterns created from the *system of regular polygons* and *fourfold system A*. However, there is sparse historical evidence for the use of these other methodologies. By contrast, and as demonstrated, the historical significance of the polygonal technique is supported by a preponderance of evidence. Nonetheless, among some of the less complex historical designs, it is not possible to know categorically whether they were produced from the

polygonal technique or an alternative methodology. There are two alternative design methodologies that have been proposed as traditional and each has merit as a vehicle for constructing geometric patterns. For the purpose of descriptive clarity, these are referred to herein as the *point-joining technique*, and the *graph paper technique*.

Since the 1970s, the point-joining technique has been advanced by a number of proponents,[68] causing it to gain support as the dominant historical design methodology

[68] –Maheronnaqsh (1976).
  –El-Said and Parman (1976).
  –Critchlow (1976).
  –Wade (1976).
  –Bakirer (1981).
  –El-Said (1993).
  –Marchant (2008), 106–123.
  –Broug (2008, 2013).

among the interested public. However, the multiple publications that advance the point-joining technique do not provide evidence for the historical use, let alone primacy, of this methodology. Much of the enthusiasm during the 1970s for the point-joining technique stemmed from its conflation with esotericism[69] wherein "the harmonious division of a circle is no more than a symbolic way of expressing *tawhid*, which is the metaphysical doctrine of Divine Unity as the source and culmination of all diversity."[70] The problem with ascribing metaphysical symbolisms to this design methodology, and indeed to the tradition of Islamic geometric design generally, is similar to that of point-joining itself: the authors provide no evidence for its historicity.[71] The general method behind point-joining constructions involves the use of a compass and straight edge: typically starting with a circle that is divided and subdivided to produce a square or regular hexagonal repeat unit, from which further divisions lead to the construction of a matrix of geometric coordinates. Lines that connect selected intersection points within this matrix will produce the completed design within its repeat unit. A fundamental feature of this technique is that each individual pattern has its own unique step-by-step construction. This is a formal process that lacks flexibility, and while it is well suited to reproducing existing designs with low to moderate complexity, it is not an especially convenient method for creating original designs. This limitation is exponentially true for creating designs with greater complexity, such as those with multiple centers of local symmetry. Even the reconstruction of preexisting patterns with particularly complex compound local symmetries via step-by-step point-joining constructions is extraordinarily cumbersome at best, and for all intents and purposes impracticable. What is more, to use this methodology to *originate* such designs begs credulity. The required independent point-joining construction for each individual pattern is in marked distinction to the inherent flexibility of the polygonal technique. With point-joining, an artist is limited by the number of patterns that have been put to memory, or that have been recorded with instructions on paper. By contrast, an artist with knowledge of the polygonal technique is able to create an unlimited number of original designs, or easily recreate existing designs as needs be. What is more, the polygonal

technique is ideally suited to creating exceptionally complex patterns with multiple centers of local symmetry.

Yet despite its limitations the point-joining technique appears to have played an important role in the history of this artistic tradition. The polygonal technique requires a high level of commitment to master, and clearly not all artists working in diverse media, and at varying degrees of geometric skill, would have received training in this design methodology. What is more, it is reasonable to assume that the transmission of the polygonal technique was formal and controlled, thereby protecting the patronal support and financial interests of the practitioners. As such, the primary role of the point-joining technique may have been as a means of providing specific instruction for individual designs to artists and craftsmen who needed access to geometric patterns, but were not privy to the methodological practices of the polygonal technique. In this way, a wide variety of geometric designs could have been introduced into the canon of general artists and craftspeople, thereby disseminating these designs into the wider cultural milieu while simultaneously protecting the interests of the specialized artists responsible for their original creation.

It is also likely that the point-joining technique occasionally provided a convenient means of scaling up patterns for their transferral to architectural surfaces. Due to the complexity limits of point-joining, this would only have been suitable with patterns of low or intermediate complexity, and artists working with more complex designs would have required alternative methods for accurately transferring scaled-up patterns for architectural locations—as per the above-referenced evidence of the polygonal technique revealed in the ceiling at Fatehpur Sikri.

Historical evidence for the point-joining technique is sparse. One rather amusing early twentieth-century anecdotal example comes from Archibald Christie who wrote:

> Oriental workers carry intricate patterns in their heads and reproduce them easily without notes or guides. There is a story that tells of an English observer, seeing a most elaborate design painted directly on a ceiling by a young craftsman, (the Englishman) sought the artist's father to congratulate him on his son's ability, but the father replied that he regarded the boy as a dolt for he knew only one pattern, but his brother was a genius—he knew three![72]

All humor aside, this story is revealing in that it relays the mnemonic practices of artists working with geometric patterns: albeit very late in the history of this tradition. While this anecdote tells us that at least some artists were reliant upon memory to recreate patterns within their limited repertoire, it also implies that such artists lacked the necessary skills that would allow them to create original designs.

[69] –El-Said and Parman (1976).
  –Critchlow (1976).
  –Burckhardt (1976).
[70] From the forward by Titus Burckhardt: El-Said and Parman (1976).
[71] The popularized claims, advanced during the 1970s, that Islamic geometric patterns are inherently associated with perennial symbolisms have been convincingly refuted as ahistorical by several scholars of note: See
  –Chorbachi (1989), 751–789.
  –Necipoğlu (1995), 73–83.

[72] Christie (1910).

However, considering the vast number of patterns from the historical record, it is unlikely that these specific point-joining constructions were held within memory alone, and it must be assumed that design scrolls and manuals were employed to a greater or lesser extent in propagating the recreation of existing designs. Regrettably few artists' scrolls (*tumar*) or bound manuscripts are known to have survived to the present, and one hopes that more will turn up with time.[73] Two are of particular importance to the question of traditional geometric design methodology: the aforementioned Topkapi Scroll and the anonymous Persian language treatise *On Similar and Complementary Interlocking Figures* in the Bibliothèque Nationale de France in Paris,[74] henceforth referred to as *Interlocking Figures*. The exceptional significance of this treatise is that the illustrations are accompanied with written step-by-step instructions for constructing the diverse range of geometric figures, including multiple geometric patterns. Except for those more complex examples that involve either conic sections or verging procedures, some of these instructions are very similar in concept to the point-joining methodology advocated since the 1970s. This is currently the only known ancient treatise that provides written instructions for constructing geometric patterns, some of which are found within the historical record. *Interlocking Figures* illustrates over 60 geometric constructions, most of which are accompanied with written instruction. Like the Topkapi Scroll, the illustrations are inked in black and red, with occasional dotted lines that provide further differentiation. The provenance of *Interlocking Figures* is uncertain and speculations for its date of origin have been based upon both linguistic analysis and comparisons with identical or near-identical geometric patterns within the architectural record.[75] Estimates for its date range between the eleventh and thirteenth centuries during either the Great Seljuk or Khwarizmshahid periods, with some portions added as late as the Timurid period when the Paris manuscript was copied. More recent research estimates its origin to circa 1300, the

later end of this spectrum.[76] The problem with comparing specific patterns from *Interlocking Figures* to architectural examples from the historical record as a means of estimating the approximate date of its original compilation is that it is impossible to know whether (1) the manuscript may have preceded and possibly influenced an architecture example, and, if so, by how long; (2) the manuscript and architectural examples were produced concurrently, possibly by the same individuals; or (3) the production of a given architectural example may have preceded and possibly influenced the manuscript, and, if so, by how long. Adding to this uncertainty is the fact that it is not known how many times the original manuscript may have been copied, and to what extent the copyists may have included examples of patterns from later dates. Nonetheless, at the very least, comparisons to the architectural record are a valuable means of contextualizing the geometric patterns in *Interlocking Figures*.

The illustrations in *Interlocking Figures* fall into several categories, including mathematical dissections of polygonal figures that can be reassembled into other figures, and can be regarded as sophisticated geometric puzzles; instructions pertaining to the construction of geometric figures such as triangles, pentagons, heptagons, and nonagons; three figures without explanatory text that appear to be *muqarnas* plan projections; and multiple examples of geometric designs ranging from the simple to moderately complex. The question naturally arises: Who created *Interlocking Figures*, and for what purpose? Alpay Özdural makes a compelling case for this treatise having possibly been compiled by a scribe as a record of meetings, or *conversazioni*, between artists and mathematicians over an unspecified period of time.[77] Gülru Necipoğlu suggests that the "anonymous author . . . seems to have been a *muhandis* with practical rather than theoretical training in geometry," and that some of the more complex mathematically precise constructions requiring an angle-bracket and conic sections were followed by instructions for simplified constructions that rely on approximations.[78] Both of these scholars place *Interlocking Figures* into context with other more widely known collaborations between medieval Muslim artists and mathematicians whereby the edification of the geometric arts was facilitated in part through the direct influence of mathematicians. Of particular note is the celebrated treatise by Abu al-Wafa al-Buzjani (940-998): *About that which the artisan needs to know about geometric constructions*. In fact, along with other works on geometry, *Interlocking Figures* is appended to a copy of this work by al-Buzjani. The general consensus among art

---

[73] The most comprehensive study of known pattern manuals and scrolls is that of Gülru Necipoğlu. See Necipoğlu (1995).

[74] MS Persan 169, fol. 180a–199a. For a thorough account of the significance of this manuscript as one of the very few historical Muslim sources of geometric analysis and instruction for Islamic geometric designs, and for its place among other historical documents concerned with the practical application of mathematics, see

–Chorbachi (1989), 751–789.
–Chorbachi (1992), 283–305.
–Necipoğlu (1995), 131–175.
–Özdural (1996), 191–211.
–Necipoğlu [ed.] (forthcoming).

[75] –Necipoğlu (1995), 168–169.
–Özdural (1996), 191–211.

---

[76] Necipoğlu [ed.] (forthcoming).

[77] Özdural (1996), 192.

[78] Necipoğlu (1995), 169.

historians is that *Interlocking Figures* was intended, at least in part, to assist artists to better understand more advanced geometric principles, and where necessary to familiarize them with approximate constructions of figures that otherwise require more complex procedures. Through this lens, the multiple geometric designs included in *Interlocking Figures* are similarly seen as instructions intended for artistic application. However, there is another interpretation of this important historical treatise. It is also possible that the focus upon geometric patterns within *Interlocking Figures* was the result of a fascination among some mathematicians to better understand the underlying geometry of an art form that was pervasive throughout their culture. Seen from this perspective, the step-by-step constructions in *Interlocking Figures* are not so much instructions for artists as exercises for students of geometry. Were this the case, these medieval constructions would be analogous to contemporary point-joining constructions promoted by multiple Western authors since the 1970s whereby people with an interest and facility with geometry and an appreciation for Islamic geometric patterns analyzed specific designs to better understand their geometric nature by creating step-by-step construction sequences.

The fact that *Interlocking Figures* is the only known historical treatise that accompanies the illustrations of geometric patterns with instructional text, coupled with the simplified instructions for creating approximate constructions that would otherwise require far greater mathematical sophistication, would appear to be a persuasive argument for the step-by-step instructions being representative of the primary methodology responsible for this geometric art form.[79] Even prior to *Interlocking Figures* becoming known to the public through the work of Wasma'a Chorbachi,[80] as mentioned, the conviction that point-joining was the preeminent design methodology employed by Muslim geometric artists had been promoted by several authors since the 1970s. More recently, the polygonal technique has become increasingly accepted as especially relevant to the development of Islamic geometric patterns—especially considering the multiple examples of historical evidence for this methodology. Despite the growth in acceptance of the polygonal technique, the historical significance of the point-jointing methodology is central to any serious study of Islamic geometric design, and all the more so in light of the constructions contained within *Interlocking Figures*.

The argument for the more exulted significance of this treatise, whereby artists were provided with necessary approximate solutions to geometric figures through collaboration with mathematicians, runs as follows: (1) there was a desire to create designs with geometric figures, such as heptagons and nonagons, that required advanced mathematical skill, such as intersecting conic sections, to produce mathematically correct constructions; (2) these mathematically correct constructions were beyond the intellectual or practical abilities of artists working in the geometric idiom; (3) and therefore mathematicians working with artists produced simplified step-by-step instructions for various geometric figures and individual patterns that would approximate true mathematical accuracy through what is described herein as point-joining constructions. The first two parts of this proposition assume that artists, in their wish to produce more complex designs employing less straightforward *n*-fold rotational symmetries, would not have conceived the very simple method of dividing the circumference of a circle or arc into a desired number of equal segments, or modular units, using a pair of dividers or compass [Fig. 295]. While this does not provide true mathematical precision, it is very fast, and no less accurate from a practical standpoint. Whether working with intersecting conic sections or the simple division of a circle's circumference into equal units, the drawing of any figure, let us say a nonagon, requires the use of tools such as dividers, straightedge, and set squares. The use of these tools can never be mathematically precise: the point of the divider will never fall at the theoretically correct intersection; the opening of the divider will never precisely conform with the precise mathematical distance; and a line between two points will never connect with absolute mathematical precision. The more steps in a handmade geometric construction, the greater the compounding error. Maintaining our example, the simple division of a circle's circumference into nine segments requires fewer steps than creating a nonagon through intersecting conic sections, and the end result is no less accurate from a practical standpoint. In addition to more complex formulae, *Interlocking Figures* indeed makes reference to this type of mathematical approximation.[81] But to assume that the presence of this divisional methodology in some of the provided instructions is an indication that artists needed to be taught this very simple procedure is disingenuous to the intelligence and innovative spirit of artists, who were, let us not forget, already well advanced in producing highly complex patterns by the time of this treatise's likely creation. This calls into question the third part of the above

---

[79] –Chorbachi (1989), 776.
   –Bulatov (1988), 52.

[80] Chorbachi (1989), 751–798.

---

[81] For example: "But we have found a technique of approximation (*taqrīb*) that, whenever we divide a right angle into nine equal parts, four parts of that angle are اجب and five parts are ب و د. And this is the limit of approximation." MS Persan 169, fol. 190a (upper right corner, diagonal text of four lines). Translation by Carl W. Ernst, Kenan Distinguished Professor of Religious Studies, The University of North Carolina at Chapel Hill.

proposition: that artists required mathematicians to produce simplified instructions for the construction of individual patterns. If artists' innate practical skills meant that they were not reliant upon mathematicians to create such geometric masterpieces as the sevenfold designs on the façade of the minaret of Mas'ud III in Ghazna, Afghanistan (1099-1115) [Figs. 280 and 281], or the design with seven- and nine-pointed stars that surrounds the *mihrab* at the Friday Mosque at Barsian, Iran (1105) [Fig. 429], then the direct contribution of mathematicians toward the growth of sophistication and maturity in this ornamental tradition becomes less significant. And if other design methodologies, such as the polygonal technique, are demonstrably superior in their ability to generate new and original designs, and if this is supported by the preponderance of historical evidence, then elevating the methodological significance of *Interlocking Figures* would appear open to question.

Many features of the anonymous *Interlocking Figures* do not support the premise that this was a manual prepared for use by artists to better equip them in their use of these construction sequences, herein referred to as point-joining, as a primary design methodology for creating new patterns. Nowhere within the text does it state that the work is intended for artists. In fact, the only references to artists within this document pertain to specific constructions used by some artists to construct rather simple designs.[82] In short, the author appears to be more influenced by artists than influence upon them. And while certainly intriguing, the large portion of this treatise dedicated to geometric dissections does not appear to be of any practical use to artists working with geometric design. Similarly, many of the instructions are of questionable relevance to artists. For example, the multiple permutations on the construction of the pentagon would have no practical value to geometric artists who it can be presumed would be very familiar with the construction of this simple figure. The inclusion of these instructions appears to corroborate a fascination with diverse geometric solutions as intellectual exercises. Significant attention is also given to the construction of the heptagon and nonagon; but as mentioned, segmenting the

circumference of a circle with a pair of dividers was a more practical way of accurately producing these polygons.

One possible reason for the preponderance of point-joining instructions in *Interlocking Figures* could have to do with the very different functions that these two design methodologies appear to have within this ornamental tradition. The polygonal technique, in both its systematic and nonsystematic variants, is predisposed to the creation of new designs. By contrast, point-joining does not conveniently lend itself to designing original patterns, but is an effective method for recreating existing designs. As proposed above, if indeed the point-joining technique was used principally for reproducing existing patterns by artists and craftspeople not otherwise trained in the very specific methodology of the polygonal technique, then it would appear reasonable to consider the possibility that the intention behind the constructions for specific geometric patterns in *Interlocking Figures* may have been to develop step-by-step instructions for such non-specialized artists and craftspeople. If this was indeed the case, *Interlocking Figures* provides important evidence of how specific geometric patterns were introduced and disseminated to artists and craftspeople throughout Muslim cultures without jeopardizing the exclusivity of methodological knowledge among the actual originators of such patterns.

Several of the geometric patterns included in *Interlocking Figures* are also found within the architectural record. Of particular interest is the presence of two notable examples from this treatise that are also found within the northeast dome chamber of the Friday Mosque at Isfahan (1088-89). Indeed, there appears to be more than a coincidental relationship between *Interlocking Figures* and this remarkable architectural monument. If the 1300 date attributed to *Interlocking Figures* is correct, the examples within the northeast dome chamber precede this treatise by approximately 200 years.[83] Figure 57 illustrates one of the most remarkable patterns from *Interlocking Figures*: the aforementioned design with sevenfold symmetry that is the only example from this treatise that includes an underlying generative tessellation typical to the polygonal technique. Considering the possibility of an earlier date of origin, Jan Hogendijk has suggested that the occurrence of this heptagonal pattern in both the anonymous treatise[84] and the northeast dome chamber of the Friday Mosque at Isfahan [Photograph 26] may indicate that the same individuals produced both during the same period, and that the presence of Omar Khayyam (1038-1141), the great Persian mathematician and poet, in Isfahan during the construction of the

---

[82] –"Some craftsmen (*ṣunnā'*) draw this problem in such a way that they take its height as seven portions and its width as six portions. The magnitude (*'uẓm*) is close." MS Persan 169, fol. 187b (four lines of upside down text at the corner of the large rectangle). Translation by Carl W. Ernst, Kenan Distinguished Professor of Religious Studies, The University of North Carolina at Chapel Hill.

–"Masters perform a test of the proportion of this problem, and Abu Bakr al-Khalil has performed the test by several methods (*wajh*, lit."face") and has achieved it. One of those [methods] is the following, which has been commented upon." MS Persan 169, fol. 189a. (bottom three lines of main text). Translation by Carl W. Ernst, Kenan Distinguished Professor of Religious Studies, The University of North Carolina at Chapel Hill.

---

[83] This 200-year discrepancy diminishes arguments for the importance of this treatise to the development of this geometric idiom.

[84] MS Persan 169, fol. 192a.

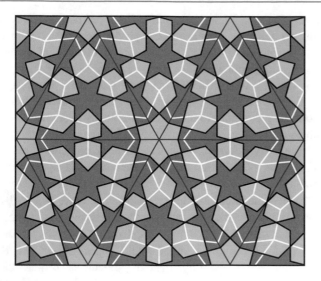

Fig. 69

northeast dome suggests the possibility that he may have been associated with this process.[85] This heptagonal pattern was used in the tympanum of one of the eight recessed arches beneath the cupola. Whether or not *Interlocking Figures* dates to this earlier period, or involved Omar Khayyam in its preparation, it would appear significant that another one of these eight arches from the northeast dome chamber employs a pattern that is almost identical to a design represented within *Interlocking Figures*.[86] This nonsystematic design is represented in Fig. 69, and is characterized by six-pointed stars placed at the vertices of the isometric grid. The only differences between the design from *Interlocking Figures* [Fig. 309b] and that from the recessed arch in Isfahan [Fig. 309a] [Photograph 27] are slightly different angles in the layout of the pattern lines, as well as the absence of regular hexagons centered at each vertex of the isometric grid. Although only small changes, the slightly adjusted pattern angles and the inclusion of the hexagons in the architectural example from Isfahan result in a significant improvement to what is already a successful design. In particular, these changes produce regular heptagons and attractive five-pointed stars within the pattern matrix. The inclusion of the regular heptagons would appear to be a willful corollary with the above-referenced heptagonal pattern in one of the neighboring recessed arches in this domed chamber. A nearly identical example of this nonsystematic design in Isfahan is also found in the Zangid doors in the portal of the Nur al-Din Bimaristan in Damascus (1154) [Fig. 309c]. The pattern in these doors has slightly more acute angles, as well as added geometric rosettes in place

of the six-pointed stars. All three of these examples are easily created from the same underlying generative tessellation. However, the illustration and written instructions in *Interlocking Figures* do not include the underlying generative tessellation. The point-joining construction sequence provided in the text of this manuscript is insufficient to complete the design, although a person familiar with this design tradition could reasonably extrapolate the complete design from the instructions provided. However, this extrapolation requires advanced knowledge of the desired end result, making the instructions unsuitable for teaching this design to anyone not already familiar with it. Be that as it may, the fact that both the heptagonal design in Fig. 57 and the nonsystematic isometric pattern in Fig. 69 were used in the Seljuk ornament of the northeast dome chamber of the Friday Mosque at Isfahan suggests the possibility that the compilers of *Interlocking Figures* were very likely familiar with this building.

There are two patterns in *Interlocking Figures* that are characterized by 10- and 12-pointed stars. With the exception of a very unsuccessful pattern with six-, seven-, and eight-pointed stars (see below), these are the only patterns represented with compound local symmetry. The first example with 10- and 12-pointed stars[87] is identical to a design from the Great Mosque of Aksaray in Turkey (1150-53) [Fig. 414]. Nonsystematic patterns that employ two seemingly incompatible regions of local symmetry within the pattern matrix were an early twelfth-century innovation; and notable twelfth-century examples include a very successful design with 7- and 9-pointed stars at the Friday Mosque at Barsian, Iran (1105) [Fig. 429], and an outstanding design with 11- and 13-pointed stars at the mausoleum of Mu'mine Khatun in Nakhichevan, Azerbaijan (1186) [Fig. 434] [Photograph 35]. As with the design from both *Interlocking Figures* and the Great Mosque of Aksaray, compound patterns with disparate local symmetries frequently rely upon more complex repetitive stratagems that transcend the more prosaic orthogonal and isometric grids. As demonstrated previously, this variety of more complex geometric design will frequently utilize either rectangular or non-regular hexagonal repeat units. Compound patterns are not a feature of Islamic geometric ornament prior to the early twelfth century. Both of these patterns with 10- and 12-pointed stars from *Interlocking Figures* repeat upon a rectangular grid, and it is worth noting that the first of the two is also represented in the Topkapi Scroll[88] wherein it is illustrated along with its underlying generative tessellation. The second design from *Interlocking Figures* with 10- and

[85] Hogendijk (2012), 37–43.
[86] MS Persan 169, fol. 193a.

[87] MS Persan 169, fol. 195b.
[88] Necipoğlu (1995), diagram no. 44.

12-pointed stars[89] is not presently known within the architectural record, although it is a beautiful design, and fully in keeping with the mature style of compound patterns. This also repeats upon a rectangular grid, but its underlying generative tessellation is completely different. While the stylistic character of this design is fully in keeping with similar patterns with compound local symmetry that were created during the period of heightened maturation, the fact that this illustration from *Interlocking Figures* is not accompanied by any additional construction lines or instructional text, and that it is near the end of the manuscript, suggests that this may have been added when the manuscript was copied at a later date. Similarly, this is likely true of the very last design in the manuscript, whose swastika aesthetic suggests a later Timurid origin.[90]

Some of the examples in *Interlocking Figures* that have been identified as geometric patterns intended for ornamental use appear to actually be mathematical exercises without ornamental utility. In particular are two varieties of motif based upon the rotational application of quadrilateral kite shapes. Great emphasis was given within *Interlocking Figures* to these two types of motif, with multiple constructions in each case. In their ideal form, both require conic sections to accurately construct the triangles that comprise these motifs, and in each case, the multiple construction sequences provide approximate solutions for their production without conic sections. Ten of the figures in *Interlocking Figures* are step-by-step point-joining instructions for constructions comprised of quadrilateral kites in fourfold rotation within a square. The kites are subdivided into secondary quadrilaterals and triangles.[91] Without their subdivision, most of these figures are similar to the rotating kite constructions that were frequently used in Islamic architectural ornament, with early examples found in the brickwork façade of the western tomb tower at Kharraqan (1093) [Fig. 27c], and on a wooden door at the Imam Ibrahim mosque in Mosul[92] (1104). The compilers of

Interlocking Figures paid considerable attention to approximate constructions of a rotating kite design with the specific geometric proportion wherein the altitude of the right triangle plus the shortest edge is equal to the hypotenuse. In the text associated with one such construction, differentiation is made between Ibn e-Heitham's method of constructing this triangle with conic sections, hyperbola and parabola, and the provided construction using a T-square.[93] In his paper that references *Interlocking Figures* Jan Hogendijk has pointed out that Ibn e-Heitham is the Persian form of Ibn al-Haytham (965-1041), an important Arab mathematician and astronomer (Alhazen) who was interested in conic sections, but whose work on this triangle is missing. Jan Hogendijk also points to the fact that Omar Khayyam (1038-1141) was also concerned with this triangle, describing it in his *treatise on the division of the quadrant*.[94] It is important to note that the fundamental constituent of a rotating kite design is a right triangle that is mirrored on its hypotenuse to create the kite motif, and that a rotating kite pattern can be made from any right triangle.[95] There are, therefore, a theoretically infinite number of rotating kite patterns—each with common 90° angles and differing pairs of acute angles.[96] It is worth noting that the proportions of the rotating kite patterns used at both Kharraqan and Mosul, and in fact almost all examples from the ornamental record, have a very simple construction that produces a specific proportion wherein the length of the edge of the central square is equal to the shortest edge of the surrounding kites. This proportion is very pleasing to the eye, but is not present in the multiple examples in *Interlocking Figures*. There are a number of ways that this visually more pleasing rotating kite motif can be easily constructed, including from a simple square or a 3 × 3 grid of nine squares [Fig. 27]. Similarly, Abu al-Wafa al-Buzjani (940-998) provided an elegant and equally simple square-based method for drawing the identical fourfold rotating motif in his *About that which the artisan needs to know about geometric constructions*.[97] *Interlocking Figures* includes five approximate constructions for the rotating kite motif created from the triangle described by Omar Khayyam. The compiler's reference to Ibn al-Haytham is a clear indication that they knew that this triangle required conic sections for a precise mathematical construction, and their reason for including the multiple approximate constructions has been proposed as a simplified approach

---

[89] MS Persan 169, fol. 196a.

[90] Several scholars have suggested the possibility that some of the illustrations in *Interlocking Figures* may date to the Timurid period. Gülru Necipoğlu has specifically referenced the final illustration, with its distinctive swastika aesthetic, as likely of Timurid origin, but goes on to mention the possibility of an earlier origin: "even this last pattern is not inconsistent with an earlier medieval repertory," Necipoğlu (1995), 180 [Part 4, note 113].

  –MS Persan 169, fol. 199a.

  –Bulatov (1988).

  –Golombek and Wilber (1988).

[91] –MS Persan 169, fols. 188a, 189b, and 19a.

  –Jan Hogendijk refers to this motif as the 12 kite pattern. See: Hogendijk (2012), 37–43.

[92] Wasma'a Khalid Chorbachi compares the rotating kite designs from the anonymous manuscript to multiple historical examples, including the door from Mosul. See Chorbachi (1989), 751–789.

[93] MS Persan 169, fol. 191a.

[94] Hogendijk (2012), 37–43.

[95] The one exception to the rule that all right triangles will produce rotating kite designs is in the case of the isosceles triangle with equal 45° acute angles. When this is mirrored along its long side it produces a square rather than a kite.

[96] Cromwell and Beltrami (2011), 84–93.

[97] Chorbachi (1989), 769.

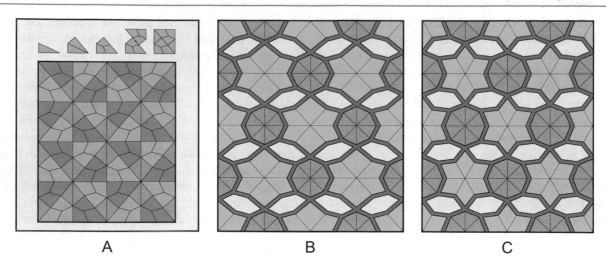

A   B   C

Fig. 70

for artists who would not have known the methodology of conic sections.[98] However, there are two problems with this assertion. Firstly, the visual character of the subdivided rotating kite designs in *Interlocking Figures* does not conform to the many examples of fourfold rotating kite designs from the architectural record. These are invariably non-subdivided. Secondly, despite assertions to the contrary,[99] the very specific geometric proportions of the rotating kite motif that results from using Omar Khayyam's elusive triangle do not appear to have been used within the architectural record. Furthermore, the supposed need to educate artists in the difficulties of creating the rotating kite motif is not supported by the fact that the examples from the architectural record are very easy to produce. It would therefore appear that the focus upon approximate constructions for these subdivided rotating kite designs with Khayyam-like proportions was more an intellectual exercise on the part of the mathematicians who compiled this section of the treatise, and less a product for use by artists in their ornamental constructions.

The other rotational kite motif that received almost equal attention in *Interlocking Figures* is comprised of two kite motifs in twofold rotational point symmetry that are placed

within a bounding rectangle. This treatise contains seven separate constructions for this motif, and as with several of the fourfold rotational kite examples, these kites are subdivided into three polygons that maintain the bilateral symmetry of the original kite. Figure 70a is constructed from the point-joining instructions given in fol. 185v from *Interlocking Figures*. As with the fourfold rotational kite motif, the basic constructive component is another triangle of specific proportion that requires conic sections to draw with mathematical precision, and like the fourfold examples, the multiple instructions for this rectangular motif make use of approximate constructions that side step the use of conic sections. This is confirmed in the written instructions wherein the author ends by stating that the triangular component is difficult to create, that it is outside Euclid's *Elements*, and otherwise requires the use of conic sections.[100] As demonstrated in the upper portion of Fig. 70a, the completed rectangle is produced from the generating triangle through mirroring the triangle on its hypotenuse to create the distinctive quadrilateral kite. This is subdivided into three secondary quadrilaterals. The subdivided kite is rotated with point symmetry such that the secondary approximate quarter octagons on adjacent sides of the two kites are contiguous. The specific proportions of the original triangle allow for the acute angles of each kite to align with the extended edges of the other, thus creating a bounding rectangle. Alpay Özdural suggests

[98] –Chorbachi (1989), 765.
    –Özdural (1996), 191–211.
    –Hogendijk (2012), 37–43.

[99] Some scholars who have written on the significance of the fourfold rotating kite designs in *Interlocking figures* have failed to differentiate between the geometric proportions of the examples from this treatise and the proportions of the examples from the architectural record. By conflating all rotating kite designs into a single complex construct requiring conic sections for mathematically accurate construction, the need for mathematicians to assist artists in the construction of simplified approximations is corroborated. See Chorbachi (1989), 751–789.

[100] "Producing a triangle such as this is difficult, and it falls outside of the Elements of Euclid. It belongs to the science of conics (*makhrūṭāt*) and it is produced by the action of moving the ruler (*misṭara*). When the height of the vertical is postulated (*mafrūḍ*) as in this example, postulated as half of segment ﺏ ﺍ, it produces the square ﻭ ﺡ ﺍ." MS Persan 169, fol. 185b. (Bottom three lines of diagonal text in upper right). Translation by Carl W. Ernst, Kenan Distinguished Professor of Religious Studies, The University of North Carolina at Chapel Hill.

that the multiple constructions for this figure were intended for artists to create a geometric design that mirrors the rectangular motif to cover the plane.[101] Figure 70b shows a widened line version of this pattern, thereby providing a more typical ornamental treatment to the line work in Fig. 70a. The deficiencies as an ornamental design are readily apparent: the six-pointed stars lack sixfold rotational symmetry, and the irregular octagons are not in conformity with the aesthetic standards of this ornamental tradition. Not surprisingly, no examples of this poorly proportioned pattern are known from the historical record. However, this basic conceptual arrangement, but with symmetrically regular six-pointed stars and regular octagons, can result in an acceptable design. Figure 70c is just such an idealized version of this otherwise unsatisfactory design, and the incorporation of regular octagons and six-pointed stars completely solves the visual imbalance of the design created from the multiple constructions in *Interlocking Figures*. This improved version is very easily created by using the exterior angles of a regular octagon to produce the angular proportions of the six-pointed stars. In light of the geometric complexity of designs produced during the period of *Interlocking Figures* estimated origin, this idealized version would have posed no intellectual challenge to a competent geometric artist of the period. While no examples of this idealized version are known in the ornamental arts of the Seljuks or the direct inheritors of their geometric traditions, this particular combination of octagons, six-pointed stars, and irregular hexagonal interstice regions was used as a generative tessellation for a design on the back wall of the previously mentioned niche in the Mamluk entry portal of the Sultan al-Nasir Hasan funerary complex in Cairo (1356-63) [Fig. 413] [Photograph 58]. Unlike almost all other geometric patterns within the architectural record, this example employs both the generative tessellation and the design itself, and is an important source of evidence for the historicity of the polygonal technique. While this unusual arrangement of octagons and six-pointed stars is responsible for the very lovely design at the Sultan al-Nasir Hasan funerary complex, it does not appear to have been used otherwise as ornament. The simplicity of construction for the idealized version in Fig. 70c is in marked contrast to the complexity of the seven constructions in *Interlocking Figures*. Had these seven constructions indeed been intended for artistic application, it is reasonable to assume that the originators of these constructions, be they mathematicians working either with or without artists, would have known that the resulting octagons and six-pointed stars would not have had regular

eightfold and sixfold symmetry. It can also be assumed that if the objective had been to create an aesthetically acceptable design, a construction sequence that provided for regularity of the octagons and six-pointed stars would have been provided—especially considering the relative simplicity of such a construction. As with the previous examples of four-fold rotational kite motifs from this treatise, it would appear that the interest in this construction was less for the purpose of informing artists of a construction sequence for producing a satisfactory geometric pattern, and more as a series of geometric exercises in their own right.

The greatest reason for calling into question the view that *Interlocking Figures* is an exposition of the primary design methodology employed during this developmental period of the geometric idiom is in the prescribed instructions for each included pattern. Many of the step-by-step constructions are rife with inaccuracies that lead to distortions, and the resulting designs are not indicative of the geometric accuracy within the vast canon of historical geometric ornament. For this reason, several of the patterns from *Interlocking Figures* fall short of their potential for creating designs that are aesthetically acceptable to this ornamental tradition. A case in point is a construction that places seven-pointed stars on the edges and near the vertices of a square repeat unit.[102] As with previous examples, this design utilizes the geometry of a fourfold rotating kite motif to structure the design upon. As clearly observed in the compilers' illustration, rather than their placement on each vertex of the square repeat, the points of each of the 4 seven-pointed stars extend beyond the vertices of the square repeat. Another failing is particularly problematic in that historical examples of geometric patterns based upon a structure of fourfold rotating kites invariably have bilateral symmetry within the kite element [Figs. 28], whereas the dotted lines that make up the kites in this example do not. What is more, some of the pattern lines that extend from the points of the seven-pointed stars are not collinear with the star, and change direction at the point of intersection. These problems are generally not in keeping with the aesthetics of this design tradition, and the inclusion of this construction in the manuscript indicates a certain naïveté on the part of the creator of this example. This design is similar in principle to an orthogonal pattern found at both the Mirjaniyya *madrasa* in Baghdad (1357), and the Amir Qijmas al-Ishaqi mosque in Cairo (1479-81) [Fig. 25c]. This also places seven-pointed stars on the edges of a square repeat unit, but has well-balanced proportions throughout. The first of the above-referenced designs with 10- and 12-pointed stars[103] is particularly revealing of the inconsistencies that result from the flawed point-joining

---

[101] Figure 70a illustrates the repetitive application of this construction, and, except for color, is identical to the prior representation by Alpay Özdural. See Özdural (1996), Fig. 7.

[102] MS Persan 169, fol. 194b.

[103] MS Persan 169, fol. 195b.

construction sequence. This example in *Interlocking Figures* is in marked contrast to the successful use of this design at the Great Mosque at Aksaray, Turkey (1150-53). As per the example of this pattern in the Topkapi Scroll, this is easily and accurately created using the polygonal technique [Fig. 414]. However, a quick study of the anonymous author's point-joining construction reveals the design to be incomplete, with a series of false starts, overdrawing, and poor angles in the unresolved region. In point of fact, the written instructions do not produce a successful design. This is a stark example of the inadequacy of the point-joining technique to accurately provide an easily followed construction for complex patterns with multiple centers of local symmetry: especially when these centers have seemingly disparate rotational symmetries such as the 10-fold and 12-fold regions within this design. The failings of this construction appear to indicate the limited scope of the author's knowledge and technical mastery of the more complex designs that were already a feature of this tradition at the likely time of the manuscript's preparation.

Other than the fivefold swastika design at the end of the treatise that was likely a Timurid addition, the only pattern with fivefold symmetry from *Interlocking Figures* is a *median* field pattern.[104] This is surprising in that fivefold patterns are an immensely important feature of this geometric art form, and were widely employed by its estimated date of origin. As with so many of the designs in this treatise, this fivefold example also fails to achieve the stylistic aesthetic to which it aspires. The greatest visual failing is in the lack of collinearity in the crossing pattern lines. With adjustments, this pattern could be made successful, and its portrayal in *Interlocking Figures* appears to be the product of someone lacking a refined understanding of the aesthetic requisites of this design tradition generally, and of the fivefold design discipline specifically.

An interesting, but ultimately disappointing, geometric pattern from *Interlocking Figures* is made up of six-, seven-, and eight-pointed stars placed into a rectangular repeat unit.[105] A cursory examination of this figure reveals several problems, most notable being the points of the seven-pointed stars not intersecting with the edges of the rectangular repeat. The lack of collinearity in the crossing pattern lines that connect the seven-pointed star with both the six- and eight-pointed stars is also problematic. A number of examples of geometric designs with sequential numbers of star types are known to the historical record: for example, a very-well-conceived pattern with five-, six-, seven-, and eight-pointed stars from the *mihrab* of the Friday Mosque

at Barsian (1105) [Fig. 332]. However, the arrangement of the sequential star forms in the example from *Interlocking Figures* appears arbitrary and contrived by comparison, and falls far short of achieving the already well-established aesthetic standards of this ornamental tradition.

Despite the presence of problematic designs in *Interlocking Figures*, there are several that are very successful, with accurate and useful point-joining instructions. Without a doubt, the most successful and remarkable design from this manuscript is the above-mentioned heptagonal design represented in Fig. 57. The design illustrated in Fig. 71a is another successful, if considerably less remarkable, pattern from *Interlocking Figures*.[106] This places six-pointed stars upon the vertices of an orthogonal grid with 90° alternating orientations much in the fashion of the historical designs in Fig. 28. Coinciding with four of the points of these 4 six-pointed stars are four of the points of an eight-pointed star centered within the square repeat unit. This central eight-pointed star is, by force, rotated out of an orthogonal alignment by 11.4254...°. Except for the fact that the 120° exterior angles of the eight-pointed stars that match those of the six-pointed stars must be inferred (no written instructions are provided), the point-joining construction for this design is complete and accurate. The oscillating orientation of the eight-pointed stars, and rotating concave octagonal shield elements of the repetitive structure (red), shares geometric properties with typical oscillating square and rotating kite designs. However, rather than the angle of declination being governed by a single polygonal element, it is the direct product of the six-pointed stars in 90° rotation. This design is a pleasing juxtaposition of six- and eightfold rotational symmetry, and its attractive qualities may well have led to its use historically, although no examples are known. Figure 71b slightly changes the design so that the proportions are determined by regular heptagons.[107] The original design suggests these heptagons within the interstices of the two star forms. By utilizing regular heptagons, the exterior obtuse angles of the six- and eight-pointed stars become the product of this polygon, as do the proportions of the concave hexagonal repetitive shield elements (red). This change is attractive in that the eye readily recognizes and appreciates the regular heptagon; but this is at the loss of the sixfold rotational symmetry of the six-pointed stars.

---

[104] MS Persan 169, fol. 193b.

[105] MS Persan 169, fol. 190b.

[106] This manuscript includes a second design (fol. 191r) that places six-pointed stars in 90° rotation around the vertices of the square repeat unit. However, the construction of the six-pointed stars is problematic in that it does not provide for the desired sixfold rotational symmetry. MS Persan 169, fol. 194a.

[107] This experimental change to the original design is the work of the author, but was inspired by an observation by Jan Hogendijk.

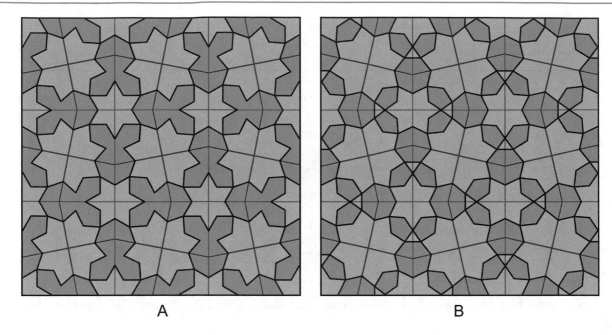

A   B

Fig. 71

One of the orthogonal designs from *Interlocking Figures* is particularly successful both aesthetically and in its point-joining construction. This is the one fourfold rotational kite figure from *Interlocking Figures* that is an actual geometric pattern, and although no examples of its use are known within the architectural record, its aesthetic character is fully in keeping with this ornamental tradition. Of particular significance is the central fourfold rotational motif comprised of a square surrounded by four rotating kites separated by four rotating chevrons, all within a bounding square [Fig. 112d]. This isolated motif is found in a number of historical patterns from Persia and Khurasan to which this example from *Interlocking Figures* is closely related. While this design is a point-joining construction, it is more easily produced using the polygonal technique. From this perspective, it is a *two-point* pattern that is created from the *system of regular polygons* through the use of the *two-uniform* $3^3.4^2$ $-3^2.4.3.4$ underlying tessellation of triangles and squares [Fig. 112c]. The point-joining instructional text that accompanies this illustration in *Interlocking Figures* states:

> Masters perform a test of the proportion of this problem, and Abu Bakr al-Khalil has performed the test by several methods (*wajh*, lit."face") and has achieved it. One of those [methods] is the following, which has been commented upon.[108]

Abu Bakr al-Khalil and his associates do not appear to have been knowledgeable of the less complex approach to constructing this pattern using underlying triangles and squares, or if they were, they cared not to reveal it. A comparison between the point-joining instructions in *Interlocking Figures* and the derivation of this design using the polygonal technique provides clear evidence of the superiority of the polygonal technique as a design methodology. This is especially true not just for its inherent simplicity, but in its greater flexibility: the ability of rearranging the polygonal modules of the *system of regular polygons* into other tessellations, thereby producing new patterns, as well as the ability to create additional patterns by applying alternative pattern lines from the other historical pattern families to each new underlying tessellation. When using the polygonal technique to create this design, the characteristic rotational motif is produced from a central square contiguously surrounded by four triangles. The underlying squares within this generative tessellation are provided with two perpendicular sets of parallel pattern lines placed on each edge, thereby identifying this as a variety of *two-point* pattern. These pattern lines extend into the adjacent underlying triangles until they meet with other extended pattern lines. An early example of a design associated with this variety of *two-point* design methodology, with the central fourfold rotational motif, was produced by Khwarizmshahid artists for the Zuzan *madrasa* in northeastern Iran (1219) [Fig. 112b] [Photograph 38]. Being that Abu Bakr al-Khalil has not yet been identified through other sources, the similarity between the example from *Interlocking Figures* and that of the Zuzan

---

[108] MS Persan 169, fol. 189a. (bottom three lines of main text). Translation by Carl W. Ernst, Kenan Distinguished Professor of Religious Studies, The University of North Carolina at Chapel Hill.

Fig. 72

madrasa raises the possibility first proposed by Alpay Özdural that portions of *Interlocking Figures* may have been produced during the Khwarizmshahid period.[109]

Another example of a successful point-joining construction from *Interlocking Figures* produces a very-well-known orthogonal design that was used widely throughout the Islamic world.[110] This *median* design is easily produced from the *fourfold system A* [Fig. 145], and was particularly popular in Khurasan during the late eleventh and early twelfth centuries. The illustration and instructions for this design in *Interlocking Figures* are an accurate, if slightly incomplete, point-joining method for its reproduction, although it is worth noting that the illustration, as drawn in this treatise, includes point-joining layout lines that obscure the actual pattern to the point of being difficult to initially identify. This presumably explains why this figure from *Interlocking Figures* has not been recognized as this particularly well-known fourfold pattern in previous studies. The inclusion of this design is significant in that it was used so widely throughout Khurasan and eastern Persia preceding the likely date of origin of this treatise.

By the time *Interlocking Figures* was written, the geometric ornamental idiom was fully mature, and the need for artists to have direct mathematical input would have been less of an aesthetic imperative than during the earlier time of Abu al-Wafa al-Buzjani (940-998). As stated, the many problems with the manuscript's constructions lead one to question the assumption of its significance to the

methodological development of this tradition. And yet, as demonstrated, this anonymous manuscript also has numerous constructions that accurately produce geometric patterns that are very acceptable, and even outstanding—as per the pattern with heptagons that is also found in the northeast dome chamber in the Friday Mosque at Isfahan.

The question of by and for whom *Interlocking Figures* was produced remains intriguing. Was it written by mathematicians for artists, or perhaps by mathematicians, inspired by and seeking to better understand the geometric work of artists? Or did the point-joining constructions of geometric patterns result from artists privy to the polygonal technique requesting assistance from mathematicians to devise step-by-step instructions that could be provided to artists and craftspeople more widely so that this art form could be adopted more pervasively in a wide range of media? These uncertainties are augmented by the lack of cohesion throughout this treatise, by the inconsistencies of mathematical sophistication, and disparities between naively conceived simplistic geometric patterns on the one hand, and well-realized construction sequences for very acceptable designs on the other. One is led to conclude that this treatise was the work of multiple individuals with variable levels of mathematical and artistic proficiency, and possibly for more than a single intent.

Recent publications that focus upon the point-joining technique include well-conceived construction sequences for numerous orthogonal and isometric designs, as well as some fivefold patterns. A typical example of a sequential point-joining construction (by author) is shown in Fig. 72. This is for a simple design that was used in a brickwork border at the tomb of Nasr ibn Ali (1012-13), the earliest of the three

---

[109] Özdural (1996).
[110] MS Persan 169, fol. 196a.

adjoining mausolea at Uzgen, Kyrgyzstan [Photograph 15]. This same design can also be created from the polygonal technique, and indeed is one of the earliest examples of a design that can be created from the *fourfold system A*. Another relatively early example of this well-known *median* design is from the mausoleum of Sultan Sanjar in Merv, Turkmenistan (1157) [Fig. 159]. As with other designs with low or moderate complexity that were very likely produced originally with the polygonal technique, this illustration demonstrates how such designs can frequently be recreated easily from point-joining methodology.

To summarize, the point-joining technique may well have been used historically as a means to recreate existing geometric patterns among artists and craftspersons working in the geometric idiom who were not privy to the more specialized and highly versatile design methodologies of the polygonal technique. While the point-joining technique is less convenient than other generative methodologies for designing original patterns of moderate complexity, and especially impractical for recreating patterns with greater complexity, it nonetheless provides a particularly useful method of recreating existing patterns of low and moderate complexity. The dissemination of specific point-joining constructions to artists and craftspeople that were not otherwise privy to some of the more esoteric design methodologies would have allowed for the widespread and repeated use of a wide variety of specific designs. In this way, point-joining constructions were likely an important contributor to the ongoing spread and consolidation of the geometric aesthetic throughout Muslim cultures.

2.5.3 The Grid Method

Some of the less complex threefold patterns can be created directly from the isometric grid. Through simple trial and error, repetitive interlocking and overlapping figures can be found that make very acceptable designs. Most of the patterns created from the isometric grid will have pattern lines that are congruent with the grid itself. More complex isometric grid designs will have pattern lines in six directions: three with the grid, and three perpendicular to the grid. Figure 73 illustrates three isometric grid designs. The first of these, Fig. 73a, is an interlocking design with three- and sixfold centers of rotational symmetry that is typical of the *p6* plane symmetry group. All of the pattern lines in this design are congruent with the grid. An example of this design is found at the Gök *madrasa* in Amasay (1266-67). The parallel pattern lines in Fig. 73b are also congruent with the grid, but interweave with one another to produce a more conventional Islamic geometric aesthetic effect. This design can also be produced very readily using the *system of regular polygons*, and a fine example was used at the Shah-i Mashhad in Gargistan, Afghanistan (1176) [Fig. 104c]. The design in Fig. 73c employs lines that are both congruent with the grid and perpendicular to the grid. This design can also be produced from the *system of regular polygons*, and the imbalance between the large and small background elements can be improved by widening the pattern lines in one direction rather than both directions. This design was used on the façade of the Mu'mine Khatun mausoleum in Nakhichevan, Azerbaijan (1186) [Fig. 101d].

The orthogonal grid can also be used to create geometric designs. At the most basic level, designs produced on this grid will maintain congruency with the orthogonal grid. Designs of this variety include the many swastika and key patterns, as well as square *Kufi* calligraphic motifs. The orthogonal nature of this variety of design made them especially relevant to the brickwork ornament championed by the Ghaznavids, Qarakhanids, Ghurids, and Seljuks. Later expressions included Timurid polychrome cut-tile mosaic. Diagonal lines were also introduced to patterns created from

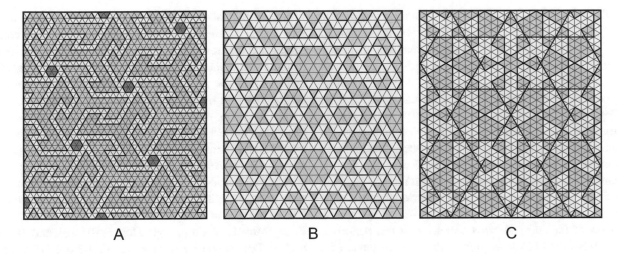

A   B   C

Fig. 73

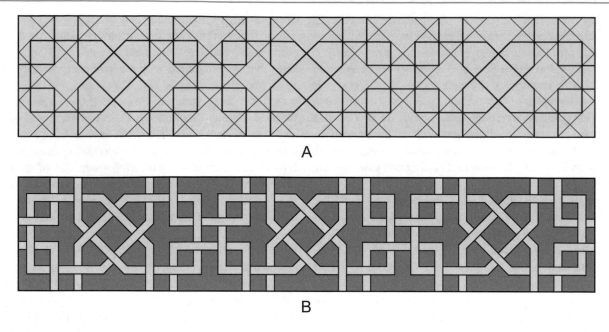

A

B

Fig. 74

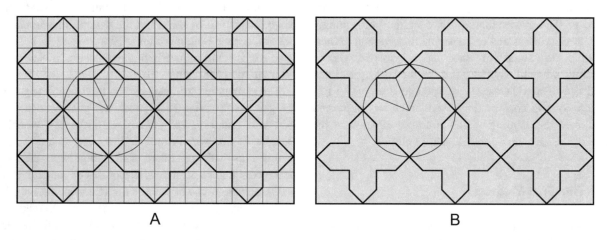

A   B

Fig. 75

the orthogonal grid, adding two more directions of pattern line. Figure 74 illustrates a border design from the minaret of Uzgen in Kyrgyzstan (twelfth century) that combines two directions of lines that are congruent with the orthogonal grid, and two directions with 45° diagonal lines. This type of design can be used to create designs with eight-pointed stars, and Fig. 75 illustrates the use of the orthogonal grid to construct the well-known star-and-cross design. Figure 75a demonstrates the problem with constructing designs with eight-pointed stars using the grid method. Because four of the points for each star are congruent with the grid, and the other four are diagonals, there are two sizes of points. The finished star does not have eightfold rotational symmetry. Examples of the star-and-cross pattern with this distortion are occasionally found in the architectural record, but almost

always from a much later date after this ornamental tradition had begun to decline. Figure 75b illustrates the correct proportion for the eight-pointed stars. More complex designs produced from the orthogonal grid frequently have the character of the *fourfold system A*. Figure 76 illustrates the orthogonal grid derivation of a widely used design, along with the same design with the correct proportions as generated from the *fourfold system A* [Fig. 145]. As mentioned above, point-joining instructions for this design were included in *Interlocking Figures*, and several examples (with correct proportions) are found in the early brickwork ornament of Khurasan, including the minaret of the Friday Mosque at Damghan, Iran (1080); the *mihrab* of the Friday Mosque at Golpayegan, Iran (1105-18); and the minaret of Daulatabad in Afghanistan (1108-09) [Photograph

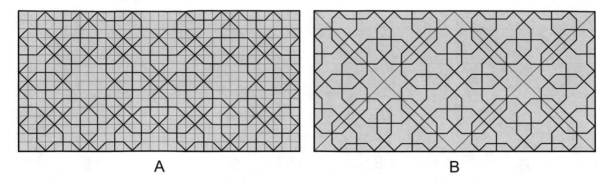

Fig. 76

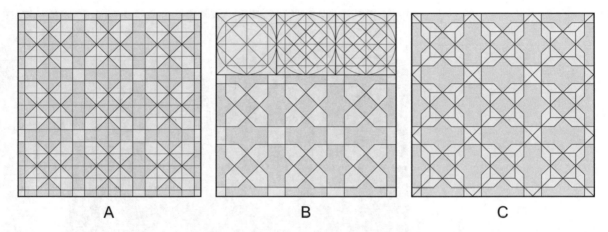

Fig. 77

20]. Working with the orthogonal grid in Fig. 76a can be a fast way of testing ideas and exploring design options. However, these designs will always have distortions that result from the difference between the length and diameter of each square cell of the orthogonal grid. As demonstrated in Fig. 76b, when using the orthogonal grid, once a design has been arrived at, it is necessary to draw it anew so that the distortions are eliminated.

As stated, there are multiple options for designing the less complex patterns in this geometric art form, and with less complex designs it is impossible to say with certainty exactly how a particular example was constructed. The same design may have been constructed one way at a given location, and another way elsewhere. Figure 77 illustrates construction solutions for a very simple fourfold design in all three of the methods discussed herein. Figure 77a demonstrates the orthogonal grid method; Fig. 77b provides a simple construction sequence for the point-joining technique; and Fig. 77c shows the generation of this pattern from an underlying tessellation associated with the polygonal technique. The benefit of the grid method is that it is a fast way to explore design options, but requires correction. The benefit of point-joining is that it is an accurate way of recreating existing designs, but lacks flexibility as a means of creating original designs. By contrast, the polygonal technique is accurate and extremely flexible, and, as pertains to the polygonal systems, very fast. What is more, the polygonal technique also provides for the generation of at least four distinctly different patterns from each underlying tessellation. By way of example, Fig. 78 illustrates designs in each of the four pattern families for the tessellation shown in Fig. 77c.

In the hands of an experienced practitioner, the orthogonal grid can be used to generate increasingly complex geometric patterns with fourfold repetitive symmetry. This especially pertains to the distinctive geometric style of Morocco and al-Andalus, and can be applied to patterns with higher order *n*-pointed stars that are multiples of 8— even up to and including 64-pointed stars. Jean-Marc Castéra has deftly demonstrated the versatility of this advanced orthogonal grid technique,[111] which he refers to as the "freehand method." His methodology takes into

---

[111] Castéra (1996).

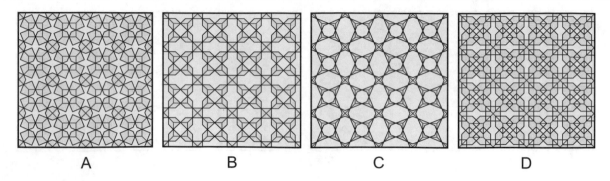

A   B   C   D

Fig. 78

Fig. 79

account the disparity between the orthogonal and diagonal coordinates, and the need to correct these approximations:

> Since we are aware of the errors brought about by these approximations, it is quite simple for us to correct them when necessary, if, for example, we wish to make an actual mosaic. In this manner, each time we create a new piece, it adopts the correct proportions, which have been geometrically deducted from those of the pieces that have been already made.[112]

Figure 79 is an example of this more complex Maghrebi form of grid method construction requiring approximate coordinates of the orthogonal grid.[113] While this method of constructing patterns is currently used in Morocco, the extent to which Maghrebi artists of the past employed this methodology is unclear. Certainly the preponderance of geometric patterns from Morocco and southern Spain is of a geometric nature that would allow for their creation in this manner. However, no ancient pattern books or scrolls from the Western regions have confirmed the historicity of this methodology, and the patterns created from the more advanced grid method can, almost always, also be created using the polygonal technique with relative ease.

2.5.4 Extended Parallel Radii

There is a category of geometric pattern that is rarely encountered, but sufficiently unusual, and indeed beautiful, as to justify methodological analysis. While no examples of *extended parallel radii* designs appear in historical scrolls or design reference books such as the *Topkapi Scroll* or

---

[112] Castéra (1996), 99. (note: this quotation is from the English edition of 1999).

[113] This photograph shows the hand of Jean-Marc Castéra using the orthogonal grid to construct the design by drawing freehand.

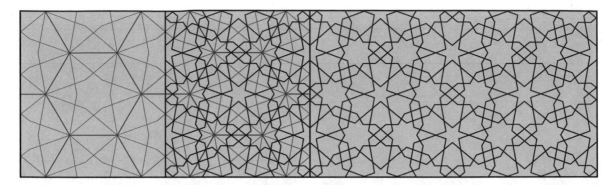

Fig. 80

Interlocking Figures, the method for constructing these patterns can be intuited by their geometric character. The essential feature of this methodology offsets the lines of a radii matrix in both directions, eliminates the original radii, and extends these parallel offsets until they meet with other extended offset lines. In its most simple form, the lines of a common grid are offset on both sides by an amount that will create a visually acceptable pattern. The very simple design in Fig. 32 can be produced in this manner, although this replicates the most basic *two-point* functionality of the polygonal technique [Fig. 96d]. As with this example, a number of early and uncomplicated geometric patterns that are easily created from the *system of regular polygons* have these parallel grid line characteristics, including a Ghurid brickwork panel from the façade of the western mausoleum at Chisht, Afghanistan (1167) [Fig. 105a]. This design is characterized by parallel offsets of the hexagonal grid and its triangular dual. An example from the synagogue in Cordoba, Spain (1316), constructed during the Nasrid period, achieves greater complexity through additional parallel offsets of lines that connect the vertices of the hexagonal and triangular grids [Fig. 105h]. When created from the polygonal technique, the distance between the parallel lines in each of these examples is determined geometrically through the lines being located at determined points within the underlying generative tessellation. By contrast, with the extended parallel radii technique the distance between the parallel lines is generally an arbitrary determination based upon the aesthetic predilections of the artist. Though less formal than other design methodologies, through trial and error, this alternative technique will nonetheless create very beautiful designs.

The more interesting extended parallel radii designs employ more complex radii matrices. As an example, Fig. 80 illustrates a design (by author) that utilizes a radii matrix with ninefold symmetry at each vertex of the regular hexagonal grid. A close inspection of this design reveals each line of the pattern to be an equally distanced parallel offset to the generative radii matrix. This is not a historical design,

although it falls within the acceptable aesthetics of this ornamental tradition. Radii matrices are a fundamental methodological component of the polygonal technique. They provide the structure upon which the generative polygonal tessellations are created. This is especially relevant to nonsystematic patterns with compound regions of *n*-fold rotational symmetries. The historical use of radii matrices is confirmed in the *Topkapi Scroll*, where they appear as incised reference lines produced with a steel stylus. Figure 81a shows a radii matrix with local regions of 8-, 10-, 12-, and 16-fold rotational symmetry set within a square repeat unit. Figure 81b illustrates an *acute* pattern produced from an underlying tessellation that can be made from this radii matrix [Figs. 404 and 405]. This *acute* pattern was used in the *iwan* of the Kemaliya *madrasa* in Konya, Turkey (1249). This same radii matrix, with its very particular combination of local symmetries, was used to create two extended parallel radii designs that date to the same approximate time and place during the Seljuk Sultanate of Rum. Figure 81c represents the extended parallel radii design from the Kaykavus hospital in Sivas, Turkey[114] (1217-18), and Fig. 81d illustrates the closely related design from the Sultan Han near Aksaray, Turkey[115] (1229). These three designs with identical symmetrical structure, but distinctly different aesthetics, all come from central Anatolia and were produced within 32 years of one another: conceivably by the same artist or artistic lineage.

2.5.5   Compass Work

The process of laying out a design with a compass or dividers was inherited by Muslim artists from their Christian counterparts who were actively engaged in the Hellenistic aesthetic that survived well into the Late Antique period.

---

[114] Schneider (1980), pattern no. 426.
[115] Schneider (1980), pattern no. 425.

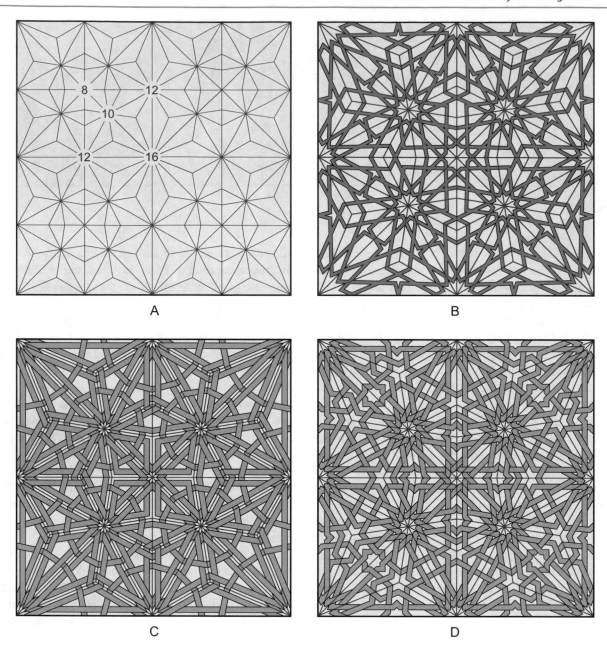

Fig. 81

The earliest examples of Islamic *compass-work* ornament are associated with several surviving Umayyad buildings: the most notable examples being several pierced stone window grilles from the Great Mosque of Damascus (706-15). While geometrically undemanding, these are significant in their application of what had previously been an ornamental device used primarily for mosaic pavements to a new expression in pierced stonework. What is more, the visual quality of these windows helped to establish interweaving geometric designs as a primary feature of Muslim aesthetics.

Compass-work ornament has its own visual character and curvilinear appeal, and it is not surprising that this artistic

practice continued among succeeding Muslim cultures, even if reduced to a role of relatively minor significance. In addition to repetitive patterns, these later expressions included nonrepetitive, stand-alone, ornamental panels primarily composed from circles within a rectangular frame. Such compass-work constructions were occasionally used as Quranic illuminations, including in the Quran produced by ibn al-Bawwab in 1001. However, this study is concerned expressly with geometric compass-work creations that have repetitive characteristics. The methodology behind these compass-work patterns is overtly apparent upon examination, and involves the drawing of circles at set points of a

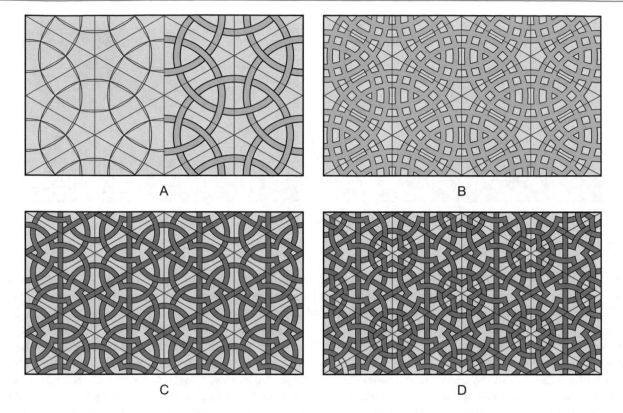

Fig. 82

given geometric grid.[116] These circles can be uninterrupted or trimmed where they intersect with other circles. Compass-work designs will sometimes incorporate s-curves within their overall matrix. Unlike patterns created from the polygonal technique, compass-work patterns will often include the generative grid along with the circular elements, thereby creating designs that include both an angular and a curvilinear quality. Figure 82a illustrates a very basic compass-work pattern made up of interweaving circles set upon the vertices of the isometric grid. The proportions of this design are easily determined by locating the center of each circle upon the vertices of the isometric grid and the radius at a point that is past the midpoint of each edge of the triangular cells that make up the isometric grid. In this illustration, the size of the circles is determined by their circumference being placed upon the vertices of the 3.4.6.4 semi-regular tessellation of triangle, squares, and hexagons. Figure 82b shows essentially the same design, but with a double-line treatment. The radii and width of the parallel circles are carefully contrived to create the distinctive network of similarly sized background elements. This compass-work design was used in the Ottoman inlaid stone ornament of the Sehzade Mehmet complex in Istanbul (1544-48), and is an excellent example of the continued

use of compass-work patterns among later Muslim cultures. Figure 82c is a representation of one of the many compass-work patterns used in the stone window grilles found in the Great Mosque of Damascus (715). This early example maintains the 3.6.3.6 semi-regular grid as part of the finished design, and the circles are located at the vertices of this grid. The circles have been trimmed where they intersect with one another, thereby opening up the design in an aesthetically pleasing fashion that also allows for greater light penetration. The trimming of these circles produces the distinctive trilobed motif at the centers of each triangular cell of the 3.6.3.6 grid. Figure 82d illustrates a slightly later Umayyad window grille from palace of Khirbat al-Mafjar in Jordon (c.743). This is identical to the previous example except for two added features: the replacement of the arcs with s-curves that create distinctive six-pointed stars at the vertices of the isometric repeat, and the small interwoven circles that surround these six-pointed stars. Given their geometric similarity, and the fact that they were produced within 30 years of one another, it is very likely that the design of the window grille from Khirbat al-Mafjar was directly influenced by the earlier example at the Great Mosque of Damascus.

Figure 83 shows an example of Tulunid compass-work ornament set on a square grid. This pattern was used on one of the arch soffits of the ibn Tulun mosque in Cairo (876-79). The circles in this example are set upon the vertices of the

---

[116] Creswell (1969), 75–80, Figs. 12 and 15.

A   B

Fig. 83

orthogonal grid, which is not incorporated into the completed design. Each circle connects with a four-lobed motif that is also placed at the vertices of the repetitive grid. Figure 83b is from the Great Mosque of Damascus, and is also comprised of circles set upon the orthogonal grid [Photograph 5]. A close examination of this design reveals an unexpected similarity with the design from the ibn Tulun. Included in the orthogonal design from the Great Mosque of Damascus are circles of the same relative size and location as those from the ibn Tulun, except that the earlier Umayyad example has double the number of circles and incorporates diagonal s-curve elements within its overall structure—thereby creating a design with far greater density. However, unlike the patterns in Figs. 82c and d, given the distance over time and territory, it is unlikely that the similarity in circular layout is the product of any direct causal influence.

2.6   Classification by Line Treatment

This final form of classification is perhaps the most obvious in that each category is readily apparent when first viewing any given design. Categorization by line treatment falls into three basic forms: (1) the basic line without widening that is almost always provided with differentiated background colors for a tiling treatment; (2) widened lines; and (3) interweaving lines. The thickness of the widened and interweaving lines is variable and determined by several criteria, including the constraints of the artistic medium; cultural conventions; geometric concordance; and aesthetic preferences of the artist [Figs. 85–88]. Differences in line treatment can greatly alter the overall appearance of a pattern, sometimes to the point of obscuring similitude between examples of the same design. While line treatment is a very basic and obvious classification, it is nonetheless important as it greatly impacts the aesthetic quality of each historical example, and as such becomes part of the descriptive analysis that accompanies any in-depth examination of this tradition, just as it is a fundamental concern to any artist engaged in working with these geometric patterns.

To conclude this discussion of classifications within Islamic geometric patterns, the wide range of diverse criteria within this tradition requires a high degree of description to fully differentiate a given example, and place it into context with the tradition as a whole. That said, the formal identification of specific and critically important aspects of this tradition allows for greater clarity and understanding of both the bold and subtle features that permeate this remarkable art form. Such formal classifications include repetitive schema; design methodology; specific pattern family; whether a pattern is nonsystematic or systematic, and if systematic, which generative system; and when relevant, the type of dual-level design. More nuanced considerations include stylistic variables such as arbitrary additive or subtractive features.

Polygonal Design Methodology

3

3.1 Background to the Polygonal Technique

Artistic methodology is an important aspect of any serious study of art. A detailed knowledge of the methods and techniques used by traditional artists, craftsmen, and architects will provide a more complete understanding of the bold trends, as well as subtle nuances within a given artistic discipline. As regards Islamic geometric patterns, the understanding of historical design methodology provides valuable insight into the initial development, refinement and maturation, and geographic distribution of this design tradition. However, such knowledge is not just relevant to historians of Islamic art and architecture. A detailed understanding of historical design methodology is especially germane to those who are involved with the incorporation of such patterns into their own creative enterprises. Yet, with the historical decline of this ornamental tradition, knowledge of the methods used to produce complex patterns was gradually lost. Even with the resurgence of interest in Islamic geometric patterns that began during the second half of the twentieth century, attempts to resuscitate this art form have been stymied due to the lack of understanding of historical methodology. This ongoing void has caused frustration among contemporary artists, designers, and architects who have had to rely upon merely copying existing designs. The loss of vitality is a great pity, for this design discipline still has much to offer. Indeed, the potential for new and original geometric patterns draws from an infinite pool that can never run dry. New designs are there for the creation, and a practical knowledge of this design methodology can be a great inspirational aid and indispensable tool for those engaged in the revival of this extraordinary artistic tradition.

In light of the above, the detailed methodological analysis provided in this chapter serves two functions: to better understand the rich diversity of historical Islamic geometric designs and to provide artists and designers with familiarity of the technical skills required for creating new and original patterns at even the most demanding levels of geometric complexity. Much can be learned from the methodological practices of the past, and there is tremendous scope for new discoveries that augment the exceptional work of past masters working within this diverse discipline.

The historical evidence for the polygonal technique establishes this as the preeminent design methodology used by Muslim artists throughout the long history of this tradition. It is therefore no surprise that the polygonal technique is especially appropriate for creating new geometric designs that conform to long-established Muslim aesthetics. The succeeding methodological analyses in this chapter examine the application of the polygonal technique to the full range of pattern types employed within this ornamental tradition. This includes each of the five historical systematic methodologies, as well as the diversity of nonsystematic geometric design varieties. Additive and subtractive variations are provided where relevant to historical examples. What is more, the specialized techniques used in creating historical dual-level patterns are extended to fulfill the precise qualifiers for self-similarity and true quasicrystallinity. This chapter concludes with an examination of the two historical varieties of domical geometric pattern application: those that employ gore segments as their repetitive device and those based upon polyhedral geometry such as the Platonic and Archimedean solids. Interspersed throughout this chapter are a small number of designs created by the author. Individually, these serve to highlight the further design potential of a given underlying polygonal tessellation, or repetitive stratagem. Collectively, these serve to touch upon the vast potential of the polygonal technique for creating new and original designs that fully conform to, or in some cases build upon, the aesthetic character of this historical art form.

The working practices associated with applying geometric patterns to the wide range of ornamental media—be it architectural, the book arts, or otherwise—are governed by their own set of practical requirements and cultural conventions. Such considerations are highly specific and

J. Bonner, *Islamic Geometric Patterns*, DOI 10.1007/978-1-4419-0217-7_3

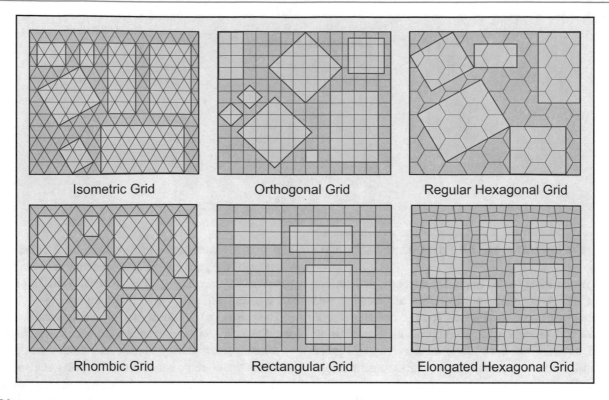

Fig. 84

generally beyond the scope of this current work. However, the conventions for panelizing, or framing, geometric patterns are fundamental to their applied use, regardless of artistic medium. The repetitive grids that are fundamental to the ability of patterns to cover the two-dimensional plane provide a wide range of proportional choices for applying designs into bounding frames. Figure 84 illustrates a number of typical framing rectangles created from repetitive grids that are proportioned upon repeat units common to Islamic geometric patterns. Any of these can be extended or reduced incrementally, and the vertical and horizontal lines of the frame generally conform to lines of symmetry within the geometric pattern.

Another secondary methodological practice involves the treatment of the pattern lines within a given design. Depending on such variables as line thickness, interweave, or use of double lines, a given design can have many contrasting aesthetic qualities. Figure 85 demonstrates diverse pattern line treatments applied to the classic fivefold *acute* pattern. Standard treatments include widened lines of variable thickness; interweaving lines of variable thickness; various forms of double-line treatments; and basic pattern with simple color differentiation applied to the pattern's cells (tiling treatment). Figure 86 illustrates the same forms of pattern line treatment applied to the classic fivefold *obtuse* pattern; Fig. 87 shows these same treatments to the pattern lines of the classic fivefold *median* design; and Fig. 88

provides typical pattern line treatments to the classic fivefold *two-point* pattern. The choice of which variety of pattern line treatment to use in a given historical example would have been determined by the aesthetic predilections of the artist as influenced to a greater or lesser degree by inherent geometric conditions, cultural aesthetic conventions, and material constraints of the designated medium.

3.1.1 Systematic Design: System of Regular Polygons

As discussed previously, the use of polygonal systems to create Islamic geometric patterns can be traced back to the formative period of this ornamental tradition, and the earliest system to be widely employed was the *system of regular polygons*. This makes use of regular triangles, squares, hexagons, and dodecagons as repetitive modules upon which pattern lines are applied. The variety of historical designs that can be created from tessellations comprised of different combinations of these polygons is surprisingly large. The most basic are of course designs that are derived from the three regular tessellations: the triangular grid, square grid, and hexagonal grid [Fig. 1]. More visually compelling and geometrically interesting patterns are created from the semi-regular, *two-uniform*, and *three-uniform* tessellations made up of these repetitive modules.

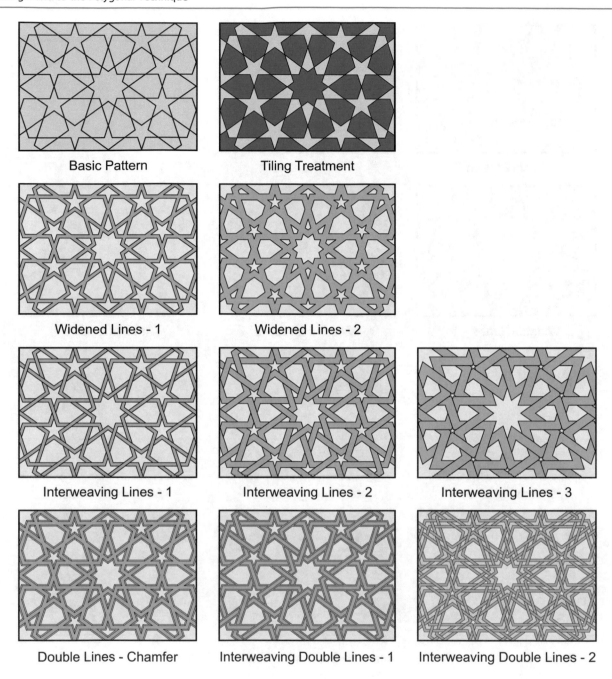

Basic Pattern

Tiling Treatment

Widened Lines - 1

Widened Lines - 2

Interweaving Lines - 1

Interweaving Lines - 2

Interweaving Lines - 3

Double Lines - Chamfer

Interweaving Double Lines - 1

Interweaving Double Lines - 2

Fig. 85

Figure 89 shows the eight semi-regular tessellations. These are characterized by a single variety of vertex with more than a single variety of polygon. The repetitive structures of five of these tessellations are isometric; two are orthogonal; and one repeats with elongated hexagons. Figure 90 illustrates 12 examples of *two-uniform* tessellations. These are characterized by two varieties of vertex. There has been some disagreement over the precise number of *two-uniform* tessellations. Depending on whether the topologies of the two varieties of vertex have consistent

or inconsistent global settings, the number of *two-uniform* tessellations can be limited or unlimited respectively.[1] The examples shown have either isometric, orthogonal, rectangular, or rhombic repeat units. Figure 91 illustrates just 4 *three-uniform* tessellations. This type of tessellation has

---

[1] A useful analysis of the different approaches to quantifying *two-uniform* tessellations has been provided by Helmer Aslaksen. See Aslaksen (2006), 533–336.

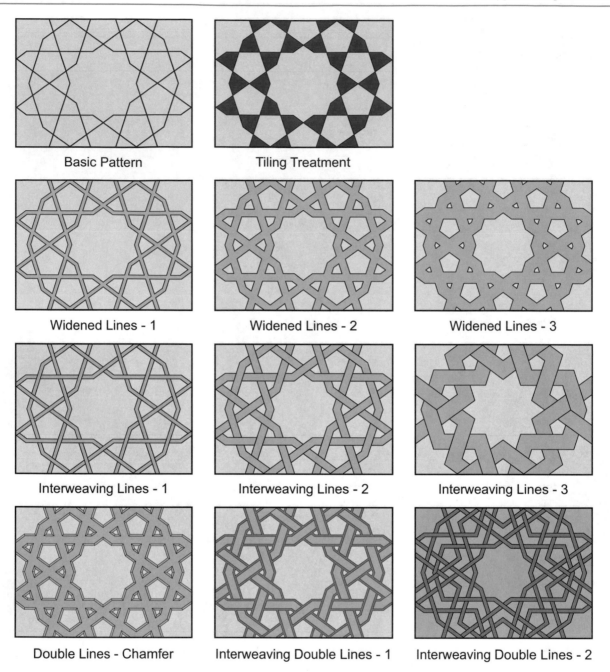

Basic Pattern

Tiling Treatment

Widened Lines - 1

Widened Lines - 2

Widened Lines - 3

Interweaving Lines - 1

Interweaving Lines - 2

Interweaving Lines - 3

Double Lines - Chamfer

Interweaving Double Lines - 1

Interweaving Double Lines - 2

Fig. 86

three types of vertex. The repetitive structures of these particular examples are either isometric or orthogonal. Geometric designs that are created from semi-regular, *two-uniform*, and *three-uniform* tessellations will invariably adhere to the same repetitive structure as the generative tessellation.

Within the historical record there is great diversity in the pattern line application to the repetitive modules that comprise the *system of regular polygons*. The chart in Fig. 92 provides the more typical pattern line applications to the triangle, square, hexagon, and dodecagon. The octagon has

been excluded from this chart due to the fact that it will only produce one tessellation with the other regular polygons: the 4.8^2 semi-regular tessellation of squares and octagons. Due to this limitation, and for the purposes of this work, the many patterns created from this tessellation are treated as a special case, and excluded from the *system of regular polygons*. The chart in Fig. 92 identifies the various pattern line applications as being *acute*, *median*, *obtuse*, or *two-point*. However, unlike the other historical design systems, or indeed the conventions for creating nonsystematic patterns, there are more that just four primary forms of pattern line

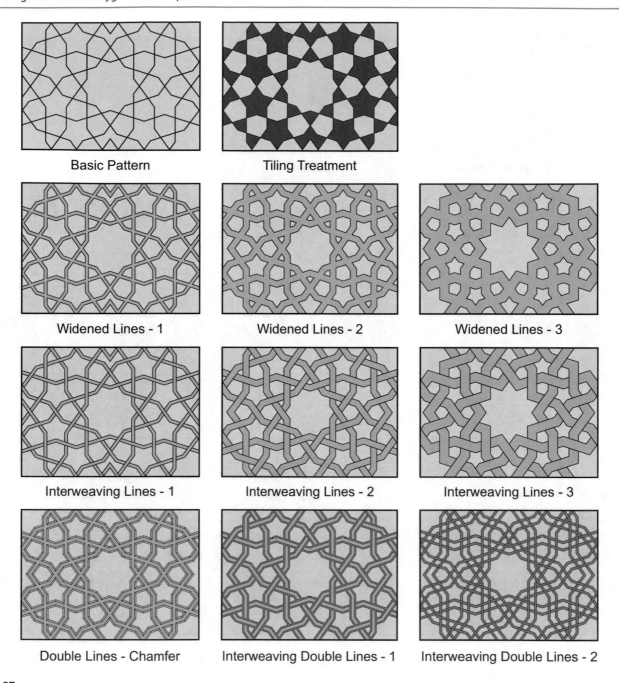

Fig. 87

application in the *system of regular polygons*. It is therefore helpful to further differentiate the types of pattern line by their angles. In this way, there are two types of *median* and *obtuse* pattern, and five types of *two-point* pattern. Figure 93 demonstrates five patterns created from different construction sequences as applied to the same $3^2.4.3.4$-$3.4.6.4$ *two-uniform* tessellation. Each of these begins by initially populating either the triangles and hexagons or just the squares with a specific set of pattern lines and extending these into the adjacent polygonal cells until they meet with other extended lines. Figure 93b illustrates two patterns that

can be constructed from the simple application of squares within the square modules. Each of these five patterns is by the author, and is not known to have been used historically. Figure 94 illustrates the same process as applied to the triangle and square cells of the $3^2.4.12$-$3.4.3.12$-3.12^2 *three-uniform* tessellation. Again, each of these four patterns is by the author and are not known to the historical record, but, like the previous five examples, is within the aesthetic scope of traditional Islamic design. Keeping in mind the innumerable tessellations that can be created from the regular polygons, the examples from Figs. 93 and 94 demonstrate

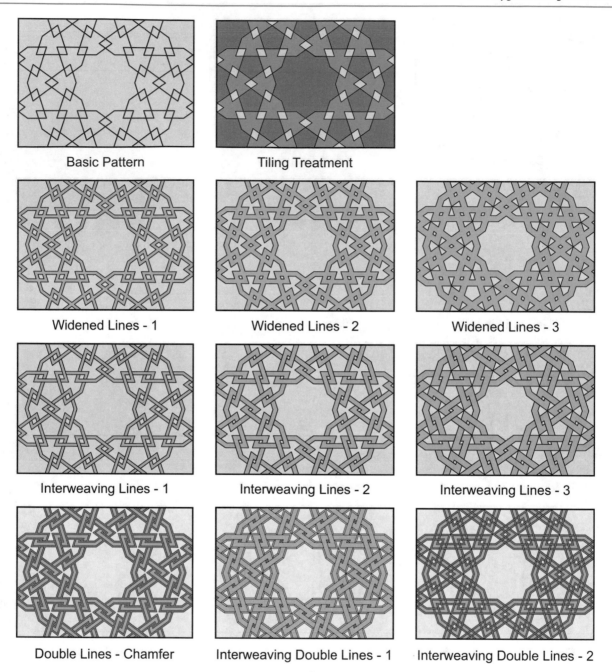

Basic Pattern   Tiling Treatment

Widened Lines - 1   Widened Lines - 2   Widened Lines - 3

Interweaving Lines - 1   Interweaving Lines - 2   Interweaving Lines - 3

Double Lines - Chamfer   Interweaving Double Lines - 1   Interweaving Double Lines - 2

Fig. 88

the vast potential for new and original patterns that are still available to the *system of regular polygons*.

As mentioned, patterns can be created by applying pattern lines to the polygonal cells of the three regular grids. Historically speaking, this is especially true of the 6^3 tessellation of regular hexagons. Figure 95 illustrates a series of patterns created from this simple tessellation. Figure 95a shows an *acute* pattern with 30° crossing pattern lines. This was used by Seljuk artists within the northeast dome chamber of the Friday Mosque at Isfahan (1088-89) [Photograph 18], and the Friday Mosque at Sin in Iran (1134), as well as

by Fatimid artists at the Sayyid Ruqayya Mashhad in Cairo (1133). Figure 95b shows the classic threefold *median* pattern with 60° crossing pattern lines. This is one of the most widely used geometric patterns throughout Muslim cultures, and two fine examples include: a pair of wooden doors at the Aljafería Palace in Zaragoza, Spain (second half of the eleventh century), and the Mamluk window grilles of the Sultan Qala'un funerary complex in Cairo (1284-85) [Photograph 55]. Figure 95c is a *median* pattern with 90° crossing pattern lines. This design is similarly ubiquitous, and a particularly impactful example from the Seljuk

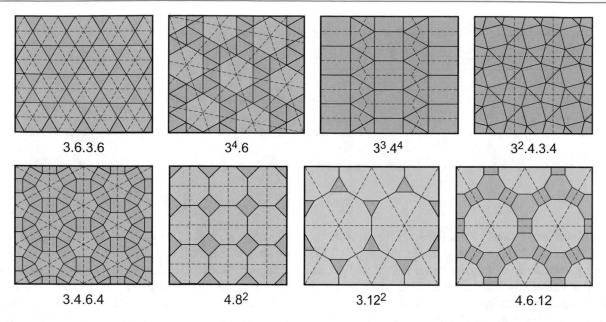

3.6.3.6 $3^4.6$ $3^3.4^4$ $3^2.4.3.4$

3.4.6.4 4.8^2 3.12^2 4.6.12

Fig. 89

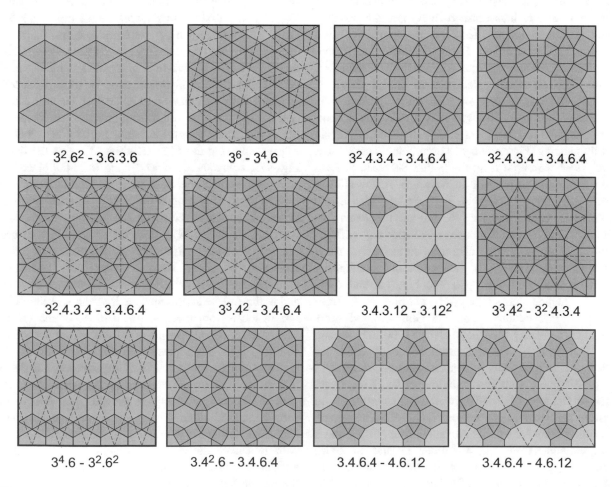

$3^2.6^2 - 3.6.3.6$ $3^6 - 3^4.6$ $3^2.4.3.4 - 3.4.6.4$ $3^2.4.3.4 - 3.4.6.4$

$3^2.4.3.4 - 3.4.6.4$ $3^3.4^2 - 3.4.6.4$ $3.4.3.12 - 3.12^2$ $3^3.4^2 - 3^2.4.3.4$

$3^4.6 - 3^2.6^2$ $3.4^2.6 - 3.4.6.4$ $3.4.6.4 - 4.6.12$ $3.4.6.4 - 4.6.12$

Fig. 90

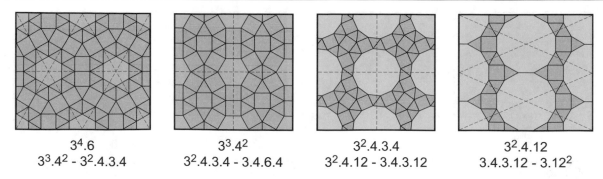

$3^4.6$
$3^3.4^2 - 3^2.4.3.4$

$3^3.4^2$
$3^2.4.3.4 - 3.4.6.4$

$3^2.4.3.4$
$3^2.4.12 - 3.4.3.12$

$3^2.4.12$
$3.4.3.12 - 3.12^2$

Fig. 91

Sultanate of Rum is found at the Sultan Han near Aksaray (1229). Figures 95d and e show *obtuse* patterns with 120° crossing pattern lines. The linear bands of Fig. 95d are, in and of themselves, the 3.6.3.6 tessellation with a widened interweaving interpretation. This design is ubiquitous throughout Muslim cultures. Figure 95e is by the author, and not known to have been used historically. Figure 95f employs pattern lines that connect to the vertices of the hexagonal grid. This design can also be produced with 120° *obtuse* pattern lines (dashed lines) that are widened to their maximum extent. This too was popularly used throughout Muslim cultures. Figure 96 illustrates 9 *two-point* patterns that are easily created from the 6^3 tessellation. Fig. 96a–f all use more typical pattern line applications, and Fig. 96g–i employ less common pattern line applications. The pattern in Fig. 96a is surprisingly uncommon and appears to have been principally used in Persian miniature painting. Figure 96b was used ubiquitously, and a relatively early example from the Seljuk Sultanate of Rum is from the Alaeddin mosque in Kırşehir, Turkey (1230). The very well known design in Fig. 96c was used at the Sabz Pushan outside Nishapur (960-85) and originates during the period of Samanid influence over this region [Photograph 11]. This design was also used on the eastern tomb tower at Kharraqan (1067-68). The pattern of superimposed hexagons in Fig. 96d was also widely used, and early examples include an Umayyad marble grill from al-Andalus (tenth century), and the Ghurid raised brick ornament at the Friday Mosque at Herat (1200). The design in Fig. 96e was equally well used and early examples include: a Fatimid window grill at the al-Azhar mosque in Cairo (970-76); an Umayyad window grille at the Great Mosque of Córdoba (987-990); two Seljuk examples from the eastern tomb tower at Kharraqan (1067-68) [Photograph 17] and the Friday Mosque at Abyaneh, Iran (1073); and a Fatimid example from the al-Aqmar mosque in Cairo (1125). The Qarakhanid design in Fig. 96f originates from the Maghak-i Attari mosque in Bukhara, Uzbekistan (1178-79), and has the further distinction of incorporating the generative hexagonal grid into the finished design. Figure

96g shows a Saminid design from the mausoleum of Arab Ata at Tim, Uzbekistan (977-78) [Photograph 12]. The earliest known use of the closely related pattern in Fig. 96h is from the eastern tomb tower at Kharraqan (1067-68). The design in Fig. 96i is Mengujekid from the Great Mosque of Divrigi in Turkey (1228-29). The three patterns in Fig. 97 are less typical *two-point* patterns associated with this same 6^3 tessellation that incorporate higher order polygons into the pattern matrix. The pattern in Fig. 97a was used by artists working for the Seljuk Sultanate of Rum for the Ahi Serafettin mosque in Ankara (1289-90). This can also be created from an alternative tessellation that employs underlying ditrigonal shield modules [Fig. 118c]. Several historical patterns were produced that are essentially variations on this design, including an early Seljuk example [Fig. 118a] from the northeast dome chamber of the Friday Mosque at Isfahan (1088-89) [Photograph 19]. When using the 6^3 tessellation to create the design in Fig. 97a, octagons are placed upon the midpoints of each edge of the generative hexagonal grid, with two corners of each octagon falling upon the polygonal edge. In this construction, the size of the octagon determines the character of the finished design. The design in Fig. 97b was used in multiple locations during the Seljuk Sultanate of Rum, including the Great Mosque at Bayburt (1220-35), the Çifte Minare *madrasa* in Erzurum (later thirteenth century), and at the Ahi Serafettin mosque in Ankara (1289-90). This design locates nonagons at the vertices of the generative hexagonal grid, and uses the grid itself within the completed pattern. Figure 97c represents a relatively common design that is comprised of a matrix of superimposed dodecagons that are located on the vertices of the isometric dual grid with the dodecagonal corners placed upon two points of each hexagonal edge. There are several nearly identical historical designs that employ superimposed dodecagons within an isometric repetitive structure, each created from a different underlying polygonal tessellation. In addition to the regular hexagonal grid of Fig. 97c, the 3.6.3.6 tessellation will also create a version of this design [Fig. 99b], as will the 3.4.6.4 tessellation [Fig. 107d]. These three design

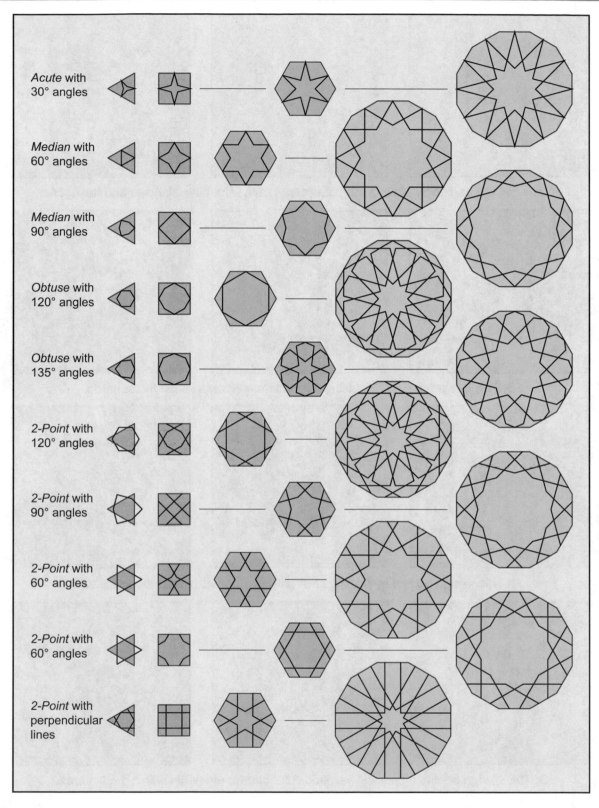

Fig. 92

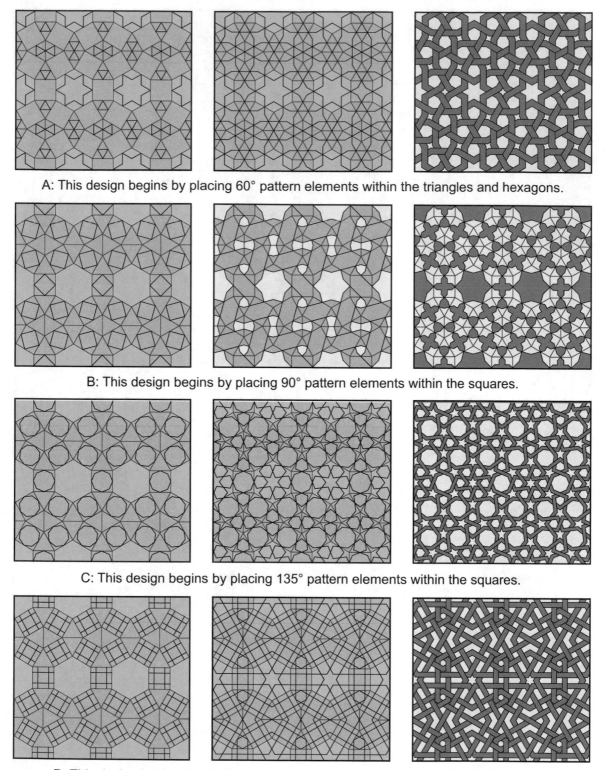

A: This design begins by placing 60° pattern elements within the triangles and hexagons.

B: This design begins by placing 90° pattern elements within the squares.

C: This design begins by placing 135° pattern elements within the squares.

D: This design begins by placing perpendicular pattern elements within the squares.

Fig. 93

variations only differ in the size of the dodecagons relative to the isometric repeat. The particular proportions of the example from Fig. 97c conform to a Fatimid window grille at the al-Azhar mosque in Cairo (970-72), as well as to a Seljuk carved stucco panel from the Friday Mosque at Forumad in Iran (twelfth century).

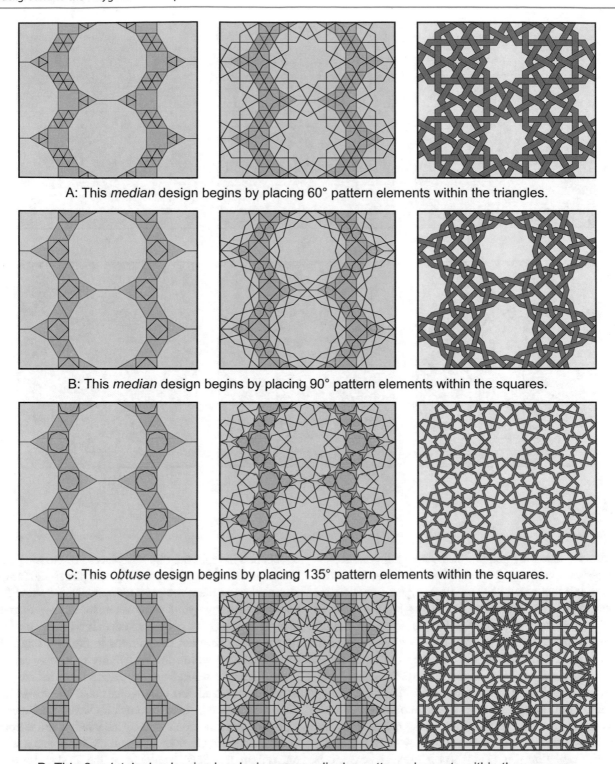

A: This *median* design begins by placing 60° pattern elements within the triangles.

B: This *median* design begins by placing 90° pattern elements within the squares.

C: This *obtuse* design begins by placing 135° pattern elements within the squares.

D: This *2-point* design begins by placing perpendicular pattern elements within the squares.

Fig. 94

A number of historical patterns were created from the 6^3 tessellation that employ two varieties of pattern line application into adjacent hexagonal cells. Figure 98 illustrates four such examples from the historical record. The blue hexagons in this figure contain standard pattern line applications, whereas their surrounding hexagons have pattern lines that are either pattern line extensions from the blue cells, or arbitrary additions within the remaining hexagonal structure.

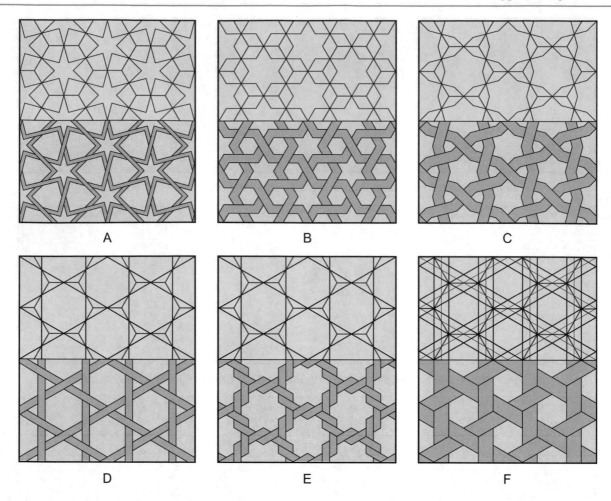

Fig. 95

The use of contrasting pattern lines within adjacent cells is relatively unusual, and generally dates from the formative period of this methodological tradition. Figure 98a was used by Ghurid artists in the decoration of the Masjid-i Jami in Herat, Afghanistan (1200); Fig. 98b is a later Ottoman example from the Great Mosque of Bursa (1396-1400) that is closely related to Fig. 98a; Fig. 98c shows a Seljuk design from the western tower at Kharraqan (1093-94); and Fig. 98d shows a Fatimid design from the *minbar* of the Haram al-Ibrahimi in Hebron, Palestine. The pattern in Fig. 98d can just as easily be created from either the 3.4.6.4 semi-regular tessellation [Fig. 106c], or the $3^4.6$-$3^3.4^2$-$3^2.4.3.4$ *three-uniform* tessellation [Fig. 114a]. As is often the case, it is not possible to know for certain which generative structure was used to produce a given historical example.

Figure 99 illustrates several historical designs that can easily be created from the 3.6.3.6 semi-regular tessellation. Figure 99a places 60° crossing pattern lines at the midpoints of each polygonal edge of the generative tessellation, and a very similar *two-point* design can be created from the 6^3

tessellation in Fig. 96e; Fig. 99b places 90° crossing pattern lines at the same locations; and Fig. 99c places 120° crossing pattern lines at these midpoints. Figures 99d–f show different varieties of *two-point* patterns that locate the pattern lines at two points on each polygonal edge. It is interesting to note the similarity between the patterns in Fig. 99c and f. Both place hexagons within the underlying triangular cells, and their differences result from the pattern lines that are chosen to penetrate into adjacent hexagonal cells. The design in Fig. 99a was frequently used throughout Muslim cultures, and early examples include an Umayyad window grille from the Great Mosque at Córdoba in Spain (987-99), and a wooden top rail in the Seljuk *minbar* of the Friday Mosque at Abyaneh, Iran (1073). The pattern in Fig. 99b is also well known, with differing proportions resulting from different polygonal extractions (e.g. Fig. 97c). An early Seljuk example of this particular variation is found at the Friday Mosque at Golpayegan in Iran (1105-18). A particularly beautiful Ilkhanid example of the well known design in Fig. 99c was used in a frontispiece of the 30-volume Quran (1313)

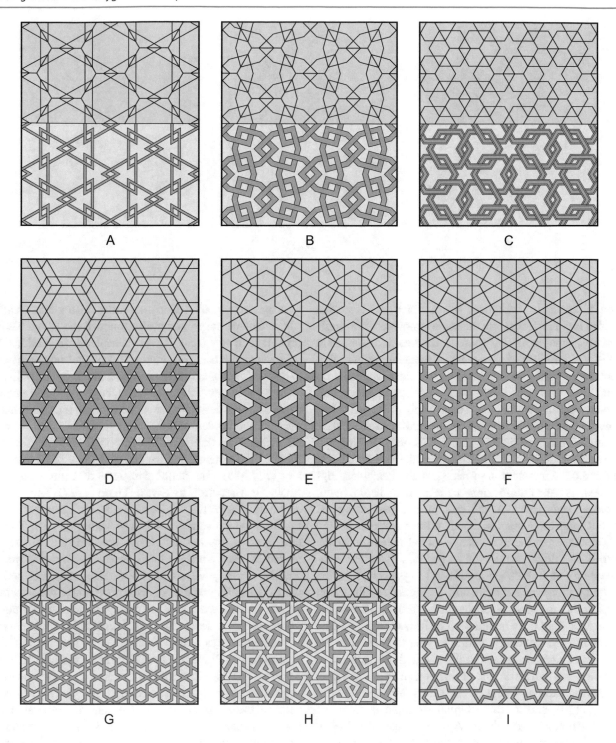

Fig. 96

commissioned by Sultan Uljaytu and calligraphed and illuminated by 'Abd Allah ibn Muhammad al-Hamadani.[2] The less common *two-point* design in Fig. 99d was used by *atabeg* artists on the sarcophagus in the mausoleum of Sultan

Duqaq in Damascus (1095-1104), and in a Ghurid *mihrab* at Lashkar-i Bazar (after 1149) [Photograph 31]. The *two-point* pattern in Fig. 99e is relatively common, and can be regarded as a variation of the design in Fig. 99a, but with different proportions within the geometric matrix. A fine *Mudéjar* example of this design is found in the raised brick ornament on the north side of the Cathedral of San Salvador at the

[2] This Ilkhanid Quran is in the National Library in Cairo: 72, pt. 19.

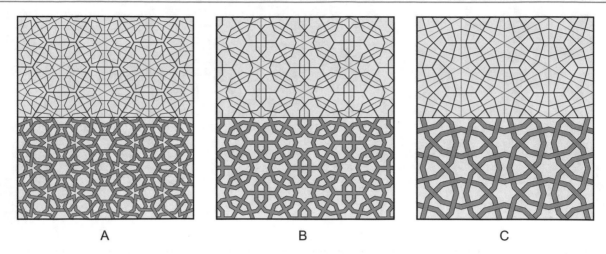

A   B   C

Fig. 97

Aljafería Palace in Zaragoza, Spain. Each of these five examples is characterized by the superimposition of a single closed polygonal motif. By contrast, the *two-point* design in Fig. 99f is comprised of a network of meandering lines that do not loop back onto themselves to close a geometric circuit. This dynamic design was used during the Seljuk Sultanate of Rum at the Ali Tusin tomb tower in Tokat, Turkey (1233-34), and in the window grilles of the Sultan Qala'un funerary complex in Cairo (1284-85) [Photograph 55]. The four designs in Fig. 100 are likewise associated with the 3.6.3.6 tessellation of triangles and hexagons. Figure 100a is an interlocking pattern that begins with six-pointed stars placed at the midpoints of the hexagonal cells, but joins these stars with single lines that connect the points (rather than the more conventional extension of the pattern lines into adjacent cells as per Fig. 99a). The rotational quality of the trilobed motif breaks symmetry with the underlying polygonal structure and distinguishes this design as conforming to the *p6* plane symmetry group, whereas the symmetry of the 3.6.3.6 tessellation on its own is *p3m1*. This is unusual in that geometric patterns generally adhere to the same plane symmetry group as their underlying generative tessellation. An early example of this design from the Seljuk Sultanate of Rum is found at the Great Mosque of Siirt, Turkey (1129), and a later Mughal example is from the tomb of I'timad ad-Dawla in Agra (c. 1628-30). Figure 100b is a *two-point* pattern with the applied pattern lines placed perpendicular to the edge. This example has the additional feature of including the generative polygonal tessellation as part of the completed pattern. The widening of the applied pattern lines follows the standard practice of equal offsets in both directions, while the widening of the hexagons within the generative tessellation is only in a single direction. This is unusual and highly effective in this circumstance: producing more evenly sized background elements than would otherwise be the case. This design

was by an Armenian Christian artist on a stone *khachkar* (fourteenth century), and is also found at the Khoja Khanate ornament of the Apak Khoja mausoleum in Kashi, China (c. seventeenth century). Figure 100c places nonagons at the centers of each generative triangle that are sized so that two of their vertices fall upon each edge of the generative triangles. This design has several historical locations, including: a thin border at the Seljuk Gunbad-i 'Alaviyan in Hamadan, Iran (late twelfth century) [Photograph 22]; the Mengujekid ornament of the Great Mosque of Divrigi, Turkey (1228-29); the Seljuk Sultanate of Rum *madrasa* of Muzaffar Barucirdi in Sivas, Tirkey (1271-72); and the Mamluk door of the Vizier al-Salih Tala'i mosque (1303). Figure 100d illustrates another *two-point* pattern. The pattern lines of this unusual design are laid out with squares placed at each vertex of the generative tessellation, and additive six-pointed stars within each hexagonal cell. As with Fig. 100b, this design also incorporates the generative tessellation as expressed by hexagons that touch corner-to-corner. This design was used in the Mamluk *madrasa* of Aqbughawiyya (1340) at the al-Azhar mosque in Cairo.

Figure 101 illustrates four historical examples of 3.6.3.6 patterns that employ both active and passive underlying polygonal cells in their creation. The active hexagonal cells in these examples are blue, and are separated by a passive hexagonal cell in each of the three orientations. The orange triangles are likewise active, and the white triangles are passive. It is interesting to note that each of the four designs places the same pattern lines within the active triangles, whereas each of the active hexagons is different. Figure 101a extends the pattern lines contained within the active hexagons and triangles until they meet within the passive hexagons. This design was used by Seljuk artists in the entry portal of the Seh Gunbad tomb tower in Orumiyeh, Iran (1180), as well as by Seljuk Sultanate of Rum artists at the Great Mosque of Niksar, Turkey (1145).

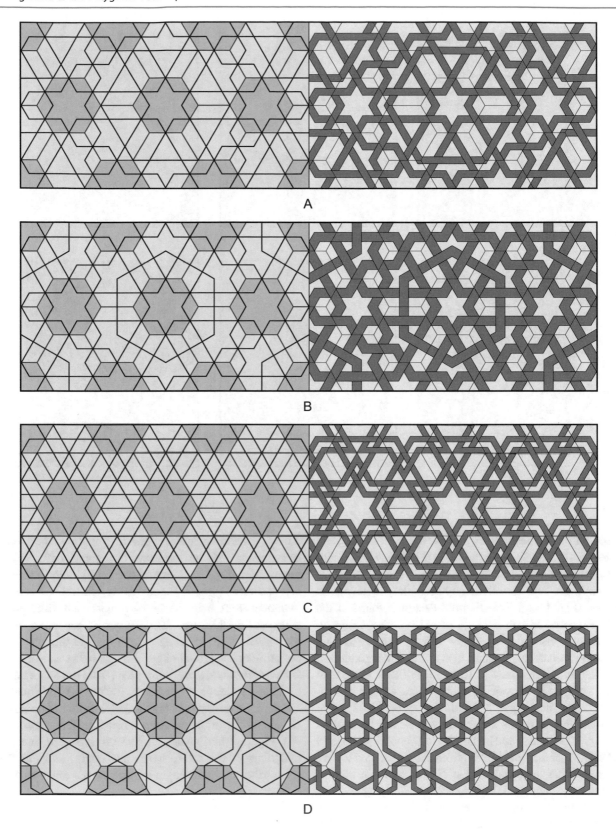

Fig. 98

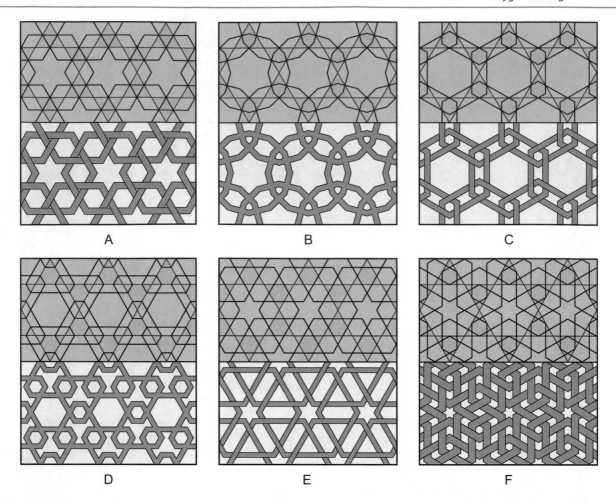

A   B   C

D   E   F

Fig. 99

A variation of this design was used as a border at the Çifte Minare *madrasa* Sivas, Turkey (1271) [Photograph 41]. Figure 101b is more conveniently created from the $3^4.6-3^3.4^2-3^2.4.3.4$ *three-uniform* tessellation of triangles, squares, and hexagons [Fig. 114c]. This Fatimid design is found at the Sayyid Ruqayya Mashhad in Cairo (1133). The design in Fig. 101c was used during the twelfth century by Zangid artist at the Bimaristan Arghun in Aleppo [Photograph 36], during the Seljuk Sutanate of Rum at the Great Mosque at Niksar, Turkey (1145), and by the Ildegizids at the mausoleum of Yusuf ibn Kathir in Nakhichevan (1161-62). Two later Anatolian examples are from the Alaeddin mosque in Konya (c. 1220), and the Huand Hatun *Complex* in Kayseri (1237). The fact that the designs from Figs. 101a and c were both used in the same building in Niksar, and that both of these patterns are created from the same underlying tessellation, is an indirect source of evidence for the use of the polygonal technique within this tradition. The closely related Ildegizid design in Fig. 101d is from the Mu'mine Khatun mausoleum in Nakhichevan, Azerbaijan (1186).

Figure 102 illustrates two patterns created from an alternative arrangement of active and passive cells from the same

3.6.3.6 generative tessellation. In this case, the isometric arrangement of the primary hexagonal cells (blue) is separated by two triangles and one centrally located secondary hexagon (grey). Figure 102a shows two historical treatments for the first of these designs: each with a different widened line thickness. This pattern is generated from the placement of 60° crossing pattern lines at the midpoints of the primary hexagonal edges. Figure 102a1 shows a design from a Ghaznavid stone relief panel from the South Palace at Lashkar-i Bazar in Afghanistan (before 1036) [Photograph 13], and the example in Fig. 102a3 is from the doors of the Zangid *minbar* at the al-Aqsa mosque in Jerusalem[3] (1168-74). Figure 102b1 places 90° crossing pattern lines at the midpoints of the active hexagonal edges, and Fig. 102b2 arbitrarily adds dodecagons into the pattern matrix, thus

---

[3] The al-Aqsa Mosque in Jerusalem is primarily a Fatimid building. However, the *minbar* was commissioned by the great Zangid ruler Nur al-Din in 1168, placing it within the sphere of Seljuk influence. See Tabbaa (2001), 86–88.

Fig. 100

creating a composition comprised of superimposed dodecagons and distinctive ditrigonal shield shapes. This design was used during the Mamluk period at the private house of Zaynab Khatun Manzil in Cairo (1468).

Figure 103 shows a pattern created from the 3.3.4.3.4 semi-regular tessellation of double triangles and oscillating squares. The size of the octagons located at the vertices

of the generative tessellation is determined by their extended lines bisecting the midpoints of each edge of the underlying square modules. This is a Khwarizmshahid design from the Zuzan *madrasa* (1219) in northeaster Iran [Photograph 39].

Figure 104 illustrates four relatively simple designs created from the 3.4.6.4 semi-regular tessellation. Figure 104a

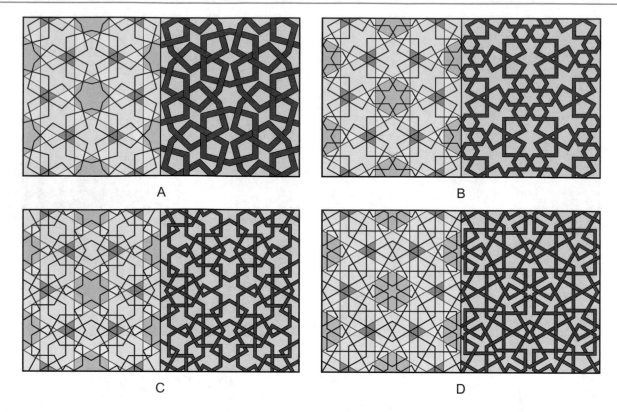

Fig. 101

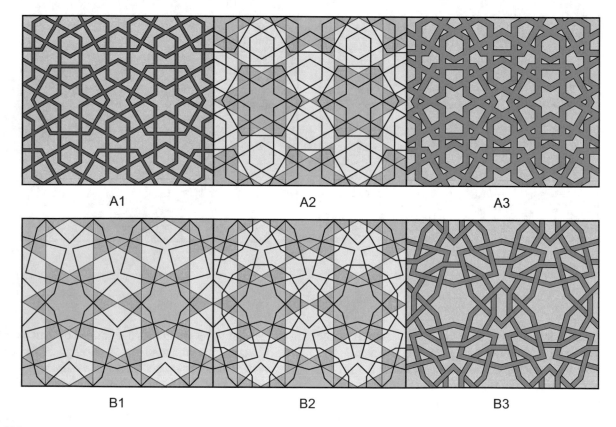

Fig. 102

Fig. 103

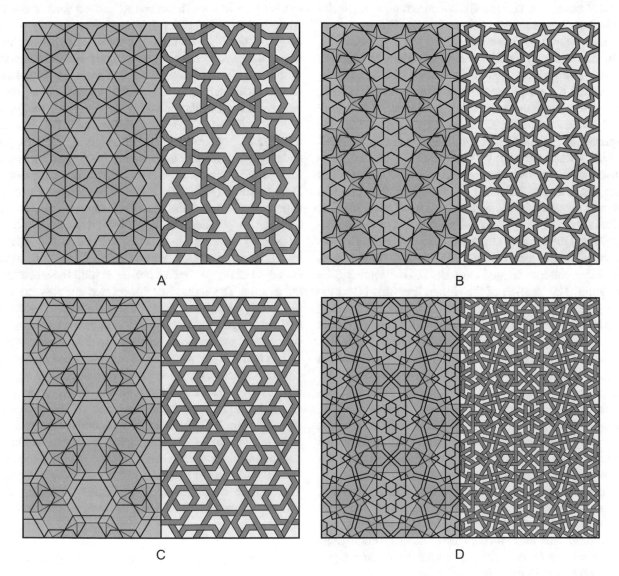

A

B

C

D

Fig. 104

employs 60° crossing pattern lines placed at the midpoints of the underlying tessellation. This is a Mamluk *median* pattern used at the Aydumur al-Bahlawan funerary complex in Cairo (1364). Figure 104b places an octagon within each underlying square module and consequently has 135° crossing pattern lines at the midpoints of each underlying polygonal edge. This *obtuse* design was used during the Seljuk Sultanate of Rum at the Gök *madrasa* and mosque in Amasya, Turkey (1266-67), as well as during the Mamluk period at the Sultan Qansuh al-Ghuri Complex in Cairo (1503-05). Figures 104c and d employ hexagons within the underlying triangles. Figure 104c is an unusual example of a hybrid *median* and *two-point* pattern wherein three sets of the pattern lines contained within the triangles (the 120° *median* pattern lines) are extended until they meet the edges of the underlying hexagonal modules, at which point the design becomes *two-point*. This layout of the applied pattern lines results in a design comprised of superimposed elongated hexagons. This is a Ghurid design from the Shah-i Mashhad in Gargistan, Afghanistan (1176). The example in Fig. 104d extends the alternative three sets of pattern lines within each underlying triangle. This pattern includes an arbitrary pattern line treatment within the underlying squares, and the six-pointed star motif is placed atypically on the corners of the underlying hexagons. The overall 3.4.6.4 generative tessellation is expressed within the completed design through emphasizing just the square module. This inventive design was used by the Qarakhanids in the anonymous southern tomb in Uzgen (1186), as well as during the Seljuk Sultanate of Rum at the Izzeddin Keykavus hospital and mausoleum in Sivas (1217-18). The atypical nature of the pattern line application of this design suggests that the former of these two historical examples was a direct influence upon the latter. Figure 105 demonstrates 9 *two-point* patterns created from the 3.4.6.4 semi-regular tessellation. The location of the pattern lines that bisect the underlying polygonal edges in Fig. 105a, b, d, g–i is determined by the 120° pattern lines of the small hexagons placed within the underlying triangular modules. The design in Fig. 105a was used by the Ghurids at the western mausoleum at Chisht, Afghanistan (1167), and by the Seljuks in the Friday Mosque at Gonabad in Iran (1212). Fig. 105b includes six-pointed stars placed at the corners of the underlying hexagons. This design is from the Qarakhanid anonymous southern tomb in Uzgen (1186). Figure 105c includes the hexagons from the underlying tessellation within the completed design. This design was used in several locations, including the Ghurid minaret of Jam (1174-75 or 1194-95), and the Chaghatayid mausoleum of Tughluq Temür in Almaliq in western China (1363). The very similar design in Fig. 105d was also in this same building [Photograph 70]. The pattern in Fig. 105e is created by placing eight-pointed stars within each underlying square module. This

design was depicted for use in a cast metal door in the *Book on the Knowledge of Ingenious Mechanical Devices* by Ismail ibn al-Razzaz al-Jazari[4] (1206). Figure 105f (by author) is a variation of Fig. 105e that mirrors every other point of the eight-pointed stars such that they become crosses—thereby creating nine-pointed stars centered on the triangular modules. Figures 105g–i are similar in that they incorporate 12-pointed stars within the underlying hexagonal modules. 12-pointed stars are typically derived from an underlying dodecagon, and their construction in these three examples is unusual. Figure 105g was used during the Seljuk Sultanate of Rum on the façade of the Usta Sagirt tomb in Ahlat, Turkey (1273); Fig. 105h shows a design from the *Mudéjar zillij* mosaic ornament at the Alcázar in Seville (1362), as well as the carved plaster ornament in the synagogue in Córdoba, Spain (1315); and Fig. 105i shows a design from the Huseyin Timur tomb in Ahlat, Turkey (1279). The widening of the pattern lines in Figs. 105a, b, e, f is offset in both directions as per standard convention, while that of Figs. 105c, d, g–i is offset in only a single direction, thereby providing a better overall balance in the size of the background elements. Figure 106 illustrates three further examples of patterns created from the 3.4.6.4 semi-regular tessellation that, to a greater or lesser extent, express the generative tessellation within the completed design. Figure 106a shows a *two-point* pattern with the generative tessellation represented as interweaving superimposed dodecagons. This was used during the Seljuk Sultanate of Rum at the Izzeddin Keykavus hospital and mausoleum in Sivas, Turkey (1217-18). Figure 106b shows a *two-point* pattern that includes the extended edges of the generative triangles to produce the conditions that provide for the 12-pointed stars. This is a Fatimid example from the al-Amri mosque in Qus, Egypt (1156). Figure 106c extends the edges of the generative hexagons until they meet inside the squares. The single line six-pointed star rosettes within the underlying hexagons are additive elements. This is also a Fatimid design from the al-Amri mosque in Qus. Figure 107 shows four patterns created from the 3.4.6.4 semi-regular tessellation that are made up of superimposed higher order polygons. Such designs have their own distinctive aesthetic. Figure 107a places octagons on the center point of the generative square modules. The size of the octagons is determined by their bisecting the midpoints of the generative triangles, thereby creating ditrigonal hexagons within each underlying triangle. This design was used during the Seljuk Sultanate of Rum at the Çifte Minare *madrasa* in Erzurum, Turkey (late thirteenth century). Figure 107b also places octagons at the same locations, but their size is determined

---

[4] Istanbul, Topkapi Sarayi Müzesi Kütüphanesi, MS A. 3472, fols. 165v–166r.

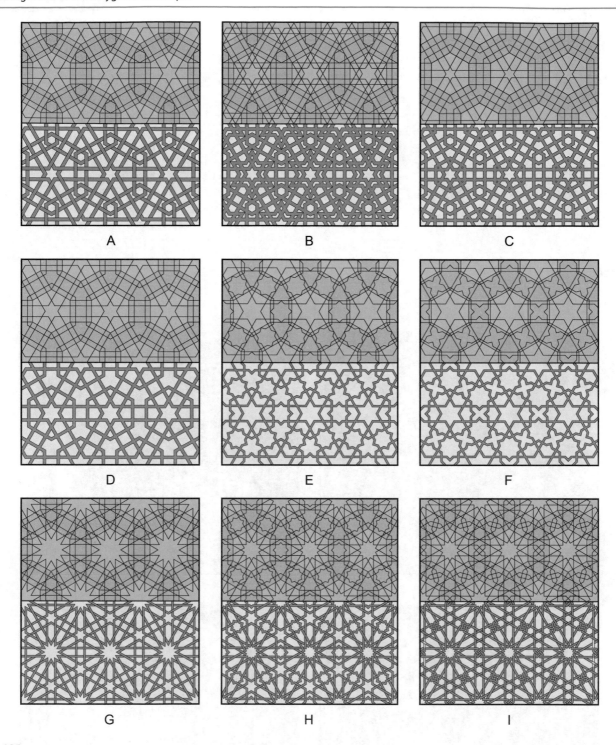

Fig. 105

by bisecting the triangular edges at 1/3 intervals. This design also incorporates hexagons whose size is determined by their midpoints being placed upon the vertices of the generative triangles. This is likewise a Seljuk Sultanate of Rum pattern, and is found at the Cincikh mosque in Aksaray, Turkey (1220-30). The size of the hexagonal pattern elements in Fig. 107c is determined by their corresponding to the size of

a regular hexagon placed within each underlying triangle [as per Fig. 105a]. However, this small hexagon is removed, and only the extended lines are kept [as per Fig. 105d]. The size of the superimposed nonagons is determined by their working together with the hexagons to create regular squares within each underlying square. This Seljuk sultanate of Rum design is from the Sungur Bey mosque in Nigde, Turkey

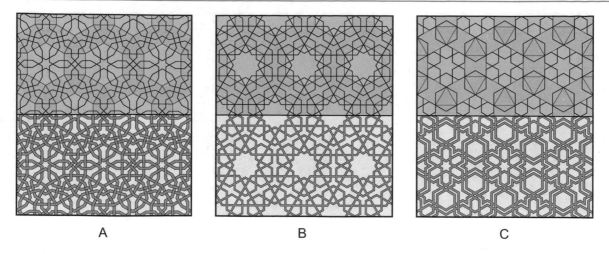

Fig. 106

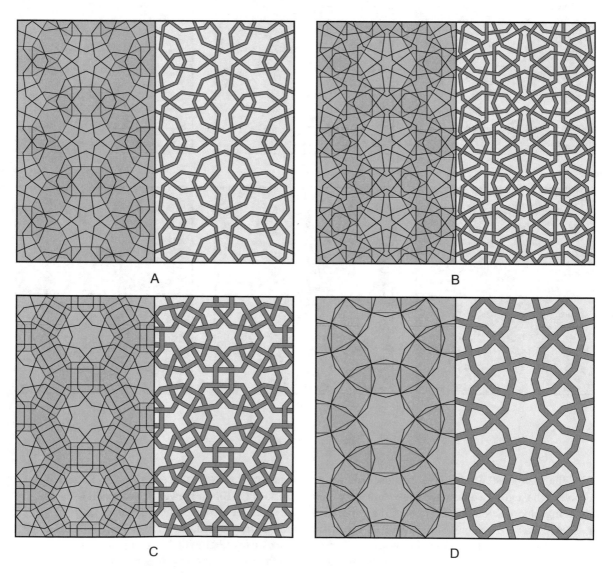

Fig. 107

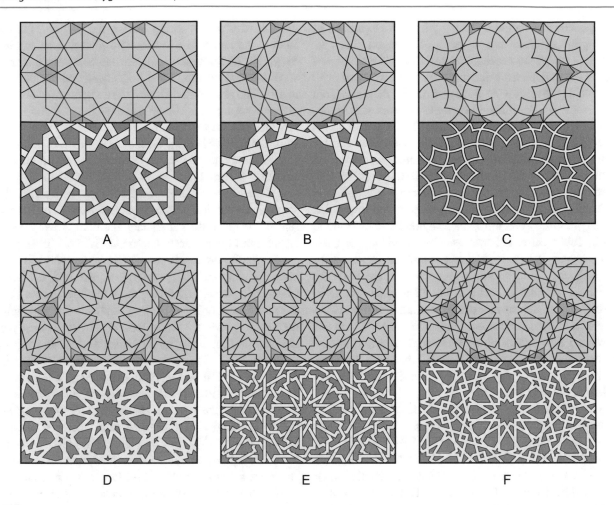

A   B   C

D   E   F

Fig. 108

(1335). Figure 107d places dodecagons at the centers of the generative hexagons: their size being determined by placing the midpoints of each dodecagonal edge upon the vertices of the generative tessellation. As stated, very similar designs comprised of superimposed dodecagons, but with slightly different proportions, can be created from different underlying polygonal structures [Figs. 97c and 99b]. The polygonal derivation in Fig. 107d is particularly successful in creating background elements that are more balanced in size and shape with one another, and conforms with a Tughluqid example from the Shah Rukn-i 'Alam in Multan, Pakistan (1320-24) [Photograph 69].

Figure 108 illustrates six historical designs that are created from the 3.12^2 semi-regular tessellation of regular triangles and dodecagons. Figure 108a is a *median* pattern that employs 60° crossing pattern lines placed at the midpoints of the underlying polygonal edges. This is a very common threefold pattern, and early Seljuk examples include the east tower at Kharraqan (1067) [Photograph 17], and the Friday Mosque of Golpayegan, Iran (1105-18). An early Fatimid example is found at the Sayyid Ruqayya

Mashhad in Cairo (1133). Multiple examples from the Seljuk Sultanate of Rum include the Great Mosque at Kayseri, Turkey (1205), and the Great Mosque at Akşehir near Konya (1213). A contemporaneous Ayyubid example is found at the Imam al-Shafi'i mausoleum in Cairo (1211). Multiple Mamluk examples include a window grille at the Ibn Tulun mosque in Cairo (1296), a door at the Vizier al-Salih Tala'i mosque in Cairo (1303), and the Amir Salar and Amir Sanjar al-Jawli complex in Cairo (1303-04). An Ilkhanid example from the same approximate period was used at the mausoleum of Uljaytu in Sultaniya (1307-13). Figure 108b shows a *median* pattern (by author) that uses 90° crossing pattern lines to create a pattern matrix comprised of superimposed dodecagons. While not known to the historical record, this design is appealing and certainly conforms to the aesthetics of this ornamental tradition. Figure 108c shows a curvilinear variation that comes from the 30-volume Quran (1313) commissioned by Sultan Uljaytu.[5] Figure 108d shows an

---

[5] This Ilkhanid Quran is in the National Library in Cairo: 72, pt. 19.

obtuse pattern that includes the typical rosette treatment to the 12-pointed star inside each dodecagon [as per Fig. 222]. One of the earliest examples of this well-known design is a Seljuk raised brick panel from the southern *iwan* of the Friday Mosque at Forumad in northeastern Iran (twelfth century). An Ilkhanid example is found at the mausoleum of Uljaytu in Sultaniya, Iran (1307-13); and locations of later Mamluk examples include the Amir Aq Sunqar funerary complex in Cairo (1346-47) [Photograph 45], and the Sultan Qansuh al-Ghuri complex in Cairo (1503-05). This design can also be made from an underlying tessellation of dodecagons separated by barrel hexagons and trapezoids [Fig. 321j]. Figure 108e utilizes a Maghrebi variation to the added 12-fold rosette. This was used by Alawid artists at the Moulay Ismail Palace in Meknès, Morocco (seventeenth century). Figure 108f shows a *two-point* pattern with the same variety of added rosette to the 12-pointed star as in Fig. 108d. Examples of this design include an illuminated page from a Quran produced in Baghdad (1303-07) that was calligraphed by Ahmad ibn al-Suhrawardi and illuminated by Muhammad ibn Aybak ibn 'Abdullah, and a Mamluk stone mosaic panel from the Amir Aq Sunqar funerary complex in Cairo (1346-47).

Figure 109 illustrates six designs created from the 4.6.12 semi-regular tessellation of squares, hexagons and dodecagons. Figure 109a shows a *median* pattern that employs 60° crossing pattern lines placed at the midpoints of the underlying polygonal edges. This is a Seljuk Sultanate of Rum pattern from the Hasbey Darül Huffazi *madrasa* in Konya (1421). Figure 109b shows an *obtuse* pattern derived from octagons placed within each square module, thus producing 135° crossing pattern lines at each midpoint of the underlying polygonal edge. This design is greatly enhanced by the underlying hexagons incorporating six-pointed star rosettes. This design can also be created from the dual of this grid (dashed lines), whereby the 5-, 6-, and 12-pointed stars are derived directly from the alternative underlying tessellation. This pattern enjoyed popularity among Mamluk artists, and examples include the Amir Sanqur al-Sa'di funerary complex in Cairo (1315); the Amir Ulmas al-Nasiri mosque and mausoleum in Cairo (1329-30); the Sultan Qansuh al-Ghuri *madrasa* (1501-03); and the Sultan Qansuh al-Ghuri Sabil Kuttab in Cairo (1503-04). Figure 109c shows an *obtuse* pattern with 120° crossing pattern lines at the underlying polygonal midpoints. This design approximates a Mamluk window grille at the Sultan Qala'un funerary complex in Cairo (1284-85). This same design can be created from an alternative underlying tessellation comprised of just triangles, squares and hexagons [Fig. 114b], with only slight differences in the proportions of the applied pattern lines. Figure 109d is unusual in that it employs the vertices of the underlying tessellation as determining coordinates for the pattern. This example is composed of just two sizes of superimposed hexagon, and is the product of artists working during the Seljuk Sultanate of Rum at the Izzeddin Keykavus hospital and mausoleum in Sivas (1217-18). Figure 109e shows a *two-point* pattern with superimposed octagons and dodecagons within the pattern matrix. The six-pointed stars at the centers of the underlying dodecagons are an arbitrary inclusion that is a sixfold corollary to the more common infill of ten-pointed stars [Fig. 224b]. This is from the Zuzan *madrasa* in northeastern Iran (1219); one of the few Khwarizmshah buildings still standing. Figure 109f shows a *two-point* pattern with the same dodecagonal motif as Fig. 109e, but with a very simple structure of parallel pattern lines that express the triangular repeat unit as well as its hexagonal dual. This is a Ghurid design from the minaret of Jam (1174-75 or 1194-95).

Figure 110 demonstrates a pattern created from the isometric $3^2.4.3.4$-$3.4.6.4$ *two-uniform* tessellation of regular triangles, squares, and hexagons. This interweaving design is created simply by offsetting, and thereby widening the lines of the generative tessellation itself. As a consequence, the shapes of the background pieces conform to the polygonal modules of the underlying tessellation. This parallel offset process can be regarded as a variety of the *two-point* design derivation that utilizes perpendicular parallel lines placed within the square modules. This atypical Seljuk design is found at the western tomb tower at Kharraqan (1093). The two designs in Fig. 111 represent a standard *two-point* pattern and variation created from the isometric $3^2.4.3.4$-$3.4.6.4$ *two-uniform* tessellation. These designs also employ the two sets of perpendicular parallel pattern lines within each underlying square module. The standard design in Fig. 111a extends the pattern lines from the underlying squares until they meet with other extended lines within the adjacent triangles and hexagons. Figure 111b replaces the six underlying triangles, six squares, and central hexagon that are located at the center of the panel with a single dodecagon. This variation produces the conditions for the central 12-pointed star. Both of these designs are found at the mausoleum of Muhammad Basharo in the village of Mazar-i Sharif in Tajikistan[6] (1342-43).

Figure 112 represents two historical *two-point* patterns that have the distinctive fourfold rotational motif created from an underlying square contiguously surrounded by four underlying triangles. The underlying generative tessellation in Fig. 122a is a $3^2.4.3.4$-$3.4.6.4$ *two-uniform* tessellation of triangles, squares, and hexagons. The resulting design in Fig. 112b is from the Khwarizmshahid *madrasa* in Zuzan, northeastern Iran (1219) [Photograph 38]. It is worth noting that as a variation, the central 12-pointed star shown in

---

[6] Not to be confused with the city of Mazar-i Sharif in Afghanistan. The village of Mazar-i Sharif, where the Mausoleum of Muhammad Basharo is sited, is approximately 25 km east of Penjikent, Tajikistan, and located in the Zarafshan River valley.

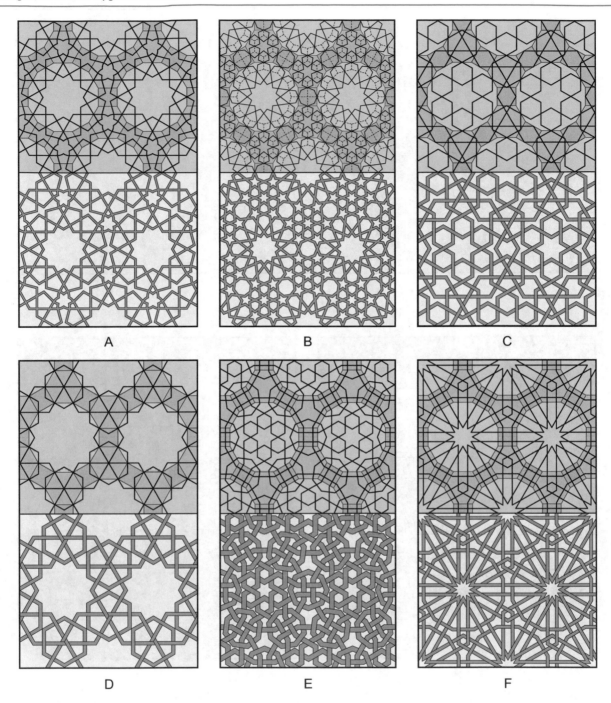

Fig. 109

Fig. 111b can also be applied to the implied central dodecagon within the underlying hexagon and surrounding triangles and squares of this underlying tessellation. The underlying generative tessellation in Fig. 112c is a $3^3.4^2$-3^2 .4.3.4 *two-uniform* tessellation of just triangles and squares. The design in Fig. 112d is from the anonymous Persian language treatise *On Similar and Complementary Interlocking Figures* in the Bibliothèque Nationale de France in Paris, but is not known within the architectural record. Both of these designs have the same derivation of perpendicular pattern line applications within the square module as the isometric design in Fig. 111a.

Perhaps the most frequently used orthogonal arrangement of polygons for creating patterns from the *system of regular polygons* is the orthogonal 3.4.3.12-3.12^2 *two-uniform* tessellation of triangles, squares and dodecagons. Figure 113 illustrates six historical designs created from this underlying tessellation. Figure 113a shows a *median* pattern with 60°

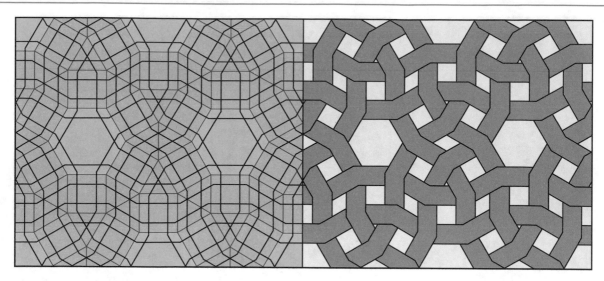

Fig. 110

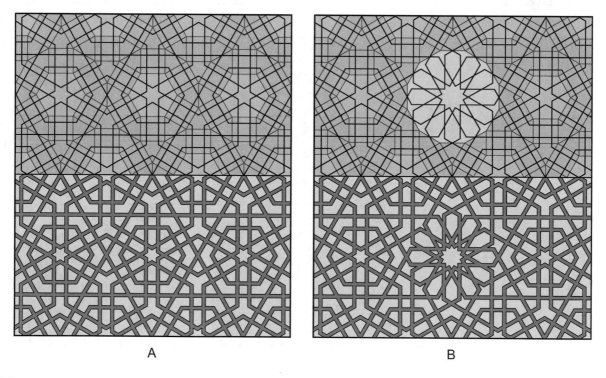

A   B

Fig. 111

crossing pattern lines that was widely used throughout Muslim cultures. An early Ayyubid example is found at the Imam al-Shafi'i mausoleum in Cairo (1211), and a later Qara Qoyunlu example was used at the Great Mosque at Van (1389-1400). Figure 113b shows a Timurid variation of this *median* pattern from the Abu'l Qasim shrine in Herat, Afghanistan (1492), that employs an arbitrary treatment to the underlying square region. Figure 113c shows an *obtuse* pattern with 120° crossing pattern lines. The proportions of

this example are consistent with the placement of regular hexagons within each underlying triangle. Multiple examples of this design exist within the architectural record. An early Zangid example was used on the wooden *minbar* from Aleppo commissioned by Nur al-Din in 1186.[7]

---

[7] This *minbar* is in the collection of the Hama National Museum in Syria.

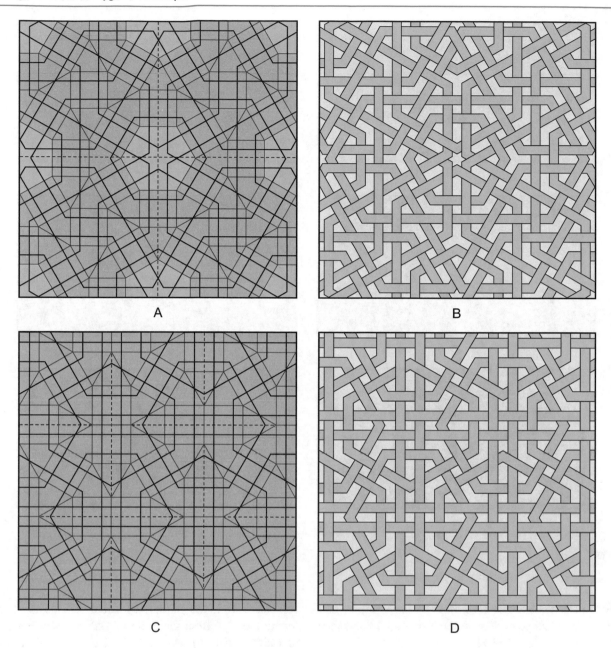

A

B

C

D

Fig. 112

Examples from the Seljuk Sultanate of Rum are found at the Sultan Han in Kayseri, Turkey (1232-36), and the Sultan Mesud tomb in Amasya, Turkey (fourteenth century). The Mamluks were particularly disposed toward this design and the many locations include: the Sultan al-Nasir Muhammad ibn Qala'un in the Cairo citadel (1295-1303); the Amir Sanqur al-Sa'di funerary complex in Cairo (1315); the Amir Altinbugha al-Maridani mosque in Cairo (1337-39); the Araq al-Silahdar mausoleum in Damascus (1349-50); an illuminated frontispiece for a Quran produced by Ya'qub ibn Khalil al-Hanafi in 1356; and an outstanding inlaid poly-chrome stone pavement in the Fort Qaytbey in Alexandria (1480s). Figure 113d shows a variation of this *obtuse* design

with small arbitrary eight-pointed stars placed within the square modules. This Nasrid variation is from the Alhambra. Figure 113e shows a *two-point* pattern with 90° crossing pattern lines. This Mamluk design is found at the Amir Qijmas al-Ishaqi mosque in Cairo (1479-81), as well as in the *minbar* of the Amir Azbak al-Yusufi complex in Cairo (1494-95) [Photograph 46]. The relative closeness in loca-tion and date invites the possibility that the same artist or atelier produced these two examples, or that the latter was copied from the former. Figure 113f is an unusual *median* pattern with an additive six-pointed star applied within alternating underlying dodecagons. This additive motif follows the fivefold convention that was occasionally

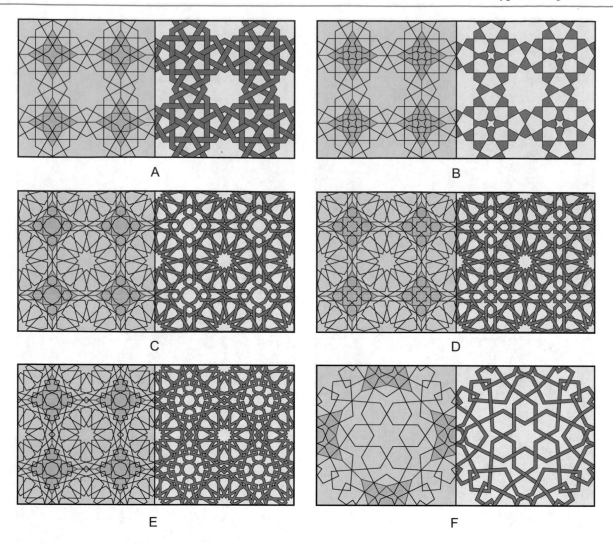

Fig. 113

applied to ten-pointed stars [Fig. 224b]. This Mamluk variation is from the stone *minbar* in the Zawiya wa-Sabil Faraj
ibn Barquq in Cairo (1400-11).

Figure 114 illustrates three closely related patterns created from the $3^4.6$-$3^3.4^2$-$3^2.4.3.4$ *three-uniform* tessellation
of triangles, squares, and hexagons. Each of these is a
median pattern with 60° crossing pattern lines. The grey
polygons in the underlying generative tessellations are passive in that they do not actively contribute to the design
process. Figure 114a can also be created from the 6^3 underlying tessellation [Fig. 98d]. This is an early design that was
used by Fatimid artists on the side panels of two wooden
minbars: that of the Haram al-Ibrahimi in Hebron, Palestine
(1094), and that of the al-Amri mosque in Qus, Egypt
(1156). Figure 114b is identical to Fig. 114a except that it
adds superimposed large hexagons into the pattern matrix.
The size and location of these are determined by 60° crossing pattern lines within the underlying triangles that are

coincident with the underlying squares. This is an early
Mamluk design from the Sultan Qala'un funerary complex
in Cairo (1284-85). As mentioned above, a design with
similar proportions can be produced from an alternative
4.6.12 tessellation [Fig. 109c]. However, the precise
proportions of the design from the Sultan Qala'un funerary
complex match the example created from the $3^4.6$-$3^3.4^2$-3^2
.4.3.4 *three-uniform* tessellation in this illustration. Figure
114c is identical to Fig. 114b except that the pattern lines
within the passive squares have been trimmed to produce the
irregular eight-pointed motifs with two perpendicular lines
of reflected symmetry. This design can also be created from
the 3.6.3.6 tessellation [Fig. 101b]. This design was used in
the spandrel of the Fatimid wooden *mihrab* from the Sayyid
Ruqayya Mashhad in Cairo (1133).[8]

_____

[8] Currently in the collection of the Islamic Museum in Cairo.

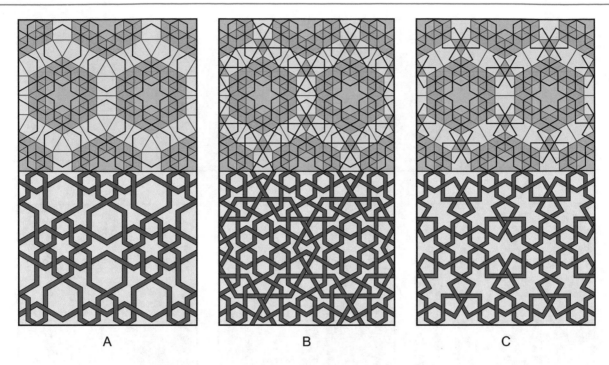

A   B   C

Fig. 114

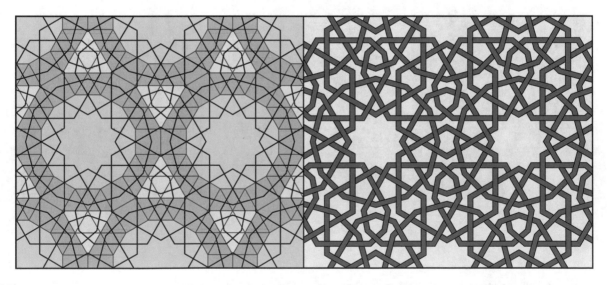

Fig. 115

Figure 115 demonstrates a pattern created from the 3^6-3^3 .4^2-3^2.4.12 *three-uniform* tessellation of triangles, squares and dodecagons. This is a *median* pattern with 60° crossing pattern lines. The grey triangles are passive, and the pattern lines contained within them are arbitrarily determined. The size of the large superimposed dodecagons within the pattern matrix is determined by their edges intersecting with the midpoints of the underlying square modules. This is a Qara Qoyunlu design from the Great Mosque at Van in Turkey (1389-1400).

Patterns created from the *system of regular polygons* occasionally employ underlying polygonal modules that are not regular. The most frequently used non-regular polygon is the distinctive ditrigonal shield module comprised of three 90° and three 150° included angles, with the angular proportions of overlapping squares in threefold rotation. Figure 116 illustrates ten examples of tessellations that employ this module. Figures 116a and b demonstrate how the ditrigon can be an interstice region within a tessellation of otherwise regular polygons. The other examples employ

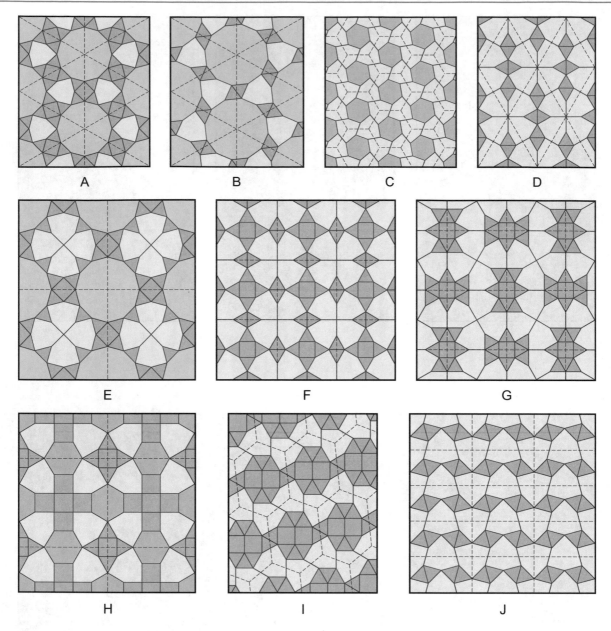

Fig. 116

the ditrigon as a tessellating entity of equal formative merit to the regular polygons within the tessellation. These ten examples demonstrate the effectiveness of this module in creating tessellations with repetitive structures that are isometric, orthogonal, rhombic, rectangular, and elongated hexagons. However, only three of these ten tessellations appear to have been used historically: Figs. 116a, e and f. Figure 117 shows the isometric *median* design created from the tessellation represented in Fig. 116a. This design was used by artists working in multiple Muslim cultures, including the Seljuk Sultanate of Rum at the Yelmaniya mosque in Cemiskezck, Turkey (1274); the Mamluks at the Aqbughawiyya *madrasa* (1340) in the al-Azhar mosque in

Cairo; and the Ottomans in their restoration of the Dome of the Rock in Jerusalem.

Figure 118 illustrates four historical patterns that can be produced from an isometric tessellation comprised of the ditrigonal modules in sixfold rotation around a six-pointed star interstice region. It is interesting to note that the underlying generative tessellation in the first three of these is itself a well-known geometric pattern created from the *system of regular polygons* [Fig. 95c]. Each of these four patterns is a variation on the same theme. Figure 118a shows the earliest such design, and possibly the prototype for later examples. This is from the Seljuk northeast dome chamber of the Friday Mosque at Isfahan (1088-89) [Photograph 19]. Figure

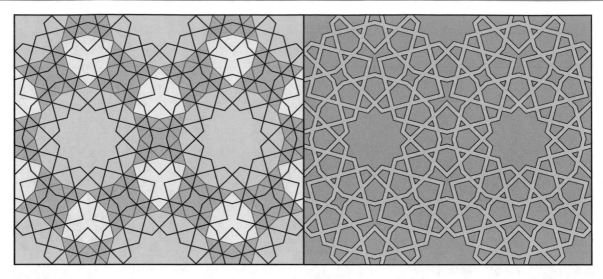

Fig. 117

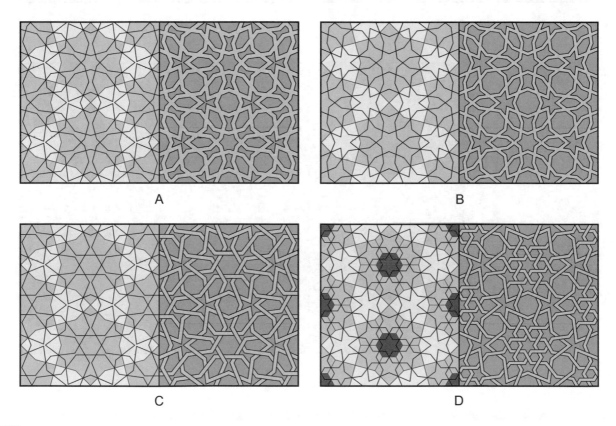

A

B

C

D

Fig. 118

188b represents essentially the same pattern, but with slightly different placement of the pattern lines. This is a Mamluk design from a window grille at the al-Anzar mosque in Cairo. Patterns created from this underlying tessellation were also produced during the Seljuk Sultanate of Rum, and Fig. 118c shows a design from the portal of the Huand Hatun

complex in Kayseri (1237), as well as the *mihrab* of the Ahi Serafettin mosque in Ankara (1289-90). The underlying tessellation of Fig. 118d further populates the interstice six-pointed stars with six pentagons that surround a central hexagon, thereby creating the conditions for the five- and six-pointed stars within these regions. This design is from

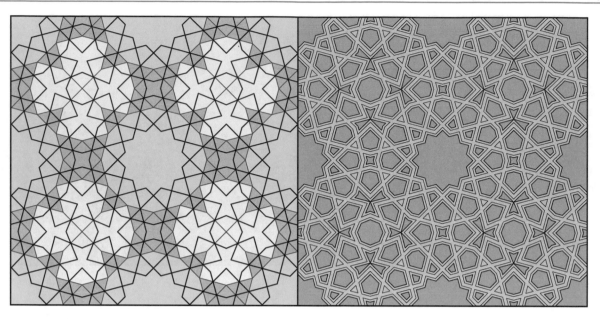

Fig. 119

Fig. 120

the *mihrab* of the Karatay *madrasa* in Antalya, Turkey (1250).

Figure 119 shows the orthogonal *median* design produced from the tessellation in Fig. 116e that includes a fourfold rotation of the underlying ditrigons that fill alternating dodecagonal cells within the underlying generative tessellation. An unusual feature of the crossing pattern lines is the predominant use of 60° angular openings, and the introduction of 45° angular openings along the coincident edges of the four ditrigonal modules. This produces the visually pleasing regular octagons at the vertex where the four

ditrigons meet. This Mamluk geometric pattern is from the side panels of the wooden *minbar* at the Vizier al-Salih Tala'i mosque in Cairo (c. 1300). The underlying generative tessellation in Fig. 120 is the third historical example from Fig. 116f, and is like that of Fig. 119 except that it places the four edge-to-edge ditrigons into each of the dodecagonal regions rather than every other one. Without the ditrigons, and associated four triangles, this would be the 3.4.3.12-3.12 [2] *two-uniform* generative tessellation [Fig. 90]. The size of the large dodecagon within the pattern matrix is determined by their transecting the midpoints of the triangle and square

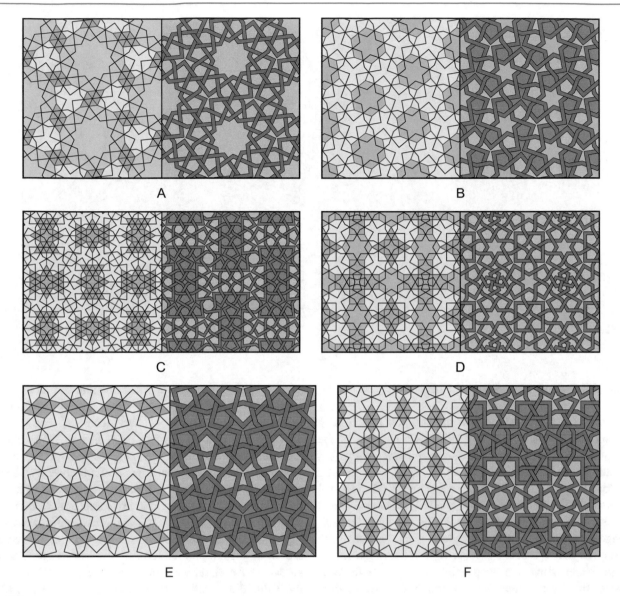

A   B

C   D

E   F

Fig. 121

modules of the generative tessellation. The earliest known use of this exceptional pattern was by Ghurid artists at both the minaret of Jam, Afghanistan (1174-75 or 1194-95), and the Shah-i Mashhad in Gargistan, Afghanistan (1176), and later examples include the work of Seljuk Sultanate of Rum artists at the Alaeddin mosque in Konya (1219-21).

The incorporation of underlying ditrigons always contributes a distinctive visual character to geometric patterns. Figure 121 illustrates six ahistorical designs (by author) from underlying generative tessellations that include this module. These examples are representative of a very large number of tessellations that can be produced with this added module; each of which will produce designs in all four of the pattern families. These six examples are all *median* patterns with 60° crossing pattern lines.

Generally, the ditrigonal shield is the only non-regular element occasionally incorporated into the *system of regular polygons*. An exception is found in a design produced during the Seljuk Sultanate of Rum, and speaks to the experimental approach to pattern making during this highly innovative period. Figure 122 illustrates a pattern created from an underlying tessellation of triangles and squares, as well as irregular pentagons that are clustered in fourfold rotation. The proportions of these pentagonal elements are determined by simply dividing the interstice regions created from the orthogonal distribution of the triangles and squares into four pieces. Unlike the ditrigonal modules, these pentagons do not tessellate in other configuration, and as such are nonsystematic. This unusual example is an orthogonal *median* design located at the tomb of Seyit Mahmut Hayrani in Aksehir near Konya (1275).

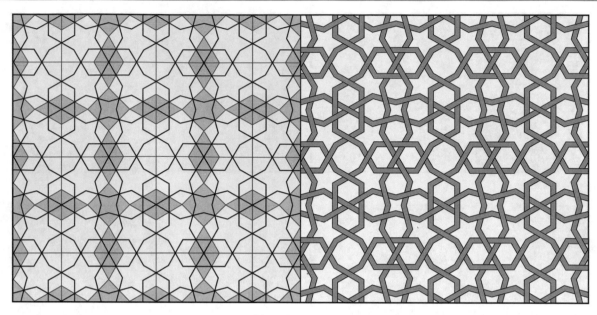

Fig. 122

3.1.2 Octagon and Square Patterns

As mentioned previously, while the octagon is a regular polygon and would appear to qualify as a member of the set of modules that comprise the *system of regular polygons*, the fact that this will only tessellate with other regular polygons in the single 4.8^2 arrangement makes this a special case that, for the purposes of greater clarity, is treated herein as a separate design category. While the singular 4.8^2 arrangement of underlying polygons is, on the one hand, limiting to the tessellating process, this arrangement of octagons and squares is responsible for more individual designs than any other single polygonal tessellation. Figure 123 demonstrates these two polygonal modules with a variety of applied pattern lines that were used to a greater or lesser extend by artists throughout Muslim cultures. As shown, multiple variations of pattern line application are possible within each of the four pattern families, and the specific examples shown are by no means exhaustive. What is more, some designs do not place pattern lines within the square modules, and others will employ alternating octagonal edges for pattern line placement.

Figure 124 illustrates the basic design from each of the four pattern families. Figure 124a shows the well-known *acute* pattern with 45° crossing pattern lines. Figure 124b shows the classic star-and-cross *median* pattern with 90° crossing pattern lines. This is one of the earliest and most widely employed Islamic geometric patterns, and is easily created from the point-to-point orthogonal arrangement of the eight-pointed stars. Figure 124c shows the *obtuse* pattern with 135° crossing pattern lines. This was also used with

great frequency throughout Muslim cultures, and can similarly be produced very easily through a corner-to-corner orthogonal arrangement of the octagons. Figure 124d shows a less well-known *two-point* pattern with 45°/135° supplementary angles of the pattern lines placed at two points upon each polygonal edge. A relatively early example of the *acute* pattern in Fig. 124a is from a panel of a metal door at the Sultan Qala'un funerary complex in Cairo (1284-85). The number of classic star-and-cross *median* designs in Fig. 124b is too numerous to elucidate herein, but some of the earliest examples include a panel from the Abbasid *minbar* at the Great Mosque of Kairouan (c. 856); one of the Tulunid arch soffits at the ibn Tulun mosque in Cairo (876-79); a Yu'firid ceiling panel from the Great Mosque of Shibam Aqyan near Kawkaban, Yemen (pre-871-72); a carved stucco panel from the No Gumbad mosque in Balkh, Afghanistan (800-50) [Photograph 10]; a Buyid carved stucco border that surrounds the *mihrab* at the Friday Mosque at Na'in, Iran (960); and a Saminid carved stucco panel from the Sabz Pushan outside Nishapur (960-85). The *obtuse* pattern in Fig. 124c has been found in the pre-Islamic Great Temple of Palmyra (c. 36), and a particularly pleasing example produced during the Seljuk Sultanate of Rum is found at the Esrefoglu Süleyman Bey in Beysehir, Turkey (1296-97). The *two-point* example in Fig. 124d is less common, and a fine example is from a Nasrid silk brocade textile at Metropolitan Museum of Art in New York[9] (fourteenth century).

---

[9] Metropolitan Museum of Art, New York, Fletcher Fund, 1929 29.22

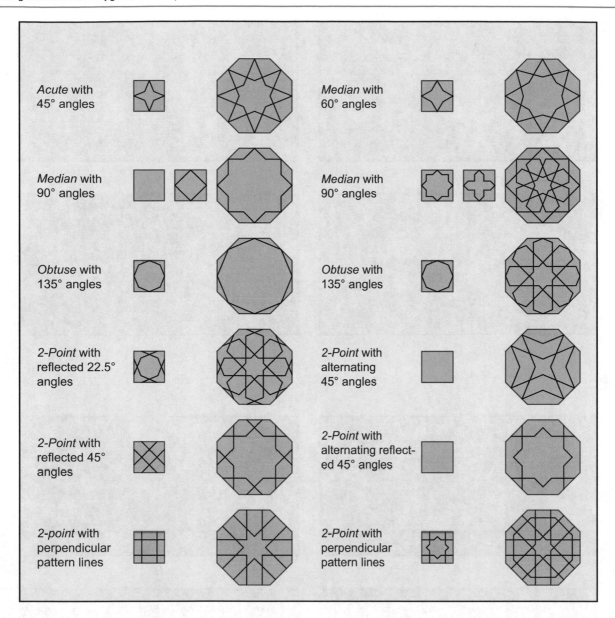

Fig. 123

Figure 125 illustrates four historical variations to the *acute* pattern created from the underlying 4.8^2 tessellation. Figure 125a removes all pattern lines from the underlying square module. This also opens up the eight-pointed stars through a simple subtractive process. This is a Seljuk variation from the minaret of Hotem Dede in Malatya, Iran (twelfth century). Figure 125b makes eight-pointed stars out of the otherwise four-pointed stars within the underlying square modules. This is an Ayyubid variation from the Sahiba *madrasa* in Damascus (1233-45). Figure 125c also places eight-pointed stars within the underlying square modules, but extends the lines of the eight-pointed stars until they meet with other extended pattern lines, thus creating a very distinctive design with two varieties of eight-pointed star. This is a Seljuk design from the Gunbad-i

Alayvian in Hamadan, Iran (late twelfth century) [Photograph 22]. Figure 125d incorporates a swastika motif into the parallel lines of Fig. 125c. The overall aesthetic results from an interlocking treatment emphasized through color contrast. This is a Seljuk Sultanate of Rum design variation from the Sirçali *madrasa* in Konya (1242-45).

Figure 126 illustrates eight variations to the *median* design of Fig. 124b. Figure 126a employs 60° crossing pattern lines rather than the more typical 90° of the classic star-and cross pattern. This is an Ayyubid variation from the Firdaws *madrasa* in Aleppo (1235-36). Figure 126b places eight-pointed stars within the square modules, and extends the pattern lines within the octagonal modules from midpoint to midpoint of the underlying polygonal edges, thereby transforming the more common eight-pointed star into an

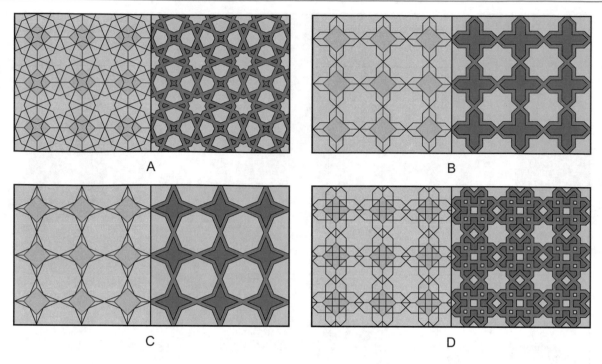

Fig. 124

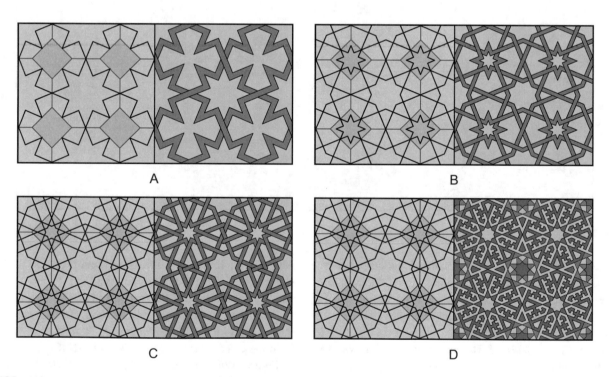

Fig. 125

octagon. This is a Nasrid variation from a wooden ceiling at the Alhambra. Figures 126c and d place a ring of squares within the octagonal modules. Figure 126c is an Umayyad variation from the mosaics at the Great Mosque of Córdoba (971), and Fig. 126b is a Timurid variation from the Ghiyathiyya *madrasa* in Khargird, Iran (1438-44). Figure

126e arbitrarily adds eight-pointed stars within the square modules and an eight-pointed star rosette within the octagons. This is a Marinid variation from the Bu 'Inaniyya *madrasa* in Fez (1350-55). Figure 126f is a frequently used subtractive pattern that radically disguises its star-and-cross origins, and historical examples include A Qarakhanid

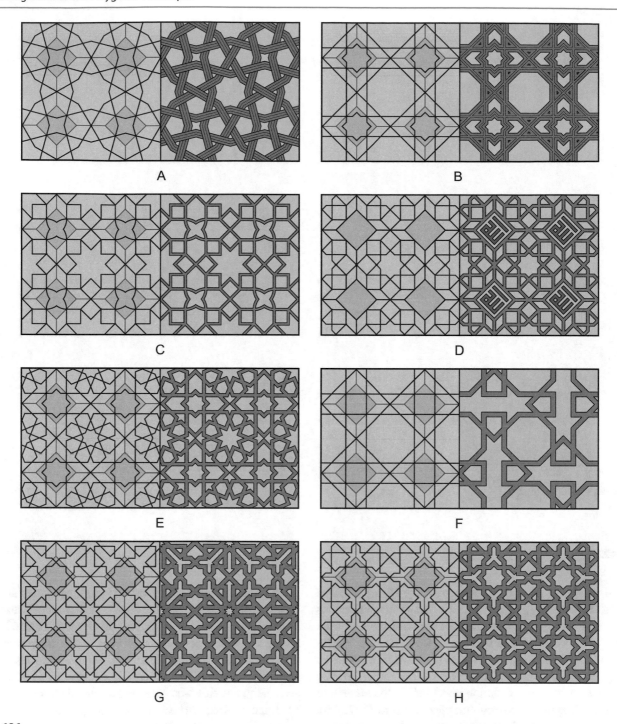

Fig. 126

variation is found at the Maghak-i Attari mosque in Bukhara (1178-79); the Zangid *minbar* at the al-Aqsa mosque (1187); and a Mamluk door (1303) at the Vizier al-Salih Tala'i mosque in Cairo. The design variations in Figs. 126g and h are typical to the Maghreb. Examples of both these variations are found at the Bu 'Inaniyya *madrasa* in Fez (1350-55).

The six designs in Fig. 127 are historical variations to the standard *obtuse* pattern in Fig. 124c. Figure 127a extends the lines of the standard *obtuse* design into the underlying square module, creating smaller octagons within the pattern matrix. This is an Ilkhanid variation from a Quranic Frontispiece dated 1304. Figure 127b shows a simple Buyid curvilinear variation from the Friday Mosque at Na'in, Iran (960). Figures 127c and d use alternating edges of the underlying octagons for pattern line placement. Figure 127c can also be created as a *two-point* pattern [Fig. 128d], and this design was used by Ibn al-Bawwab in his celebrated Baghdad

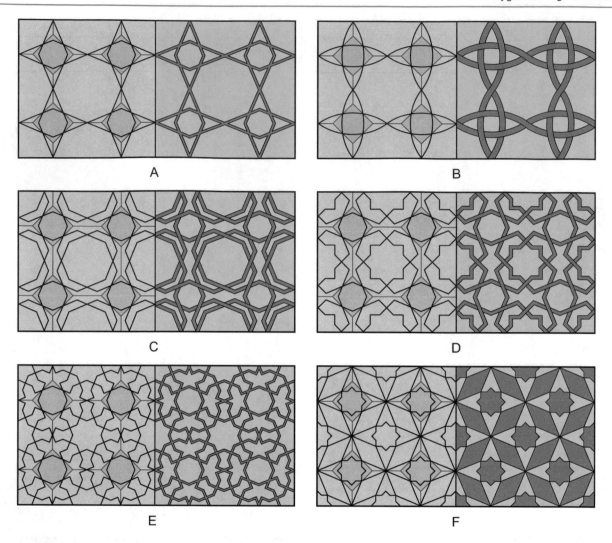

Fig. 127

Quran produced in 1001 [Photograph 6]. Figure 127d is similar to that of Fig. 127c in its use of alternating midpoints of the underlying octagons. This appears to have first been used by Fatimid artist in a wooden *mihrab* from the mausoleum of Sayyidah Nafisah in Cairo (1138-46), and later by Mengujekid artists at the Great Mosque of Divrigi, Turkey (1228-29). A Mamluk variation was used on the minaret at the Sultan Qaytbay funerary complex in Cairo (1472-74). Figure 127e employs bilateral concave octagonal motifs that are common to the *fourfold system A*. This is a Seljuk variation from an unattributed stucco panel at the Tehran Museum. Figure 127f places arbitrary eight-pointed stars within each underlying square module, and reflects four of the octagonal angles within the underlying octagons to produce large four-pointed stars. This variation comes from the Mughal pavement at the Taj Mahal in India (1632-53).

Figure 128 illustrates eight historical *two-point* patterns created from the 4.8^2 tessellation of squares and octagons. Figure 128a uses alternating underlying polygonal edges for

pattern line placement, with 45°/135° supplementary angles. This is a Maghrebi design that appears to have first been used by the Almohads at the al-Kutubiyya mosque in Marrakech, Morocco (twelfth century), and later in the Nasrid carved stucco ornament of the Alhambra. Figure 128b also uses alternating edges of the underlying octagon for pattern line placement, and this was also used at the Alhambra. Figures 128c and d place two sets of 45° crossing pattern lines at each of the two points of alternating underlying octagonal edges. The earliest known use of the pattern in Fig. 128c is from a Ghurid carved stucco panels from Lashkar-i Bazar[10] in Afghanistan (after 1149), and a later Seljuk Sultanate of Rum example is from the Karatay *madrasa* (1251-55). Figure 128d shows an alternative method for creating the design used by Ibn al-Bawwab as per Fig. 127c above

---

[10] In the collection of the National Museum of Afghanistan in Kabul, Afghanistan. See Crane, Howard and Trousdale (1972), 215–226.

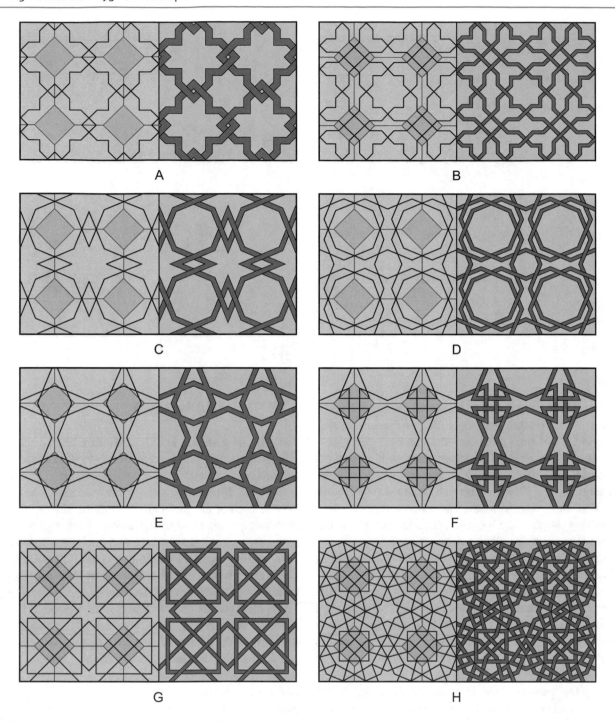

Fig. 128

[Photograph 6]. Figures 128e and f also employ alternating edges, and differ only in the treatment of the underlying square module. Both variations are Mengujekid: Fig. 128e being from the Divrigi hospital in Turkey (1228-29), and Fig. 128f from the nearby Great Mosque of Divrigi (1228-29). Figure 128g employs pattern lines that are perpendicular to the underlying polygonal edges. This design is also from the Great Mosque of Divrigi, and it is possible that these three examples were created by the same artist. Figure 128h is derived from an assortment of pattern line features that all transect the two points of each underlying polygonal edge. These include squares, double sets of 60° crossing pattern lines, and large dodecagons. The success of this pattern is the result of offsetting the pattern lines in a single direction rather than the more common practice of widening lines equally in both directions. This exceptionally successful Ghurid design is from the raised brick ornament of the Friday Mosque at Herat (1200).

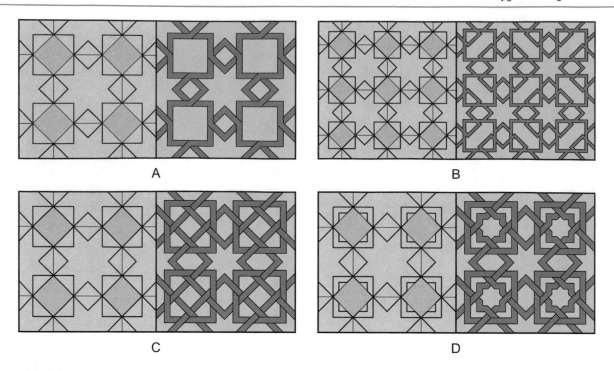

Fig. 129

Figure 129 represents four historical design variations that orient their pattern lines upon the vertices of the 4.8^2 tessellation rather than points on the underlying polygonal edges. While this alternative design practice is atypical, it will occasionally produce patterns that are very successful. The design in Fig. 129a is comprised of two sizes of square as well as eight-pointed stars. The earliest occurrence of this pattern is on one of the Tulunid carved stucco arch soffits at the ibn Tulun mosque in Cairo (876-79) [Photograph 9]. Later locations include the Almohad al-Kutubiyya mosque in Marrakech, Morocco (twelfth century), and the Nasrid Alhambra. Figure 129b is very similar except for the discontinuous lines that break with standard methodological practices. However, it is nonetheless handsome in its overall composition. This is an Artuqid pattern from the Great Mosque of Dunaysir in Kiziltepe, Turkey (1204). Figures 129c and d achieve their interweave by widening the pattern lines in a single direction. These designs can also be created from the *two-point* process [Fig. 128g], in which case the pattern lines are widened in both directions until the three lines meet at a single point. A variation of the design in Fig. 129c (with scallops incorporated into the pattern matrix) was used in a window grille by the Umayyads in Córdoba (980-90). Later Ghurid examples include the raised brick ornaments of the minaret of Jam in Afghanistan (1174-75 or 1194-95), and the western mausoleum at Chisht, Afghanistan (1167). The very similar design in Fig. 129d was used in the *Mudéjar* ornament at the Alcazar in Seville (1364-66), and by the Nasrids at the Alhambra.

3.1.3 Fourfold System A

The *fourfold system A* is comprised of a limited number of polygonal modules that can tessellate in an unlimited number of combinations. Figure 130 illustrates the polygonal modules of this system, along with their associated pattern lines in each of the four pattern families. It is important to note that the nine polygonal modules represented in this figure are not exhaustive, and that additional modules are occasionally employed within this system. These secondary modules are often derived via interstice regions through the process of tessellating with the otherwise standard modules. The applied pattern lines in this system, as well as the other historical systems that follow, are generally more formalized than that of the *system of regular polygons*. This is due in part to the fact that the modules of the *system of regular polygons* can tessellate both isometrically and orthogonally, eliciting a broad range of associated angular openings for the crossing pattern lines of the various pattern families: e.g., 30°, 45°, 60°, 90°, 120°, and 135° [Fig. 92]. By contrast, the standardized angular openings of the *fourfold system A* employ fewer angles: 45°, 90° and 135°. Another reason for the greater diversity of pattern line application to the *system of regular polygons* is the earlier provenance of this system, with earlier examples being produced during a period of greater exploratory experimentation and prior to the generalized standardization that came with the maturity of this ornamental tradition.

There are three edge lengths among the polygons of the *fourfold system A*. This forces the polygonal modules to

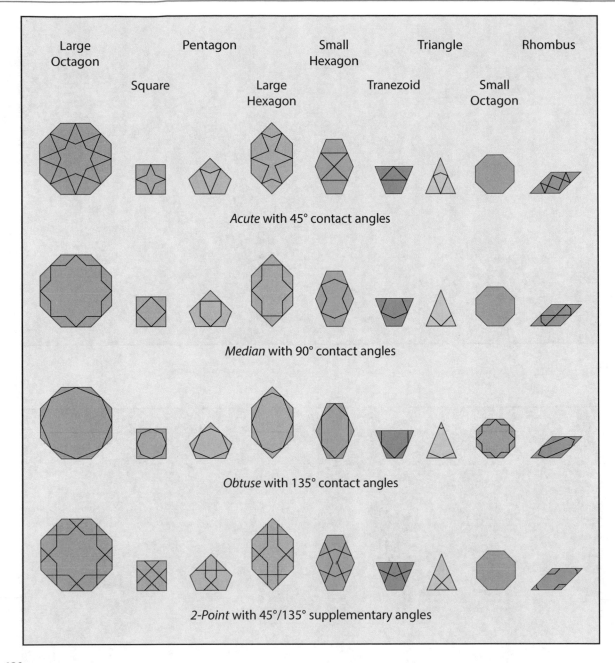

Fig. 130

associate with one another based upon edge determinants. It is worth noting that the only regular polygons in the *fourfold system A* are the square and two sizes of octagon. The applied pattern lines of the larger octagon and square modules are identical to four examples from the previous section detailing patterns created from the 4.8^2 tessellation. Indeed, such patterns can be equally regarded as part of the *system of regular polygons* or the *fourfold system-A*. This overlap is all the more reason for this 4.8^2 variety of design to be given its own categorization.

Figure 131 demonstrates methods for constructing the polygonal modules of the *fourfold system A* using the large octagon as the foundation from which each additional module is derived. Figure 132 provides the proportional relationships of the three edge lengths as they relate to the foundational octagon. The image on the right represents the edge lengths as nested squares. The indicated proportional relationship of 1:1.4142... between the short and long edges is $\sqrt{2}$.

Figure 133 illustrates the seven rotational combinations of the triangle from the *fourfold system A*, with applied pattern lines from the *acute* family. The outer long edges of the rotated clusters of triangles must be coincident with the long edges, be they other triangles in 180° rotation

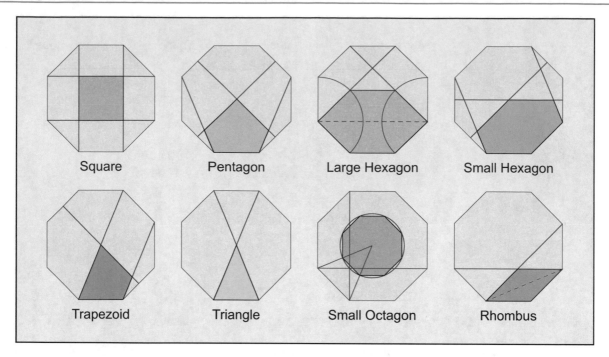

Fig. 131

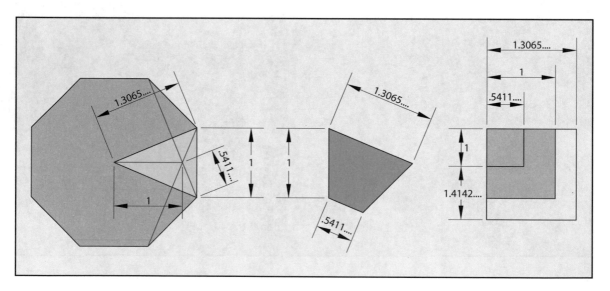

Fig. 132

(as per the cluster of four triangles), the long edges of the trapezoids, or a square interstice module (as per the cluster of six triangles). The demonstrated arrangements of clustered triangles in 45° rotational increments, with coincident triangles, trapezoids, or square, create pattern motifs (in this case associated with the *acute* family) that are well known to this ornamental tradition.

Figure 134 demonstrates the dualing characteristic between the underlying tessellations created from the *four-fold system A*. A remarkably feature of this system provides for each tessellation to have a dual relationship with another

tessellation comprised of polygonal modules from this same system. As a consequence, it is possible to create a specific pattern from either of the two dualing underlying tessellations. In this example, the two tessellations are identical in every respect except that their locations shift between the rectangular repeat unit (solid black line) and the rectangular dual of the repeat unit (dashed black line). In examples such as this, the dual tessellations are reciprocals. Figure 135 illustrates three patterns (by author) from the *acute*, *median* and *obtuse* families that are created from the dualing underlying reciprocal tessellations from Fig. 134. Although made

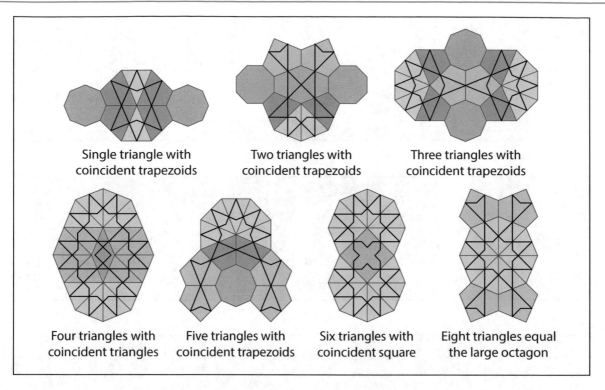

Single triangle with coincident trapezoids

Two triangles with coincident trapezoids

Three triangles with coincident trapezoids

Four triangles with coincident triangles

Five triangles with coincident trapezoids

Six triangles with coincident square

Eight triangles equal the large octagon

Fig. 133

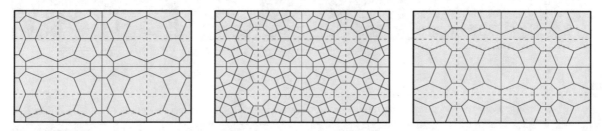

Fig. 134

from dual tessellations, the *obtuse* pattern in Fig. 135a and the *acute* patterns in Fig. 135b are identical: the *obtuse* pattern being created from 135° angular openings (where the bisector of the angle is perpendicular to the polygonal edge), and the *acute* pattern being produced from 45° angular openings. Similarly, the *acute* pattern in Fig. 135e and the *obtuse* pattern in Fig. 135f are also identical. What is more, the *acute* pattern in Fig. 135b is the reciprocal of the *acute* pattern in Fig. 135e; and the *obtuse* pattern in Fig. 135a is the reciprocal of the *obtuse* pattern in Fig. 135f. *Acute* and *obtuse* patterns created from the *fourfold system A* are demonstrably correlated, and without knowing the specific underlying tessellation of a given historical example, patterns produced from 45° and 135° angular openings can equally be regarded as either *acute* or *obtuse*. By contrast, the 90° crossing pattern lines of the *median* patterns in Figs. 135c and d produce the same design with the same relative location in both dualing tessellations.

Designs created from tessellations made up of just a single underlying polygonal module are always highly repetitive, and lacking primary star forms. These invariably qualify as field patterns. Figure 136 illustrates four such patterns created from just the small hexagon of the *fourfold system A*. Figures 136a and c demonstrate the reciprocal relationship between *acute* and *obtuse* patterns common to this system; and in this case the orientation of the pattern is rotated 90°. Figure 136b shows a well-known *median* pattern, with an early Anatolian Seljuk example used at the Great Mosque of Niksar in Turkey (1145). The *two-point* pattern in Fig. 136d is by the author, and not known to the historical record. The six field patterns in Fig. 137 are created from the underlying tessellations comprised of just the large hexagon from this system. Figure 137a shows an *acute* design (by author) with 45° crossing pattern lines, while the variation in Fig. 137b (by author) incorporates *two-point* lines upon the horizontal polygonal edges into the otherwise *acute*

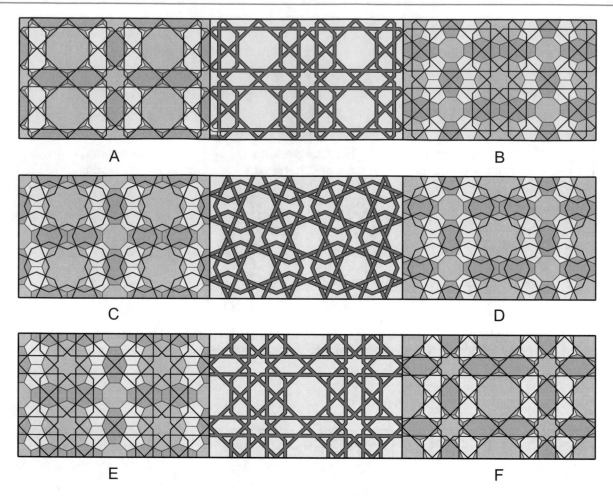

A   B

C   D

E   F

Fig. 135

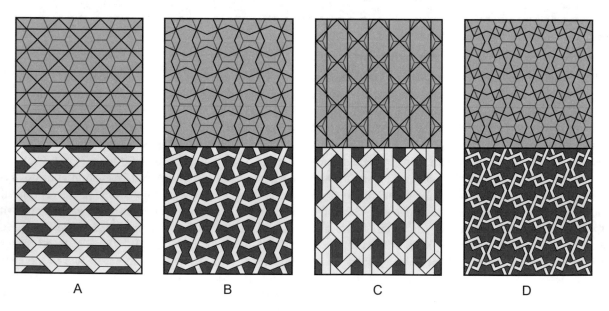

A   B   C   D

Fig. 136

Fig. 137

design. Figure 137c shows a conventional *median* design (by author) with 90° crossing pattern lines, while Fig. 137d shows a historical *median* pattern that employs 60° crossing patterns lines. The use of 60° angular openings within the *fourfold system A* is unusual. This is a Seljuk example from the Kwaja Atabek mausoleum in Kerman, Iran (1100-1150). Figure 137e shows an *obtuse* pattern (by author), and Fig. 137f shows a *two-point* pattern (by author).

The majority of field patterns associated with the *fourfold system A* were produced from underlying tessellations comprised of two or more polygonal modules. Figure 138

shows six field patterns created from just two modules: the square and large hexagon. Figure 138a shows a Seljuk *acute* pattern from the *mihrab* of the Ibrahim mosque at Salihin in Aleppo (1112). This was also used by the Saltukids on the Tepsi minaret in Ezurum, Turkey (1124-32). The *acute* variation in Fig. 138b is a Ghaznavid design from the Ribat-i Mahi near Mashhad, Iran (1019-20) [Photograph 14]. This arbitrarily changes the pattern lines associated with the underlying square module. Figure 138c shows a very-well-known *median* design that was used in one of its earliest locations at the eastern tomb tower at Kharraqan

A

B

C

D

E

F

Fig. 138

(1067-68). The *median* variation in Fig. 138d is a Qara Qoyunlu design from the Great Mosque of Van (1389-1400). Figure 138e shows an *obtuse* design (by author) that is a matrix of superimposed octagons. And the *obtuse* variation in Fig. 138f is a Qarakhanid design from the Maghak-i Attari mosque in Bukhara, Uzbekistan (1178-79). This design is entirely comprised of superimposed dodecagons. Figure 139 shows a further variation of the

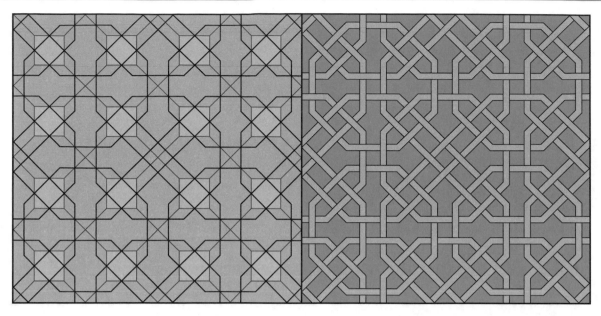

Fig. 139

Fig. 140

median pattern from Fig. 138c. This design has both varieties of applied pattern from Figs. 138c and d within a single construction. This was produced during the Seljuk Sultanate of Rum and is from the *mihrab* of the Great Mosque of Erzurum (1179). The underlying polygonal tessellation in Fig. 140 is comprised of large hexagons and pentagons, and the repeat unit for this *median* pattern is either of the two hexagons illustrated. These are duals of one another. The resulting pattern is a network of superimposed non-regular octagons. This was produced during the Seljuk Sultanate of Rum and is found at multiple locations in Turkey, including the Sultan Han in Kayseri (1232-36), and the Esrefoglu Süleyman Bey in Beysehir, Turkey (1296-97). The field pattern in Fig. 141 employs four generative polygonal modules: the large hexagons, the

small hexagon, the pentagon, and a small rhombic interstice element. The earliest example of this *median* design is from the Ildegizid mausoleum of Yusuf ibn Kathir in Nakhichevan (1161-62), and two other examples were used during the Seljuk Sultanate of Rum: at the Haunt Hatun in Kayseri (1238), and the Haci Kiliç mosque and *madrasa* in Kayseri (1249). Figure 142 demonstrates how local regions of an otherwise repetitive field pattern can be changed to include eight-pointed stars. Selected modules from the underlying tessellation in Fig. 139 have been replaced with pentagons and the large octagon so that eight-pointed stars are introduced into the pattern matrix. This example is also the product of artists working in the Seljuk Sultanate of Rum, and is from a wooden balustrade at the Esrefoglu Süleyman Bey in Beysehir, Turkey (1296-97).

Fig. 141

Fig. 142

There are a large number of patterns created from the *fourfold system A* that have as their basis an orthogonal arrangement of underlying large hexagons. Figure 143 demonstrates how this layout of the large hexagons produces eight-pointed star interstice regions. In point of fact, this underlying tessellation is essentially the same as the classic star and cross design produced from the 4.8^2 generative tessellation [Fig. 124b]. This eight-pointed star interstice region produces very acceptable pattern elements in each of the four pattern families. The two examples in this

A   B

Fig. 143

illustration were both produced during the Seljuk Sultanate of Rum. Figure 143a represents a *median* design from the Great Mosque of Erzurum, Turkey (1179). Ghurid artists employed this same design just 18 years later at the eastern mausoleum at Chisht, Afghanistan (1197). The variation in Fig. 143b is from the Gök *madrasa* and mosque in Amasya, Turkey (1266-67). The patterns in Fig. 144 employ the same underlying tessellation. Figure 144a shows a *median* pattern with atypical 60° crossing pattern lines. This design was used during the Muzaffarid period at the Friday Mosque at Kerman (1349). This design is similar to a Chaghatayid or Sufid example from the Task-Kala caravanserai in Konye-Urgench, Turkmenistan (fourteenth century) [Fig. 147b]. Figure 144b shows an *obtuse* pattern (by author) with an additive eightfold rosette within each eight-pointed star.

This design is geometrically similar to a Qarakhanid design from the Maghak-i Attari mosque in Bukhara, Uzbekistan (1178-79) [Photograph 16], in that both include superimposed octagons set upon the vertices and centers of their respective square repeats. However, the octagons in Fig. 144b are all of the same size, whereas those in the Qarakhanid example are of two different sizes [Fig. 151]. Figure 145 illustrates two underlying generative tessellations for the same *median* pattern. This is a very-well-known design employed widely throughout Muslim cultures. The underlying tessellation on the right side of this figure fills the large eight-pointed star interstice regions of the previous examples with underlying large octagons surrounded by eight pentagons as per Fig. 142. The earliest occurrences of this design are from the raised brick

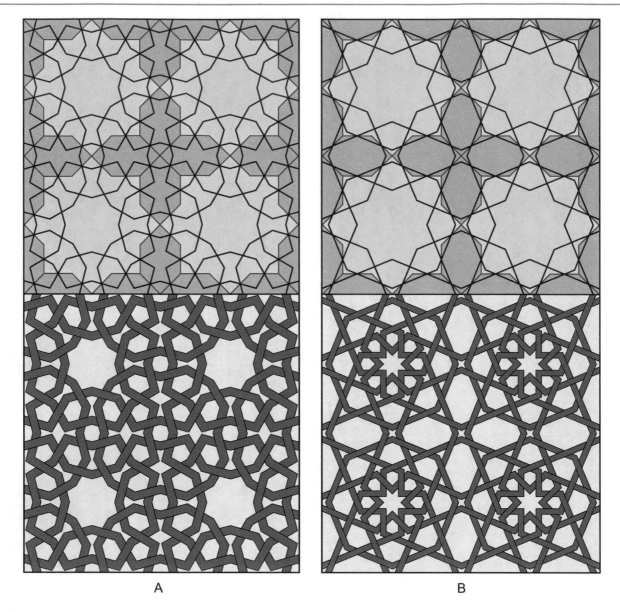

A   B

Fig. 144

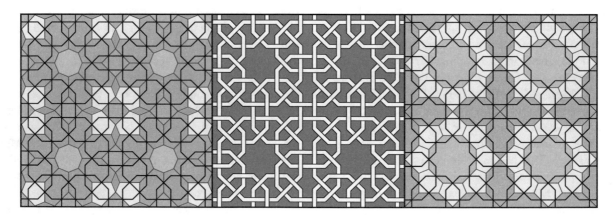

Fig. 145

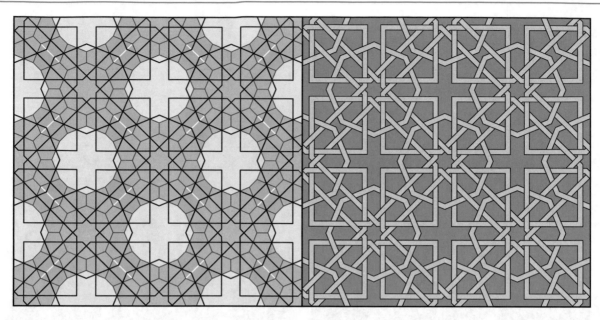

Fig. 146

ornaments of the Seljuks and Ghurids in Khurasan, including the minaret of the Friday Mosque at Damghan, Iran (1080); an example that is incorporated into a *Kufi* inscription at the Friday Mosque at Golpayegan, Iran (1105-18); the minaret of Daulatabad in Afghanistan (1108-09) [Photograph 20]; the minaret of the Friday Mosque at Saveh, Iran (1110); the Friday Mosque at Sangan-e Pa'in, Iran (late twelfth century); and the minaret of Jam in central Afghanistan (1174-75 or 1194-95). This is also illustrated in the anonymous treatise *On Similar and Complementary Interlocking Figures*.[11] Later examples of this ever-popular design include the Shaybanid polychromatic brick ornament at the Shir Dar *madrasa* in Registan Square, Samarkand (1619-36) [Photograph 72]; the Mughal stone mosaics of the mausoleum of Akbar in Sikandra, India (1613); and the Ottoman ornament of the Bayt Ghazalah private residence in Aleppo (seventeenth century). The underlying tessellation in Fig. 146 also incorporates large octagons surrounded by eight pentagons. In this case the ring of pentagons is separated by small hexagons located at the midpoints of the repetitive edges. This produces a non-regular interstice dodecagon at the centers of the square repeat into which the design's cruciform element is located. The angular openings of the crossing pattern lines in Fig. 146 are of two types: 45° and 67.5°. This is a Seljuk Sultanate of Rum design from the Çifte *madrasa* in Kayseri (1205).

The placement of eight underlying squares in rotation is a device that was used with some frequency for creating designs from the *fourfold system A*. The pattern lines in

Fig. 147a are almost identical to those of Fig. 146, but produced from an underlying tessellation that places eight squares and eight rhombi in eightfold rotation at the vertices of the square repeat unit. Indeed, both these two designs can be produced from either underlying tessellation. The design in Fig. 147a is unusual in that it employs three varieties of crossing pattern line placed at the midpoints of the underlying polygonal edges: 45°, 67.5°, and 90°. This is a Seljuk pattern that surrounds the portal of the Friday Mosque at Gonabad (1212). The design in Fig. 147b is also unusual. The underlying tessellation does not make use of the eight rhombi in rotation, and these areas now become eight-pointed star interstice regions. The application of the pattern lines within these regions of the underlying tessellation is unconventional. The entire pattern matrix is determined by the placement of eight regular hexagons around the interstice eight-pointed stars that are centered on the vertices of the square repeat unit (dashed lines). This results in the hexagons being located upon the vertices of the 4.8^2 tessellation of octagons and squares (white lines). These hexagons are aligned to the indicated sets of eight radii at every other underlying interstice eight-pointed star. While the proportion of the resulting eight-pointed stars within these regions is not ideal, the extended lines of the regular hexagons serendipitously produce regular octagons within the alternating underlying eight-pointed stars at the centers of the repeat units. The size and placement of each hexagon are determined by placing two of the vertices of each hexagon at the midpoints of the underlying squares. Extending the lines of the hexagons to intersection points of other extended lines completes this highly unusual design. This is either a Sufid or Chaghatayid pattern that is found on the arch soffit of the

[11] MS Persan 169, fol. 196a.

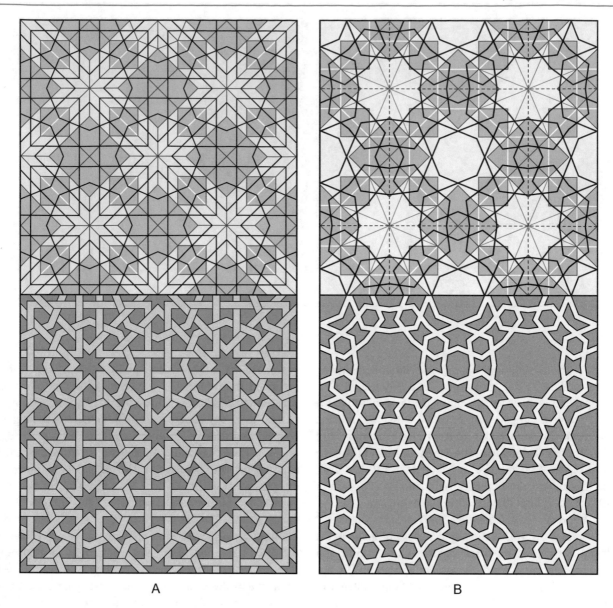

A   B

Fig. 147

entry *iwan* at the Task-Kala caravanserai in Konye-Urgench, Turkmenistan (fourteenth c). The *median* pattern in Fig. 148 has shared characteristics with the *median* design in Fig. 139, but with eight-pointed stars and nested octagons at the nodal centers. This design fills the large eight-pointed star interstice region in Figs. 143 and 144 with the eight squares and rhombi in rotation as per the previous examples, and the resulting *median* pattern is, in fact, a variation of the design in Fig. 143a, the only difference being the added small octagons that result from the ring of eight underlying squares. This Seljuk Sultanate of Rum design is located at the Alay Han near Aksaray (1155-97). Figure 149 demonstrates the construction of a pattern that utilizes the ring of eight squares that adhere to the Maghrebi stylistic conventions. Figure 149a employs the standard applied

pattern lines associated with the *median* family to each of the underlying polygonal modules except for the squares. These applied pattern lines place notches in the standard applied squares so that they become crosses. This design (by author) is an adequate example of a *median* pattern in the western style produced from the *fourfold system A*. However, Fig. 149b demonstrates how further arbitrary modifications to the cruciform pattern lines in the underlying square modules will create a more interesting variation. This example is Nasrid and was used in the Palace of Myrtles at the Alhambra (1370). Both of these treatments to the applied pattern lines associated with the underlying squares were occasional features of the Maghrebi use of this design system. Figure 150 illustrates two patterns created from underlying tessellations wherein the rings of eight squares are

Fig. 148

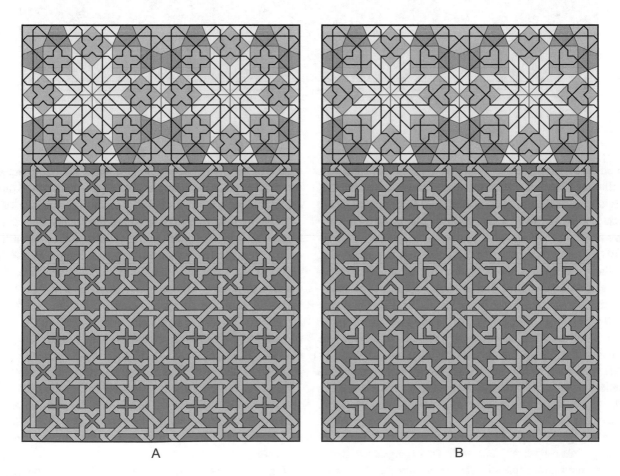

A   B

Fig. 149

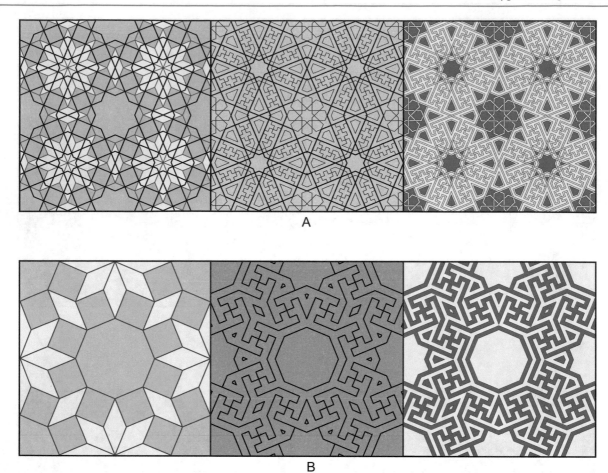

Fig. 150

rotated 22.5° from the previous examples. This produces a tessellation of squares, rhombi, and large octagons. Each of the two historical examples created from this tessellation are provided with additive swastika motifs incorporated into the pattern matrix. Figure 150a can also be created through modifying the *acute* pattern produced from the underlying tessellation of squares and octagons [Fig. 125d]. This example was produced during Seljuk Sultanate of Rum and is from the Karatay *madrasa* in Konya (1251-55). The swastika design in Fig. 150b is distinctive in that it incorporates the generative tessellation itself as the basis for the completed design, and is derived by applying the swastika motifs into each of the square modules of the generative tessellation. This is a Timurid design from a marble column at the Gawhar Shad *madrasa* and mausoleum in Herat, Afghanistan (1417-38).

The design in Fig. 151 incorporates two varieties of crossing pattern line: 45° and 90°. These angles are determined by the pattern line distribution within the underlying triangular modules. These two types of crossing pattern line produce both the small four-pointed stars that are typical to *acute* patterns within this system, and the large eight-pointed

stars that are typical to *median* patterns. By being equal to the short edge of the triangles, the edge length of the underlying squares does not conform to the typical size of the square modules from this system. However, this smaller square module works well in the context of the two varieties of crossing pattern line. Similarly, the length of the long edges of the triangular module determines the size of the underlying octagon, and this is also atypical to the proportions of the octagonal modules within the *fourfold system A*. This is a Qarakhanid pattern used on the façade of the Maghak-i Attari mosque in Bukhara, Uzbekistan (1178-79) [Photograph 16]. Figure 152 illustrates another pattern created from this unusual underlying tessellation. This is a *two-point* pattern produced during the Seljuk Sultanate of Rum that is from the Sirçali *madrasa* in Konya, Turkey (1242-45). The distinctive pattern line treatment within the underlying square element is an arbitrary inclusion.

Figure 153 shows an orthogonal design from the *fourfold system A* with standard *median* pattern line application to the underlying polygons. This design was used by the Seljuks in the Malik mosque in Kerman (eleventh to twelfth century),

Fig. 151

Fig. 152

Fig. 153

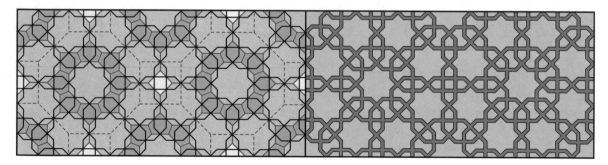

Fig. 154

as well as the Mengujekids in the mausoleum of Behram Shah in Kemah, Turkey (1228).

Most patterns created from the *fourfold system A* repeat upon the orthogonal grid. While the example in Fig. 154 is no exception, this design utilizes the 4.8^2 arrangement of squares and octagons as its repetitive schema. Although relatively uncommon, the use of multiple repetitive elements—in this case squares and octagons—within a single design was also applied to the *fourfold system B* and the *fivefold system*. These forms of hybrid repeat provide further means of creating ever-greater pattern complexity and visual interest. The underlying generative octagons in this example are placed edge to edge at each vertex of the 4.8^2 grid. Octagons, pentagons, and squares provide further polygonal infill of the generative tessellation. Several examples of this design are known to the historical record, including a Seljuk Sultanate of Rum stone border at the Zeynebiye *madrasa* in Hani, Turkey (eleventh to twelfth centuries); the Mamluk minaret of the Attar mosque in Tripoli, Lebanon (1350); the

border surrounding the early Ottoman *mihrab* in the Hatuniye *madrasa* in Karaman, Turkey (1382); and the Mughal tomb of I'timad al-Daula in Agra (1622-28) [Photograph 73]. Figure 155 illustrates another orthogonal *median* design created from this system that uses the 4.8^2 arrangement of squares and octagons as its repetitive schema. Except for the applied pattern lines into the underlying octagons that rest upon the vertices of the repetitive octagon, this design is much the same as that of Fig. 154. The applied pattern lines within these modules run parallel to the octagons of the underlying generative tessellation. This is a Qarakhanid design from the stone façade of the Maghak-i Attari mosque in Bukhara, Uzbekistan (1178-79). Figure 156 shows another *median* design with a hybrid repetitive structure. This example employs oscillating squares and rhombi in a highly eccentric repetitive structure. The included angles of the rhombi are 64.4712...° and 115.5288...°, and do not align with the inherent geometry of either the underlying tessellation or the pattern itself. Due

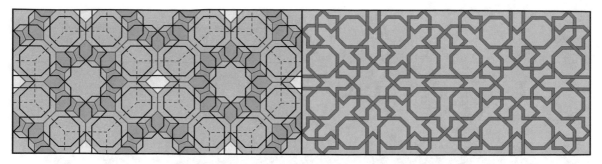

Fig. 155

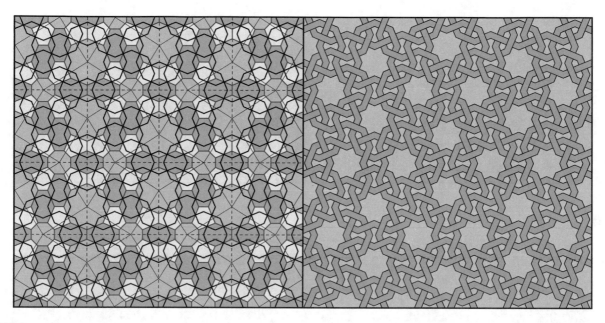

Fig. 156

to this, the eight-pointed stars that are located on the vertices of the square and rhombic grid are not in alignment with the grid itself. As with other oscillating square patterns, this design is orthogonal, but with a comparatively large amount of geometric information within each square repeat unit. This remarkable design is testament to the inventive skills of geometric artists working in the Seljuk Sultanate of Rum. This example comes from the *mihrab* of the Huand Hatun complex in Kayseri, Turkey (1237-38), and is also represented in the Topkapi Scroll.[12] The unusual character and repetitive complexity of this design suggest the possibility that the compiler of the Topkapi Scroll may have had either direct or indirect knowledge of the example in Kayseri.

The square is by far the most common repeat unit for patterns with fourfold symmetry. However rhombic repeat units that have 45° and 135° included angles were also occasionally employed. Figure 157 depicts three designs with this repetitive structure that are created from the *four-fold system A*. The underlying tessellation for these examples places four large octagons at the vertices of each rhombus. These are separated along the sides of each rhombus by a pentagon. Figure 157a represents a Seljuk Sultanate of Rum *acute* design from the Haund Hatun complex in Kayseri (1237-38), along with a variation of this design with additive swastikas from the Topkapi Scroll.[13] Like the design from Fig. 156, the fact that this example is also found at both the Huand Hatun complex in Kayseri and the Topkapi Scroll suggests a possible connection between these two historical sources. Figure 157b is a *median* design that can be created from either an underlying tessellation of just large octagons and pentagons, or one of small octagons, pentagons, and small hexagons. This is a Timurid design from the Bibi Khanum in Samarkand, Uzbekistan (1398-

---

[12] Necipoğlu (1995), diagram no. 61.

[13] Necipoğlu (1995), diagram no. 67.

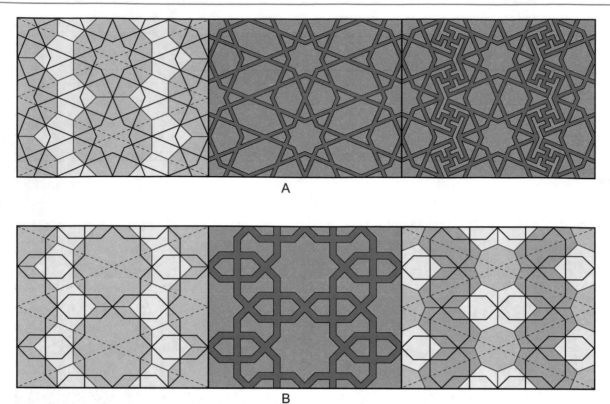

A

B

Fig. 157

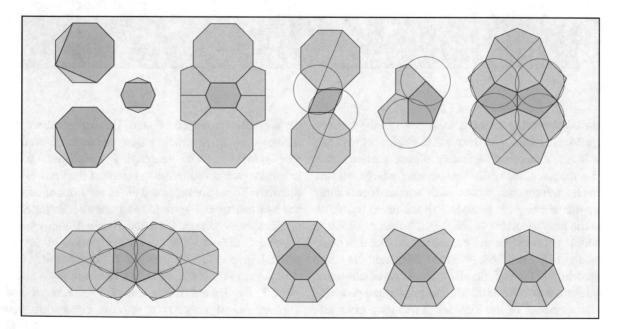

Fig. 158

1404), and a magnificent Ilkhanid additive variation was used in the mausoleum of Uljaytu in Sultaniya, Iran (1305-1313) [Fig. 66].

Figure 158 demonstrates several secondary polygonal modules from the *fourfold system A* that were occasionally used in creating patterns. These are derived primarily through either truncation of the large and small octagons, or as interstice regions in tessellations of the standard polygonal modules. These additional modules were only used in the eastern regions to a limited extent, and are more

Fig. 159

Fig. 160

commonly associated with *fourfold system A* designs of the Maghreb. Figure 159 illustrates a historical *median* pattern that incorporates one of the secondary hexagons into the underlying generative tessellation. This particularly early geometric pattern is Qarakhanid and is found on the mausoleum of Nasr ibn Ali—the middle of the three contiguous mausolea at Uzgen, Kyrgyzstan (1012-13) [Photograph 15]. This is an early date for the use of an underlying tessellation with this level of relative sophistication, and it is very likely that this example from Uzgen may have been created from the orthogonal grid method. This design was also used by Seljuk artists at the Sultan Sanjar mausoleum in Merv, Turkmenistan (1157). Figure 160 shows another Qarakhanid design that employs a secondary module within its underlying tessellation. This generative tessellation includes both the standard pentagon from this system, as well as one of the secondary pentagons from Fig. 158. The widened interweaving pattern line is arrived at through offsetting the basic pattern lines in only a single direction rather than the far more common practice of offsetting equally in both directions. This will often have a pronounced effect on the visual balance of a design. This highly

Fig. 161

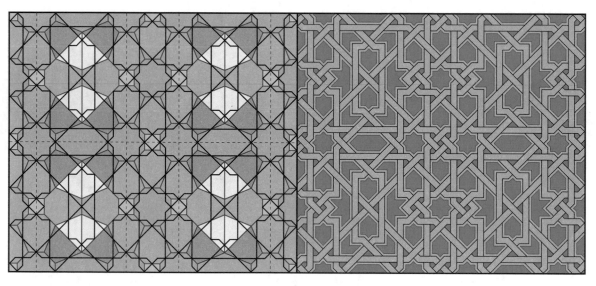

Fig. 162

successful *acute* design is from the anonymous southern mausoleum of the three mausolea at Uzgen (1186). Figure 161 illustrates a *median* pattern created from the *fourfold system A* that employs an underlying tessellation comprised of octagons, truncated octagons, and interstice eight-pointed stars. The unusual character of this historical example is due to their being two varieties of interweaving widened lines: those produced from offsetting the singular pattern lines in both directions equally and those that offset the pattern lines in one direction only. This unusual practice is represented in the middle panel of this figure. The single pattern line offsets are double the width of the twin offsets, thereby making all the widened lines the same thickness. This inconsistent line

widening methodology results in obscuring the eight-pointed stars within the underlying octagons. This is a Zangid design from several window grilles at the Nur al-Din Bimaristan in Damascus (1154) [Photograph 37]. Figure 162 shows a Nasrid *median* design that utilizes two distinct rectangular repeat units, either of which can be used on its own to fill the two-dimensional plane, thereby qualifying it as a hybrid design. This example demonstrates the more complex results achieved through the use of secondary underlying polygonal modules combined with a more flexible and arbitrary approach to the application of the pattern lines to the underlying tessellation. This more refined use to the *fourfold system A* provided for greater artistic license that the more

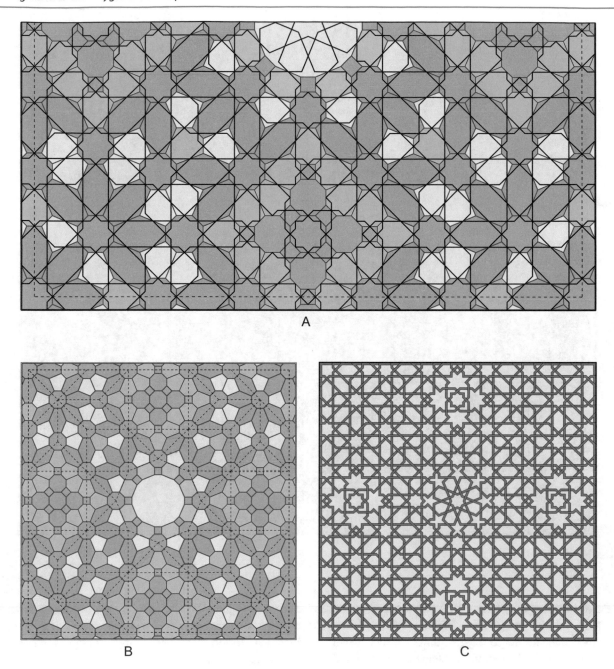

Fig. 163

formulaic approach to using design systems. This approach to the *fourfold system A* in particular was typical in the western regions under the patronage of the Nasrids of Spain and the Marinids of Morocco. This example is from a wooden door at the Alhambra in Granada, Spain (1370) [Photograph 62]. The design in Fig. 163 is also a very successful *median* design with inclusions of the secondary modules from this system. Figure 163a details the more varied approach to the pattern line application typical of the Maghrebi tradition. Of particular note is the treatment within the cluster of five underlying octagons at the central periphery of the repeat. Also of note

are the parallel pattern lines that create a bordering frame around the periphery of the panel. This is a common framing device in the western geometric tradition. As shown, this outer frame of parallel pattern lines typically extends beyond the line of symmetry created by the edges of the repeat unit (dashed line). Figure 163b illustrates how the underlying tessellation for this design includes multiple secondary polygonal modules, including two varieties of truncated octagon. The central 16-sided element is an interstice of the squares and smaller truncated octagons. This element allows for the central eight-pointed star that is rotated out of alignment with

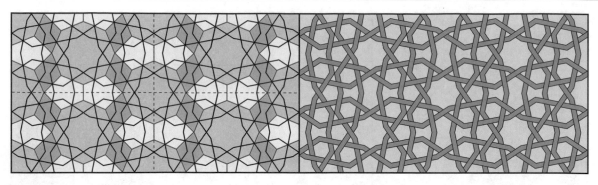

Fig. 164

Fig. 165

the rest of the design by 22.5°. This very appealing feature appears to be unique to this design. Figure 163b also illustrates how the large repeat unit of this panel is comprised of multiple repetitive elements that include a central octagon, large and small squares, half squares, rhombi, and half rhombi. This feature identifies this as a hybrid design, and all but the central octagon can be used on their own to create successful patterns. Figure 163c represents the interweaving line version of this design as per the original from the Hall of Ambassadors at the Alhambra. Figure 164 shows a Timurid *median* pattern with crossing pattern lines set at 67.4031. . . .. The underlying pentagons and rhombi are secondary polygonal modules that are less typical to this system. This is from a cut-tile mosaic panel at the Shah-i Zinda funerary complex in Samarkand, Uzbekistan (1386) [Photograph 74].

The Maghreb was the only region that incorporated 16-pointed stars into designs associated with the *fourfold*

system A. Figure 165 shows a *Mudéjar median* pattern with a large percentage of secondary underlying polygonal modules within the underlying polygonal tessellation, including 16-gons surrounded by elongated pentagons. These contribute significantly to the overall complexity of the design. As with other Maghrebi examples created from this system, changes to the standard pattern line application to particular polygonal modules add to the distinctive aesthetic character of this design. This design is from the Synagogue del Tránsito in Toledo (1360). The *acute* pattern in Fig. 166 is from a Nasrid window grille at the Alhambra. This hybrid design is comprised of two distinct repetitive elements (dashed line), either of which can be used on its own. This also employs a wide assortment of primary and secondary underlying polygonal modules, nonstandard pattern line application, and a central 16-pointed star within the upper square repeat. Figure 167 illustrates a Nasrid *acute* design from a *zillij* panel from the Alhambra. Like the

Fig. 166

Fig. 167

previous example, this is created from primary and secondary underlying polygonal modules, nonstandard pattern line placement, and 16-pointed stars in the Maghrebi style. Figure 168 shows another *acute* design from a window grille at the Alhambra that employs these same methodological characteristics. This design has a particularly large repeat unit.

3.1.4 Fourfold System B

There are fewer polygonal modules that make up the *fourfold system B* than that of the *fourfold system A*. Yet even the comparatively small number will tessellate in innumerable combinations, and will produce a tremendous diversity of geometric designs. As such, the examples in this section are

Fig. 168

representative of the historical record, but only scratch the surface of the design potential offered by this methodological system. Figure 169 illustrates the five primary polygonal modules that comprise the *fourfold system B*, along with the standard applied pattern lines in each of the four pattern families. As with other historical systems, additional polygonal modules are occasionally employed within this system, and these are generally the product of interstice regions in tessellations made up of otherwise standard polygonal modules. Exceptions to this form of secondary module are the additional polygonal modules that allow for the incorporation of 16-pointed stars into the pattern matrix. These are very similar to the equivalent modules from the *fourfold system A*. The angular openings of the *acute* crossing pattern lines in the *fourfold system B* are 45°, 70.5288...° for *median* designs, and 112.5° for *obtuse* designs. There are two varieties of applied pattern line for the large hexagons of the *acute* family. Type A has the standard 45° crossing pattern lines located upon the midpoints of the hexagonal edges. Type B introduces a pair of parallel lines that allow for the creation of octagons within the pattern matrix as per Fig. 172. With the exception of the octagonal module, the *two-point* patterns created from this system generally have applied pattern lines that are parallel to the lines from the *obtuse* family, but set upon two points of each polygonal edge. As shown, the *two-point* applied pattern lines to the octagons generally employ 45° supplementary angles, while those of the other modules generally have 33.75° and 146.25° supplementary angles. There are just two edge lengths among the polygonal modules of the *fourfold system B*. This requires the modules to tessellate with one another in conformity with edge length. The proportions and simple constructions for each

of these modules is illustrated in Fig. 170. The pentagon is easily constructed from two octagons or alternatively from the 16-gon; the rhombus and both varieties of hexagon are interstice regions created from tessellating with the octagons and pentagons.

Figure 171 demonstrates the origin of the seemingly unusual angular openings for the crossing pattern lines of the *median* and *obtuse* families. Figure 171a shows the octagonal origin for the 70.5288...° angle of the crossing pattern lines for the *median* family. A line that connects the midpoint of the octagon with the third sequential vertex determines this angle. Figure 171b illustrates the 112.5° angular opening of the *obtuse* family that results from connecting the midpoints of the short edges of the pentagon.

Figure 172 illustrates the functionality of the two varieties of *acute* pattern line application to the large hexagons. Figure 172a shows the standard *acute* pattern line application to the frequently found configuration of two large hexagons and pentagon. Figure 172b demonstrates how the addition of two parallel pattern lines results in the creation of regular octagons within the pattern matrix. This variation was widely incorporated into the *acute* patterns created from the *fourfold system B*.

Figure 173 illustrates designs from each of the four pattern families as applied to the most basic of the underlying generative tessellations produced from the *fourfold system B*. The dual of the underlying tessellation (dashed line) is the 4.8^2 semi-regular tessellation, and each of these patterns can be alternatively created from this generative tessellation. The *acute* pattern in Fig. 173a is the most ubiquitous design produced from this system, and examples are found throughout the Islamic world. The earliest extant example appears to be from the arch spandrel of the Seljuk *mihrab* in the Friday

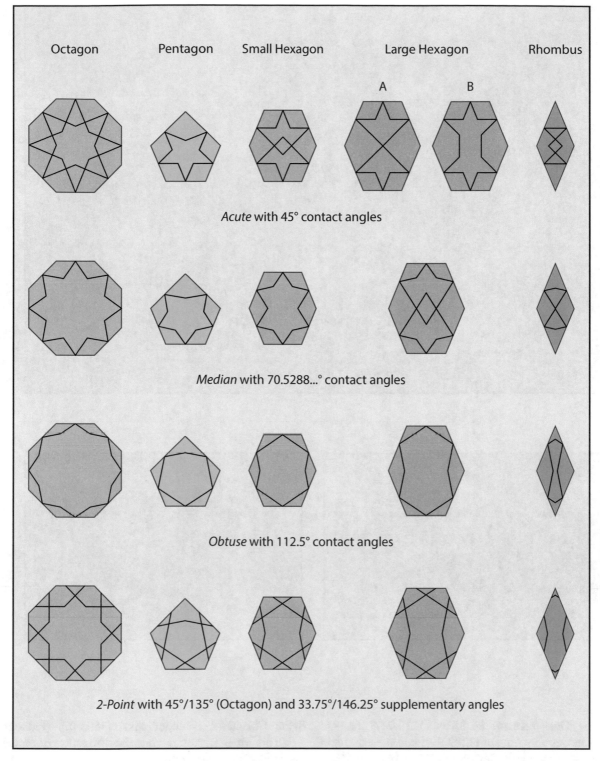

Fig. 169

Mosque at Sin, Iran (1134). Other early locations were predominantly in Iraq and the Levant, and include the base of the Zangid minaret of the Great Mosque of Nur al-Din in Mosul, Iraq (1170-72); two side panels in the Ayyubid stone *mihrab* in the Zahiriyya *madrasa* in Aleppo, Syria (1217); a wooden soffit at the Farafra *khanqah* (Dayfa Khatun) in Aleppo (1237-38); and the back wall of the arched recess in the Ayyubid wooden *mihrab* (1245-46) of the Halawiyya mosque in Aleppo. Anatolian examples from the Sultanate of Rum include the Donar Kunbet tomb tower in Kayseri

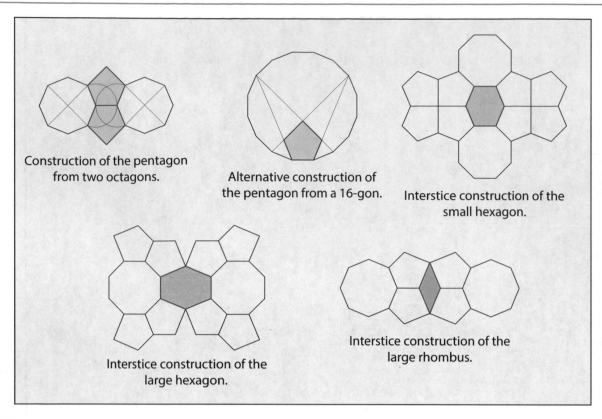

Construction of the pentagon from two octagons.

Alternative construction of the pentagon from a 16-gon.

Interstice construction of the small hexagon.

Interstice construction of the large hexagon.

Interstice construction of the large rhombus.

Fig. 170

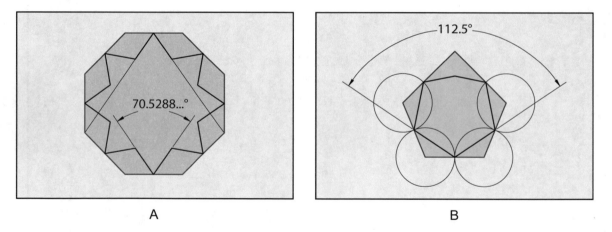

A

B

Fig. 171

(1276); the Gök *madrasa* in Sivas (1270-71); the Ahi Serafettin mosque in Ankara (1289-90); and the Esrefoglu Süleyman Bey mosque in Beysehir (1297). An early Mamluk example was used in the window grilles of the Sultan Qala'un funerary complex in Cairo (1284-85) [Photograph 55]. There are also many fine examples from the eastern regions produced after the Mongol destruction. These include a cut-tile mosaic panel from the Abdulla Ansari complex in Gazargah near Herat, Afghanistan (1425-27) [Photograph 76], and a marble *jali* screen from the tomb of

Salim Chishti at Fatehpur Sikri (1605-07) [Photograph 77]. This Mughal example has the distinction of including the underlying generative tessellation as part of the completed screen, thereby providing valuable evidence of the historicity of the polygonal technique. Somewhat surprisingly, the *median* pattern in Fig. 173b (by author) does not appear to have been used historically. The *obtuse* pattern in Fig. 173c and the *two-point* pattern in Fig. 173d both enjoyed limited historical use. The *obtuse* pattern was used by the Ayyubids in an inlaid stone panel at the Firdaws

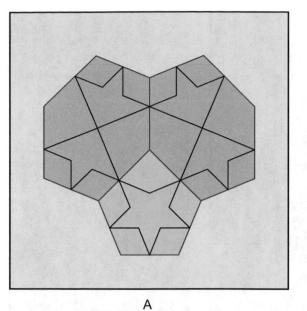
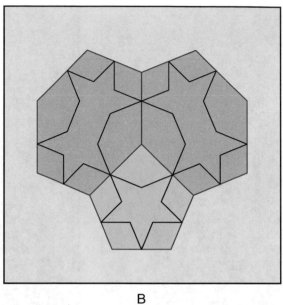

A

B

Fig. 172

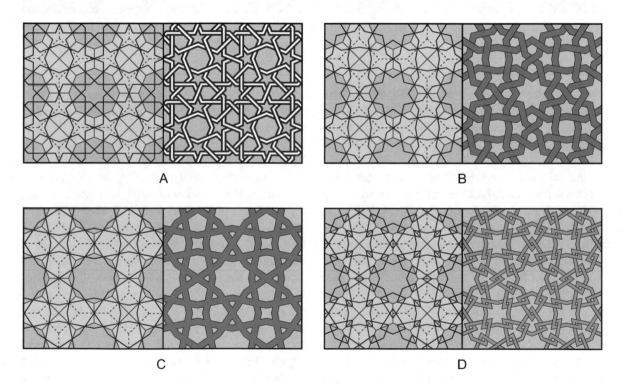

A

B

C

D

Fig. 173

madrasa in Aleppo (1235-36), while the *two-point* design was used by Mamluk artists in the stone ceiling of the Ashrafiyya *madrasa* in Jerusalem (1482). Figure 174 shows three additional *two-point* designs created from the same underlying tessellation introduced in Fig. 173. All three designs are variations on the *two-point* pattern in Fig. 173d, and each includes a square centered on the vertex

where all four pentagons meet. Figure 174a shows a Mamluk variation that is found at both the Aqbughawiyya *madrasa* (1340) in the al-Azhar mosque in Cairo and the Sultan al-Mu'ayyad Shaykh complex in Cairo (1415-22). Figure 174b shows a Ghurid design from the Friday Mosque at Herat, Afghanistan (1200) [Photograph 32]. Figure 174c shows a Seljuk Sultanate of Rum variation from the

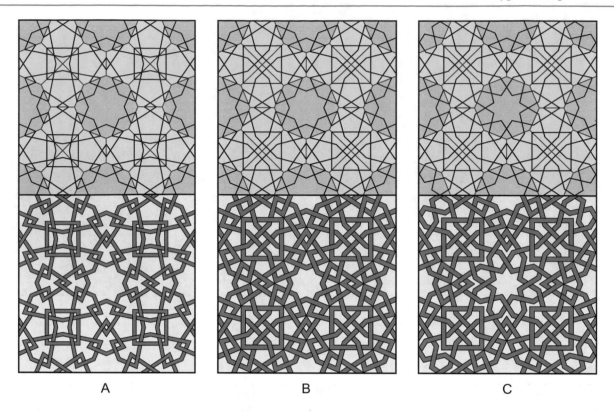

A   B   C

Fig. 174

Bimarhane hospital in Amasya, Turkey (1308-09). Figures 174b and c are similar in their inclusion of superimposed dodecagons into their pattern matrices, but Fig. 174c differs in its arbitrary treatment of the alternating eight-pointed stars.

There are a large number of orthogonal geometric patterns created from underlying tessellations in which the octagons are separated by hexagons along the edges of the square repeat unit. A feature of this variety of underlying tessellation is the four clustered pentagons at the center of each square repeat unit. It is interesting to note that both varieties of hexagon from the *fourfold system B* were used within such configurations of underlying modules. Figure 175 illustrates the *acute*, *median*, and *obtuse* designs created from an underlying tessellation that employs the small hexagon from the *fourfold system B*. While fully acceptable to the aesthetics of this tradition, the *acute* design in Fig. 175a was not generally used. Rather, the historical *acute* designs associated with this configuration of underlying polygonal modules tend to employ the large hexagon as per Fig. 177a. This is a subtle difference that is not readily apparent to casual observation. The *acute* design in Fig. 175b is an arbitrary variation of the standard *acute* pattern. This extends designated lines within the pentagons to allow for an eight-pointed star at the center of each repeat unit. This variation was used in several locations in the Maghreb including: the entry portal of the Sidi Bou Medina mosque

in Tlemcen, Algeria (1346); a wooden door from the Bu 'Inaniyya *madrasa* in Fez (1350-55); and a *zillij* panel from the Alcazar in Seville (fourteenth century). The double-line treatment of this illustration conforms with the example from Tlemcen. Figure 175c is a Muzaffarid *median* design from the Friday Mosque at Kerman (1349). The *obtuse* pattern in Fig. 175d was used in several locations: by Ayyubid artists at the portal of the Palace of Malik al-Zahir at the Aleppo citadel (before 1193), by Mamluk artists at the Taybarsiyya *madrasa* (1309) in the al-Azhar mosque in Cairo, and by Ottoman artists in the Great Mosque at Bursa (1396-1400). Figure 176 illustrates three historical *two-point* patterns created from the same underlying tessellation as that of Fig. 175, and as with the 3 *two-point* examples in Fig. 174, each of these incorporates a square centered at the vertex of the four clustered pentagons. Figure 176a shows a Mamluk design from the Sidi Madyan mosque in Cairo (1465); Fig. 176b shows a Mamluk design from the Sultan al-Mu'ayyad Shaykh complex in Cairo (1415-22); and Fig. 176c shows a much earlier Qarakhanid example from the southern anonymous tomb among the three contiguous mausolea in Uzgen, Kyrgyzstan (1186). While following the same conceptual layout, the underlying tessellation in Fig. 177 employs the large hexagon rather than the small hexagon from the *fourfold system B*. As a consequence, the size and proportion of the pentagons are different from the standard pentagon from this system, and

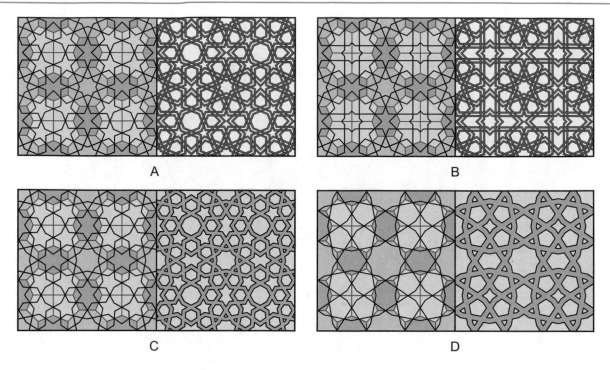

Fig. 175

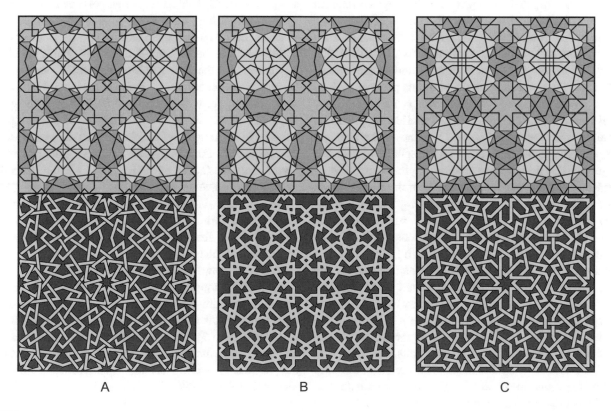

Fig. 176

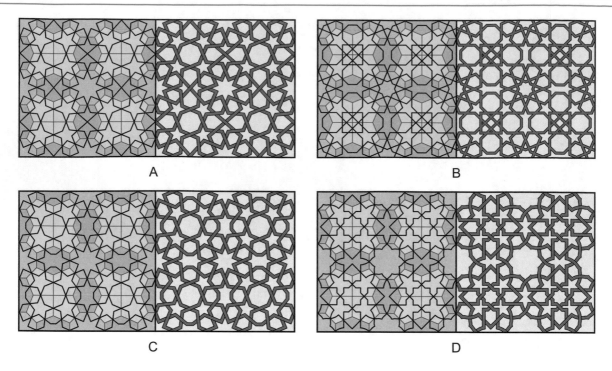

Fig. 177

are derived by simply dividing the interstice region produced from the octagons and hexagons into four quadrants. This new interstice pentagon is not found in other historical patterns, and its unique association with this tessellation disqualifies it from inclusion with the standard modules of this system. The singular application of this pentagonal form did not preclude this tessellation from broad appeal, and indeed patterns created from this tessellation were used by several Muslim cultures. Figure 177a shows the standard *acute* design produced from this tessellation. Multiple examples of this design were used during the Seljuk Sultanate of Rum, and one of the earliest is from the Sultan Han in Aksaray, Turkey (1229). A roughly contemporaneous example by Ayyubid artists was used for a balustrade in the minaret of the Aqsab mosque in Damascus (1234). Figure 177b shows a variation of the *acute* design in Fig. 177a that effectually introduces octagons into the pattern matrix, and thereby changes the pattern lines associated with the cluster of four pentagons. This design was used by Zangid artists in the wooden *mihrab* of the Lower Maqam Ibrahim in the citadel of Aleppo (1168), as well as on the *minbar* of the al-Aqsa mosque in Jerusalem (1187). The *median* pattern in Fig. 177c was used by the Ilkhanids in the arch over the entry portal of the round tower in Maragha, Iran (thirteenth century). The design in Fig. 177d is unusual for its inclusion of curvilinear lines into the pattern matrix. This is a Nasrid variation from a ceiling at the Alhambra. Figure 178 illustrates three historical designs that are not created from the *fourfold system B*, but whose underlying tessellation

shares the conceptual arrangement of octagons separated by hexagons along each edge of the square repeat unit. Because of this conceptual similarity, these designs are included within this section. The hexagon in this tessellation is regular, and the four resulting interstitial pentagons are proportioned accordingly. Figure 178a shows a Mamluk *acute* design from the Vizier al-Salih Tala'i mosque in Cairo (1303). Figure 178b shows a Mamluk *median* pattern from a carved stone relief panel at the Altinbugha mosque in Aleppo (1318). And Fig. 178c shows an Ildegizid *median* design from the façade of the Mu'mine Khatun mausoleum in Nakhichevan, Azerbaijan (1186).

As with the *fourfold system A*, geometric patterns that employ the hybrid repetitive structure of the 4.8^2 grid were also produced from the *fourfold system B*. Figure 179 illustrates two historical examples of this variety of design. Figure 179a shows an *acute* pattern that was relatively well known throughout Muslim cultures. The rings of eight octagons that are centered on each vertex of the square repeat unit result from the utilization of the type B large hexagon [Fig. 169]. This is a distinct visual characteristic of all designs that include this feature. It is interesting to note that the design within just the repetitive square region is identical to that of Fig. 173a. The earliest use of this design appears to date from the Seljuk Sultanate of Rum and is found at the Izzeddin Keykavus hospital and mausoleum in Sivas (1217-18). Gerd Schneider's excellent research on the Islamic geometric patterns of the Seljuk Sultanate of Rum identified no less than ten examples of this design produced

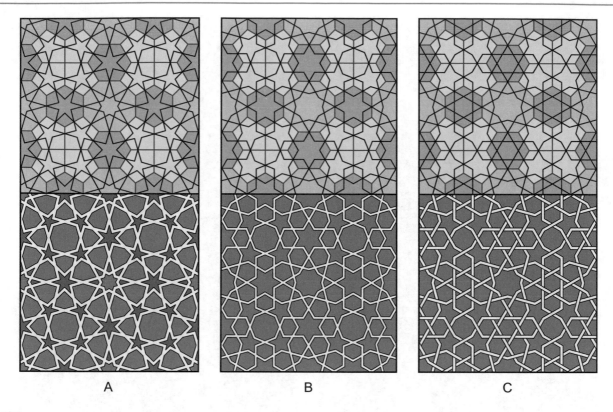

Fig. 178

in Anatolia during this period.[14] A Kartid example of this design was used in the painted fresco ornament of the mausoleum of Shaykh Ahmed-i Jam at Torbat-i Jam in northeastern Iran (1442-45) [Photograph 75]. There are also several Mamluk examples of this design, including the side panel of a *minbar* (1296) for the ibn Tulun mosque; a Quranic Illumination from Hebron[15] (1369); and an Armenian *katshkerim* carved stone relief panel set into the exterior walls of the Cathedral of St. James in Jerusalem (thirteenth to fourteenth centuries). It is worth noting that during the period of the Mamluks and the Seljuk Sultanate of Rum, the Armenian Christians occasionally adopted Islamic geometric design into their carved stone decoration in both Armenia and, to a lesser extent, in Jerusalem. Like the thirteenth- and fourteenth-century Coptic Christians in Cairo and the Catholics and Jews in al-Andalus, the Armenian Church was not averse to adopting the prevailing geometric aesthetics of their Muslim neighbors. The *obtuse* pattern in Fig. 179b is produced from the same underlying tessellation. This was created during the Seljuk Sultanate of Rum and is found at the Hudavent tomb in Nidge (1312). Figure 180 illustrates two further designs created from the same underlying tessellation

set within the 4.8² hybrid repetitive structure. Figure 180a shows a *median* design (by author), and Fig. 180b shows a *two-point* design (by author). These two examples demonstrate the efficacy of this system in creating original patterns that adhere to the aesthetics of this ornamental tradition.

As with the *fourfold system A*, several examples of patterns that employ the rhombic repeat unit with 45° and 135° included angles were produced from the *fourfold system B*. Figure 181 illustrates an *acute* pattern that uses this repeat. This pattern includes the type B large hexagons within its underlying tessellation. This produces the distinctive bands of zigzag octagons throughout the pattern matrix. The earliest example of this pattern appears to date from the Seljuk Sultanate of Rum: located at the Izzeddin Keykavus hospital and mausoleum in Sivas, Turkey (1217-18). Artists working under the auspices of this dynasty included multiple subsequent examples of this design in their work.[16] A later Timurid example of this design comes from the carved stucco ornament of the mausoleum of Amir Burunduq in the Shah-i Zinda funerary complex in Samarkand, Uzbekistan (1390). Figure 182 illustrates an *acute* design that utilizes two separate hybrid repeat units (dashed lines) within its overall rectangular repeat: a square and rhombus. On its own, the design within the square repeat is identical to

---

[14] Schneider (1980), pattern no. 320.

[15] Mamluk *mashaf*: Quranic manuscript No. 16; Islamic Museum, al-Aqsa Mosque, al-Haram al-Sharif, Jerusalem.

[16] Schneider (1980), six historical examples of pattern no. 322.

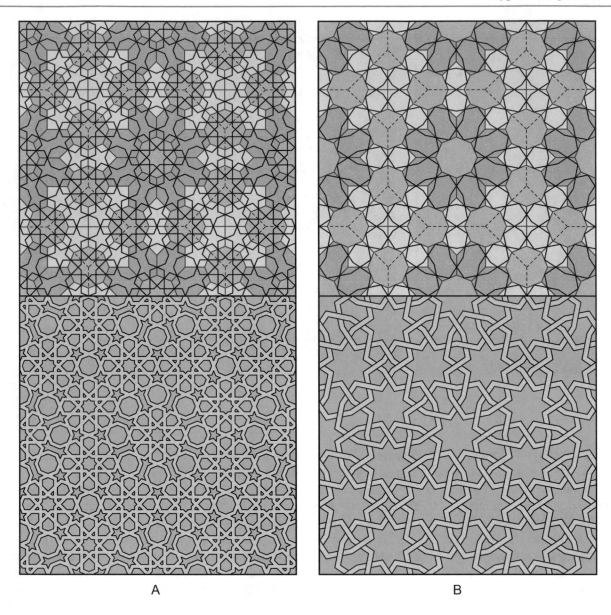

A   B

Fig. 179

that of Fig. 173a; and the design within the rhombic repeat is the same as the design in Fig. 181. These work together by virtue of the identical edge configuration in the underlying tessellation between that of the square and the short edge of the half rhombic triangle. This ingenious hybrid design is the work of Ildegizid artists working on the Mu'mine Khatun mausoleum in Nakhichevan, Azerbaijan (1186), and a later example is found at the Abbasid Palace of the Qal'a in Baghdad (c. 1230) [Photograph 34]. It is interesting to note that this twelfth-century use of the rhombic component of this hybrid design predates the earliest known use of this rhombus as a single repeat unit at the Izzeddin Keykavus hospital and mausoleum by some 32 years. Figure 183 shows a *two-point* pattern that also uses the rhombic repeat with 45° and 135° included angles. In addition to octagons, small

hexagons, and pentagons, the underlying tessellation for this design includes the rhombic module. The use of this module is relatively unusual. This design was used by Mamluk artists in two carved stone panels above the *mihrab* at the al-Mar'a mosque in Cairo (1468).

The historical use of the *fourfold system B* was less inclined to include secondary interstice elements within the underlying tessellations as the *fourfold system A*. However, the few noteworthy examples help to emphasize the design flexibility of systematic methodologies generally, and the potential for interesting original designs that can still be created using the *fourfold system B*. Figure 184 illustrates an *obtuse* pattern created from an underlying tessellation comprised of octagons, pentagons, large hexagons, and two interstice elements: squares and rhombi. This rhombus has

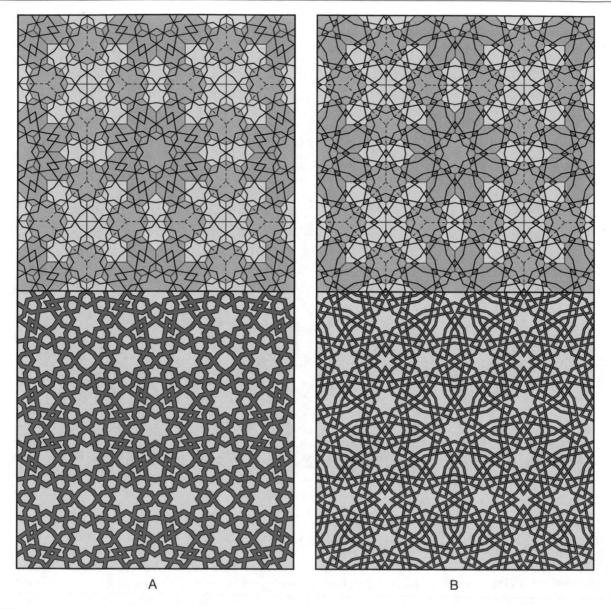

Fig. 180

the same included angles as the standard rhombus module [Fig. 169], but the edge length corresponds to the shorter edges of the polygonal modules rather than to the longer edge. Another unusual feature of this design is the variable angular openings of the crossing pattern lines. In addition to the standard 112.5° of the *obtuse* family, this example includes 90° and 104.1776 . . .° angled crossing pattern lines. This latter angular opening is a product of incorporating regular octagons into the region of extended pattern lines from the *obtuse* eight-pointed stars. This is a Mamluk design from a stone lintel at the Sultan Qaytbay Sabil in Jerusalem (1482).

Patterns that incorporate 16-pointed stars are less common to the *fourfold system B* than to the *fourfold system A*. However, those that were produced are very successful, and

much like those of their *fourfold system A* counterparts. Figure 185 illustrates two designs created from the same underlying tessellation with the additional secondary polygonal modules that allow for the incorporation of 16-pointed stars into the pattern matrix. These designs employ the same 4.8^2 hybrid repetitive structure as the examples from Figs. 179 and 180 (dashed lines). The 16-fold ring of pentagons within the underlying tessellation have variable proportions, with those in the *two-point* pattern being closer to the proportions of a regular pentagon. Figure 185a shows a Nasrid *acute* pattern from the *zillij* mosaics at the Alhambra [Photograph 63]. This was used subsequently at the Sa'dian tombs in Marrakesh, Morocco (sixteenth century), and in an Alawid stucco ceiling of the Moulay Ishmail mausoleum in Meknès, Morocco (seventeenth century). Figure 185b

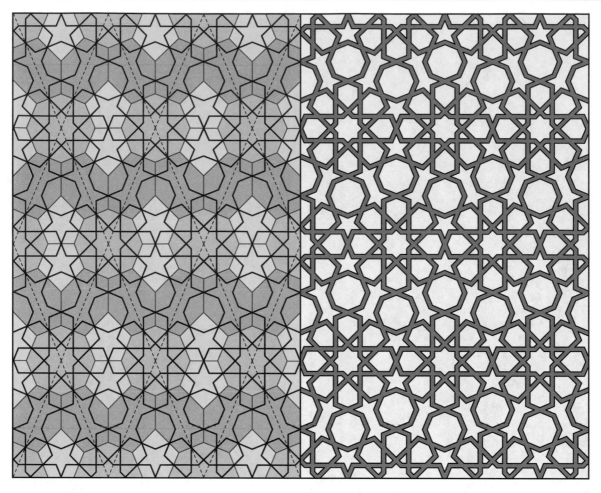

Fig. 181

shows a *two-point* design from the Imam al-Shafi'i mauso-leum in Cairo (1211). This is a rare example of a fourfold systematic design with incorporated 16-pointed stars from a region other than the Maghreb. Figure 186 shows an *acute* design created from this system that incorporates 8- and 16-pointed stars into a pattern matrix that repeats with both a rhombic grid (dash lines) and hexagonal grid (white). The artist who created this design resolved the pattern lines associated with the disproportionately long pentagons that separate the 16-gons in the underlying tessellation very nicely. This is a Marinid design from a wooden ceiling at the Bu Inaniyya *madrasa* in Fes, Morocco (1350-55) [Photograph 64].

3.1.5 Fivefold System

The *fivefold system* is immensely important to the history of Islamic art and architecture. More than any of the other systematic design methodologies, the *fivefold system* received significant innovations in pattern line variations,

ever-greater design complexity, and repetitive geometric structures. This was also the most broadly dispersed meth-odological design system throughout the Islamic world, with the greatest diversity and the largest number of representa-tive examples. The recognizable qualities of patterns created from this system were a significant contributing factor to the cohesive aesthetic of Islamic geometric art throughout the length and breadth of this tradition. The *fivefold system* has significantly more polygonal modules than the previously discussed design systems, and the greater the number of polygonal components, the greater the diversity of tessellating potential within a given system. Even with the large number of historical designs created from this system, there is still tremendous potential for creating original patterns. This extends to all of the varieties of design that the *fivefold system* was applied to, including field patterns; designs that repeat with a single rhombic, rectangular, or hexagonal repetitive cell; patterns that employ hybrid repet-itive stratagems with more than a single repetitive cell; designs that transition between more than one scale of polygonal module; designs that incorporate 20-pointed

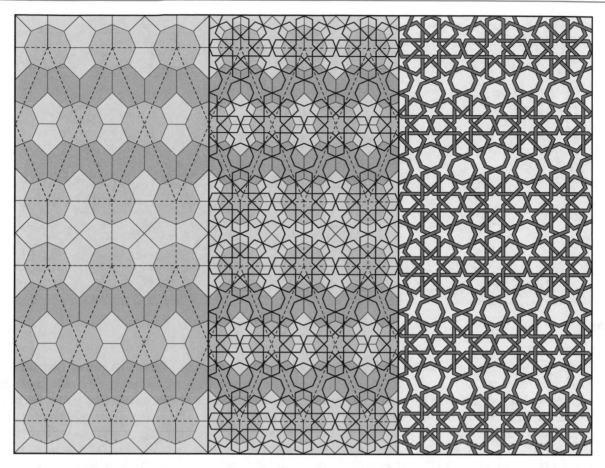

Fig. 182

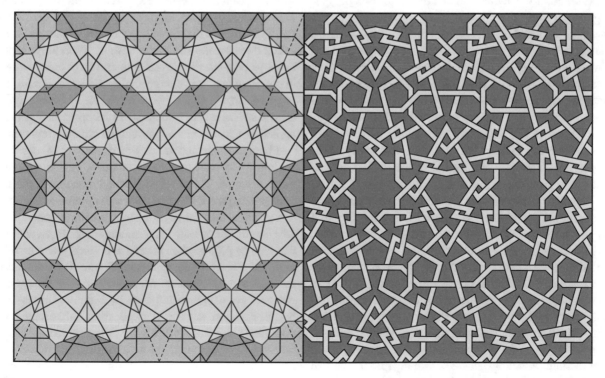

Fig. 183

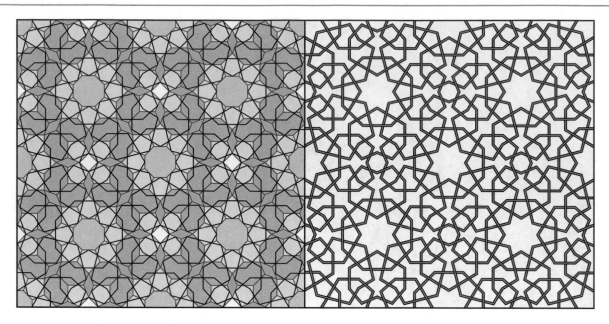

Fig. 184

stars into the pattern matrix; and dual-level designs with self-similar characteristics.

The angular openings of the crossing pattern lines in the *fivefold system* all relate directly to the geometry of the decagon. Figure 187 demonstrates how the angular openings of the *acute* family are 36°, those of the *median* family are 72°, and those of the *obtuse* family are 108°. By convention, the applied pattern lines of the *two-point* family follow the angles of the *obtuse* family, but allied to two points on each polygonal edge rather than the midpoint. The supplementary angles are therefore 36° and 144°. The application of the angles associated with each pattern family is more standardized than in the *system of regular polygons*, as well as in both fourfold systems. Yet there are significantly more conventions for arbitrary design modification—sometime additive, and sometimes subtractive—that were applied to the *fivefold system*, and led to even greater overall design diversity.

As with both fourfold systems, there are two edge lengths in the polygonal modules that comprise the *fivefold system*. The individual polygons in this system may have just the short or long edge lengths, or may be a combination of both. These edge lengths have a *φ* (*phi*) proportional relationship, and indeed the *golden ratio* is inherent to the *fivefold system*. Figure 187 illustrates the modules from this system that have only the short edge length. Being that the fundamental module to this system is the decagon, the shorter edge length of these modules can be regarded as the primary edge length of this system. This figure also demonstrates the standard pattern lines in each of the four pattern families applied to each of these modules. The visual character of the applied pattern lines of certain modules (shaded) within specific pattern families is less acceptable to the aesthetics of this tradition.

Specifically, the *acute* pattern lines created from the concave and long hexagons are generally less acceptable design features within this system and are rarely found in historical examples. Figure 188 provides the additional polygonal modules that have both edge lengths, as well as those that have only the long edge lengths. The *median* pattern lines applied to the two conjoined triangular modules (1/5 of the dodecagon) that make the quadrangular kite shape are occasionally given the specialized treatment shown in this figure. The shading over the triangles in the *obtuse* and *two-point* families indicates that these two varieties of pattern line application do not make successful designs. With the exception of the large pentagon, large rhombus and large concave hexagon in the *acute* family, the large pentagons, large rhombi, and the large concave hexagons were rarely used historically. When they were, there is considerable variation in their pattern line application, and the examples shown are only representative.

Figure 189 demonstrates the simplicity of constructing the more common polygonal modules of this system from the dodecagon. The short edge length equals the sides of the decagon and the long edge equals the radius of the decagon, and as stated these have a *φ* proportional relationship. Figure 190 demonstrates how some modules of the *fivefold system* can also be produced as interstice regions when tessellating with the decagon and pentagon. Figure 190a demonstrates the concave hexagon as a product of four decagons in a rhombic arrangement and Fig. 190b demonstrates how the barrel hexagon can be derived from an arrangement of decagons and pentagons. Figure 191 demonstrates the creation of new modules through decagonal mirroring and decagonal union. Figure 191a illustrates how the thin rhombus, the long hexagon, and the octagon can be created from

A

B

Fig. 185

reflected decagons. These three examples also show how each can be produced as interstice regions of a tessellation of other modules from the system. Figure 191a further illustrates how the wide rhombus can be created from two superimposed pentagons. Figure 191b demonstrates how the union of decagons can create new larger modules: in this case, three types of conjoined decagons. This figure also shows how the barrel hexagon can be created from superimposed pentagons. Patterns created from the larger conjoined decagons are relatively unusual. As geometric

figures these twinned decagons conform to Johannes Kepler's "fused decagon pairs" or "monsters,"[17] although their earliest use as a design vehicle predates Kepler by over a century. Figure 192 provides examples of pattern line application to two varieties of conjoined decagons in each of the four pattern families. The conjoined decagons in Fig. 192a have three edges of superimposition, while those in

---

[17] Kepler (1619).

Fig. 186

Fig. 192b have two. The application of pattern lines to larger polygonal modules such as these is open to greater artistic license: for example, the rosette infill of the *obtuse* and *two-point* pattern lines in Fig. 192a is an arbitrarily derived motif rather than purely the product of the pattern lines as it relates to the polygonal edge conditions (as per the *obtuse* and *two-point* pattern lines in Fig. 192b). Figure 193 shows several additional examples of decagonal truncations with their applied pattern lines. It is worth noting that the applied pattern lines in each of the four families do not necessarily work well with every truncation. The pattern lines in the examples shown in this figure have a synergistic relationship between the midpoints of the decagonal edges and the longer truncated edges. Truncated decagons were rarely employed traditionally, but offer a further range of potentiality for contemporary designers. Figure 194 illustrates multiple tessellating configurations that employ truncated decagons. The upper eight examples are radial, and the lower four are linear. These varieties of radial and linear combinations of truncated decagons are generally ahistorical, and are included because they nonetheless have valid potential for creating contemporary designs that fall within the aesthetics of this design tradition.[18] Figure 195 demonstrates the φ

proportional relationship between the short and long edges of several representative polygonal modules that make up the *fivefold system*, including the decagon and pentagon. These golden ratio proportions are an inherent aspect of both the underlying polygonal tessellations, and their applied pattern lines.

Figure 196 illustrates eight examples of tessellating configurations that employ the triangle module from the *fivefold system*. This triangle is 1/10 of the decagon, and the examples illustrated can be thought of as partial decagons. These are analogous to the use of the assembled 1/8 triangular modules from the *fourfold system A* [Fig. 133]. As shown in this figure, the historical use of contiguous 1/10 triangles in a rotational assemblage is almost always associated with *acute* patterns. While very successful with the applied pattern lines in this family, these arrangements are generally less successful when applied to the other three pattern families. Of particular note are the pattern elements that are created at the acute triangular vertices. With the exception of the two 7/10 arrangements, these are relatively common features of more complex *acute* patterns created from this system [Figs. 254–256 and Fig. 266]. The features of the 8/10 arrangement are the same as that of the *acute* example in the conjoined decagons of Fig. 192b.

The *fivefold system* has greater nuance than either of the fourfold systems. Just as there are fivefold polygonal modules that are less acceptable to the applied pattern lines

---

[18] The contemporary American designer Marc Pelletier has used underlying tessellations with radial assemblies of truncated decagons to produce a series of very successful dual-level designs.

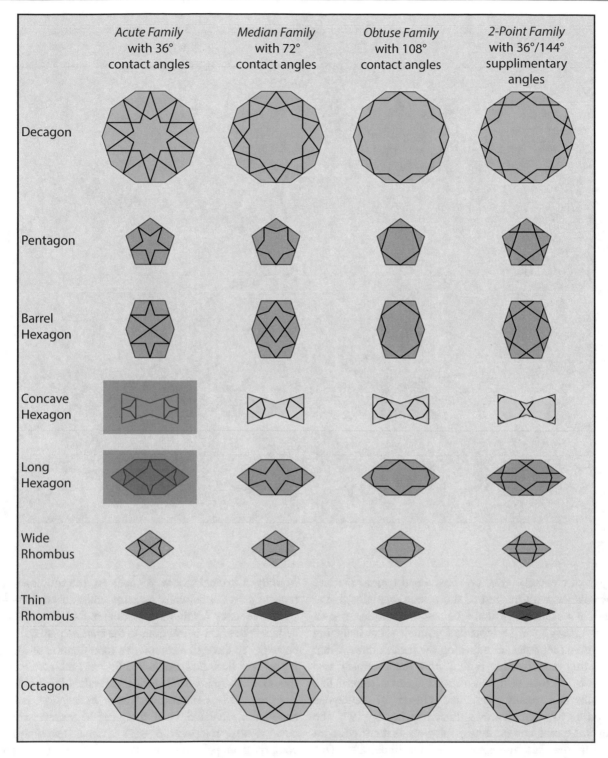

Fig. 187

of specific pattern families [Figs. 187 and 188], it is also important to note that the standard pattern line applications do not always work well in all modular configurations. Figure 197 demonstrates the pattern lines in each of the four pattern families applied to an assembly of six pentagons surrounding a thin rhombus. This configuration can be seen to create very successful motifs within the *obtuse* and *two-point* families, but unacceptable features in both the *acute* and *median* families. The problems with the pattern lines within these two latter families are the multiple small

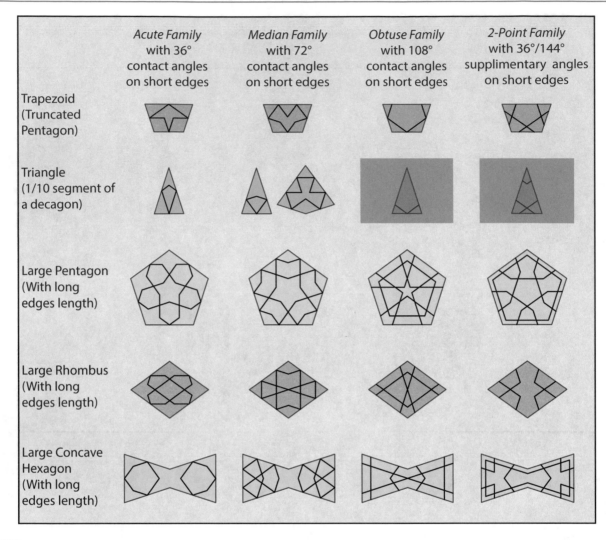

Fig. 188

background elements, creating constrained regions that are out of balance with the rest of the design, and outside the aesthetic expectations of this ornamental tradition. In such cases, the design can be remedied through either changing the polygonal modules, or adjusting the pattern lines within the existing tessellation. Figure 198 demonstrates two methods of correcting the unacceptable *acute* pattern line conditions that result from this cluster of pentagons surrounding the thin rhombus illustrated in Fig. 197. The first of these replaces the four pentagons that are edge-to-edge with the thin rhombus with four trapezoids. This provides for the use of the large wide rhombus within the newly created interstice region. The second replaces all six pentagons with six trapezoids, and fills the void with the large concave hexagon. Each of these are very successful corrections, and well known to the historical record. Figure 199 illustrates the same corrective measures within the *median* family. The solution that employs four trapezoids is not particularly attractive and, not surprisingly, is absent

from the historical record. By contrast, the solution with six trapezoids is aesthetically pleasing and, indeed, found with some frequency within the historical record. Figure 199 also demonstrates how corrections to the *median* pattern lines can be achieved through a subtractive modification of the crossing pattern lines that are shared by both the pentagons and rhombus. There is also historical precedent for this solution.

Figure 200 demonstrates how underlying polygonal tessellations created from the *fivefold system* will often have a dual relationship with another tessellation also comprised of polygonal modules from this same system. This is analogous to many of the underlying tessellations of the *fourfold system A* [Fig. 134]. Both tessellations in Fig. 200 will create patterns in each of the four pattern families, and frequently a single geometric pattern can be created from either with equal ease. It is therefore not always possible to know with certainty the underlying polygonal tessellation that was used to create a specific historical example. The dual relationship between the tessellations in Fig. 200a

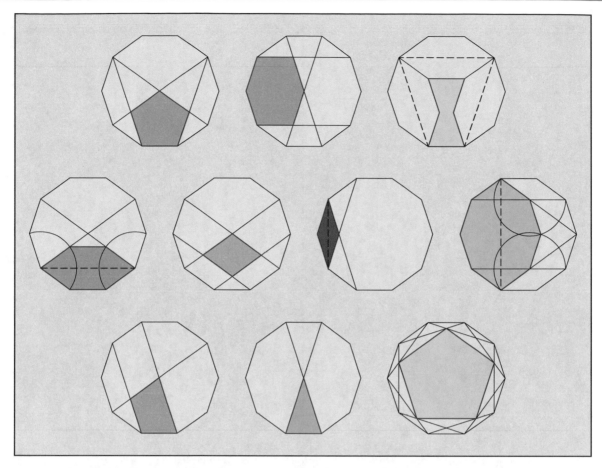

Fig. 189

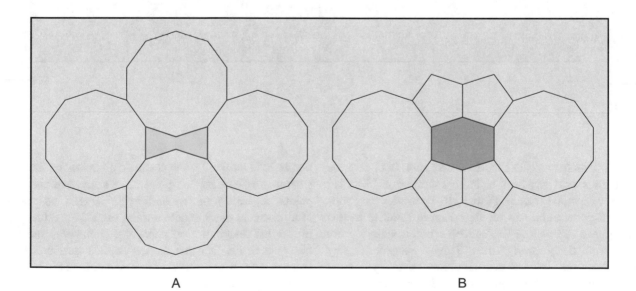

A   B

Fig. 190

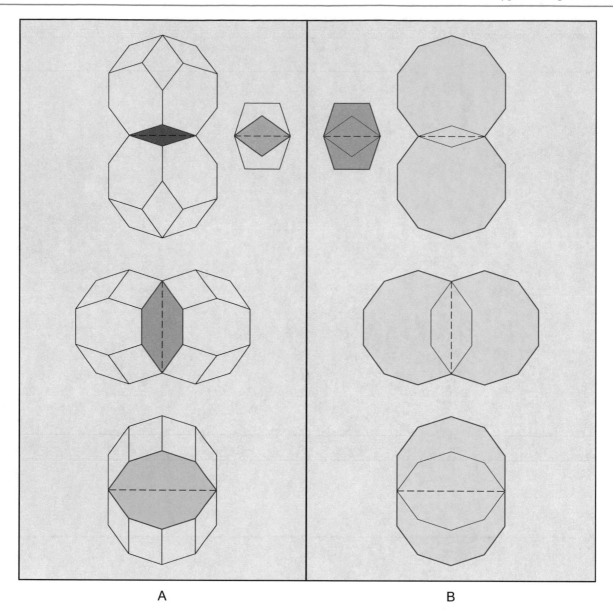

Fig. 191

and c is demonstrated in Fig. 200b. As said, both of these tessellations will produce a large number of historical designs with equal facility. Figure 201 illustrates the application of pattern lines in all four pattern families to this tessellation. The pattern in Fig. 201a is the classic *acute* design from the *fivefold system*. This is characterized by 36° angular openings at the midpoints of the polygonal edges of the tessellation comprised of decagons, pentagons and barrel hexagons. When examined from the perspective of the generative tessellation of decagons and concave hexagons, the angular opening for this design is 144°. However, this change in the angular opening does not change the categorization of the pattern family. The fivefold *acute*

family is identified by the aesthetic character of the design wherein the five- and ten-pointed stars are governed by 36° points, regardless of the underlying tessellation that was used to create them. Another more favorable feature of the underlying tessellation of decagons, pentagons, and barrel hexagons is the fact that the ten-pointed stars are directly generated from the underlying tessellation. This is not the case with the tessellation of decagons and concave hexagons, wherein the ten-pointed stars are produced through an additive process. This limitation of the underlying tessellation of just decagons and concave hexagons is also the case with the *median* pattern from Fig. 201b. This design has 72° angular openings at the midpoints of the

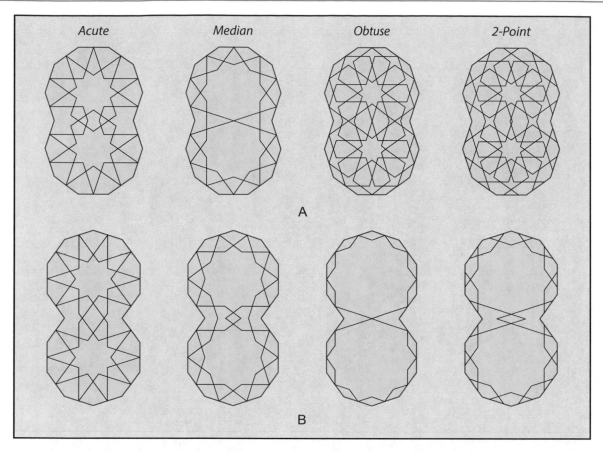

Fig. 192

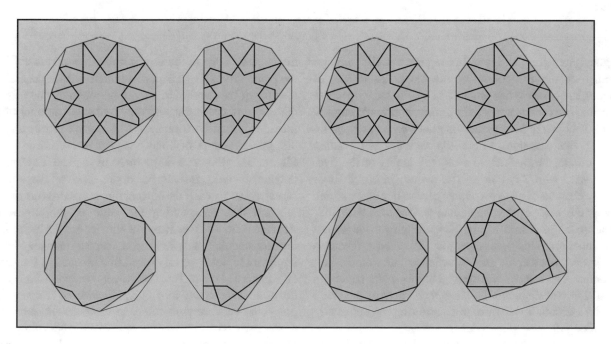

Fig. 193

Fig. 194

decagons, pentagons, and barrel hexagons, and 108° angular openings at the midpoints of the dual tessellation of decagons and concave hexagons. The 108° angular openings are standard to the *obtuse* pattern family, and would suggest that this design can be regarded as either *median* or *obtuse*. However, this example is rightfully regarded as a *median* pattern due to the distinctive aesthetic quality of the five-pointed stars with 72° points. The pattern in Fig. 201c is similarly characterized as an *obtuse* pattern due to the strong aesthetic character of the pentagonal pattern lines with 108° angular openings placed within the underlying pentagons of the tessellation. This is despite the 72° angular openings of the pattern lines in the dual tessellation that would otherwise define it as a *median* pattern. The *two-point* pattern in Fig. 201d has the distinctive *two-point* characteristics in both generative tessellations, albeit with differing supplementary angles along the polygonal edges.

A key methodological criterion of the polygonal technique relies upon the population of each edge of the pattern's repeat unit with a specific configuration of polygonal modules, and associated pattern lines, so that they match

the opposite edge of the repeat unit when this is applied to the plane through translation symmetry. As a general rule, the greater the number of polygonal modules that rest upon the edges of the repeat unit the greater the amount of geometric information contained within each repeat unit, and the greater degree of complexity in the resulting pattern. Figure 202 illustrates a number of typical examples of polygonal configurations for application to the edges of repeat units. In each case decagons are placed at the ends of each line representing the edge of the repeat unit. Arrangements such as these are equally applicable to rhombic, rectangular, and hexagonal repeat units, and—it is important to note—are also used in constructing the secondary pattern layout for dual-level designs. Figure 203 demonstrates the step-by-step application of two very basic polygonal edge configurations to rhombic and rectangular repeat units. The upper row employs the edge configuration of two decagons separated by two edge-to-edge pentagons to produce a rhombic repeat unit with 72° and 108° included angles. The central interstice region is conveniently the barrel hexagon from the *fivefold system*. The next row uses

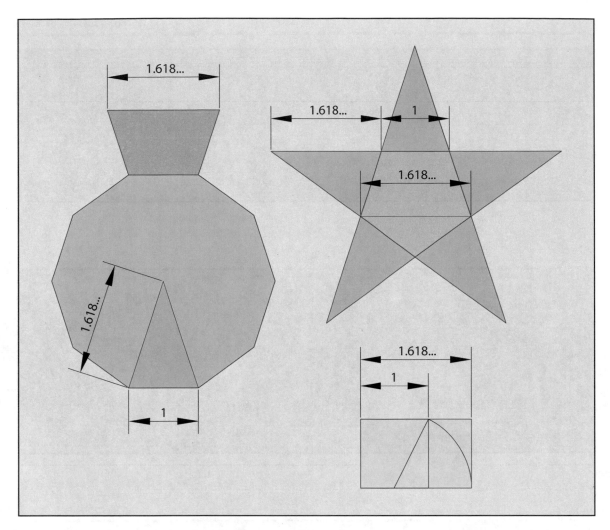

Fig. 195

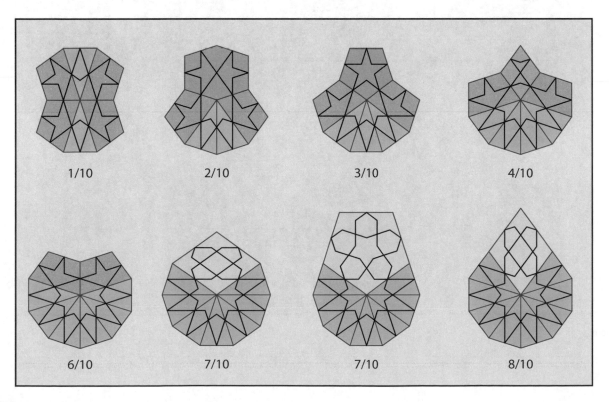

1/10	2/10	3/10	4/10
6/10	7/10	7/10	8/10

Fig. 196

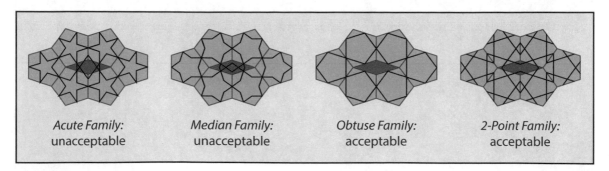

Acute Family:
unacceptable

Median Family:
unacceptable

Obtuse Family:
acceptable

2-Point Family:
acceptable

Fig. 197

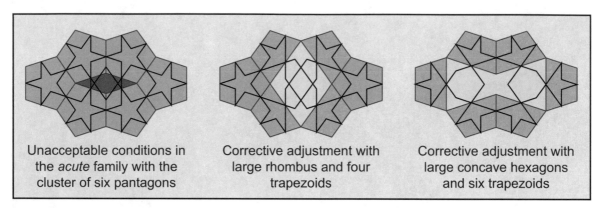

Unacceptable conditions in
the *acute* family with the
cluster of six pantagons

Corrective adjustment with
large rhombus and four
trapezoids

Corrective adjustment with
large concave hexagons
and six trapezoids

Fig. 198

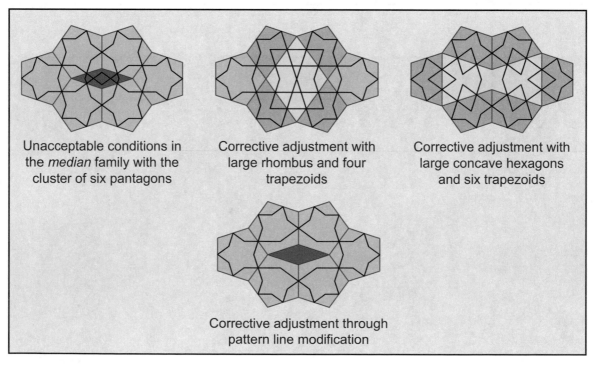

Unacceptable conditions in
the *median* family with the
cluster of six pantagons

Corrective adjustment with
large rhombus and four
trapezoids

Corrective adjustment with
large concave hexagons
and six trapezoids

Corrective adjustment through
pattern line modification

Fig. 199

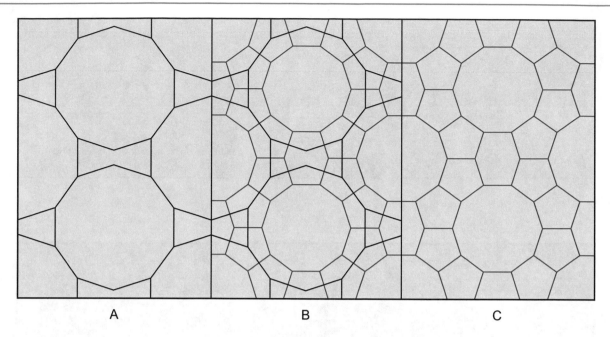

Fig. 200

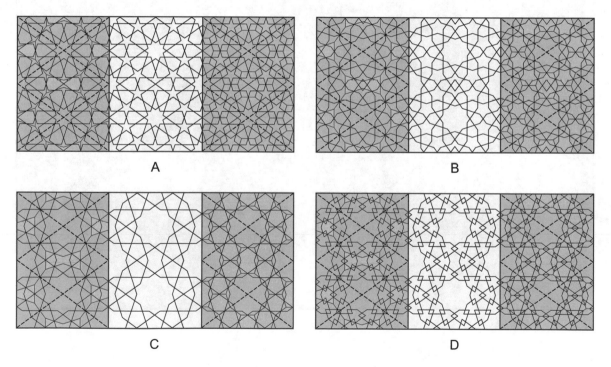

Fig. 201

the same edge configuration for the short edge of a rectangular repeat unit and two decagons separated by the barrel hexagon for the long edge. The central interstice region is once again easily filled with modules from the *fivefold system*. The third row uses two decagons separated by a barrel hexagon as the edge configuration of a rhombic repeat unit

with the 72° and 108° included angles, with an infill of pentagons and a thin rhombus in the central interstice region. And the bottom row uses this same edge configuration for a rhombic repeat unit with 36° and 144° included angles, with two wide rhombi filling the remaining interstice regions. Each of these four repeat units produces patterns that are

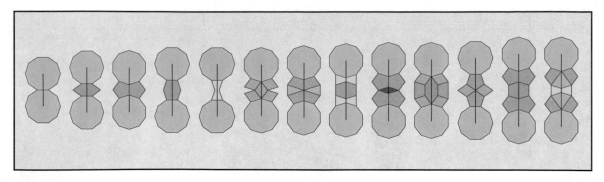

Fig. 202

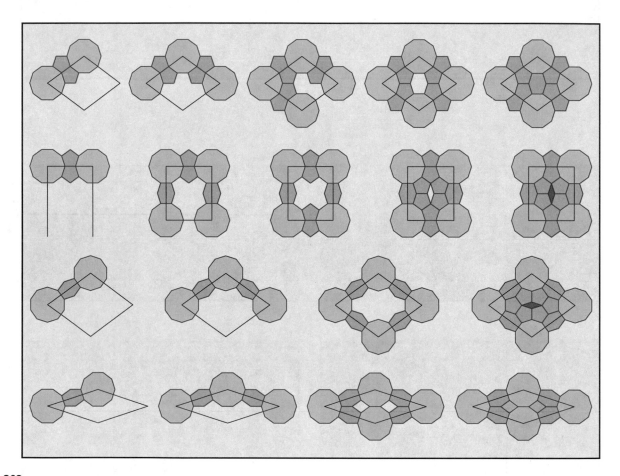

Fig. 203

well known to the historical record. Figure 204 illustrates a step-by-step construction of a more complex edge configuration for a rhombic repeat unit with 72° and 108° included angles. The more polygonal information along each edge of the repeat unit, the larger the resulting interstice region; and the larger the interstice regions, the more options for secondary polygonal infill. The final image in this figure suggests an infill with conjoined decagons, and this can be a very successful solution to populating the interstice region, but other options are possible. Figure 205 shows eight alternative polygonal arrangements for populating this interstice region. Figure 205a illustrates four infill treatments that have two axis or reflected symmetry. This is typical, but not mandatory with this tradition, and each of these infill configurations will create significantly different geometric patterns. By contrast, the polygonal infill configurations in Fig. 205b have twofold rotational point symmetry. While rare, this is occasionally found within the historical record [Fig. 266].

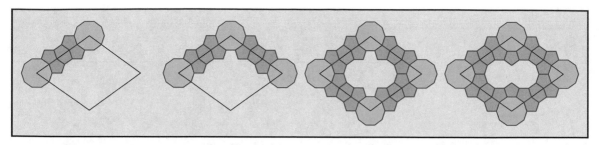

Fig. 204

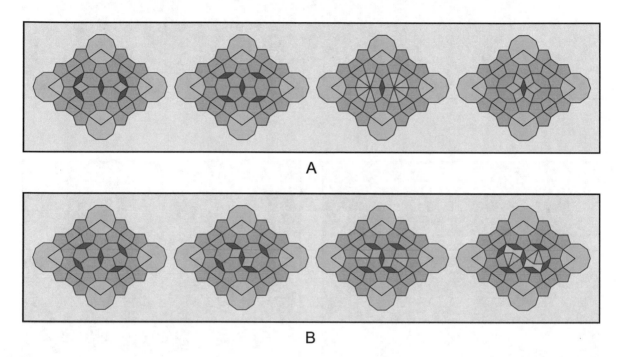

A

B

Fig. 205

Interstice regions within generative tessellations will often create design challenges that engender creative solutions. A case in point is a fivefold pattern from the Ghaznavid minaret of Mas'ud III in Ghazni, Afghanistan (1099-1115).[19] Figure 206 shows how this design is derived from a tessellation of corner-to-corner touching decagons placed upon a rhombic grid with concave decagonal interstice regions separating the decagons. Atypically, the decagonal tessellation is maintained as part of the completed design. This exceptional fivefold pattern originates from the period when fivefold patterns were first being introduced into Islamic ornament, and the use of this decagonal

tessellation is more experimental, and less akin to the standard conventions of the *fivefold system*. This experimental approach is seen in the non-regular seven-pointed stars that result from the treatment of the pattern lines within the interstice region.

In contrast to both fourfold systems, the greater number of secondary polygonal modules within the *fivefold system* provides for their significantly greater design potential in creating field patterns. These exclude the decagon within their underlying polygonal structures, and hence lack the ten-pointed stars that are otherwise a standard feature of designs produced from this system. The earliest fivefold field pattern appears to be Seljuk: used in the interior ornament of the Khwaja Atabek mausoleum in Kerman (1100-1150) [Fig. 211]. Indeed, this variety of fivefold pattern was especially popular among Seljuk artists, with the greatest number being used as border designs during the Seljuk Sultanate of Rum. The absence of decagons in the

---

[19] The fivefold panel from the minaret of Mas'ud III presented herein is in very poor repair, and available photographs are of low quality. For these reasons the reconstruction of this example may differ slightly from the actual design in Ghazni.

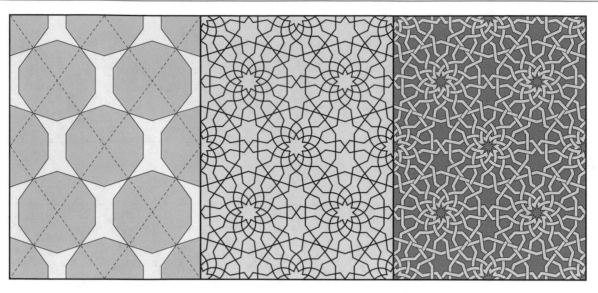

Fig. 206

A   B

Fig. 207

underlying tessellations of fivefold field patterns generally precludes the use of rhombic repeat units, and field patterns created from the *fivefold system* most commonly repeat upon either a rectangular or hexagonal grid. Figure 207 shows a *two-point* field pattern from the Great Mosque at Malatya in Turkey (1237-38) that repeats upon a rectangular grid. This is an unusual example in that the majority of fivefold field patterns are either of the *median* or *obtuse* pattern families. The underlying generative tessellation that produces this design is comprised of pentagons, barrel hexagons, and long hexagons. Figure 208 represents an *obtuse* field pattern from the Haci Kilic *madrasa* in Kayseri, Turkey (1275). This repeats upon a rectangular grid and the underlying generative tessellation is made up of pentagons, barrel hexagons,

long hexagons, wide rhombi, and thin rhombi. This historical example utilizes only the highlighted isolated linear region as a border design. Figure 209 shows a *median* field pattern from the hospital (1205) associated with the Çifte *madrasa* in Kayseri, Turkey. This repeats upon a rectangular grid, and is created from an underlying generative tessellation comprised of concave hexagons, trapezoids that are half-concave hexagons, and wide rhombi. The short edges of the four clustered trapezoids define the small thin rhombus that is not typical to the *fivefold system*. Anomalies such as this abound in the application of the *fivefold system*, and this cluster of four half-concave hexagons around a small thin rhombus is an unusual tessellating condition that works very well in the *median* family: creating the distinctive

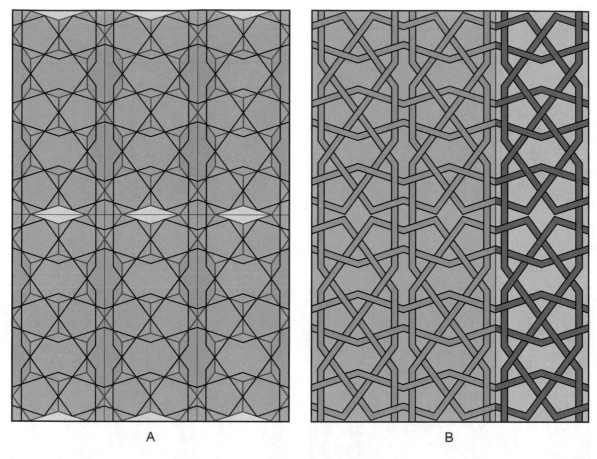

Fig. 208

Fig. 209

12-sided motif located at the vertex of each rectangular repeat. Figure 210 shows a field pattern from the Huand Hatun complex in Kayseri, Turkey (1237), that has an unusually broad rectangular repeat. The underlying generative tessellation has concave hexagons, long hexagons, and wide rhombi placed in fivefold rotational symmetry. Figure 211 shows the field pattern from the above-mentioned Khwaja Atabek mausoleum in Kerman, Iran, and later at the Sahib Ata mosque in Konya, Turkey (1258), and the *minbar* of the Esrefoglu mosque in Beysehir,

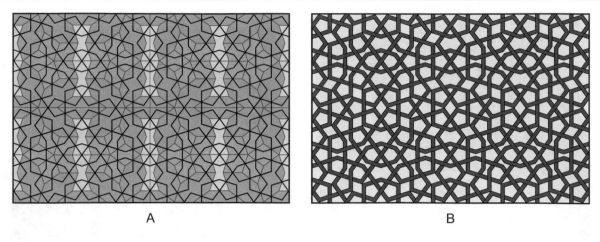

A   B

Fig. 210

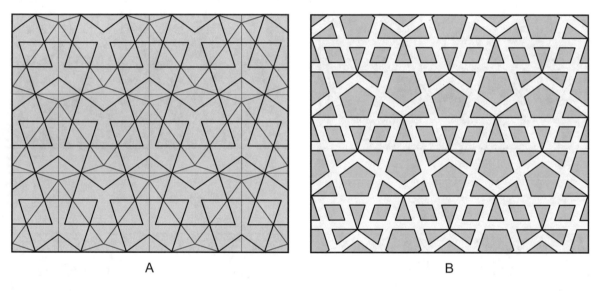

A   B

Fig. 211

Turkey (1296-97). This *median* pattern repeats rectilinearly, although it simultaneously repeats with a 6-sided chevron devise [Fig. 20]. The underlying generative tessellation is comprised of kite-shaped quadrilaterals with bilateral symmetry, each made up of two adjacent 1/10 decagonal triangles from this system [Fig. 188]. As repetitive elements, these quadrilaterals alternate in 180° rotation to fill the plane. This quadrangular module was used with some frequency in *median* patterns. However, this design is unique in that the underlying generative tessellation is comprised solely of this module.

Most historical examples of fivefold field patterns repeat upon a hexagonal grid. Figure 212 shows a Mamluk *median* design from the Amir Ghanim al-Bahlawan funerary complex in Cairo (1478). The underlying generative tessellation utilizes the same arrangement of linearly alternating quadrilaterals as the field pattern from the Khwaja Atabek

mausoleum in Kerman, but also includes an adjacent linear band of edge-to-edge concave hexagons separating the quadrilaterals. Figure 213b shows an *obtuse* field pattern found in several locations in Anatolia, including Sitte Melik tomb in Divrigi, Turkey (1196); the Alaeddin mosque in Konya, Turkey (1219-21); and the Mama Hatun tomb tower in Tercan, Turkey (thirteenth century). This design can be produced with equal facility by either of the two dual grids. Figure 213a employs pentagons, barrel hexagons, and thin rhombi, and the angular opening of the crossing pattern lines is 108°: the angle associated with *obtuse* patterns. The alternative underlying generative tessellation in Fig. 213c employs concave hexagons and long hexagons, with 72° angular openings. This angle is a standard feature of *median* patterns. However, as with other patterns with these characteristics, this design is regarded as an *obtuse* pattern due to the visually dominant pentagonal motif. Figure 214

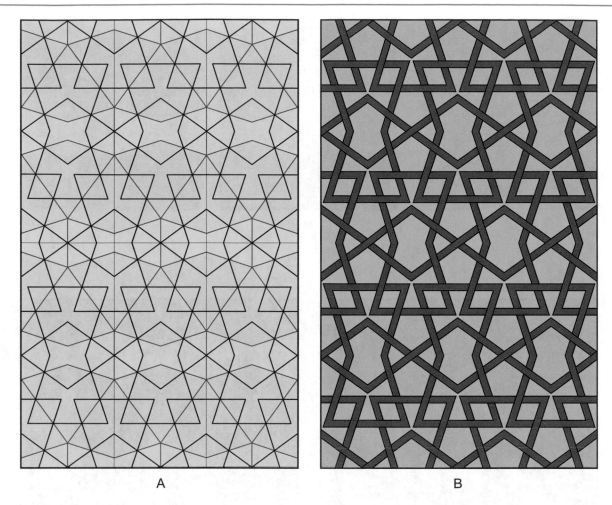

Fig. 212

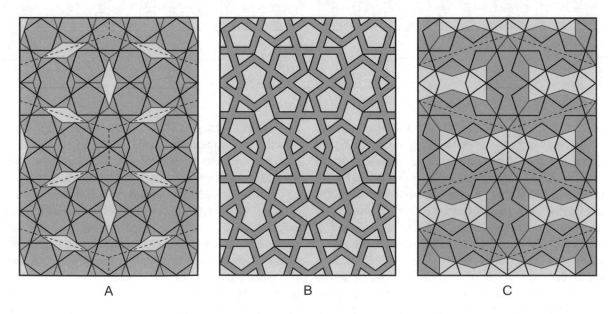

Fig. 213

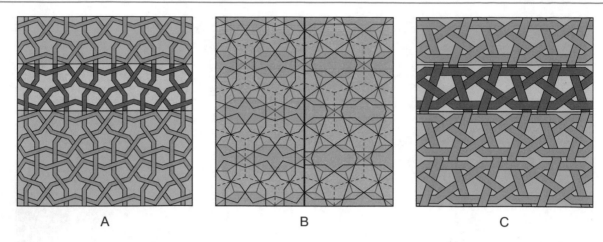

Fig. 214

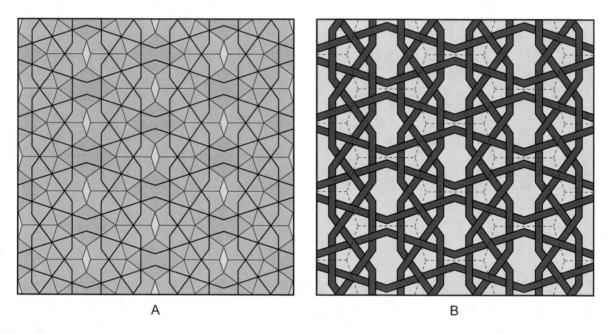

Fig. 215

illustrates two field patterns created from the same underlying tessellation of just pentagons and long hexagons. Figure 214a shows a *median* design from the Alaeddin mosque in Konya (1219-21). As highlighted, this was used as a border design. Figure 214c shows an *obtuse* design from the Çifte Minare *madrasa* in Erzurum (late thirteenth century) that was also used as a border. Figure 215 illustrates a *median* field pattern also from the Çifte Minare *madrasa* in Erzurum. The underlying generative tessellation of this pattern incorporates the same layout of four half-concave hexagons placed around a small thin rhombus as found in the previously cited example from the hospital in Kayseri (Fig. 209). Figure 216 shows a *median* field pattern from the Sultan Han

at Kayseri, Turkey (1232-36). The underlying generative tessellation utilizes the same kite shaped quadrilateral modules, with the same *median* pattern line application, as the patterns in Figs. 211 and 212. The *median* pattern in Fig. 217 is from the exterior cut-tile mosaic façade of the Great Mosque in Malatya, Turkey (1237-38). The underlying tessellation in Fig. 217a incorporates the cluster of four trapezoids (half-concave hexagons) that are edge to edge with a small thin rhombus: a feature of Figs. 209 and 215. The *median* pattern in Fig. 218 also incorporates this underlying configuration. This contemporaneous design is from the Huand Hatun in Kayseri (1237), and has an unusually long hexagonal repeat unit. The underlying generative

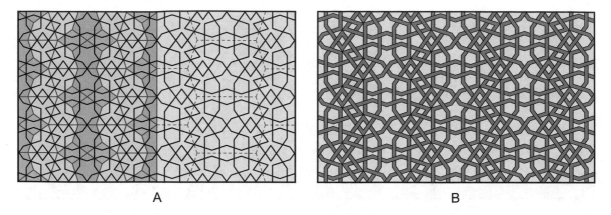

Fig. 216

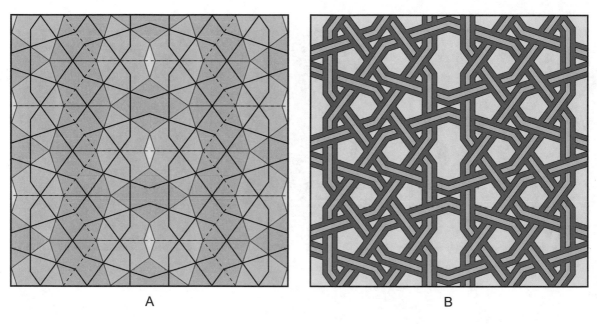

Fig. 217

Fig. 218

Fig. 219

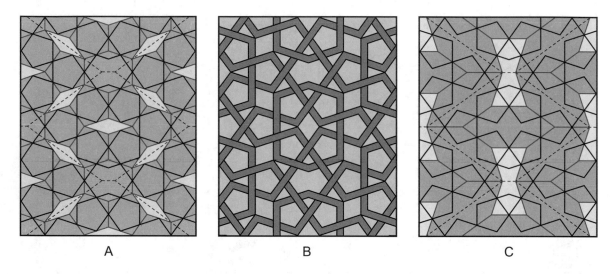

Fig. 220

tessellation of this *median* field pattern utilizes the same arrangement of four half-concave hexagons placed around a small thin rhombus seen in Figs. 209, 215, and 217, but separates these with alternating kite-shaped quadrangles as per Figs. 211, 212, and 216. The *median* pattern in Fig. 219 is from the Hekim Bey mosque in Konya (1270-80). This is created from an underlying tessellation of pentagons, long hexagons, and wide rhombi. Figure 220b shows an *obtuse* field pattern from the Great Mosque in Malatya (1237-38). This design was also used by Chaghatayid artists at the Bayan Quli Khan mausoleum in Bukhara (1357-58). As typical with this family of pattern, it can be created from either of two underlying tessellations. Figure 220a employs pentagons, barrel hexagons, and thin rhombi, and the angular opening of the crossing pattern lines is 108°— the angle associated with *obtuse* patterns. The alternative underlying generative tessellation in Fig. 220c employs concave hexagons and long hexagons, with 72° angular

openings—the angle associated with *median* patterns. However, for purposes of clarity of identification, the visually dominant pentagonal motif, with its 108° angles, identifies this as an *obtuse* pattern.

3.1.6 Fivefold System: Pattern Variation and Modification

There is considerable historical variation in the application of pattern lines to the decagons in the *fivefold system*. Figure 221 illustrates standard pattern line variations in each of the four pattern families as applied to the decagon. The decision as to what form of primary ten-pointed star to incorporate into a given design is a balance between an artist's and potentially client's personal predilections, regional and cultural design conventions, and constraints of the end material over the design process. Generally, the *acute* and *median*

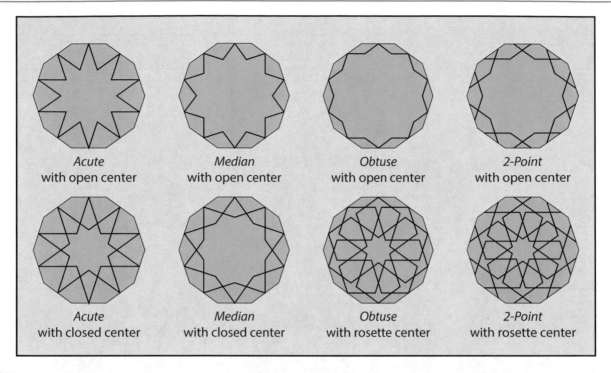

Acute with open center	*Median* with open center	*Obtuse* with open center	*2-Point* with open center
Acute with closed center	*Median* with closed center	*Obtuse* with rosette center	*2-Point* with rosette center

Fig. 221

pattern lines that are applied to the underlying decagon conform to the basic alternatives offered in this illustration. By contrast, the decagons of the *obtuse* and *two-point* families lend themselves to a greater amount of stylistic variation that is not governed by the inherent geometry of the underlying decagon itself. The most common variety of arbitrary additive elaboration of *obtuse* and *two-point* patterns introduces a tenfold star rosette into the otherwise open ten-pointed stars. Figure 222 demonstrates a simple process for creating the tenfold geometric rosette within both the *obtuse* and *two-point* families. This involves the following: (A) draw a radius from the center of the decagon to the midpoint of an edge; (B) extend pattern lines that are perpendicular to the radius as shown; (C) bisect angle and draw a circle as shown; (D) copy and rotate the bisected angle to the decagon's radius; (E) trim the dart motif, draw two new radii as shown, and extend the rotated pattern lines to the new radii; and (F) copy and rotate this motif ten times around the center of the decagon.

Subtractive processes can also be applied to the *median* crossing pattern lines of the standard ten-pointed star created from the decagon surrounded by pentagons and barrel hexagons. Figure 223 demonstrates the most widely used variety of *median* pattern modification—both among patterns created from the *fivefold system* as well as with nonsystematic designs. This form of modification to the primary stars was especially popular with the Mamluk artists of Egypt. Figures 223a through d illustrate how the pattern lines can be removed, and replaced with a central rosette

comprised of a ten-pointed star and ten darts. This motif is typically associated with fivefold patterns in the *acute* family. This process involves the removal of the 72° crossing pattern lines located on the decagonal edges in Fig. 223b; the extension of the remaining pattern lines in Fig. 223c; and the incorporation of a central ten-pointed star rosette as per Fig. 223d. Figure 223e shows the standard *median* pattern without this modification, and Fig. 223f shows this same pattern with the modification. The application of this modification to each decagonal region radically transforms the character of the original, and the replacement of the five-pointed stars with two darts and two shield shapes provides this modified pattern with the characteristics of the *acute* family. In fact, Fig. 223g shows how the modified design from Fig. 223f is identical to an *acute* pattern that is more efficiently produced from its own underlying tessellation comprised of decagons, barrel hexagons, trapezoids, and large concave hexagons. This alternative method for creating the same design makes use of the corrective adjustment to the underlying tessellation that employs six trapezoids demonstrated in Fig. 198.

Two varieties of *obtuse* modification to the ten-pointed stars created from the decagon are illustrated in Fig. 224. This modification was popular among Seljuk artists in both Persia and Anatolia. Figure 224a transforms the region of the decagon from a ten-pointed star into a fivefold motif with a pentagon at the center of the decagon. The less common modification in Fig. 224b replaces the ten-pointed star with a five-pointed star with 72° points as per the *median* family.

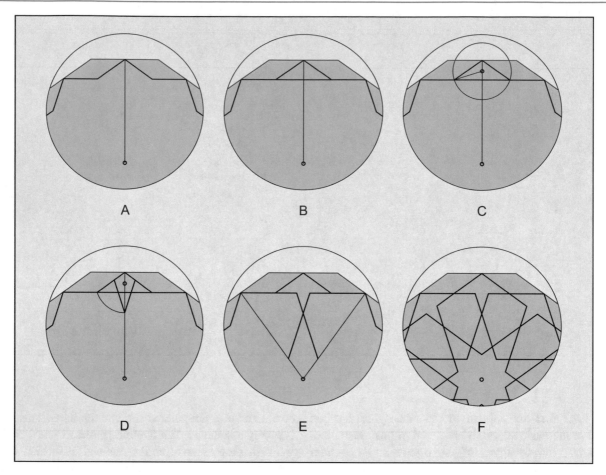

Fig. 222

When applied to every ten-pointed star within a given design, each of these modifications transforms the design into a field pattern.

There are multiple conventions for the arbitrary treatment of the ten-pointed stars in the *two-point* family. Figure 225 illustrates three examples. Figure 225a shows the standard *two-point* design with its mirrored pattern lines along each decagonal edge. Figure 225b extends, rather than mirrors, the lines within the pentagons to create the tenfold geometric rosette. Figure 225c extends the standard pattern lines of Fig. 225a into a tenfold geometric rosette of smaller scale than that of Fig. 225b. And Fig. 225d mirrors two of the lines within each underlying pentagon to create a distinctive ring of ten rhombi. Each of these variations was used historically.

As applied to otherwise complete patterns, arbitrary modifications to the standard pattern lines can have a pronounced effect on the visual character of a geometric design. Pattern modification is, therefore, an extremely important, if frequently overlooked, aspect of this ornamental tradition. The conventions for pattern modification are often regional and cultural, and find their most expansive representations within designs created from the *fivefold system*. What is more, specific pattern modifications established within the

fivefold system almost certainly served as analogs for similar application to designs created from other systematic and nonsystematic expressions of the polygonal technique. As mentioned above, *acute* patterns created from the *fivefold system* received less variational attention than those produced from the other pattern families. Figure 226d illustrates and outstanding exception to this general rule. This Ilkhanid example is found at the mausoleum of Uljaytu in Sultaniya (1307-13), and is based upon the classic *acute* pattern created from the underlying tessellation of decagons, pentagons and barrel hexagons placed upon a rhombic repeat shown in Fig. 226a. The equal width of the two colored foreground and background elements in this design from Sultaniya are akin to the *ablaq* ornament of the Ayyubids and Mamluks. The standard *acute* pattern in Fig. 226b is characterized by the 36° angular openings of the pattern lines placed at the midpoints of the underlying polygonal edges. Figure 226c illustrates this frequently encountered interpretation with widened interweaving lines, and an early example of this classic fivefold pattern was used on the intrados of the Ghurid ceremonial arch at Bust (1149) [Photograph 33]. Seljuk artists used a widened line version as a border within the *iwan* of the Friday Mosque at Gonabad (1212) [Photograph 23].

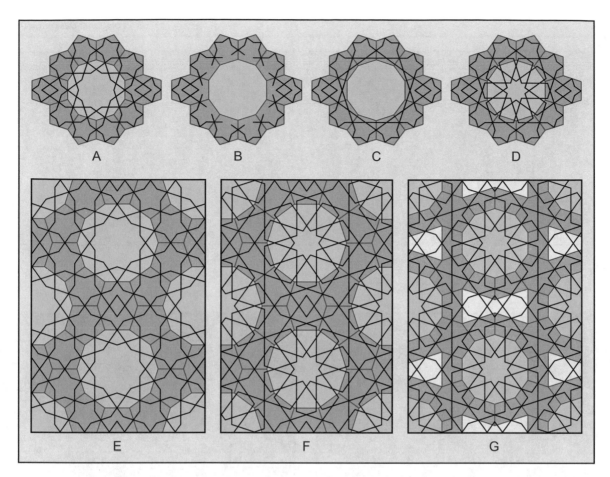

A   B   C   D

E   F   G

Fig. 223

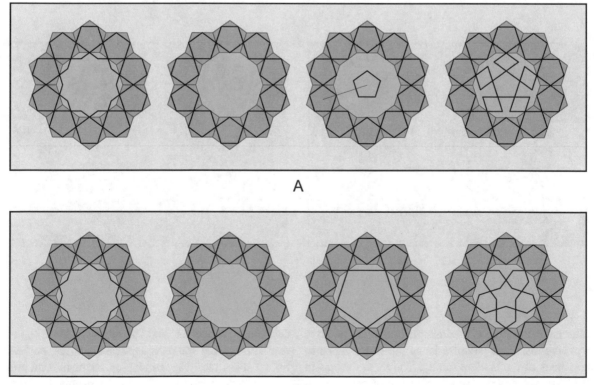

A

B

Fig. 224

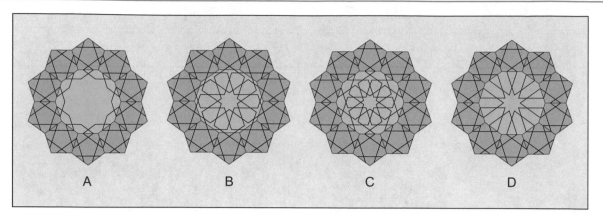

Fig. 225

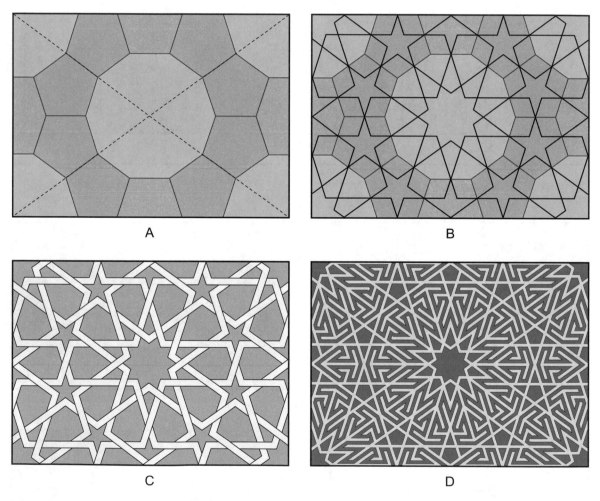

Fig. 226

As referenced, there are a considerable number of additive and subtractive historical modifications that were applied to the pattern lines of the *median* family. The three designs in Fig. 227 illustrate a progressive modification of the standard *median* pattern created from the rhombic tessellation of

decagons, pentagons and barrel hexagons. Figure 227a demonstrates the standard application of the *median* pattern lines to this underlying tessellation. As mentioned previously, this is characterized by 72° angular openings of the pattern lines at each underlying polygonal midpoint. Figure 227b

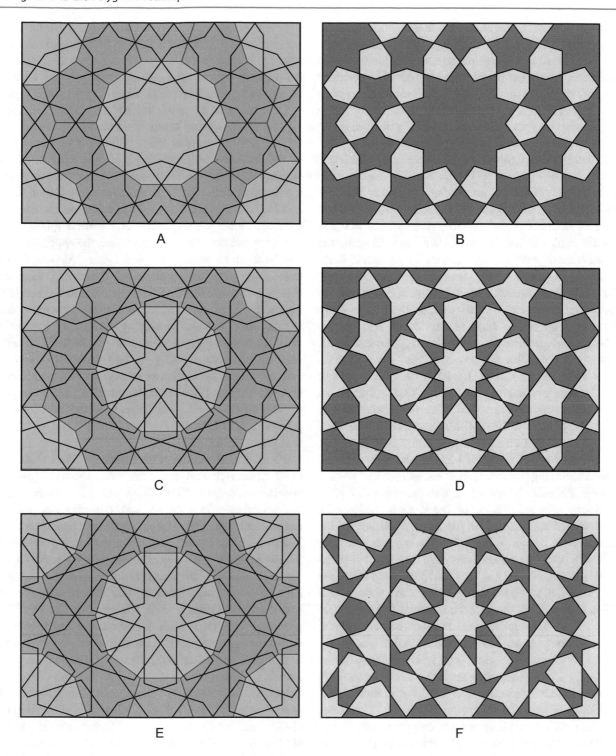

Fig. 227

represents the commonly found tiled interpretation of this design, including in the exterior cut-tile mosaic of the Bibi Khanum mosque in Samarkand (1399-1405). Widened line interpretations of this design were also well known and include the main façade of the Sultan al-Ghuri *madrasa* in Cairo (1503-05). This standard line treatment has the open center within the ten-pointed stars. Figures 227c and d

illustrate the historical variation to the ten-pointed stars that is relatively common among *median* patterns [Fig. 223d]. This example arbitrarily applies this modification to alternating decagonal regions, thereby creating a ten-pointed star rosette at just the center of the design, and not at the corners. Historical examples with this alternating modification are uncommon, but this was used by Qara Qoyunlu artists in a cut-tile mosaic border at the Imamzada Darb-i Imam in Isfahan (1453). Figures 227e and f apply this same modification to each underlying decagonal region. This has the effect of eliminating two of the points from each of the five-pointed stars in the original design. This modification maintains only a tenuous association with the underlying tessellation, effectively transforming the standard *median* pattern into an *acute* pattern that is well known to the historical record. Indeed, this same design can be produced more efficiently from either an underlying tessellation of edge-to-edge decagons [Fig. 232f] or underlying decagons that are separated by barrel hexagons [Fig. 232h]. This is a good example of both the occasional interchangeability between pattern families, and the methodological uncertainty that is often associated with specific geometric designs. Early examples of this design are found on the exterior façade of the Ildegizid mausoleum of Mu'mine Khatun in Nakhichevan, Azerbaijan (1186), and on the Zangid entry door at the Awn al-Din Meshhad in Mosul, Iraq (1248). This design was particularly popular among Mamluk artists in Cairo, and examples are found at the Sultan Qala'un funerary complex in Cairo (1284-85), the Amir Sanqur al-Sa'di funerary complex in Cairo (1315), the Hasan Sadaqah mausoleum in Cairo (1315-21), the Sultan Qaytbay funerary complex in Cairo (1472-74), the Qadi Abu Bakr Muzhir complex in Cairo (1479-80), and the Amir Azbek al-Yusufi complex in Cairo (1494-95) [Photograph 46]. The process of removing two points from each of the five-pointed stars in Fig. 227e can be replicated with the removal of other combinations of points. Each of the designs in Fig. 228 truncates two of the points from the five-pointed stars of the standard *median* pattern in Fig. 227a, thereby removing two of the five 72° crossing pattern lines from the midpoints of the underlying pentagonal edges. This modification in Fig. 228a is the same as that presented in Fig. 227e, with a widened line treatment in Fig. 228b. Figure 228c truncates two alternative points from the five-pointed stars: those associated with the edge-to-edge pentagons, but not the shared edges of the pentagons and barrel hexagons. The resulting interweaving pattern in Fig. 228d was used in the late Abbasid mausoleum of 'Umar al-Suhrawardi in Baghdad (early thirteenth century). Figures 228e and f remove yet another pair of the 72° crossing pattern lines. While certainly acceptable to the aesthetics of this tradition, somewhat surprisingly, this design (by author) does not appear to be historical.

Arbitrary modifications to fivefold *obtuse* patterns are also relatively common. Figure 229a illustrates the standard *obtuse* pattern along with its underlying generative tessellation of decagons, pentagons and barrel hexagons. Figure 229b shows an interweaving line version of this very-well-known standard pattern. This design is 1 of the 3 fivefold examples that were used in the Seljuk northeast dome chamber in the Friday Mosque at Isfahan (1088-89) [Photograph

21], and is, as such, one of the earliest extant fivefold patterns known to the historical record. This example from Isfahan is the earliest known design to include the ten-pointed star rosettes within each of the underlying decagonal regions [Fig. 221]. Figures 229c and d represent a historical example with the arbitrary additive treatment applied to alternating ten-pointed stars. This modification introduces a pentagonal motif within the ten-pointed stars in the fashion illustrated in Fig. 224a. This design with alternating decagonal infill was used in the frontispiece of a Mamluk Quran (1369) commissioned by Sultan Sha'ban (r. 1363-77). Figures 229e and f represent the classic *obtuse* pattern with the same additive modification applied to each of the ten-pointed stars. As mentioned, the application of the modification to each ten-pointed star has the effect of transforming the pattern into a field pattern. This modified pattern is found at the Haci Kilic *madrasa* in Kayseri (1275). Figure 230 shows a variation of the classic *obtuse* design created from this same underlying tessellation. Figures 230b and c illustrate the application of fivefold rotational features into the underlying pentagons and barrel hexagons. This example shares the aesthetics of the more common fourfold swastika designs [Figs. 150a, b and 157a], and examples of this design are found at the Friday Mosque at Yezd (1324) [Photograph 80], in the Topkapi Scroll,[20] and at the Imam mosque in Isfahan (1611-38).

Figure 231 illustrates modifications to the standard *two-point* pattern created from the same rhombic underlying tessellation. Figure 231a shows the relationship between the underlying tessellation and the standard *two-point* design, and Fig. 231b shows a widened interweaving line treatment of this pattern. In this illustration, the widths of the interweaving lines are maximized so that the acute angles of the superimposed kite elements touch one another. Figure 231c arbitrarily introduces the star rosette into alternating underlying decagonal regions, whereas Fig. 231d applies the geometric rosette to each decagonal region. Figure 231d shows the most frequently encountered version of the classic fivefold design, and an early example is found in the tympanum of an arch over the entry of the Gunbad-i Alayvian in Hamadan[21] (1150-1200) [Photograph 22]. Later Mamluk examples include the Sultan Qaytbay Sabil-Kuttab in Cairo (1479), and the Qadi Abu Bakr Muzhir complex in Cairo (1479-80). An especially attractive Mamluk use of this design is from a stone mosaic panel in the Metropolitan

---

[20] Necipoğlu (1995), diagram no. 8.

[21] Based upon stylistic features of the architectural form, Herzfeld and Wilber both identified the Gunbad-i Alayvian as early Ilkhanid. However, the architectural ornament is more characteristic of the Seljuks.
– Herzfeld (1922), 186–199.
– Wilber (1955), 151–152.

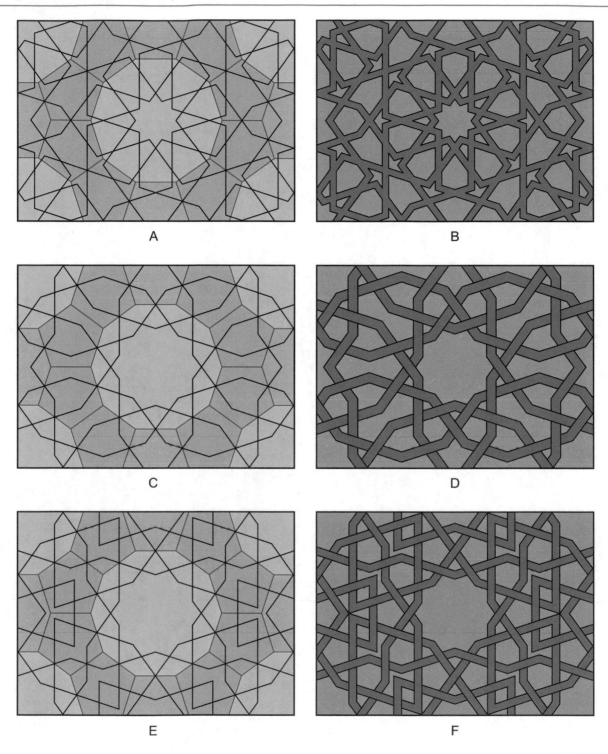

A   B

C   D

E   F

Fig. 228

Museum of Art in New York City[22] [Photograph 47]. Figures 231e and f represent a modification that changes the angle of

declination of the pattern lines within the underlying decagons to be less acute than the standard design in Figure 231a. This design modification was open to stylistic variation [Fig. 225d], and the particular proportions of this illustration were used by artists during the late Abbasid period at the Mustansiriya in Baghdad (1227-34).

---

[22] Metropolitan Museum of Art, New York: gift of the Hagop Kevorkian Fund, 1970; dado panel [Egypt] (1970.327.8).

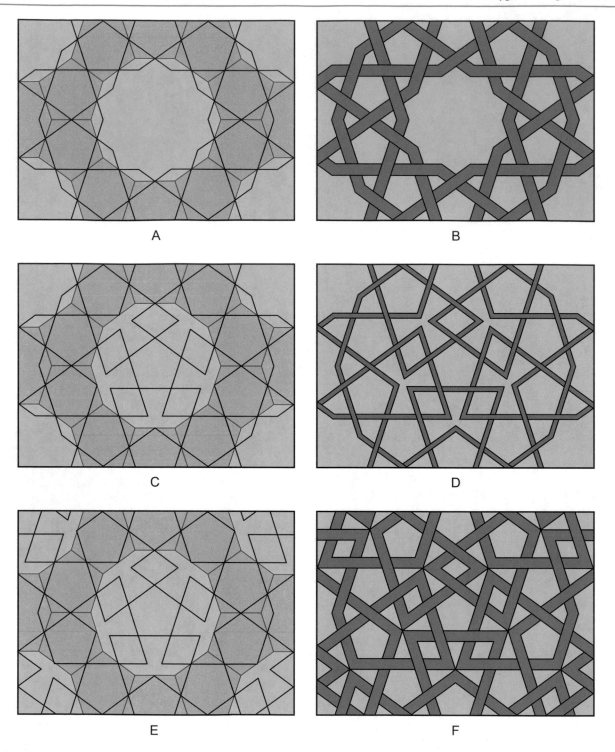

Fig. 229

3.1.7   Fivefold System: Wide Rhombic Repeat Unit

Patterns created from the *fivefold system* make use of several varieties of repetitive stratagem. As with the previous examples of the standard fivefold designs in each of the four pattern families, the most common repeat unit is the wide rhombus with 72° and 108° included angles. The most basic underlying tessellation that employs this repeat unit is comprised of just decagons and concave hexagons. Figure

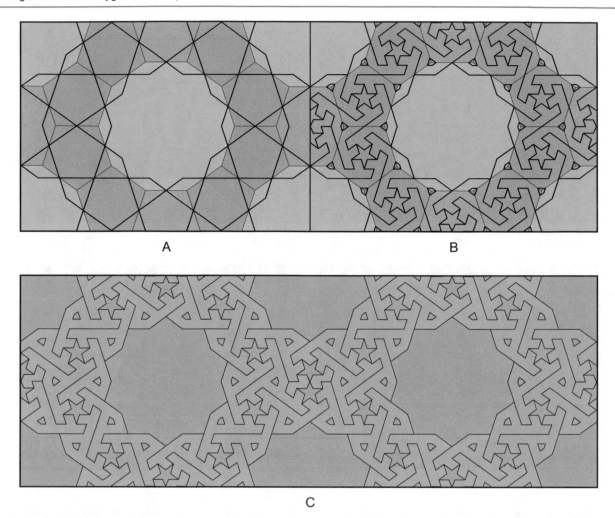

Fig. 230

232 illustrates the designs produced from this underlying tessellation in each of the four pattern families. Figure 232a shows an *acute* pattern that is not historical. The two small elements within the concave hexagon are out of balance with the rest of the design. The widened line version in Fig. 232b, while interesting, does not comport with the aesthetics of the *fivefold system*. Figure 232c employs the 72° angular openings of the *median* family. However, the visually dominant features of the resulting design in Fig. 232d are the 108° angles of the pentagons and concave octagonal shield shapes that are characteristic of the *obtuse* family. Indeed, Fig. 232e demonstrates how this same design—the classic *obtuse* pattern—is created from the underlying tessellation of decagons, pentagons, and barrel hexagons as per Fig. 229a. As demonstrated earlier, the well-known *acute* pattern in Fig. 232g can be produced with either 108° angular openings, as per Fig. 232f, or with 36° angular openings in the alternative underlying tessellation in Fig. 232h. Interestingly, and as demonstrated, this can also be created as an arbitrary modification of the standard

median pattern [Fig. 223]. The derivation in Fig. 232f requires an additive geometric rosette, whereas that of Fig. 232h produces this feature automatically from each midpoint of the alternative underlying polygonal edges. (Historical examples of this design are provided in the text associated with Fig. 227f.) The classic *two-point* pattern in Fig. 232j can also be made from the underlying tessellation of just decagons and concave hexagons, although the pattern lines have more contact points with the underlying tessellation in Fig. 232k. (Historical examples of this design are provided in the text associated with Fig. 231a.)

The five designs in Fig. 233 repeat upon the same rhombic grid. As illustrated, the two upper patterns can be created from either of two reciprocal underlying polygonal tessellations, and follow the dual qualities demonstrated in Fig. 200. The underlying tessellation in Fig. 233a is comprised of decagons, long hexagons, and concave hexagons, and the applied pattern lines have 72° angular openings placed at the midpoints of each polygonal edge. This angular opening is ordinarily associated with the

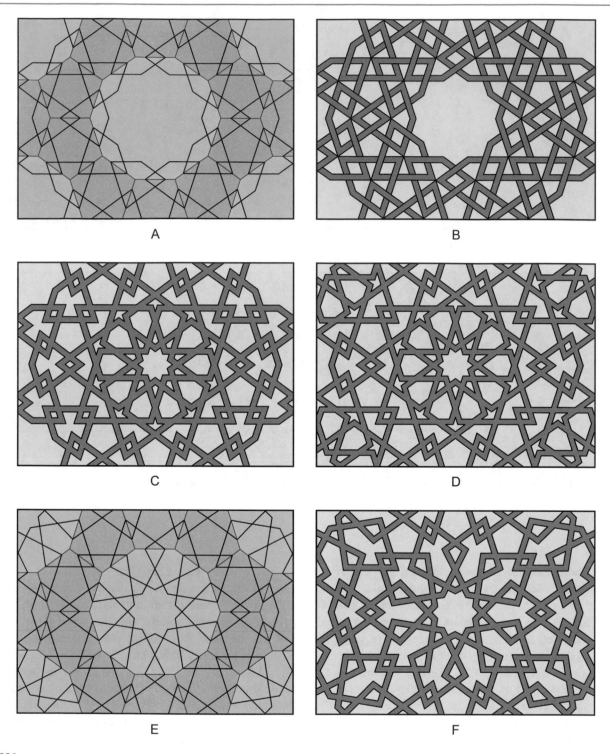

Fig. 231

median pattern family. However, the visual character of the resulting design in Fig. 233b is identifiably *obtuse*, as per the 108° angular opening of the pattern lines applied to the underlying decagons, pentagons, barrel hexagons, and thin rhombi in Fig. 233c. This is a very-well-known design, and two relatively early examples include a late Abbasid carved

stucco arch spandrel at the Palace of the Qal'a in Baghdad (c. 1220), and a Seljuk Sultanate of Rum courtyard portal at the Agzikara Han in Turkey (1242-43). Mamluk examples include the stone mosaic in the *mihrab* niche of the Sultan Qala'un funerary complex in Cairo (1284-85), and a frontispiece from the Quran commissioned by Sultan Faraj ibn

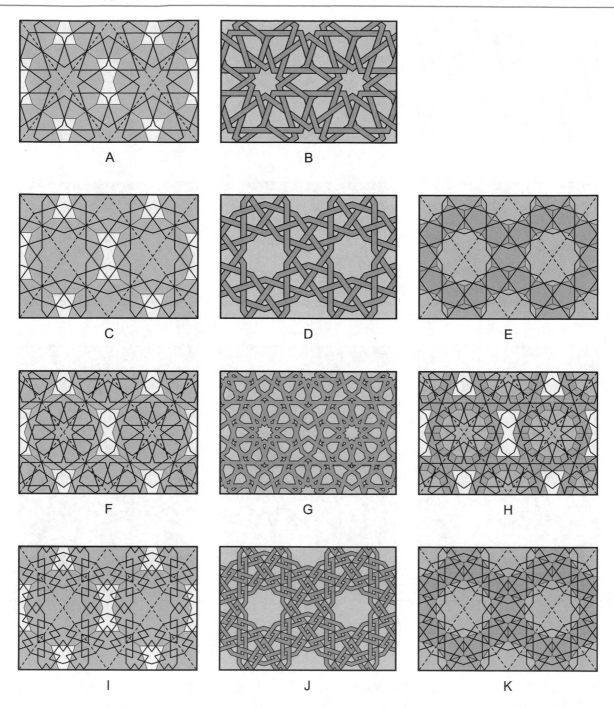

Fig. 232

[23] British Library, London, BL Or. MS 848, ff. 1v-2.

Barquq[23] (1399-1411) [Photograph 48]. A Muzaffarid cut-tile mosaic example is found in the main portico of the Friday Mosque at Kerman (1349). These same two reciprocal underlying tessellations can also be used to create the *two-point* pattern in Fig. 233e. The angles of declination of

the pattern lines located at each underlying polygonal edge in Fig. 233d have 72° while those of Fig. 233f have 36°. This *two-point* design was used by Mamluk artists in the *mihrab* at the Sultan Qansuh al-Ghuri complex in Cairo (1503-05). As demonstrated in Fig. 197, the arrangement of six underlying pentagons surrounding the thin rhombi in Fig. 233c and f, while well suited to both the *obtuse* and *two-point* families, do not generate acceptable pattern features within

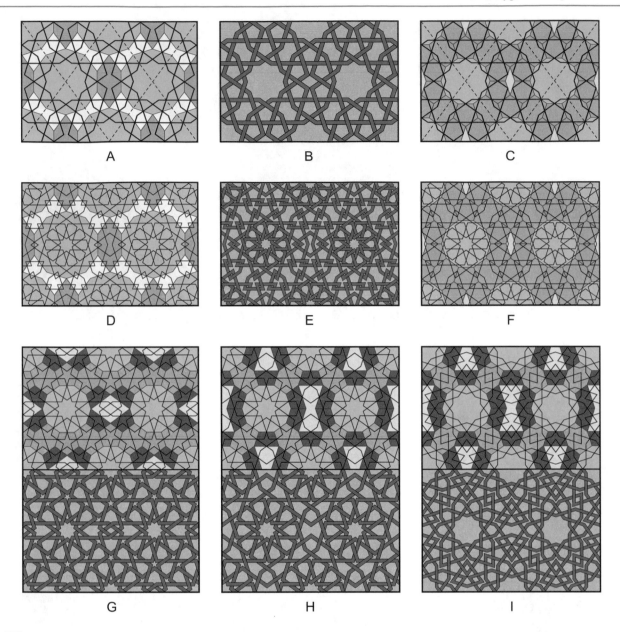

Fig. 233

the *acute* and *median* families. Figure 233g illustrates an *acute* pattern (by author) created from the modified underlying tessellation that replaces four of the six pentagons with trapezoids, thereby creating a large central wide rhombus that is contiguous with the long edges of the trapezoids. The resulting *acute* pattern does not appear to have been used historically, which is surprising in that this method of modifying the underlying tessellation was relatively well known [e.g. Fig. 245d], and this rather basic example results in a very attractive pattern. Figure 233h shows an *acute* design that employs the modification to the underlying tessellation that replaces all six of the pentagons that surround

the thin rhombi with trapezoids, leaving a large concave hexagon as an interstice region at the center of the repeat. As mentioned previously, this well-known pattern can also be created from the underlying tessellation of just edge-to-edge decagons and concave hexagons shown in Fig. 232f, as well as through the modification of *median* pattern lines as per Fig. 223f. (Historical examples of this fine design are provided in the text associated with Fig. 227e.) The *median* pattern in Fig. 233i is also created from the same underlying tessellation as the *acute* pattern in Fig. 233h. This design employs two points upon each of the long edges of the underlying tessellation, and while a *median* pattern, it has

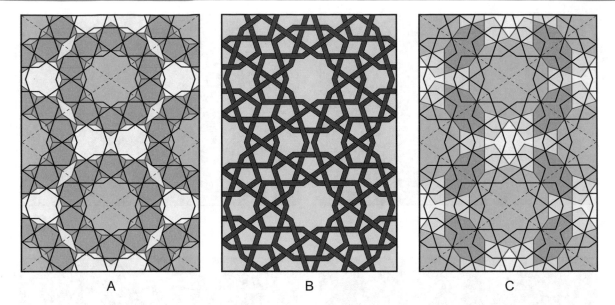

A   B   C

Fig. 234

distinctive characteristics of the *two-point* family. This fine example was used by Timurid artists at the Ghiyathiyya *madrasa* in Khargird, Iran (1438-44).

Figure 234 illustrates an *obtuse* design from the Seljuk Sultanate of Rum that is found at the Gök *madrasa* in Tokat (1275-80). This design repeats on a rhombic grid and can also be created from two different underlying tessellations. Figure 234a shows the underlying tessellation comprised of decagons surrounded by a ring of ten pentagons, with thin rhombi and large irregular decagonal interstice regions. The 108° angular opening of the pattern lines at each midpoint of the polygonal edge creates the pentagons contained within underlying pentagons that characterize the *obtuse* family. Figure 234c demonstrates how this same *obtuse* design can be produced from an alternative tessellation comprised of decagons, long hexagons, half-concave hexagons, and interstice regions with applied 72° angular openings. The visual appeal of Fig. 234b is augmented by the influence of the interstice regions upon the completed design. Figure 235 illustrates two versions of another *obtuse* design that repeats upon a rhombic grid and can be created from either of the two underlying tessellations. Figure 235c shows a Mamluk interlocking tiled version from a Quranic frontispiece (1313) illuminated by Aydoğdu bin Abdullah al-Badri and Ali bin Muhammad al-Rassam.[24] This is a relatively rare example of a geometric design produced by known artists. Figure 235d shows a Seljuk Sultanate of Rum interweaving version from the Huand Hatun mosque complex in Kayseri (1237) in which the pattern lines have been widened to the maximum

amount, allowing the lines to touch corner to corner with other widened lines. Figure 235a demonstrates the relationship between this pattern to the underlying generative tessellation of decagons, pentagons, barrel hexagons, and thin rhombi, whereas Fig. 235b shows the same design created from underlying decagons, long hexagons, and concave hexagons. Figure 236 shows a *median* pattern that repeats upon the same rhombic grid and is created from an atypical underlying tessellation comprised of decagons, pentagons, wide rhombi, and triangular elements that are half of a wide rhombus. Unusually, the edge length of the decagons is equal to the newly created long edge of the triangular modules. Typically, the edges of the underlying pentagons that surround the decagon are coincident with the decagonal edges. In this case the decagons and pentagons meet at their vertices. The experimental spirit exhibited in this pattern provided the *fivefold system*, and indeed other design systems, with increased originality and diversity. This design was used by Kartid artists at the Shamsiya *madrasa* in Yazd (1329-30), as well as by Timurid artists at the Ulugh Beg *madrasa* in Samarkand (1417-20). The *median* pattern in Fig. 237 also orientates the underlying pentagons such that their vertices face the centers of the ten-pointed stars. The underlying tessellation of this design is comprised of pentagons, long hexagons, wide rhombi, atypical octagons, and atypical ten-pointed star polygons. The underlying octagons (pink) in this figure can be substituted with a combination of two wide rhombi, four half-concave hexagons, and a central thin rhombi as per the underlying tessellation in Fig. 209a. The 10-pointed star interstice region at each corner of the repeat unit is essentially the same as the half wide rhombi and decagons from Fig. 236, except for the treatment of the applied pattern lines. This design is from the Seljuk

---

[24] This Quran is in the collection of the Museum of Turkish and Islamic Arts; Sultanahmet, Istanbul, Turkey: Museum Inventory Number 450.

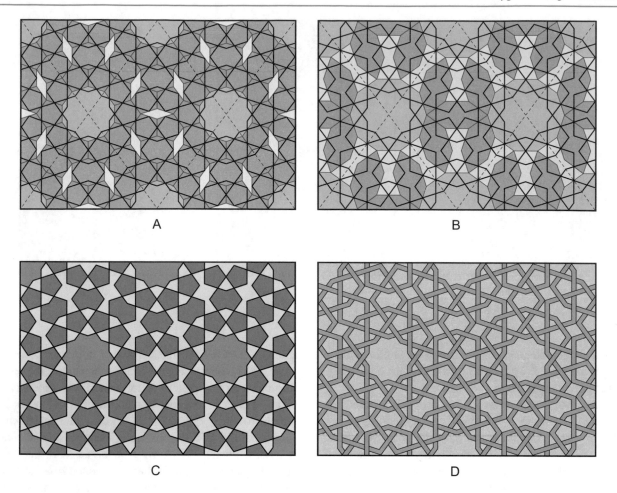

A   B

C   D

Fig. 235

Sultanate of Rum and is found at the Sultan Han in Kayseri (1232-36) [Photograph 42].

There are two historical designs created from the *fivefold system* that repeat upon the rhombic grid with 72° and 108° angles that have particularly large amounts of geometric information within their respective repeats. The underlying tessellations of both of these examples share the same basic structure,[25] but differ in the application of secondary underlying polygonal modules, and infill treatments of the underlying decagons. Both of these examples are the work of Seljuk artists: one from the Seljuk Sultanate of Rum, and the other from the Great Seljuks in Persia; and both are masterpieces of geometric design. The *obtuse* pattern in Fig. 238 is from the cut-tile mosaic ornament of the Karatay *madrasa* in Konya (1251-52). Figure 238a illustrates the concept behind the underlying generative tessellation. This is based upon the well-known rhombic tessellation of

decagons separated by two edge-to-edge pentagons, with barrel hexagons completing the coverage (black) [Fig. 200c]. Edge-to-edge underlying decagons are placed at each vertex of this initial polygonal structure. This produces a distinctive ring of ten decagons at the vertices of the repetitive rhombic grid. Figure 238b applies further polygonal infill into the central regions of the initial pentagons and barrel hexagons, as well as into arbitrarily selected secondary decagons. It is worth noting that the large interstice region at each repetitive vertex that was not provided with further polygonal infill was a purposeful exclusion. Figure 238c applies the pattern lines into the underlying tessellation in Fig. 238b. There are two additive features to this pattern line application: the ten-pointed stars and rosettes at the vertices of the repetitive grid, and the underlying decagons that remained unfilled in Fig. 238b. The central motive of the former introduces the rosette and ten darts in the same basic formula as demonstrated in Fig. 222. The treatment of the pattern lines within the open underlying decagons follows the Seljuk practice detailed in Fig. 224a. The end result in Fig. 238d is an exceptionally well-worked-out geometric design of considerable

---

[25] The author is indebted to both Emil Makovicky and Jean-Marc Castéra for independently discovering the geometric similarity between these 2 fivefold patterns. See Castéra (2016).

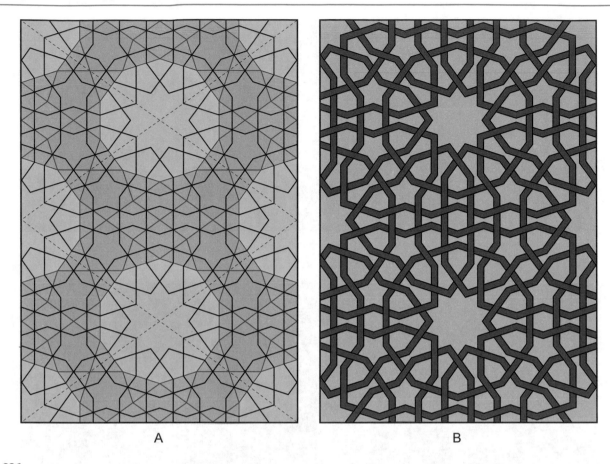

A   B

Fig. 236

complexity. Figure 239 illustrates the repetitive layout and underlying generative tessellation of the remarkable raised brick design on the façade of the Gunbad-i Qabud in Maragha (1196-97) [Photograph 24]. This tomb tower is decagonal in plan, and the geometric pattern created from this underlying tessellation is applied continuously across nine of the ten sides: the tenth side being the portico. What is more, in a visual tour de force, and feat of artistic dexterity, the pattern flows uninterrupted across the ten engaged columns at each corner of the tomb tower. Like the example from the Karatay *madrasa* in Konya, Fig. 239a shows how the starting point of this design is the identical placement of edge-to-edge decagons at each vertex of the standard tessellation of decagons, pentagons, and barrel hexagons. Figure 239b fills in this decagonal network with long hexagons, concave hexagons, and wide rhombi from the *fivefold system*. Three of the secondary decagons located on the vertices of the primary barrel hexagon are not provided with secondary infill, nor are the decagons that are placed upon the vertices of the rhombic repetitive grid at the base of the tessellation in Fig. 239c. Keeping the three decagons unfilled provides this tessellation with reflection symmetry along the long vertical axis of each rhombic repeat unit, but not along the shorter horizontal axis. This type of break in symmetry is

very unusual within this tradition. The highlighted lower region in Fig. 239c represents the portion of this tessellation that was used in Maragha. It is noteworthy that the artist who devised this design chose to cut the pattern off above the horizontal line of symmetry. This is another unusual form of symmetry break. Figure 239d illustrates the underlying tessellation in four of the nine uninterrupted linear repetitive cells that span nine of the ten sides of this tomb tower. Whereas the artist kept the set of decagons that rest upon the repetitive rhombic vertices along the lower edge of the design, the similarly placed decagons in the upper set of repetitive vertices have been filled in with long hexagons and concave hexagons. This is yet another break in symmetry. With this further infill of these decagons, the rhombic repeat in Fig. 239c no longer has translation symmetry, and to achieve translation symmetry with the tessellation in Fig. 239d one would have to mirror the tessellation upon the upper horizontal line of symmetry—thus achieving a rectangular repeat unit with clear dodecagons at each corner. Figure 239e shows the underlying tessellation throughout all nine of the ten sides that received this pattern. The grey zones represent the regions of the design that wrap around the ten engaged columns at each corner of the tower. These regions are half circles in plan. Figure 240a illustrates the

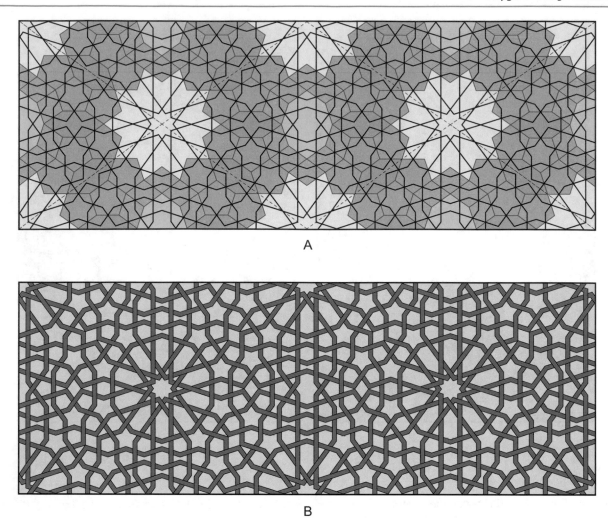

Fig. 237

median pattern as applied to 4/9 of the underlying generative tessellation of the design from the Gunbad-i Qabud. Each of the unfilled underlying decagons from Fig. 239d has been provided with the relatively common Seljuk fivefold rotational motif [Fig. 224a]. This arbitrary additive modification affectively transforms this to a field pattern, and the fact that the fivefold symmetry of these modifications does not align with the vertical reflective symmetry of the multiple 1/9 divisions is another break in symmetry. In fact, this entire geometric construction can be thought of as an exercise in symmetry breaking. Figure 240b illustrates the same 4/9 segment of this design without the underlying generative tessellation. It is worth noting that the design from the Gunbad-i Qabud includes an additive secondary level of pattern in the background of the primary design [Fig. 67] [Photograph 24]. This provides a further level of complexity and visual interest, and is an early outlier of the dual-level design aesthetic that developed in the same general region during the fifteenth century.

3.1.8 Fivefold System: Thin Rhombic Repeat Unit

Although less common, the diverse types of repeat units employed within the *fivefold system* also include the rhombus with 36° and 144° included angles [Fig. 5b]. Figure 241b illustrates an *obtuse* pattern that was used by artists working during the Seljuk Sultanate of Rum at the Muzaffar Buruciya *madrasa* in Sivas (1271-72), as well as by Mamluk artists at the Amir Altinbugha al-Maridani mosque in Cairo (1337-39). Figure 241a shows how this design can be produced from an underlying tessellation of decagons, pentagons, barrel hexagons, and thin rhombi. This derivation employs the 108° angular opening of the crossing pattern lines associated with the *obtuse* family. Figure 241c demonstrates how this same design can be created from a tessellation of decagons, long hexagons, and concave hexagons, with 72° angular openings. Figure 242 illustrates an *acute* design with atypical irregular pentagons incorporated into the underlying

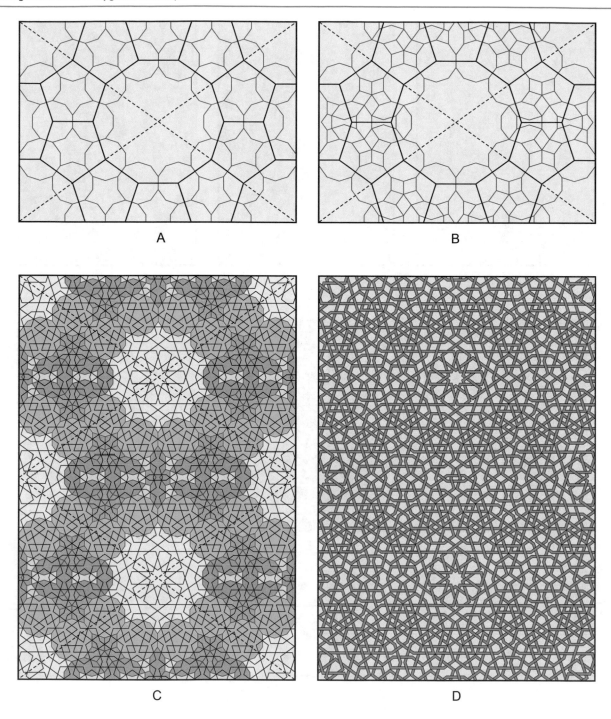

Fig. 238

generative tessellation. The earliest known use of these unusual underlying pentagons, and the distinctive pattern motif they create within the *acute* family, is from one of the recessed arches in the upper portion of the northeast dome chamber in the Friday Mosque at Isfahan (1088-89) [Fig. 261] [Photograph 25]. The two nonconforming edge lengths of these pentagons have a φ proportional relationship to the standard edge length, and share the same longer edge

length as the adjacent scaled-up wide rhombi. The extension of the *acute* pattern lines from the irregular pentagons into the adjacent wide rhombi creates crossing pattern lines with 72° angular openings. This change from 36° angular openings to 72° angular openings constitutes a regional change from the *acute* family to the *median* family. Transitions between pattern families within a single design are unusual, and invariably rely on manipulations in the

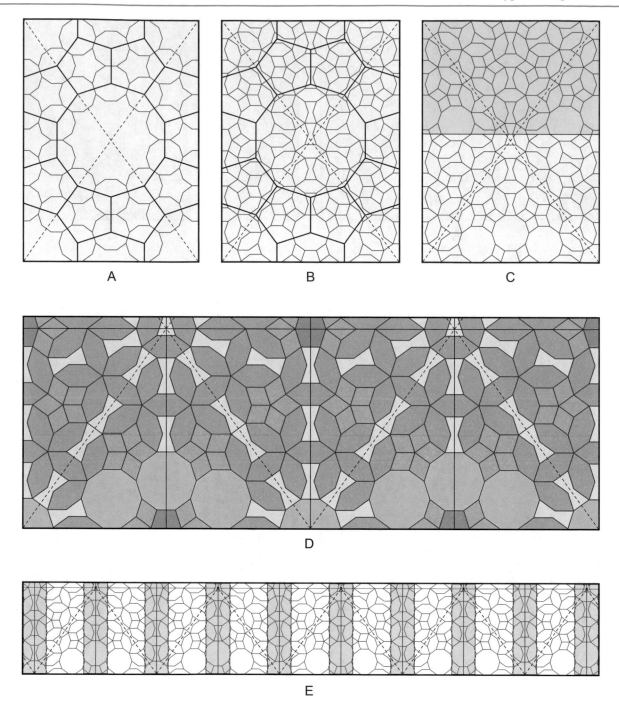

Fig. 239

scale of the underlying polygonal modules [Figs. 269 and 270]. This design is relatively well known, with early examples being produced contemporaneously by artists during the Seljuk Sultanate of Rum at the Huand Hatun *madrasa* in Kayseri (1237) and by Kartid artist at the Turbat-i Shaykh Ahmad-i Jam in Torbat-i Jam, northeastern Iran (1236). Later Mamluk examples include the *minbar* of the *khanqah* and mosque of Sultan al-Ashraf Barsbay (1432-

33), and a *minbar* door in the collection of the Victoria and Albert Museum in London.[26] The 36° acute angles of this repeat unit allow the rhombus to become a 1/10 segment of a

---

[26] This *minbar* door panel is a nineteenth century copy, produced in Egypt, of a Mamluk original of uncertain origin. Victoria and Albert Museum, London, England, collection code FWK.

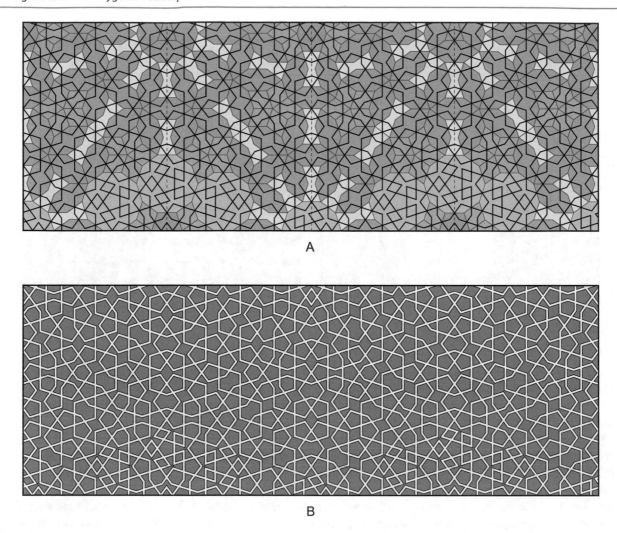

A

B

Fig. 240

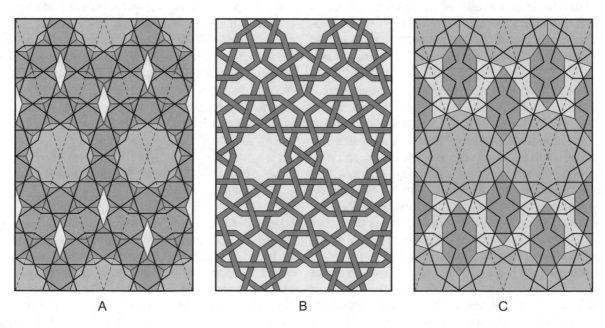

A   B   C

Fig. 241

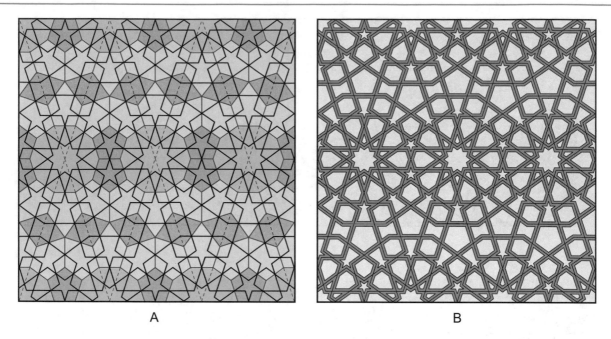

A   B

Fig. 242

decagon when cut in half between the two vertices at the obtuse included angles. This 1/10 triangle can then be provided with tenfold rotation symmetry to create a decagonal radial design. The pierced marble radial design on the sides of the Ottoman *minbar* at the Selimiya complex in Edirne, Turkey (1568-74), employs a design that rotates and copies the half repeat unit in Fig. 242 in just such fashion. The patterns in Fig. 243 employ an underlying tessellation with contiguous truncated decagons placed upon each vertex of the rhombic repeat unit. This arrangement can also be achieved through overlapping the decagons [Fig. 191b]. This creates a series of continuous linear bands comprised of truncated decagons that run through the vertices of this rhombic grid. The design demonstrated in Fig. 243a is a *two-point* design that is unusual in that it utilizes double sets of 72° crossing pattern lines placed at two locations on each of the edges of the underlying pentagons. This provides for 54° angles of declination rather than either 72° or 36° that are far more common to *two-point* patterns created from the *fivefold system*. This feature creates the large and small five-pointed stars that are characteristic of the *median* family at each of the underlying pentagons. The precise placement of the 72° crossing pattern lines on the underlying pentagonal edges is so contrived as to allow for the selected extended lines of the smaller five-pointed stars to intersect with the midpoints of the long edges of the truncated decagons. This in turn allows for the creation of the ten-pointed star rosette at the center of each truncated decagon. Figure 243b demonstrates the aesthetic success of this highly unusual and ingenious design. This pattern was used as a border in the late Abbasid main entry portal of the

Mustansiriyah *madrasa* in Baghdad (1227-34). It is worth noting that the generative schema of a much later Ottoman design from a door of the Sultan Bayezid II Kulliyesi in Istanbul (1501-06) is remarkably similar to this design from Baghdad [Fig. 270]. Figure 243d illustrates a Mamluk *two-point* design that is produced from the same underlying tessellation as the *two-point* design from the Mustansiriyah *madrasa* in Baghdad. Figure 243c demonstrates the application of the standard *two-point* pattern lines with 36° angles of declination. This *two-point* design is from the Amir Qijmas al-Ishaqi mosque in Cairo (1479-81). Figure 244 illustrates a more complex Mamluk *two-point* pattern that also repeats upon the same acute rhombic grid. This exceptional example is from a *minbar* (1468-96) commissioned by Sultan Qaytbay that is in the collection of the Victoria and Albert museum in London. This design is created from an underlying tessellation of decagons, pentagons, barrel hexagons, and thin rhombi. Star rosettes have been arbitrarily added into the ten-pointed stars located at the vertices of the rhombic repeat.

3.1.9   Fivefold System: Rectangular Repeat Units

Many patterns created from the *fivefold system* repeat upon a rectangular grid. Figure 245 features six such designs. Figure 245a shows an *obtuse* pattern that can be easily created from the rectangular tessellation of decagons, pentagons, barrel hexagons, and thin rhombi. The first known use of this popular pattern is the work of Qarakhanid artists, and is found in the rear portico of the Maghak-i Attari mosque in

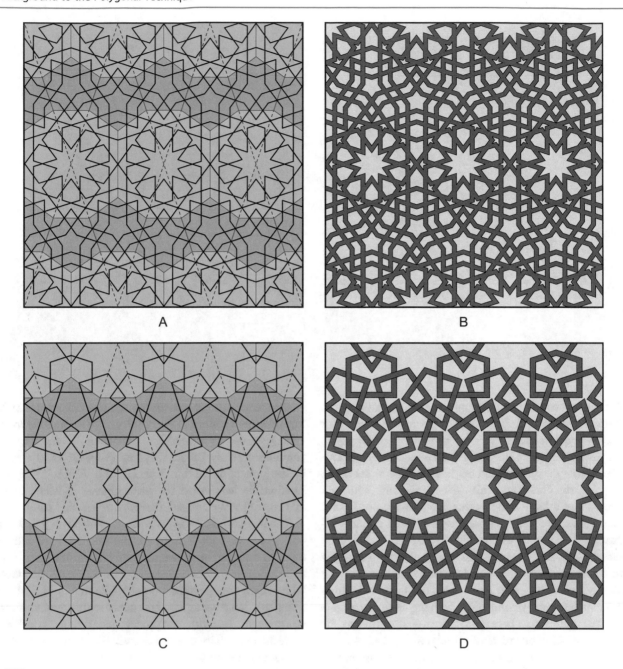

A

B

C

D

Fig. 243

Bukhara (1178-79). Later examples include the Sultan Han in Aksary, Turkey (1229); the Friday Mosque at Ashtarjan, Iran (1315-16); the Shah-i Zinda funerary complex in Samarkand, Uzbekistan (1386); and the Abdulla Ansari complex in Gazargah, Afghanistan (1425-27). Figure 245b shows a *two-point* pattern from the Mughal mausoleum of Humayun in Delhi (1562-72) [Photograph 79]. Like the illustration, this Mughal example keeps the regions within the ten-pointed stars open. By contrast, the *two-point* pattern in Fig. 245c incorporates the arbitrary design modification that places ten-pointed star rosettes within the central underlying decagonal regions. This particular variety of star

rosette modification is relatively uncommon [Fig. 225c]. This variation is a Mamluk design from the *mihrab* of the Qadi Abu Bakr Muzhir complex in Cairo (1479-80). As with other designs created from the *fivefold system*, each of the designs in Fig. 245a–c can also be created from an alternative underlying tessellation, in this case comprised of decagons, long hexagons, and concave hexagons (not shown). As shown, the underlying thin rhombi surrounded by six pentagons in Fig. 245a–c work exceedingly well with the *obtuse* and *two-point* pattern families. However, this arrangement of underlying polygons does not work nicely with *acute* and *median* patterns. The *acute* pattern in Fig.

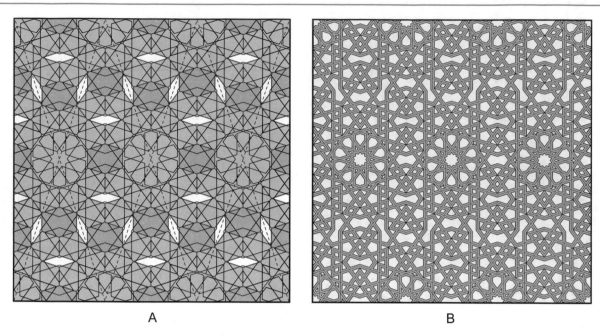

A   B

Fig. 244

245d is created from an underlying tessellation that replaces the thin rhombus and four of the pentagons with four trapezoids and a central large wide rhombus. This design is from a *zillij* panel at the Bu 'Inaniyya *madrasa* in Fez (1350-55). The *acute* design (by author) in Fig. 245e replaces the underlying rhombus and all six pentagons with six trapezoids and a large concave hexagon. While not known to the historical record, this is very similar to several historical examples that make use of this alteration of the underlying tessellation, for example, that of Fig. 233h. Figure 245f shows a *median* design produced from the same underlying tessellation as in Fig. 245e. This is also an ahistorical design (by author) that is fully compatible with the aesthetics of this ornamental tradition. These two types of alteration to the underlying tessellation are demonstrated in Fig. 198. As with many *obtuse* designs created from the *fivefold system*, the *obtuse* design in Fig. 246 can be created from two different underlying tessellations. Figure 246a employs an underlying tessellation comprised of decagons, pentagons, barrel hexagons, and thin rhombi, while that of Fig. 246c has decagons, long hexagons, and concave hexagons. This is an Ilkhanid pattern from the mausoleum of Uljaytu in Sultaniya, Iran (1307-13). Figure 247b shows a rectangular *obtuse* pattern produced by artists during the Seljuk Sultanate of Rum. Once again, this design can be created from two different underlying tessellations. Figure 247a employs an underlying tessellation comprised of decagons, pentagons, barrel hexagons, and thin rhombi, while that of Fig. 247c has decagons, long hexagons, and concave hexagons. This pattern is found at the Sirçali *madrasa* in Konya (1242-45), as

well as at the Ahi Serafettin mosque in Ankara (1289-90). The *two-point* pattern in Fig. 248b was used on three roughly contemporaneous Mamluk wooden *minbars*: one commissioned by Sultan Qaytbey (1468-96) and currently in the collection of the Victoria and Albert Museum in London; the *minbar* of the Amir Qijmas al-Ishaqi mosque in Cairo (1479-81); and the *minbar* of the Amir Azbak al-Yusufi complex in Cairo (1494-95) [Photograph 46]. This *two-point* design can be created from either of the two underlying tessellations. Figure 248a employs an underlying tessellation comprised of decagons, pentagons, barrel hexagons, and thin rhombi, while that of Fig. 248c has decagons, long hexagons, and concave hexagons. As with other examples with reciprocal underlying tessellations, these are essentially duals of one another. The pattern within the underlying decagonal regions of the rectangular repeat employs ten-pointed star rosettes [Fig. 225b]. The *obtuse* pattern in Fig. 249b has an unusually large amount of geometric information within each rectangular repeat unit. This was created during the Seljuk Sultanate of Rum for the Yusuf ben Yakub *madrasa* in Cay, Turkey (1278). Figure 249a shows the construction from an underlying tessellation comprised of decagons, pentagons, barrel hexagons, and thin rhombi, while Fig. 249c employs decagons, long hexagons, concave hexagons, conjoined long hexagons, and half-concave hexagons. The pattern lines at the center of the conjoined long hexagons do not rest conveniently upon the edges of this underlying polygon. However, the pattern lines in this same region of the underlying tessellation in Fig. 249a sit precisely upon the midpoints of the underlying polygonal

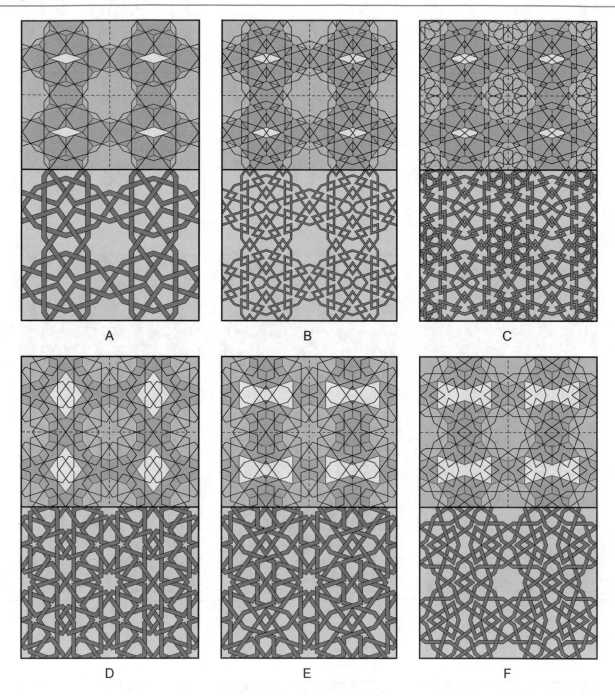

Fig. 245

edges. It therefore appears more likely that the artist responsible for this design employed the construction illustrated in Fig. 249a. The *acute* design in Fig. 250b is from a Mamluk metal window grille at the al-Azhar mosque in Cairo. This design is created from an underlying tessellation of decagons, pentagons, barrel hexagons, and long hexagons. This arrangement places decagons at each vertex of the rectangular repeat, as well as at the center of each repeat. As mentioned previously, not all polygonal modules within

the *fivefold system* work nicely in each of the four pattern families. As highlighted previously, a case in point is the long hexagon within the *acute* family [Fig. 187]. The artist responsible for this design sought to ameliorate the constrained conditions of the *acute* pattern lines within the long hexagon by adjusting the angles of the pattern lines as they enter the long hexagon, resulting in these pattern lines being noncollinear. This is an example of the willful departure from convention to achieve a more pleasing design.

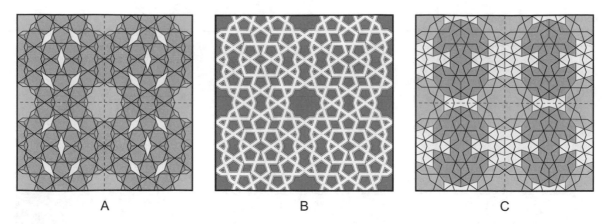

A

B

C

Fig. 246

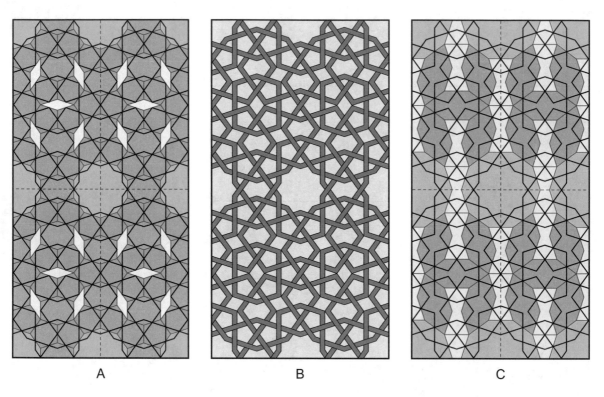

A

B

C

Fig. 247

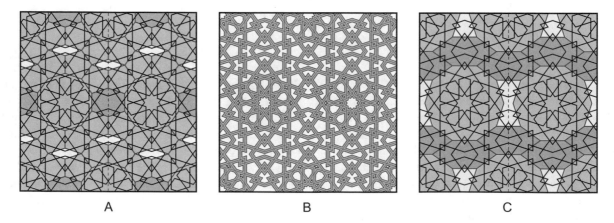

A

B

C

Fig. 248

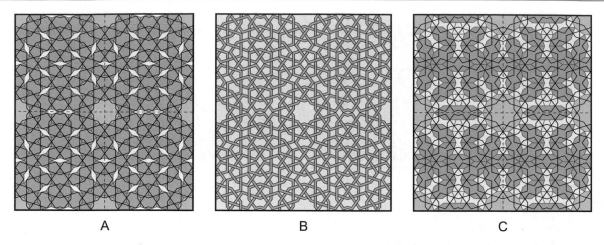

A B C

Fig. 249

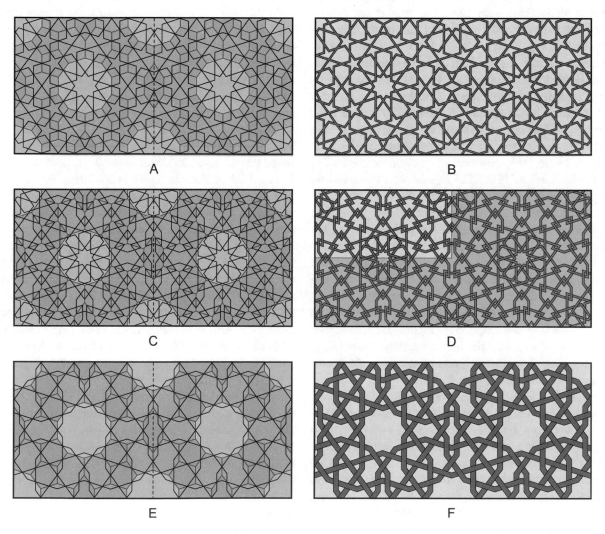

A B

C D

E F

Fig. 250

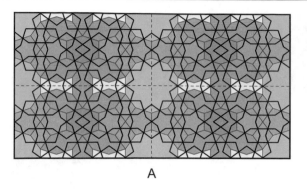

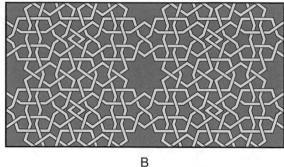

A   B

Fig. 251

However, a general rule of this ornamental tradition is for crossing pattern lines to remain collinear at the point were they intersect with one another. While at first glance this appears as an acceptable modification, upon closer inspection, the noncollinearity is somewhat problematic. The *two-point* pattern in Fig. 250d is created from the same underlying tessellation. In Fig. 250c, four of the parallel lines within the long hexagons are much closer together when compared with the pattern density of the rest of the design, and this would appear to be a problem. However, the Mamluk artist who produced this *two-point* pattern widened the line in only one direction: outward from the adjacent parallel line, thereby avoiding the problem of pattern density. The highlighted region in Fig. 250d represents a carved stone relief panel from the Sultan Qaytbay Sabil in Jerusalem (1482). This relief panel further reduces any remaining sense of constraint within the parallel pattern lines of the long hexagon by only using a quarter of this module within the finish relief panel. Figure 250e illustrates another use of this same underlying tessellation by Mamluk artists. The resulting *obtuse* pattern in Fig. 250f was used on the railing of the stone *minbar* of the Sultan Barquq mausoleum in Cairo (1384-86) [Photograph 57]. Figure 251a illustrates the derivation of a rectangular *median* pattern created from an underlying tessellation comprised of decagons, pentagons, barrel hexagons, long hexagons, concave hexagons, and wide rhombi. This combination of underlying polygons is somewhat unusual in that it combines the underlying long hexagons and concave hexagons with underlying pentagons and barrel hexagons. Within the *fivefold system*, the 72° crossing pattern lines associated with the pentagons and barrel hexagons create distinctive features of the *median* pattern family, particularly the five-pointed stars. However, within the long hexagon and concave hexagon the 72° crossing pattern lines produce features that are characteristic of the *obtuse* family, such as the pentagons, concave decagons, and shields associated with these modules. The combination of these elements within a single underlying tessellation therefore produces features that are both *median* and *obtuse*.

Figure 251b shows an interweaving version of this design that was produced by artists during the Seljuk Sultanate of Rum for the Külük mosque in Kayseri (1280-90). Figure 252b illustrates a Mamluk *two-point* pattern from the Amir Qijmas al-Ishaqi mosque in Cairo (1479-81). Figure 252a demonstrates how this pattern is created from an underlying tessellation that places decagons at the vertices and center of the rectangular repeat unit, and pentagons, barrel hexagons, and thin rhombi in the connective polygonal field. The pattern lines in Fig. 252b have been widened to their maximum outward expansion so that the outer corners of the independent kites, rhombi, and concave hexagons meet at a single point. Figure 253 illustrates a Timurid *median* pattern from the Imam Reza complex in Mashhad, Iran (1405-18). The underlying generative tessellation in Fig. 253a is unusual in that it has two sizes of decagon. The edge length of the smaller decagons is standard to this system by virtue if its edge lengths equaling those of the barrel hexagons and wide rhombi. The long edge of the triangular element determines the size of the larger decagons, and these atypical modules are simply wide rhombi that have been divided into half. This larger variety of underlying decagon is also seen in the earlier Kartid example from Yazd [Fig. 236]. Figure 254b shows a Timurid *acute* design from the Shah-i Zinda complex in Samarkand (fourteenth century). Figure 254a shows how the cluster of underlying trapezoids and triangles produces the distinctive eight-sided shield motif located at the midpoint of the longer side of the rectangular repeat unit. This arrangement of trapezoids and triangles is especially relevant to the *acute* pattern family [Fig. 196]. Figure 254d illustrates another pattern that includes this distinctive motif, albeit in a much more complex geometric structure. Figure 254c shows how the same arrangement of underlying polygons responsible for this feature is similarly placed at the midpoints of the long edge of the rectangular repeat unit. This design includes features that are derived from two configurations of contiguous triangles: one associated with 2/10 of the decagon, and the other with 6/10 [Fig. 196]. This example is considerably more complex and was produced by

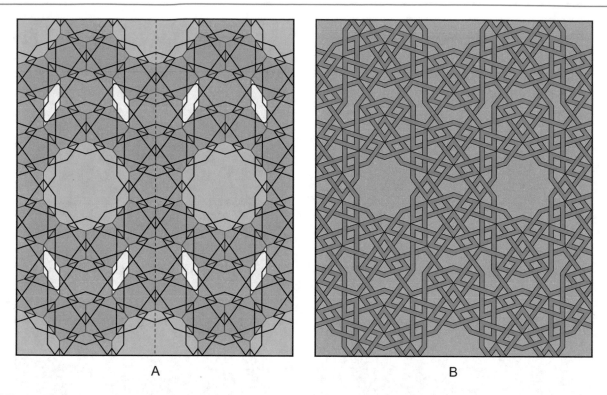

A

B

Fig. 252

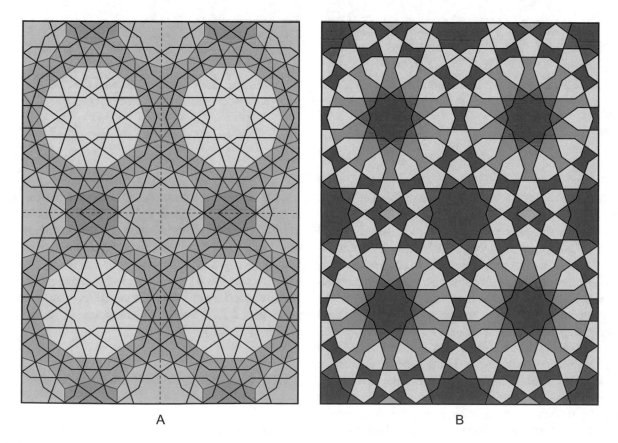

A

B

Fig. 253

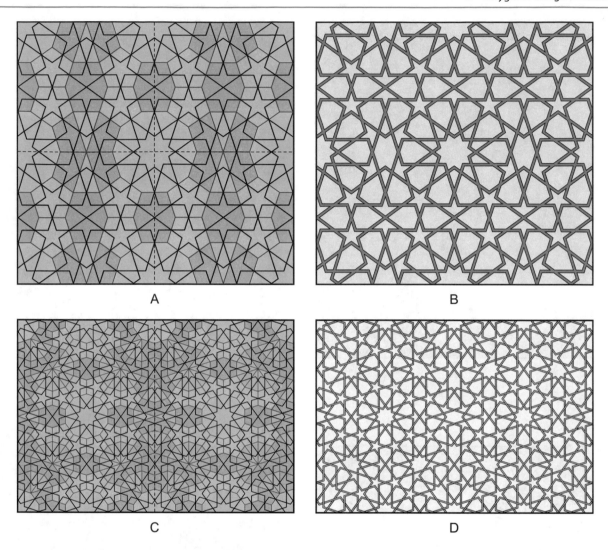

A

B

C

D

Fig. 254

Muzaffarid artists for the Friday Mosque at Kerman (1349). Figure 255b shows an *acute* pattern that was used by Mamluk artists on the *minbar* (1300) of the al-Salih Tala'i mosque in Cairo, as well as by Mughal artists at the mausoleum of Akbar in Sikandra, India (c. 1612). Figure 255a shows how this design is derived from an underlying tessellation that places decagons at the vertices and center of the rectangular repeat, with connective pentagons, barrel hexagons, trapezoids, and triangles from the *fivefold system*. The repetitive structure of this design is such that the geometric information contained within the rectangular repeat unit is identical to that of its dual. Like the previous example, this design also employs groupings of two edge-to-edge underlying triangles that combine to equal a 2/10 portion of the decagon. Figure 256b shows a Shaybanid *acute* design from a door of the Kukeldash *madrasa* (1568-69) in the Lab-i Hauz complex in Bukhara [Photograph 78]. A later

Janid Khanate example was used at the Bala Hauz mosque in Bukhara (1712). Figure 256a shows how this is created from an underlying tessellation comprised of decagons, pentagons, barrel hexagons, trapezoids, and triangles. Each of the long edges of the trapezoids is contiguous with a long edge of a triangle. This example includes distinctive pattern motifs that are created from the underlying partial decagons made up of 2/10 and 3/10 groupings of the triangular modules [Fig. 196].

3.1.10 Fivefold System: Hexagonal Repeat Units

Most patterns created from the *fivefold system* that repeat upon a hexagonal grid are field patterns [Figs. 212–220]. However, patterns with ten-pointed stars can also use

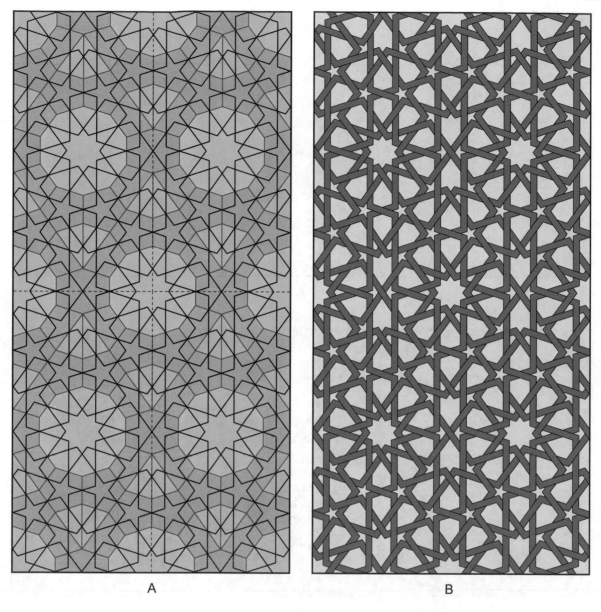

A   B

Fig. 255

hexagonal repeat units, although very few are known to the historical record. Figure 257a illustrates a particularly successful example of a *median* design that employs a hexagonal repeat unit that places the underlying decagons upon the midpoints of the long edges of the repeat unit. This is unusual in that the primary underlying polygonal modules are ordinarily located at the vertices of the repetitive grid. The distribution of ten-pointed stars is the same as patterns that repeat on a rhombic grid with 36° and 144° included angles [Figs. 241–244], and technically this design can be said to repeat on this same rhombic grid. However, for the purposes of this study and for reasons of clarity, repeat units are identified as having reflected symmetry in the pattern lines located along their edges. The tenfold radial symmetry

of the ten-pointed stars in this design does not align with the neighboring ten-pointed stars that are separated with an underlying wide rhombus module. Rather, the pattern lines between these ten-pointed stars are askew from one another and lack reflected symmetry. Again, for the purposes of this study, their lack of reflected symmetry therefore precludes this design from being categorized as repeating on a rhombic grid. This skewed feature between the ten-pointed stars is relatively uncommon and creates a pleasing visual tension within the design. It is worth noting that this interesting design can also repeat with a rectangular cell (not shown) that contains the area of two hexagonal repetitive cells. While the design in Fig. 257b is very successful, the artist from the Seljuk Sultanate of Rum who created this pattern

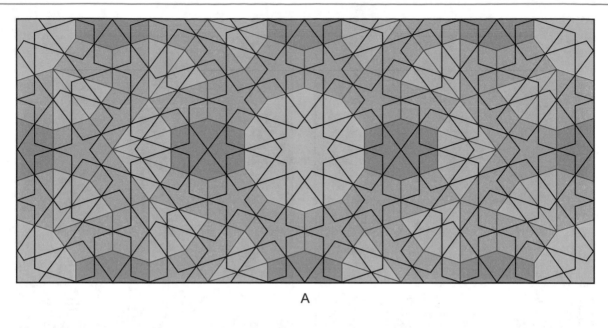

A

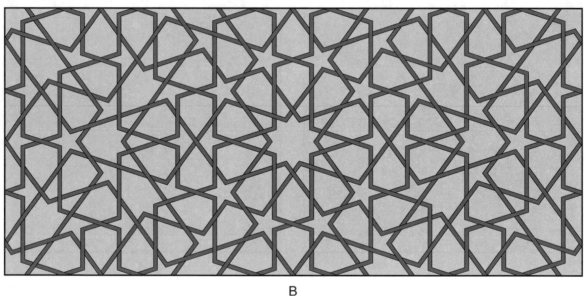

B

Fig. 256

arbitrarily filled each ten-pointed star with the relatively common modification that introduces five-pointed stars [Fig. 224b]. This results in the very balanced field pattern in Fig. 257c from the Huand Hatun in Kayseri (1237). Figure 258b shows a Mamluk *two-point* pattern that repeats upon a hexagonal grid. The underlying tessellation shown in Fig. 258a places the decagons at each vertex of the repeat unit, with edge-to-edge decagons at the two short edges of the repeat, and decagons separated by two mirrored pentagons along the other four longer edges. The other underlying polygonal modules within this generative tessellation are pentagons, wide rhombi, and an unusual stellated interstice region at the center of the repeat unit. The small highlighted rectangular region in Fig. 258b represents the isolated region

of this design that was used in the entry portal of the Ashrafiyya *madrasa* in Jerusalem (1482). By only using this limited region of the overall design, the artist successfully circumvented the problem of the tightly constrained parallel pattern lines that result from the application of *two-point* pattern lines to the underlying wide rhombi [Fig. 187]. This rather clever solution is conceptually identical to that employed in the stone relief panel from the Sultan Qaytbay Sabil in Jerusalem [Fig. 250d]. The fact that both these idiosyncratic Mamluk examples are from Jerusalem and were produced in the same year strongly indicates the likelihood of their being produced by the same artist or atelier. Figure 259b illustrates an unusual Ilkhanid design from the Gunbad-i Gaffariyya in Maragha, Iran (1328).

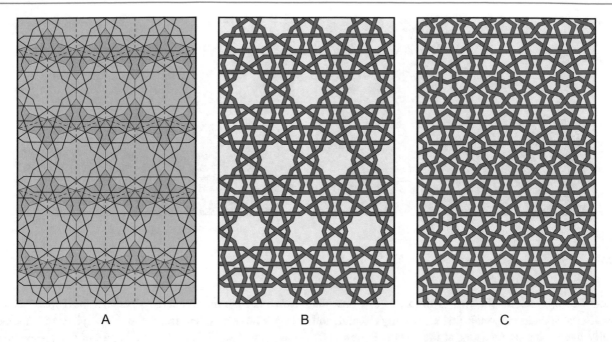

A   B   C

Fig. 257

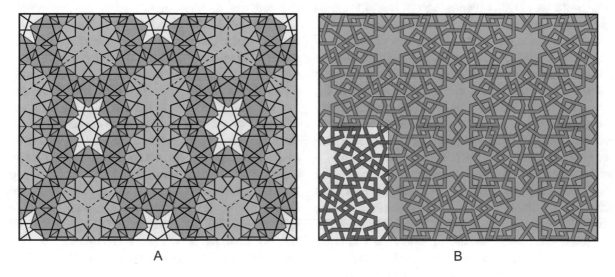

A   B

Fig. 258

Rather than placing the pattern lines upon either the midpoints or two points of the underlying polygonal edges, this design utilizes the vertices of the underlying tessellation. As such, it does not conform to any of the four pattern families. The underlying generative tessellation is comprised of a linear band of truncated decagons and pentagons, and is identical to the underlying tessellations of the two designs in Fig. 243. Whereas the earlier examples repeat with the thin rhombic repeat associated with this system, the atypical application of pattern lines in this example are more closely associated with the hexagonal dual of the rhombic repeat. The linear band of partial decagons at the top and bottom of

the panel have a different set of applied pattern lines than the central set of overlapping decagons, and this breaks the translation symmetry of the hexagonal grid. To add to the eccentricity of this design, the widened line treatment combines interweaving and interlocking qualities.

3.1.11 Fivefold System: Radial Designs

Of all the historical design systems, the *fivefold system* was the most widely used for creating patterns with rotation symmetry. This variety of design was afforded four distinct

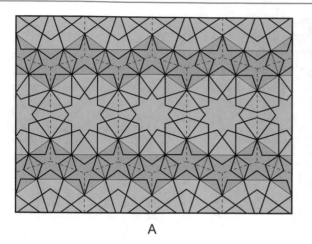

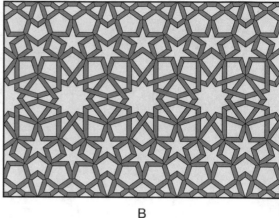

A   B

Fig. 259

manners of application. The first is the most obvious: the production of standalone panels that are usually circular, and typically have a ten-pointed star at the center. Figure 260b shows just such a design, produced during the Khwarizmshahid period and located at the Zuzan *madrasa* in northeastern Iran (1219) [Photograph 40]. As illustrated and like so many designs created from the *fivefold system*, this example can be easily created from either of two underlying tessellations. A better known example of a stand-alone tenfold radial design is the aforementioned pierced marble circular panel on the sides of the Ottoman *minbar* at the Selimiya complex in Edirne, Turkey (1568-74). Each 1/10 radial segment of this panel is half of the rhombic repeat unit that was first used in several locations by artists during the Seljuk Sultanate of Rum [Fig. 242]. Radial designs with five- or tenfold symmetry were occasionally used on the flat soffits of *muqarnas* constructions. This of course requires the shape of the soffit to also have compatible rotation symmetry: for example, pentagons, decagons, five-pointed stars, or ten-pointed stars. The most geometrically interesting examples of fivefold radial pattern making are in the secondary infill of the primary design elements in dual-level designs. Again, the primary elements receiving the secondary infill will themselves have fivefold rotation symmetry, including pentagons, decagons, five-pointed stars, and ten-pointed stars. Marinid examples of such fivefold rotational elements within dual-level designs are found at both the Bu'Inaniyya *madrasa* (1350-55) and the al-'Attarin *madrasa* in Fez, Morocco (1323) [Figs. 474 and 476]. Timurid, Qara Qoyunlu, and Safavid artists created several dual-level designs that contain regions with either fivefold or tenfold rotation symmetry, including examples from the Imamzada Darb-i Imam in Isfahan (1453) [Fig. 451] [Photograph 97], and the Madar-i Shah in Isfahan (1706-1714) [Figs. 453 and 468]. The single most significant group of fivefold dual-level designs with regions of rotation

symmetry are the five examples from the Topkapi Scroll.[27] Two fine examples from the Topkapi Scroll include a ten-pointed star [Fig. 22a] that is part of a dual-level design comprised of an *obtuse* primary pattern and a *median* secondary pattern[28], and a second example that is self-similar in that both the primary and secondary patterns are from the same *median* family [Fig. 449]. The last variety of radial pattern created from the *fivefold system* involved the application of geometric designs onto the surface of domes. Figure 21 illustrates an example from the Samosa Mahal at Fatehpur Sikri, India[29] (seventeenth century) that used 8/10 of a two-dimensional radial design to create a conical form that was then used on the eight gore segments of a dome. A small amount of distortion invariably results from this method of applying geometric designs to the surfaces of domes. The Topkapi Scroll depicts several gore segments that are presumably intended for application to domes,[30] including the design created from the *fivefold system* represented in Fig. 260e. The underlying generative tessellation for this *acute* design is comprised of decagons, pentagons, barrel hexagons, and clustered triangles.

3.1.12 Fivefold System: Hybrid Designs

As with the *fourfold system A* and *fourfold system B*, the *fivefold system* was also used to create hybrid designs that contain more than a single repetitive cell within their overall composition. Hybrid designs bring greater complexity to a given design, and their composition requires greater

---

[27] Necipoğlu (1995), diagram nos. 28, 29, 31, 32, and 34.

[28] Necipoğlu (1995), diagram no. 29.

[29] Hankin (1925a), Figs. 45–50.

[30] Necipoğlu (1995), diagram nos. 4a, 10b, and 90a.

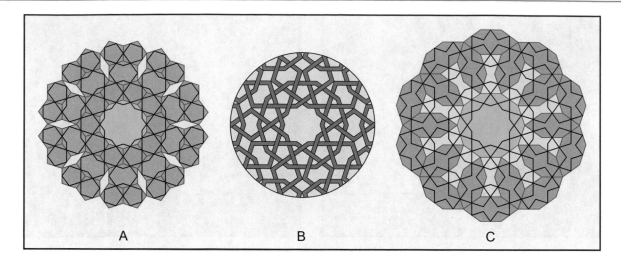

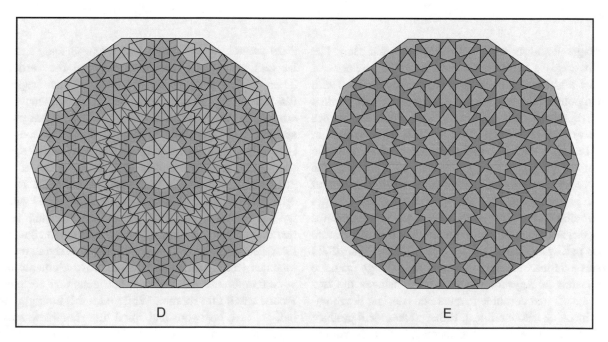

Fig. 260

geometric aptitude. Examples of hybrid designs generally date to the period when this ornamental tradition was reaching full maturity during the thirteenth century, with notable twelfth-century exceptions such as the *fourfold system A* design at the Maghak-i Attari mosque in Bukhara (1178-79) [Fig. 155], and the *fourfold system B* design at the Mu'mine Khatun in Nakhichevan, Azerbaijan (1186) [Fig. 182]. However, it is remarkable that the earliest known hybrid design predates these examples by a century. Among the different patterns that were included in the upper recessed arches in the northeast dome chamber of the Friday Mosque at Isfahan (1088-89) is the fivefold hybrid design in Fig. 261b that is composed of at least two separate repetitive cells [Photograph 25]. This is all the more remarkable in

that along with two other examples from the northeast dome chamber [Figs. 229a and 496] these are the earliest extant fivefold Islamic geometric designs known to the historical record. This hybrid design has a central pentagonal region with fivefold rotational symmetry that is attached to rhombic cells on either side of the pentagon. These rhombi have 72° and 108° included angles, and the pattern lines associated with each is the classic *acute* pattern [Fig. 226b]. Because the boundary of the arch limits the amount of geometric design, only portions of the rhombic cells are contained within the arch. Due to this limitation, the structure of the repetitive cells is somewhat ambiguous and must be inferred—especially where the pattern moves outward from the central arched region. What is certain is the use of

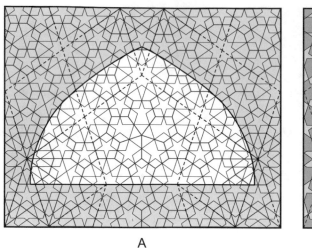

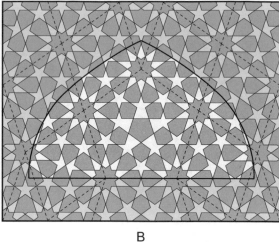

A   B

Fig. 261

the pentagonal and rhombic cells within the arch itself. The 36° leftover space between the rhombic cell at the apex of the arch is a 1/10 segment of the circle, and easily filled with an added point of the ten-pointed star, or conceivably with a second rhombus with 36° and 144° included angles (as shown). The underlying tessellation is demonstrated in Fig. 261a. This employs five irregular pentagons at the center of the pentagonal repetitive cell. This produces the distinctive arrangement of pattern lines wherein a central pentagon is surrounded by 5 nine-sided motifs derived from the five-pointed star. This pattern line configuration shares the properties of a popular design from the Huand Hatun *madrasa* in Kayseri, Turkey (1237) [Fig. 242b]. Figure 262d illustrates a Seljuk Sultanate of Rum hybrid design from the Huand Hatun in Kayseri. Figure 262a illustrates the two repetitive cells that combine to create this design: the rhombus and hexagon (dashed lines). Either of these will produce very good patterns when used on its own. In fact, the pattern within just the rhombus is the very-well-known classic *obtuse* design. It would appear significant that the Huand Hatun also contains a remarkable hybrid pattern created from the *fourfold system A* that also has just two repetitive elements [Fig. 156]. Without question, the artists responsible for the geometric ornament of this building were exceptionally skilled, with a sophisticated understanding of hybrid design methodology. Figure 262b represents an interweaving treatment of the standard *median* pattern created from the underlying tessellation in Fig. 262a.[31] Figure

262b shows how this hybrid design repeats upon a rectangular grid that places ten-pointed stars at the vertices and 2 ten-pointed stars within the field of each repeat unit (dashed lines). The pattern in Fig. 262d includes an arbitrary modification of the standard design that places an additive fivefold motif within each of the ten-pointed stars [Fig. 224a], affectively transforming the design with ten-pointed stars into a field pattern. As indicated in Fig. 262c and d, the artist's decisions regarding the rotational orientation of the fivefold additive modifications within the ten-pointed stars changes the otherwise rectangular repeat to one that is twice as long as the unmodified design (dashed lines). In addition to the decagons, the underlying polygonal modules within the hexagonal repetitive elements include wide rhombi and half-concave hexagons that are clustered around small thin rhombi. While rare, this configuration of half-concave hexagons and small thin rhombi was used in other locations during the Seljuk Sultanate of Rum [Figs. 209, 215, 217 and 218]. Figure 263c shows a more complex fivefold hybrid *obtuse* design from the Izzeddin Keykavus hospital and mausoleum in Sivas (1217). Figure 263a shows the four repetitive cells that make up this design. These include a large rhombus with 72° and 108° included angles, a smaller rhombus with the same proportion, a rhombus with 36° and 144° included angles with the same edge length as the larger wider rhombus, and a triangle that is half the thinner rhombus. The short edge of the triangle is the

---

[31] The fact that the rhombic repetitive cells are populated with *obtuse* pattern lines, and the hexagonal regions are referred to herein as *median*, is less confusing than it might appear. As demonstrated in Fig. 232c, e, the classic *obtuse* pattern can be created from either of the two underlying tessellations, but the 108° crossing pattern lines that produce the pentagons located inside the underlying pentagons identify

this as an *obtuse* design. However, the pattern within the hexagonal region is overtly of the *median* family, and by sharing the 108°, 72°, 108°, and 72° angles at each intersection of the pattern, they work seamlessly with one another. It is logical to categorize this overall hybrid design as *median* in that the hexagonal regions (with their distinct *median* pattern characteristics) comprise a significantly larger area of the total design than the rhombic repeat units.

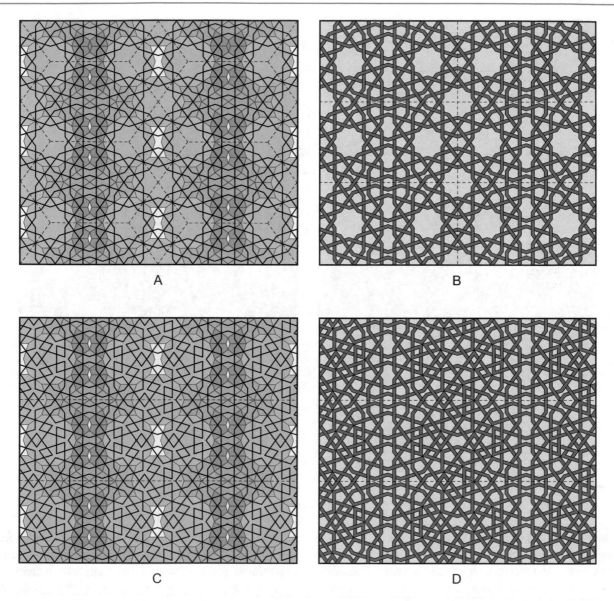

Fig. 262

same length as the edge of the smaller wide rhombus. The small wide rhombus is identical to that of Fig. 262a and on its own produces the classic fivefold *obtuse* design. As with all repetitive cells used within this tradition, it is required that the underlying polygonal structure and the resulting applied pattern lines of each repetitive edge must match all other edges of equal length. In this case, there are two edge lengths. The short lengths have underlying edge-to-edge decagons placed on each vertex, and the longer edge lengths are populated with two underlying decagons separated by a concave hexagon placed in its long orientation. It is worth noting that the underlying polygonal modules that generate this design could also be the dual of this tessellation,

comprised of decagons, pentagons, barrel hexagons, and thin rhombi [Fig. 200]. Figure 263b illustrates the hybrid tessellation of these four repetitive elements, and Fig. 263c shows how this combination has translation symmetry with a rectangular repeat. Figure 264c shows 1 of 2 fivefold hybrid designs from the Karatay Han near Kayseri (1235-41). As shown in Fig. 264a, this *acute* pattern is comprised of four repetitive cells: the barrel hexagon, thin rhombus, wide rhombus, and a triangle that is half a wide rhombus. There are two edge lengths in this group, each with its own underlying polygonal configuration. On its own, the wide rhombus makes the classic *acute* pattern. Figure 264b indicates an interstice region within with underlying tessellation of the

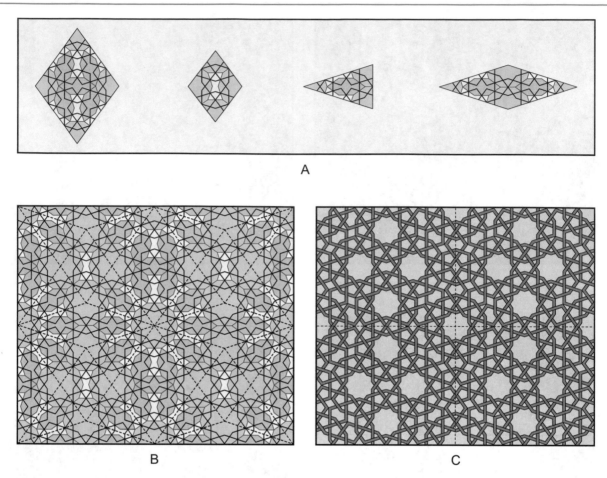

Fig. 263

repetitive barrel hexagon. This creates an interesting, if not completely successful, feature in the derived pattern. Had the artist employed the 3/10 cluster of three triangles [Fig. 196], this would be a more satisfactory design. However, the use of clustered triangles, each being 1/10 of a decagon, to create visually appealing *acute* patterns did not come into use until the fourteenth century, and for all their geometric skill and artistry, it appears that this seemingly simple innovation was not known to the thirteenth century artists working on this building. Figure 264c shows how the arrangement of the four repetitive elements combine together in a rectangular repeat unit with translation symmetry. Figure 265c shows a second hybrid *acute* design from the Karatay Han near Kayseri that was doubtless created by the same artist as the previous example. Figure 265a illustrates the six repetitive cells that comprise this design. The first four of these are identical to those of Fig. 264. There are three edge lengths in this group, each with its own underlying polygonal configuration. The barrel hexagon repetitive cell employs the same atypical, and not wholly satisfactory, interstice region as the design from Fig. 264. The two underlying long hexagons (green) in the second repetitive hexagonal cell produce pattern lines that

are not generally acceptable within the *acute* family [Fig. 187]. The 36° crossing pattern lines produce constrained regions within the applied pattern lines associated with this particular hexagon. This is one of the only historical examples of an *acute* design that employs this unsatisfactory motif. The panel in Fig. 265c represents the full rectangular repeat unit that this particular combination of repetitive elements creates. Figure 266 illustrates a Mughal hybrid pattern from a stone mosaic panel on the façade of the I'timad al-Daula in Agra, India (1622-28). This type of pattern was not typical to the Mughal period, but it is nonetheless very successful. The three repetitive cells are simply the rhombus with 72° and 108° included angles, a triangle that is half of the rhombus, and a rectangle with a short edge length that is equal to the long edge of the triangle. Once again, the pattern lines within the rhombi are the classic *acute* design. The pattern lines within the rectangle provide this example with its most distinctive feature: the rotational point symmetry at the center of the rectangular cell. This is created from an arrangement of 12 edge-to-edge underlying triangles that form an *s*-curve with point symmetry at its center. The *acute* pattern in Fig. 267b was produced by Mamluk artists

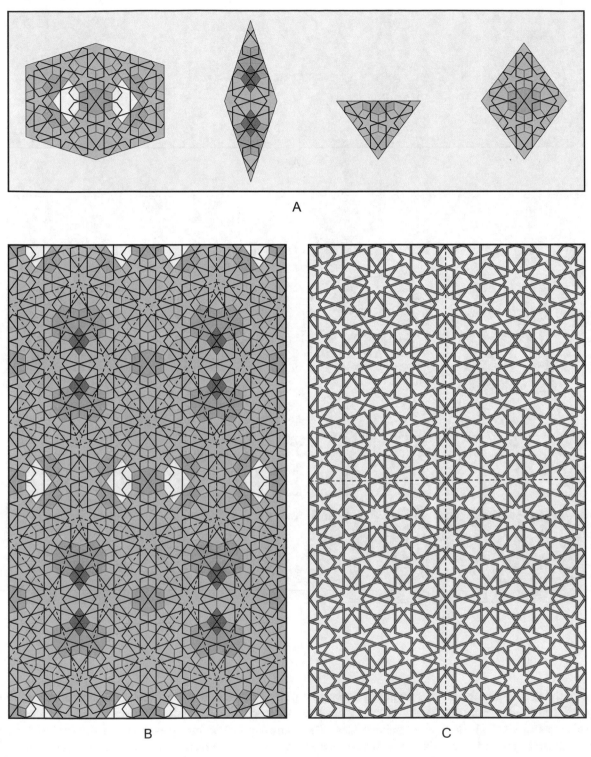

A

B   C

Fig. 264

for the Qadi Abu Bakr ibn Muzhir in Cairo (1479-80). This exceptional fivefold design is one of the very few Islamic geometric patterns to incorporate 20-pointed stars into a matrix of 10-pointed stars. The overall repeat unit with translation symmetry is the wide rhombus with 72° and 108° included angles and 20-pointed stars at each vertex (dashed line). However, Fig. 267a demonstrates how the repetitive structure of this design is comprised of an arrangement of repetitive decagonal and concave hexagonal cells. Some of the vertices of these decagons and concave hexagons have

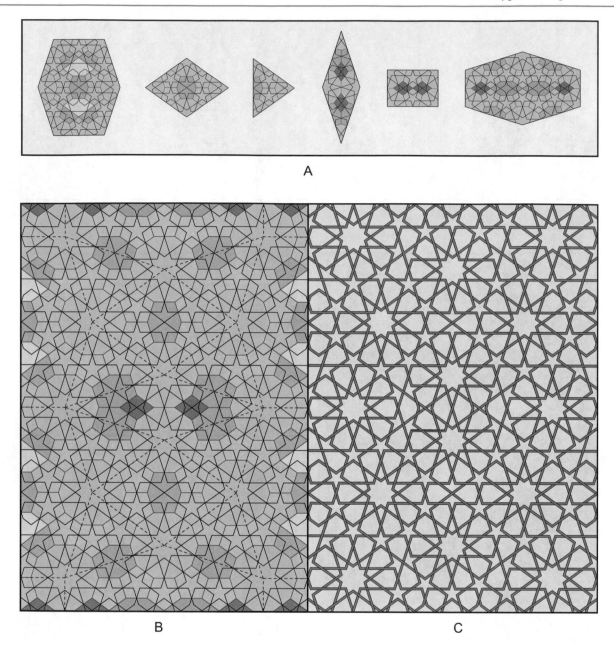

Fig. 265

partial underlying decagons made up of six adjacent triangles that equal 6/10 of a decagon [Fig. 196]. The center of each decagonal repetitive cell houses a 20-pointed star. The incorporation of 20-pointed stars within the *fivefold system* is analogous to the incorporation of 16-pointed stars within the *fourfold system A* and *fourfold system B* [Figs. 165–168, and 185–186]. Figure 268b illustrates a Marinid *acute* hybrid pattern with 10- and 20-pointed stars that is conceptually similar to the previous Mamluk example. This outstanding design is from the Bu 'Inaniyya *madrasa* in Fez, Morocco (1350-55). Figure 268a shows how the hybrid repetitive cells of this example are made up of decagons and rhombi with

72° and 108° included angles. Figure 268b indicates the rectangular repeat unit with translation symmetry. The rectangular dual of this repeat unit (not shown) has the identical geometric information as the repeat unit itself, which is to say that the pattern is self-dualing. As with the Mamluk example, the 10-pointed stars are located at the vertices of the hybrid structure, and the 20-pointed stars are placed at the center of each repetitive decagonal cell. Once again, on its own, the rhombic repetitive cell is the repeat unit for the classic fivefold *acute* pattern [Fig. 226b]. This design from the Bu 'Inaniyya *madrasa* is arguably one of the most beautiful fivefold *acute* patterns from this ornamental tradition.

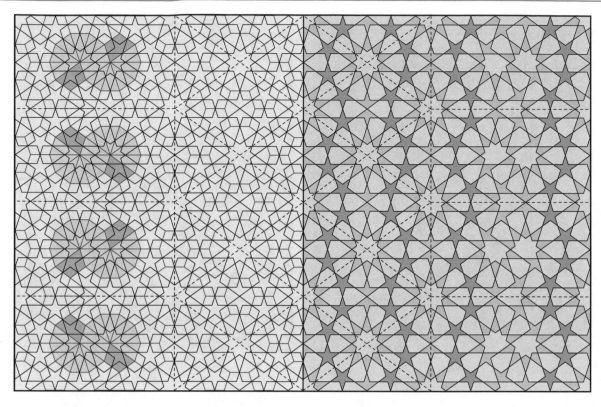

Fig. 266

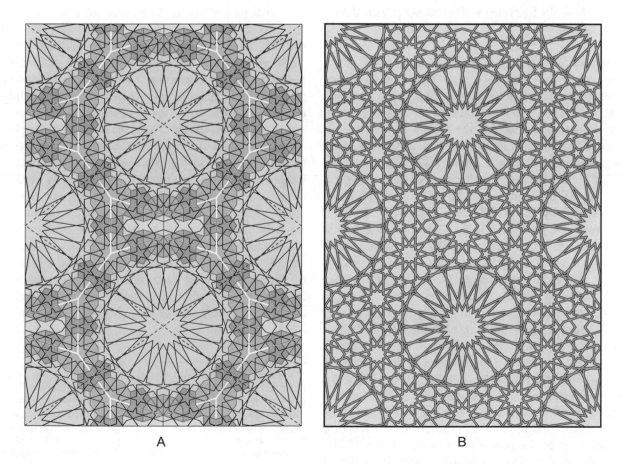

A

B

Fig. 267

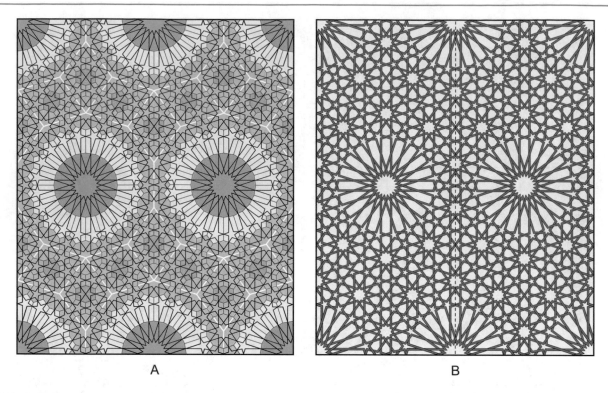

A   B

Fig. 268

3.1.13 Fivefold System: Patterns with Variable Scale

A very unusual and very rare use of the *fivefold system* derives the pattern lines from two scales of underlying polygonal modules within a single underlying tessellation. The design in Fig. 269b is from a door produced during the Seljuk Sultanate of Rum for the Hekim Bey mosque in Konya (1270-80), and currently in the Ince Minare Medrese History Museum in Konya. Figure 269a shows how the underlying tessellation transitions between polygonal modules of two different scales. The smaller scale polygons, as represented by the smaller decagons and surrounding pentagons, have pattern lines from the *acute* family. As these *acute* pattern lines extend into the adjacent larger scale wide rhombi, decagons, long hexagons, and concave hexagons, the pattern lines naturally convert to the *median* family. The proportional relationship between these two scales of underlying polygon is 1.3764. . .. This is a product of φ: the inherent proportional relationships within the *fivefold system*. If the small decagonal and pentagonal edges are taken as 1, the length of a line that connects any two nonadjacent corners of the pentagon is 1.6180. . . [Fig. 195]; and if the short diagonal of a rhombus with 72° and 108° included angles is 1.6180. . . then the edge of this rhombus is 1.3764. . .. The pattern lines within the larger underlying decagonal modules are provided with an arbitrary modification that disguises the ten-pointed star in a similar fashion as in Fig. 224b. The introduced five-pointed stars at the center

of each large underlying decagon are identical in size and shape to the five-pointed stars associated with the smaller scaled pentagons. This provides visual similitude between the *acute* region and the *obtuse* region, and helps to unify the design. Filling the large ten-pointed stars with this motif also provides the design with a more balanced density throughout. This design is one of the few historical examples of this variety of pattern manipulation, which is surprising in that the incorporation of variable scaled pattern elements offers tremendous innovative appeal. However, the aesthetics of this ornamental tradition generally seeks to achieve an overall balance in design density, and the fact that regions of reduced scale in the underlying generative tessellation will typically cause concomitant regions of greater density within the resulting pattern matrix would likely have been an aesthetic impediment to the development of this category of design. Traditional aesthetics aside, it is possible for diminishing scale within a geometric design to be visually appealing and intellectually satisfying[32] [Figs. 484 and 485],

---

[32] The author has produced several fivefold designs that scale the polygonal elements of the underlying generative tessellation by a factor of *phi*. These include recursive self-similar examples with quasicrystal characteristics. The artist Joe Bartholomew has also produced a number of patterns that employ this unusual scaling feature. The designs of both the author and Joe Bartholomew have the added characteristic of diminishing the scale of the polygonal edges in a continual, and theoretically infinite, sequence: whereas the historical examples only make use of two scale lengths.

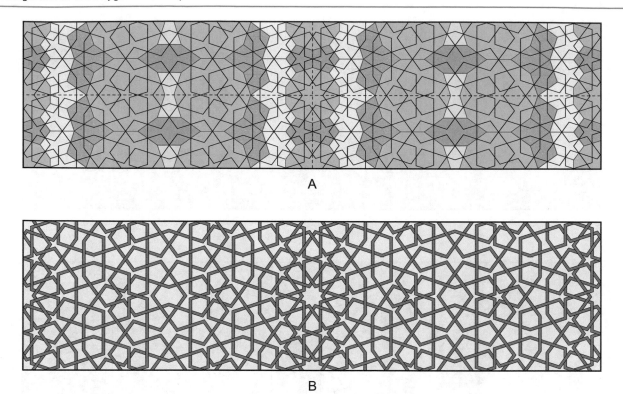

A

B

Fig. 269

albeit very deferent from historical examples. Figure 270c shows a design from the Sultan Bayezid II Kulliyesi in Istanbul (1501-06) that is also created from an underlying tessellation with variable scaled polygonal modules, and also transitions between two different pattern families. Figure 270b demonstrates how the smaller scale polygons have pattern lines from the *acute* family, while the larger scale polygons are populated with the *two-point* family, with two sets of crossing pattern lines with 72° angular openings. This creates the less acute five-pointed stars with points of 72° that are more commonly associated with the *median* family, thereby introducing the characteristics of a third pattern family into this one design. Figure 270a demonstrates how the proportional relationship between both scales of the underlying polygonal modules is derived from a ring of ten smaller pentagons fitting precisely within the larger scaled decagon. The black lines in Fig. 270a illustrate the truncation lines that create the trapezoids in Fig. 270b. This allows for the dart-shaped *acute* pattern lines within the smaller scaled trapezoids to extend into the larger scaled pentagons, thereby determining the placement of the two points upon each edge for the larger scale *two-point* pattern lines. The proportion of the two edges lengths is 1.9021.... As with the example from the Hekim Bey mosque, this is a product of φ. If the sides of the small underlying decagons and pentagons are taken as 1, then the distance between two outer points in the smaller ring of ten edge-to-edge pentagons that surround the decagon is 1.9021, the length of the large pentagonal

edges. This also equals the distance between two consecutive corners of the small underlying decagon. The scale of the larger underlying polygons is almost double that of the smaller. By doubling the amount of pattern lines application to the larger underlying polygons through the use of *two-point* application with lines that continue in both directions beyond the underlying polygonal edges, the artist who designed this outstanding pattern successfully balanced the overall design density within both regions of variable scale. Both in terms of its visual splendor and geometric ingenuity, as well as for the excellent quality of the woodworking, this example of Ottoman geometric ornament is nothing short of a masterpiece.

3.1.14 Sevenfold System

Patterns created from the *sevenfold system* are characterized by the presence of heptagons, 7-pointed stars, and 14-pointed stars. These are rare, with every historical example known to the author being included within this study.[33]

---

[33] Much of the material in this section on sevenfold geometric design, including many of the illustrations, was first presented to the public at the 2012 Bridges Conference, and first published in their Conference Proceedings. See:
–Bonner and Pelletier (2012).
–Pelletier and Bonner (2012).

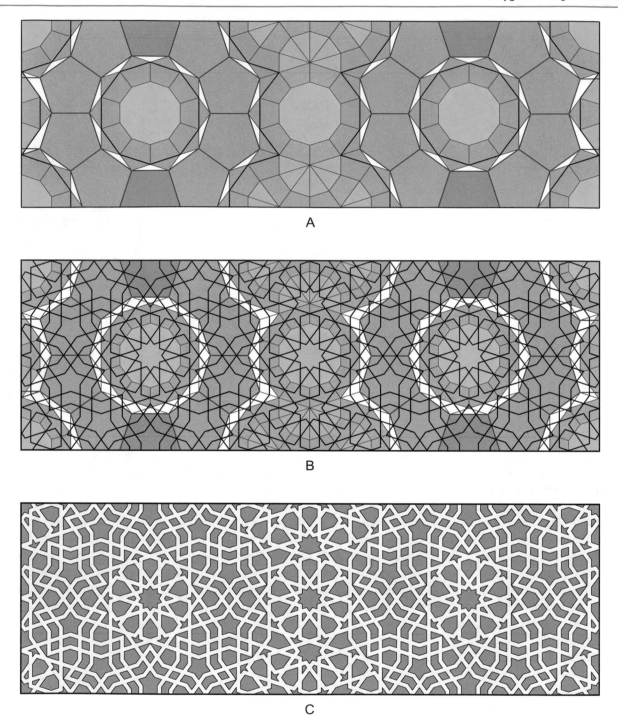

Fig. 270

As with the other historical design systems, the *sevenfold system* employs a limited set of underlying polygonal modules to which pattern lines are applied in each of the four pattern families. Considering their beauty, the rarity of this variety of design would not appear to be due to any aesthetic predilection against their appearance. Rather, one must conclude that knowledge of this system was held and passed on to only a very few select artists within those Muslim cultures that included such designs within their

ornamental canon. Considering their broad spread over time and territory, it appears likely that the discovery of the *sevenfold system* may have occurred independently in several locations rather than as a continuum of inherited knowledge. Considering the paucity of historical examples, it is impossible to know for certain to what extent artists working with sevenfold patterns were aware of this as a design system *per se*, or were merely applying the methodology of the polygonal technique to sevenfold geometry to

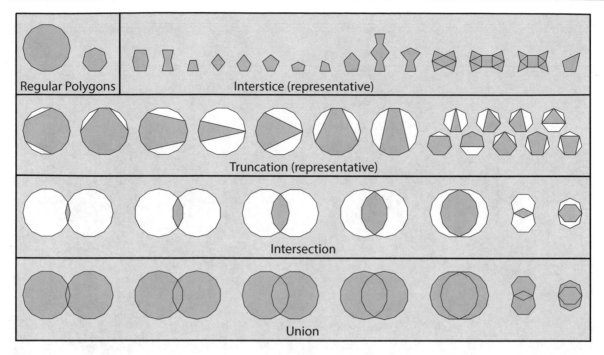

Fig. 271

arrive at stand-alone patterns without realizing the systematic potential of the underlying polygons within the generative tessellations they produced. This historical ambiguity in no way diminishes the fact that each of the historical examples can be created from the limited set of underlying generative polygons that comprise the *sevenfold system*, nor the fact that this system has extraordinary potential for creating countless original designs for contemporary artists.

As with the *fivefold system*, the earliest example of a pattern created from underlying polygonal modules of the *sevenfold system* is from the Seljuk work on the northeast dome chamber of the Friday Mosque at Isfahan (1088-89). Soon after this example, two Ghaznavid sevenfold patterns were incorporated into the exterior façade of the minaret of Mas'ud III in Gazna, Afghanistan (1099-1115). Following this, there is a hiatus of roughly a century before several rather simple sevenfold patterns were produced in Anatolia during the Seljuk Sultanate of Rum. It is possible that all three of these locations may have been isolated developments of sevenfold pattern making. The added sophistication of incorporated 14-pointed stars did not transpire until another century had passed. Mamluk artists appear to have independently developed the more fully expanded set of underlying polygonal modules that comprise this system in the early fourteenth century. Indeed, the majority of historical examples of sevenfold patterns with 14-pointed stars are Mamluk, with a few notable Ottoman and Timurid examples that were presumably influenced by their earlier Mamluk precursors.

Figure 271 shows the five types of underlying generative polygons that comprise the *sevenfold system*. These consist

of the heptagon and tetradecagon, the two regular polygons native to this system; those that result as interstice regions through tessellating with other polygonal modules from this system; those that result from truncating the heptagon or tetradecagon; those that result from the intersection of the heptagon or tetradecagon; and those that result from the union of the heptagon or tetradecagon. The examples illustrated in this figure are not exhaustive, and there are many more interstice and truncation modules than shown. It is also worth noting that not all of the polygonal modules shown were used historically, and that the *sevenfold system* offers tremendous innovative opportunity to contemporary artists seeking to create original geometric designs.

Figure 272 illustrates the pattern line applications to the two regular polygons of the *sevenfold system*. The angular openings of the pattern lines are determined by drawing lines that connect the midpoint of the sides of the tetradecagon, ranging from adjacent sides [14-*s*1] through six sequential sides [14-*s*6].[34] The *two-point* pattern lines are shown as a 14-*s*4 edge-to-edge sequence, but other *two-point* sequences are also possible. With the six possible midpoint-to-midpoint line sequences [14-*s*1–14-*s*6], the *acute*, *median*, and *obtuse* pattern family assignments are less specific than with the other pattern systems. As such, the designation of the pattern family in the *sevenfold system* is generally descriptive of the aesthetic character of a given pattern rather than the specific angular opening employed. The line sequence

---

[34] This method of defining star forms roughly follows the nomenclature of A.J. Lee. See: Lee (1995), 182–197.

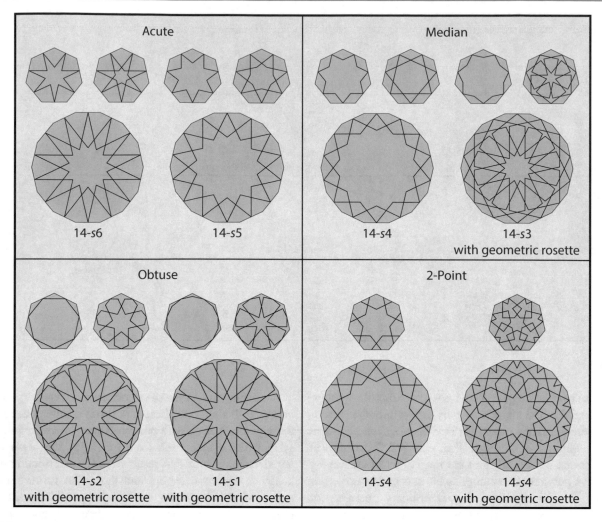

Fig. 272

nomenclature [14-s1–14-s6] is, therefore, necessary for accurately identifying the precise character of any given pattern created from this system. The *median*, *obtuse*, and *two-point* tetradecagons are provided with additive geometric rosettes that are typical to this ornamental tradition. However, other varieties of rosette, and other treatments of the 14-pointed stars are also possible, and these are only meant as representative examples. Figure 273 shows the interstice polygonal modules with associated pattern lines in each of the four pattern families. Again, this is only a representative sample of the interstice modules that are generated from this system. Unlike Fig. 272, only one midpoint-to-midpoint line sequence is shown for each family. These are the more commonly employed within the limited number of historical designs generated from this system. Figure 274 demonstrates the pattern lines applied onto the truncated polygonal modules. Again, this is only a representative sample of the truncation modules that are generated from this system, and only one of the midpoint-to-midpoint line sequences is shown for each family. Truncated tetradecagons were not a feature of historical

methodological practices. However, when used in rotation with matching truncated edges, they can provide a positive design contribution.[35] Note: those truncated tetradecagons that have no applied pattern lines do not make acceptable design features within the 14-s2 *obtuse* and 14-s4 *two-point* families. Figure 275 shows the polygonal modules that are derived from intersections of the heptagon and tetradecagon. Only the more visually acceptable midpoint-to-midpoint line sequences are shown for each family. The two modules without pattern lines do not generate acceptable features within the *obtuse* family. Figure 276 illustrates the polygonal modules that are derived from the union of both heptagons and tetradecagons along with their associated pattern lines in each of the four pattern families. Several of the conjoined tetradecagons have no pattern lines as these polygonal modules do not work well with the particular variety of line sequence.

---

[35] Pelletier and Bonner (2012).

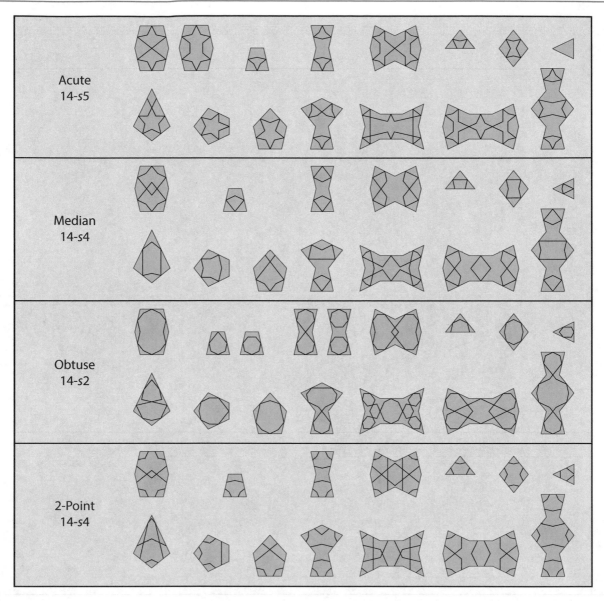

Fig. 273

The geometric properties of the *sevenfold system* are governed by the inherent proportional ratios of the heptagon. Figure 277 demonstrates how the heptagonal edge, taken as 1, relates to *ρ (rho)* as the length that connects two consecutive edges (1.80193774...), and *σ (delta)* as the length that connects three consecutive edges (2.24697960...). These ratios are analogous to the *φ (phi)* proportional ratio of the golden section (1.61803398...) that is inherent within the pentagon and which functions analogously as the proportional determinant within the *fivefold system*. Figure 278 provides several examples of linear arrangements of tetradecagons and various secondary modules of the *sevenfold system*. This demonstrates how the sevenfold proportions of the heptagon determine the

tessellating properties created from the polygonal modules of this generative system. Each interval of two tetradecagons (either overlapping, edge to edge, or separated by secondary polygonal modules) can be used on its own as an edge configuration for both rhombic and rectangular repeat units. The proportions of the linear arrangements of three tetradecagons as shown in this figure are especially relevant to the design of the smaller scaled secondary pattern when creating dual-level designs from the *sevenfold system*.

Figure 279b illustrates an *acute* field pattern that is one of the motifs in the group of arches in the upper portion of the square base of the northeast dome in the Friday Mosque at Isfahan [Photograph 26]. The only other known example of this design is from the anonymous treatise *On Similar and*

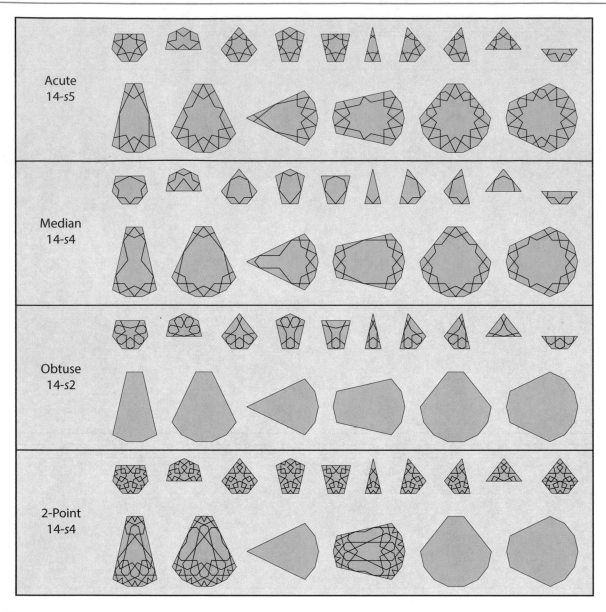

Fig. 274

Complementary Interlocking Figures. This treatise is estimated to date to circa 1300,[36] and the fact that these are the only known examples of this very distinctive design suggests a direct causal influence between them. As detailed in the previous chapter, it is highly significant that the underlying generative tessellation in Fig. 279a is represented along with the *acute* pattern in the illustration in this treatise. Considering that this is the earliest known sevenfold pattern, it is somewhat surprising that the polygonal modules that comprise the underlying tessellation do not include the heptagon. Significantly, the absence of the heptagon suggests

that the artist who created this design may have understood the systematic potential of the two varieties of underlying hexagon. The underlying hexagons in Fig. 279c are simply created from the intersection of two heptagons. Figure 279d shows how the underlying barrel hexagons can be derived as an interstice region of an arrangement of heptagons and overlapping heptagons. Figure 279e produces the same interstice regions, but with heptagons and hexagons from Fig. 279c; and Fig. 279f produces the barrel hexagon from an interstice region of an arrangement of just the hexagons from Fig. 279c. This last example is the arrangement that was used to produce the design in the northeast dome chamber. These polygonal arrangements demonstrate the determinant nature of the heptagon, and the artist who conceived

[36] Necipoğlu [ed.] (Forthcoming).

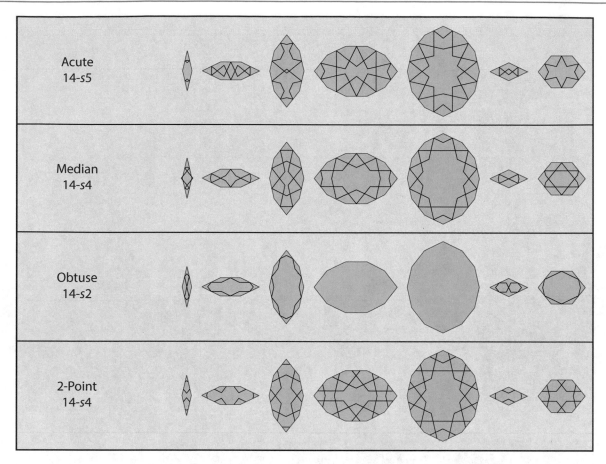

Fig. 275

this design could not have derived the two underlying hexagonal modules without starting with the heptagon. This artist would have therefore been aware of at least three polygonal components, and it is plausible that this artist would have known that these underlying modules could be arranged into other generative tessellations, thereby producing other sevenfold patterns. If this artist was also responsible for the two patterns created from the *fivefold system* that are in other arches in this same area of the northeast dome chamber—a supposition that would appear most likely—then we can conclude that this artist was knowledgeable of systematic design methodology more generally. It would therefore seem entirely reasonable for this artist to seek a means to also produce sevenfold patterns systematically. Figure 279a shows how this design repeats upon either of the two dualing hexagonal repetitive grids. With the lack of any primary star forms, this example falls into the category of field pattern.

The exterior façade of the minaret of Mas'ud III in Ghazni, Afghanistan (1099-1115), includes two designs with sevenfold symmetry.[37] Along with the single example

from the northeast dome in the Friday Mosque at Isfahan these are among the earliest examples of complex sevenfold pattern making to have ever been produced.

Figure 280 illustrates the construction sequence for one of the two Ghaznavid sevenfold designs from the façade of the minaret of Mas'ud III in Ghazni, Afghanistan (1099-1115). Figure 280a shows the underlying generative tessellation comprised of edge-to-edge heptagons that repeat upon an elongated hexagonal grid. The interstice regions in this configuration are filled with edge-to-edge irregular pentagons. Figure 280b demonstrates the first step in the placement of the pattern lines. These are unusual in that they are set upon the vertices of the generative grid rather than the midpoints of each heptagonal edge. This pattern was created during the developmental period that preceded the methodological codification of the polygonal technique, and is an excellent example of the artistic experimentation prevalent under the Ghaznavid patronage. Figure 280c shows the completion of the pattern through the incorporation of secondary pattern lines, and the modification of the primary seven-pointed

---

[37] The 2 sevenfold panels from the minaret of Mas'ud III presented herein are in very poor repair, and available photographs are of low

quality. For these reasons the reconstructions of these two examples may differ slightly from the actual designs in Ghazni.

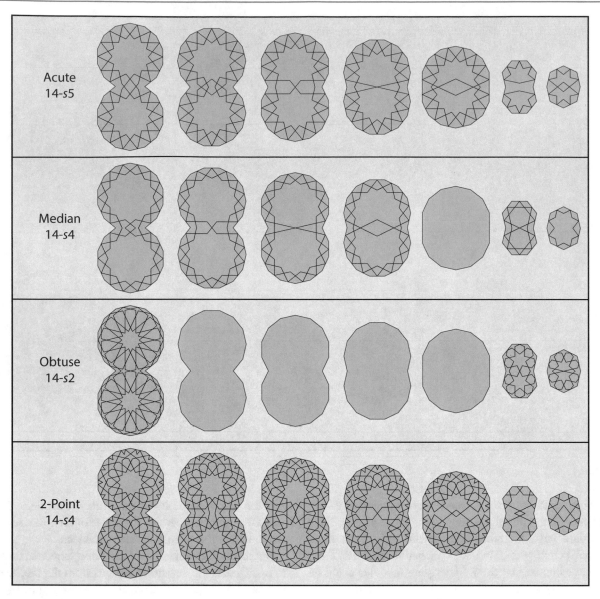

Fig. 276

stars from Fig. 280b to include a seven-pointed star rosette placed at each vertex of the hexagonal repeat. The secondary pattern lines in Fig. 280c elegantly employ 60° angles placed at two points of each polygonal edge. These extend into the underlying pentagons to create the distinctive five-pointed stars. Figure 280d is a representation of the raised brick panel with widened interweaving lines from the façade of this monument. Figure 281 shows the construction sequence for the second sevenfold pattern from the minaret of Mas'ud III in Ghazni. Figure 281a indicates how both the repeat unit and underlying generative tessellation are the same as the previous example from the same building. Figure 281b demonstrates the first step in the placement of the pattern lines. These follow the more conventional approach of placing the crossing pattern lines at the midpoints of the

underlying polygons. Figure 281c shows the completion of the pattern through the incorporation of a relatively complex network of secondary pattern lines. The overlapping kite elements provide the aesthetics of the *two-point* pattern family. Figure 281d shows an approximation of this raised brick design.

Figure 282 illustrates four patterns created from the same underlying tessellation of edge-to-edge heptagons that was used in the Ghaznavid patterns in Figs. 280 and 281. Each of these four designs utilizes the midpoints of the underlying polygonal edges for pattern line application, and they differ significantly from the two Ghaznavid designs in that the pattern line application is more standardized: wholly determined by the underlying tessellation, without the inclusion of secondary pattern lines. The historical patterns in this

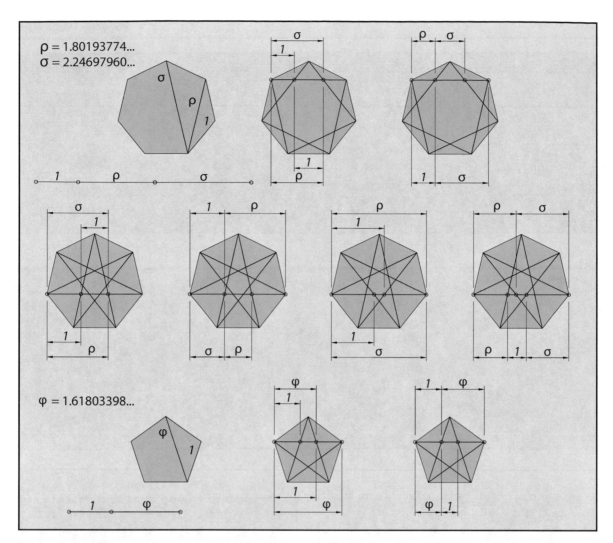

ρ = 1.80193774...
σ = 2.24697960...

φ = 1.61803398...

Fig. 277

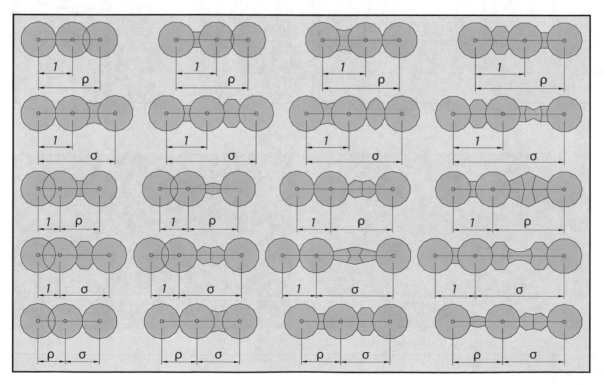

Fig. 278

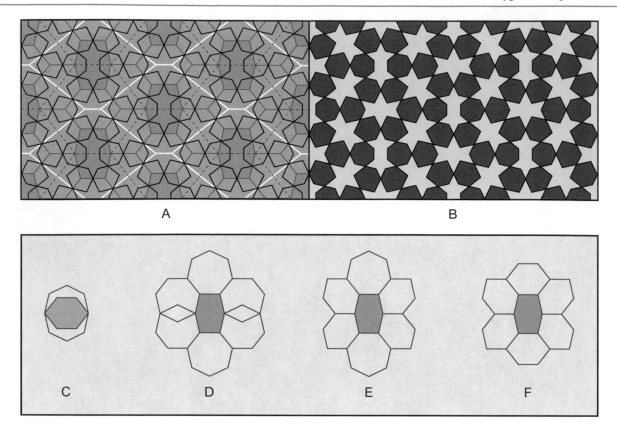

Fig. 279

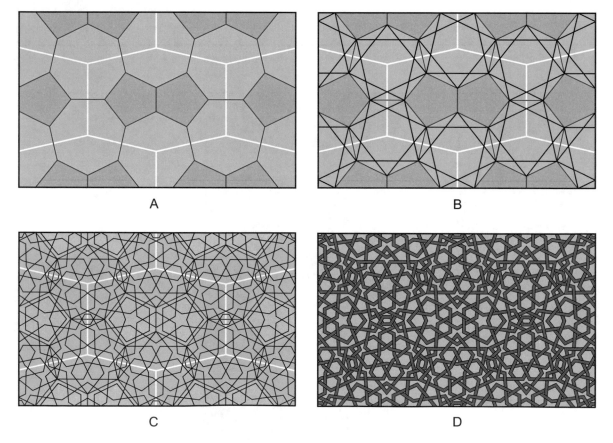

Fig. 280

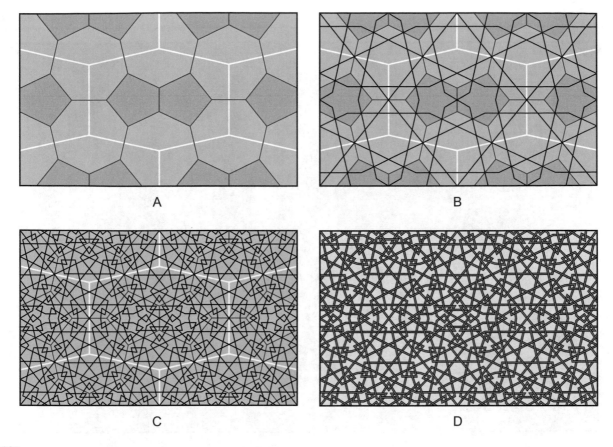

A

B

C

D

Fig. 281

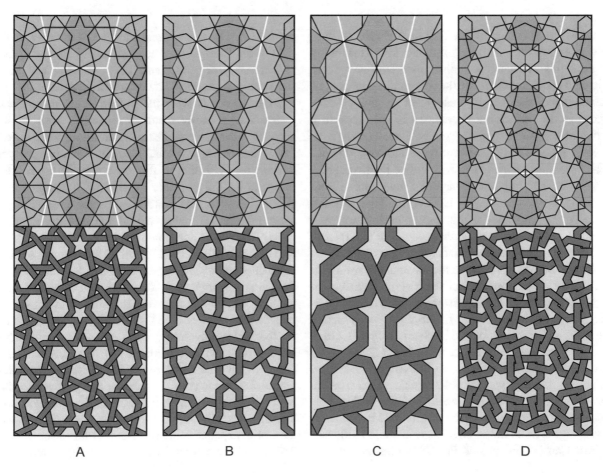

A

B

C

D

Fig. 282

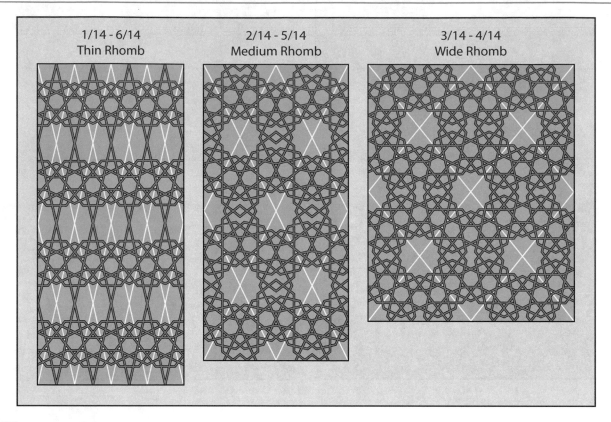

Fig. 283

figure were produced during the Seljuk Sultanate of Rum roughly a century later than the examples from Ghazni, by which time this ornamental tradition had reached greater maturity and codification. While the pattern line applications are standardized, the simple deployment of edge-to-edge heptagons in the underlying tessellation is not necessarily indicative of a systematic methodology. Certainly, the heptagon and pentagonal interstice elements are both modules from the *sevenfold system*. However, unlike the two underlying hexagons and implicit heptagon of the example from the northeast dome chamber in Isfahan, the heptagon and interstice pentagons of these four examples will not rearrange into other tessellations on their own. For this reason, despite the fact that these two modules are members of the larger family of polygons contained within the *sevenfold system*, the artists responsible for the historical designs in this figure were not necessarily aware of the otherwise systematic nature of the sevenfold patterns they constructed. Figure 282a is an *acute* pattern [14-s5] from the Great Mosque of Dunaysir in Kiziltepe, Turkey (1200-04), as well as at the Alaeddin mosque in Nidge (1223). Figure 282b shows a *median* pattern [14-s4] that, on its own, is not known within the historical record, but was the basis for the more complex design from the minaret of Mas'ud III in Fig. 281b. Figure 282c shows an *obtuse* pattern [14-s2] from the Eğirdir Han (1229-36), and Fig. 282d illustrates a *two-*

point pattern from the Great Mosque of Malatya in Turkey (1237-38). These three Anatolian examples were produced within a 38-year period, and it is possible that all three are the work of a single person or artistic lineage.

Mamluk artists were the first to develop the *sevenfold system* into its fully mature expression; with 14-pointed stars, considerable complexity resulting from the large number of underlying polygonal modules, and diverse repetitive stratagems. With the notable exception of the rectangular design at the Sultan al-Mu'ayyad Shaikh complex in Cairo (1412-22), all of the historical designs created from this system during its mature expression repeat upon rhombic grids. There are three rhombi that are the product of sevenfold symmetry [Fig. 10]. Figure 283 shows grids made up of these three rhombi along with applied 14-s4 *median* patterns (by author). Only the medium rhombus with 2/14 and 5/14 included angles and the wide rhombus with 3/14 and 4/14 included angles were used historically as repeat units. Unlike the fourfold and fivefold systems, none of the known historical sevenfold designs made use of hybrid repetitive cells. However, the methodology of the polygonal technique is also well suited to providing greater design diversity through hybrid repetitive constructions with the *sevenfold system*. This is especially relevant to contemporary artists with an interest in expanding the repertoire of the *sevenfold system*. To this end, the patterns in each of these three rhombic repeat

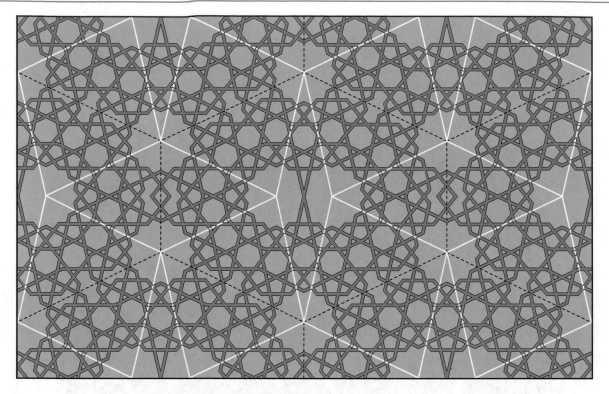

Fig. 284

units in Fig. 283 have identical edge configurations, allowing them to also be used in combination with one another. The design in Fig. 284 employs all three of the rhombi from Fig. 283. This hybrid 14-*s4 median* design (by author) has overall hexagonal translation symmetry (dashed lines). The design in Fig. 285 uses the same three rhombi in an array that tessellates the tetradecagon with 14-fold radial symmetry. It is interesting to note that these three rhombi can also be used to cover the plane non-periodically.

The earliest known example of a design created from the more mature expression of the *sevenfold system* is from one of the stone lintels in the south elevation of the Qawtawiyya *madrasa* in Tripoli, Lebanon[38] (1316-26). This design, along with its underlying generative tessellation is illustrated in Fig. 286a. This is a 14-*s2 obtuse* pattern in which the underlying tetradecagons are filled with a star rosette that follows the convention of the *fivefold system* [Fig. 222]. Figure 286b employs the same generative tessellation but with an added ring of 14 trapezoids within the tetradecagons. As with the pentagons in this tessellation, these trapezoids are also truncated heptagons [Fig. 271]. This is a 14-*s4 median* pattern that was used in a carved stucco panel at the Amir

Burunduq mausoleum at the Shah-i Zinda complex in Samarkand (1390-1420), as well as in a carved stone panel in the exterior façade of the Amir Qijmas al-Ishaqi mosque in Cairo (1479-81). Figure 286c demonstrates how the same pattern can be created from the dual of the tessellation in Fig. 286b, in which case this design can be categorized as a 14-*s4 acute* pattern. In both cases, the polygonal modules of their respective tessellations are members of the *sevenfold system*. Both of the designs in Fig. 286 repeat upon the medium rhombic grid with 2/14 and 5/14 included angles [Fig. 10b]. Figure 287 illustrates the only other historical design known to repeat upon this rhombic grid. Figure 287a shows the 14-*s2 obtuse* pattern along with the underlying generative tessellation comprised of tetradecagons, concave hexagons, and edge-to-edge triangles. Figure 287b shows a widened line version of this pattern that was used in a side panel of the wooden *minbar* in the Sultan Barsbay complex at the northern cemetery in Cairo[39] (1432). Figure 287c shows a subtractive variation of this design that removes the pattern lines from one of the underlying triangular modules. This earlier version was used on an Ottoman wooden door of the Bayezid Pasa mosque in Amasya, Turkey (1414-19).

---

[38] This pattern was first illustrated in Tripoli, the Old City: Monument Survey-Mosques and Madrasas: A Sourcebook of Maps and Architectural Drawings. Beirut: American University of Beirut, Department of Architecture. Saliba [ed.] (1994).

[39] This design is also found in the Coptic entry door at the Hanging Church (al-Mu'allaqa) in Cairo. However, this door appears to have been produced during a more recent restoration.

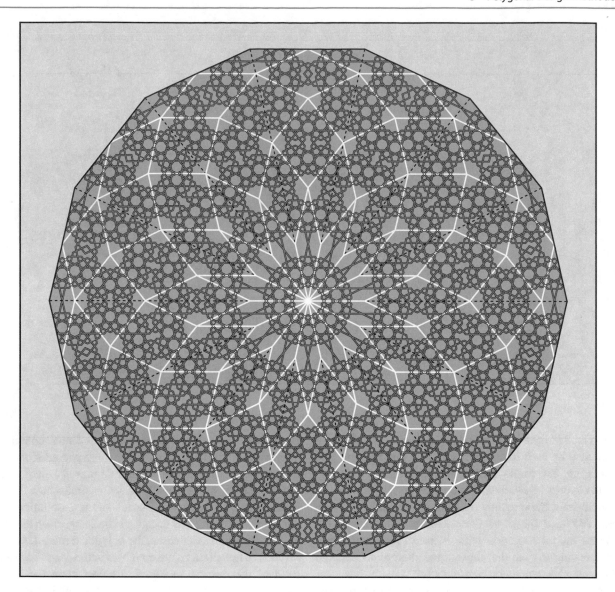

Fig. 285

Following its rise to full maturity, the majority of designs created from the *sevenfold system* repeat upon a rhombic grid with 3/14 and 4/14 included angles [Fig 10a]. The design with prominent 14-pointed stars separated by twin 5-pointed stars in Fig. 288 repeats upon this grid. This is from a Mamluk door at the Sultan Qansuh al-Ghuri complex in Cairo, Egypt (1503-05) [Photograph 49]. Figure 288a illustrates the derivation of this pattern from an underlying tessellation of edge-to-edge tetradecagons and interstice concave decagons. The pattern lines associated with this tessellation are categorized as 14-*s*1 of the *obtuse* family with an additive 14-fold rosette within each tetradecagon. Figure 288b employs an alternative generative tessellation of tetradecagons, pentagons, trapezoids, and concave hexagons; and the associated pattern lines in this derivation fall into the 14-*s*6 *acute* family. As is often the case, these alternative tessellations have a dual relationship. The

additive process for creating the central 14-pointed stars in Fig. 288a is very straightforward, and one has to assume would have been well within the skill set of any artist work at this level of sophistication, and the original designer of this very successful design is as likely to have used one of these underlying tessellations as the other. Figure 289c illustrates an *acute* design from the ceiling of the courtyard cistern at the Suleymaniya mosque in Istanbul (1550-58) [Photograph 81]. This same basic design was used in the courtyard of a house that belonged to a Christian trader in Aleppo (1757), albeit with a different arbitrary treatment applied to the centers of the 14-pointed stars. Figure 289a illustrates how this pattern repeats upon the rhombic grid with 3/14 and 4/14 included angles, and is generated from an underlying tessellation comprised of tetradecagons placed at each repetitive vertex separated by a barrel hexagon on each repetitive edge, trapezoids and central concave decagonal interstice regions.

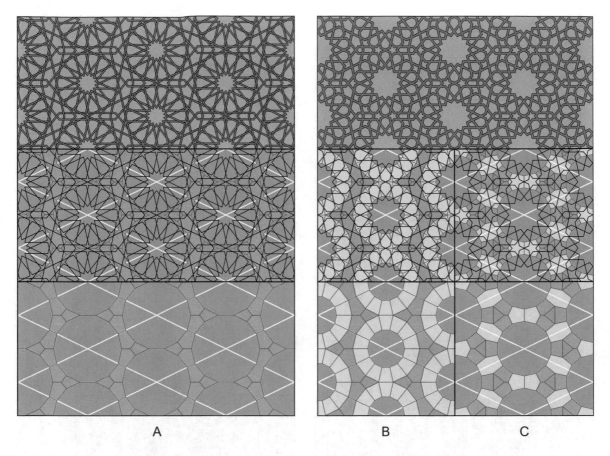

A   B   C

Fig. 286

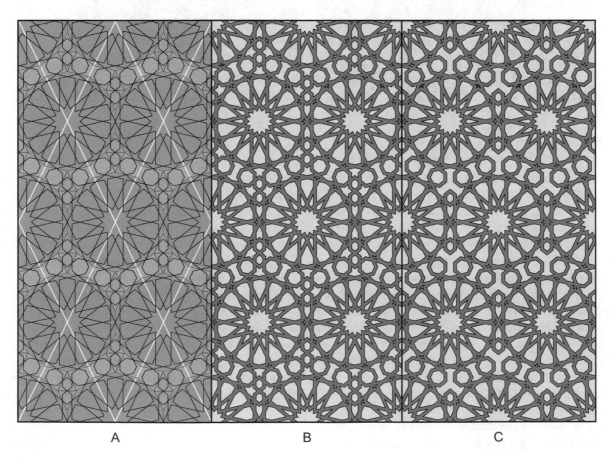

A   B   C

Fig. 287

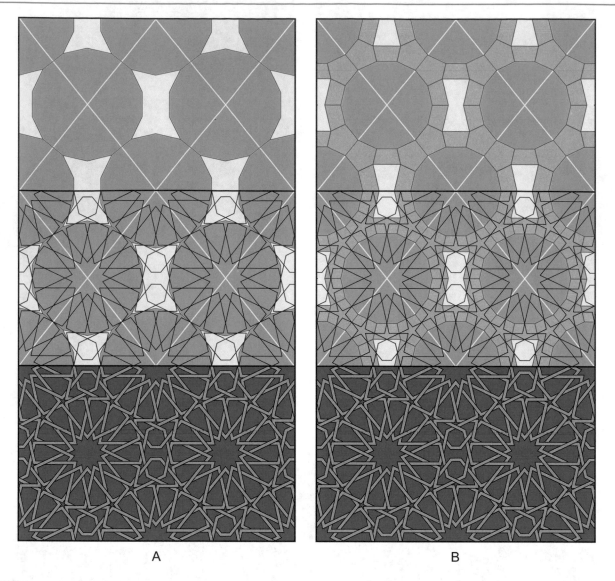

A   B

Fig. 288

Of course the barrel hexagons are the equivalent of two contiguous trapezoids, and the outer long edges of this arrangement of trapezoids are identical to that of Fig. 288a. Figure 289b shows how the vertices of the two kite motifs located within the interstice regions are fixed upon the vertices of the underlying trapezoids. This is an unusual feature, but highly effective in these circumstances. Figure 289c includes the design modifications at the center of the 14-pointed stars that are present in the example from the Suleymaniya. This involves extending every other line of the central 14-pointed star so that a 7-pointed star is created, and further filling each 7-pointed star with a ring of pentagons surrounding a central heptagon. These seven pentagons actually have a functional purpose: they enclose nodules that spray water from the ceiling into the pool of water within the cistern. The underlying generative tessellation in Fig. 290a is identical to that of Fig. 289a except that the interstice regions

have been filled with two varieties of triangle and a central rectangle. Other than this region with introduced triangles and rectangle, the applied pattern lines in Fig. 290b are identical to the design in Fig. 289b, and remain in the 14-*s*6 *acute* family. The pattern lines associated with the central rectangle produce an octagon that is almost regular. This design is from the *minbar* doors of the Haram al-Ibrahimi in Hebron, Palestine. This *minbar* was produced by Fatimid artists for the Mashhad Nabi Hussein in al-Majdal Asqalan, Palestine[40] (1191-92), and moved to its current location in Hebron by Ṣalāḥ ad-Dīn a century later. However, the pattern with 14-pointed stars on the *minbar* doors is distinctly Mamluk, and clearly a later addition, as is the pattern with 12-pointed stars on the back panel of the

---

[40] Now Ashkelon, Israel.

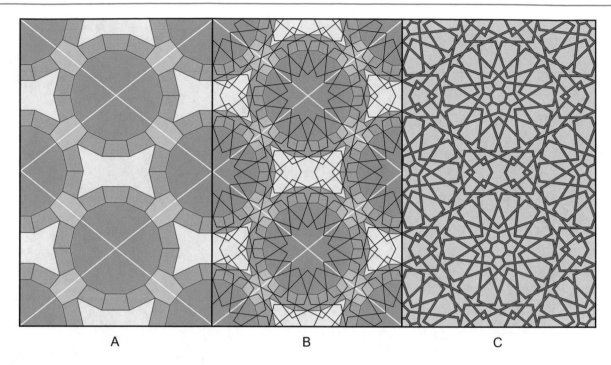

Fig. 289

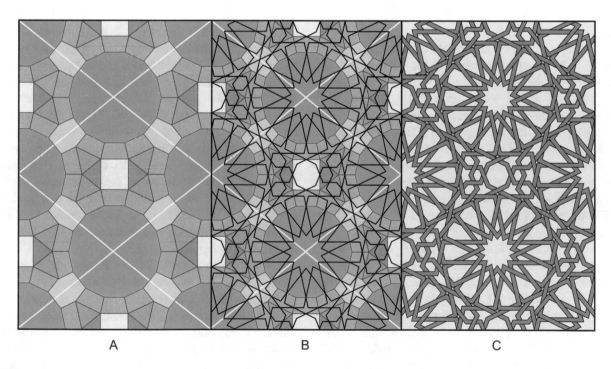

Fig. 290

minbar's platform. Figure 290c employs and interweaving line as per the historical example from Hebron. The underlying tessellation in Fig. 291a provides an alternative polygonal infill of the concave decagonal interstice regions from Fig. 289a. This infill is comprised of four kite-shaped quadrilaterals and an irregular hexagon at the center. This design was recorded by Bourgoin,[41] but its location is unattributed. Considering that Bourgoin was working in Egypt, Syria, and the Levant, this design is presumably Mamluk,

---

[41] Bourgoin (1879), pl. 165.

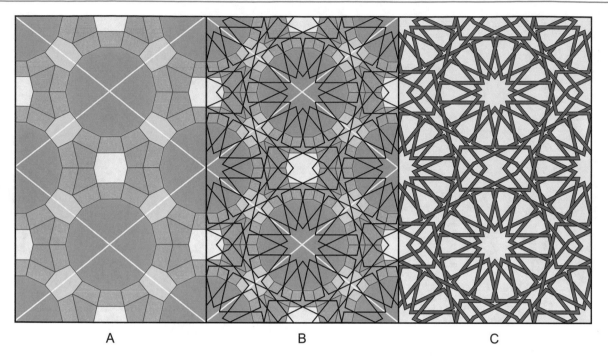

A   B   C

Fig. 291

but could also be Ottoman. Figure 291b demonstrates the placement of the pattern lines onto the underlying polygonal edges, and other than the region of the underlying kites and hexagon, this pattern is the same as Figs. 289 and 290. As such, it is also in the 14-*s*6 *acute* family. Figure 291c represents the completed design with interweaving lines as per Mamluk and Ottoman convention. The two designs in Fig. 292 are produced from the same underlying tessellation of tetradecagons separated by concave hexagons, with conjoined heptagons in the center of each repeat. The repeat unit for this design is the same rhombus with 3/14 and 4/14 included angles. Figure 292a illustrates a 14-*s*2 *obtuse* pattern from the *minbar* door at the ʿAbd al-Ghani al-Fakhri mosque in Cairo (1418). The underlying conjoined heptagons are responsible for the two point-to-point seven-pointed stars that are a distinctive feature of this design. This design was also collected by Bourgoin,[42] but its location is similarly unattributed; and Bourgoin's work was the likely source for Ernest Hanbury Hankin's analysis that includes the underlying generative tessellation, but remains unattributed.[43] Figure 292b shows a 14-*s*4 *2-point* design that was used on the wooden congregational Quran stand in the Sultan Qansuh al-Ghuri complex in Cairo, Egypt (1503-05). This is the only known *two-point* pattern created from the sevenfold system during the period of full maturity. The

design in Fig. 293 is 1 of only 2 known sevenfold examples that originate in the eastern regions during the period of full maturity, the other being the example in Fig. 286b. This also utilizes the rhombic repeat unit with 3/14 and 4/14 included angles, and is found in the Timurid shrine complex of Imam Reza in Mashhad, Iran (1405-18). Figure 293a shows the underlying generative tessellation comprised of tetradecagons separated by long hexagons, with pentagons and shorter hexagons within the center of each rhombic repeat unit. Figure 293b shows the derivation of the 14-*s*3 *median* pattern. This is readily apparent as analogous to the *median* designs created from the *fivefold system*. Figure 293c shows the completed pattern with interweaving lines as per the historical example.

The design in Figure 294 is the only known historical sevenfold example that repeats upon a rectangular grid. The proportions of each rectangular repeat unit appear at first glance to be a square, but are actually not quite equilateral. This beautiful design is from the side panels of the minbar in the Sultan al-Muʾayyad Shaykh complex in Cairo (1412-22) [Photograph 50]. Figure 294a illustrates the unusual non-aligned edges of the adjacent tetradecagons where they are separated by two mirrored triangles. Figure 294b demonstrates how the pattern lines in this region have an interesting skewed dynamic that results from this nonaligned triangular configuration. This is a 14-*s*2 *obtuse* pattern with a 14-fold star rosette in the center of each tetradecagon. Figure 294c represents the interweaving design as per the Mamluk historical example.

---

[42] Bourgoin (1879), pl. 167.
[43] Hankin 1925a, pl. 35.

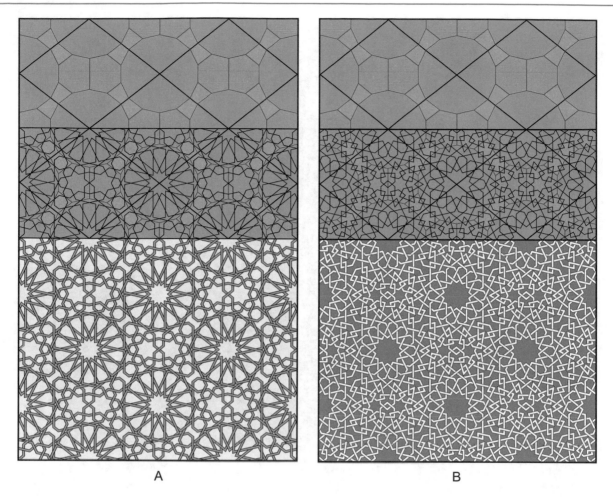

Fig. 292

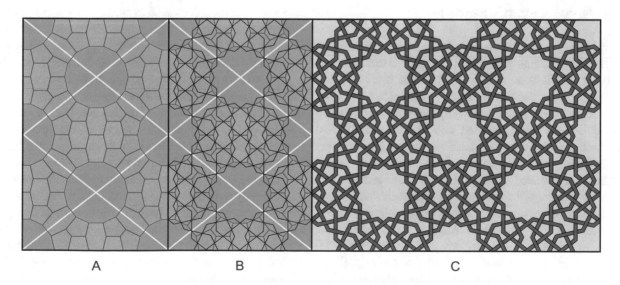

Fig. 293

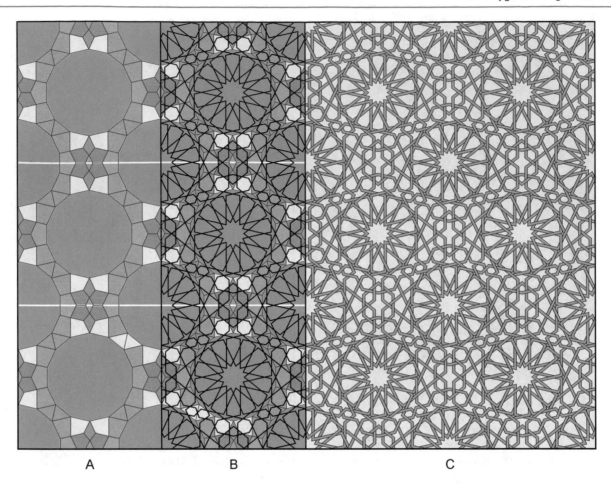

Fig. 294

A   B   C

3.2   Nonsystematic Patterns

As mentioned previously, patterns created from the polygonal technique fall into two broad categories: those whose
underlying tessellations are systematic, and those that are
nonsystematic. As expounded in the previous section, systematic design methodology utilizes a limited set of polygonal modules that are assembled into diverse tessellations
onto which associated pattern lines are placed. By contrast,
nonsystematic patterns are derived from underlying
tessellations comprised of polygons that are specific to the
tessellation and will not reassemble into additional
arrangements. Both types of underlying tessellation allow
for the creation of designs in each of the four pattern
families. Nonsystematic designs vary in complexity between
those with only a single variety of primary star form, to those
that include multiple regions of local symmetry expressed as
multiple star forms within a single pattern. Indeed, it was
through the nonsystematic use of the polygonal technique
that Muslim artists were able to produce the extraordinary
patterns with such unusual combinations as 7- and 9-pointed

stars, 9- and 11-pointed stars, 11- and 13-pointed stars, and
sequential combinations such as 9-, 10-, 11-, and 12-pointed
stars. Nonsystematic design methodology exploits the full
range of repetitive strategies, including orthogonal and isometric grids, rhombic grids, rectangular grids, and both
regular and irregular hexagonal grids. Without wanting to
diminish the remarkable achievements in the historical use
of the five design systems, the creative energy spent on
developing the nonsystematic design methodology resulted
in many of the most geometrically sophisticated and innovative patterns known to this ornamental tradition.

Nonsystematic patterns frequently include the more common primary stars with 6, 8, 10, and 12 points. These are also
standard features among systematic design methodology: 6-
and 12-pointed stars to the *system of regular polygons*,
8-pointed stars to both the *fourfold system A* and the *fourfold
system B*, and 10-pointed stars to the *fivefold system*. Patterns
with less accessible star forms, for example, those with 9, 11,
13, and 15 points are invariably nonsystematic. In examining
diverse design methodologies, the previous chapter details
why the polygonal technique is the only traditional methodology that allows for the creation of patterns with these more

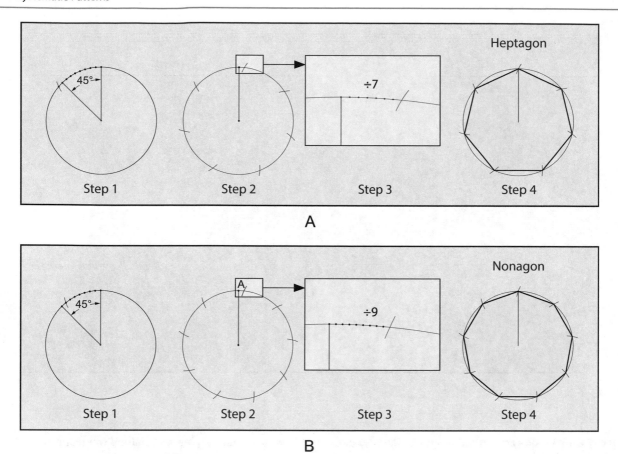

Fig. 295

enigmatic symmetries. More specifically, the previous chapter examines the question of whether Muslim geometric artists were dependant upon mathematicians for instructions in the use of conic sections and/or approximate constructions for accurately producing the higher order regular polygons that enable the creation of these types of stars. Historians of Islamic art and mathematics have tended to overlook a very basic, yet very accurate method of creating these otherwise problematic polygons. Figure 295 demonstrates a simple, fast, and effective method of drawing polygons that otherwise requires complex mathematical procedures to construct. This method approximates the regular polygon through dividing the circumference of a circle into the requisite number of segments with a compass or pair of dividers. Step 1 of Fig. 295a demonstrates the drawing of the heptagon by dividing a 1/8 segment of a circle (45°) into eight approximately equal parts (halves, quarters, eighths). The compass is then set at a length that is approximately 1/8 larger than the 45° segment. Step 2 places marks progressively around the circle seven times. The last mark will have a slight shortfall of the vertical starting point. Step 3 divides the shortfall into seven approximate equal parts so that the compass setting is increased by the 1/7 division. This is then used to re-remark the circle from the initial vertical position. Step 4 creates the heptagon by connecting the new marked divisions. Figure 295b demonstrates the drawing of the nonagon in the same fashion, except that the first setting of the compass is decreased by 1/8 rather than increased. If, during Steps 2 and 3, the compass falls beyond the vertical starting point, decrease the compass setting by an amount equal to a division of the long-fall by the number of sides of the intended polygon. This same approximate method can be used with equal ease to draw higher order polygons such as those with 11, 13, 14, 15, etc. sides. This approximate technique is as accurate from a practical standpoint as using conic sections. By way of example, if one is making an enneagon with conic sections, while the end result will be theoretically precise, the actual drawing will only be as accurate as one's drawing skills and equipment allow. People are not computers, and such drawings will inevitably have inaccuracies, and these inaccuracies will be no less that those resulting from the approximate technique outlined above. With practice, in two or three incremental steps, one can quickly divide the circle into less tangible divisions that are, for all intents and purposes, functionally accurate.

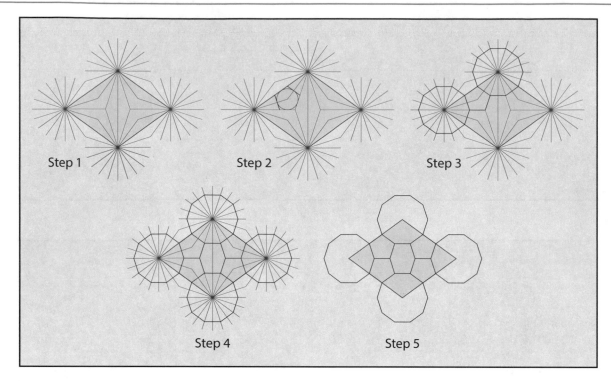

Fig. 296

Nonsystematic design methodology has three sequential phases: the construction of a radii matrix; the making of the underlying generative tessellation; and the extraction of the geometric pattern. Radii matrices provide a very effective means of constructing the underlying generative tessellations of both simple and complex designs, and evidence from the Topkapi Scroll indicates that these were used to set up the underlying tessellations for both nonsystematic[44] and systematic designs.[45] The radii matrices present in the Topkapi Scroll are invariably un-inked lines (dead lines) scribed into the surface of the paper with a steel stylus. In the case of systematic designs, as explained previously, the underlying tessellations are comprised of modular polygonal elements, with associated pattern lines, that are assembled into different combinations. However, in laying out a given combination for placement of a design on a wall or on paper (such as a Quranic frontispiece or the Topkapi Scroll), radii matrices provide a very effective means of accurately drawing the underlying tessellation, whether it be systematic or nonsystematic.

The radii matrix establishes the regions of local symmetry within a given design. For nonsystematic patterns of low complexity this might be the 12-pointed stars on the vertices of either an isometric or orthogonal repetitive grid. With more complex patterns, these regions of local symmetry allow for the placement of different stars with *n*-fold symmetry at the repetitive vertices, centers of the repeat, midpoints of the repetitive edges, and/or within the field of the repeat unit. Figure 296 demonstrates the use of a radii matrix to construct the well-known fivefold underlying tessellation that repeats upon a rhombic grid. The construction of tessellations from radii matrices typically begins with the establishment of the pentagons, followed by the primary polygons with *n*-fold local symmetry (in this case decagons), followed by interstice regions (in this case the barrel hexagon). The radii matrix in Fig. 296 is associated with the *fivefold system*. Although this is systematic, the proportional regularity inherent within the *fivefold system* provides a useful demonstration of the ideal relationship between the radii matrix and its resulting tessellation. In this example, the pentagons are regular, and the edge lengths of all the polygonal elements are the same. When this methodology is applied to nonsystematic pattern generation the pentagonal proportions and edge lengths invariably become irregular. However, as a general rule, the closer they are to the ideal—as exemplified by the characteristics of fivefold

---

[44] Examples from the Topkapi Scroll that use radii matrices for drawing nonsystematic designs include: diagram nos. 30 (13 and 16-pointed stars), 35 (8 and 12-pointed stars), 39 (10 and 12-pointed stars in the rectangular portion), 42 (9 and 11-pointed stars), 44 (10 and 12-pointed stars), and 63 (12-pointed stars on the isometric grid).

[45] Examples from the Topkapi Scroll that use radii matrices for drawing systematic designs include: diagram nos. 33, 53, 54, 55, 64, 73, and 90a from the *fivefold system*; and nos. 39 (square portion) and 57 from the *fourfold system B*.

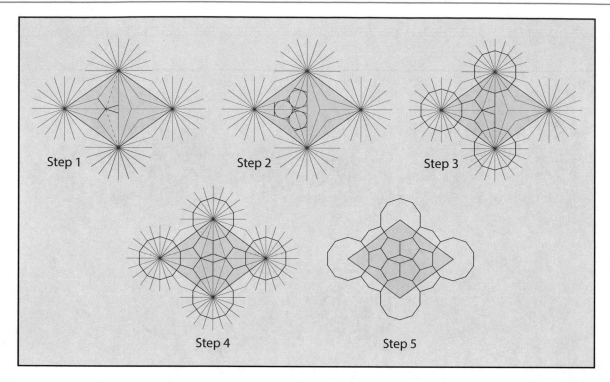

Fig. 297

symmetry—the better the quality of the geometric pattern that is produced from the generative tessellation. Conversely, the greater the disproportion within the generative polygons the less likelihood of creating a successful pattern. It is important to note that in nonsystematic pattern making, while the pentagons, hexagons, and other elements within the polygonal matrix that separate the primary polygons are not regular, the primary polygons with *n*-fold local symmetry are always regular. This provides for the regularity of the *n*-pointed stars that characterize this design tradition. Step 1 of Fig. 296 shows an array of 20 radii placed at each vertex of the standard wide rhombus associated with fivefold symmetry, with the lines extended into the rhombus until they meet with other extended lines. This methodology is often assisted by placing twice the number of radii (in this case 20) as there are number of sides to the primary polygon (in this case 10). In Step 2 a circle is drawn that is tangent to the red radii, and lines are drawn that intersect the circle and are perpendicular to the two blue radii that meet at the center of the circle. Step 3 shows the regular pentagon that these two lines create, as well as the two decagons created by rotating each line ten times around their respective rhombic vertex. Step 4 mirrors the pentagons and decagons, thereby creating the barrel hexagon at the center of the rhombus. Step 5 shows the completed underlying tessellation without the radii. Note: The edges of the pentagons are congruent with the red radii.

Just as a single underlying tessellation will create multiple geometric patterns, one of the potent features of radii

matrix design methodology is the ability of a single radii matrix to generate more than one underlying polygonal tessellation. This provides for a relatively large number of geometric patterns that can be created from a single radii matrix. Figure 297 shows the construction of another well-known fivefold underlying tessellation that is created from the same radii matrix as that of Fig. 296. The difference between the constructions of these two underlying tessellations is in the position of the pentagons relative to the 20 radii at each rhombic vertex. In this example, the pentagons have congruent edges with the blue radii rather than the red radii. In each case, the first objective is to create the pentagons, and from the pentagons, the decagons. Step 1 shows how this is achieved by drawing the lines that connect the vertices of the red and blue radii, and mirroring these on the indicated red radii (dashed lines). Step 2 demonstrates how the pentagons are established by following the same method as the previous example: by introducing circles that are tangent to the relevant radii and applying lines that are perpendicular to the radii. Step 3 shows how the pentagonal edges create the decagons, and Step 4 produces the thin rhombus, and barrel hexagons on each repetitive edge through mirroring the pentagons and decagons. Two features of this tessellation are the primary polygons (in this case decagons) being separated along the edge of the repeat unit by barrel hexagons, and the cluster of six pentagons surrounding the thin rhombus. Each of these common fivefold features is also encountered frequently in

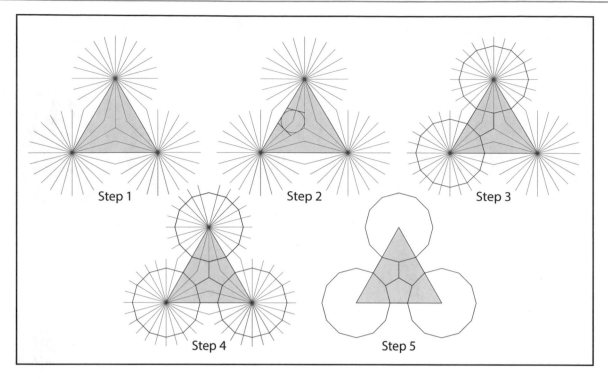

Fig. 298

the application of this radii matrix based design methodology to nonsystematic pattern generation, although the pentagons will not be regular and the edge lengths of the polygons will not all be identical.

The most basic nonsystematic patterns place just one variety of primary star at the vertices of the repetitive grid. The radii matrices for such designs are simple geometric structures, as are the underlying tessellations that they will produce. Figure 298 illustrates an isometric radii matrix that creates underlying tessellations with dodecagons at each vertex of the triangular repeat. Three irregular pentagons clustered at the center of the triangle separate these dodecagons. Step 1 places an array of 24 radii at each corner of the triangle. Step 2 places a circle that is tangent to the three red radii, and places two lines that are perpendicular to the blue radii and intersect the circle and blue radii. Step 3 identifies the irregular pentagon that is implicit in Step 2, and uses the two lines from Step 2 to produce the two dodecagons. Step 4 mirrors these elements to complete the tessellation, and Step 5 illustrates the tessellation without the radii. Figure 299 shows another underlying tessellation created from the same isometric radii matrix as the example in Fig. 298. The dodecagons in this example are separated by barrel hexagons, with a similar cluster of three irregular pentagons at the center of the repeat. As established in Steps 1, 2, and 3, the pentagons in this underlying tessellation are tangent with the blue radii. Figure 300a illustrates the *acute*, *median*, *obtuse*, and *two-point* patterns created

from the underlying tessellation in Fig. 298, while the four patterns in Fig. 300b were created from the underlying tessellation in Fig. 299. The tremendous potency of this methodological practice is demonstrated by the fact that all eight of these designs are created from two underlying tessellations that in turn are produced from just a single radii matrix. Numerous examples of the *acute* pattern in Fig. 300a are known to the historical record, and locations include the Great Mosque of Niksar in Turkey (1145); the Izzeddin Keykavus hospital and mausoleum in Sivas (1217); the Abbasid Palace of the Qal'a in Baghdad (c. 1220); the Great Mosque of Divrigi (1228-29); the mausoleum of Uljaytu in Sultaniya, Iran (1307-13) [Photograph 82]; the Friday Mosque at Varamin, Iran (1326); the Amir Qawsun mosque in Cairo[46] (1329-1330) [Photograph 52]; the *madrasa* al-Mirjaniyya in Baghdad (1357); the Taşkın Paşa mosque in Damsa Köy bei Ürgüp, Turkey[47] (c. mid-fourteenth century); the Khatuniyya *madrasa* in Tripoli, Lebanon (1373-74); and the Amir Mahmud al-Ustadar complex in Cairo (1394-95). The *median* design in Fig. 300a was used at the Mustansiriyah *madrasa* in

---

[46] This example is from a pair of *minbar* doors is in the collection of the Metropolitan Museum of Art in New York City: accession number 91.1.2064.

[47] This example is from a Karamanid wooden *mihrab* currently in the collection of the Ethnography Museum of Ankara: inventory no. 11541.

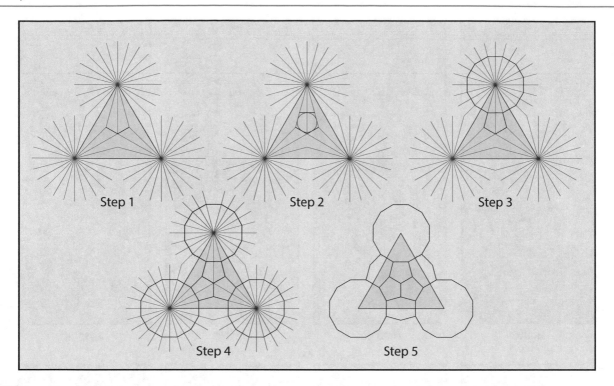

Fig. 299

Baghdad (1227-34), as well as at the mausoleum of Uljaytu at Sultaniya, Iran (1313-14). The *obtuse* design in Fig. 300a (by author) is not known to the historical record, but is similar to a well-known and superior pattern created from the *system of regular polygons* [Fig. 108a]. An example of the *two-point* pattern in Figure 300a is found at the Ribat Ahmad ibn Sulayman al-Rifa'i in Cairo (1291). The *acute* design in Fig. 300b was used in multiple locations, including the exterior ornament of the Abbasid Palace of the Qal'a in Baghdad (c. 1220), a pair of Mamluk doors at the al-Azhar mosque, as well as two representations in the Topkapi Scroll.[48] It is interesting to note that one of the examples from the Abbasid Palace of the Qal'a is a hybrid construction that combines the triangular repetitive cell of this example with a square repeat that shares geometric information along the shared repetitive edges [Figs. 23d–f], and that this same hybrid use of the triangular and square repetitive cells is also shown in diagram number 35 of the Topkapi Scroll. While use of the *median* pattern in Fig. 300b (by author) is not known historically, this design meets the aesthetic criteria of this tradition. The *obtuse* pattern in Fig. 300b was used in the Aq Qoyunlu ornament of the Friday Mosque at Isfahan (1475), and the *two-point* pattern in this figure was used in several Mamluk locations, including a large wooden door in

the Sultan al-Mu'ayyad Shaykh complex in Cairo (1415-22); the portal of the Ribat Khawand Zaynab in Cairo (1456); and a carved stone lintel in the Ashrafiyya *madrasa* in Jerusalem (1482).

Figure 301 demonstrates the construction sequence of four additional underlying tessellations that can also be created from the radii matrix introduced in Fig. 298. The upper two have the same polygonal configuration on each edge of the triangular repeat, whereas the two lower examples have two identical edges and one unique edge. Figure 302 shows the applied pattern lines from each of the four pattern families to the additional underlying tessellations created in Fig. 301. The patterns in the far left column are *acute* designs, those in the central left column are *median* designs, those in the central right column are *obtuse* designs, and those in the far right column are *two-point* designs. None of these 16 patterns are known to the historical record, and some are more acceptable to traditional aesthetic conventions than others. These 16 designs (by author), in addition to the eight patterns illustrated in Fig. 300, demonstrate the high level of generative potential of just a single radii matrix.

Figure 303 demonstrates the variety of isometric tessellations that can be created from a single radii matrix comprised of two centers of local symmetry. This radii matrix places 24 radii at each corner of the triangular repeat, and 18 radii at the center of the repeat. Each of the five tessellations places dodecagons at the triangular corners and nonagons at the center of each repeat. These, in turn, produce 12- and 9-pointed stars, respectively. As demonstrated

---

[48] Necipoğlu (1995), diagram nos. 35 (triangular portion with subtractive variation at the center of the triangle) and 63 (also shown with the same subtractive variation).

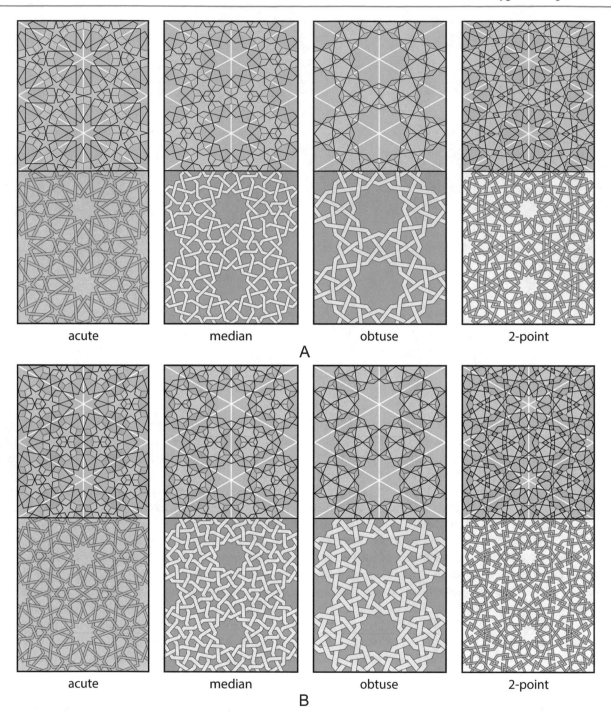

Fig. 300

in the previous examples, each of these five tessellations will generate distinct patterns from each of the four pattern families (not shown).

Figure 304 demonstrates the variety of orthogonal tessellations that can be created from a single radii matrix comprised of two centers of local symmetry. This radii matrix places 24 radii at each corner of the square repeat, and 16 radii at the center of the repeat. Each of the five tessellations places

dodecagons at the corners of the square and octagons at the center of each repeat. These, in turn, produce 12- and 8-pointed stars, respectively. Each of these five tessellations will generate patterns from each of the four pattern families (not shown). Only the tessellations in the top two rows (A and B) are known to have been used historically to generate geometric patterns.

Figure 305 demonstrates the variety of rectangular tessellations that can be created from a single radii matrix

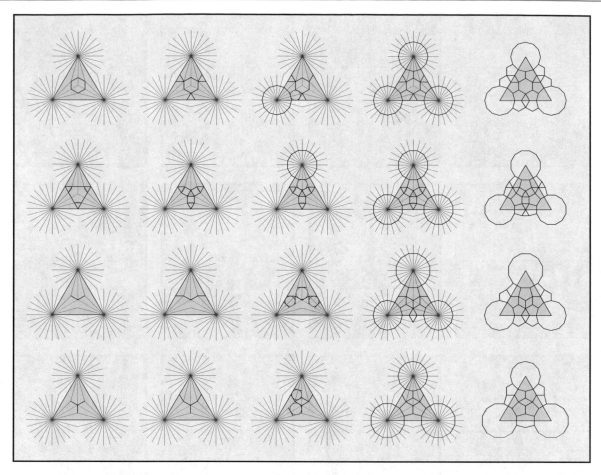

Fig. 301

comprised of two centers of local symmetry. This radii matrix places 24 radii at each corner of the rectangular repeat, and 20 radii at the center of the repeat. Each of the five tessellations places dodecagons at the corners of the rectangle and decagons at the center of each repeat. These, in turn, produce 12- and 10-pointed stars, respectively. Each of these five tessellations will create patterns from each of the four pattern families (not shown). Only the underlying tessellation in the top row is known to have been used historically [Fig. 414].

Figure 306 shows three square tessellations that can be created from a single radii matrix comprised of four centers of local symmetry. This radii matrix places 24 radii at the corner of the square repeat, 8 radii at the center of the repeat, 20 radii at the midpoints of the repeat, and 18 radii within the field of the polygonal matrix. These regions of local symmetry correspond to dodecagons, octagons (or regions with fourfold symmetry), decagons, and nonagons respectively. These, in turn, produce complex patterns with 12-, 10-, 9-, and 8-pointed stars (or octagons). As in the previous figures, each of these three tessellations will generate four distinct patterns; one from each of the pattern families (not shown). Only the upper tessellation is known to the historical record [Fig. 400].

Figure 307a illustrates the classic *acute* pattern created from the *fivefold system* along with a highlighted detail. The five- and ten-pointed stars have five- and tenfold rotational symmetry, respectively, and the applied pattern lines uniformly bisect the midpoints of each underlying polygonal edge. This uniformity is also seen in the standardized 36° angular opening of the crossing pattern lines at each midpoint of the underlying polygonal edge. By contrast, Fig. 307b shows how nonsystematic patterns do not share this inherent uniformity. While both patterns share the same basic configuration of pentagons and barrel hexagons that surround their respective primary polygons, the underlying pentagons in Fig. 307b are not regular, and the polygonal edge lengths are not all equal. As such, the five-pointed stars do not have rotational symmetry. As mentioned previously, when creating a nonsystematic design, the general objective is to create pattern elements that are as close as possible to the ideal proportions exemplified in the *fivefold system*. In order to achieve this, the placement of the crossing pattern lines upon the polygonal edges of the underlying tessellation will not always be located precisely at the midpoints, but may have to move up or down the polygonal edge in order to produce better looking design proportions. This is

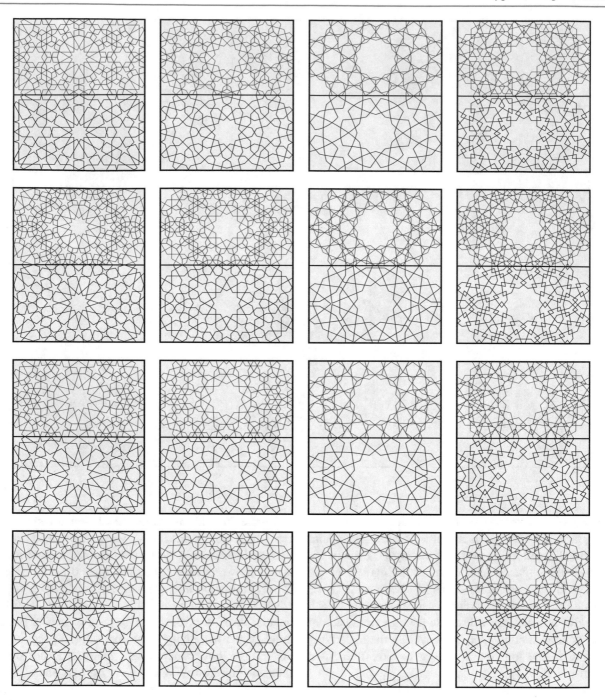

Fig. 302

demonstrated by the pattern line placements within the green squares in Fig. 307b. Similarly, in order to achieve a more balanced effect, it is also occasionally necessary to move the intersection of the crossing pattern lines slightly off of the underlying polygonal edge, as shown in the crossing pattern lines within the green circles in Fig. 307b. What is more, it is often necessary for the angles of the crossing pattern lines to slightly vary from location to location. The precise placement and angles of the pattern lines are subtle aesthetic decisions

made by the artist. The one consistent area of uniformity is in the rotational symmetry of the primary star forms.

As nonsystematic patterns become more complex they are more likely to contain design elements that are asymmetrical. Figure 308 is an orthogonal design with 12- and 16-pointed stars, as well as 7-pointed stars within the pattern matrix [Fig. 396b]. Many of the constituent shapes that make up this design are asymmetrical, including the shapes that have been highlighted in blue and green. The aesthetics of this tradition

Fig. 303

are highly reliant upon symmetry, and patterns with a greater preponderance of asymmetrical components are generally less likely to be pleasing to the eye. However, the visual discord of a given asymmetrical pattern element is rectified through reflection. In this way, asymmetrical elements are paired with identical elements through reflection; thereby providing the symmetry that is fundamental to the visual appeal of this tradition. The blue truncated stars in Fig. 308a demonstrate how a reflected pair can be immediately adjacent to the line of reflection, while the blue truncated stars in Fig. 308b are separated by a similar element that has bilateral reflected symmetry. This is also the case with the green hexagons in Fig. 308a. As a general rule, the closer together the reflected pairs, the more successful the design, but many factors play into the aesthetic success of particularly complex designs, and there are many exceptions to this general rule.

3.2.1 Isometric Designs with a Single Region of Local Symmetry

The level of complexity of patterns created from each of the systematic design methodologies is largely a product of the ratio between the number of secondary connective polygons and the single variety of primary polygons that comprise the underlying tessellation. While this general principle is also true of nonsystematic patterns, the number of different primary *n*-sided polygons within a single underlying generative tessellation is a further variable in determining complexity. In this way, nonsystematic patterns created from an underlying tessellation with primary polygons of a single variety that are connected by a minimal number of secondary polygons will be the least complex, whereas those produced from multiple varieties of primary polygon that are

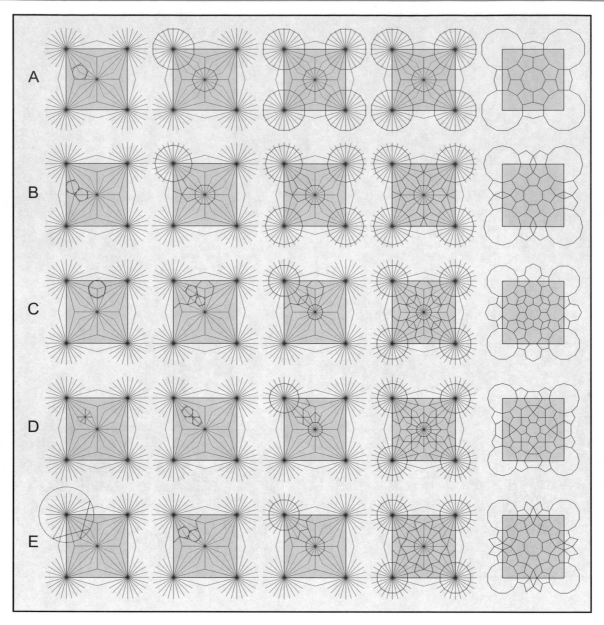

Fig. 304

connected by a large number of secondary polygons will be the most complex.

Complexity and beauty should not be conflated. The three patterns in Fig. 309 are not particularly complex, but each is well balanced and imbued with visual interest. Each is created from the same nonsystematic underlying tessellation comprised of two edge-to-edge regular pentagons placed at the midpoints of each edge of the triangular repetitive cell such that the outer corners of the twinned pentagons touch. This creates two interstice regions: one that is a six-pointed star located at the vertices of the isometric grid, and the other a shield-shaped ditrigon located at the center of each triangle. The angular openings

of the applied pattern lines in Fig. 309a are determined in part by the strategic placement of regular hexagons centered on each vertex of the isometric grid. Each edge of these hexagons contributes to the formation of regular heptagons placed at the outer points of each underlying six-pointed star. This Seljuk design, with its distinctive heptagons, is from the upper arches in the base of the northeast dome chamber in the Friday Mosque at Isfahan (1088-89) [Photograph 27]. Figure 309b shows more or less the same design, but with slightly different angular openings within the applied pattern lines, and without the hexagon centered on the vertices of the isometric grid. The lack of these hexagons disallows the heptagonal motif, and transforms the ring of

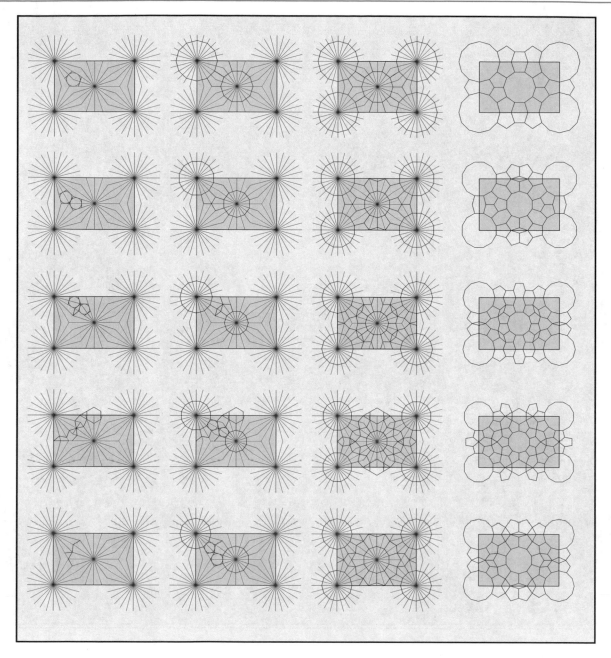

Fig. 305

6 five-pointed stars from the previous pattern to the ring of 6 ten-sided elongated motifs that surround the central six-pointed star. This design is from the anonymous treatise *On Similar and Complementary Interlocking Figures* in the Bibliothèque Nationale de France in Paris.[49] The close relationship between the designs in Figs. 309a and b indicates a likely connection between the geometric patterns in the northeast dome chamber in Isfahan and this anonymous treatise. Figure 309c is very similar to the

pattern in Fig. 309a accept that the angular openings of the crossing pattern lines are more acute, and the six-pointed stars have been arbitrarily modified with a sixfold star rosette. This is a Zangid design from the Nur al-Din Bimaristan in Damascus (1154).

A distinct group of nonsystematic geometric designs are created from underlying tessellations that place nonagons at the vertices of a regular hexagonal grid. This variety of geometric design is most frequently found in the ornament of the Seljuk Sultanate of Rum. Figure 310 illustrates the least complex example of such a generative tessellation, with the nonagons in edge-to-edge hexagonal contact. This

---

[49] MS Persan 169, fol. 193a.

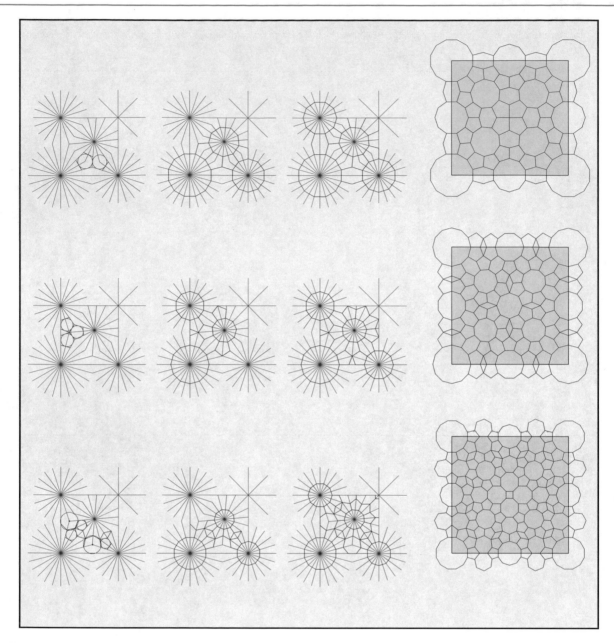

Fig. 306

arrangement of underlying nonagons creates six-pointed star interstice regions at the centers of the hexagonal repeat units. This very simple underlying tessellation produces an *acute* pattern with six-pointed stars within the interstice regions and nine-pointed stars at the vertices of the hexagonal repetitive grid. This design was produced during the Seljuk Sultanate of Rum and is found in the triangular pendentives that support the dome at the Alaeddin mosque in Konya (1218-28). The design in Fig. 311 breaks with the repetitive convention of placing the underlying nonagons at the vertices of the hexagonal grid. Figure 311a demonstrates how the nonagons are placed at the vertices of a rhombic grid with 60° and 120° included angles. The polygonal connective

matrix is comprised of triangles and shield-shaped ditrigons. Figure 311b shows how this creates an unusual *median* pattern comprised of nine-pointed stars that all share the same directional orientation, and it is due to this orientation that the translation symmetry is rhombic rather than hexagonal. Ignoring the arbitrary chirality of the interweaving pattern lines in Fig. 311c, this pattern adheres to the *p3m1* plane symmetry group, and is one of the more interesting examples of an Islamic star pattern based upon this relatively uncommon symmetry group. Artist from the Seljuk Sultanate of Rum incorporated this design into the mosaic ornament of the Great Mosque of Malatya (1237-38). The three designs in Fig. 312 are created from an underlying

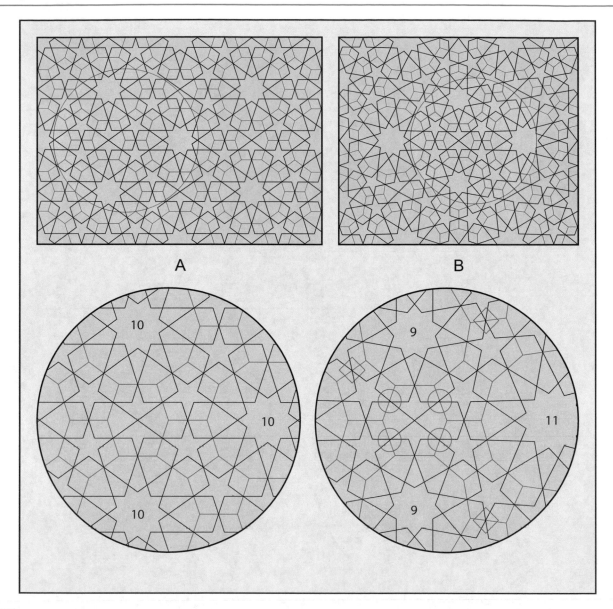

Fig. 307

tessellation that separates the nonagons at each hexagonal vertex with a ring of pentagons. This arrangement produces underlying six-pointed star interstice regions that are identical to those in Fig. 309. Each of these three examples is an *acute* pattern, and they only differ in the angular openings of their crossing pattern lines, and the applied pattern lines associated with the underlying six-point star interstice region. Figure 312a shows a Mamluk design from the Sultan Qaytbay complex in Cairo (1472-74) [Photograph 51]. Figure 312b is found at the Alay Han near Aksaray (1155-92) [Photograph 43], the Huand Hatun in Kayseri, Turkey (1238), as well as the Agzikara Han near Aksaray, Turkey (1242-43). The angular openings of the crossing pattern lines in this example are determined by the incorporation of regular heptagons (yellow). These heptagons are achieved

in an identical fashion as the design in Fig. 309a from the northeast dome chamber in the Friday Mosque at Isfahan (1088-89). This unusual heptagonal design feature is unique to these two examples, and it would appear likely that they share a causative agent rather than having a fully independent origin. Yet their respective origins span 150 years over a distance of approximately 2000 km. The occurrence of this identical design feature over such disparate time and location may have resulted from the use of inherited pattern scrolls. The design in Fig. 312c is from one of the triangular pendentives at the Alaeddin mosque in Konya. This design eliminates the central hexagon within the pattern matrix, producing a motif that is conceptually identical to that of the comparable region in Fig. 309b. The *obtuse* pattern in Fig. 313a is created from the same underlying tessellation as

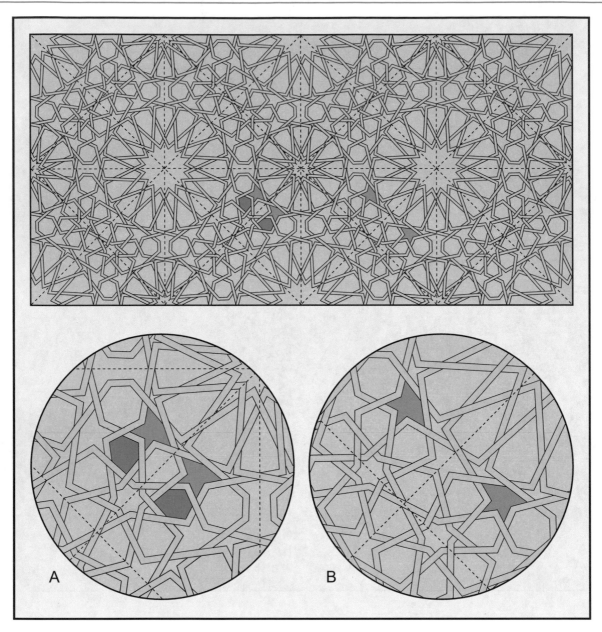

Fig. 308

the examples in Fig. 312. This has an arbitrary pattern line treatment within the central six-pointed star interstice region. This *obtuse* design was used by Shaybanid artists at the Kukeltash *madrasa* in Bukhara (1568-69) [Photograph 83], and the Tilla Kari *madrasa* in Samarkand (1646-60), and by the Janids at the Nadir Diwan Beg *madrasa* and *khanqah* in Bukhara (1622). Figure 313b shows a very successful *two-point* pattern also created from this underlying tessellation. This is from a Mamluk stone mosaic panel in the entry portal of the Ashrafiyya *madrasa* in Jerusalem (1482). Figure 313c shows an *acute* pattern created from a variation of this underlying tessellation that clusters six contiguous barrel hexagons around a central regular

hexagon. The applied pattern lines associated with the underlying barrel hexagons and pentagons are a corollary of similar *acute* pattern features in the *fourfold system B* [Fig. 172b]. However, unlike the fourfold example, the generated octagons are not regular—although they appear to be. This example is also from the Seljuk Sultanate of Rum, and is found at the Izzeddin Kaykavus hospital and mausoleum in Sivas (1217). Figure 314 illustrates the construction sequence for the radii matrix that creates the underlying tessellation responsible for the patterns in Figs. 312 and 313a and b. Step 1 places 18 radii at each vertex of the regular hexagonal repeat unit. Step 2 establishes the edges of the nonagons, as well as the separating pentagon with the

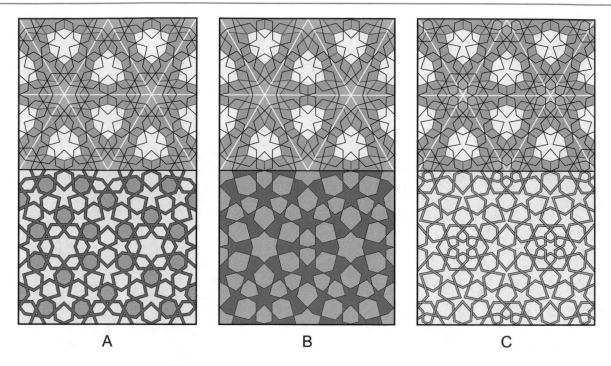

A   B   C

Fig. 309

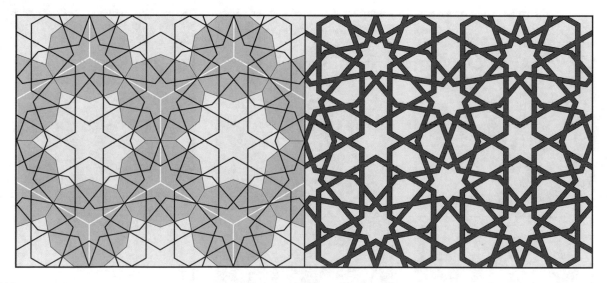

Fig. 310

placement of a circle that is tangent to the red radii and centered on the vertex of the blue radii. Step 3 completes the nonagons and pentagon. Step 4 rotates these around the hexagon. Step 5 creates the six pentagons that surround the central six-pointed star, and Step 6 shows the complete underlying tessellation.

Figure 315 illustrates another geometric pattern with nine-pointed stars placed upon the vertices of the regular hexagonal grid. Barrel hexagons rather than the mirrored pentagons in Figs. 312 and 313 separate the nonagons within

the underlying generative tessellation. This underlying tessellation has a large irregular dodecagonal interstice region at the center of each hexagonal repeat, and the applied pattern lines into this region are partially determined by the arbitrary placement of regular octagons within the pattern matrix. This pattern was produced by artists during the Seljuk Sultanate of Rum and is from the Gök *madrasa* and mosque in Amasya, Turkey (1266-67). Figure 316 demonstrates how the same radii matrix as that of Fig. 314 also produces the underlying tessellation for the design of Fig. 315. Step 1

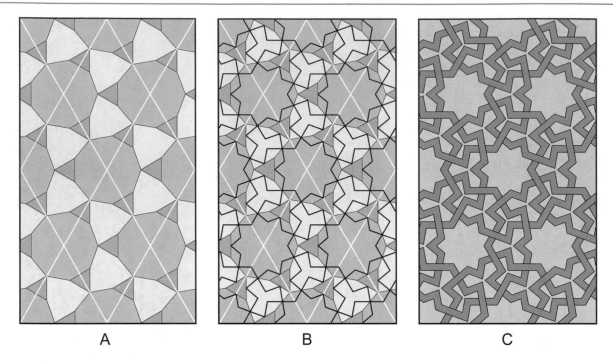

Fig. 311

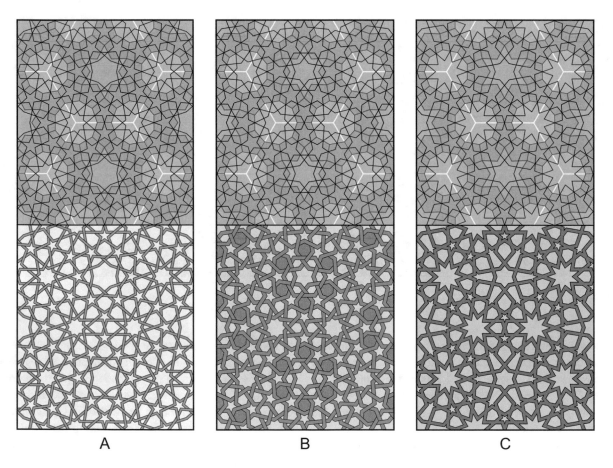

Fig. 312

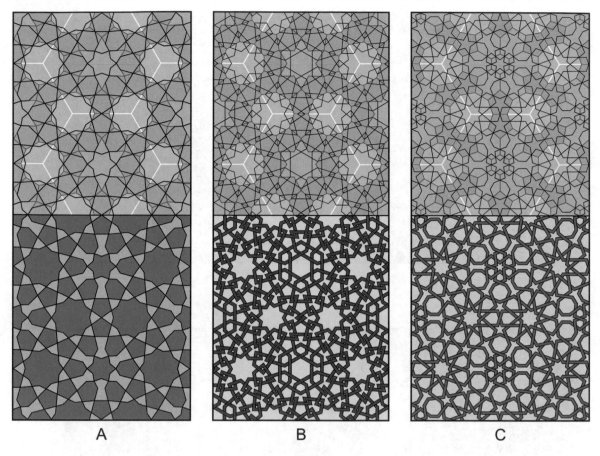

A B C

Fig. 313

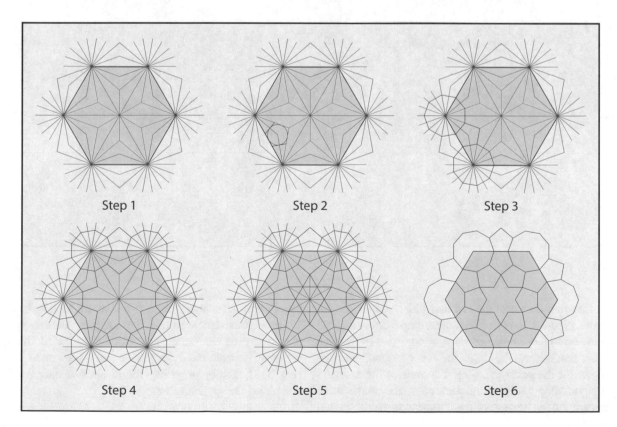

Step 1 Step 2 Step 3

Step 4 Step 5 Step 6

Fig. 314

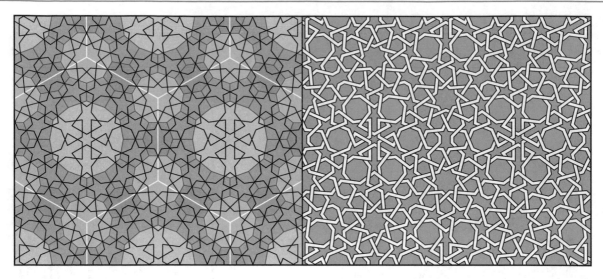

Fig. 315

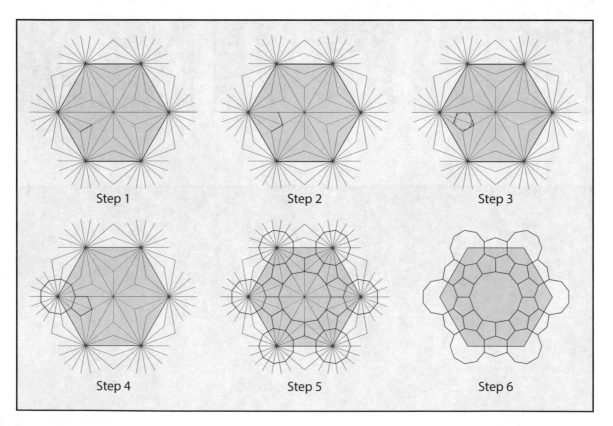

Fig. 316

draws a line that connects the red and blue radii as shown. Step 2 mirrors this line on the red radius. Step 3 draws a circle that is tangent to these reflected lines, and tangent to the blue radii. This establishes the edge for the nonagon, and the proportions for the pentagon. Step 4 completes the nonagon and pentagon. Step 5 mirrors the pentagons, and rotates these elements around the hexagon. Step 6 illustrates the completed

tessellation. A notable feature of this tessellation is the ring of 12 pentagons that surround the irregular dodecagon. The irregularity of the dodecagon can be corrected through a different constructive sequence of the radii matrix [Fig. 348], thereby providing for 12-pointed stars within the completed design [Figs. 346 and 347].

Fig. 317

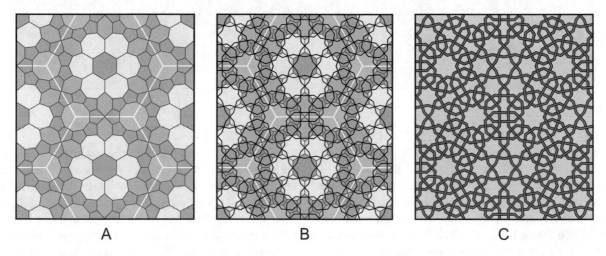

A   B   C

Fig. 318

As in the *fivefold system*, the underlying pentagons of nonsystematic patterns can be truncated into trapezoids that produce the dart motif within the *acute* pattern family. The design in Fig. 317 (by author) is just such a variation of the *acute* example in Fig. 312a. This underlying tessellation truncates the set of six pentagons that define the underlying six-pointed star so that a large underlying hexagonal interstice region is created. The lines of the dart motif inside each underlying trapezoid extend into the underlying hexagon to make the six-pointed star rosette at the center of each repeat unit. The construction of the central six-pointed star follows the standard practice for creating additive star rosettes [Fig. 222], with the point of rotation being the vertex of the underlying pentagon, two trapezoids, and large hexagon. It is somewhat surprising that this design modification does not appear to have been used historically.

Figure 318 shows a *median* pattern from the Great Mosque of Malatya (1237-38) that places nine-pointed stars at the vertices of the regular hexagonal grid, and six- and seven-pointed stars within the pattern matrix. Figure 318a illustrates the unusual characteristic of six edge-to-edge irregular heptagons in rotation around a central hexagon. The noticeably larger scale of the central hexagon and surrounding heptagons, relative to the length of the polygonal edges and size of the nonagons is unusual, and is responsible for the noticeable change in the design density between the peripheral and central regions of each repeat unit. The denser region of his geometric pattern shares some of the visual characteristics of *median* patterns from the *fourfold system A* [Fig. 145]. This is due to the similarity of the pentagonal and elongated hexagonal cells within the underlying tessellation and the application of approximate 90°

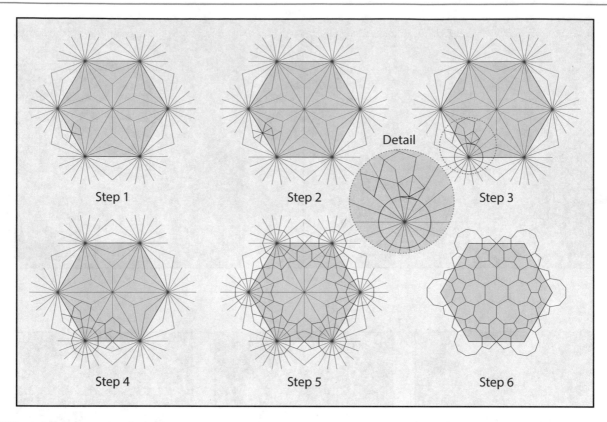

Fig. 319

crossing pattern lines. Interestingly, the less dense region shares the qualities of the classic *median* pattern with 90° crossing pattern lines created from the underlying tessellation of just regular hexagons [Fig. 95c]. Figure 319 demonstrates the construction of the underlying tessellation used to create the design in Fig. 318. This utilizes the same radii matrix as the previous examples that allows for the placement of nonagons on the vertices of the hexagonal grid. Step 1 of this more complex tessellation places intersecting perpendicular lines at the midpoint of one of the repetitive edges. In Step 2, lines are added that are perpendicular to the repetitive edge, and connect the crossing lines of Step 1 to the nearby red radii, thereby creating a hexagonal region. Step 3 mirrors one of the lines from Step 2, and uses these mirrored lines to establish the radius of a circle, which in turn, locates the edge of the nonagon. Step 4 mirrors the information from Step 3, and Step 5 rotates this information around the hexagonal repeat. Step 6 completes the tessellation by introducing irregular heptagons surrounding the central hexagon. It is important to note that while this specific construction results in a polygonal tessellation that successfully creates the *median* design from the Great Mosque of Malatya, as patterns and underlying tessellations become more complex, a degree of conjecture is required in establishing a construction sequence that allows for the creation of the underlying tessellations from radii matrices. The step-by-step examples provided herein are therefore not

to be regarded as necessarily those used by artists of the past, but rather workable procedures that accurately replicate the underlying structures that provide for the design of the many examples of particularly complex patterns. Allowance should be made for other construction sequences that utilize radii matrices and arrive at the same polygonal result.

The designs in Fig. 300 demonstrate the wide diversity of nonsystematic patterns with 12-pointed stars as the single primary star form. A modified version of the underlying tessellation that creates these patterns produces the *acute* pattern in Fig. 320. This modification results from the truncation of the three edge-to-edge pentagons at the center of each triangular repeat. These truncated pentagons become three trapezoids that are contiguous with a central equilateral triangle. This *acute* design was used widely by many Muslim cultures. Figure 320a illustrates the truncation of the three pentagons from the underlying tessellation pictured in Fig. 300a, and Fig. 320b demonstrates the application of *acute* pattern lines into this modified underlying tessellation. As with other *acute* patterns, these trapezoids produce the distinctive dart motif. Note: the pattern lines of the obtuse angle of each dart are not located precisely on the midpoint of the truncating polygonal edge, but slightly inside the trapezoid. Different historical examples of this pattern treat this condition differently. The pattern line application of this example is determined by the decision to have parallel lines within the 12-pointed stars, and a regular hexagon within the

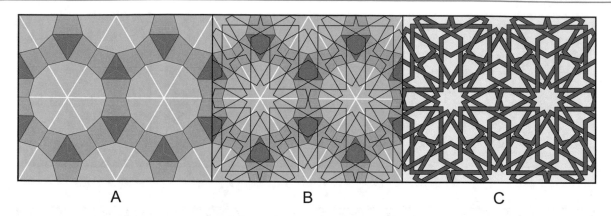

Fig. 320

underlying triangles. Early examples of this very popular design include the archivolt of the Zangid *mihrab* in the Upper Maqam Ibrahim at the citadel of Aleppo (c.1214); the vertical side panel of the Mengujekid *minbar* in the Great Mosque of Divrigi (1228-29); and the Seljuk Sultanate of Rum carved stone ornament at the Çifte Minare *madrasa* in Sivas, Turkey (1271). The seven designs in Fig. 321b–h are derived from another isometric underlying tessellation with a similar arrangement of three edge-to-edge pentagons at the center of each repetitive triangle, but with a barrel hexagon that separates each dodecagon [Fig. 299]. The three designs in Fig. 321j–l are produced from the underlying tessellation in Fig. 321i that truncates the three central pentagons in the same fashion as the example in Fig. 320. The *acute* pattern in Fig. 321b was used in multiple locations, including the kiosk of the Keybudadiya at Kayseri (1224-26); the late Abbasid Palace of the Qal'a in Baghdad (c. 1220) [Photograph 28]; and a Mamluk door at the Al-Azhar mosque in Cairo. A tiled variation of this *acute* design was used together with a square repetitive cell with matching edge conditions to create a very attractive type B dual-level design from the Topkapi Scroll[50] [Figs. 458 and 459]. The design in Fig. 321c is the standard *median* pattern, and that of Fig. 321d is a modification that introduces a star rosette place within the 12-pointed star in the manner that was especially popular among Mamluk artists. Surprisingly, neither of these *median* designs (by author) appears to have been used historically. The *obtuse* pattern in Fig. 321e was used as the triangular component of a Timurid type B dual-level design from the Friday Mosque at Isfahan. This dual-level design combines this triangular pattern with the square pattern in Fig. 379f. The example in Fig. 321f modifies this isometric *obtuse* design in Fig. 321e by adding a 6-pointed star motif into the otherwise 12-pointed stars, and subtracting the small hexagonal pattern feature at the center of each triangular

repeat, thereby opening up the most congested region of the original. The overall effect of these arbitrary modifications is a field pattern with more evenly balanced pattern elements. This design was used at the Imamzada Darb-i Imam in Isfahan (1453). This same essential design, without the removed central hexagons, was used in the stone window grilles of the Sultan Qala'un funerary complex in Cairo (1284-85). Figure 321g shows the standard *two-point* pattern (by author) that does not appear to have been used historically; and the design in Fig. 321h employs the arbitrary modification of the 12-pointed stars through the introduction of the 12-pointed star rosette. As mentioned, this form of modification was particularly popular among Mamluk artists, and indeed, this design is found in several Mamluk locations, including: a panel from the *minbar* stair rail, as well as one of the interior wooden doors at the Sultan al-Mu'ayyad Shaykh complex in Cairo (1415-22); the portal of the Ribat Khawand Zaynab in Cairo (1456); and a stone lintel for a window at the Ashrafiyya *madrasa* in Jerusalem (1482). The modified underlying tessellation in Fig. 321i allows for the production of the well-known design in Fig. 321j, and examples include one of the Ilkhanid ceiling vaults of the mausoleum of Uljaytu in Sultaniya, Iran (1307-13); a Mamluk stone mosaic panel from the Amir Aq Sunqar funerary complex in Cairo (1346-47); and the doors of a Mamluk cupboard at the Sultan Qansuh al-Ghuri complex in Cairo (1503-05). Figure 321k shows the standard *median* pattern (by author) created from this modified underlying tessellation, and although this design is not known to the historical record, the variation in Fig. 321l was used in the late Abbasid tomb tower of Umar al-Suhrawardi in Baghdad (1234). The *acute* pattern from this modified underlying tessellation can also be created from the 3.12^2 underlying tessellation of triangles and dodecagons [Fig. 108d].

Figure 322 illustrates the derivation of a Timurid *median* pattern with 18-pointed stars placed at the vertices of the isometric grid and an arbitrary 6-pointed star at the center of each triangular repeat. Ordinarily, the outward extension of

---

[50] Necipoğlu (1995), diagram no. 38.

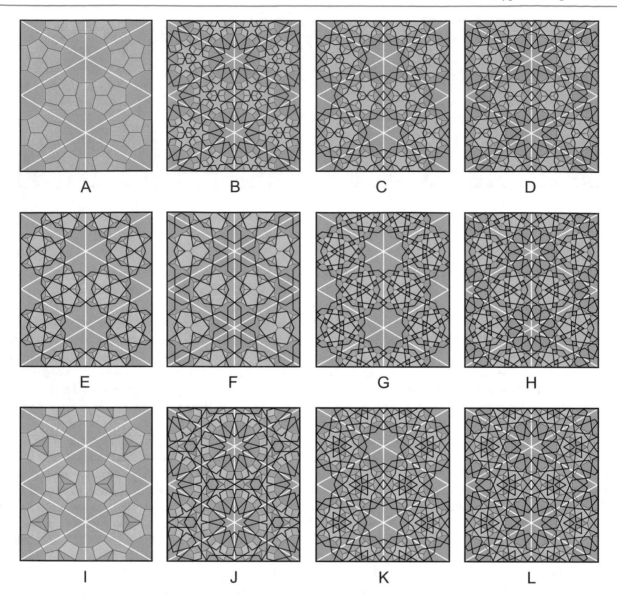

Fig. 321

the lines from the tips of the six-pointed stars would be collinear. As a widened line or interweaving line treatment, this configuration would be less acceptable to the aesthetics of this design tradition. However, by treating this example as a tiling, the unconventional character of the noncollinearity becomes a distinctive and appealing feature within the design. This isometric pattern is from the Abdulla Ansari complex in Gazargah near Herat, Afghanistan (1425-27) [Photograph 84]. Patterns with 18-pointed stars are surprisingly uncommon, and the construction of this example is relatively simple. The underlying tessellation in Fig. 322a is made up of 18-gons placed at each vertex of the triangular repeat. These are separated by barrel hexagons along the edges of the repeat. The central region of the triangular repeat is an interstice six-pointed star that is divided into

six quadrilateral kite shapes with bilateral symmetry. The angular openings of the applied *median* pattern lines in Fig. 322b are determined by their 18-*s*6 placements within the 18-gons. The underlying tessellation that creates this *median* pattern is well suited to the other three pattern families. Figure 323 illustrates these three additional designs (by author). Figure 323a shows an *acute* pattern, Fig. 323b illustrates an *obtuse* pattern, and Fig. 323c shows a *two-point* pattern. This underlying tessellation can be modified to produce yet more patterns. The three nonhistorical *median* patterns (by author) in Fig. 324 are created by changing the central sixfold region of the underlying tessellation in the previous examples. Figure 324a incorporates an underlying central ditrigonal element surrounded by six pentagons, Fig. 324b replaces the six pentagons with six trapezoids and a

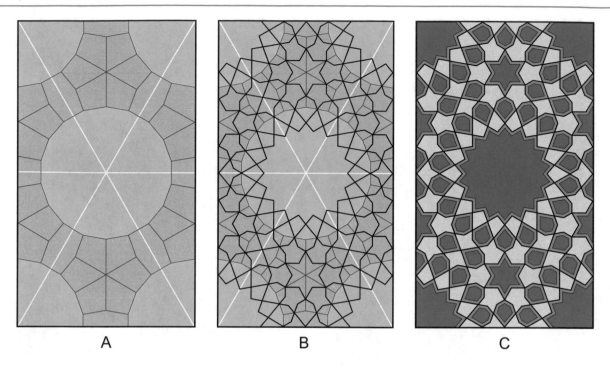

Fig. 322

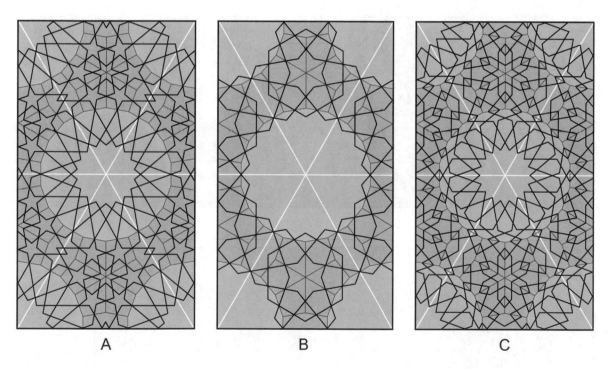

Fig. 323

central triangle, and Fig. 324c places six pentagons around a central regular hexagon. Figures 324a and b have tight regions in the pattern matrix that are problematic, but the design in Fig. 324c is very acceptable. Figure 325a demonstrates the simple construction for the radii matrix that produces the *median* pattern in Fig. 322. A

determination for the size of the 18-gon is shown in the detail, wherein an angle between a radius and an applied line that is perpendicular to the edge of the triangular repeat is bisected, thus establishing the placement for the edge of both the 18-gon and barrel hexagon. This construction produces a polygonal matrix between the 18-gons with edges that are

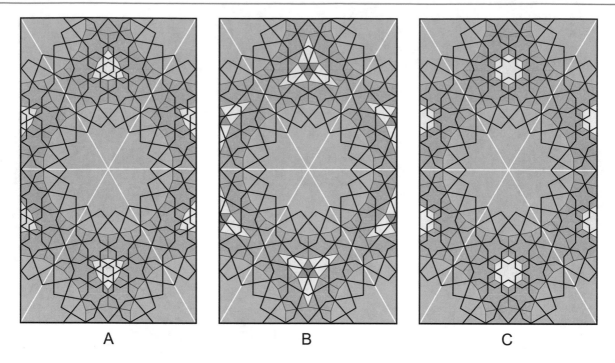

Fig. 324

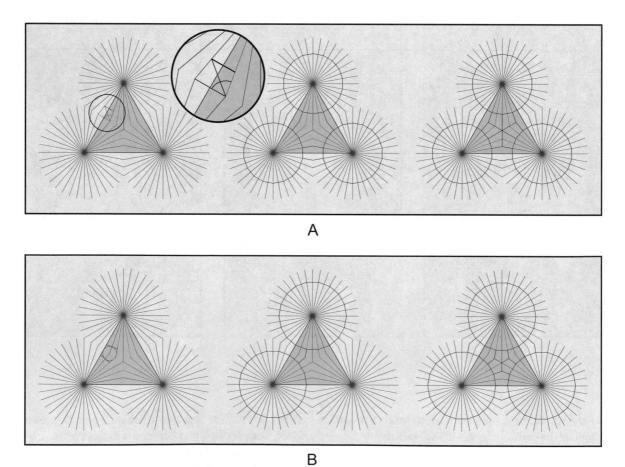

Fig. 325

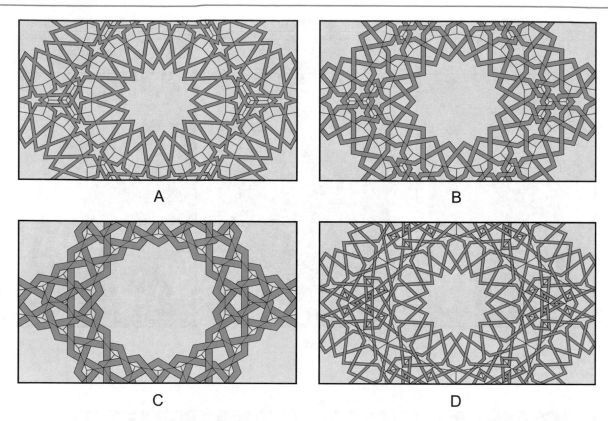

A

B

C

D

Fig. 326

congruent with the blue radii. Figure 325b introduces an alternative underlying tessellation created from this same radii matrix. This follows the previously demonstrated construction of pentagons created from circles that are tangent with the radii. The resulting ring of pentagons that surround each 18-gon are tangent with the red radii. Although this underlying tessellation does not appear to have been used within the historical record, it is fully in keeping with the expected characteristics for creating good patterns, and, as demonstrated in Fig. 326, will produce acceptable patterns in each of the four pattern families (by author). Figure 326a shows the *acute* pattern created from this new underlying tessellation; Fig. 326b shows the *median* pattern; Fig. 326c shows the *obtuse* pattern; and Fig. 326d shows the *two-point* pattern.

Figure 327 demonstrates the origins of a particularly beautiful nonsystematic *two-point* design with 24-pointed stars placed at the vertices of the isometric grid. Figure 327a shows how the polygonal matrix surrounding the underlying 24-gons is comprised of rings of 24 pentagons, two varieties of hexagon, and heptagons. The *two-point* pattern line application is testament to the ingenuity of the artist who created this design. The pattern lines within each of the clustered hexagons are particularly interesting. Each is provided with two perpendicular axis of reflected symmetry rather than the more conventional sixfold rotational symmetry commonly found within regular hexagons. This design is the product of

artists working under the Seljuk Sultanate of Rum. The earliest known example is from a carved stone relief panel at the Nalinci mosque in Konya (1255-65). A later cut-tile mosaic example is from the *mihrab* niche at the Esrefoglu Süleyman Bey mosque in Beysehir, Turkey (1296-97) [Photograph 44]. An additional example is from a thirteenth-century stone fragment found in the Alaeddin Hill excavations in Konya.[51] Figure 328 shows three patterns (by author) created from the same underlying tessellation as shown in Fig. 327a. Although not known to the historical record, each of these is very acceptable to the aesthetics of this ornamental tradition. Figure 328a shows an *acute* design, Fig. 328b shows a *median* design, and Fig. 328c shows an *obtuse* design with *median* characteristics. The radii matrix in Fig. 329 provides the means for creating the underlying tessellation for the patterns in Figs. 237 and 238. This radii matrix places 48 radii at each vertex of the triangular repeat. Step 1 identifies an internal angle of 135°, and draws a line from the center of the triangle to the midpoint of the triangular edge. The 135° angle is close to the included angles of a heptagon. Based upon this observation, Step 2 draws 14 radii at the designated intersection of the red radii. Step 3 places a regular heptagon centered at the same

---

[51] This fragment on display at the Karatay *madrasa* Museum in Konya.

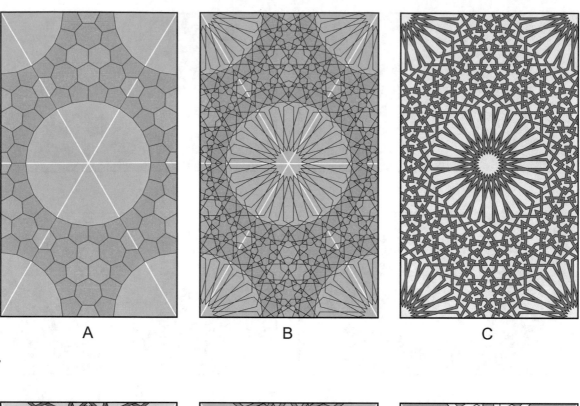

Fig. 327

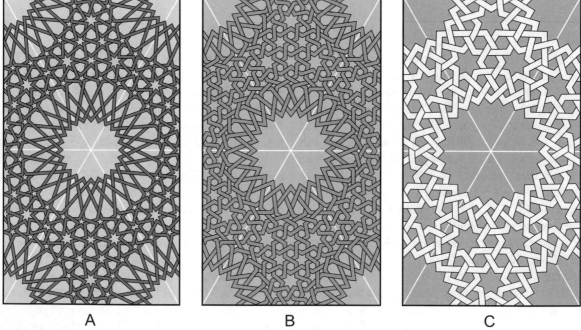

Fig. 328

intersection. The edge that is parallel to the triangular edge of the repeat rests upon the intersection of the two 14-fold radii and the two 48-fold red radii. The fact that the edges of the heptagon do not quite align with the red and blue radii is acceptable. Step 4 mirrors the lines of the heptagon as shown, and adds a line connecting the heptagon with the center of the triangular repeat. Step 5 places circles that are tangent with the red radii, and draws lines that are perpendicular to the red radii where they intersect with the circles. Step 6 uses these lines to determine the size of the 24-gon and the ring of pentagons. Step 7 rotates these elements throughout the triangular repeat, and Step 8 is the complete

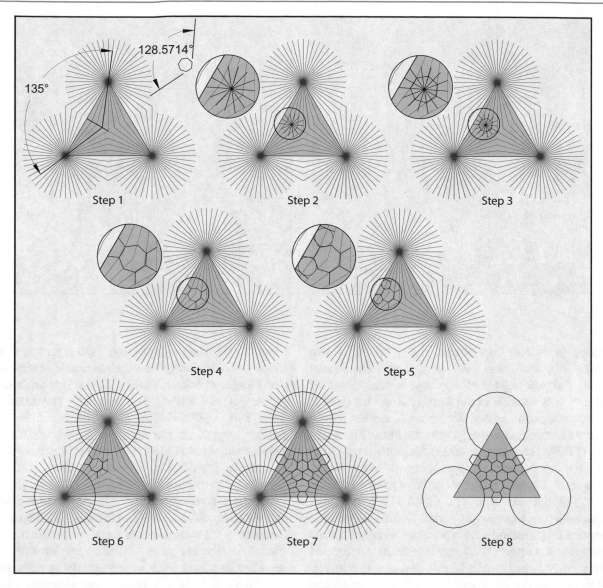

Fig. 329

underlying tessellation. As mentioned previously, this specific step-by-step sequence is certainly not the only method of creating the underlying tessellation from the radii matrix, and as such is presented as a representative rather than a definitive example.

3.2.2 Orthogonal Designs with a Single Region of Local Symmetry

Nonsystematic orthogonal patterns with single regions of primary local symmetry are also well known to the historical record. Whereas the *n*-fold primary stars of such isometric patterns will invariably be divisible by three, those of orthogonal designs will be divisible by four. Just as with isometric examples, nonsystematic orthogonal patterns with

only a single variety of local symmetry are diverse in both the types of stars and levels of complexity. Among the least complex of this variety of design are those created from simple underlying tessellations that include different combinations of octagons, irregular pentagons, squares, and triangles. Figure 330 illustrates a design that is produced from what is perhaps the least complex of such underlying tessellations. This is a Ghurid *acute* design from the Shah-i Mashhad *madrasa* at Gargistan in the remote Badghis Province of northwestern Afghanistan (1176). This very simple field pattern is characterized by octagons, four-pointed stars, and unusually elongated five-pointed stars. Other than the inclusion of the five-pointed stars, this is similar in concept to a simple design created from the 4.8^2 arrangement of underlying octagons and squares [Fig. 124c]. Figure 331 includes two designs where the octagonal elements in the

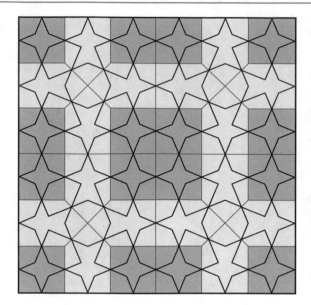

Fig. 330

underlying generative tessellations are separated by two irregular pentagons. These pairs of pentagons touch point to point and are separated by edge-to-edge irregular triangles. The orientation of the pentagons and triangles in these two examples is rotated 90° from one another. Figure 331a shows a Seljuk *acute* design from the Friday Mosque at Barsian (1105). The octagons within the pattern matrix are regular, but the heptagons are only approximate. Figure 331b shows an Ilkhanid *two-point* additive pattern from the tomb of Uljaytu in Sultaniya (1313-14) [Photograph 68]. The underlying generative tessellation for this design is determined by placing four equilateral triangles around each octagon, and squares with edge lengths that are equal to the octagons and equilateral triangles placed vertex-to-vertex with the triangles. The irregular triangles and pentagons are a product of this polygonal arrangement. The Ilkhanids were particularly disposed toward achieving greater geometric complexity through additive pattern elements. While almost all of their additive patterns were derived from one or another of the generative systems, this example is unusual in that the initial pattern is nonsystematic, albeit relatively simple. Figure 332 illustrates three designs created from the same underlying tessellation that places octagons at the vertices of the orthogonal grid, separated by two edge-to-edge regular hexagons along each edge of the repeat. The interior region of this polygonal arrangement is filled with irregular pentagons and heptagons, with a square at the center of the repeat. The seven-pointed stars of the *median* pattern in Fig. 332a have bifold symmetry and are distinctly non-regular. Several historical *median* designs were produced from this underlying tessellation, the earliest of which is a tiled version from the Seljuk ornament within the *muqarnas* of the *mihrab* in the Friday Mosque at Barsian

in Iran (1105). The design in Fig. 332b is an interweaving version produced during the Seljuk Sultanate of Rum for the Great Mosque of Niksar, Turkey (1145). This was also used by Qara Qoyunlu artists in the portal of the Great Mosque in Van in Turkey (1389-1400). The variation in Fig. 332c is from the Ildegizid exterior façade of the Mu'min Khatun in Nakhichevan, Azerbaijan (1186). This differs from the earlier example from Naksar in the continuous pattern line treatment of the six-pointed stars located along the edges of the square repeat. The design in Fig. 332e is from the Amir Sarghitmish *madrasa* in Cairo (1356). Figure 332d demonstrates the unusual character of the pattern line application. With the exception of the underlying square, star rosette motifs inhabit each of the underlying polygons. As applied to the underlying heptagons and pentagons, the outer petals of the star rosettes provide a vehicle for overcoming the irregularity of these two underlying polygons. This allows for the central five-pointed stars inside the underlying pentagons, and the seven-pointed stars inside the underlying heptagons to have accurate rotation symmetry. By contrast, the multiple five- and six-pointed stars located at the vertices of the underlying tessellation in Fig. 332d are noticeably irregular. The separation of the underlying octagons with regular hexagons is similar in concept to the nonsystematic underlying tessellation in Fig. 178.

The two designs in Fig. 333 are somewhat unusual in that while their underlying tessellation has two primary polygons, the derived patterns only utilizes one of these for the primary star form. In each case, the underlying dodecagons are used to create the 12-pointed stars, while the underlying octagons contribute less overtly to the completed designs. The pattern in Fig. 333a is Ayyubid, and was used on the wooden *mihrab* (1245-46) of the

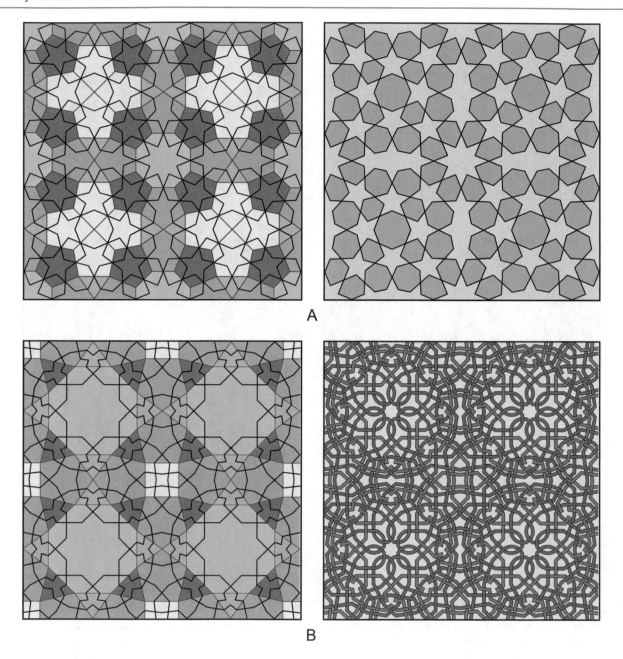

A

B

Fig. 331

Halawiyya mosque and *madrasa* in Aleppo. This makes use of 60° crossing pattern lines placed within the dodecagons in a 12-*s*4 arrangement that conforms to the *median* pattern family. The lines of the 12-pointed star are extended into the octagon and interstice regions until they meet with other extended lines. This pattern is then provided with additive six-pointed stars within the dodecagons, and additive four-pointed stars within the octagons. The design in Fig. 333b employs the same 12-pointed star within the dodecagons, but only extends the pattern lines into the interstice region. To great aesthetic effect, this design places octagons at the center of each underlying octagon. The size of these added

octagons is determined by a 1/4 division of four of the underlying octagonal edges, as per *two-point* patterns. However, the size of these elements may also have been an arbitrary decision based upon visual effect. This design was used in two late Abbasid building in Baghdad: the Palace of the Qal'a (c. 1220) and the Mustansiriyah *madrasa* (1227-34).

As with nonsystematic isometric patterns, there are a relatively large number of orthogonal designs that employ 12-pointed stars as the single primary star form. The underlying tessellation in Fig. 334a is made up of dodecagons and squares in an orthogonal arrangement, with interstice

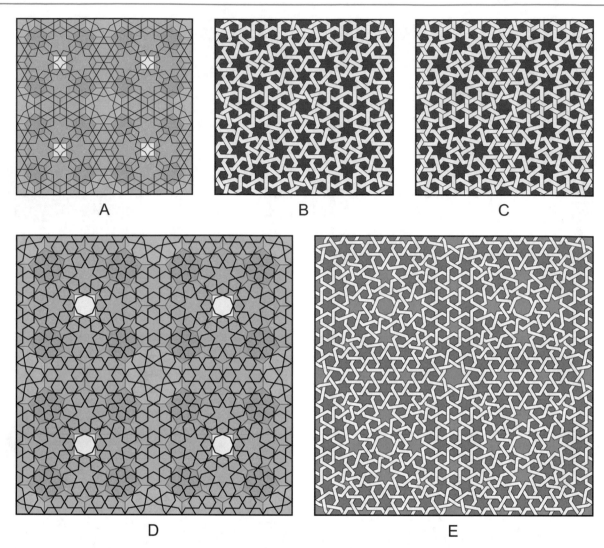

A   B   C

D   E

Fig. 332

regions that are divided into four pentagons. The *acute* design created from this tessellation has 45° crossing pattern lines placed at the midpoints of each underlying dodecagonal and square edge. The size of the regular octagon located at the center of the square repeat is determined by its extended lines being collinear with the crossing pattern lines that extend from the underlying squares into the adjacent pentagons. The sets of 45° crossing pattern lines within each dodecagon in Fig. 334a do not produce parallel lines within the 12-pointed stars. Rather, the artist responsible for this design chose to use the 30° angular openings in Fig. 334b. This subtle adjustment produces 12-pointed stars with parallel lines. The resulting Mamluk design represented in Fig. 334c was used for a pierced stone window grille at the Sultan Qala'un in Cairo (1284-85) [Photograph 55]. In addition to aesthetic preference, it is possible that the artist preferred the smaller sized 12-pointed stars as this would provide greater structural integrity within the pierced stone, as well as a more uniform light diffusion. Figure 334b shows

how the crossing pattern lines of the 12-pointed stars have been moved inward from their dodecagonal edges, resulting in an elongation to the points of the five-pointed and four-pointed stars that share these edges. Moving the crossing pattern lines inward from the underlying polygonal edges is an arbitrary process aimed at achieving either an aesthetic or practical result. In this case, whereas the use 30° crossing pattern lines provides both the visual appeal of parallel lines within the 12-pointed stars and a reduced central 12-pointed star, their slight movement inward from the dodecagonal midpoints elongates the points of the 4- and 5-pointed stars in a somewhat atypical fashion, with the fourfold symmetry of the 4-pointed stars being sacrificed. Figure 335 demonstrates the derivation of four *acute* patterns created from an underlying tessellation comprised of dodecagons and two varieties of pentagon. The pattern in Fig. 335a is from an arched tympanum in a Seljuk gate in the Friday Mosque at Isfahan (after 1121-22). This uses 30° crossing pattern lines at each underlying polygonal edge. However,

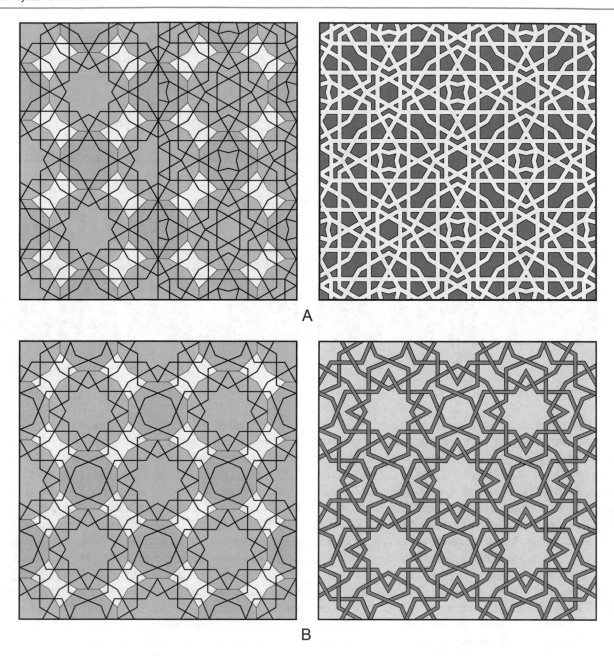

A

B

Fig. 333

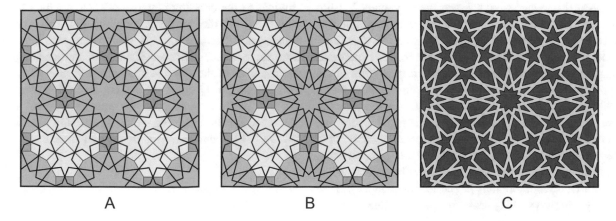

A   B   C

Fig. 334

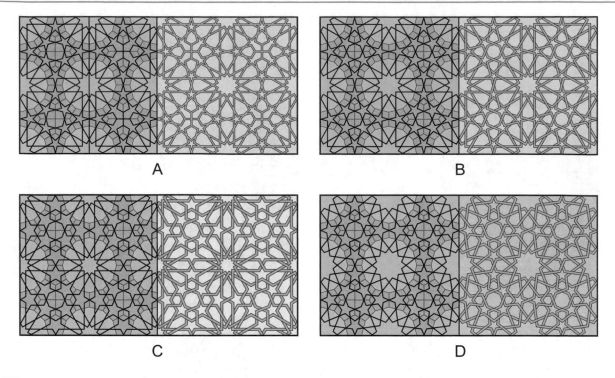

Fig. 335

this angular opening creates a box-like octagon where the four underlying pentagons meet at the center of each repeat unit. This box-like octagon is not especially appealing and is significantly larger that the other pattern elements in this design. The artist who created this design modified the octagons with a stellated cruciform motif that is atypical to this tradition, but nicely overcomes the visual imbalance of the box-like octagon. This example of pattern adjustment is illustrative of the experimental approach to design that was taking place during the formative period when this example was produced, and it is perhaps not coincidental that these ten-sided atypical stellated motifs are very similar to the nine-sided stellated motifs used in the fivefold hybrid design in the nearby northeast dome chamber [Fig. 261]. The *acute* design in Fig. 335b replaces the boxlike octagon in Fig. 335a with a regular octagon with 45° crossing pattern lines. This is a very-well-known design that was used by many Muslim cultures in many locations. A particularly early example was used during the Seljuk Sultanate of Rum at the Great Mosque of Siirt (1129). The variation in Fig. 335c increases the angular opening of selected crossing pattern lines of the five-pointed stars from 30° to 60°. As with the example in Fig. 334, this example also moves the 30° angled crossing pattern lines of the 12-pointed stars inward from the midpoints of the underlying dodecagons, thereby extending the length of the points in the adjacent five-pointed stars. This example is from the Hall of the Ambassadors at the Alhambra (fourteenth century). The example in Fig. 335d uniformly employs 45° crossing pattern lines at the

midpoints of each underlying polygonal edge. This design is from a frontispiece of the 30 volume Quran written and illuminated by 'Abd Allah ibn Muhammad al-Hamadani[52] (1313) at the likely behest of the Ilkhanid Sultan Uljaytu.[53] These examples of nuanced pattern line application result in the distinct visual character of each of these four otherwise very similar designs. Figure 336 illustrates four additional designs created from the same underlying tessellation as that of the previous example. Figure 336a shows a *median* pattern that places central octagons with 45° crossing pattern lines into a pattern matrix that is otherwise comprised of 60° crossing pattern lines. This is an Artuqid design from the *mihrab* of the Great Mosque of Silvan (1152-57). Figure 336b shows a *median* pattern that uses 60° crossing pattern lines uniformly, but includes the customary *median* modification that changes the quality of the primary stars through eliminating the crossing pattern lines that are located at the midpoints of the primary polygon—in this case the dodecagons [Fig. 223]. This design can also be created from the 3.4.3.12-3.12² tessellation of decagons, squares, and triangles of the *system of regular polygons* [Fig. 113c]. This is a relatively common design, and a fine Muzaffarid example is from the Friday Mosque at Kerman (1349). The *obtuse* pattern in Fig. 336c is visually similar to a *median* pattern that can also be created from the 3.4.3.12-

[52] Cairo, National Library, 72, part 22.

[53] Lings (1976), 119.

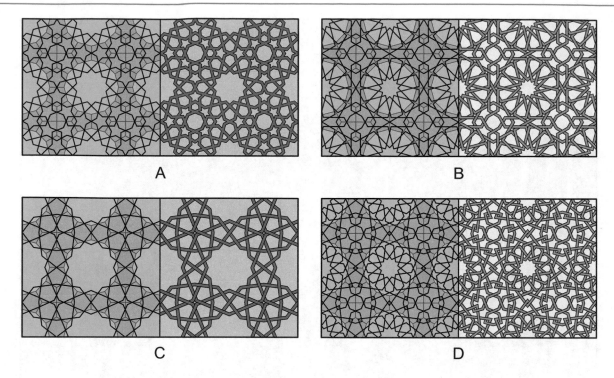

Fig. 336

Fig. 337

3.12^2 tessellation [Fig. 113a]: the difference being in the small variation between the angles of the pattern lines. While the angular variation is small, the visual results are very apparent: the design in Fig. 336c includes regular superimposed octagons, whereas the example from the *system of regular polygons* sacrifices these octagons in favor of the regular squares within the pattern. The superimposed octagons are an appealing feature and it is somewhat surprising that the *median* design in Fig. 336c (by author)

does not appear to have been used historically. Figure 336d shows a *two-point* pattern that incorporates an octagon at the center of the repeat unit where four clustered underlying pentagons meet. This example also employs one of the more common variations to the primary stars, in this case 12-pointed [Fig. 225b]. This pattern was used in the *minbar* of the Amir Qijmas al-Ishaqi mosque in Cairo (1479-81). Figure 337 illustrates the derivation of a particularly popular *acute* design that was used throughout Muslim cultures. This

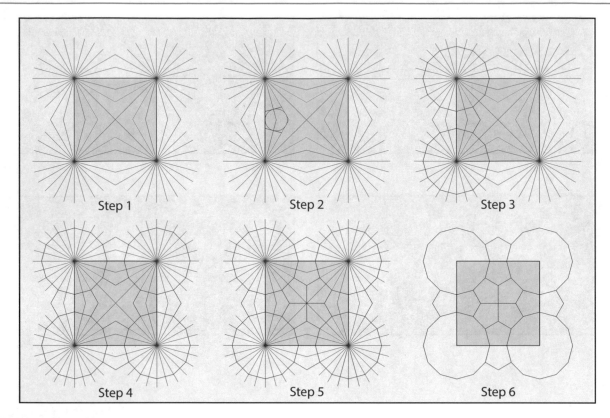

Fig. 338

is created from a modification of the underlying tessellation in Figs. 335 and 336 that truncates the four pentagons at the center of the repeat unit, turning them into trapezoids. This produces a square at the center of the repeat unit that is contiguous with the long edge of each trapezoid. One of the earliest uses of this design in on the Ildegizid façade of the Mu'mine Khatun mausoleum in Nakhichevan, Azerbaijan (1186). Later examples include the door of the Zangid entry portal at the otherwise Mamluk Bimaristan Arghun in Aleppo; the Mamluk entry doors and incised stonework at the Zahiriyya *madrasa* and mausoleum of Sultan al-Zahir Baybars in Damascus (1277-81); a pair of bronze doors from the Seljuk *atabeg* of Cizre, Turkey (thirteenth century); the *minbar* at the Mamluk funerary complex of Sultan al-Zahir Barquq in Cairo (1384-86); and the *minbar* doors at Amir Taghribardi funerary complex in Cairo (1440). Figure 338 demonstrates a construction sequence for the underlying tessellation in Figs. 335 and 336. Step 1 places 24 radii at each vertex of the square repeat unit. Step 2 draws a circle that is tangent to the red radii. This circle is used to determine the edges of the dodecagons, as well as the separating pentagon. Step 3 completes the dodecagons and pentagon. Step 4 rotates these throughout the square repeat unit. Note: the underlying tessellation in Fig. 337 can be produced at this stage by simply connecting the four inward facing points of the four pentagons, thereby

creating four trapezoids surrounding a central square. Step 5 determines the four clustered pentagons at the center of the repeat, and Step 6 shows the completed tessellation.

The underlying tessellation in Fig. 339 is unusual in that it surrounds each dodecagon with a ring of trapezoids. In this case, these trapezoids are not modifications of previously placed pentagons. Although the proportions differ, the arrangement of four trapezoids and two triangles located at the midpoints of each square repeat is conceptually identical to the 1/10 arrangement of similar modules from the *fivefold system* [Fig. 196]. In both cases, this arrangement lends itself to the *acute* pattern family, and the similarity in the resulting concave octagonal pattern motif centered in this arrangement is readily apparent when compared to certain historical fivefold designs, for example, a Timurid design from the Shah-i Zinda in Samarkand [Fig. 254b]. The underlying tessellation for the design in Fig. 339 is essentially corner-to-corner dodecagons placed upon the vertices of the orthogonal grid with four equilateral triangles placed at the center of the square repeat unit. This creates the elongated rhombic interstice regions shown as two mirrored triangles. While the *acute* pattern that this tessellation creates is attractive, the irregular eight-pointed star at the center of the repeat unit is improved by adding four more points, thereby making this into a regular 12-pointed star as per Fig. 340a. As mentioned previously, the geometric design tradition in the Maghreb

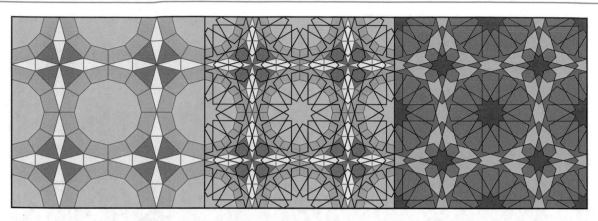

Fig. 339

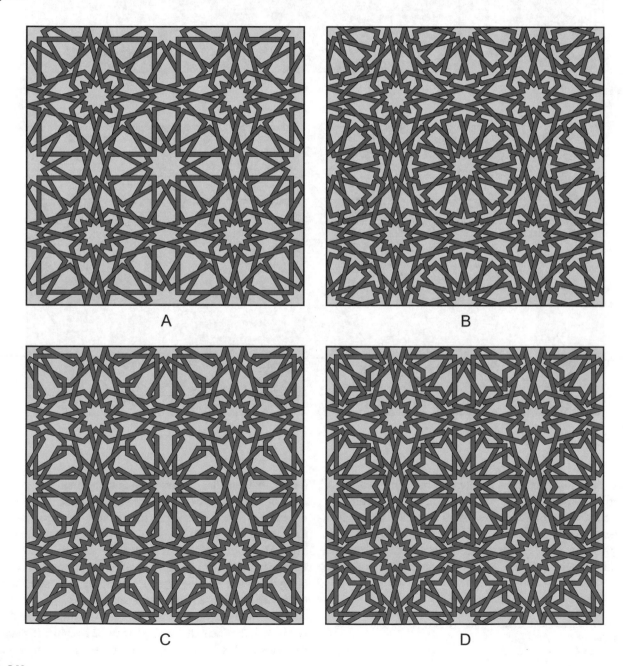

A

B

C

D

Fig. 340

Fig. 341

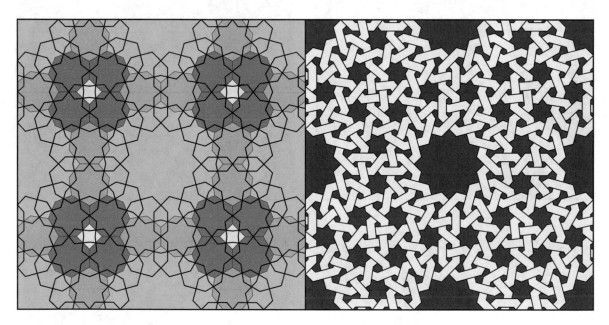

Fig. 342

placed particular emphasis on the arbitrary modification of the primary star forms.[54] In addition to the arbitrarily modified non-regular 8-pointed stars that have been given four more points so that they become 12-pointed stars, the examples in Fig. 340b–d illustrate three distinctive variations to the primary 12-pointed stars. Each of these

four variations has its own distinct aesthetic merit, and each of these four examples was used in the *zillij* cut-tile mosaics of the Alhambra (fourteenth century). In fact, each was used within the single *zillij* panel represented in Fig. 341. The incorporation of multiple varieties of modified 12-pointed stars into a single panel is an effective means of bringing broad-ranging design variation into what would otherwise be a more repetitive aesthetic.

The underlying tessellation in Fig. 342 is unusual in that it is comprised principally of rings with 12 irregular heptagons that surround a 12-pointed star interstice region at each

---

[54] The work of Jean-Marc Castéra explores these design conventions in great depth.
      –Castéra (1996).

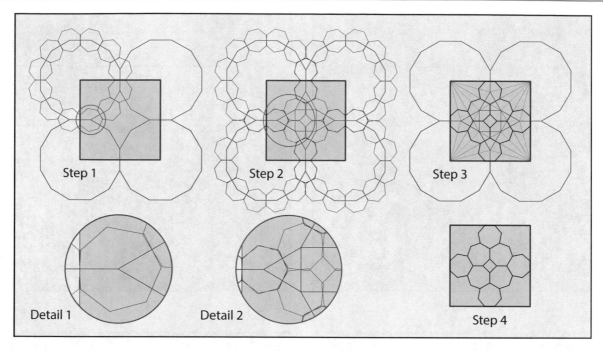

Step 1

Step 2

Step 3

Detail 1

Detail 2

Step 4

Fig. 343

corner of the orthogonal grid. This is an Ilkhanid *median* design from the mausoleum of Uljaytu in Sultaniya, Iran (1307-13) [Photograph 87]. The heptagons in the underlying tessellation are of two varieties, and they produce a very distinctive design characterized by twelve 7-pointed stars surrounding 12-pointed stars. The four underlying heptagons at the center of each repeat unit create a motif of four point-to-point seven-pointed stars. Figure 343 demonstrates a construction of the underlying tessellation for this design. This involves the placement of 12 regular heptagons in 12-fold rotation upon the vertices of a dodecagon as shown, as per Step 1. Detail 1 shows how the midpoints of the heptagonal edges intersect with the midpoints of the dodecagonal edges, but the edges of adjacent heptagons overlap rather than being contiguous. Step 2 places this ring of heptagons at each corner of the square repeat unit. This conceptually completes the tessellation. However, the nonaligned edges of the adjacent heptagons need to be corrected to make the pattern line application easier. Detail 2 shows how the edges of the heptagons are not aligned with one another. This is rectified by simply making all the irregular heptagonal edges that intersect with the formative dodecagons, triangles, and squares perpendicular with these formative edges. Step 3 shows the new set of heptagons that are slightly irregular, and have edges that are aligned with the red radii of the radii matrix. Step 4 shows the completed tessellation. This interesting tessellation will create very acceptable designs in each of the four pattern families, although no other examples are known to the historical record.

The *acute* design in Fig. 344d is from a Mamluk frontispiece of an illuminated Quran (1369) commissioned by Sultan Sha'ban.[55] This places 16-pointed stars at the vertices of the orthogonal grid and octagons at the center of each square repeat unit. This underlying tessellation includes an unusual feature of alternating pentagons and barrel hexagons that surround the 16-gons. Figure 344a illustrates a standard *acute* pattern line application to the underlying tessellation. This places a four-pointed star within the underlying square at the center of the repeat unit. The Mamluk artist who created this design chose to alter this central region. Figure 344b arbitrarily introduces an octagon surrounded by 4 eight-sided mushroom-shaped elements at the center of each repeat unit. This Quranic illumination places the central repetitive regions at the corners of the panel, allowing only 1/4 of each central region to be represented. For reasons that remain unclear, the artist chose to mirror the 1/4 segment of the octagon at each corner of the panel, thereby reversing the direction of the foot of each mushroom shape. However, this does not work well when repeating the design with translation symmetry, and has been ignored for the purposes of this study. This artist introduced the further modification of the primary 12-pointed stars shown in Fig. 344c. This conforms to the common practice for modifying the primary stars of *median* designs [Fig. 223], although this is applied to an *acute* rather than a *median* pattern. The design in Fig. 345

---

[55] Cairo National Library; 7, ff. IV-2r.

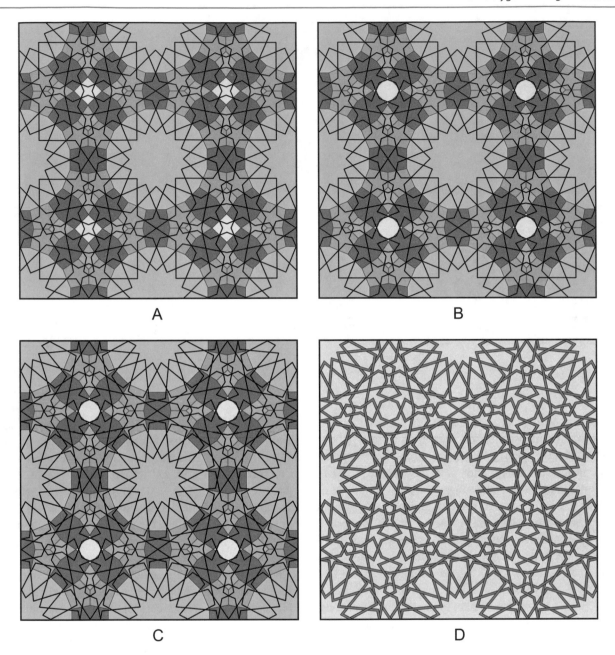

A

B

C

D

Fig. 344

is from the same Mamluk illuminated Quran commissioned by Sultan Sha'ban,[56] and this example also places 16-pointed stars onto the vertices of the orthogonal grid. Like the underlying tessellation of Fig. 344, there are alternating barrel hexagons arranged around the 16-gon, but in this case, the barrel hexagons are rotated 90°, and separated by an unusual octagonal interstice region. Such oddly proportioned interstice regions are often problematic when applying the pattern lines. However, in this case the bilateral pattern lines work very nicely within this atypical

feature. The similarity between this example and the previous pattern is not surprising in that both of these designs with 16-pointed stars were presumably the work of the same artist. Figure 345a illustrates the standard *acute* pattern with 45° crossing pattern lines. As with the previous example from the same Quran, Fig. 345b modifies the primary 12-pointed stars in the manner that was particularly popular among Mamluk artists [Fig. 223]. Again, this modification is typically applied to *median* patterns, but in this case has been applied to the *acute* pattern. The angles of the resulting dart motifs within this modification are consequently more acute than normally found with *median* patterns. Figure 345c shows four repeat units of this design in a two-color tiling

---

[56] Cairo National Library; 54, ff. IV-2r.

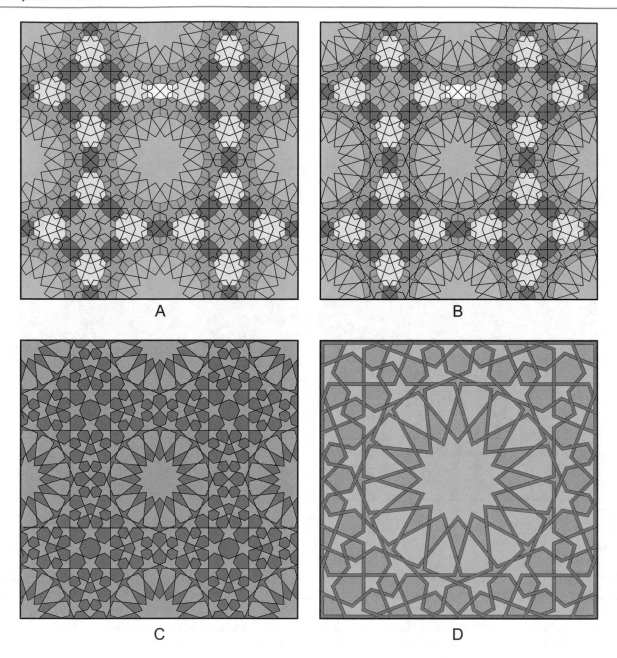

A

B

C

D

Fig. 345

expression. Figure 345d is a representation of the design as it was used in the Quranic illumination, albeit with different colorization and no floral infill.

3.2.3 Isometric Designs with Multiple Regions of Local Symmetry

There is a wide diversity of nonsystematic patterns that repeat upon an isometric grid and have more than one region of primary local symmetry. These invariably place objects with sixfold symmetry at the vertices of the triangular grid,

and objects with threefold symmetry at the vertices of the dual-hexagonal grid. Additional introduced regions of local symmetry may include the midpoints along the repetitive edges, which will always have twofold point symmetry (i.e., n-pointed stars with even numbers), and occasionally locations within the field of the design. The n-fold symmetry of local regions within the field is only limited by the inherent geometry of a given construction.

There are only a few nonsystematic underlying tessellations that were used historically to create designs in each of the four pattern families. The underlying tessellation that produces the patterns in Figs. 346 and 347 is perhaps the

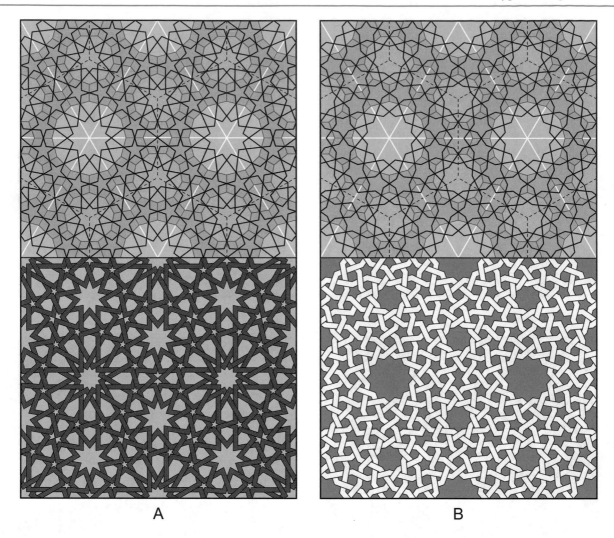

A   B

Fig. 346

most prolifically used, and the designs created from each of the pattern families are exceptionally beautiful. This underlying tessellation places dodecagons at the vertices of the triangular grid and nonagons at the vertices of the dual-hexagonal grid. A ring of pentagons surrounds each dodecagon, and the nonagons are separated by barrel hexagons. This configuration of polygons is the least complex of several historical underlying tessellations that were used to create patterns with 9- and 12-pointed stars. Figure 346a demonstrates the derivation of the *acute* pattern created from this underlying tessellation. This pattern was used widely throughout Muslim cultures, and the interweaving example in this figure represents the interpretation used on two illuminated facing pages of a Moroccan Quran written in 1568[57] [Photograph 65]. Early examples of this nonsystematic *acute* design include a Mengujekid carved stone

border from the east portal of the Great Mosque and hospital of Divrigi (1228-29); a Mamluk bronze door from the Sultan al-Zahir Baybars *madrasa* in Cairo[58] (1262-63); and the Mamluk ornament of the mausoleum of Fatima Khatun in Cairo (1283-84). Perhaps the most unexpected location for the use of this *acute* pattern is from a *Mudéjar* door in the Cathedral of Santo Domingo in Cusco, Peru (1559-1654). Figure 346b illustrates the *median* design created from the same underlying tessellation. Unlike the *acute* pattern, this design is uncommon, and a fine historical example is found at the mausoleum of Uljaytu in Sultaniya, Iran (1307-13). This is depicted at a much later date within one of the Tashkent scroll fragments (sixteenth or seventeenth century) at the Institute of Oriental Studies in Tashkent, Uzbekistan. Figure 347a shows an *obtuse* pattern created from this same underlying tessellation, and the earliest known use of this

---

[57] London, British Library, Or. 1405, ff. 399v-400r.

[58] This Mamluk bi-fold door presently serves as the entry door of the French Embassy in Egypt.

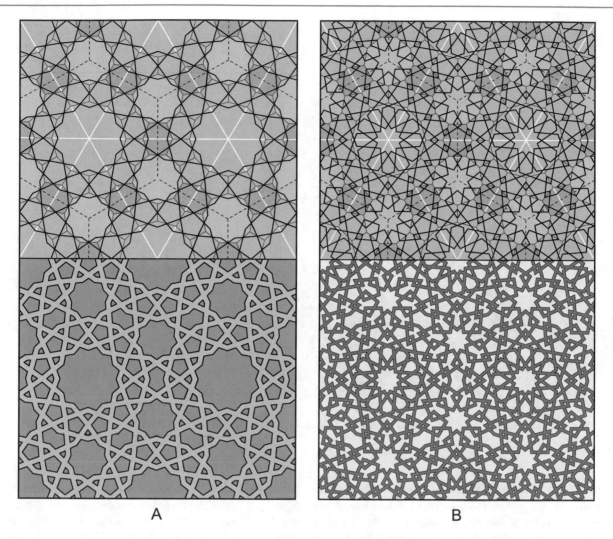

A   B

Fig. 347

design is in the carved stucco ornament that frames the arched Fatimid *mihrab* at the al-Azhar mosque in Cairo. This was likely produced during the first half of the twelfth century as part of the extensive renovations of the Fatimid caliph al-Hafiz li-Din Allah[59] (r. 1131-49). As such, this example is significant in that it appears to be the earliest extant nonsystematic pattern with more than one variety of primary star to originate from outside those regions under direct Seljuk artistic influence. Another early example of this *obtuse* pattern is found in the *mihrab* of the Great Mosque at Aksehir in Turkey (1213). Locations of later examples include the Qara Qoyunlu ornament of the Great Mosque of Van (1389-1400); a Timurid cut-tile mosaic border from the Abdulla Ansari complex in Gazargah, Afghanistan (1425-27) [Photograph 85]; a pierced stone screen from the minaret balcony of the Amir Ghanim al-Bahlawan funerary

complex in Cairo (1478); and the Shaybanid cut-tile mosaic arch spandrel in the entry portal of the Kalyan mosque in Bukhara (1514). The *two-point* pattern in Fig. 347b is also created from this same tessellation, and employs the most frequently used modification to the 12-pointed stars [Fig. 223], and the distinctive variation to the pattern lines of the 9-pointed stars [Fig. 225d]. The angles of the pattern lines within the nine-pointed stars are derived from the *two-point* 9-*s*3 pattern line application. The earliest known example of this design is Ilkhanid: from the mausoleum of Uljaytu in Sultaniya, Iran (1307-13). Several Mamluk examples include a panel from the Mamluk *minbar* at the Qadi Abu Bakr Muzhir complex in Cairo (1479-80); an incised stone relief panel in the entry portal of the Qanibay Amir Akhur funerary complex in Cairo (1503-04); and a contemporaneous rectangular side panel from the *minbar* of the Sultan Qansuh al-Ghuri complex (1503-05) [Photograph 53]. Figure 348 demonstrates a construction of the underlying tessellation that produces the patterns in Figs. 346 and

---

[59] Rabbit (1996), 45–76.

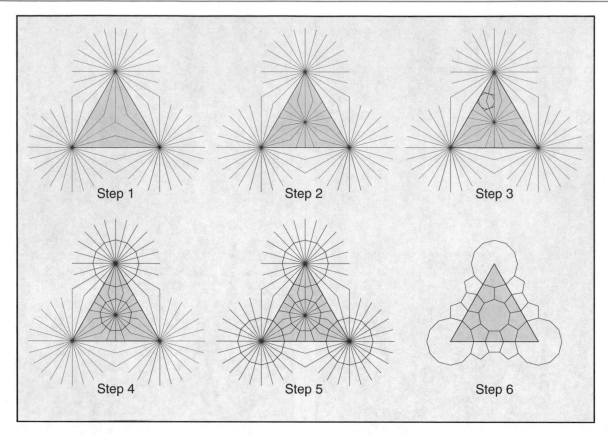

Fig. 348

347. This begins with the placement of 24 radii at the vertices of an equilateral triangle as shown in Step 1. Step 2 introduces 18 radii at the center of the triangle, and completes the radii matrix that produces this underlying tessellation. Step 3 determines the sides of the nonagon and dodecagon, as well as the proportions of the irregular pentagons. This follows the standard procedure described earlier. These three polygonal elements are completed in Step 4. Step 5 mirrors and rotates these elements throughout the triangle, and Step 6 illustrates the completed underlying tessellation.

Figure 349 shows how the same radii matrix with 24-fold and 18-fold centers of local symmetry can be used to create another underlying tessellation that also produces fine patterns with 12- and 9-pointed stars. As in the *fivefold system*, the arrangement of six pentagons surrounding the thin rhombus makes this alternative underlying tessellation especially appropriate for *obtuse* and *two-point* patterns [Fig. 197]. Step 1 draws a line that connects the vertices of the red and blue radii as shown. Step 2 reflects this line upon the edge of the triangle, and upon one of the red central radii. Step 3 uses these lines to determine the edges of the nonagons and dodecagons, as well as two varieties of pentagon. Step 4 completes the nonagon and dodecagon, as well as the pentagons. Step 5 mirrors and rotates these elements throughout the triangle, thereby determining the barrel

hexagons and small rhombi along each triangular edge. Step 6 illustrates the completed underlying tessellation. Figure 350a illustrates the *obtuse* pattern (by author) created from this underlying tessellation, and Fig. 350b illustrates the *two-point* pattern (by author) produced from this same underlying tessellation, and the 12-pointed stars have been modified in the standard fashion [Fig. 223]. While both of these patterns are aesthetically acceptable to this tradition, neither appears to have been used historically. As with fivefold examples with the cluster of six pentagons surrounding a thin rhombus, the underlying tessellation in Fig. 350 requires modification for the production of successful *acute* patterns. As explained previously [Fig. 198], there are two varieties of such modification. The *acute* design (by author) in Fig. 351a is created from the truncation of just four of the six clustered pentagons. This transforms the four pentagons into trapezoids that then define a wide rhombic interstice region. The *acute* design (by author) in Fig. 351b modifies the cluster of six pentagons by truncating all six pentagons such that they become six trapezoids surrounding a concave hexagonal interstice region. The pattern line application within the trapezoids of both these designs maintains the distinctive dart motif common to *acute* patterns. Although aesthetically acceptable, neither of these modified patterns is known to the historical record.

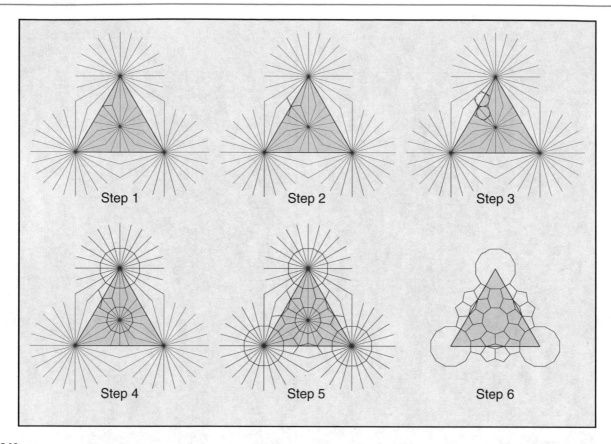

Step 1 Step 2 Step 3

Step 4 Step 5 Step 6

Fig. 349

Figure 352 illustrates the derivation of another *obtuse* pattern with 12-pointed stars placed at the vertices of the isometric grid and 9-pointed stars at the vertices of the dual-hexagonal grid. The underlying polygonal elements and applied pattern lines have shared characteristics with many designs created from the *fivefold system* [Fig. 233a]. What would otherwise be 12-pointed stars within this design have been given a sixfold additive modification that is conceptually the same as the far more common fivefold modification found in *obtuse* patterns from the *fivefold system* [Fig. 224a]. As with analogous patterns created from similar underlying polygonal modules from the *fivefold system*, this pattern can also be created from an alternative underlying tessellation of dodecagons, pentagons, barrel hexagons, and thin rhombi (not shown). This *obtuse* design was produced during the Seljuk Sultanate of Rum, and is from the Sultan Han in Aksaray, Turkey (1229). Figure 353 demonstrates how the construction of the underlying tessellation for this pattern can be produced from a radii matrix that begins with Step 1 wherein a line that intersects the third vertices of the nonagon is extended outside the nonagon to a length that is equal to the sides of the nonagon. Step 2 mirrors this process as shown and places a line (red) at the ends of these extended lines that will become an edge of the triangular repeat. Step 3 mirrors the two extended lines to make the

hexagonal element, and rotates the edge three times around the nonagon to make an equilateral triangle that is centered on the nonagon. Step 4 rotates the elongated hexagons to each side of the triangle. Step 5 uses the acute points of the elongated hexagons to determine the size and position of the dodecagons. And Step 6 shows the completed triangular repeat for the underlying tessellation.

Figure 354 details the construction of yet another *obtuse* pattern with 12-pointed stars on the vertices of the isometric grid and 9-pointed stars at the vertices of the dual hexagonal grid. Like the pattern in Fig. 352, this pattern can be made from either of two underlying tessellations. Figure 354a illustrates the pattern derivation from an underlying tessellation of dodecagons, nonagons, pentagons, barrel hexagons, and thin rhombi. Figure 354b uses a dual underlying tessellation of dodecagons, nonagons, elongated hexagons, and concave hexagons to create the same design. Figure 354c demonstrates the dual relationship between these alternative tessellations. This is analogous to the dual relationship in many examples from the *fivefold system* [Fig. 200]. This design is also from the Seljuk Sultanate of Rum, and comes from the Susuz Han in the village of Susuzköy, Turkey (1246).

Figure 355 shows the construction of a Mamluk *median* design from a window lintel in the Qartawiyya *madrasa* in

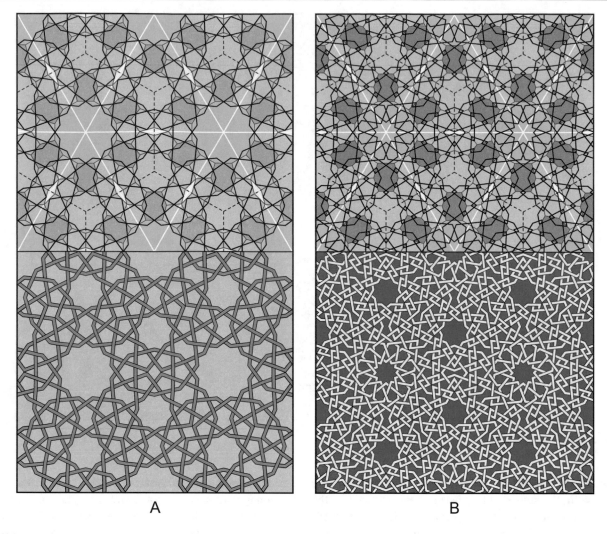

A   B

Fig. 350

Tripoli, Lebanon (1392). Figure 355b shows the standard *median* derivation for this pattern, and Fig. 355a demonstrates how this design is created from an underlying tessellation comprised of 15-gons, dodecagons, pentagons and barrel hexagons. The design in Fig. 355d is the historical example that modifies the 15-pointed stars as per the common fivefold convention [Fig. 223]. These two patterns place 12-pointed stars on the vertices of the isometric grid, and 15-pointed stars at the vertices of the dual hexagonal grid. Figure 355a shows how the pattern lines within the pentagons and barrel hexagons that surround the 15-gon are alternating five-pointed stars and darts. In Fig. 355c the pattern lines within these polygonal elements eliminate the points that extend into the 15-gons, and extend the other lines to create the ring of 15 darts that characterize this modification. Figure 356 demonstrates a construction of the underlying tessellation that produces this pattern. This process begins with the placement of 24 radii at the vertices of an equilateral triangle and 30 radii at the center of the triangular repeat as shown in Step

1. Step 2 places a line that connects the red and blue vertices as shown. Step 3 mirrors this line along the edge of the repeat. Step 4 determines the sides of the dodecagon and 15-gon as well as the proportions of two irregular pentagons. Step 5 completes these elements. Step 6 mirrors one of the pentagons, determines a third pentagon by mirroring one of the sides of the 15-gon and mirroring the adjacent radii, and establishes the barrel hexagon from the blue radii. Step 7 mirrors these elements as shown. Step 8 rotates these elements three times around the center of the triangular repeat. And Step 9 shows the completed underlying tessellation.

Figure 357 demonstrates the derivation of an *acute* pattern with 18-pointed stars placed upon the vertices of the isometric grid and 9-pointed stars upon the vertices of the dual-hexagonal grid. This design also has regular octagons at the midpoints of the edges of both the triangular and hexagonal repetitive cells, where these two grids intersect. This is a Mamluk design from the Amir Qijmas al-Ishaqi mosque in Cairo (1479-81). Figure 358 demonstrates a construction of

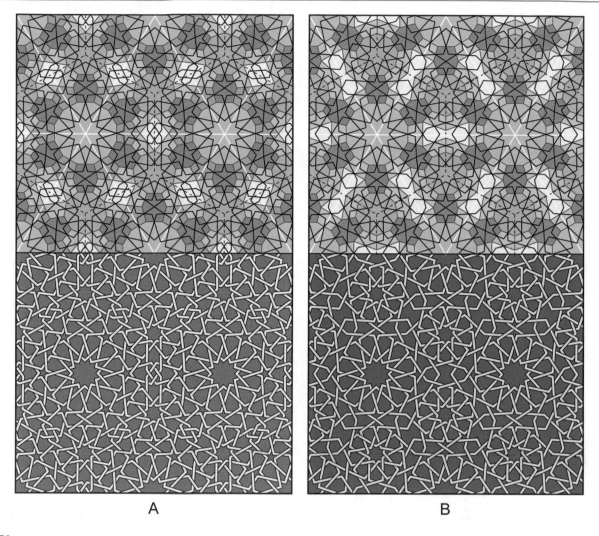

A   B

Fig. 351

the underlying tessellation that produces this pattern. Step 1 locates 36 radii at the vertices of an equilateral triangle and 18 radii at the center of the triangular repeat. Step 2 establishes the size of the 18-gon and nonagon, as well as the pentagon that separates the 18-gon and nonagon. Step 3 completes the 18-gon and nonagon, as well as the separating pentagon. Step 4 establishes two more pentagons placed at two midpoints of the triangular repetitive edges. Step 5 mirrors these new pentagons along the sides of the triangular repeat, and rotates the polygonal elements throughout the triangular repeat. Step 6 illustrates the completed underlying tessellation.

Figure 359 shows an *acute* design with 18-pointed stars at the vertices of the isometric grid and 12-pointed stars at the vertices of the dual hexagonal grid. This pattern can be created from either of two underlying tessellations. Figure 359a is a simple edge-to-edge configuration of 18-gons and dodecagons, while the underlying tessellation of Fig. 359d is comprised of the same primary polygons, but with a

connective polygonal matrix that includes pentagons, trapezoids, and interstice concave hexagons. Figures 359c and f are identical except for scale. This Mamluk design is from the Sultan al-Zahir Barquq *madrasa* and *khanqah* in Cairo (1384-1386) [Photograph 54]. Figure 360 shows a construction of the underlying tessellation that produces the pattern from Fig. 359. This begins with the placement of 36 radii at the vertices of an equilateral triangle and 24 radii at the center of the triangular repeat. Step 1 illustrates this radii matrix. Step 2 determines the sides of the dodecagon and 18-gon, as well as the proportions of the irregular pentagon that separates these primary polygons. This also places a line that connects the end points of blue and red radii that is a precursor for another pentagon. Step 3 completes these polygonal elements, as well as establishes the proportions for the additional pentagons and thin rhombus by mirroring the precursor line from the previous step. Step 4 mirrors these additional pentagons. Step 5 rotates these polygonal elements throughout the triangular repeat;

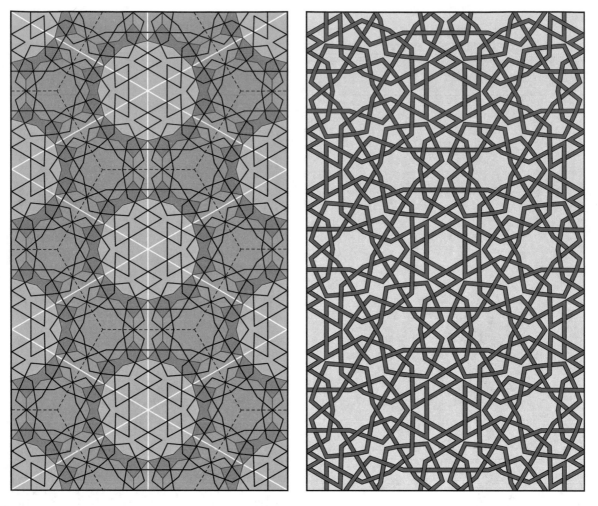

Fig. 352

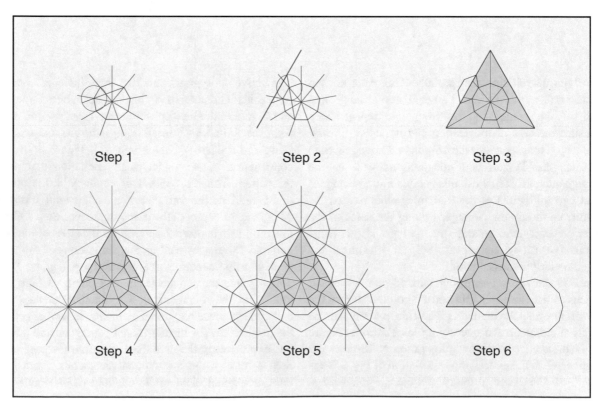

Step 1   Step 2   Step 3

Step 4   Step 5   Step 6

Fig. 353

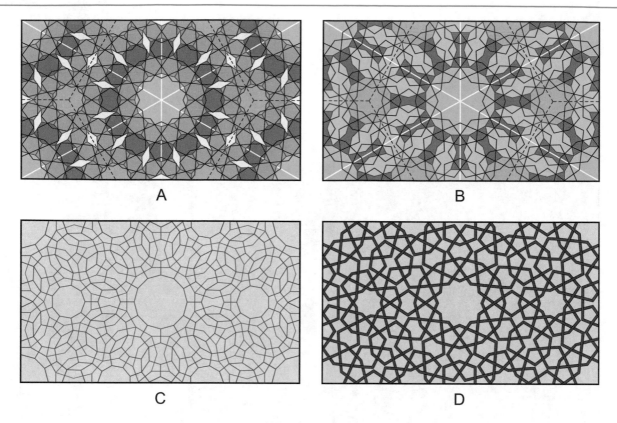

Fig. 354

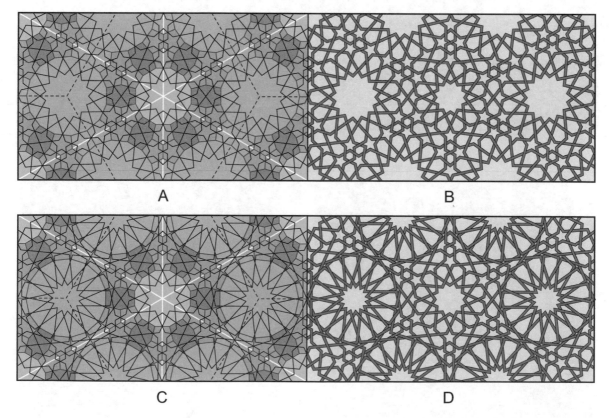

Fig. 355

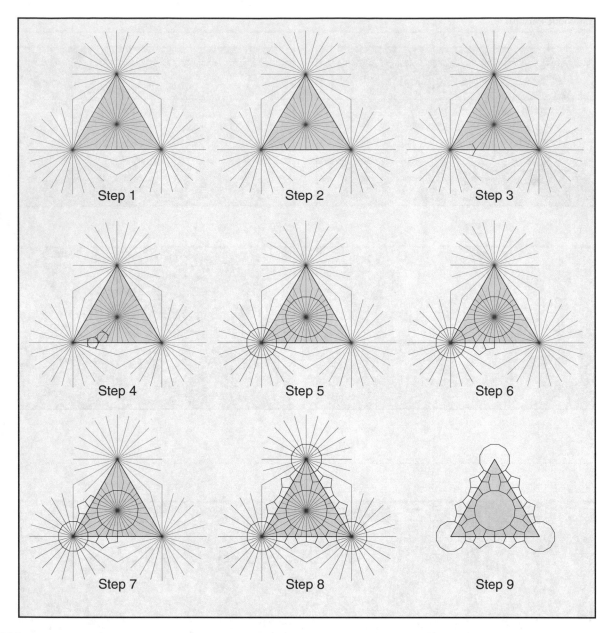

Step 1　　　　　　Step 2　　　　　　Step 3

Step 4　　　　　　Step 5　　　　　　Step 6

Step 7　　　　　　Step 8　　　　　　Step 9

Fig. 356

Fig. 357

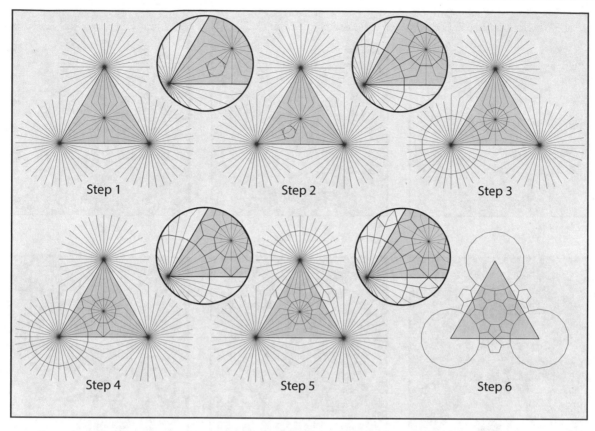

Step 1

Step 2

Step 3

Step 4

Step 5

Step 6

Fig. 358

and Step 6 completes the underlying tessellation. As with previous examples, the arrangement of six clustered pentagons surrounding a thin rhombus that occurs along each edge of the triangle is not suitable for the production of *acute* patterns. The *acute* design in Fig. 359 makes use of the modification to these clustered pentagons as per established convention [Fig. 198] wherein all six pentagons are truncated to form six trapezoids with a concave hexagonal interstice region.

As mentioned, some isometric designs with more than one variety of primary star place one of these stars at the vertices of the triangular grid and the other at the midpoints of the edges of the triangular cell. The *acute* design in Fig. 361 places 12-pointed stars at the vertices of the isometric grid, and 8-pointed stars at these midpoints. The application of the pattern lines allows for the introduction of nonagons at the centers of the triangular repetitive cells, which are also the vertices of the dual-hexagonal grid. This design is from the Seljuk Sultanate of Rum, and was used on one of the many gravestones at Ahlat near the shores of Lake Van in eastern Turkey (thirteenth to fifteenth centuries). The construction of the underlying tessellation that creates this design is shown in Fig. 362. This begins with setting up the radii matrix. Step 1 places 24 radii at the vertices of the triangular repeat. Step 2 introduces 16 radii at the midpoints of the edges of the triangular repeat. This completes the radii matrix. Step 3 determines the size and placement of the dodecagons and octagons within the underlying tessellation. This also identifies the proportions of the pentagons that surround the dodecagons. Step 4 completes the dodecagon, octagon, and pentagon. Step 5 mirrors the pentagon. Step 6 rotates these polygonal elements throughout the triangular repeat. Step 7 fills the interstice regions at the center of the triangle from Step 6 with three irregular hexagons with edges that are congruent with the red radii. And Step 8 illustrates the completed underlying tessellation. Figure 363 shows the construction of an *obtuse* pattern that places 12-pointed stars at the vertices of the isometric grid and 10-pointed stars at the midpoints of each edge of the triangular repeat units. A measure of the success of this fine Qara Qoyunlu design is that it appears less complex than it actually is. This was used in the raised brick ornament of the Great Mosque at Van in eastern Turkey (1389-1400) [Photograph 86]. The similarity and close proximity of this example to the previous design in Fig. 361, albeit with ten- rather than eight-pointed stars, suggest the possibility that they may have been produced by the same individual or atelier, or at the least, that one may have inspired the origin

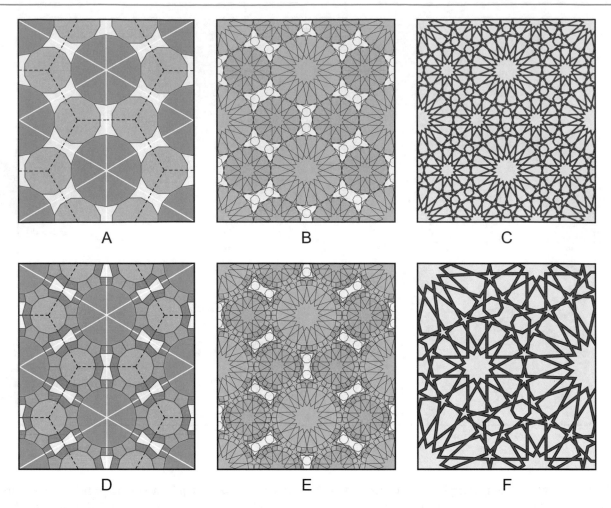

Fig. 359

of the other. The construction of the underlying tessellation that produces this design is demonstrated in Fig. 364. This begins with the construction of the radii matrix. Step 1 places 24 radii at the vertices of the triangular repeat. Step 2 introduces 20 radii at the midpoints of the edges of the triangular repeat, thus completing the radii matrix. Step 3 determines the size and placement of the dodecagons and decagons within the underlying tessellation. This also identifies the proportions of the pentagons that surround the dodecagons. Step 4 completes the dodecagon, decagon, and pentagons. Step 5 mirrors the pentagon. Step 6 rotates these polygonal elements throughout the triangular repeat. Step 7 fills the interstice regions at the center of the triangle from Step 6 with three clustered pentagons at the center of the triangle surrounded by three irregular hexagons. And Step 8 illustrates the completed underlying tessellation.

The *acute* pattern in Fig. 365 has three regions of local symmetry, with 24-pointed stars at the vertices of the isometric grid, 12-pointed stars at the vertices of the dual hexagons grid, and 8-pointed stars at the midpoints of the

edges of the triangular repetitive cells. The crossing pattern lines of the 24-pointed stars and 12-pointed stars have been arbitrarily moved slightly inward from the midpoints of their associated underlying polygonal edges toward the centers of their respective primary polygons. This allows the parallel pattern lines within the underlying octagons to be replicated within the underlying dodecagons, and closely simulated within the 24-gons. Were the crossing pattern line associated with the 24- and 12-pointed stars placed on the midpoints of their respective primary polygons, the sizes of the central 24- and 12-pointed stars would be substantially increased, thereby changing the aesthetic character of the finished design. This is a Seljuk Sultanate of Rum pattern from the *mihrab* niche at the Great Mosque of Ermenek in Turkey (1303). Figure 366 demonstrates a construction of the underlying tessellation for this design, beginning with the construction of the radii matrix. Step 1 places 48 radii at the vertices of the triangular repeat. Step 2 introduces 24 at the center of the repeat. Step 3 places 16 radii at the midpoints of the edges of the repeat. This completes the radii matrix.

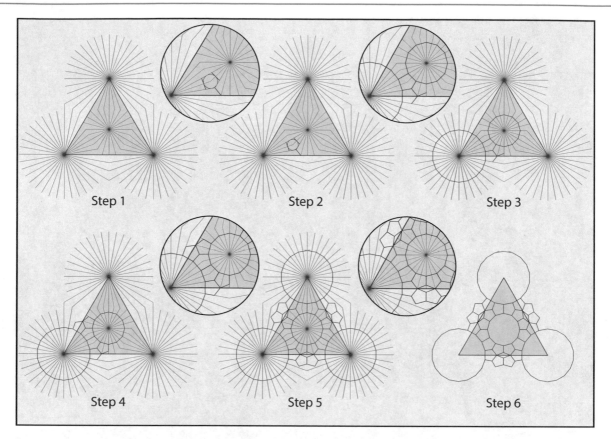

Fig. 360

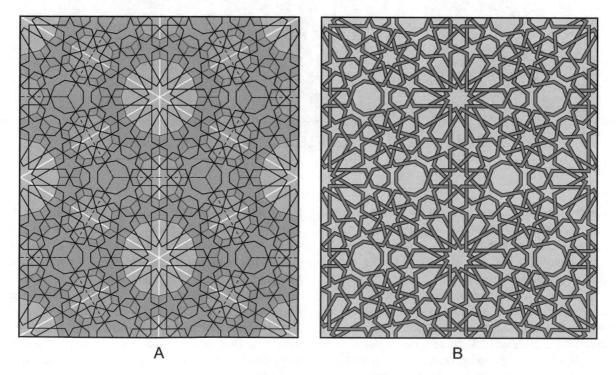

Fig. 361

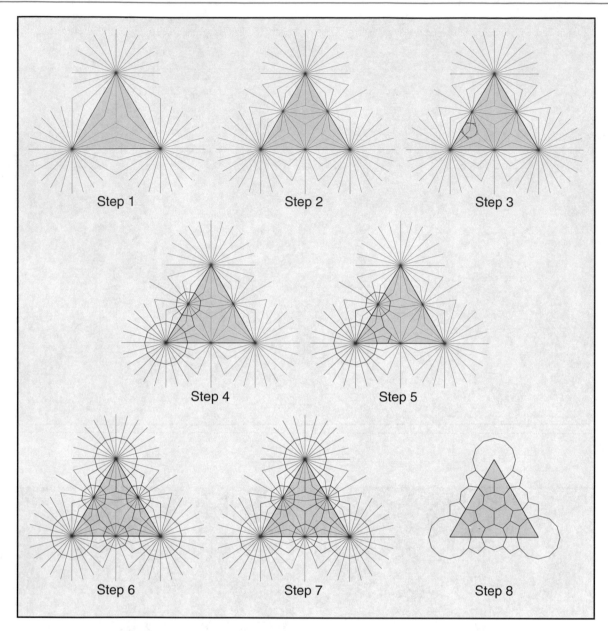

Step 1   Step 2   Step 3

Step 4   Step 5

Step 6   Step 7   Step 8

Fig. 362

Step 4 determines the size and placement of the dodecagons and octagons within the underlying tessellation, as well as the pentagon that separates these two polygons. This also places a 24-gon of a suitable size to allow for a well-proportioned pentagon between the 24-gon and the octagon. Step 5 emphasizes the pentagons the separate the 24-gon, dodecagon, and octagon. Step 6 mirrors these polygons. Step 7 rotates these throughout the triangular repeat; and Step 8 illustrates the completed underlying tessellation.

The design in Fig. 367 incorporates a fourth region of local symmetry located within the field of the pattern matrix. This remarkable *acute* design has 9-, 10-, 11-, and 12-pointed stars, and is arguably the most complex

nonsystematic isometric design from this ornamental tradition. Figure 367a shows how the underlying tessellation for this design places dodecagons at the vertices of the isometric grid (blue), nonagons at the vertices of the dual hexagonal grid (green), decagons at the midpoints of the edges of the triangular cells (purple), and hendecagons within the field of the polygonal matrix (grey). These primary polygons allow for the introduction of the four varieties of local symmetry to the completed design. The four primary polygons are surrounded by a connective matrix of pentagons and barrel hexagons. This masterpiece of geometric art was produced during the Seljuk Sultanate of Rum, and two examples are found within the architectural record: the carved stone

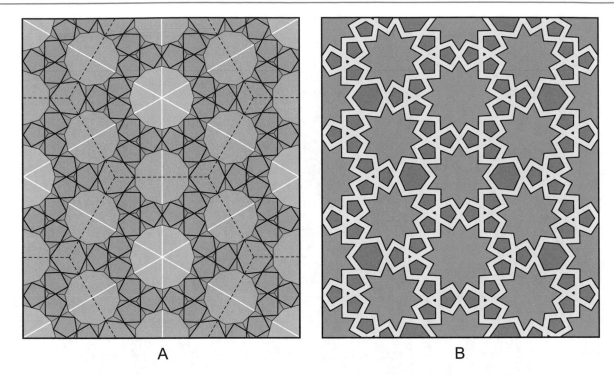

A   B

Fig. 363

ornament in the courtyard portal of the Seri Han near Avanos (1230-35), and the carved stone ornament of the entry to the mosque at the Karatay Han near Kayseri (1235-41). These contemporaneous examples are only 65 km apart, and are almost certainly the work of the same artist or atelier. The three designs with 9-, 10-, 11-, and 12-pointed stars in Fig. 368 (by author) are created from the same underlying tessellation as the previous *acute* pattern. Figure 368a shows a *median* pattern, Fig. 368b shows an *obtuse* pattern, and Fig. 368c shows a *two-point* pattern. Each of these is acceptable to the aesthetics of this tradition, but are unknown to the historical record. Collectively, these demonstrate the efficacy of the polygonal technique in creating highly complex designs in each of the four pattern families from a single underlying tessellation. The construction of the underlying tessellation for these designs is significantly more complex than previous examples. Figure 369 details the creation of just the radii matrix. This requires subtle geometric adjustments during the construction to allow for the incorporation of the regions with 11-fold radial symmetry. Step 1 places 24 radii at the vertices of the triangular repeat. Step 2 adds 18 radii to the center of the triangle. Step 3 incorporates 20 radii at the midpoint of each edge of the triangle. The maker of this pattern could have finished the radii matrix at this point—moving directly to the establishment of an underlying tessellation that will create patterns with 9-, 10-, and 12-pointed stars. However, a close examination of the radii matrix in Step 3 reveals the potentiality of the further incorporation of local 11-fold symmetry. The

introduction of the 11-fold region begins with Step 4, which highlights the 114° and 66° angles of the radii from Step 3. As further shown, these are very close to the 114.5454...° and 65.4545...° angles found in 22-fold rotational symmetry, and indicate that regions with 11-fold symmetry might successfully be incorporated into this radii matrix. Step 5 places 22 radii at the intersection of the radii from the 24-, 20-, and 18-fold centers. Detail 1 shows how the line of radius from the 20-fold center does not quite intersect the 22-fold center; and conversely, the line of radius from the 22-fold center does not quite intersect with the 20-fold center. Detail 2 introduces a vertical line half way between these two centers. Detail 3 moves the 22 radii vertically by the small amount so that the two lines of radius from the 20- and 22-fold centers intersect. This detail also trims these lines. These two radii now appear collinear, but actually have a slight angle off 180°. Step 6 rotates the region with 22-fold symmetry into the other two positions within the triangle. The small dots indicate the radii intersections that appear collinear. Figure 370 demonstrates the construction of the underlying tessellation from the radii matrix in Fig. 369. This follows established procedure of using tangent circles to determine the pentagons and primary higher order polygons. Step 1 determines the edge size and location for the nonagon, decagon, and hendecagon through placing circles at the vertices of the red radii that are tangent to the blue radii. Step 2 completes these three primary polygons. Step 3 emphasizes the various pentagons and barrel hexagons that were also created from the tangent

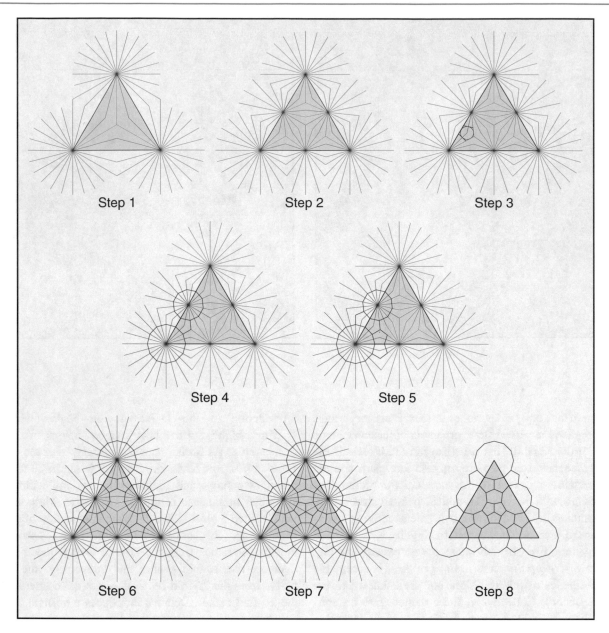

Fig. 364

circles that separate the primary polygons of Step 2. Step 3 also determines the dodecagon by drawing a line that connects red and blue radii, thereby creating a pentagon, and drawing a circle that is tangent with this pentagon and the blue radii. This circle is centered on the edge of the triangular repeat unit. Step 4 completes the dodecagon, and emphasizes the additional pentagons and barrel hexagons implicit within the radii matrix. This step also mirrors the polygonal elements. Step 5 rotates the polygonal elements from Step 5 throughout the triangular repeat; and Step 6 shows the completed underlying tessellation.

Clearly, Muslim artists were immensely innovative in developing repetitive strategies that allowed for the use of multiple centers of local symmetry within a single pattern. However, they by no means exhausted the pool of repetitive polygonal stratagems that provide for the creation of visually compelling geometric designs. An exploring mind can discover new forms of underlying polygonal repetitive stratagems that are effective, but do not appear within the historical record. These can provide contemporary artists with tremendous potentiality for the creation of original nonsystematic geometric designs, often imbued with considerable complexity. Within the isometric family, there are at least two contemporary repetitive stratagems that are particularly interesting, and worthy of detail. Both of these will produce a very large number of very successful patterns, and

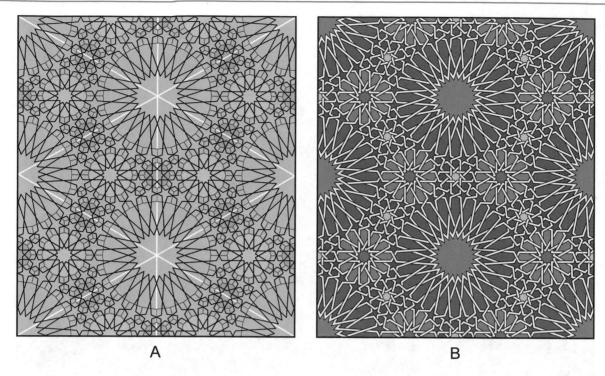

A

B

Fig. 365

can be regarded as doorways into uncharted creative territory.

Figure 371 illustrates the principle behind the first of these contemporary methods for producing underlying generative tessellations. This type of polygonal construction is governed by the use of a central polygon with sides that are multiples of three (e.g., triangle, hexagon, nonagon, dodecagon). Onto this central polygon are placed three edge-to-edge polygons in threefold rotation around the central polygon. Lines are drawn that bisect these three new polygons. These three lines are extended until they meet, thus creating an equilateral triangle. The triangle and polygonal elements are mirrored to create a rhombus that is then rotated three times to create a regular hexagonal repeat unit with translation symmetry. Patterns based upon this repetitive schema are of the *p31m* plane symmetry group. In addition to the author, both Peter Cromwell in the United Kingdom, and Goossen Karssenberg of the Netherlands independently discovered this method of constructing polygonal matrices.[60] The examples in Fig. 371 are just eight such

tessellations, but very many more are possible, and each of these has the potential for creating very successful patterns in each of the four pattern families. Examples A, B, C, and D all use the hexagon as the central polygon, while examples E, F, G, and H use the nonagon at the center. Example A adds three pentagons to the hexagon, creating an interstice region of three clustered irregular heptagons. Example B adds heptagons to the central hexagons. This serendipitously provides for the incorporation of dodecagons into the generative tessellation. Example C adds octagons to the central hexagons, and example D adds nonagons to the hexagons. The inclusion of the nonagons allows for the secondary placement of additional hexagons at the vertices of the repetitive grid. This tessellation makes particularly nice geometric patterns. Example E adds heptagons to the central nonagons. This tessellation also produces very nice patterns. Example F adds octagons to the central nonagons, and examples G and H both add decagons. The arrangement of nonagons and decagons in Example H allows for the serendipitous addition of 15-gons at the vertices of the

---

[60] Peter Cromwell provides two designs of his own creation that subscribe to this repetitive principle. He explains that the application of this triangular repeat is inspired by an analogous square design within the Topkapi Scroll. Goossen Karssenberg in the Netherlands independently discovered this same methodology. He is a teacher of mathematics, and a specialist in Islamic geometric design. He sent the author the arrangement of heptagons surrounding central nonagons as per Fig. 371e in the spring of 2014. Parenthetically, the author created

several designs in the 1990s that employ this same repetitive strategy, for example, Fig. 371d. In 2008 the author also applied this strategy to the triangles of the octahedron to produce a spherical design with regularly distributed seven-pointed stars. However, it was not until receiving the letter from Goossen Karssenberg that the author realized of the potential for applying this methodological principle more broadly to produce a diverse variety of underlying tessellations such as those represented within Fig. 371. See Cromwell (2010), 73–85.

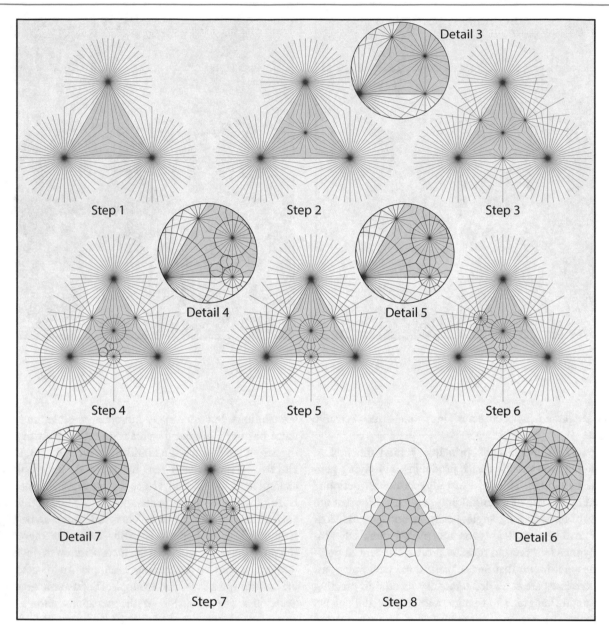

Fig. 366

rhombic repeat. Figure 372 shows the construction of a nonhistorical *median* pattern (by author) comprised of seven- and nine-pointed stars created from the tessellation in Fig. 371e.[61] This has shared visual characteristics to the *median* patterns created from the *fourfold system A* [for examples, Figs. 145 and 154]. Indeed, the success of this design is predictable through the *principle of adjacent numbers* wherein a star form that works well on its own (in this

case the eight-pointed star) indicates the likely success of patterns that use stars that are one numeric step above and below this star form. This is a useful principle for predicting patterns with unexpected combinations of local symmetry—in this case seven- and nine-pointed stars. Figure 373a illustrates a more complex form of this repetitive strategy that combines central dodecagons with three edge-to-edge hendecagons (11-gons) in threefold rotation around it. As with the examples from Fig. 371, the elements from the initial triangle have been mirrored to form a rhombic cell that is then rotated three times to produce the hexagonal repeat unit with translation symmetry. Figure 373b illustrates the serendipitous addition of tridecagons

---

[61] This *median* design was created by the author from the underlying tessellation produced by Goossen Karssenberg, and sent to the author in the spring of 2014.

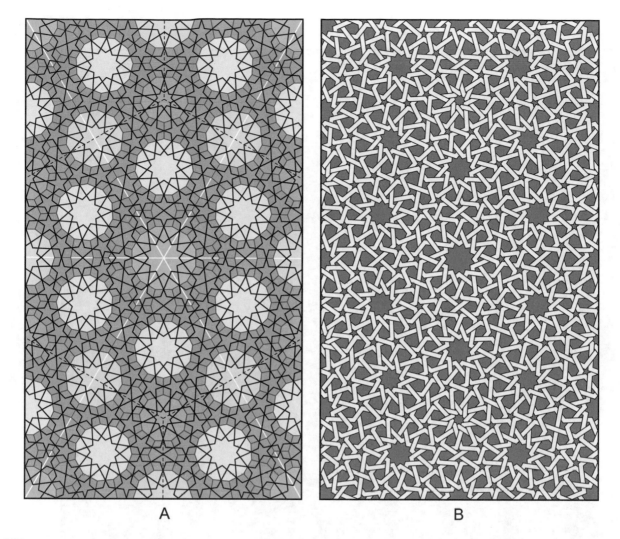

A

B

Fig. 367

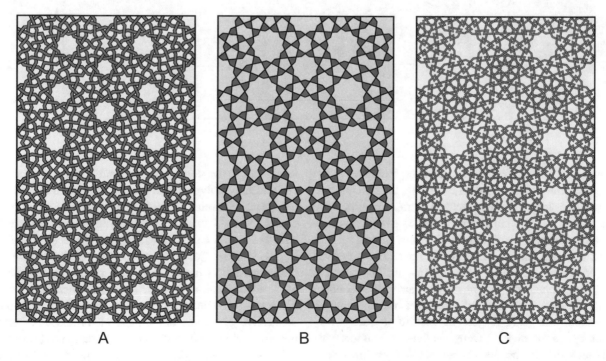

A

B

C

Fig. 368

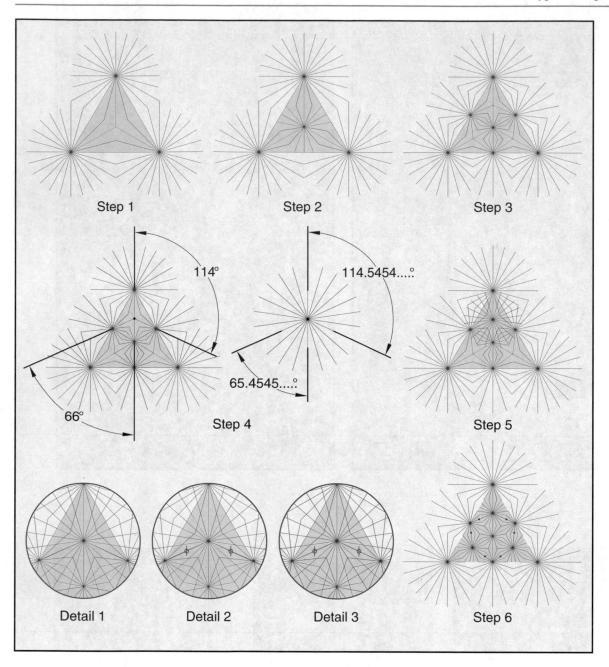

Step 1   Step 2   Step 3

114°

114.5454.....°

66°   65.4545.....°   Step 5

Step 4

Detail 1   Detail 2   Detail 3   Step 6

Fig. 369

(13-gons) into this configuration. However, the edges of the tridecagons are not quite congruent with their neighbors. This configuration of hendecagons, dodecagons, and tridecagons (11-, 12-, and 13-gons) can be used on its own to create patterns, but has limitations. However, as demonstrated previously, greater design potential is derived from underlying generative tessellations that separate the primary polygons with a connective matrix of secondary polygons. Figures 373c and d demonstrate how the polygonal tessellation in Fig. 373b can be used to produce a very satisfactory radii matrix from which new underlying tessellations can be created. Lines are simply introduced

that connect the centers to the vertices and midpoints of each primary polygon, and extend these radii outward until they meet with other extended radii. Figure 373d shows the completed radii matrix. Figure 373e shows an underlying tessellation that can be created from this radii matrix, and Fig. 373f shows just the underlying tessellation within its hexagonal repeat unit. The initial non-congruence between the tridecagons and their adjacent hendecagons and dodecagons is eliminated through the matrix of secondary connective polygons, allowing each of the three primary polygons within this underlying tessellation to have regular *n*-fold symmetry. Figure 374 is an *acute* pattern (by author)

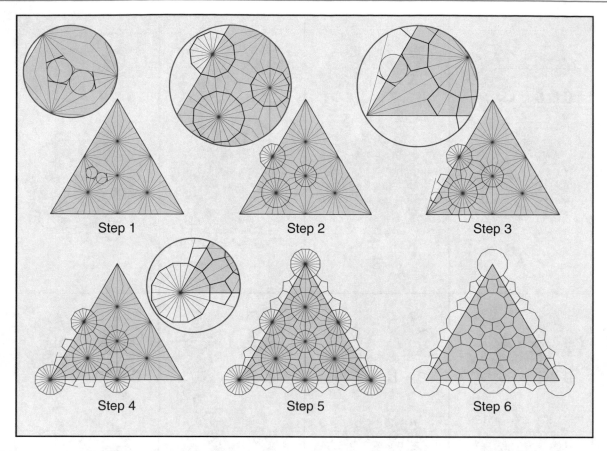

Step 1

Step 2

Step 3

Step 4

Step 5

Step 6

Fig. 370

created from the underlying tessellation constructed in Fig. 373. This design is comprised of 11-, 12-, and 13-pointed stars, with octagons also incorporated into the pattern matrix. While not historical, this *acute* pattern comports with the aesthetics of this tradition, especially as practiced in Anatolia during the Seljuk Sultanate of Rum. This seemingly eccentric combination of star forms is less mysterious when compared with the 3.12^2 semi-regular isometric tessellation of dodecagons and triangles [Fig. 89]. This places six edge-to-edge dodecagons around each individual dodecagon. The underlying tessellation for the pattern in Fig. 374 can be interpreted as having replaced these six surrounding dodecagons with alternating hendecagons (green) and tridecagons (yellow). This is another form of the *principle of adjacent numbers* where the original polygonal figure, in this case the dodecagon, is maintained, and the surrounding dodecagons are replaced with polygons that have alternating plus-one and minus-one number of sides. The underlying tessellation used to produce this pattern will also make very successful *median*, *obtuse*, and *two-point* patterns.

Figure 375 demonstrates the principle behind the second hexagonally based repetitive stratagem that, although ahistorical, nonetheless produces a very wide variety of

underlying generative tessellations with tremendous potential for contemporary geometric artists. This method of creating nonsystematic patterns with unusual combinations of local symmetry makes use of two varieties of irregular hexagonal cell (red) within a single repetitive construction. The blue hexagons in Fig. 375 are centered on the isometric grid, while the beige hexagons are centered upon the dual hexagonal grid. The distribution of these two types of hexagon is similar to the active and passive underlying hexagonal grid in the patterns of Fig. 98. However, the included angles of both varieties of hexagon in Fig. 375 have been adjusted to conform to the angles associated with specific polygons: thus allowing for the incorporation of unusual combinations of primary polygons, with their associated local symmetries, into an underlying generative tessellation. The angular proportions of Fig. 375a are derived from a combination of regular pentagons and heptagons at the vertices of the irregular hexagonal grid, and regular hexagons at the centers of the beige hexagons. This will produce patterns with five-, six-, and seven-pointed stars. The alternating pentagons and heptagons around each hexagon are another example of the *principle of adjacent numbers*. The lack of precise edge-to-edge alignment in this configuration of polygons will cause problems when extracting patterns. As in Fig. 373, this can

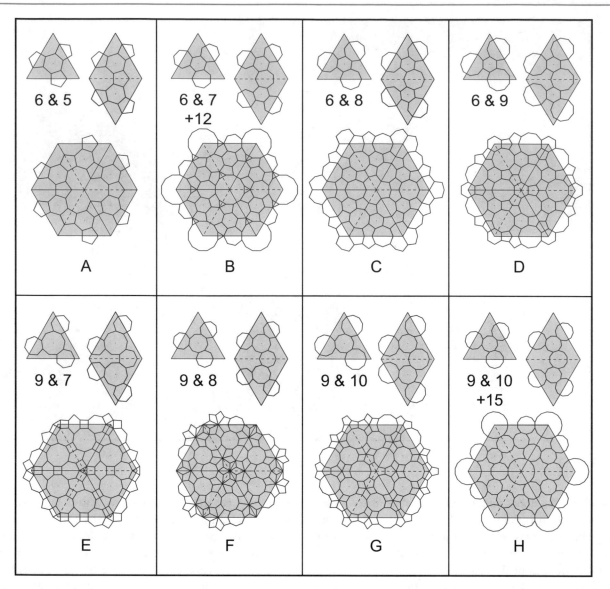

Fig. 371

be overcome by using the regular polygons as a layout for a radii matrix that can then be used to create a new underlying tessellation with connecting irregular polygons separating the regular primary polygons. The proportions of the hexagonal grid in Fig. 375b are based upon an arrangement of tetradecagons (14-gons) and decagons at the vertices of the hexagonal grid, with dodecagons at the centers of the blue hexagons. This will make patterns with 10-, 12-, and 14-pointed stars. The included angles of what would otherwise be the blue hexagons in Fig. 375c are 60° and 180°: affectively turning the hexagon into an equilateral triangle. This places octagons and dodecagons into an edge-to-edge configuration that allows for patterns with 8- and 12-pointed stars. Figure 375d places octagons and decagons at the vertices of the hexagonal grid, and hexagons at the centers of the blue hexagons. This will produce patterns with six-,

eight-, and ten-pointed stars. Figure 375e places hendecagons (11-gons) and tridecagons (13-gons) at the vertices of the hexagonal grid, with dodecagons at the centers of the brown hexagons. As with the example in Fig. 374, this will produce patterns with 11-, 12-, and 13-pointed stars, although with a completely different repetitive schema. Figure 375f places heptagons and octagons onto the vertices of the hexagonal grid, allowing for patterns with seven- and eight-pointed stars. Figure 375g places nonagons and decagons at the vertices of the hexagonal grid, allowing for patterns with nine- and ten-pointed stars. And Fig. 375h places decagons and hendecagons at the vertices of the hexagonal grid, with nonagons at the centers of the brown hexagons. This will produce patterns with 9-, 10-, and 11-pointed stars, again with a demonstration of the *principle of adjacent numbers* with the decagon being the

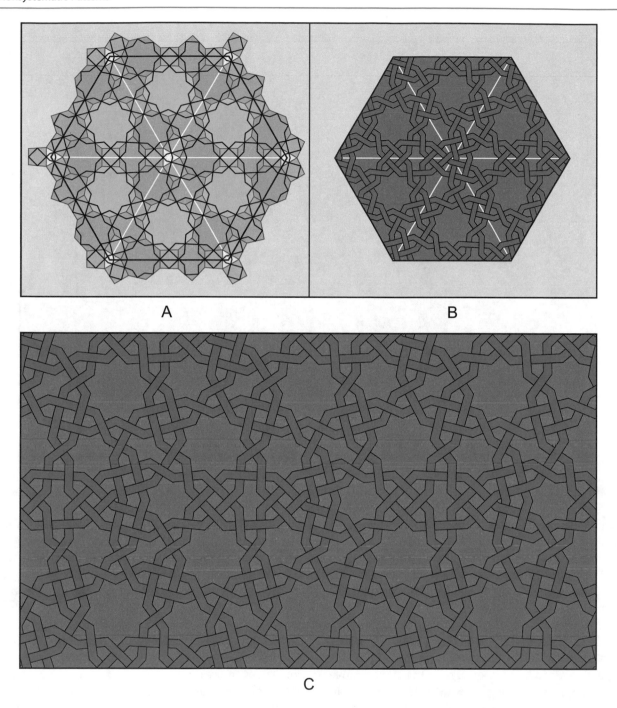

Fig. 372

central polygonal in the numeric chain. As with the examples illustrated in Fig. 371, this eccentric method of providing for atypical and unexpected regions of local symmetry within an otherwise isometric structure offers tremendous potential for creating new patterns to contemporary geometric artists. As mentioned in reference to Fig. 375a above, these polygonal arrangements can be used as layouts for radii matrices that can then be used to create very successful underlying tessellations with primary *n*-fold

polygons separated by a connective matrix of irregular pentagons, barrel hexagons, and other case-sensitive polygons. By way of example, Fig. 376 further develops the arrangement of hendecagons, decagons, and nonagons from Fig. 375h into an underlying tessellation suitable for creating a variety of patterns. Figure 376a illustrates the isometric grid along with its dual hexagonal grid. Each included angle within the hexagonal grid can be represented as a 1/3 division of the circle. In Fig. 376b the included

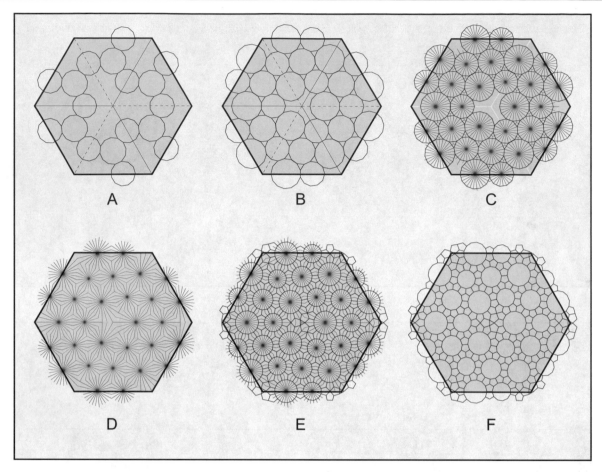

Fig. 373

angles of the two varieties of hexagon now correspond to alternating 10- and 11-fold divisions of the circle. It is important to note that the 10- and 11-fold radii do not actually align, and that the edges of both varieties of hexagon are made up of two noncollinear lines that intersect at the small black dots. What appear to be hexagons are, therefore, actually dodecagons. The nonaligning of the edges is a standard feature of this variety of geometric structure, and all of the examples from Fig. 375 (with the exception of C) have this anomaly. From the perspective of geometric precision this might be regarded as a fault, but for the purposes of creating geometric designs this anomaly is of no consequence. Figure 376c places 10 and 11 radii at the vertices of this (pseudo) hexagonal grid, and 9 radii at the center of each beige (pseudo) hexagon. Again, the black dots indicate the intersections of the noncollinear radii. From this radii matrix the underlying tessellation in Fig. 376d is easily created using the standard conventions for creating such tessellations. An important feature of the matrix of pentagons and barrel hexagons that separate the primary polygons is that their being located at the points of intersection of the not-quite-collinear radii overcomes any problems associated with the nonaligned radii, thereby allowing for

the primary polygons (in this case hendecagons, decagons, and nonagons) to be regular. Figure 377a illustrates the *obtuse* pattern (by author) created from the underlying tessellation created in the previous figure. Figure 377b demonstrates how this same pattern can be created from the nesting of primary polygons as per Fig. 375h. As explained, the radii associated with these primary polygons are not quite aligned, and therefore the polygonal edges are not able to be in precise edge-to-edge contact. This produces difficulties in laying out the pattern lines. By contrast, the matrix of pentagons and barrel hexagons in Fig. 377a allows for the regularity of the primary polygons to not be in conflict with one another; making the application of the pattern lines less problematic. Figure 377c demonstrates the dual relationship between these two underlying tessellations. This is analogous to many other examples, especially in the *fivefold system* [Fig. 200]. Figure 377d shows a very acceptable interweaving version of the *obtuse* design that can be created from either of these dual underlying tessellations. Figure 378 demonstrates the derivation of the *acute* pattern (by author) that is produced from the underlying tessellation in Fig. 376d. This design combines 9-, 10-, and 11-pointed stars in a balanced arrangement that

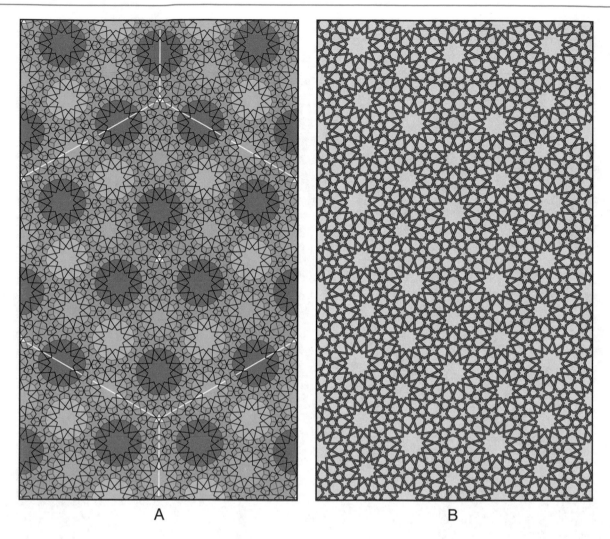

Fig. 374

is fully in keeping with the aesthetics of this ornamental tradition. Just as decagons tessellate on their own (with concave hexagonal interstice regions as per Fig. 200a), so also will decagons successfully tessellate with their numerically adjacent nonagons and hendecagons. This is an expression of the *principle of adjacent numbers*. The *median* and *two-point* patterns that this underlying tessellation creates are also very acceptable (not shown).

3.2.4 Orthogonal Designs with Multiple Regions of Local Symmetry

The most common variety of orthogonal pattern with multiple regions of local symmetry employs 12-pointed stars at the vertices of the orthogonal grid and 8-pointed stars at the center of each square repeat unit. Several underlying tessellations were used that produce such patterns, but the most common surrounds each underlying octagon with a

ring of pentagons, and separates each dodecagon with a barrel hexagon. The simplicity of constructing this underlying tessellation from a radii matrix is demonstrated in Fig. 304a. Evidence for the use of this radii matrix in producing the *median* pattern created from this underlying tessellation is found in the Topkapi Scroll.[62] Patterns that can be created from this underlying tessellation can also be made from a tessellation of edge-to-edge dodecagons and octagons, with concave hexagonal interstice regions. As with other examples of a single design that can be created from either of two generative tessellations, these underlying tessellations have a dual-like relationship. Figures 379–382 illustrate the derivation of patterns from each of the four pattern families from both of these two closely related

---

[62] Necipoğlu (1995), diagram no. 35 is a hybrid design with both square and triangular repetitive cells. The radii matrix and underlying tessellation in the square element alone are the same as those of Fig. 304a.

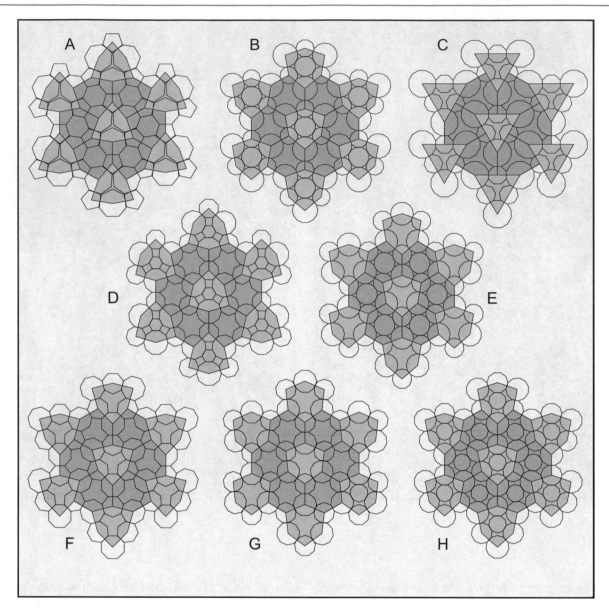

Fig. 375

underlying tessellations. The three *acute* patterns with 8- and 12-pointed stars in Fig. 379 demonstrate how subtle differences in the angles of the applied pattern lines can change the overall character of what is otherwise the same design. The Mengujekid example in Fig. 379b is from the Kale mosque in Divrigi, Turkey (1180-81), and has wider interweaving lines and slightly more acute angles in the five-pointed stars. This example also uses a truncated pattern line treatment within the underlying barrel hexagons. All of these features produce a visual quality that is distinct from the version of this *acute* pattern in Fig. 379e. Many examples of this *acute* design are found throughout the Islamic world, although the specific proportions will vary from example to example. In addition to line thickness, these variations are caused by subtle differences in the angles of the crossing

acute pattern lines. Notable examples of this *acute* design include a Seljuk border pattern in the entry of the Friday Mosque at Gonabad (1212) [Photograph 23], and two late Abbasid exemplars: one from the Palace of the Qal'a in Baghdad (c. 1220), and the other from the Mustansariyya in Baghdad (1227-34). A Mamluk example of this design was used as a side panel of the *minbar* of the Amir Azbak al-Yusufi complex in Cairo (1494-95) [Photograph 46]. The slightly less acute angles of the five-pointed stars combined with narrower widened lines creates a more open design with background elements that are proportionally larger than those of Fig. 379b. A tiling version of this design was used within a recessed arch tympanum in the southern corner of the southeast *iwan* in the Friday Mosque at Isfahan [Photograph 29]. This is stylistically similar to several of the Seljuk

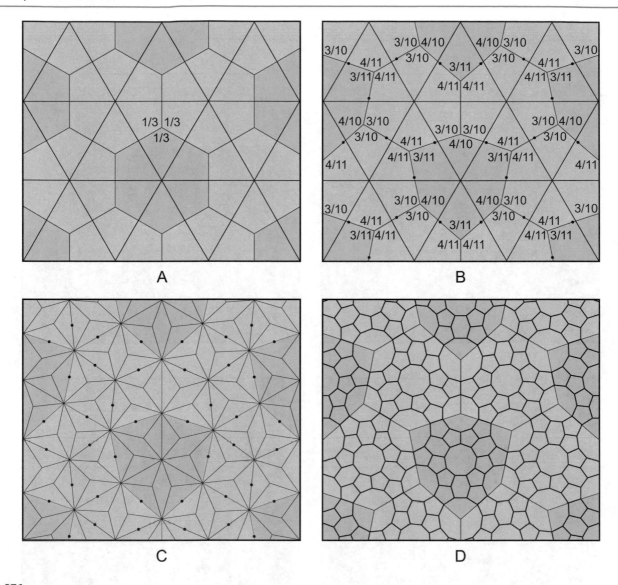

Fig. 376

designs in the nearby northeast dome chamber (1088-89), and its provenance appears to be Seljuk from the late eleventh or early twelfth century. The pattern lines of the 12-pointed stars of Fig. 379h are comprised of 12 parallel sets. This pulls the crossing pattern lines inward from the midpoints of the underlying dodecagon in Fig. 379i. This was a common convention among artists in the Maghreb and North Africa, and the Ottoman example illustrated here comes from the Great Mosque of Sfax in Tunisia (eighteenth century). An earlier example from the Seljuk Sultanate of Rum was used in the minaret of the Great Mosque of Siirt in Turkey (1129). While both underlying tessellations in Fig. 379 will make all three variations of the *acute* design, the tessellation with the ring of pentagons and barrel hexagons provide contact points for all of the key pattern line placements. By contrast, the 8- and 12-pointed star rosettes within the alternative tessellation of edge-to-edge octagons

and dodecagons require an additive process that is not directly the product of the underlying generative tessellation. Figure 380 demonstrates the construction of two *median* patterns from these two underlying tessellations. Figure 380b shows the standard *median* derivation without design modification. An Ayyubid carved stone example of this design is found in the city walls of the Bab Antakeya in Aleppo (1245-47); an Ilkhanid example is found in the portal of the Gunbad-i Gaffariyya in Maragha, Iran (1328); and Timurid examples are found in the cut-tile mosaics at both the Bibi Khanum in Samarkand (1398-1404), and the Friday Mosque at Herat (fifteenth century). The design in Fig. 380e was employed in the triangular side panels of the *minbar* at the Sultan Mu'ayyad mosque in Cairo (1415-21). This example uses the arbitrary modification that was typically applied to *median* patterns [Fig. 223] and was especially popular among Mamluk artists. Another feature of this

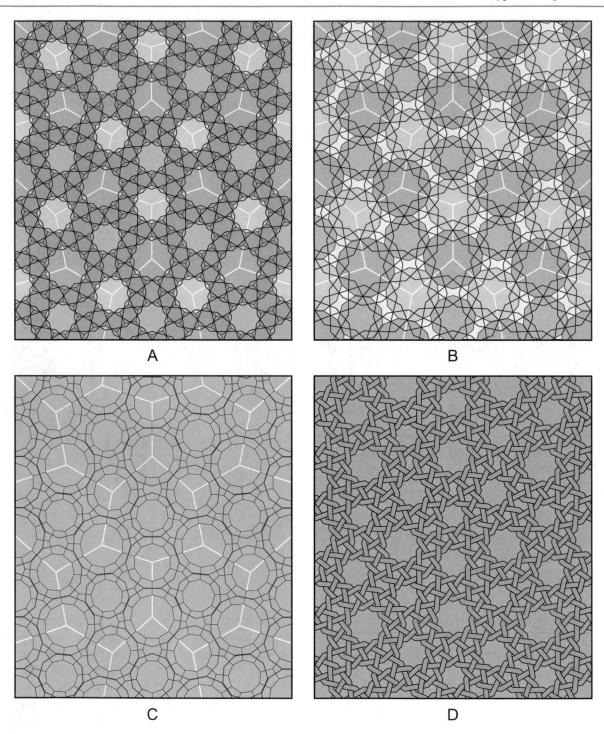

Fig. 377

design is the arbitrary incorporation of heptagonal elements within the pattern matrix. Although occasionally found in the geometric patterns of other Muslim cultures, for example those produced during the Seljuk Sultanate of Rum, arbitrary inclusions such as these heptagons are a relatively common feature within the Mamluk geometric tradition. As with the *acute* patterns created from the underlying tessellation of

edge-to-edge dodecagons and octagons, not all of the crossing pattern lines of the *median* 12-pointed stars are fixed upon the midpoints of the dodecagonal and octagonal edges in Fig. 380a and d. For this reason, the alternative underlying tessellation that includes pentagons and barrel hexagons is generally more convenient for producing these *median* designs. Figure 381 illustrates two *obtuse* patterns that can

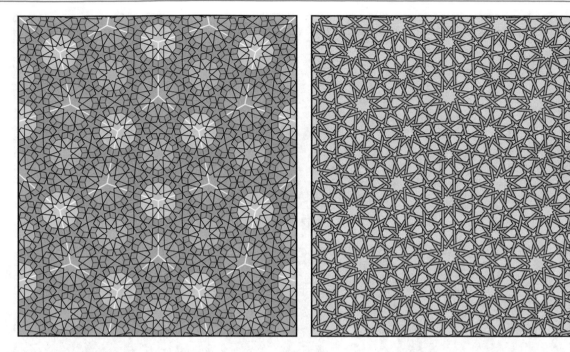

Fig. 378

be created from the same two underlying tessellations. Unlike the previous examples, all of the pattern lines of the 8- and 12-pointed stars in the *obtuse* example in Fig. 381a are directly associated with the midpoints of the underlying dodecagonal edges. For this reason each of the two underlying tessellations is equally expedient for generating these *obtuse* designs. Mamluk examples of the design in Fig. 381b are found in the blind arches that surround the exterior drum of the dome at the Hasan Sadaqah mausoleum in Cairo (1315-21), and in a window grill on the drum of the dome at the contemporaneous Amir Sanqur al-Sa'di funerary complex in Cairo (1315). Ilkhanid examples from roughly the same period include a vaulted ceiling panel from the mausoleum of Uljaytu in Sultaniya, Iran (1307-13), and a cut-tile mosaic border at the Gunbad-i Gaffariyya in Maragha, Iran (1328). Later eastern examples include a Qarjar cut-tile mosaic arch over the reconstructed entry door at the Aramgah-i Ni'mat Allah Vali in Mahan, Iran (nineteenth century); a cut-tile mosaic panel from the Qarjar restorations of the Malik mosque in Kerman, Iran; and a stone mosaic panel created by Mughal artists at the tomb of Akbar in Sikandra, India (1612). The *obtuse* design in Fig. 381e is also from the mausoleum of Uljaytu. This example includes a modification to the decagonal region that disguises the 12-pointed star through an arbitrarily added sixfold motif that is similar in concept to the well-known fivefold modification [Fig. 224a]. Figure 382b shows a Mamluk *two-point* pattern also produced from either of these underlying tessellations, but the example with the underlying pentagons and barrel hexagons has more contact points with the

crossing pattern lines. As with the *median* pattern in Fig. 380e, this design incorporates the same variety of modification to the 12-pointed stars [Fig. 223]. This example is from the triangular side panels of the *minbar* at the Princess Asal Bay mosque in Fayyum, Egypt (1497-99).

As with the previous examples, the *two-point* pattern in Fig. 383 also places 12-pointed stars at the vertices of the orthogonal grid and 8-pointed stars at the center of each square repeat unit. However, the underlying tessellation for this design separates the dodecagons and octagons with a barrel hexagon rather than the two pentagons from the previous example. This underlying tessellation also contains the characteristic cluster of six pentagons surrounding a thin rhombus that is a frequent feature of fivefold underlying tessellations. As detailed previously, this cluster of six pentagons is well suited to producing *obtuse* and *two-point* patterns. However, for *acute* and *median* patterns, the pentagons require truncation for the creation of well-composed patterns. This *two-point* pattern is of Mamluk origin, and comes from the Sultan Qaytbay funerary complex in Cairo (1472-74). Figure 304b shows how the underlying tessellation that produces this *two-point* pattern is constructed from the same radii matrix as the previous set of patterns with 8- and 12-pointed stars.

The two *median* designs in Fig. 384 also place 12-pointed stars at the vertices of the orthogonal grid and 8-pointed stars at the center of the square repeat unit. The underlying tessellations responsible for these two designs are similar to that of Fig. 304a, except for the ring of eight rhombi replacing the ring of eight pentagons. The proportions of

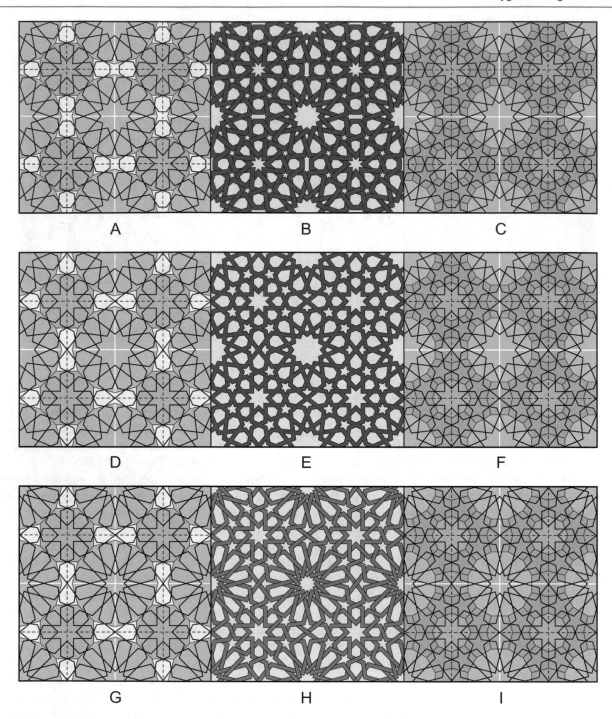

Fig. 379

the eight-pointed stars in Fig. 384a are determined by mirroring the lines at the edges of the underlying octagons, whereas the eight-pointed stars in Fig. 384b are the product of the pattern lines within the rhombi continuing toward the center of the repeat until they meet with other continued lines. The example in Fig. 384a is In'juid from the tympanum in the east portal of the Friday Mosque at Shiraz (1351), and the design in Fig. 384b is Muzaffarid from the Friday

Mosque at Yazd (1365). The relative closeness in time and proximity between these two very similar *median* designs suggests the possibility of their being produced by the same artist or atelier. The design in Fig. 384a was also used in the Timurid cut-tile ornament of the Gur-i Amir complex in Samarkand (1403-04). The arrangement of underlying polygons that surround the region with eightfold local symmetry in the two designs in this figure are analogous to a type

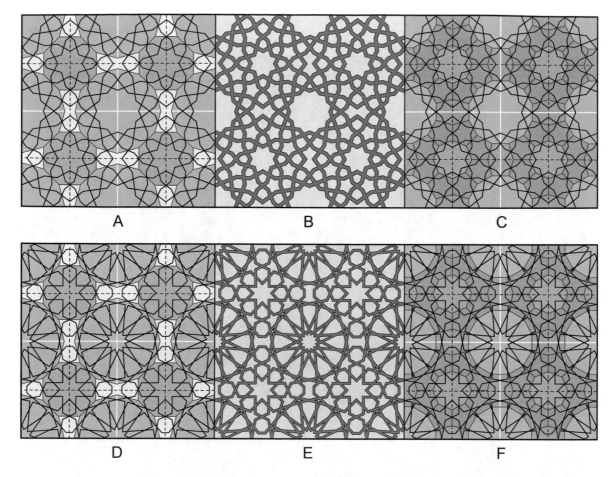

A   B   C

D   E   F

Fig. 380

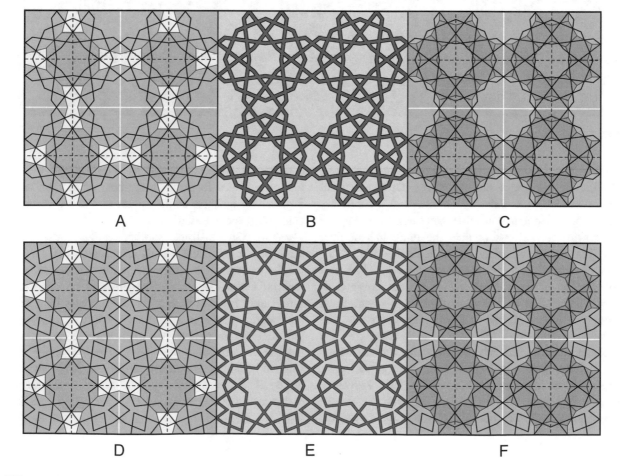

A   B   C

D   E   F

Fig. 381

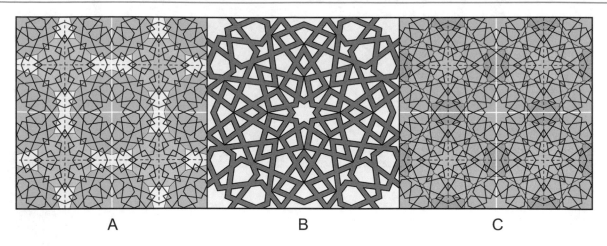

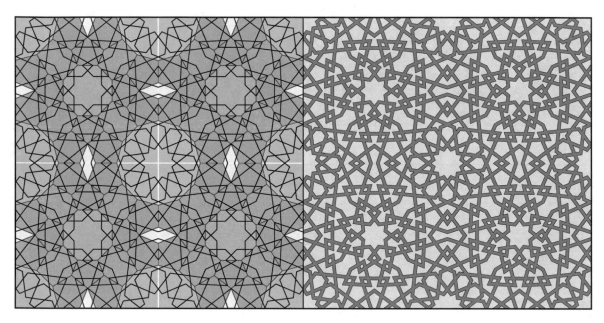

A   B   C

Fig. 382

Fig. 383

of pattern created from the *fivefold system* [Figs. 236, 237, and 253]. This employs an underlying arrangement of pentagons and/or wide rhombi that surround the primary centers of local symmetry such that their vertices are aligned with the primary radii rather than the midpoints of their edges. This leaves either an *n*-fold interstice star at the primary centers of local symmetry, as per Fig. 384b, or a ring of triangles surrounding the *n*-fold primary polygon, as per Fig. 384a. Most of the designs with this variety of polygonal arrangement, be it fivefold or otherwise, are from the eastern regions, and date to the period after the Mongol conquest. Figure 385 demonstrates a construction for the underlying tessellation that produces these designs. Step 1 starts with the radii matrix from Fig. 304a. However, rather than the central region being filled with the ring of pentagons surrounding an octagon, this step shows just the

dodecagons and barrel hexagons. Step 2 simply mirrors the indicated angles to produce the ring of rhombi; and Step 3 places an octagon at the center of the repeat unit, thus completing the tessellation.

Figure 386 illustrates another *median* pattern with the same placement of the 8- and 12-pointed stars. As with the previous example, this also employs the unusual arrangement of vertex-to-vertex pentagons that surround both the 8-fold and 12-fold centers of radial symmetry. This produces the underlying triangular regions that surround each primary polygon, and the pattern line application along the edges of the primary polygons employs two points rather than the singular midpoints that are typical of *median* patterns. The cluster of four contiguous rhombi produces a design feature that is also rather unusual, and occasionally found in some fivefold patterns. This nonsystematic Timurid design is from

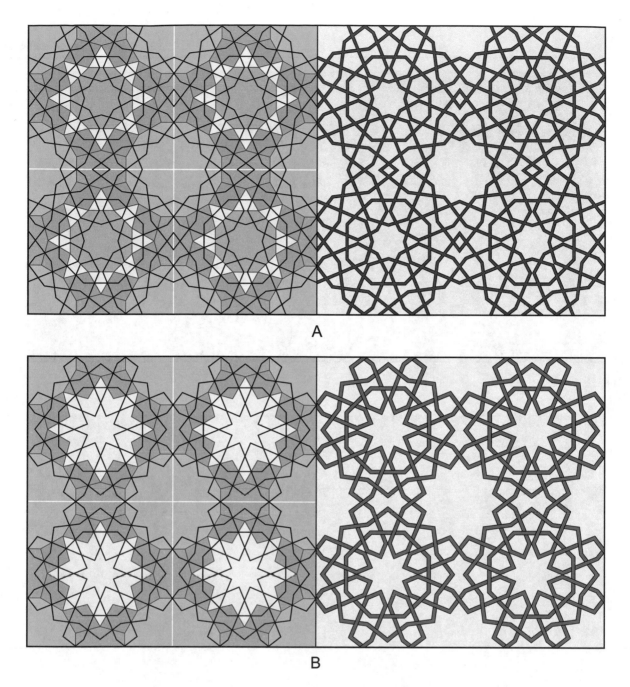

Fig. 384

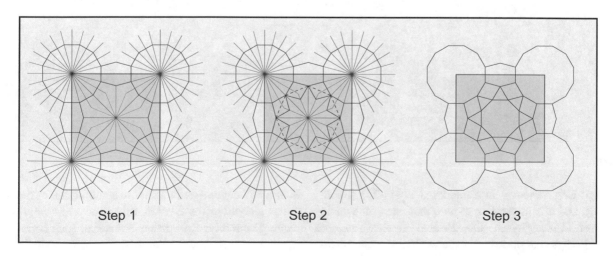

Step 1 Step 2 Step 3

Fig. 385

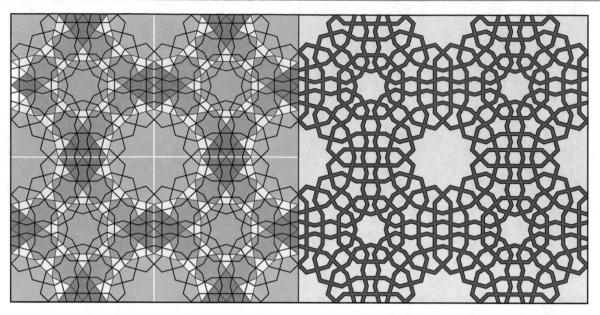

Fig. 386

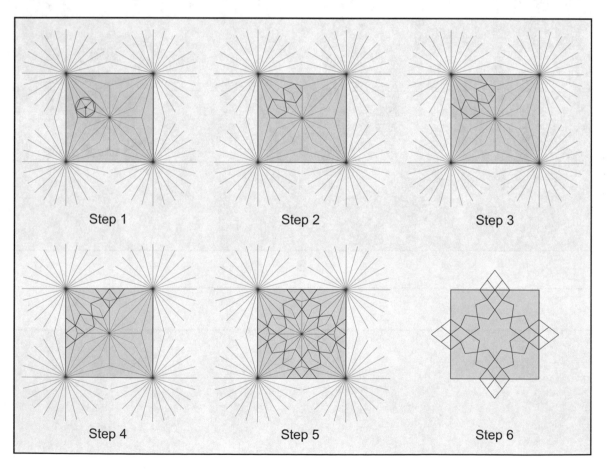

Fig. 387

the Ulugh Beg *madrasa* in Samarkand (1417-20) [Photograph 88], and it is interesting to note that one of the analogous fivefold designs is also located at this *madrasa* [Fig. 236]. These are presumably the work of the same artist.

Figure 387 demonstrates a simple construction of the underlying generative tessellation using the radii matrix technique. The radii matrix places 24 radii at each corner of the square repeat unit, and 16 radii at the center. Step 1 draws a

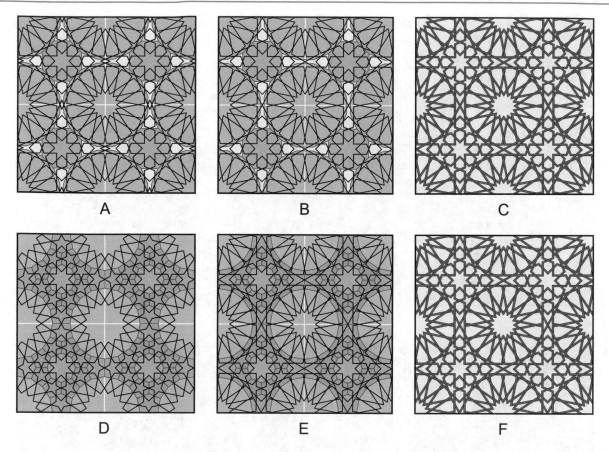

A B C

D E F

Fig. 388

circle at the indicated intersection of the red radii that is tangent to all three of the blue radii. A pentagon is drawn inside the circle with vertices at the intersect points of the red radii and circle, as well as the points at which the circle is tangent with the blue radii. Step 2 mirrors this pentagon. Step 3 mirrors a single edge of each pentagon; establishing one quarter of a 12-pointed star. Step 4 mirrors the lines from Step 3 to create the rhombi. Step 5 rotates these elements throughout the square repeat. And Step 6 shows the completed tessellation.

Figure 388 shows two alternative constructions for an *acute* design that places 16-pointed stars on the vertices of the orthogonal grid and 8-pointed stars at the center of each square repeat unit. The underlying tessellation in Fig. 388a is comprised of edge-to-edge 16-gons and octagons, with elongated concave octagonal interstice regions separating the adjacent 16-gons. The pattern lines associated with the two adjacent parallel edges of the 16-gons overlap to create a small rhombus at the center of the interstice region. This is an atypical feature that is not generally in keeping with this ornamental tradition. The historical example of this design changes the angles of the pattern lines associated with these two edges of the underlying 16-gon such that they create the more familiar dart shape shown in Fig. 388b. This results in

the new dart shapes being larger and differently proportioned than the adjacent darts that surround the 16-pointed stars. Figures 388d and e illustrate an alternative method of creating this design that uses an underlying tessellation with the commonly found feature of a ring of pentagons. The example in Fig. 388d shows a reasonably successful *acute* pattern, although the proportion of the five-pointed stars created from the large pentagons (yellow) is not ideal. This less desirable feature is resolved in Fig. 388e through the modification of the 16-pointed star rosette using the common Mamluk method [Fig. 223]. The end result of this modification is very successful, and the Mamluk design pictured in this figure is from the Sultan al-Mu'ayyad Shaykh complex in Cairo (1412-22) [Photograph 60]. Figure 389 demonstrates two additional designs produced from the same underlying tessellation with 16-gons and octagons as the previous example. Figure 389a shows an *acute* pattern with the shared points of the 5- and 16-pointed stars being moved off of the midpoints of the 16-gon inward toward the vertices of the square repeat unit. This elongates the five-pointed stars, changing the overall aesthetic effect in a manner that is most frequently found in the Maghreb. This example was used in a *Mudéjar* window grille in the ibn Shushen Synagogue of Toledo (1180), and now known as

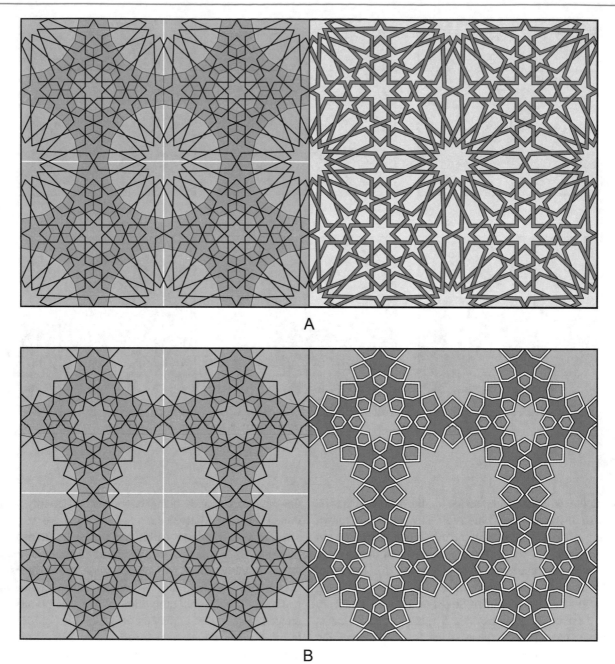

A

B

Fig. 389

the Santa Maria la Blanca. Another example of this design was used for an illuminated frontispiece of the Quran (1310) commissioned by the Ilkhanid Sultan Uljaytu.[63] Figure 389b shows a Timurid *median* pattern created from the same underlying tessellation that is found in the entry *iwan* of the Ulugh Beg *madrasa* in Samarkand (1417-20) [Photograph 89]. Figure 390 illustrates a construction of the

underlying tessellation from Fig. 389. The radii matrix places 32 radii at each corner of the square repeat unit, and 16 radii at the center. Step 1 shows the standard placement of a circle at the intersection of the chosen radii that is tangent with the red radii, and lines that are perpendicular to the blue radii that intersect with the circle and blue radii. This establishes the edge of the 16-gon, the edge of the octagons, and the separating pentagon. Step 2 shows all three of these features. Step 3 mirrors the pentagon. Step 4 rotates these features throughout the square repeat, identifying the second variety of pentagon in the process. These are located at the

[63] Calligraphed by 'Ali ibn Muhammad al-Husayni in Mosul (1310). British Library, Or. 4945, ff. IV-2r.

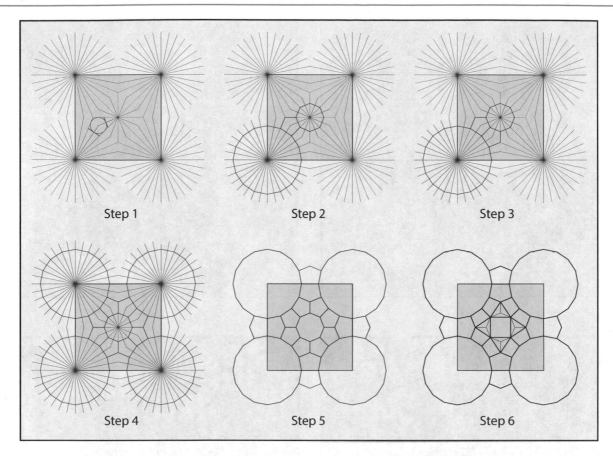

Fig. 390

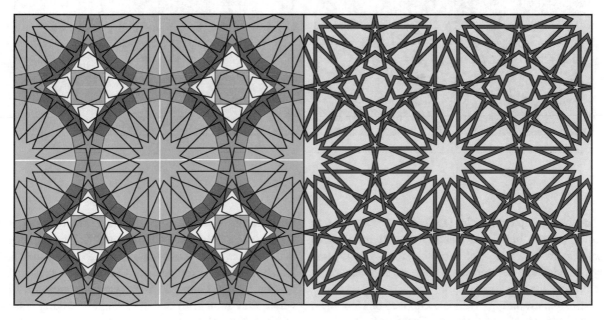

Fig. 391

midpoints of the repetitive edges. Step 5 shows the completed tessellation. Step 6 shows a modification to the ring of eight pentagons that produces the design in Fig. 391. This modification transforms the tessellation, and the derived patterns, from having 8- and 12-fold local symmetries to 4- and 12-fold symmetries, thereby eliminating the eight-pointed stars from the finished pattern. This was used for a Mamluk *acute* design from the Sultan

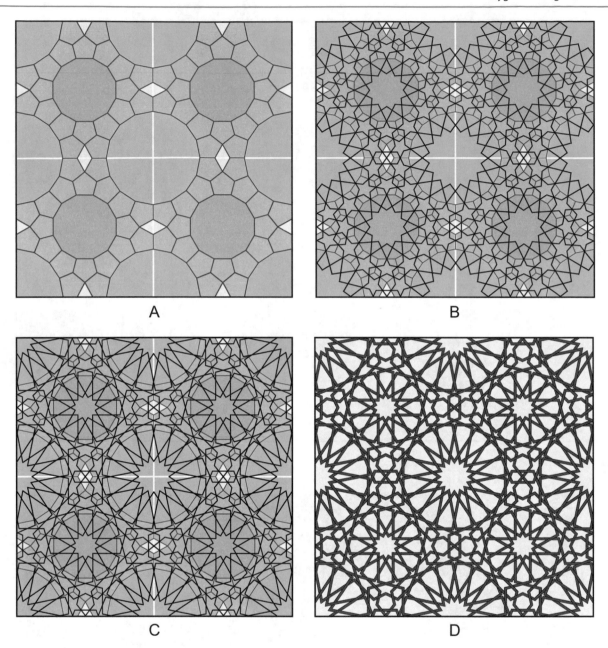

Fig. 392

Hasan funerary complex in Cairo (1356-63). The application of crossing pattern lines to the central underlying squares foregoes the more acute angles of the *acute* family, and thereby provides for the inclusion of the regular octagon in this location.

The *acute* pattern in Fig. 392d is comprised of 16-pointed stars at the vertices of the orthogonal grid and 12-pointed stars at the centers of each repeat unit. This design is a modification of the standard *acute* pattern shown in Fig. 392b that follows the common convention, especially among Mamluk artists, for changing *acute* and *median* patterns [Fig. 223]. Applying this modification to both

varieties of primary star radically changes the overall appearance of the design. This modified pattern was used in the bronze entry doors that first graced the Sultan al-Nasir Hasan funerary complex in Cairo (1356-63), but were moved to the Sultan al-Mu'ayyad complex in 1416-17, where they remain to this day. The underlying tessellation in Fig. 393 employs a modification to the tessellation in Fig. 392a wherein the six pentagons that surround each thin rhombus are truncated into trapezoids. This follows the convention for modifying star rosettes of the *acute* family that was established within the *fivefold system* [Fig. 198]. This modification changes what would otherwise

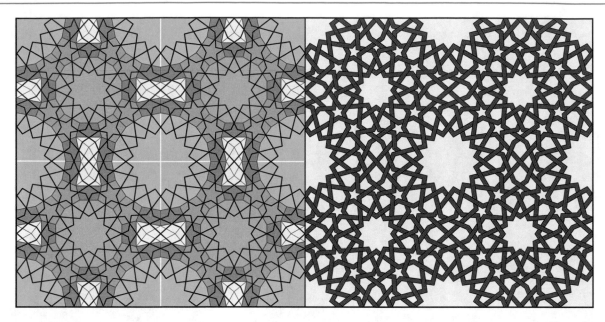

Fig. 393

be 6 five-pointed stars into 6 dart motifs. The earliest known example of this *acute* pattern was produced by Seljuk artists for a panel within the *muqarnas* hood of the *mihrab* in the Friday Mosque at Barsian (1105). This is a surprisingly early date for such a sophisticated design. Rather than using widened or interweaving lines (as per the illustration), the tiled expression of this example from Barsian follows the Seljuk geometric aesthetic exemplified by the roughly contemporaneous geometric designs in the nearby northeast dome chamber of the Friday Mosque at Isfahan (1088-89). The interweaving expression pictured in Fig. 393 was used by Mamluk artists in several locations, including: a window grill in the mosque of Altinbugha al-Maridani in Cairo (1337-39) [Photograph 56]; a curvilinear variation from the bronze entry doors of the Sultan al-Nasir Hasan funerary complex in Cairo (1356-63); and the triangular side panel of the wooden *minbar* (1468-96) commissioned by Sultan Qaytbay, and currently on display at the Victoria and Albert Museum in London. Figure 394 demonstrates a construction of the underlying tessellation with 16-gons and dodecagons from Figs. 392 and 393. This makes use of a radii matrix with 32 radii placed at the corners of the square repeat unit, and 24 radii at the centers. Step 1 places a circle at the intersection of two blue radii that is tangent to the red radii. Lines have been drawn that are perpendicular to the blue radii and intersect with the circle and blue radii. Step 2 uses these lines to produce the 16-gon and dodecagon, as well as the separating pentagon. Step 3 mirrors the pentagon as well as draws two additional pentagons that connect the end points of projected red and blue radii. Step 4 creates a third set of pentagons along the repetitive edges by mirroring

lines from Step 3. Step 5 rotates these elements throughout the square repeat. And Step 6 shows the completed tessellation, along with the potential truncated cluster of six pentagons for the design from Fig. 393. The *two-point* pattern in Fig. 395 also employs 16-pointed stars at the corners of the square repeat unit and 12-pointed stars at the center of each repeat. This design was used in several Mamluk locations, including the triangular side panel of the stone *minbar* at the Sultan al-Zahir Barquq in Cairo (1384-86) [Photograph 57]; the triangular side panel of the wooden *minbar* at the Amir Qijmas al-Ishaqi mosque in Cairo (1479-81); and the triangular side panel of the wooden *minbar* at the Sultan Qansuh al-Ghuri complex in Cairo (1503-05). The underlying tessellation separates the 16-gons and dodecagons with a barrel hexagon, and separates the adjacent 16-gons with an arrangement of ten pentagons surrounding two mirrored irregular heptagons. The precise placement of the pattern lines of the *two-point* family is inherently flexible, and in this case those associated with the heptagons have been carefully placed so that they produce twin seven-pointed stars that have true sevenfold rotational symmetry. Figure 396 illustrates two *acute* patterns that are also created from the underlying tessellation in Fig. 395. The standard *acute* pattern (by author) that is created from this underlying tessellation is represented in Fig. 396a, but there are no known examples of this design in the historical record. The design in Fig. 396b is a Mamluk example that modifies the 12- and 16-pointed stars using the technique demonstrated in Fig. 223 wherein the surrounding five-pointed stars are transformed into darts. The pattern lines associated with the barrel hexagons in

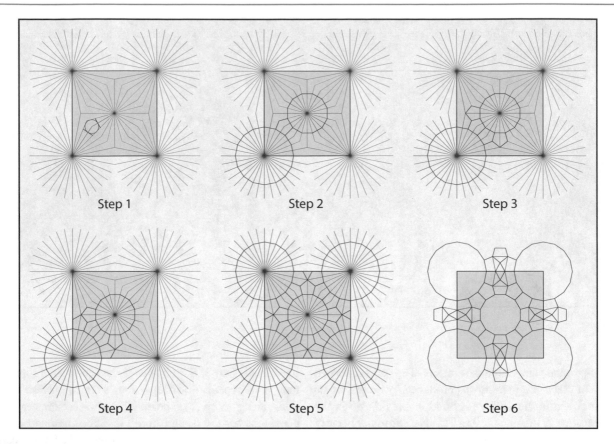

Fig. 394

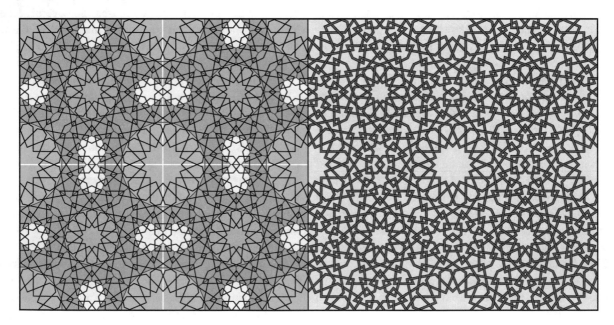

Fig. 395

Fig. 396b have been arbitrarily adjusted to produce two nearly regular heptagons within the pattern matrix. This form of pattern adjustment was frequently employed by Mamluk artists, and was occasionally used elsewhere. This Mamluk *acute* design is from a carved stone panel at the base of the minaret at the Mughulbay Taz mosque in Cairo (1466). Figure 397 demonstrates how the construction of the underlying tessellation used in creating the patterns in

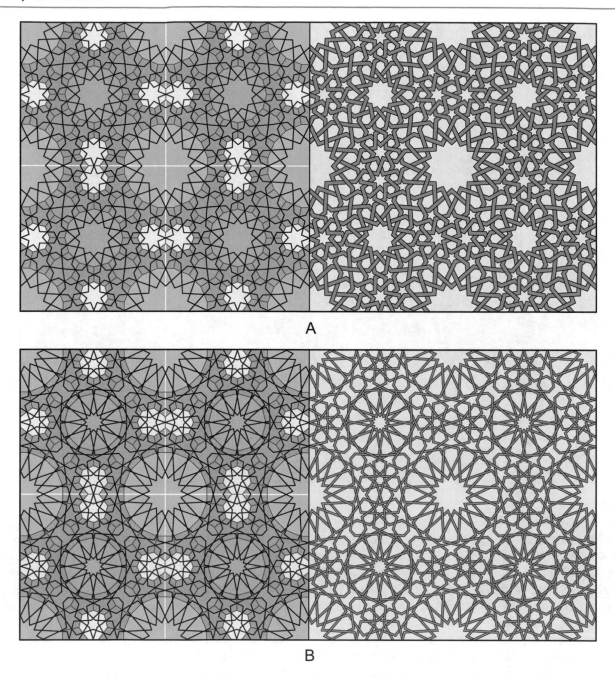

A

B

Fig. 396

Figs. 395 and 396 employs the same radii matrix as that of Fig. 394. This is made up of 32 radii placed at the corners, and 24 radii at the centers of the square repeat unit. However, the underlying tessellation shown in Fig. 397 employs an entirely different derivation from this radii matrix. In creating the underlying tessellation, Step 1 draws an almost regular heptagon at the midpoint of the edge of the square repeat. The size is determined by the vertex of a regular heptagon intersecting the blue radii of 32-fold symmetry, as per the detail. The detail also shows how two of the heptagonal vertices do not quite fall upon the vertices of the pair of intersecting red radii. These nonaligned heptagonal vertices are therefore moved so that they rest on the intersection of the red radii, making the heptagon slightly irregular. Step 2 draws two circles; one centered on a red radius of 32-fold symmetry and they other on a red radius of 24-fold symmetry. These circles are tangent to the blue radii. Lines that are perpendicular to these red radii are drawn that determine the size of the 16-gon and dodecagon. Step 3 shows these primary polygons as well as the two pentagons

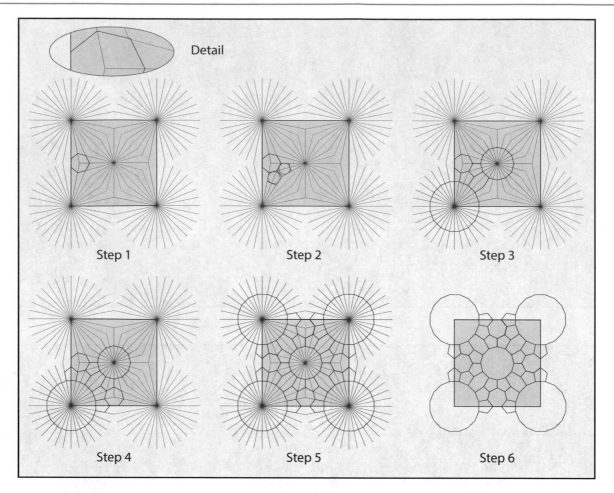

Fig. 397

and a barrel hexagon between the 16-gon and dodecagon. Step 4 mirrors these elements and introduces pentagons along the edges of the repeat that are determined by mirroring the edge of the heptagon. Step 5 rotates these elements throughout the square repeat. And Step 6 shows the completed tessellation.

It is rare for orthogonal patterns with just two primary star forms to not locate these on the vertices and center of the square repeat unit. Figure 398 shows an unusual *acute* design that places 16-pointed stars at the vertices of the orthogonal grid and four 13-pointed stars within the field of each square repeat unit. Patterns that place primary star forms within the field of the repeat ordinarily utilize the more typical locations first; for example, the vertices of the orthogonal grid and orthogonal dual grid, and the midpoints of the edges of each repeat unit. It is interesting to note that the centers of the 13-pointed stars appear to fall upon the vertices of the semi-regular 4.8^2 tessellation of squares and octagons [Fig. 89]. However, close examination reveals that there are two different distances between the locations of the 13-pointed stars, thereby disqualifying this from adhering to the 4.8^2 semi-regular symmetry. This interesting design is

from the Topkapi Scroll, but is not otherwise known to the historical record.[64] A construction of the underlying tessellation for this eccentric *acute* pattern is demonstrated in Fig. 399. Step 1 places 32 radii at each vertex of the square repeat unit. The four pairs of emphasized radii (black) have internal angles of 135°. This angle is close to the 138.4615...° found within a 13-fold division of a circle, indicating that a 13-pointed star can be placed at these locations. Step 2 introduces 26 radii at these four locations. The small black dots indicate the intersection points for the 32- and 26-fold radii that do not quite align. Therefore, the red radii connecting these centers are not quite collinear, although they appear so. Step 3 determines the edges of the tridecagons, as well as the separating pentagon using the standard formula. Step 4 draws the two completed tridecagons as well as the two separating pentagons. Step 5 uses intersect points within the radii matrix to produce further pentagons. This step also determines the edge of the 16-gon using a circle in tangent with the blue radii. Step 6 mirrors these elements. Step 7

---

[64] Necipoğlu (1995), diagram no. 30.

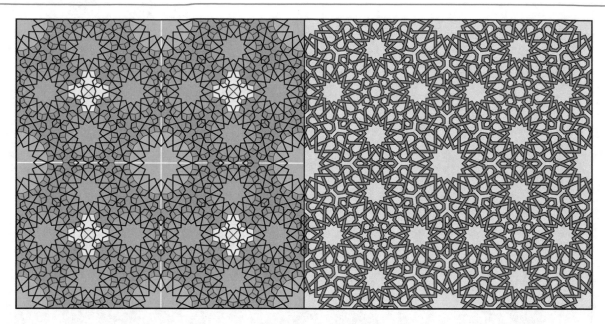

Fig. 398

completes one quadrant by making the final variety of pentagon along the edges of the repeat. Step 8 rotates these elements throughout the repeat. And Step 9 completes this rather remarkable tessellation. It is worth noting that the un-inked scribed "dead lines" of this drawing in the Topkapi Scroll show both the radii matrix and the underlying tessellation as the generative schema employed in the design of this pattern.

There are relatively few historical orthogonal patterns that employ three primary star forms within their overall make up. The *acute* pattern in Fig. 400 is one of the most complex orthogonal designs found within Islamic geometric art. This is comprised of 12-pointed stars placed at the vertices of the orthogonal grid, octagons at the centers of each repeat unit, 10-pointed stars at the midpoints of the repetitive edges, and 9-pointed stars within the field of the repeat. The nine-pointed stars rest upon the diagonals of the square repeat. As with so many of the particularly complex Islamic geometric designs, this is the product of artists working during the Seljuk Sultanate of Rum, and is found at several Anatolian locations, including the Kayseri hospital (1205-06); the Agzikara Han near Akseri (1236-46); and the Çifte Kumbet in Kaysari (1247). Figure 401 shows a construction for the underlying tessellation that creates this pattern. Once again, this begins with the creation of a radii matrix. Step 1 places 24 radii next to 20 radii such that they share a horizontal radius (red). The 45° diagonal radius of the 24 radii is extended to meet the extended vertical radius of the 20 radii to establish the fundamental domain (blue) of the eventual design. The 81° angle of this construction is

close to the 80° angle of a ninefold division of the circle, indicating the potential location of a nine-pointed star. Step 2 places 18 radii at this location. This is aligned with the 45° diagonal radius from the 24 radii. Note: Detail 1 shows how the relevant radius from these 18 radii does not connect with the center of the 20 radii, and is not quite parallel with the closest radius from the 20 radii. Step 3 begins the process of correcting this situation by copying the 45° diagonal radius from the 24 radii to a location midway between the 18 and 20 radii. As shown in the Detail 2, Step 4 trims the two nonaligned radii with this copied diagonal line. Step 5 moves the group of 18 radii along the copied diagonal line so that the two trimmed radii meet. This intersection is indicated with the black dot. While these two radii are not actually collinear, for the purposes of this design process they function as thought they are. Step 6 introduced 8 radii at the 45° upper corner of the fundamental domain. Step 7 determines the edges of the three primary polygons, as well as the two separating pentagons. Step 8 draws the dodecagon, decagons, and nonagon, as well as the connecting pentagons and barrel hexagon. Step 9 mirrors these elements and the fundamental domain. Step 10 rotates the elements from Step 9 four times, and Step 11 shows the completed tessellation.

The *acute* design in Fig. 402 is from one of the side panels of the *minbar* at the Sultan Qaytbay funerary complex in Cairo (1472-74). This Mamluk example places 16-pointed stars at the vertices of the orthogonal grid, 12-pointed stars at the center of each square repeat unit, and 10-pointed stars at the midpoints of each edge of the repeat. In keeping with Mamluk aesthetic practices, the pattern lines within the

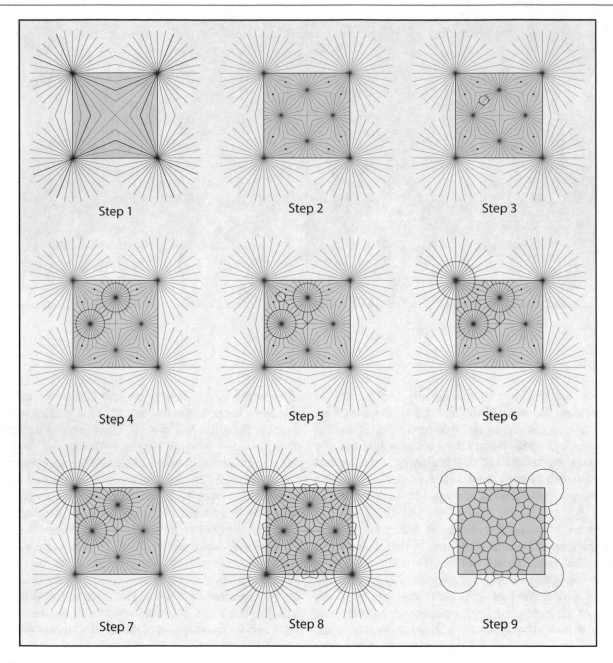

Fig. 399

concave hexagons have been adjusted to produce twin heptagons. With its three regions of local symmetry, this is one of the more complex orthogonal geometric patterns created by Mamluk artists. Figure 403 shows a construction for the underlying tessellation used for creating the pattern in Fig. 402. Step 1 illustrates a radii matrix with 32 radii at the corners of the square repeat, 24 radii at the center of the repeat, and 20 radii at the midpoint of the edges of the repeat. Step 2 determines the edges of the primary polygons, as well as two of the pentagons using the standard formula of a circle in tangent with the red radii and applied perpendicular lines. Step 3 draws the 16-gon, dodecagons and decagon, as well as the pentagons and barrel hexagon. These pentagons and barrel hexagon are determined from the intersections of the primary polygonal edges and once the primary polygons have been established, are implicit within the radii matrix. Step 4 mirrors these elements, and adds two pentagons to complete one quadrant. The large circle provides for regularity in the edge lengths of the new pentagons. Step 5 rotates these elements throughout the square repeat. And Step 6 completes the tessellation. Note: the truncation of the clustered six pentagons, and resulting concave hexagons is also indicated. This variety of modification [Fig. 198] is required for the *acute* pattern in Fig. 402.

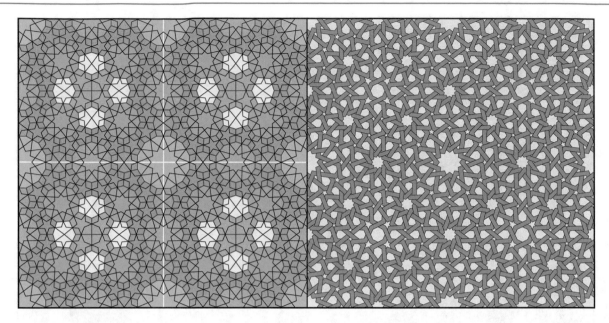

Fig. 400

The design in Fig. 404b places 16-pointed stars at the vertices of the orthogonal grid, 8-pointed stars at the center of each square repeat unit, 12-pointed stars at the midpoints of each repeat unit, and 10-pointed stars on the repetitive diagonals within the field. This *acute* pattern is from the Kemaliya *madrasa* in Konya (1249), and shares its regions of local symmetry with two other examples from the Seljuk Sultanate of Rum; although these others are created from the *extended parallel radii* design methodology rather than the polygonal technique [Fig. 81]. Each of the three primary star forms in this example is comprised of sets of parallel lines, all of equal width. This is a distinctive feature more commonly seen in the geometric art of the Maghreb. As in other examples that have this feature, the crossing pattern lines of the primary stars will not necessarily fall upon the midpoints of the primary polygonal edges, but will be moved inward toward the polygonal centers. This allows for the uniformity in the width of the parallel pattern lines that make up the primary stars. As shown in the upper left panel in Fig. 404a, the width of the parallel lines are established within the decagonal region, and copied into the 16-gons and dodecagons. The eight-pointed stars are atypical in that they are not produced directly from an underlying octagon, but through extending the lines within the ten-pointed stars toward the center of the square repeat unit, and rotating these four sets of diagonal parallel pattern lines by 45°. Figure 405 illustrates a construction for the underlying tessellation used for making the design from Fig. 404. Step 1 shows one quadrant of a radii matrix with 32 radii in the corner of the repeat unit (upper left), and 24 radii at the midpoints of the repeat (upper right and lower left). The two black radii have an included angle of 150° which is relatively close to the

144° found in a tenfold division of a circle. Step 2 introduces 20 radii at this point. The radii between the 20- and 24-fold centers are not aligned. To correct this, the black radius between the 32- and 20-fold centers is copied to points that are midway between the two 20- and 24-fold centers. Step 3 trims the radii with these copied lines. Step 4 moves the 20 radii along the black diagonal lines until the trimmed radii intersect. The black dots indicate the intersections of these radii, and what functions as a line of radius between the 24- and 20-fold centers is actually two intersecting noncollinear radii. Step 5 determines the edges of the primary polygons as well as the separating pentagons. Step 6 draws the 16-gon, dodecagon and decagons. Step 7 mirrors the dodecagon and fills in the separating matrix of pentagons. Step 8 finishes the quadrant, and Step 9 shows the completed tessellation in a full repeat. Note: The four large pentagons at the center of the repeat are out of scale to the other pentagonal elements. This would ordinarily cause problems. However, by applying pattern lines to these regions as illustrated in Fig. 404, this problem is elegantly overcome through the introduction of the eight-pointed star.

Figure 406 demonstrates a highly versatile method of creating orthogonal patterns with multiple regions of local symmetry. This repetitive schema is an orthogonal corollary to the isometric method illustrated in Figs. 371 through 374. However, whereas the isometric methodology has no historical precedent, several patterns that conform to this orthogonal repetitive structure are known historically. The construction of tessellations with this technique requires a central polygon with sides that are multiples of four. Four *n*-sided polygons are placed edge-to-edge in rotation around the central polygon, and these are divided in half to

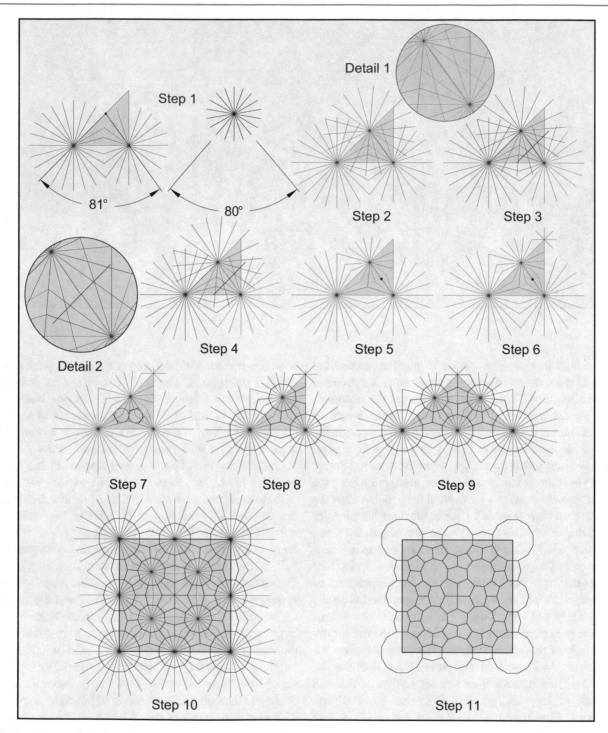

Fig. 401

determine the square repetitive unit. This unit is reflected on its vertical and horizontal edges to create the larger square repeat unit with translation symmetry. Unless an interweaving line introduces chirality to the pattern lines, this variety of pattern always conforms to the *cmm* plane symmetry group. The shaded squares in Fig. 406 indicate the *oscillating square* feature of this variety of design. In this regard, they conform to the historical examples illustrated in Figs. 23 through 25. However, the application of the polygons, and secondary infill of further polygons provides for greater complexity than most historical *oscillating square* patterns. Figure 406a uses a central octagon surrounded by four heptagons. Figure 406b adds nonagons to the central octagon, and Fig. 406c uses four hendecagons

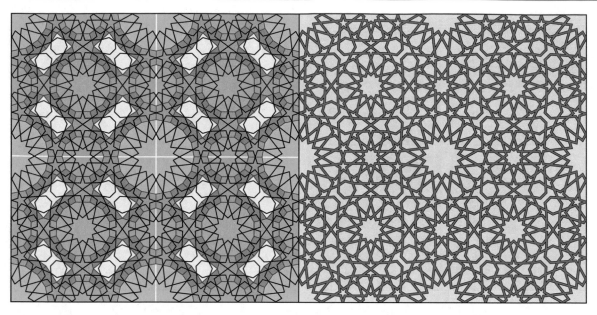

Fig. 402

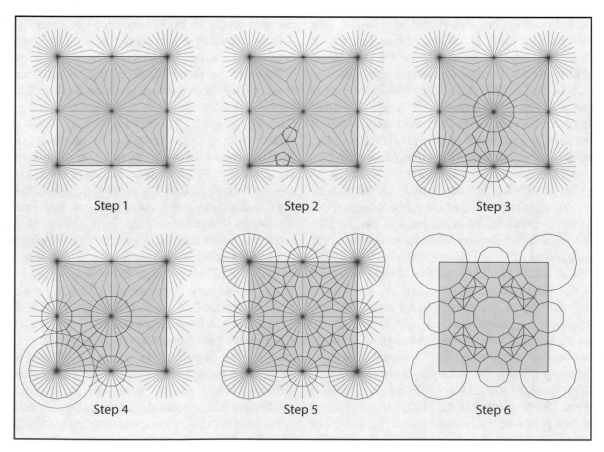

Fig. 403

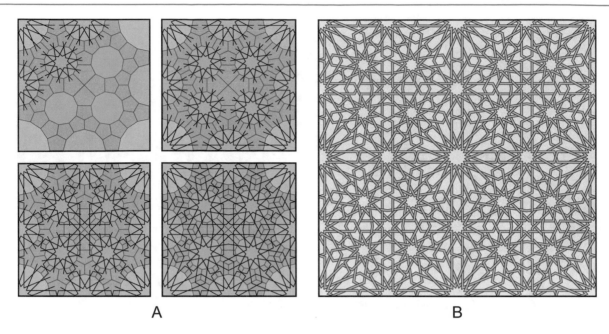

Fig. 404

(11-gons) surrounding the central octagon. Figure 406d places heptagons around the central dodecagon. Figure 406e places four octagons around the central dodecagon. Figure 406f shows four nonagons around the central dodecagon. Figure 406g adds four dodecagons around the central dodecagon. And Fig. 406h places four tridecagons (13-gons) around the central dodecagon. Each of these has been further developed with the subdivision of the interstice regions into secondary polygons, and each is well suited to producing very acceptable geometric designs. Figure 407 illustrates the design potential of using just the central polygon and the four surrounding polygons for creating patterns. In these two examples the interstice region is passive, and has not been filled with secondary polygons. With the exception of the secondary polygons, the underlying tessellation in Fig. 407a is the same as that of Fig. 406a, and that of Fig. 407c is the same as Fig. 406c. The *median* pattern in Fig. 407b is comprised of octagons and seven-pointed stars. This was used historically in many locations, and is one of the better-known patterns with seven-pointed stars, although clearly not created from the *sevenfold system*. The application of the octagon within the underlying octagon is unusual, and certainly an eight-pointed star could have been used in this location. The proportions of the octagon are determined by using two points on every other underlying octagonal edge. In this case, the two points are determined by dividing the edge into quarters. Historical examples of this design include: a Jalayirid arch spandrel at the Mirjaniyya *madrasa* in Baghdad (1357), and a small carved stone relief panel in the Mamluk *iwan* of the Amir Qijmas al-Ishaqi mosque in Cairo (1479-81). The pattern in Fig. 407d (by author) uses

the same design process of only using the primary polygons, with the interstice regions being passive, and favoring an octagon within the underlying octagons over an eight-pointed star. The main difference between these two designs is that the historical example combines 7-pointed stars with octagons and the other uses 11-pointed stars and octagons. The two designs in Fig. 408 (by author) demonstrate the efficacy of using secondary polygonal elements that fill the interstice regions between the two primary polygons. The *acute* pattern in Fig. 408b is comprised of 8- and 12-pointed stars and is created from the underlying tessellation of central dodecagons and surrounding octagons. The placement of the pattern lines allow for the incorporation of octagons within the pattern matrix. The underlying tessellation is the same as Fig. 406e. The *obtuse* pattern in Fig. 408d is made up of 9- and 12-pointed stars. The underlying tessellation in Fig. 408c, and the resulting *obtuse* pattern, are analogous to examples created from the *fivefold system* [Fig. 235b].

Just as in with the analogous isometric examples, Fig. 409 demonstrates how the orthogonal arrangement of a central polygon with edges that are a multiple of four, and four surrounding *n*-sided polygons can be used to make radii matrices from which very acceptable underlying tessellations can be created. A radii matrix has been added to the tessellation of dodecagons and nonagons in the upper left panel of Fig. 409a. The radii can be seen to converge on a point that allows for another radial center with 20-fold symmetry. This indicates that a ten-pointed star can be placed at these locations. The upper right panel shows just the radii matrix. The lower left panel of Fig. 409a illustrates a tessellation of primary dodecagons, decagons and

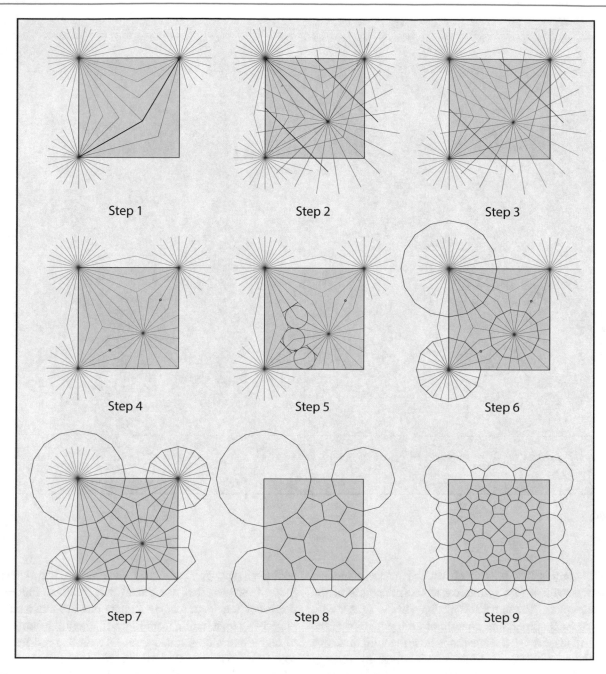

Step 1 Step 2 Step 3

Step 4 Step 5 Step 6

Step 7 Step 8 Step 9

Fig. 405

nonagons that are separated by a polygonal matrix of pentagons and barrel hexagons. And the lower right panel shows just the tessellation and repetitive structure. Note: the decagons placed at the two adjacent centers with 20-fold symmetry are conjoined, and this underlying feature can produce attractive patterns [Fig. 192]. The *acute* pattern with 9-, 10-, and 12-pointed stars in Fig. 409b (by author) is very acceptable to the aesthetics of this ornamental tradition.

Figure 410 illustrates another approach to producing underlying tessellations with the same repetitive schema as

the examples in Figs. 406 through 409. This technique dispenses with the use of the primary polygons as the first step in establishing a radii matrix (as per Fig. 409), and starts directly with the production of the radii matrix. Similarly, rather than focusing on the reflected square as the repetitive element during the design process, this technique begins with the concave octagonal shield shape that is always associated with this type of repetitive structures. As indicated in the examples from Figs. 406 through 409, the shield-shaped repetitive cells require 90° rotation to cover the plane. The shield shapes can be thought of as a square

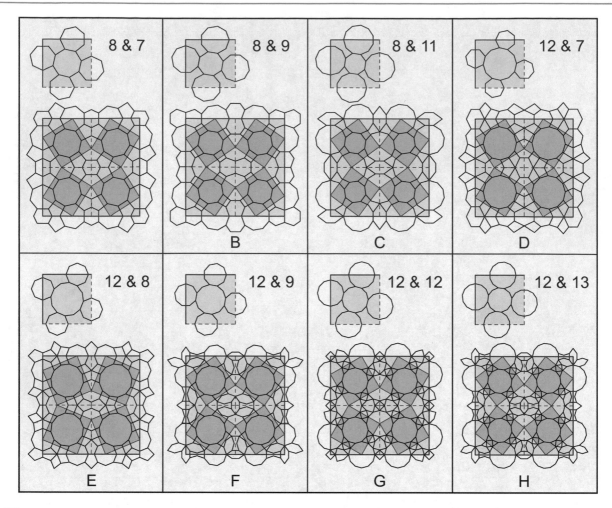

Fig. 406

with 90° corners that have been rotated to accommodate symmetries with *n*-fold symmetry at the midpoints of the otherwise square. The upper left of Fig. 410 is a square with 16 radii at each corner, and 16 radii at each midpoint of the square's edges. Step 1 demonstrates how the midpoints can accommodate a different *n*-fold radial center by rotating each corner an amount that conforms to the introduced symmetry—in this case 18-fold. The radii at the four corners of the square must always be a multiple of four to maintain the right angle, but the introduced *n*-fold symmetry at the erstwhile midpoints is entirely flexible. Step 2 extends the radii of the 16- and 18-fold centers until they intersect, thus completing the radii matrix. Step 3 determines the size of the octagon and nonagon in the standard method using circles and perpendicular lines. Step 4 draws the primary polygons, as well as the separating barrel hexagon and pentagons. Step 5 mirrors these elements around the periphery of the shield shape. Step 6 mirrors the pentagons to the other edges of each nonagon. Step 7 fills the remaining space with

additional pentagons, triangles, and a central barrel hexagon. Step 8 shows the completed tessellation. Figure 411 illustrates four patterns (by author) that are made from the underlying tessellation from Fig. 410. Each is a combination of eight- and nine-pointed stars. The dashed red lines indicate reflection symmetry, and the black dots indicate points of rotational symmetry. The dashed black lines represent the shield shapes in fourfold rotation around each rotation point. As mentioned, these designs have the same repetitive structure as the design from Figs. 406 through 409. Figure 411a shows an *acute* pattern. Figure 411b shows an *obtuse* pattern. Figure 411c shows a *median* pattern, and Fig. 411d shows a *two-point* pattern.

These two approaches to creating complex orthogonal geometric patterns with multiple regions of local symmetry that conform to the aesthetic standards of this artistic tradition offer tremendous scope for contemporary artists and designers working in this discipline.

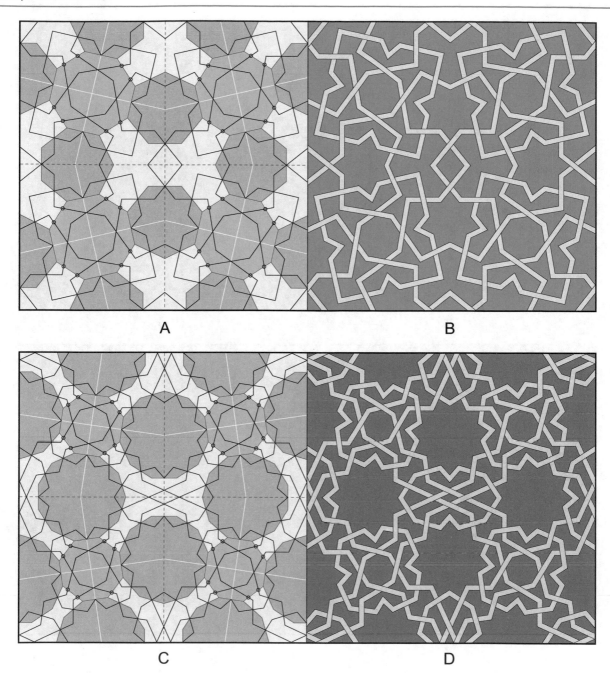

Fig. 407

3.2.5 Rectangular Designs with Multiple Regions of Local Symmetry

Nonsystematic designs with rectangular repeat units range from the rather simple to the very complex. The application of multiple regions of local symmetry follows several conventions that are specific to this repetitive schema. The *n*-fold symmetry located at the corners of the rectangular repeats unit is, perforce, divisible by two. Similarly, if the center of the rectangular repeat unit is also populated with a region of local symmetry, this will likewise be divisible by

2—the center being a vertex of the rectangular dual grid. If there are primary stars located upon the edges of the repeat unit, these can have either even or odd number of points, as can primary stars located within the field of the repeat unit.

Figure 412 is an *acute* field design that, while relatively simple in its geometric composition, is nonetheless appealing to the eye. This places octagons at the vertices of a rectangular grid, as well as at the vertices of the identical dual grid, and the geometric information contained within the repeat unit is identical to that of its dual. The underlying tessellation is comprised of just two polygonal elements:

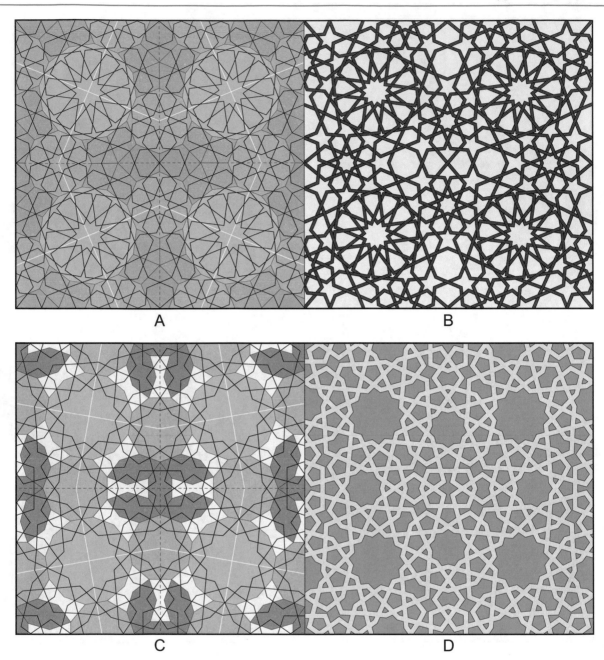

Fig. 408

irregular pentagons and irregular hexagons, and the pattern adheres to the *cmm* plane symmetry group. This is a Mamluk design from an incised stone border in the entry portal of the Sultan al-Nasir Hasan funerary complex in Cairo (1356-63) [Photograph 58].

The design in Fig. 413 is also from the entry portal of the Sultan al-Nasir Hasan funerary complex in Cairo [Photograph 58]. This pattern places eight-pointed stars at the corners of a rectangular repeat unit and regular hexagons at the midpoints of each repetitive edge. This also places an eight-pointed star at the center of the repeat, and like the

previous example the geometric information contained within each repeat is identical to that of the dual repeat. Also like the previous example this design conforms to the *cmm* plane symmetry group. These regions of eightfold and sixfold local symmetry surround a central distorted hourglass motif that is produced from underlying hexagons that are unusual in that they have 180° point symmetry rather than reflection symmetry (black circles). Figure 413b is a representation of the design found in a recessed niche in the entry *iwan* of this complex in Cairo. This is notable in that it is a rare architectural example of a geometric pattern that

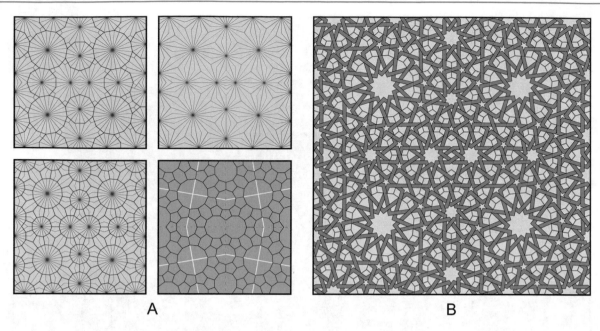

A   B

Fig. 409

includes its underlying generative tessellation as part of the overall ornament. It is worth noting that except for the pattern lines that make up the hexagons, the pattern lines are not collinear as they cross the midpoints of the underlying polygons. Changing the angular openings of the pattern lines at the underlying polygonal midpoints is typically avoided, and tends to only be used whenever the pattern elements would not otherwise be well balanced. Such exceptions to the norm are far more common among nonsystematic designs than in patterns created from any of the generative systems.

Figure 414 illustrates one of the most elegant nonsystematic designs based upon a rectangular repetitive grid. This is an *acute* pattern that places 12-pointed stars at the vertices of a rectangular grid, and 10-pointed stars at the vertices of the rectangular dual grid. The 10- and 12-pointed stars are separated by a matrix of 5-pointed stars and mirrored dart motifs. Figure 414a illustrates how these are derived from an underlying tessellation comprised of decagons, dodecagons, pentagons and barrel hexagons. As mentioned previously, the regularity of the pentagons and barrel hexagons from the *fivefold system* provide ideal conditions for creating patterns in all four families, and the success of this well-balanced pattern is in part the result of the relative regularity of the underlying pentagons—which is to say their closeness to the proportions of the regular pentagon and barrel hexagon of the *fivefold system*. This design was created by artists during the Seljuk Sultanate of Rum where it was used in the Great Mosque at Aksaray in Turkey (1150-53). It also appears in the anonymous manuscript, *On Similar and Complementary Interlocking*

Figures,[65] as well as the Topkapi Scroll.[66] As discussed previously, portions of the anonymous manuscript appear to have been directly influenced by Seljuk geometric ornament, and the rarity of this design suggest the possibility of a link between this manuscript and the example from Aksaray. In addition to the *acute* pattern in Fig. 414b, the underlying tessellation in Fig. 414a will also produce very acceptable designs in each of the other three pattern families: Fig. 415a shows the *median* pattern that this underlying tessellation produces; Fig. 415b shows the *obtuse* pattern; and Fig. 415c shows the *two-point* pattern. Although not used historically, each of these three patterns (by author) conforms to the aesthetics of this ornamental tradition. Figure 416 demonstrates a construction of the underlying tessellation used for creating the designs in Figs. 414 and 415. Step 1 draws 24 radii within two edge-to-edge dodecagons. Step 2 places a decagon, with 20 radii, in a corner-to-corner arrangement with the two dodecagons. Step 3 mirrors the dodecagons on the vertical radii of the decagon. This establishes the rectangular repeat unit. It is worth noting that this arrangement of dodecagons and decagons can be used for creating patterns, although the non-congruent edges are problematic. Step 4 illustrates the radii matrix that results from the previous steps. Step 5 establishes the edges of the dodecagons and decagons through the standard method using a circle that is tangent to the radii. Step 6 draws the dodecagon and decagon, as well as the two separating

---

[65] MS Persan 169, fol. 195b.

[66] Necipoğlu (1995), diagram no. 44.

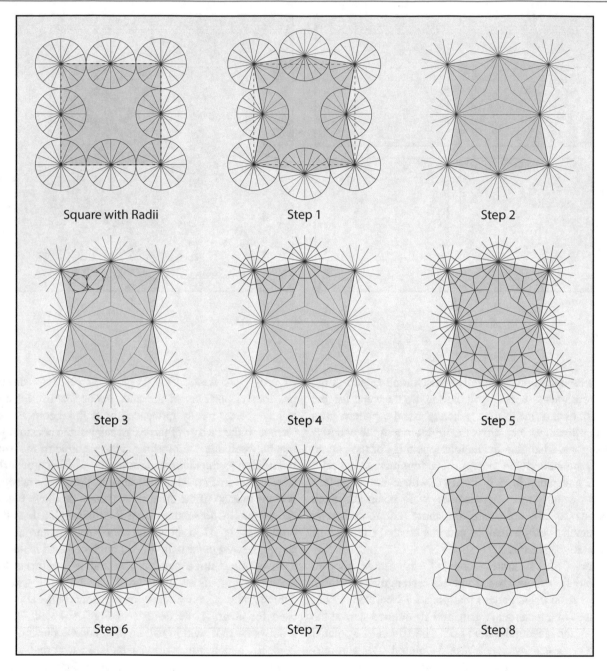

Fig. 410

pentagons. Step 7 mirrors these elements throughout the rectangle. Step 8 mirrors the trapezoids at the midpoints of the long edge of the rectangle, thus producing the barrel hexagons; and Step 9 shows the completed tessellation. It is worth noting that the Topkapi Scroll's depiction of the *acute* design in Fig. 414 includes the underlying tessellation as red lines, and the radii matrix as scribed "dead lines," and the proportional relationships within these features of the example from the Topkapi Scroll comport with those contained with Fig. 416.

The *acute* pattern in Fig. 417 is from the entry door of the Abd al-Ghani al-Fakhri mosque in Cairo (1418). This places 10-pointed stars upon the vertices of a rectangular repeat unit, 10-pointed stars at the midpoints of the long edge of the repeat unit, and two 11-pointed stars within the field of the repeat unit. The six trapezoids that have coincident edges with the concave hexagons in Fig. 417a are created by truncating the six pentagons that surround a thin rhombus. As explained earlier, the configuration of six pentagons surrounding a thin rhombus is not suited to the *acute* pattern family, but, as this example demonstrates, their truncation

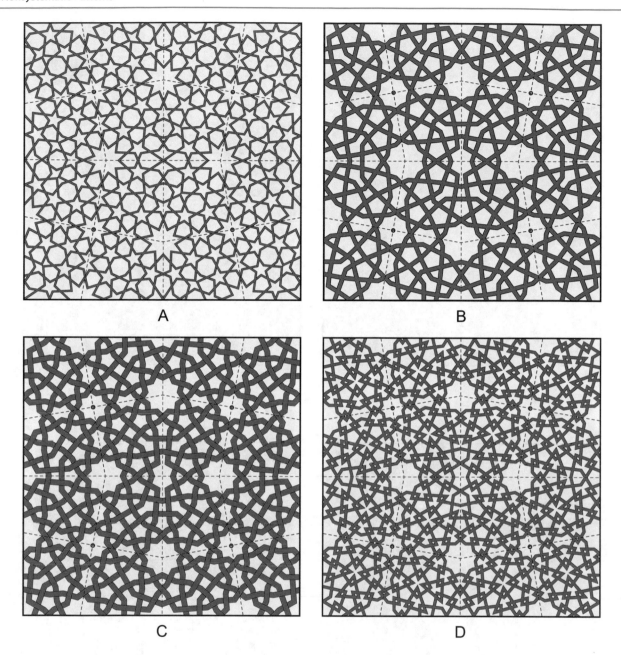

A

B

C

D

Fig. 411

Fig. 412

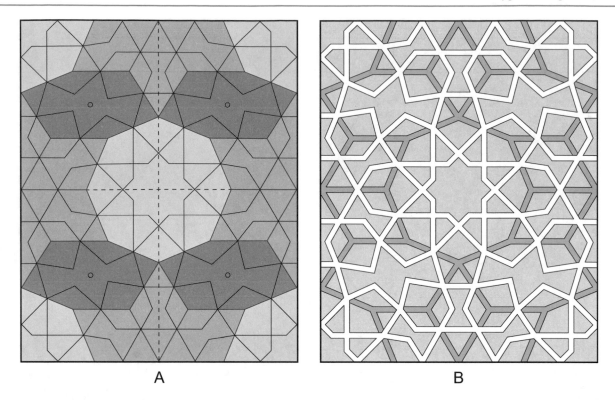

A   B

Fig. 413

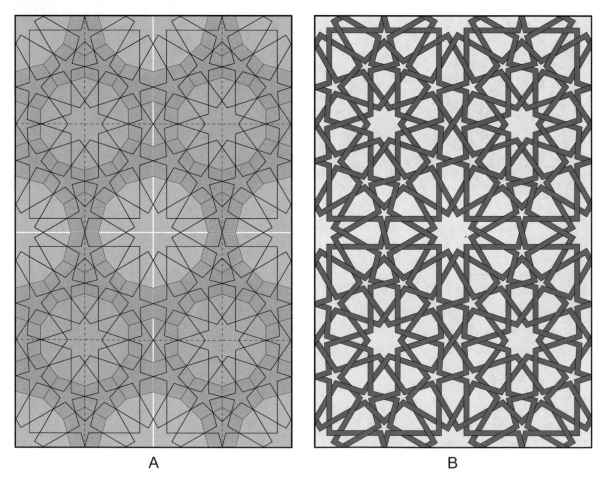

A   B

Fig. 414

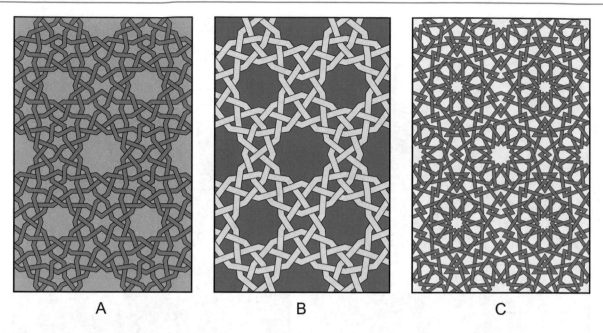

A   B   C

Fig. 415

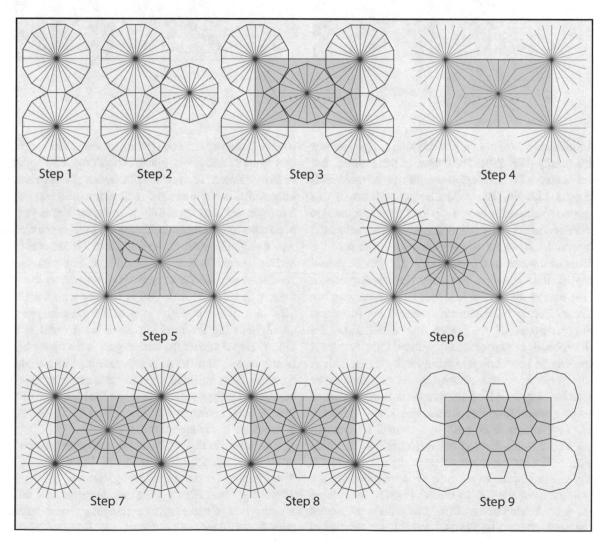

Step 1   Step 2   Step 3   Step 4

Step 5   Step 6

Step 7   Step 8   Step 9

Fig. 416

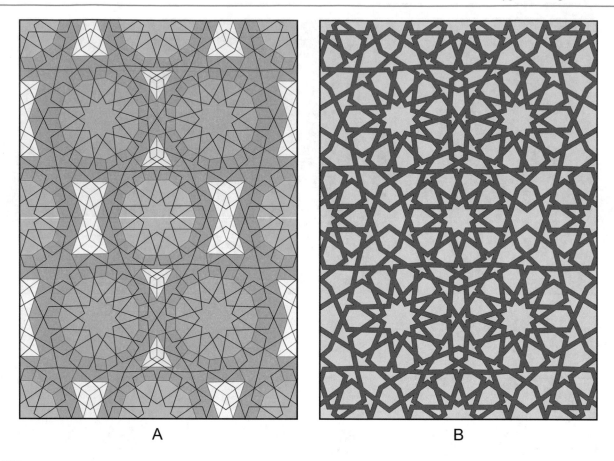

A   B

Fig. 417

into six trapezoids allows for very acceptable design features within this family [Fig. 198]. The cluster of three underlying pentagons has also been transformed into three trapezoids surrounding a triangle. The particular proportions of this design cause the trapezoid that is divided by a vertical line of reflective symmetry to be proportionally narrower that the other trapezoids. In order to provide visual balance, the pattern lines of the dart motif associated with this trapezoid extend beyond the midpoints. Similarly, the crossing pattern lines of the dart motifs in the adjacent barrel hexagons are not collinear. Generally, the more eccentric the distortion within the polygons of the underlying tessellation, the greater likelihood of the crossing pattern lines requiring noncollinearity to produce more acceptable results. This is one of the more eccentric geometric designs created by Mamluk artists. Figure 418 demonstrates a construction of the underlying tessellation that creates this Mamluk design. Step 1 places two sets of 20 radii in a vertical orientation. The angle of the two indicated extended radii is 72°. Step 2 places 22 radii near the intersection of the extended radii from Step 1. The choice of 22 radii is determined by the relative closeness of the 72° to the 65.4545...° associated with an 11-fold division of the circle. The precise placement allows the radii connecting the 20- and 22-fold centers to

intersect midway between the rotational centers, as indicated by the upper and lower black dots. These radii appear more or less collinear. The precise placement also provides for the intersection of the three red radii at the third black dot. Step 3 mirrors the radii from Step 2 on the indicated vertical and horizontal axes from Step 2. This determines the rectangular repeat unit. The multiple small black dots indicate the intersection points where the radii are not collinear. This completes the radii matrix. Step 4 determines the edges of a decagon using a circle that is tangent with the blue radii. This also determines the size of the hendecagon through simply drawing the circle centered on the end of the blue radius. Step 5 copies the dodecagon to the upper left corner of the repeat, and draws a separating pentagon and barrel hexagon. Step 6 fills in one quadrant of the repeat with the connecting polygonal matrix using the blue lines of the radii matrix. Step 7 mirrors this throughout the repeat. And Step 8 shows the completed tessellation.

The design in Fig. 419b is from a *Mudéjar* stucco window grille from Synagogue Transito in Toledo, Spain (1360). This design is a complex arrangement of 6-, 8-, 14-, and 18-pointed stars. Figure 419a demonstrates how this design is comprised of two distinct rectangular repeat patterns: one considerably longer than the other. Either of these can be

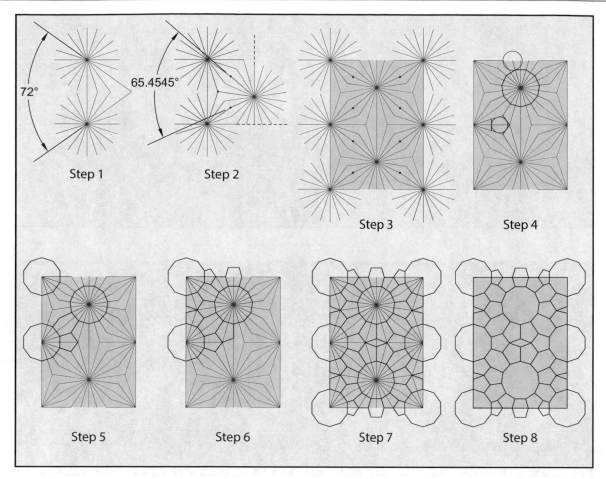

Fig. 418

used independently as a repeat unit, and in this respect, this example can be considered a hybrid design. In this case the shared arrangement of underlying polygons on the short edges of the two repetitive rectangles (18-gons separated by squares) allows them to be used together. Both repetitive rectangles place 18-pointed stars on each corner, and 8-pointed stars within the field of the repeats. The long rectangle also incorporates 14-pointed stars at the center of each repeat unit. Figure 420 shows four stages for a construction of the underlying tessellation for the design from Fig. 419. In the interest of brevity, this sequence is somewhat truncated compared to previous examples. Stage 1 is the radii matrix with 36 radii at each of the corners of both repetitive rectangles, 28 radii at the center of the larger repetitive cell, and 16 radii at appropriate locations throughout the radii matrix. Noncollinearity within the radii matrix is indicated by the black dots. Stage 2 determines the sizes and edge locations of the primary polygons. This standard process makes use of circles placed at intersections within the radii matrix that are tangent to adjacent radii. Stage 3 uses the radii matrix to identify the remaining secondary polygonal elements, including the pentagons, barrel

hexagons, and small rectangles. And Stage 4 is the completed tessellation. As mentioned previously, the underlying tessellations of both rectangular repeat units can also be used on their own for pattern generation, although neither are known to have been used on their own within the historical record.

The *acute* pattern in Fig. 421 places 12-pointed stars on the corners of the rectangular repeat unit, 8-pointed stars at the midpoints of the long edges of the repeat, and two 9-pointed stars within the field of each repeat unit. This pattern also incorporates octagons within the pattern matrix. This is a Mengujekid design from the Great Mosque of Divrigi in Turkey (1228-29). Figure 422 provides a construction of the underlying tessellation used for creating this design. Step 1 illustrates two sets of radii in a vertical orientation, one with 24 radii, the other with 16 radii. Step 2 extends two of the blue radii until they meet. The angle between these extended radii is 52.5° and this is close to the 60° that are 3/18. This provides the opportunity for incorporating a nine-pointed star at this location. Step 3 places 18 radii at the indicated point of intersection from Step 2. These are orientated with the extended line of radius

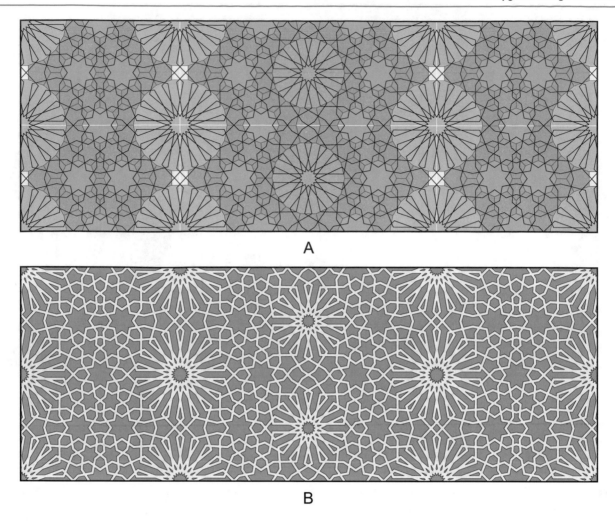

Fig. 419

from the 24 radii in Step 2. However, these 18 radii do not align with the relevant lines of radius from the 16 radii. This problem is solved by moving the 18 radii a requisite distance along the black line of radius between the 24 and 18 radii. This distance is determined by copying the black line of radius to the intersection of 24 and 16 red radii, indicated with a black dot. Step 4 trims the red and blue radii that cross the copied black radius from Step 3. Step 5 moves the 18-fold set of radii from the end of the trimmed blue radius to the end of the trimmed blue radius from the 16 radii. These two radii are not collinear but function as such during the process of extracting the underlying tessellation. A black dot indicates the intersection of the two blue radii that are almost collinear. Moving the 18 radii along the line of radius that connects the 24 and 18 radii also connects three of the red radii from these regions of local symmetry. This also provides the necessary conditions within the radii matrix for extracting the underlying tessellation. Step 6 mirrors the radii on the vertical and horizontal dashed red lines indicated in Step 5 to complete the radii matrix. Step 7 determines the

edges of the dodecagon, nonagon, and octagon using the standard method of drawing circles that are tangent to the red radii. Step 8 draws the three primary polygons. Step 9 mirrors the primary polygons throughout the repeat. Step 10 fills in the secondary polygonal elements using the radii matrix; and Step 11 illustrates the completed underlying tessellation.

Figure 423 is a representation of an *acute* design with 10-pointed stars at the corners of the rectangular repeat unit, 8-pointed stars at the midpoints of the long edge of the repeat, and two 11-pointed stars within the field of each repeat unit. Figure 423a shows the significant disparity between the sizes of the underlying pentagons. This causes the five-pointed stars and adjacent pattern elements in Fig. 423b to vary in size considerably. This, in turn, produces an undulating density throughout the design that is generally less desirable. The discrepancies in the underlying pentagons and barrel hexagons create five- and six-pointed stars that have unsatisfactory characteristics, as are the irregular octagons within the pattern matrix. Another undesirable

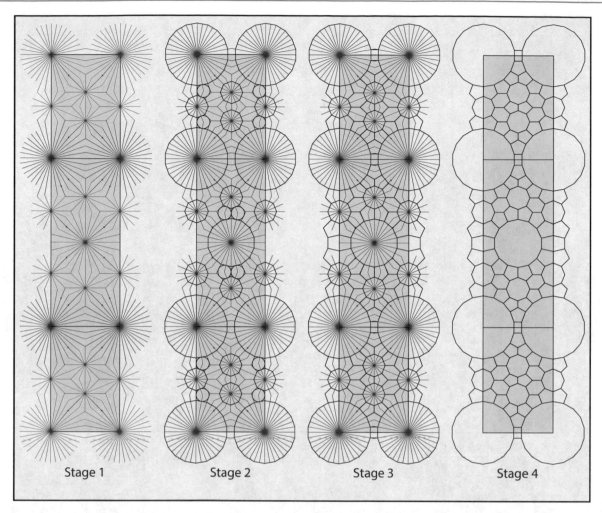

Stage 1   Stage 2   Stage 3   Stage 4

Fig. 420

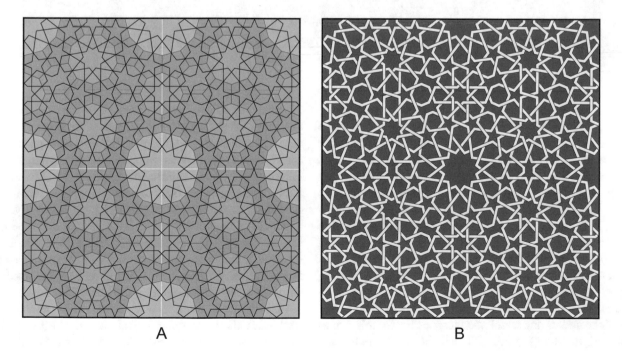

A										B

Fig. 421

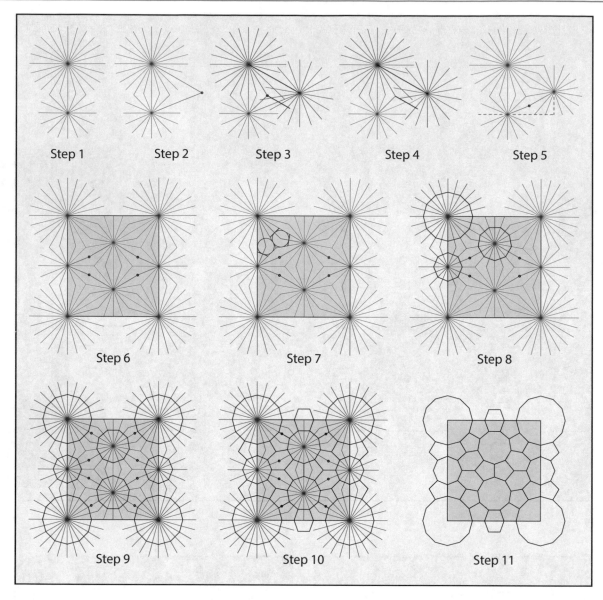

Fig. 422

result of the discrepancies in the sizes of the underlying pentagons is the evident distortion in the proportions of the star rosettes associated with the 11-pointed stars. As a result of these many problems this example lacks elegance and is not a particularly successful geometric design. Were it not for its interesting provenance, these problematic features would have precluded inclusion within this study. This design is from a stone *khachkar* produced by the Christian monk Momik in Noravank, Armenia, during the thirteenth century. He had presumably received training in this geometric art form from artists working in the neighboring Seljuk Sultanate of Rum, although he does not appear to have mastered this discipline. Figure 424 shows a construction of the underlying tessellation responsible for this Armenian design. Step 1 places 20 radii above 16 radii, and

22 radii at the point of intersection between the two extended blue radii. Step 2 moves the 22 radii to a location that approximates equal conditions in the intersecting radii that connect the 22-fold center with the 20- and 16-fold centers. The two black lines indicate the collinear ideal, while the adjacent red and blue radii indicate the approximate equal conditions between these regions of local symmetry. It is important to note that the fact that these red and blue radii are so far from being collinear indicates the strong likelihood for there being problems in the finished tessellation, as well as any derived patterns. This is the basic flaw that results in all of the many problems with the completed design. Step 3 mirrors these radii to create the radii matrix, and establishes the rectangular repeat. Again, the black lines indicate the collinear ideal. Step 4 places the primary

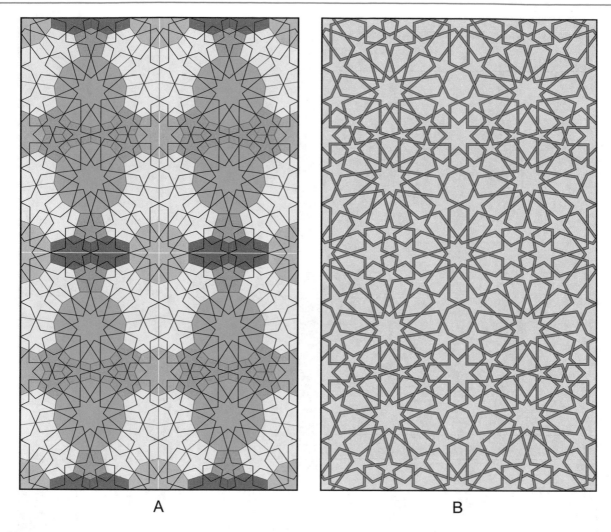

A   B

Fig. 423

polygons into the radii matrix. Due to the unusually high degree of noncollinearity within the radii matrix, the standard method of circles in tangent with radii does not work well in this case. Rather, the size of the primary polygons has been arrived at through trial and error, with the objective being the least amount of discrepancy between the pentagonal elements. Step 5 fills in the pentagons along the lines of radius. Step 6 adds further pentagons and barrel hexagons. And Step 7 completes the tessellation. The differences between the size and nonconformity of the pentagons (and pseudo pentagons) are readily apparent, and as mentioned produces unavoidable irregularities in any patterns that are created from this underlying tessellation.

The rectangular *acute* pattern in Fig. 425 is of a type characterized by linear bands of different primary star forms. In this case the vertical orientation creates an alternating series of 12-, 11-, 10-, 11-, 12-, 11-, 10-, 11-, and 12-pointed stars, etc. The proportion of the rectangular repeat unit for this pattern is unusually long, with 12-pointed

stars at the vertices of the repeat unit; 10-pointed stars at the midpoints of the long edge of the repeat; and two 11-pointed stars within the field of the repeat, each located at the approximate centers between the 12- and 10-pointed stars. Figure 425a illustrates the cluster of three pentagons (brown) within the underlying tessellation, as well as an oscillating band of elongated pentagons (pink). These two types of pentagon create very different types of five-pointed stars, and the resulting five-pointed stars in Fig. 425b reflect this discrepancy. As with the pattern in Fig. 423, this discrepancy creates variation within the overall density of the design. However, this example is more cohesive, balanced, and subtle than the previous example, and the variable density is less problematic. This design was created during the Seljuk Sultanate of Rum and is found in the portal of the Erkilet Kiosk near Kayseri, Turkey (1241). Figure 426 illustrates a construction of the underlying tessellation for this pattern. Step 1 shows two sets of 24 radii, one above the other, with extended red radii that has a 60° angle between

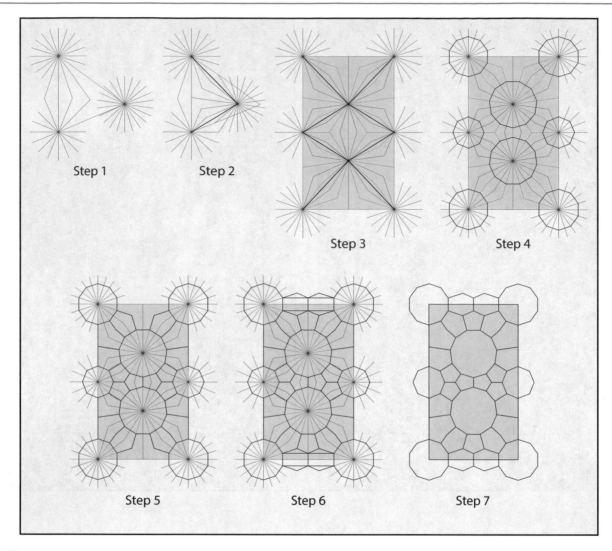

Fig. 424

them. This is close enough to the 65.4545...° angle of the 2/11 division of a circle to provide for an 11-pointed star at this approximate location. Step 2 places an array of 22 radii at the intersection of the two extended red radii from Step 1. This is orientated such that the opening between two of the radii is horizontally aligned, thereby allowing the 22 radii to dialogue equally with the two sets of 24 radii. The fact that the two red radii with the 65.4545...° angle between them are not aligned with the 60° radii from the two sets of 24 radii means that the 22 radii must be moved horizontally to allow these nonaligned radii to intersect at a point that is midway between the 24 and 22 radii centers. To achieve this, two black horizontal lines are drawn at these locations. Step 3 trims the red radii that cross the black lines. Step 4 moves the 22-fold center horizontally so that the two sets of trimmed radii meet at the point indicated by the black dot. These radii appear more or less collinear. Step 5 places two

sets of 20 radii such that three blue radii meet at the indicated location (central black dot), thus establishing half of the rectangular repeat. The orientation of these 20 radii is also horizontally aligned. The upper and lower black dots indicate the intersection of what appears to be collinear blue radii. Step 6 determines the edges of the primary polygons through the standard method of circles in tangent with radii. Step 7 draws the dodecagons, hendecagons, and decagons. Step 8 fills in the polygonal matrix using the lines of the radii matrix. And Step 9 shows half of the completed tessellation. This requires mirroring along the indicated axis for the full repeat unit. Like the pattern in Fig. 425, the *acute* pattern in Fig. 427 is also comprised of 10-, 11-, and 12-pointed stars in a 12, 11, 10, 11, 12, 11, 10, 11, and 12 columnar arrangement. Similarly, this design has an especially long repeat unit with 12-pointed stars at each corner, 10-pointed stars at the midpoints of the long edges of the repeat, and 11-pointed

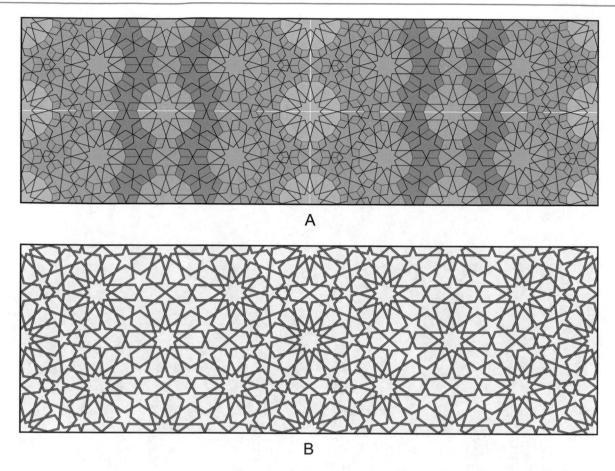

A

B

Fig. 425

stars between the 10- and 12-pointed stars. These two historical examples share the same repetitive and numeric schema, but the actual patterns are quite distinct from one another. Figure 427a shows the typical six trapezoids surrounding the concave hexagon, and as detailed previously, this arrangement of polygons is particularly well suited to the *acute* pattern family [Fig. 198]. This is a Zangid design that was reported by Ernst Herzfeld[67] to have come from a door at the Lower Maqam Ibrahim in the citadel of Aleppo (c. 1230), but is now missing. In noting the highly idiosyncratic conceptual similarities between this Zangid design and the example from the Erkilet Kiosk, and the fact that both are roughly contemporaneous and in relatively close proximity of less than 500 km, it would appear possible that the former of these two geometric designs had a direct influence upon the latter, possibly being created by the same artist or artistic lineage. A construction for the underlying tessellation for the Zangid pattern in Fig. 427 is shown in Fig. 428. As is sometimes the case, a useful starting point is an arrangement of polygons that define the regions of local symmetry. Step 1 shows half of the rectangular repeat unit, with corner-to-corner dodecagons placed at the two corners of the repeat. A hendecagon (11-gon) is placed such that its size and location are determined by two of the corners touching a corner of each dodecagon. Decagons have been placed at the other two corners of the half repeat. These are scaled and located such that their edges are as close as possible to being congruent with the edges of the hendecagon while maintaining their center points upon the same horizontal level as the center points of the dodecagons. Step 2 places radii within each of these polygons. Step 3 shows the finished radii matrix. It is worth noting that this radii matrix is identical to that shown in Fig. 226, and the fact that this can be constructed in more than a single manner is an indication of the flexibility within this design methodology. Step 4 places a new set of dodecagons, hendecagons, and decagons within the set of polygons from Step 1. These are scaled such that a ring of well-proportioned trapezoids surrounds each of the primary polygons. Step 5 shows the completed tessellation.

[67] Herzfeld (1954-56), Fig. 56.

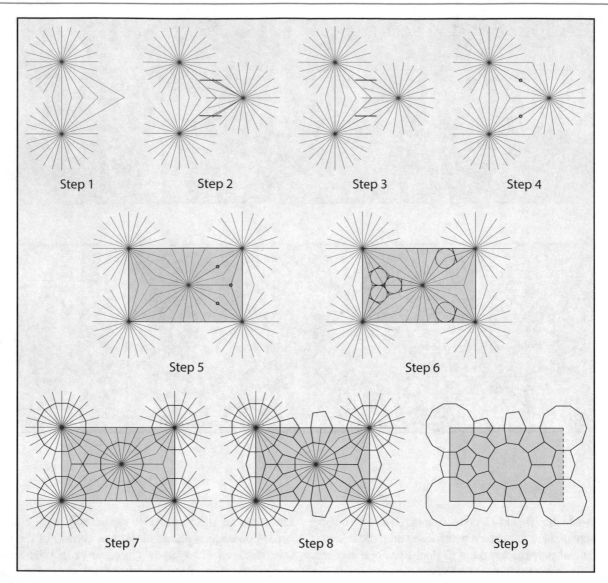

Fig. 426

3.2.6 Hexagonal Designs with Multiple Regions of Local Symmetry

There are relatively few historical nonsystematic patterns with multiple regions of local symmetry that repeat upon a hexagonal grid. Yet several of the examples that adhere to this repetitive schema are among the most geometrically interesting designs within this ornamental tradition. In particular is small set of patterns that employ the *principle of adjacent numbers* wherein the two regions of local symmetry are one numeric step above and one numeric step below an *n*-fold symmetry that works conveniently on its own to fill the plane. In this way, just as eight-pointed stars are predisposed to work on their own, so also should patterns with seven- and nine-pointed stars. Similarly, patterns with just 10-pointed stars suggest the possibility for those with 9-

and 11-pointed stars, and those comprised of 12-pointed stars anticipate patterns with 11- and 13-pointed stars. Each of the following historical examples that employ the *principle of adjacent numbers* repeats upon an elongated hexagonal grid that places one of the two types of primary star at each of the vertices of the repetitive grid. The dual of these grids are also comprised of elongated hexagons, the vertices of which have the other primary star form. The proportion of each grid is determined by the *n*-fold symmetry of the primary stars, and the dual grids are perpendicularly orientated from one another.

Figure 429d represents a Seljuk border that surrounds the *mihrab* of the Friday Mosque at Barsian, near Isfahan (1105). This is a *median* pattern that juxtaposes seven- and nine-pointed stars that repeat upon either of two perpendicular elongated hexagonal grids indicated in Fig. 429e.

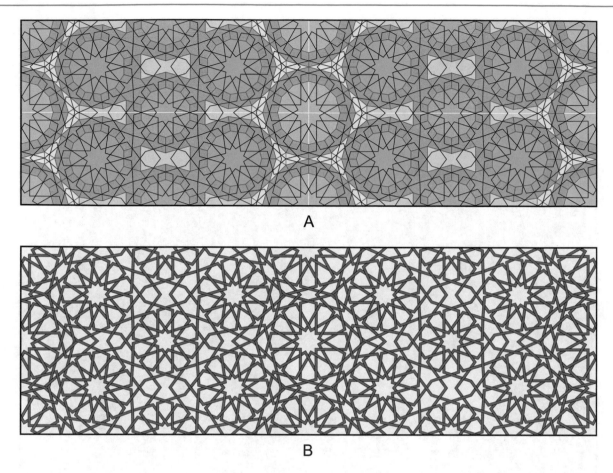

A

B

Fig. 427

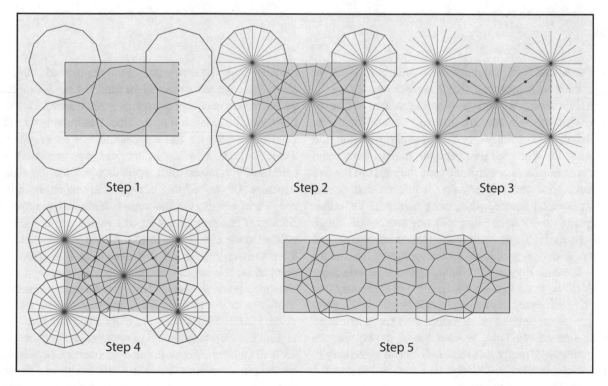

Step 1

Step 2

Step 3

Step 4

Step 5

Fig. 428

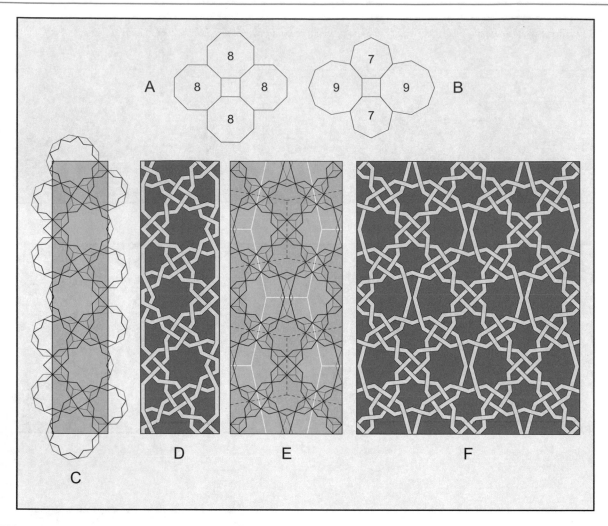

Fig. 429

Figures 429a and b demonstrate the *principle of adjacent numbers* wherein a polygon that tessellates easily on its own—in this case the octagon—indicates the ability of polygons that are one numeric step above and below to also tessellate successfully—in this case the nonagon and heptagon. Note: While the four octagons that cluster around the central square have congruent edges, the edges of the two regular nonagons and two regular heptagons that cluster around the central square, while congruent with the edges of the square, do not quite align with one another (although they are close). Figure 429c shows the linear arrangement of these two underlying polygons that produces the border design at Barsian. Figure 429d demonstrates how extending the width of the border past the linear line of symmetry that divides the underlying heptagons in half disguises the presence of the seven-pointed stars. Rather than being truncated along the line of symmetry at the edge of the border, the pattern continues slightly past the linear line of symmetry to create a wider border, and designated pattern lines at the edges of the border are woven back into the design. In this

example from Barsian, some of the unresolved pattern lines along the edge are extended beyond the border to become part of the adjacent *kufi* calligraphy (not shown). Figure 429e illustrates the extension of the linear underlying tessellation from Fig. 429c to a full tessellation that covers the plane. This illustration also shows the two perpendicularly oriented hexagonal dual grids that provide the repetitive structure. The underlying nonagons in this arrangement are conjoined, and unlike the superb historical example from Barsian, the pattern lines in this region are not full nine-pointed stars. The design in Fig. 429f (by author) creates an interweaving version of the design in Fig. 429e. While acceptable, this would be more pleasing were it to have full nine-pointed stars. This design (sans interweave) conforms to the *pmm* plane symmetry group, and shares visual characteristics with the classic star and cross pattern created from the tessellation of octagons and squares. Figure 430a demonstrates how the arrangement of heptagons and overlapping nonagons from Fig. 429e can be used to produce a tessellation wherein the primary polygons are surrounded

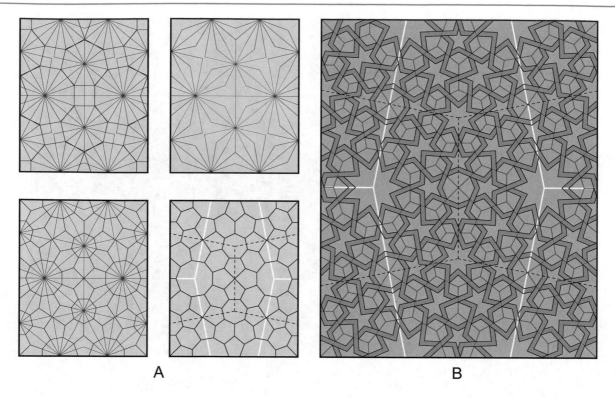

Fig. 430

by a polygonal matrix of pentagon and barrel hexagons. This further development has the advantage of full nonagons that do not overlap, thereby allowing for non-interrupted nine-pointed stars within any resulting patterns. Figure 430a shows the sequence of producing this secondary new underlying tessellation. This sequence begins with the creation of a radii matrix, from which the polygonal matrix is easily derived. Figure 430b shows an *acute* pattern (by author) created from this tessellation. The four clustered pentagons distributed throughout the underlying tessellation provides for the octagons within the pattern matrix. These octagons and the pattern lines surrounding the octagons have shared qualities with some *acute* designs created from *the fourfold system B* [Fig. 177a]. The repetitive structure of this design is identical to that of Fig. 429f, with two varieties of perpendicularly oriented hexagonal grids.

The *acute* pattern in Fig. 431 has the same dual hexagonal repetitive structure as the examples from Figs. 429 and 430. However, in this case the included angles of the hexagonal repeat units are derived from 9- and 11-fold local symmetries. The horizontally oriented hexagonal grid has 11-pointed stars at the vertices, and the vertically orientated dual-hexagonal grid has 9-pointed stars at the vertices. This remarkable pattern is from the Topkapi Scroll,[68] but is currently unknown to the architectural record. As per the

principle of adjacent numbers, the occurrence of 9- and 11-pointed stars is presupposed by the convenience of 10-pointed stars for covering the plane. What is more, the layout of the 5-pointed stars and facing darts that separate the 9- and 11-pointed stars bears remarkable similarity to that of the classic fivefold *acute* design [Fig. 226]. Figure 432 illustrates a construction of the underlying tessellation responsible for this design from the Topkapi scroll. Figure 432a shows how nonagons and hendecagons can be arranged around a concave hexagon in a very similar manner as decagons around a concave hexagon. Unlike the decagons, the edges of the nonagons and hendecagons are not quite contiguous. The ability of these polygons to work together in this fashion is an expression of the *principle of adjacent numbers* wherein the 9- and 11-sided polygons are able to cover the plane in a similar manner as the 10-sided polygon is. Figure 432b illustrates this two-dimensional coverage, along with the radii matrix that this arrangement facilitates. Figure 432c shows just the radii matrix, along with black dots that indicate locations were the radii are not quite collinear. Figure 432d illustrates the radii matrix along with a polygonal tessellation that is created from the radii matrix. This is produced using the standard formula of circles in tangent with the radii that has been detailed previously. Figure 432e illustrates this new tessellation with hendecagons at the vertices of the horizontally oriented hexagonal grid, and nonagons at the vertices of the vertically orientated dual hexagonal grid. Figure 432f shows this

---

[68] Necipoğlu (1995), diagram no. 42.

Fig. 431

underlying tessellation along with the *acute* pattern that is produced from this tessellation. The shaded region is the minimal rectangular repetitive cell, and represents the portion of this design illustrated in the Topkapi Scroll. It is worth mentioning that along with the pattern itself, the illustration in the Topkapi Scroll accompanies the design with both the radii matrix and the underlying tessellation. Figure 433 shows an alternative construction sequence for the underlying tessellation for the *acute* pattern from the Topkapi Scroll in Fig. 431. Step 1 is an array of 22 radii. Step 2 mirrors the 22 radii. Step 3 mirrors the 22 radii again. Step 4 shows half of the hexagonal repeat unit along with the extended radii from each corner. The indicated angle is 40.9091...°. This is close to the 40° angle associated with a ninefold division of the circle, and indicates that a nine-pointed star should be able to work at this location. Step 5 places 18 radii at the intersection of the four blue radii from Step 4. These radii can be seen to be out of alignment with the four blue radii that originate at each corner and meet at the center of the trapezoid. The black horizontal line indicates the small amount of horizontal movement of the 18-fold center to make these blue radii intersect. Step 6 shows the results of this move. The radii that connect the 22-fold and 18-fold centers are not quite collinear, but function as though they are. The intersections are indicated in black dots. Step 7 determines the edges of the nonagon and hendecagon through the placement of a circle that is tangent with the blue radii. Step 8 draws these primary polygons, as well as the two separating pentagons.

Step 9 mirrors the hendecagons to each corner of the half repeat, and fills in the half repeat with further pentagons and barrel hexagons. And Step 10 is the completed tessellation. It is worth mentioning that this underlying tessellation will make very acceptable designs in all three of the other pattern families.

The pattern in Fig. 434 is from the Mu'mine Khatun mausoleum in Nakhichevan, Azerbaijan (1186). This remarkable Ildegizid pattern is comprised of 11- and 13-pointed stars [Photograph 35]. Figure 434a illustrates how the underlying hendecagons are located at the vertices of a horizontally oriented hexagonal grid, and the tridecagons are placed on the vertices of a vertically oriented dual hexagonal grid. Both of these hexagonal grids have translation symmetry and can be regarded as the repeat unit. The turquoise hendecagons and tridecagons that surround the stars in Fig. 434b are an arbitrary addition on the part of the artist. The overall balance of this design is aesthetically pleasing, and this non-challenging quality belies the considerable geometric complexity that underlies this construction. The *principle of adjacent numbers* is also a feature of this Ildegizid pattern. Figure 435a demonstrates how the square and triangular arrangements of the dodecagon have analogous arrangements with the hendecagon and tridecagon. Of course the proportions of the square and regular triangles change according to the angles associated with the 11- and 13-fold division of the circle. Figure 435b places hendecagons and tridecagons into an arrangement wherein their edges are very close to being contiguous.

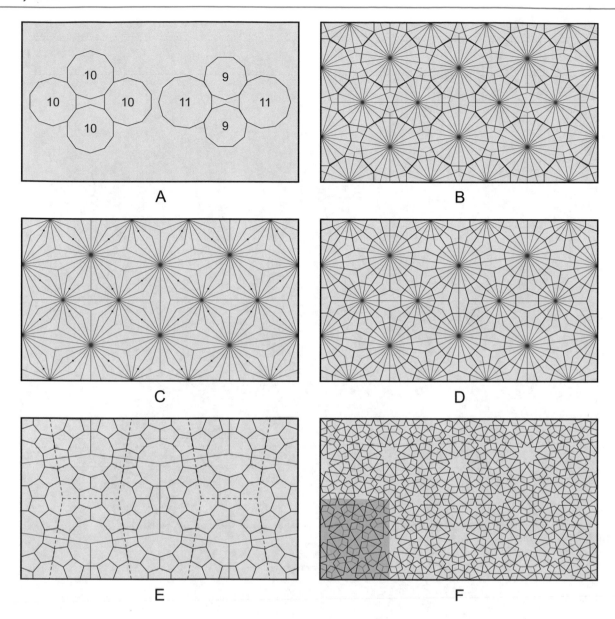

Fig. 432

This also places a radii matrix of 22- and 26-fold rotational centers within these figures. Figure 435c shows the radii matrix alone, with added black dots that indicate the intersection of radii that appear collinear. Figure 435d shows the new tessellation with connecting pentagons, along with the generative radii matrix. Figure 435e shows just the new tessellation along with the two perpendicular dual hexagonal grids. And Fig. 435f shows the underlying tessellation with the *acute* pattern from Nakhichevan. Figure 436 illustrates the analogous relationship between the underlying tessellation for the pattern in Fig. 434 and a hybrid tessellation of dodecagons placed upon the vertices of a grid comprised of squares and equilateral triangles. The tessellation in Fig. 436a has vertical and horizontal lines of reflected symmetry (dashed lines), and the full rectangular panel has translation symmetry. As a repeat, this rectangle has twice the area of either of the hexagonal repeat units. The point of intersection of the red diagonal lines, where four pentagons meet, has 180° rotation symmetry. The square panel in Fig. 436b also has two lines of reflected symmetry, and also has translation symmetry. Rather than 180° rotation symmetry, the analogous location were four pentagons meet, and the red diagonal lines intersect has 90° rotation symmetry. The visual similarity between these two tessellations is especially apparent in the respective arrangements of the matrix of pentagons that separate the primary polygons. The underlying tessellation for the pattern in Fig. 434 will also make very acceptable designs in the other three pattern families.

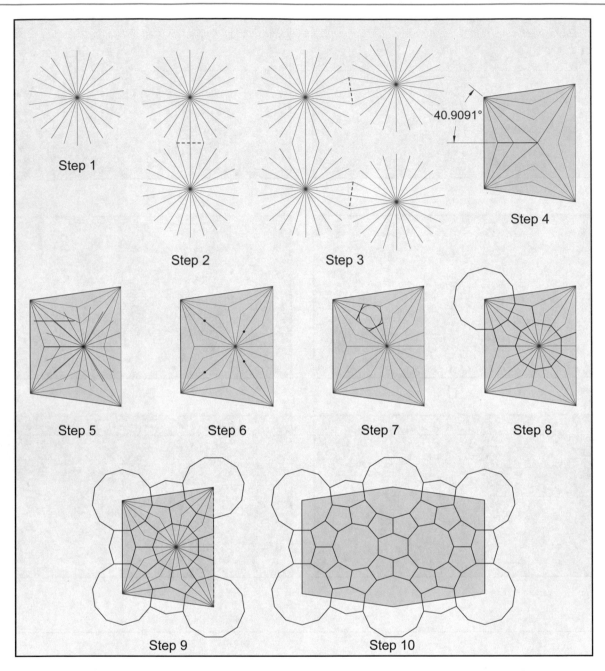

Fig. 433

Each of the three designs (by author) in Fig. 437 is comprised of 11- and 13-pointed stars, and repeats on either of the two hexagonal grids. Figure 437a shows a *median* pattern that has been modified in the common technique often found in Mamluk designs [Fig. 223]. Figure 437b shows an *obtuse* pattern. And Fig. 437c shows a *two-point* pattern with a standard variation to the primary stars [Fig. 225b]. Figure 438 demonstrates a construction of the underlying tessellation for the pattern in Fig. 434. Step 1 mirrors 22 radii so they have a vertical orientation. Step 2 mirrors these 22-fold centers as shown, thereby identifying

the trapezoid (blue) that is half the hexagonal repeat unit. The indicated angle is 81.8181...°. This is close to the 83.0769...° associated with 6/26 of a 26-fold division of a circle. Step 3 places 26 radii at the intersection of the extended blue radii from Step 2. Four of the 26-fold blue radii do not quite align with the associated 22-fold radii. To correct this non-alignment two black horizontal lines are placed for moving the 26-fold center and for trimming the blue radii. The slight movement of the 26-fold center must be horizontal so that the final resting place remains central within the repeat. Step 4 shows the radii matrix after moving

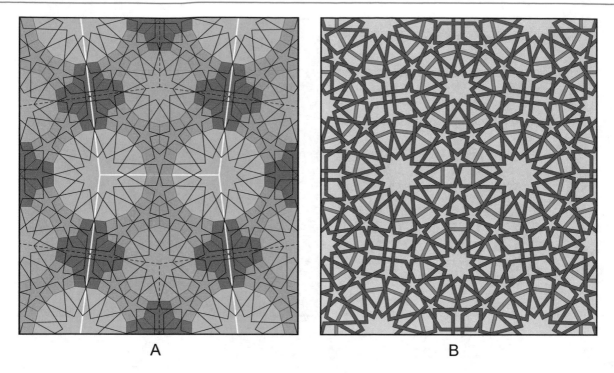

Fig. 434

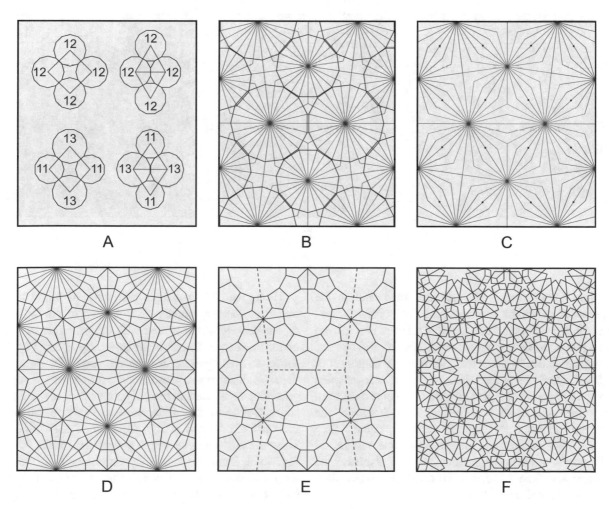

Fig. 435

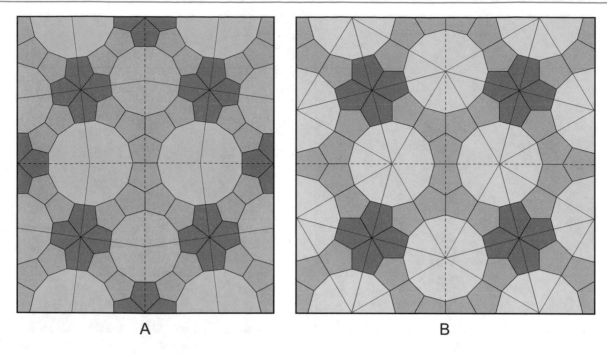

Fig. 436

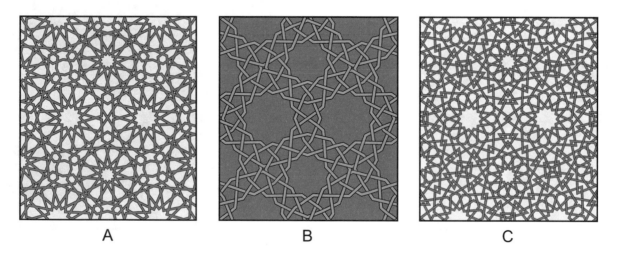

Fig. 437

the 26-fold center so that the trimmed blue radii of the 26-fold center meets the trimmed blue radii of the 22-fold centers. The black dots indicate the intersections of the trimmed blue radii that connect the 22- and 26-fold centers. These appear collinear, but actually have slight angles off of 180°. Step 5 determines the edges of the hendecagons and tridecagon using a circle that is tangent to the blue radii. Step 6 draws the two primary polygons. Step 7 mirrors the hendecagons into the other three corner locations. Step 8 uses the lines of the radii matrix to fill the half repeat with the matrix of connecting pentagons. Step 9 mirrors the half repeat; and Step 10 shows the completed tessellation within its hexagonal repeat unit.

Hexagonal nonsystematic patterns with more than one region of local symmetry do not have to conform to the repetitive schema associated with the principle of adjacent numbers—as per the previous three examples. Figure 439 shows an *acute* pattern that repeats upon a hexagonal grid, with 9-pointed stars placed at the vertices of the repetitive grid, 12-pointed stars at the midpoints of the aligned parallel edges, 8-pointed stars at the centers of each repeat unit, and four 10-pointed stars within the field of each repeat unit. There are also octagons within the pattern matrix. As seen in Fig. 439a, there is substantial irregularity in the sizes and shapes of the polygons that make up the connecting matrix, and this transfers to the significant disparity and

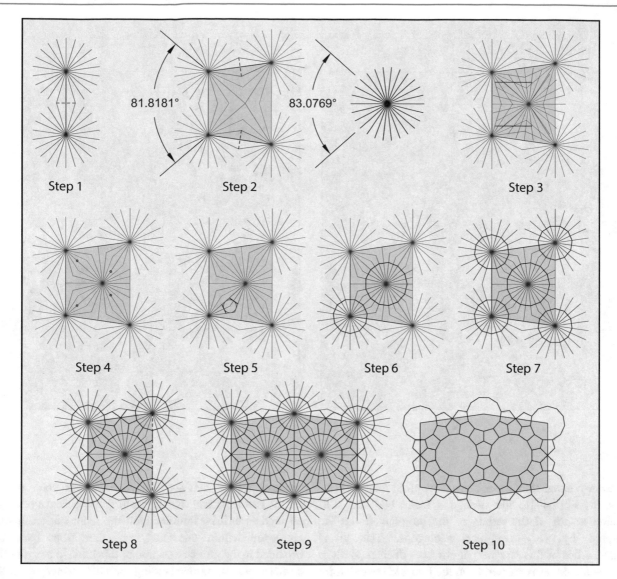

Fig. 438

inconsistency within the pattern matrix of Fig. 439b. This design originates from the Seljuk Sultanate of Rum and is found in the courtyard portal of the Karatay Han near Kayseri, Turkey (1235-41).

3.2.7 Radial Designs with Multiple Regions of Local Symmetry

Nonsystematic geometric patterns with radial symmetry are rare within the ornamental arts of Muslim cultures. The *obtuse* pattern in Fig. 440 dates from the Safavid period, and is from one of the star shaped soffits within the *muqarnas* of the southeast *iwan* (fifteenth to sixteenth century) at the Friday Mosque in Isfahan. The soffit is a seven-pointed star, and the *obtuse* design contained within this soffit has sevenfold rotational symmetry. A 14-pointed star

is located at the center of the 7-pointed star, and 11-pointed stars are placed at the inside angles of the 7-pointed star. The furthest angle of the seven-pointed star has a partial nine-pointed star. Figure 440 demonstrates how this design can be created from either of two underlying tessellations. In a similar fashion as the two tessellations in Fig. 200, these underlying tessellations have a dual relationship.

Figure 441 shows a design also from a soffit in the *muqarnas* semidome of the southeast *iwan* in the Friday Mosque at Isfahan [Photograph 90]. This is a radial design with tenfold rotation symmetry set within a ten-pointed star. This *median* pattern places a ten-pointed star at the radial center, and 10 ten-pointed stars at the outer obtuse angles of the ten-pointed star panel. Ten 7-pointed stars surround the central 10-pointed star, and partial 7-pointed stars separate the 10-pointed partial stars at the periphery of the panel. The heptagons that form the ring in Fig. 441a are not regular, and

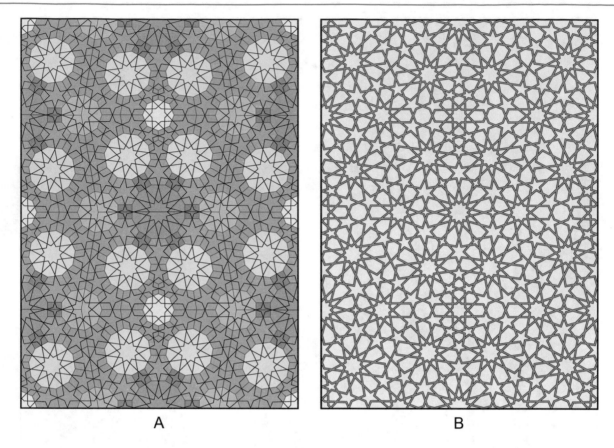

A   B

Fig. 439

the seven-pointed stars do not have sevenfold rotation symmetry. In Fig. 441a, the decagons located at the 108° included angles of the points of the ten-pointed star are separated by very irregular heptagons. The visual characteristics of this *median* pattern have distinct parallels with *median* patterns created from the *fourfold system A*, for example, Figs. 154 and 159.

3.3   Dual-Level Designs

The tradition of dual-level geometric patterns with self-similar characteristics is one of the most striking developments in the long and multifaceted history of Islamic geometric ornament. The secondary application of a smaller scale geometric design into an otherwise independent larger scale primary pattern was highly innovative, and the development of dual-level patterns was the last great geometric and aesthetic advance of this artistic tradition. The multi-level quality of this variety of historical Islamic geometric design has characteristics that invite comparison to several areas of contemporary mathematics and crystallography. What is more, and as per the examples at the end of this section, the methodological practices that were employed by the originators of these dual-level designs can be used to

produce designs that accurately conform to some of these recent mathematical and crystallographic discoveries. Not all readers will be familiar with the technical terminology employed within this field, and definitions have been provided in the glossary section of this book to help facilitate a more precise understanding of such concepts as self-similarly, aperiodicity, quasiperiodicity, quasicrystallinity, subdivision rules, and Penrose matching rules.

The mature style of dual-level design developed in the Maghreb during the fourteenth century, and approximately a century later in Persia, Khurasan, and Transoxiana. Precursors to the fully mature style are found in several locations in the eastern regions, including the exterior façade of the Gunbad-i Qabud in Maragha, Iran (1196-97) [Fig. 67] [Photograph 24]; the minaret of the Yakutiye *madrasa* in Erzurum, Turkey (1310); and the mausoleum of Uljaytu in Sultaniya (1307-13) [Photograph 96]. In the Maghreb, the incorporation of dual-level designs was implemented as a fully mature tradition from its onset during the fourteenth century, indicating that it may have been innovated by a single person or atelier rather than part of an ongoing developmental process. The mature style of dual-level design in both the eastern and western regions is characterized by the almost universal use of three-, four-, and fivefold polygonal systems. While appearing to have considerable complexity,

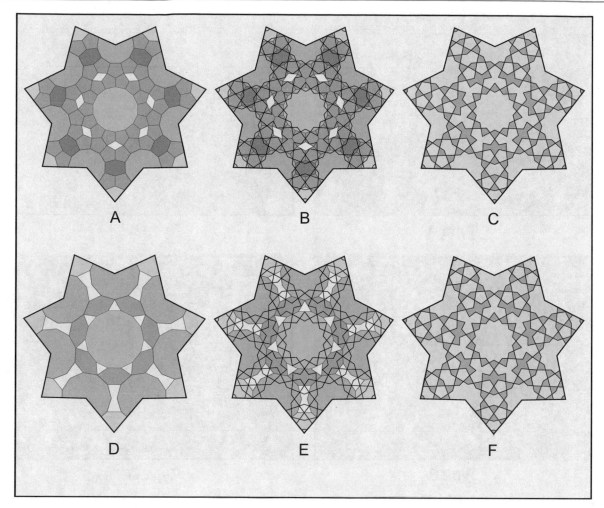

Fig. 440

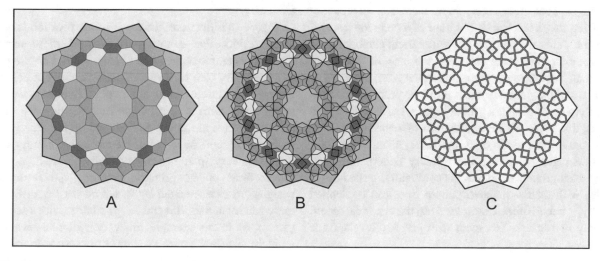

Fig. 441

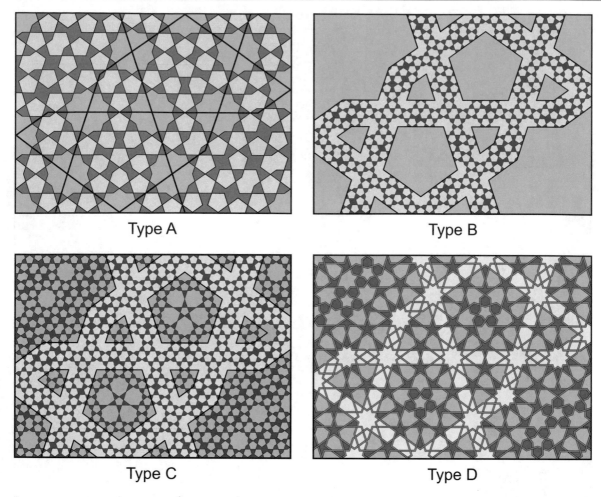

Type A

Type B

Type C

Type D

Fig. 442

the systematic basis behind their design methodology is surprisingly accessible, and their increased complexity stems from the geometric elaboration of a common systematic theme rather than increased symmetrical elaboration.

In creating dual-level designs, the use of polygonal systems such as the *system of regular polygons*, the *fourfold system A*, and the *fivefold system* provides proportional continuity between a design's primary and secondary levels. In practice, this involves the application of scaled-down secondary underlying polygonal modules, along with their associated pattern lines, to the primary pattern lines and primary background elements. Theoretically, scaled-down modules with their associated pattern lines can be applied infinitely,[69] but all of the examples from the historical record have only two levels. Yet, even with just two levels, these

designs occasionally fulfill the mathematical criteria for self-similarity.

Figure 442 illustrates the four historical varieties of dual-level design. For comparison, each of these examples (by author) employs the same classic fivefold *obtuse* pattern as the primary pattern. Type A dual-level designs emphasize the primary pattern with a single plain line upon which the secondary pattern is constructed. Both the primary and secondary patterns are uninterrupted, and fill the repetitive cell to the full extent. As with this example, the primary stars of the secondary pattern are typically located upon the intersections of the primary pattern. Type B dual-level designs are characterized by widening the lines of the primary pattern and infilling the widened lines with a secondary pattern. As in this example, this variety of dual-level design typically places the primary stars of the secondary pattern at the vertices of the widened primary pattern. The background regions are either left plain (as in this case) or provided with a floral or occasionally calligraphic design. As with type B designs, type C dual-level designs also widen the primary pattern and apply the secondary pattern into the widened

[69] Jean-Marc Castéra has illustrated this infinite recursive principle very effectively in animation and with several published consecutive plates from this animation. Castéra (1996), 277.

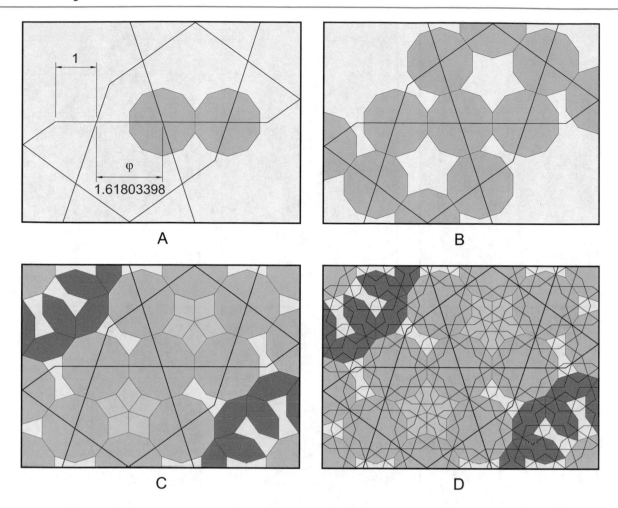

Fig. 443

lines. However, this variety of dual-level design extends the secondary pattern to fully cover the repetitive cell. The bounding lines of the primary pattern are maintained, but the differentiation between the primary and secondary patterns is emphasized through the use of color. Type D dual-level designs use color as the only method of differentiating the primary and secondary patterns. In this variety of dual-level design the primary pattern is exposed through picking out the appropriate background regions of the secondary pattern. Without the color differentiation, type D dual-level designs would appear as particularly complex single-level geometric designs. Types A, B, and C are native to Persia, Khurasan, and Transoxiana, with types A and B being most common and type C being comparatively rare. Type D dual-level designs are exclusively found in Morocco and al-Andalus.

Figure 443 demonstrates the type A application of the secondary pattern to the primary pattern through detailing the example illustrated in Fig. 442. This example employs the *fivefold system*, but the same basic methodology also works with each of the other historical design systems. As

shown, the primary pattern in this example is comprised of two intervals that have the *phi* [φ] proportions of the golden ratio. Figure 443a determines the scale of the secondary pattern by centering two edge-to-edge decagons at each end of the longer interval. As shown in Fig. 447, other arrangements of underlying polygons can be used, but the key to this design methodology is placing the scaled-down polygonal modules so that they fit within the intervals of the primary pattern. Figure 443b applies the scaled-down decagons to other intersections throughout the primary pattern. Figure 443c fills in the remaining background with additional modules from the same system to complete the secondary underlying tessellation. And Fig. 443d applies pattern lines associated with the *median* family to complete the design (as per Fig. 442: type A). Although less common, some historical type A dual-level designs place the primary polygons of the secondary pattern at the intersections of the primary pattern as well as the vertices of the primary underlying tessellation. Figure 444a places two edge-to-edge decagons such that one is placed upon the primary pattern (black) and the other is placed on the primary underlying

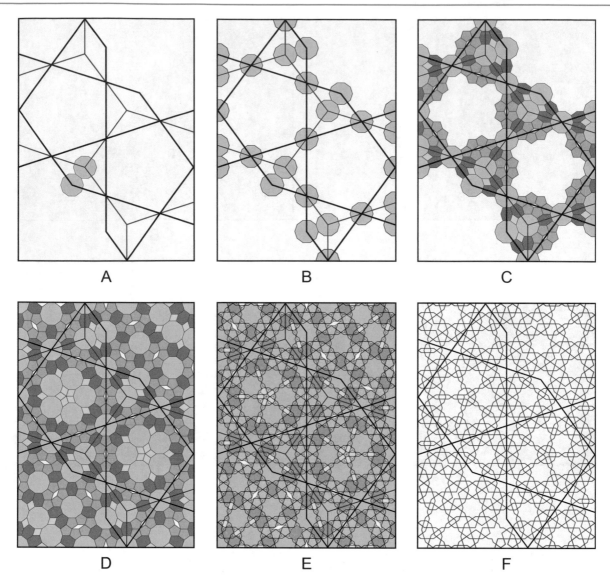

Fig. 444

tessellation (red). These two locations are the closest points between these two varieties of vertex. Figure 444b places these scaled-down decagons at all the intersections of the primary pattern and all the vertices of the primary underlying tessellation. Figure 444c populates the edges of the primary pattern and the edges of the primary underlying tessellation with further polygonal modules from the same system. Figure 444d fills in the remaining background to complete the secondary underlying tessellation. Figure 444e illustrates the application of the secondary *obtuse* pattern to the secondary underlying tessellation; and Fig. 444f shows just the pattern lines that make up the dual-level design (by author) created from this tessellation.

   A primary concern in creating type B and type C dual-level designs is determining the proportions of the widened primary design. By widening the lines so that the proportions

adhere to the proportions of the generative system—in this case the *fivefold system*—the widened lines will allow for the scaled-down modules from the same system to fit precisely into the widened region. Figure 445a demonstrates how an appropriate proportion for the widened lines can be determined easily by employing modules from the *fivefold system*: in this case the wide rhombus and the trapezoid placed in a pentagonal configuration. The use of these modules insures that the intervals within the widened lines have the *phi* proportions of the golden section that are required for adding the scaled-down polygonal modules when working with the *fivefold system*. Figure 445b places decagons at the intersections of the widened lines, and the scale of these decagons is determined by their edge-to-edge placement throughout this network. Figure 445c shows the type B variation with the trimmed away secondary modules so

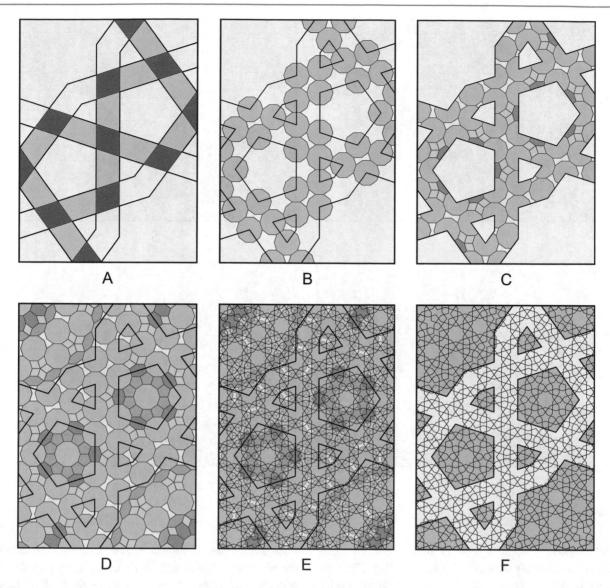

A

B

C

D

E

F

Fig. 445

that only the widened lines are populated with both the underlying tessellation and the applied pattern lines of the *median* family. Figure 445d through f are type C dual-level designs. Figure 445d shows the same widened primary lines and the same applied secondary modules as Fig. 445c. However, the secondary underlying tessellation is extended throughout the repetitive cell, thereby infilling the background regions with additional polygonal modules. Figure 445e applies the *median* pattern lines to this tessellation; and Fig. 445f illustrates the completed type C dual-level design. It is interesting to note the pattern line conditions within the decagons that are located within the widened primary design, but not centered on a vertex of the primary design. These are near to the inside corner of the primary ten-pointed stars. The fact that these secondary decagons are not centered on the primary vertices, and that the point at which their edges cross the lines of the primary design are not the

midpoints of the secondary decagon (as is the case with the decagons that are placed on the vertices of the primary widened design), might lead one to expect that the applied secondary pattern lines inside these decagons would be arbitrary, and would not necessarily work well with the primary widened pattern lines. However, a careful look at the applied pattern lines to these secondary decagons in Fig. 445f reveals that the interior points of the secondary ten-pointed stars rest precisely upon the lines of the primary design. This type of concordance in intersection points between the applied pattern lines of multiple levels of design is a remarkable, albeit standard feature of the recursive use of these design systems, and is the result of the proportional continuum that is inherent throughout the multiple levels of scaled-down polygonal modules and their resulting pattern lines.

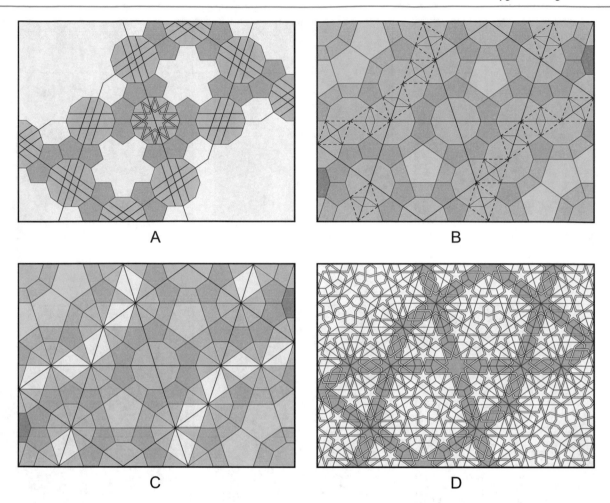

Fig. 446

The method for creating type D dual-level design from polygonal systems is essentially the same as that of the type A designs, with one important proviso: the rotational orientation of the primary polygons for the secondary pattern—in this case scaled-down decagons—must allow for the creation of secondary pattern elements that express the primary pattern through color differentiation. Figure 446a illustrates how this is achieved by placing the decagons such that two sets of the applied secondary pattern lines that make up the ten-pointed stars run parallel with the directions of the primary pattern. The requirement of secondary pattern lines that are parallel with the primary pattern lines also makes the choice of pattern family critical to this process. Rotational orientation and suitable pattern family are required regardless of the polygonal system that is being used. Once the initial application of suitably oriented secondary polygonal modules has been applied to the vertices of the primary pattern, the infill of the remaining areas is completed. Figure 446b shows a completed secondary tessellation with overlapping decagons. Figure 446c shows how

these overlapping decagons can be transformed to allow for the placement of multiple 1/10 triangular segments of the decagon and large rhombi. These configurations produce good results with the *acute* family [Fig. 196]. Figure 446d shows the *acute* secondary pattern, along with the primary pattern that is differentiated through the use of color within the secondary background elements.

Figure 447 illustrates a selection of alternative secondary polygonal arrangements for populating the same pentagonal region that represents the primary pattern lines. This simple process involves first determining the edge configuration, followed by filling in the remaining interior of the primary pentagon with further polygonal modules. Invariably, the scale factor between levels is governed by the secondary polygonal edge configuration, and the precise scale will always be an expression of the inherent proportions of the generative system. The examples in this figure are produced from the *fivefold system*, and the consequent scale factors are expressions of *phi*. The column on the left side of this illustration indicates two edges of the primary underlying generative

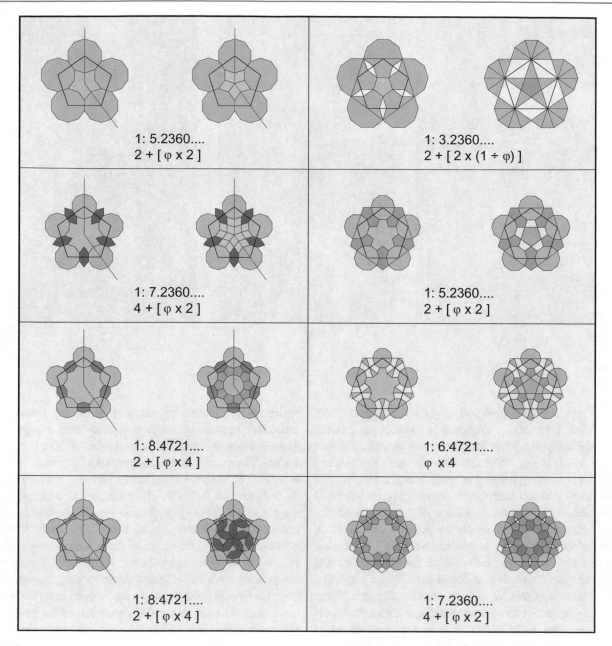

Fig. 447

decagon in green, and the primary pentagonal pattern lines of the *obtuse* family in black. The provided scaling ratio is between the length of the secondary polygonal edges (red) and the primary decagonal edges (green). The edges of the primary polygonal modules in the column on the right are the green pentagons, and the primary pattern lines are the black pentagons that contact the midpoints of the green pentagons. Once again, the provided scaling ratio is between the length of the secondary polygonal edges (red) and the primary decagonal edges (green). In all eight of these examples the scaling ration is indicated as an expression of *phi*.

3.3.1 Historical Examples of Type A Dual-Level Designs

Figure 448 is a dual-level design from a Mughal pierced marble *jali* window grille at the I'timad al-Daula in Agra, India (1622-28). Figure 448a illustrates the two generative tessellations of regular hexagons that produce both levels of design in Fig. 448b. Taken on their own, the design of both the primary and secondary levels is the classic six-pointed star *median* design created from the *system of regular polygons* [Fig. 95b]. The proportional scale between the

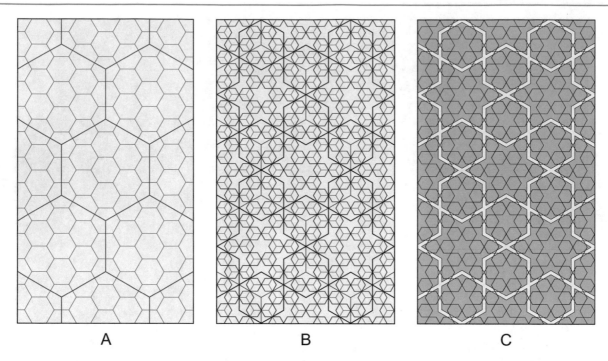

A   B   C

Fig. 448

primary and secondary hexagonal grids corresponds with
$2 \times \sqrt{3}$, or 1:3.4610.... Figure 448b applies six-pointed
stars to the midpoints of each primary and secondary underlying hexagon. Figure 448c illustrates the dual-level pattern
without its underlying generative tessellation. Peter Cromwell
has observed that this example is the only known historical
Islamic dual-level geometric design with scale invariance
wherein the primary and secondary patterns are identical,
but for their scale and 90° rotational orientations.[70] Geometric
structures with scale invariance such as this Mughal example
from India have a high degree of self-similar symmetry. The
scale invariance of this example applies equally to both the
underlying generative tessellation and the resulting dual-level
design. This exactitude is distinct from the looser forms of
self-similarity of other historical dual-level designs wherein
the primary and secondary patterns are both created from
polygonal modules contained within the same modular system, and where both the primary and secondary patterns are of
the same pattern family, but where the two respective patterns
are not otherwise identical. Still less precise are those historical dual-level patterns were the primary and secondary
patterns are created from the same generative system, but
the primary design and secondary design are produced from
different pattern families.

Of the many dual-level designs illustrated in the Topkapi
scroll, number 31[71] has scale invariance within a limited

region of the overall repeat unit. Both the primary and
secondary patterns of the type A dual-level design in Fig.
449 are of the *median* family, and created from the *fivefold
system*. The overall rectangular repeat of the primary design
is recursively iterated at the center of the secondary pattern.
As indicated in Fig. 450d, this rectangular arrangement of
4 ten-pointed stars is placed upon four of the vertices of the
central hexagon in the primary design. The only difference
between the scaled-down use of the primary design is in the
treatment of the ten-pointed stars: the scaled-down
ten-pointed stars in the example from the Topkapi Scroll
have ten kites in rotation, while the primary stars are without
kites. As indicated, this scaled-down repetitive rectangle is
also used to the right and left of the central location. Aside
from these iterative regions, the secondary pattern is not
totally the same as the primary pattern, and the self-
similarity conforms to the use of *median* patterns created
from the same methodological system at two scales. Note:
The example from the Topkapi Scroll is black lines only, and
the color in this illustration has been added for visual clarity.
Figure 450 demonstrates the method of creating the secondary pattern by applying scaled-down polygonal modules
from the same *fivefold system* to key locations of the primary
pattern. These scaled-down polygonal modules are present
in the scribed "dead lines" of the Topkapi Scroll. Figure
450a shows a rectangular repeat unit with the underlying
tessellation of decagons, pentagons and wide rhombi and the
primary *median* pattern created from this tessellation. Figure
450b places decagons with applied pattern lines of the
median family upon intersections of the primary pattern.

[70] Cromwell (2016).
[71] Necipoğlu (1995), diagram no. 31.

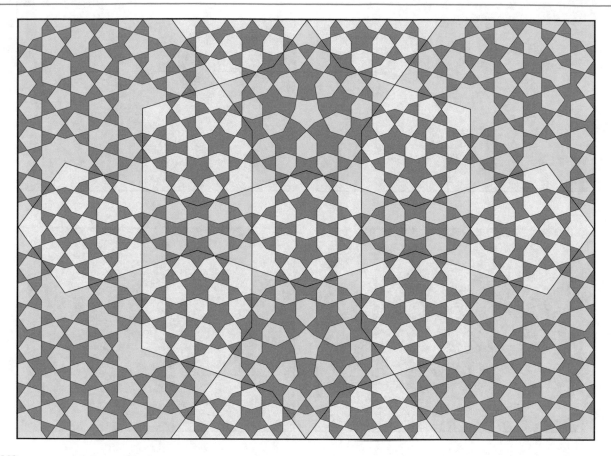

Fig. 449

These secondary decagons are scaled such that they fit in an edge-to-edge arrangement at the 72° and 108° angles of the primary pattern. The leftover regions are filled in Fig. 450c. The inherent *phi* proportions of the *fivefold system* insure that the scaled-down polygonal modules will seamlessly tessellate the remaining areas between the secondary decagons. The scale factor for this dual-level design is 1:5.2360.... and the relation to *phi* is $2 + [(1 + \sqrt{5} \div 2) \times 2]$, or more simply as $2 + [\varphi \times 2]$. As mentioned, the secondary pattern and secondary tessellation in each of the three shaded rectangles at the center of Fig. 450d are identical to the rectangular repeat, underlying tessellation and primary pattern in Fig. 450a (except for the addition of the kite elements within the ten-pointed stars).

Figure 451 represents a Qara Qoyunlu cut-tile mosaic type A dual-level design in one of the blind arches at the Imamzada Darb-i Imam in Isfahan (1453) [Photograph 97]. This design is repeated within an arched spandrel at the Imamzada Darb-i Imam, the only difference being the colorization, and slight variations within the secondary infill of the primary ten-pointed stars. The primary pattern in this example is the classic fivefold *obtuse* design [Fig. 229a], and the secondary pattern is also from the *obtuse* family. While both the primary and secondary patterns are created from the *fivefold system*, the primary pattern is not replicated within

the secondary pattern. For this reason, there is no scale invariance present in this example. Therefore, the more loosely defined self-similarity in this dual-level design is a product of its using polygonal modules from the same *fivefold system* at both scales, as well as the same pattern family at both scales. The secondary pattern of this historical example has some anomalous properties that warrant examination. Specifically, the secondary pattern contained with the pentagons of the primary pattern have neither reflected nor rotational symmetry. Rather, the polygonal infill of the pentagonal regions eschews these more conventional forms of symmetry in favor of a pattern that, at first glance, appears to have rotation symmetry, but breaks this around the periphery of the pentagonal infill. As pointed out by Peter Cromwell,[72] the secondary pattern within each of the primary pentagons has the exact same anomalous pattern, but with different rotational orientations within each respective pentagon. It has been suggested that the non-rotational symmetry of the secondary design contained within the primary pentagons was a mistake on the part of the artist, perhaps introduced

---

[72] In previous writings concerning this dual-level pattern [Bonner 2003] the author has misidentified the secondary pattern within the primary pentagons as having rotational symmetry. See: Cromwell (2016).

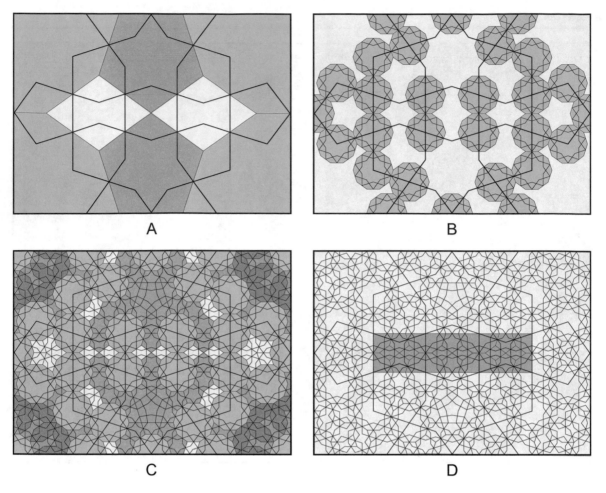

A

B

C

D

Fig. 450

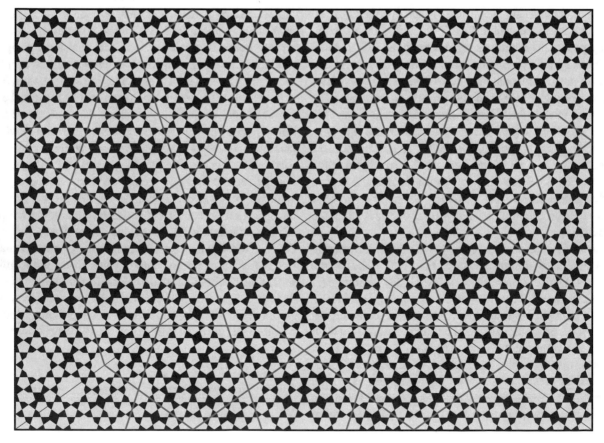

Fig. 451

while the original panel was being repaired.[73] However, the occurrence of this unusual feature within each of the primary pentagonal elements would appear to indicate a willful intent. This is further confirmed by the fact that the primary pentagonal regions of the dual-level design from the nearby arch spandrel at the Imamzada Darb-i Imam also employs the same unusual arrangement of secondary design elements within the primary pentagonal regions. To add to this refutation, both of these dual-level examples at the Imamzada Darb-i Imam are in excellent condition and show no signs of having been repaired. The anomalous secondary pattern treatment within the pentagonal regions interrupts the rotational symmetry that would otherwise be a standard feature of these two examples. Additionally, this anomalous treatment of the secondary pattern within the primary pentagonal regions interrupts the reflection symmetry that would otherwise be present, thereby changing the plane symmetry group for both these dual-level designs to *p1* rather than *pmm*. In 2007 Peter Lu and Paul Steinhardt claimed to have discovered "nearly perfect" quasicrystalline tilings in the ornamental mosaics at the Imamzada Darb-i Imam.[74] Despite the significant media attention such claims generated, there are serious problems with their arguments. Their claim that "the Darb-i Imam tessellation is not embedded in a periodic framework and can, in principle, be extended into an infinite quasiperiodic pattern" is contradicted by the fact that this dual-level design has very obvious translation symmetry. As indicated by the red diagonal lines in Fig. 451, this dual-level design repeats upon a rhombic grid, and it is immaterial that the secondary pattern is comparatively complex. Seeking to demonstrate quasicrystalline aperiodicity by only examining an isolated region of the secondary pattern, and comparing this to Penrose tilings with subdivision rules, ignores the fact that the isolated region is unquestionably part of a larger periodic structure.[75] Their article created considerable

debate over the merits of their claim of quasicrystallinity within the design at the Darb-i Imam. A persuasive counter argument has been made by Peter Cromwell who details the flaws in their claim to have identified Penrose subdivision rules within the example from the Darb-i Imam.[76] Lu and Steinhardt are correct in identifying the self-similar characteristics between the primary and secondary patterns, but they were not the first to discover recursive self-similarity within this tradition,[77] or even at the Imamzada Darb-i Imam.[78] Nor were they the first, as claimed, to recognize how the secondary pattern is the product of a set of underlying polygonal modules, or the correlation between the use of these modules in both this design from Isfahan and examples from the Topkapi Scroll.[79] Similarly, these authors are correct in identifying the potential of the *fivefold system* (although they do not use this prior terminology) for making Islamic geometric patterns that are true quasiperiodic structures devoid of translation symmetry. Here again, they were not the first to identify this remarkable capability of the *fivefold system*.[80]

Figure 452a shows the classic fivefold *obtuse* design along with its underlying generative tessellation. Figure 452b applies two secondary underlying tessellations to the primary pattern in Fig. 452a. As is common within the *fivefold system* [Fig. 200], these have a dual relationship and either can be used to create the secondary *obtuse* pattern from this historical example. Figure 452c demonstrates the method of determining the size of the scaled-down secondary polygonal modules: edge-to-edge decagons are placed such that their centers rest upon the closest interval in the primary design. Decagons are then placed at each intersection of the primary design, and these are connected with concave hexagons along the relevant primary pattern lines.

[73] Lu and Steinhardt (2007a).

[74] "the tessellation approach was combined with self-similar transformations to construct nearly perfect quasi-crystalline Penrose patterns, five centuries before their discovery in the West." Lu and Steinhardt (2007a). Note: the crystallographer Emil Makovicky and the artist and mathematician Jean-Marc Castéra made prior claims to the discovery of quasicrystallinity among examples of Islamic geometric design.
– Makovicky (1992).
– Castéra (1999a).

[75] Lu and Steinhardt are mistaken in their attribution of aperiodicity to the arch spandrel pattern at the Darb-i Imam ("the Darb-i Imam shrine stands out as the only example found thus far that is not part of a periodic pattern and that instead displays a self-similar subdivision into smaller tiles that can be continued ad infinitum to obtain an infinite perfectly quasi-crystalline pattern" Lu and Steinhardt (2007b)). As demonstrated in Fig. 451, this example has clearly evident translation symmetry. Further, Lu and Steinhardt are critical of Makovicky for using an isolated region of a much larger periodic structure in Makovicky's arguing for the quasi-crystallinity of the fivefold pattern

from the exterior façade of the Gunbad-i Kabud in Maragha. Yet by not recognizing the overall periodic structure of the pattern from the Darb-i Imam and only considering an isolated region within its overall repetitive cell, Lu and Steinhardt inadvertently avail themselves to the very argument they make against Makovicky. Ironically, this is despite their explicitly stating "The identification of tilings as Penrose or quasi-crystalline (or periodic for that matter) must, by definition, be based on the symmetries and properties of the overall pattern, not just isolated fragments" Lu and Steinhardt (2007b).
– Makovicky (1992).
– Lu and Steinhardt (2007a), 1108.
– Lu and Steinhardt (2007b).

[76] Cromwell (2015).

[77] – Makovicky (1992).
– Castéra (1996).
– Bonner (2003).

[78] Bonner (2003).

[79] – Bonner (2003).
– Saltzman (2008).

[80] –Makovicky (1992).
– Bonner (2003), 11–12: with recursive non-periodic fivefold examples provided in the illustrations.

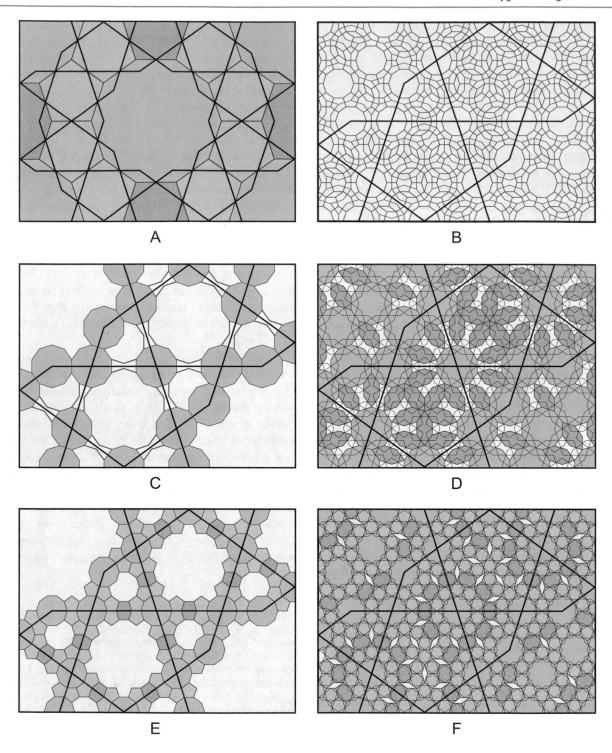

A

B

C

D

E

F

Fig. 452

Figure 452d completes the underlying tessellation by applying further decagons, concave hexagons and long hexagons into the unfilled background regions from Fig. 452c. The initial layout of the alternative underlying secondary tessellation that produces the secondary pattern is demonstrated in Fig. 452e. This places decagons that are separated by two contiguous pentagons at the shortest interval of the primary design. The continued population of the primary pattern lines with scaled-down polygonal modules simply copies the secondary decagons and pentagons to each vertex of the primary pattern, and the gaps are conveniently filled with barrel hexagons from the *fivefold system*. Figure 452f fills the undeveloped areas of Fig. 452e with additional pentagons, barrel hexagons and thin rhombi of the *fivefold*

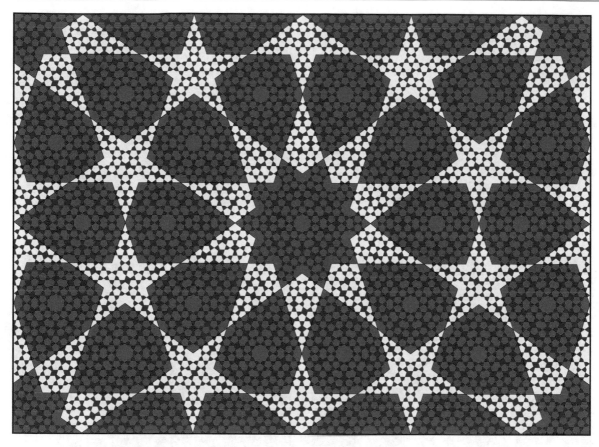

Fig. 453

system. The scale factor for this dual-level design is 1:8.4721... which can be expressed as $2 + [\varphi \times 4]$, or as $4 + [\sqrt{5} \times 2]$.

Figure 453 shows a Safavid cut-tile mosaic type A dual-level design from the Madar-i Shah in Isfahan (1706-14) that is also created from the *fivefold system*. The primary pattern is the classic *acute* design found throughout the Islamic world, and the secondary level is an *obtuse* pattern. This is a looser form of self-similarity that applies only to the design methodology wherein the same underlying polygonal modules from the *fivefold system* are employed recursively at differing scales, but does not apply to the visual result wherein two distinct pattern families are used at different scales. Needless to say, the use of two different pattern families at both levels precludes the possibility of scale invariance within the design. Figure 454a shows the classic fivefold *acute* design along with its underlying generative tessellation. Figure 454b shows two secondary grids with dual characteristics, either of which can be used to create the secondary *obtuse* pattern from this dual-level design. Figure 454c illustrates the applied secondary decagons, concave hexagons, and long hexagonal modules to the intersections of the primary design, as well as to the vertices of the underlying tessellation for the primary design. The scale of

the secondary modules is determined by making the centers of two edge-to-edge decagons equal the shortest distance between intersections in the primary design. Figure 454d shows the primary and secondary designs along with the primary and secondary underlying tessellation. Figure 454e shows the alternative underlying tessellation for producing the secondary *obtuse* pattern. This is comprised of decagons, pentagons, barrel hexagons, and thin rhombi. The decagons are likewise placed at the intersections of the primary design as well as the vertices of the primary underlying tessellation. The scale of the secondary modules is determined by a configuration of two decagons separated by two pentagons placed at the shortest interval of the primary design. Figure 454f illustrates the completed design along with the primary and secondary underlying tessellations. The scale factor between the two levels is 1:13.7082... which can be expressed as $4 + [\varphi \times 6]$, or as $7 + [\sqrt{5} \times 3]$.

3.3.2 Historical Examples of Type B Dual-Level Designs

Figure 455a illustrates a Janid type B dual-level design from the Nadir Divan Beg *madrasa* in Bukhara (1622-23)

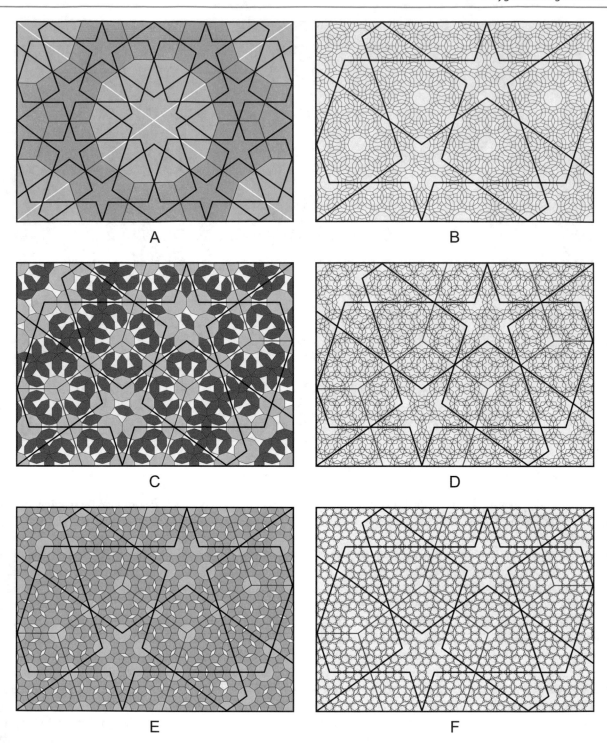

A

B

C

D

E

F

Fig. 454

[Photograph 100]. As stated, this variety of dual-level design widens the primary pattern lines and fills this widened region with the secondary design. Figure 455b illustrates the simplicity of the primary design: a linear band of hexagons and triangles, with the widened lines determined by the governing isometric grid. The secondary design is from the *obtuse* family, and is comprised of six- and nine-pointed stars, with the six-pointed stars placed at the vertices of the

governing isometric grid. As shown, the design and width of the widened lines of the primary pattern are easily derived from the isometric grid, but can also be identified as the 3.6.3.6 semi-regular tessellation [Fig. 89]. The secondary pattern is nonsystematic, and works within this structure by virtue of the triangular repetitive cells. Figure 456a shows the secondary underlying tessellation of nonagons surrounded by pentagons, with six-pointed star interstice

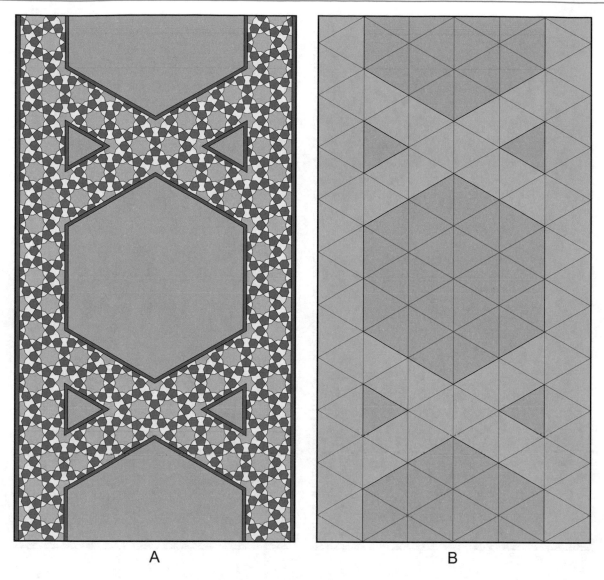

A   B

Fig. 455

regions at each vertex of the repetitive isometric grid. Figure 456b shows the secondary *obtuse* pattern created from the underlying tessellation. The pattern lines within each underlying six-pointed star interstice region are an arbitrary treatment that is not determined by the underlying tessellation. The secondary design on its own is essentially identical to several single-level examples from the same region, including an earlier Shaybanid pattern at the Kukeltash *madrasa* in Bukhara (1568-69) that may have served as an inspiration for this dual design [Fig. 313a].

Figure 457a represents one of the two very similar Timurid type B dual-level designs from the Friday Mosque at Varzaneh near Isfahan (1442-44). As per the previous example, Figure 457b demonstrates how the primary widened lines can be easily derived from the isometric grid. The primary design is one of the most basic threefold patterns,

and can be easily created from the isometric grid (as shown), or from an underlying tessellation of just hexagons [Fig. 95b]. Figure 457b illustrates the underlying tessellation that creates the secondary pattern. This is comprised of dodecagons and triangles with the dodecagons placed at the vertices of the isometric grid. This secondary pattern was frequently used as a single level design [Fig. 108a]. The use of the isometric grid as an underlying structure creates, by necessity, different thicknesses in the outer vertical and horizontal borders.

Figure 458a illustrates an exceptional type B dual-level design from the Topkapi Scroll.[81] While the background regions conform to a 3.6.3.6 arrangement of hexagons and

---

[81] Necipoğlu (1995), diagram no. 38.

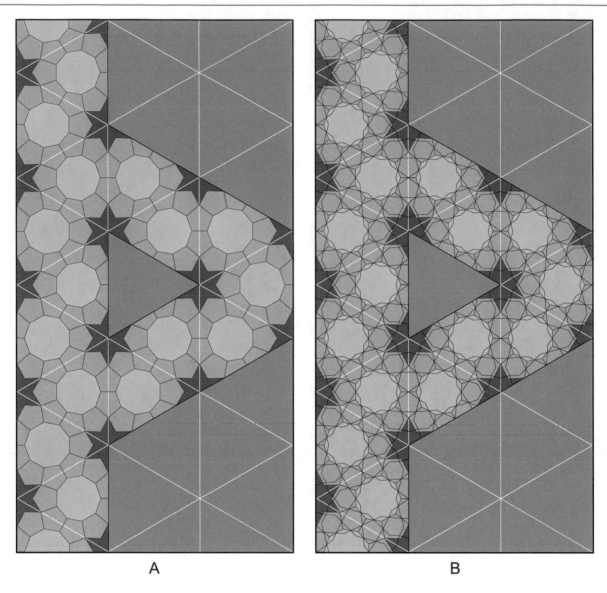

A   B

Fig. 456

triangles [Fig. 89], the widened lines of the primary design actually conform to the arrangement of hexagons, squares, and triangles in the $3^2.4.3.4$-$3.4.6.4$ *two-uniform* tessellation [Fig. 90]. By selecting the triangles and squares to create the widened line, the artist was able to incorporate a secondary pattern that has an identical edge configuration in the underlying tessellations for these two repetitive elements. This was a very clever contrivance that more than makes up for the noncollinearity of the widened pattern lines associated with the neighboring primary hexagonal elements. Figure 458b illustrates the application of the underlying tessellation and the resulting *acute* pattern within the square and triangular repetitive cells. These place the dodecagons at each repetitive vertex and separate each with a barrel hexagon. The background regions in the primary design could have also been filled with the triangular repeat, but keeping these

areas open creates the dual-level dynamic. The combined use of the square and triangular repetitive cells qualifies the secondary pattern as a hybrid design, and but for the arbitrary treatment of the pattern lines at the center of the triangular repeat unit, the hybrid use of these two repetitive cells is provided a second representation within the Topkapi Scroll, as a design suitable for an arch tympanum rather than as a dual-level design [Figs. 23d–f].[82] Figure 459a shows the application of the secondary underlying tessellation to the squares and triangles of the primary design. Of particular interest is the seamless incorporation of the rectilinear border in what is otherwise a $3^2.4.3.4$-$3.4.6.4$ tessellation. Figure 459b applies the secondary *median* pattern lines to

---

[82] Necipoğlu (1995), diagram no. 35.

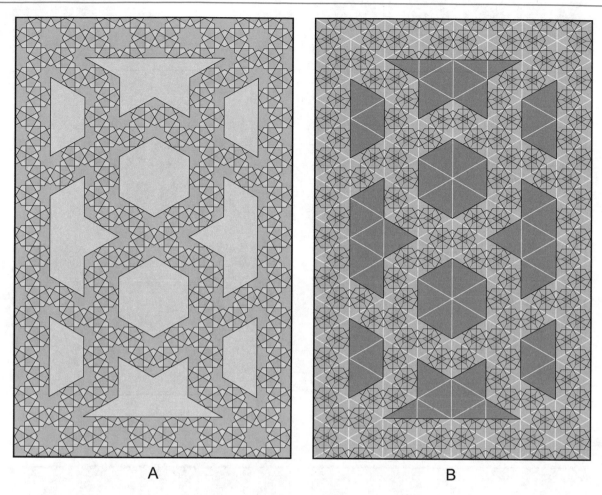

A   B

Fig. 457

the underlying tessellation of Fig. 459a. Once again, the secondary pattern in this dual-level design is nonsystematic and works by virtue of the hybrid use of the square and triangular repetitive cells.

Figure 460a illustrates a Timurid type B dual-level design from the Gawhar Shad mausoleum in Mashhad (1416-18). Both the primary and secondary levels are *median* designs produced from the *fourfold system A*. Figure 460b shows the underlying tessellation that produces the secondary design along with the applied pattern lines of the *median* family. Much of the secondary design is the classic star and cross pattern created from the semi-regular 4.8^2 tessellation of squares and octagons [Fig. 124b]. Figure 461a shows the primary *median* pattern and its underlying generative tessellation. This is a field pattern comprised of superimposed octagons, and is one of several historical designs created from this simple underlying tessellation of large hexagons and squares [Fig. 138c]. Figure 461b demonstrates a method for widening the primary pattern lines that gives the proportions used in this historical example. Each edge of the superimposed octagons has eight applied squares, with the centers of the outer two squares placed upon the corners

of the octagon. Figure 461c shows the widened line version of the primary design. Figure 461d places the secondary polygonal modules, along with their associated *median* pattern lines into the widened primary pattern. This pattern is only self-similar in its use of modules from the *fourfold system A* at two scales, and its employment of the *median* pattern family at both levels. However, the primary pattern has no eight-pointed stars, whereas the secondary pattern does; and the secondary pattern is not a widened line pattern. Therefore, this example is only loosely self-similar. The proportional scale between the two levels is 1:10.2426... which can be expressed as $6 + [\sqrt{2} \times 3]$.

Figure 462a represents an Aq Qoyonlu[83] cut-tile mosaic panel from the Friday Mosque at Isfahan (1475) [Photograph 99]. Figure 462b demonstrates how the widened primary design of this type B dual-level example is a simple assembly of squares and triangles that produces octagons, four-pointed stars, and concave octagons as background

---

[83] This may have been produced during the sixteenth century under Safavid patronage. See Necipoğlu (1995), 37.

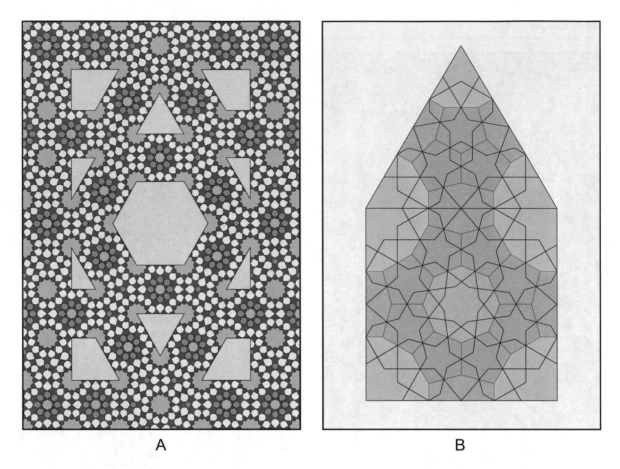

A   B

Fig. 458

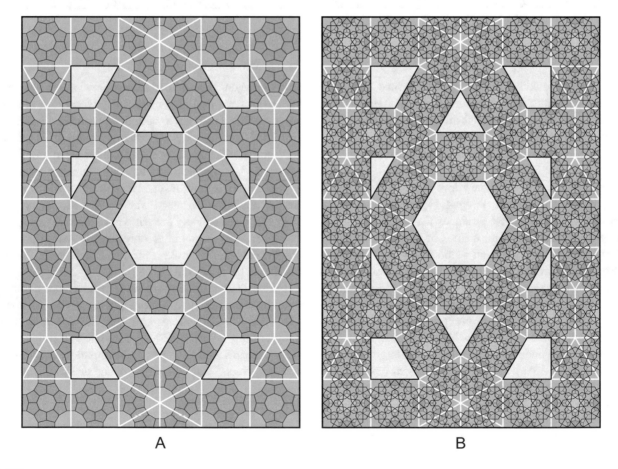

A   B

Fig. 459

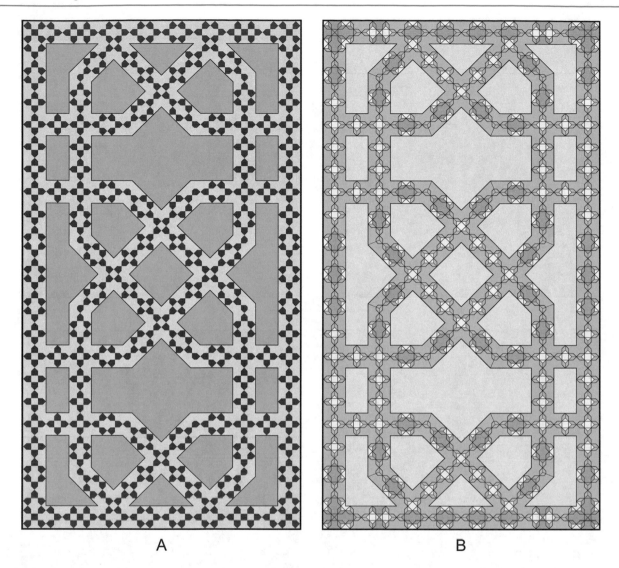

A   B

Fig. 460

elements. Each triangle is a 1/8 segment of an octagon. Figure 462c details the application of the secondary polygonal modules to the primary squares and triangles, along with their associated *median* pattern lines. These modules are from the *fourfold system A*. The hybrid use of two distinct repetitive cells is similar in concept to the example in Figure 458. On their own, the square cells produce the ubiquitous star and cross pattern, while the triangular cells produce a very pleasing design that was used previously in several locations, including: the Haund Hatun complex in Kayseri (1237-38) [Fig. 157b]; the mausoleum of Uljaytu in Sultaniya, Iran (1305-1313) [Fig. 66]; and at the Bibi Khanum in Samarkand, Uzbekistan (1398-1404). Figure 462d shows the application of the secondary underlying polygonal modules and associated pattern lines to the complete panel.

Figure 463 represents a Qara Qoyunlu widened line dual-level design from the Imamzada Darb-i Imam in Isfahan

(1453-54) [Photograph 98]. This design is created from the *fivefold system*, with the primary pattern being an unusual hybrid of *acute* and *median* widened pattern lines, and the secondary pattern being from the *obtuse* family. This is one of the most remarkable architectural examples of type B pattern making, and is the work of Sayyid Mahmud-i Naqash. (Note: The outer border is not represented.) Figure 464a illustrates the underlying tessellation along with the applied pattern lines for this example from the Darb-i Imam. The applied pattern lines are not collinear where they cross the midpoints of the underlying decagons. It is at these locations that the pattern lines change from *acute* to *median*. This is an unusual feature that ordinarily creates discontinuity within a given design, and only occasionally found within this ornamental tradition. In this widened line dual-level example, the noncollinearity is both eccentric and visually pleasing. Figure 464b shows how the proportions of the

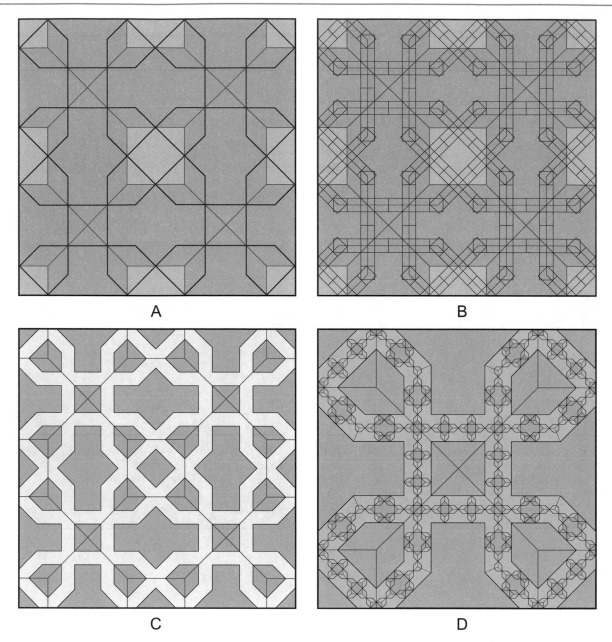

Fig. 461

widened line are determined by the wide rhombus with 72°
and 108° included angles. Figure 464c illustrates the application of the secondary underlying polygonal modules to the
three regions that make up the widened line configuration in
Fig. 464b. This also shows the secondary underlying
polygons with the associated *obtuse* pattern lines. Figure
464d applies the three regions with secondary polygons
and applied pattern lines to the widened lines of the primary
pattern. The proportional scale factor between the primary
and secondary levels is 1:14.3261… which can be expressed
as 3 + [φ × 7], or as 4 + [φ × 5] + √5.

Figure 465 illustrates another type B dual-level design
from the Topkapi Scroll.[84] This example uses the *acute*
family in the primary level and the *obtuse* family within
the widened lines of the secondary level. The patterns within
both levels are produced from the *fivefold system*. Although
this does not appear to have been used architecturally, this is
arguably the most complex and successful type B dual-level
design from the historical record. Figure 466a illustrates the
primary pattern with its underlying generative tessellation.

---

[84] Necipoğlu (1995), diagram no. 49.

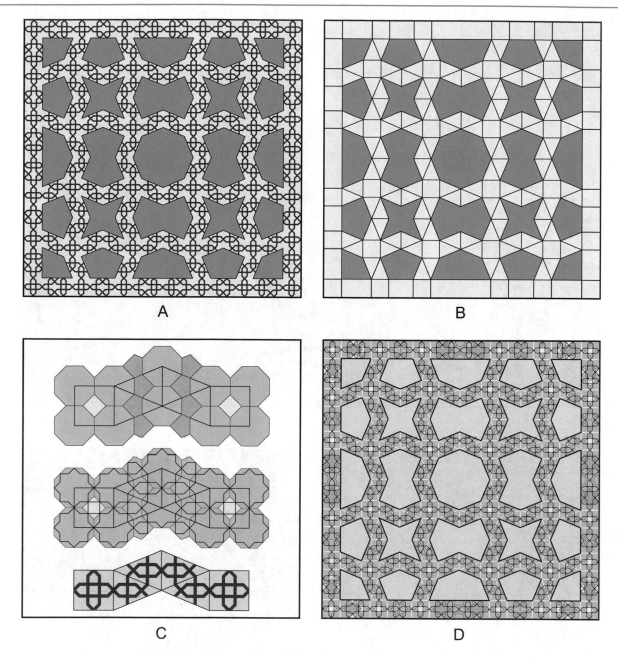

Fig. 462

Figure 466b shows how the proportions within the widened lines of the primary pattern are determined by a series of rhombi, triangles, trapezoids and pentagon that all relate to the *fivefold system*. Figure 466c places the underlying polygonal modules onto the widened line segments, as well as shows the polygons with their associated *obtuse* pattern lines. Figure 466d shows the widened line primary pattern with the secondary generative polygons and associated pattern lines. As with many fivefold patterns, the *median* secondary pattern can also be constructed from an alternative grid with dual characteristics (not shown). The scale factor for this remarkable dual-level design is 1:17.9442... or as an expression of *phi* as 5 + [$\varphi \times 8$], or as 9 + [$\sqrt{5} \times 4$].

3.3.3 Historical Examples of Type C Dual-Level Designs

Like type B dual-level designs, the type C design methodology involves the widening of the basic primary pattern, but extends the secondary pattern beyond the region of the widened lines so that it fills the entire plane. Differentiation between the primary and secondary patterns is provided in two ways: by emphasizing the widened lines themselves, as well as through color contrast between the region of the widened lines and its background. This is the least common variety of dual-level design with only a handful of examples known to the historical record. Figure 467a illustrates a

Fig. 463

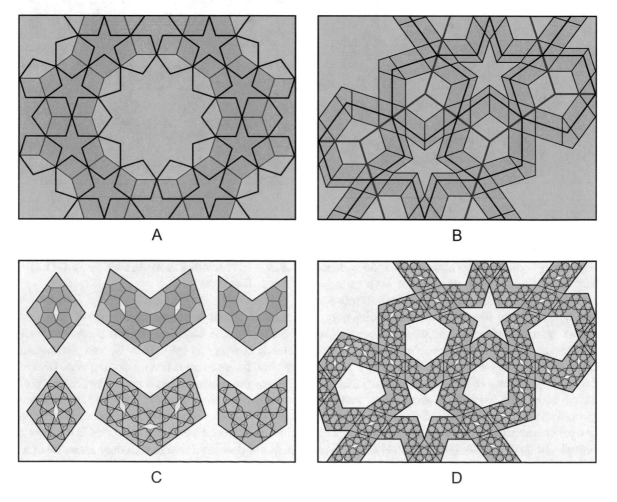

A

B

C

D

Fig. 464

Fig. 465

Shaybanid type C dual-level design from a wooden ceiling at the Khwaja Zayn al-Din mosque and *khanqah* in Bukhara (c. 1500-1550). The widened line primary design is the classic threefold *median* pattern with six-pointed stars used repeatedly throughout the Islamic world [Fig. 95b]. Figure 467b shows how the proportions of the widened line are simply derived from the isometric grid (red). The secondary pattern repeats upon the isometric grid, and the differentiation between the primary and secondary levels in the wooden ceiling is achieved through relief rather than color. The secondary pattern places six-pointed stars at the vertices of the isometric grid. These are separated by hexagons located at the center of each triangular cell. Were it not for the inclusion of the widened lines within the overall pattern matrix the secondary pattern would function independently as ornamental surface coverage. This is a typical feature of type C dual-level designs.

Figure 468 represents one of the more complex historical type C dual-level designs. This Safavid example is from the Madar-i Shah in Isfahan (1706-14). Both levels are *obtuse* patterns created from the *fivefold system*, providing this design with a higher level of self-similarity than many historical dual-level designs. Figure 469a illustrates the underlying tessellation of edge-to-edge decagons and concave hexagonal interstice regions that is one of the most common generative tessellations within the *fivefold system*. The *obtuse* primary pattern lines are also represented. Figure 469b widens the *obtuse* primary pattern lines by using rhombi and triangles associated with fivefold proportions. This method of widening is very similar to that shown in Fig. 466b, although in this example the widening is applied to an

obtuse rather than an *acute* pattern. Figure 469c places the secondary decagons onto the vertices of the widened primary pattern. The scale of the decagons is determined by applying the same arrangement of decagons and concave hexagons to the rhombic regions as is used in the primary underlying tessellation wherein the four decagons surround a concave hexagon. This feature, and the fact that both the primary and secondary patterns are of the same pattern family, provides scale invariance between the small rhombic regions of the secondary pattern and rhombic repeat of the primary design. Figure 469c also shows the full infill of the secondary polygonal modules. The additional decagons within the primary ten-pointed stars are located at the vertices of the array of rhombi located within each of the primary ten-pointed stars in Fig. 469b. Figure 469d applies the associated *obtuse* pattern lines to the secondary underlying polygons. It is worth noting that both the primary and secondary *obtuse* patterns can be derived with equal ease from an alternative underlying tessellation with dual characteristics comprised of decagons, pentagons, barrel hexagons and thin rhombi [Fig. 200]. The scale factor between the primary and secondary levels is 1:15.3262... or as an expression of *phi* as $4 + [\varphi \times 7]$, or as $6 + [\varphi \times 3] + [\sqrt{5} \times 2]$.

3.3.4 Historical Examples of Type D Dual-Level Designs

Type D dual-level designs are similar to the type C variety in that the secondary pattern has complete surface coverage

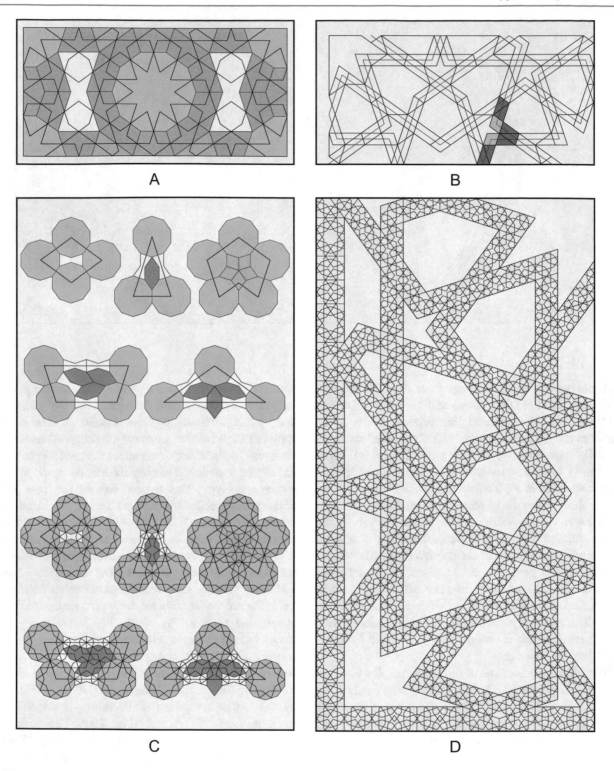

Fig. 466

and the primary pattern is expressed as a widened line. However, the visual characteristics of these two types of dual-level design are very distinct from one another. Type D designs rely upon the stars of the secondary pattern having parallel lines that have a collinear orientation with those of

neighboring stars. This creates a channel of secondary background elements that are provided their own color, thereby differentiating the widened line from the rest of the design. Without this color, the design would appear as a standard, albeit rather complex, geometric design. Figure 470a

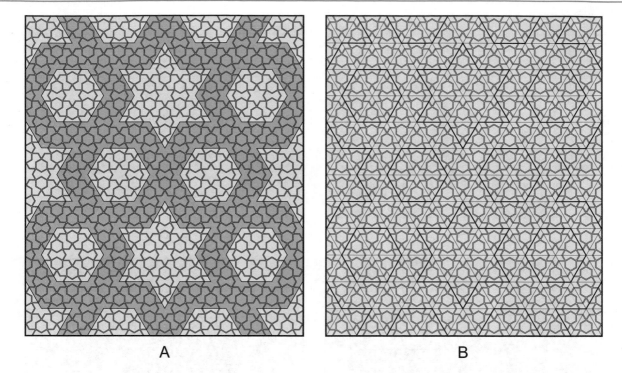

Fig. 467

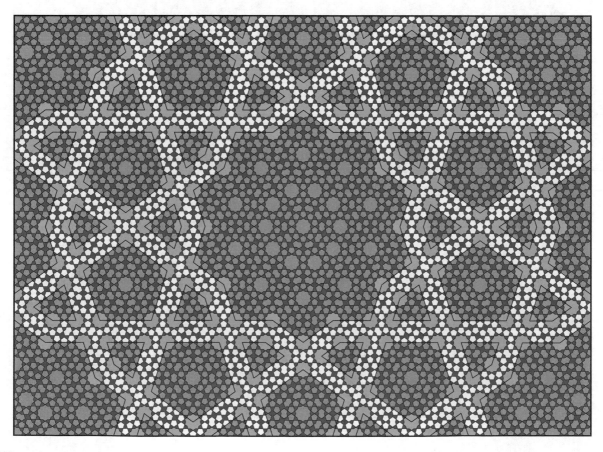

Fig. 468

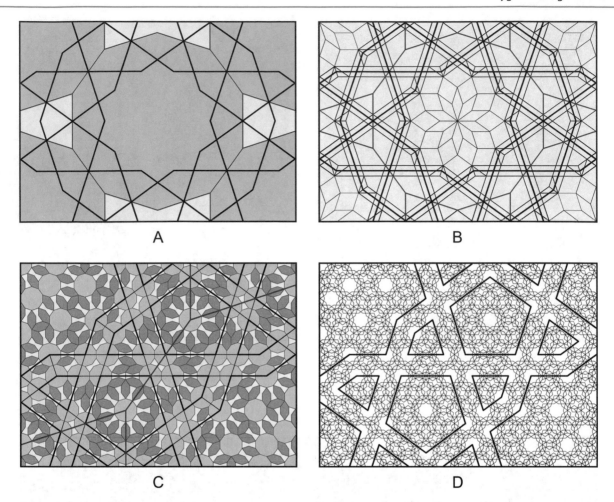

A

B

C

D

Fig. 469

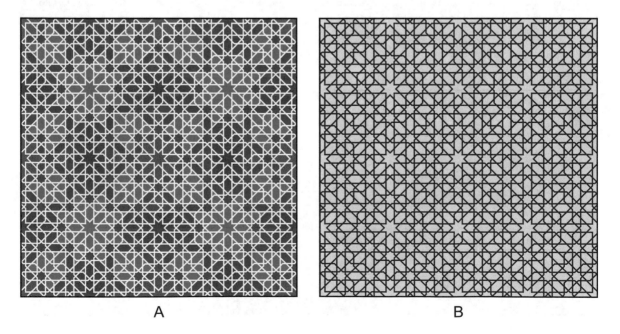

A

B

Fig. 470

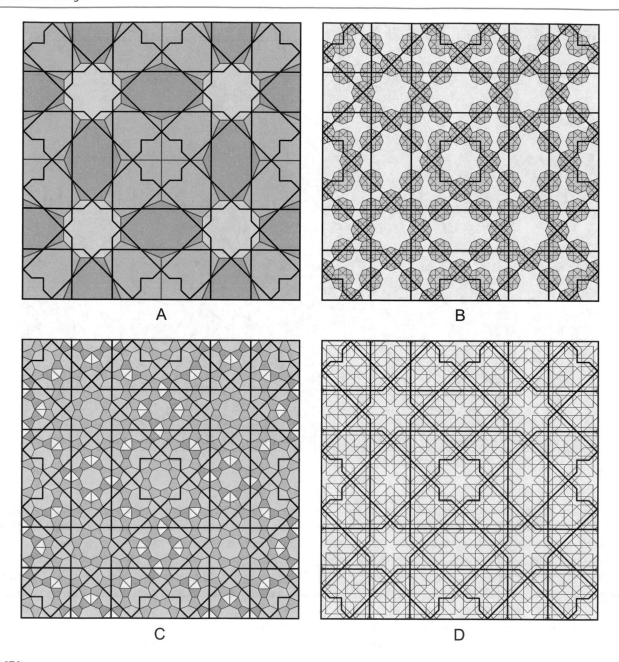

Fig. 471

represents a *Mudéjar* type D dual-level design from the Alcazar in Seville (1364-66) [Photograph 101]. Like the majority of dual-level designs from the Maghreb, both the primary and secondary patterns are associated with the *four-fold system A*. To demonstrate the reliance upon color to differentiate the primary pattern in this variety of dual-level design, the geometric pattern in Fig. 470b is the secondary design alone. As is visually apparent, it is next to impossible to ascertain the primary design within this overall pattern matrix. Figure 471a illustrates the underlying tessellation along with the associated pattern lines of the *obtuse* family. Figure 471b places secondary octagons at the intersections of the primary pattern. These are scaled such that a ring of

eight edge-to-edge octagons fits onto each primary eight-pointed star. This illustration demonstrates how the lines of each small secondary eight-pointed star are parallel to their immediately adjacent lines of the primary pattern. Figure 471c fills in the background regions from Fig. 471b with additional polygonal modules from the *fourfold system A*. Figure 471d shows the primary pattern as single lines, more or less in the style of type A dual-level designs. The proportional scale factor between the primary and secondary levels is 1:4.8284... which can also be expressed as $2 + (\sqrt{2} \times 2)$.

Figure 472 represents a Nasrid *zillij* mosaic dual-level panel from the Alhambra [Photograph 102]. The primary

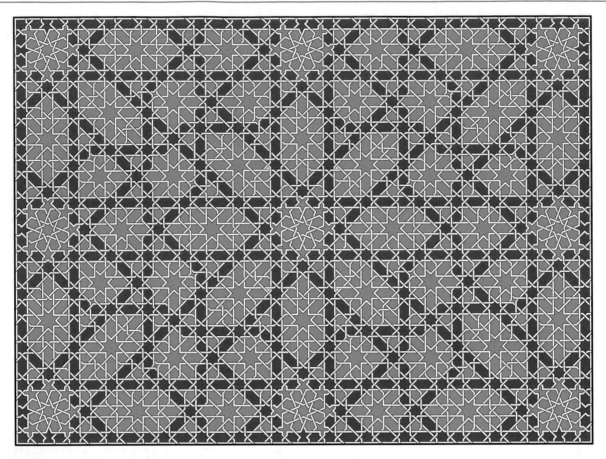

Fig. 472

and secondary patterns are also created from the *fourfold system A*. Figure 473a shows the derivation of the primary *acute* pattern from an underlying tessellation of octagons, pentagons and triangles. This pattern repeats upon a rectangular grid. Figure 473b places octagons at the intersections of the primary pattern. As with the example from the Alcazar in Fig. 470, the size of the octagons is determined by their eightfold edge-to-edge placement around the primary eight-pointed stars. The associated pattern lines are also from the *acute* family, giving this dual-level design a strong self-similar characteristic. Figure 473c places further octagons (blue) at key locations within the primary pattern. The locations of this new set of octagons are determined by the underlying tessellation (red lines) of the primary design. Figure 473d fills in the remaining background with additional polygonal modules from the *fourfold system A*. And like the example from the Alcazar, the proportional scale factor between the primary and secondary levels is $2 + (\sqrt{2} \times 2)$, or 1:4.82842712.

Figure 474 represents a Marinid type D dual-level design from the Bu'Inaniyya *madrasa* in Fez (1350-55). This is one of the few dual-level designs from the Maghreb that is created from the *fivefold system*. Rather than a conventional

geometric pattern, the primary design is a polygonal tessellation of decagons that are placed edge to edge in the horizontal orientation, and corner to corner in the vertical orientation, and separated by eight-sided interstice regions. The secondary pattern is from the *acute* family, and the discrepancy between the visual quality and methodological origin between the primary and secondary patterns means that this dual-level design does not possess self-similarity. It is nevertheless a very appealing design. The arrangement of decagons in the primary pattern produces an overall repeat unit that is rectangular. Figure 475a illustrates the aforementioned arrangements of decagons touching edge-to-edge horizontally, and corner-to-corner vertically. Figure 475b subdivides this decagonal arrangement into smaller polygonal components. Figure 475c applies scaled-down polygonal modules from the *fivefold system*, along with their associated *acute* pattern lines, to the four varieties of polygonal component from Fig. 475b. Each of these four polygonal components can be used to created patterns on their own, and qualifies this secondary pattern as a hybrid design, much like the group of repetitive fivefold elements in several Anatolian Seljuk hybrid patterns [Figs. 263–265]. Figure 475d applies the secondary elements and associated pattern

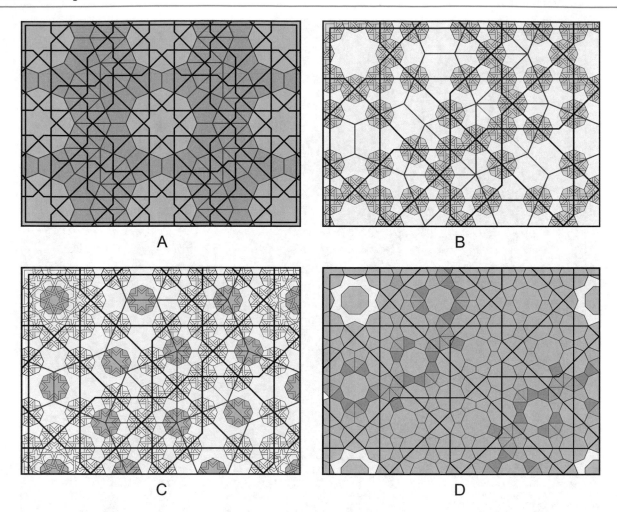

A

B

C

D

Fig. 473

lines from Fig. 475c to the overall structure in Fig. 475b. The scale factor between the primary decagons in Fig. 475a and the secondary underlying decagons in Fig. 475d (blue) is 1: 4.2360... or as $1 + [\varphi \times 2]$, or as $2 + \sqrt{5}$.

Figure 476 shows another Marinid design that makes use of the same arrangement of decagons as the example in Fig. 474. This design is from the al-'Attarin *madrasa* in Fez (1323) [Photograph 103], and, like the previous example, the secondary level is an *acute* pattern created from the *fivefold system*. The obvious difference between these two Marinid dual-level designs is the level of complexity of the secondary pattern. As with the dual-level design at the Darb-i Imam [Figs. 451 and 452], and the highly complex pattern that shrouds the Gunbad-i Qabud in Maragha [Figs. 239 and 240], it has been argued that this dual-level design from the al-'Attarin *madrasa* is quasicrystalline.[85] While the isolated region of local tenfold symmetry contained within each primary decagon undoubtedly has shared characteristics

with decagonal quasicrystals, the fact remains that this decagonal region is only a subset of design elements that exists as part of a larger pattern matrix that employs a rectangular repeat with translation symmetry—the antithesis of quasicrystallinity. Figure 477a illustrates the origin of the ten-pointed stars at the centers of each primary decagon. Figure 477b divides the structure into smaller repetitive components that are essentially the same as the previous example from the Bu'Inaniyya *madrasa*. Figure 477c places modules from the *fivefold system* into the construction from Fig. 477b. The scale of these secondary modules is determined by an arrangement of three linear decagons separated by two sets of mirrored pentagons, with the distance between the centers of the outer two decagons equaling the edge length of the primary decagons. Figure 477d places *acute* pattern lines into the secondary polygonal modules. The similarities between the previous example from the Bu'Inaniyya *madrasa* and this example from the al-'Attarin *madrasa*, and their closeness in proximity and date suggest the strong possibility that they were designed by the same individual or are the product of the same artistic lineage. The

---

[85] Ajlouni (2012).

Fig. 474

scale factor between the primary decagons in Fig. 477a and the secondary underlying decagons in Fig. 477c (blue) is precisely double that of the previous example, which is to say: 1:8.4721… or as 2 + [φ × 4], or as 4 + [$\sqrt{5}$ × 2].

3.3.5 Potential for New Multilevel Designs

In both the east and west, the number of overall examples of dual-level designs is relatively small. This is somewhat surprising considering their visual and intellectual distinction. This relative rarity is presumably due to the highly specialized design methodology required of this tradition,

and a consequent paucity of specialists familiar with this dual-level discipline. It is certainly not the case that this rarity is in any way the result of an exhaustion of the creative potential that this methodology offers. On the contrary, multilevel design methodology offers contemporary artists an unlimited capacity for highly innovative original designs that expand upon the remarkable examples of the past. As detailed above, historical dual-level designs were of just four varieties, and utilized just four of the design systems: the *system of regular polygons*, the *fourfold system A*, the *fourfold system B*, and the *fivefold system*. In addition, all historical examples repeat with conventional translation symmetry. Even constraining oneself to these features,

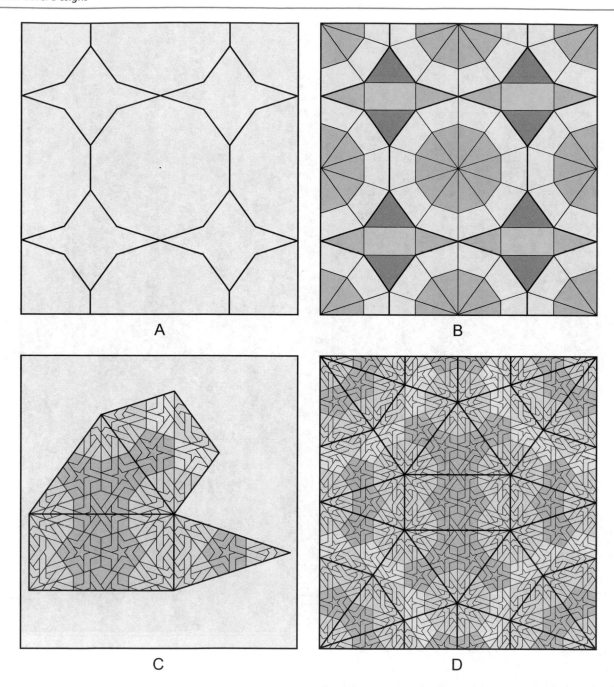

Fig. 475

there is limitless potential for the creation of designs with great beauty and originality. However, contemporary explorations in multilevel design allow for significant innovation beyond these methodological constraints, including: new varieties of design beyond the types A, B, C, and D; examples that employ more than just two recursive levels of design; multilevel designs that employ the *sevenfold system*; self-similar designs with fully realized scale invariance; and aperiodic multilevel designs that are truly quasicrystalline. Undoubtedly, this is an area of design that has extraordinary potential to contemporary geometric artists.

As shown, the primary patterns in the two historical fivefold type D examples from Fez are a grid of decagons rather than a conventional geometric pattern. By contrast, as per Figs. 470 and 472, artists in the Maghreb typically used conventional geometric patterns for the primary pattern in their fourfold type D dual-level constructions. Although the *fivefold system* was not used in this way historically, it will make very acceptable type D dual-level designs. The design in Fig. 478 (by author) uses the classic fivefold *obtuse* pattern for the primary level, and an *acute* pattern for the secondary level. Figure 479a shows how the secondary

Fig. 476

pattern is created from the placement of underlying generative decagons at the intersections in the primary pattern
lines, as well as at the center of the primary ten-pointed
star. The size of these secondary decagons is determined
by an arrangement of two decagons separated by mirrored
pentagons, with the distance between the centers of the
decagons being equal to the edge of the pentagons within
the primary pattern. Figure 479b adds the secondary *acute*
pattern lines to this underlying secondary tessellation. The
proportional scale between the two levels is 1:5.2360... and
as a function of *phi* can be expressed as $2 + [\varphi \times 2]$, or as
$3 + \sqrt{5}$. This is the same scale factor as one of the dual-level
designs in the Topkapi Scroll [Fig. 449].

As mentioned, though no historical examples are known,
the modules that comprise the *fivefold system* can be used to
create patterns with five- and tenfold local symmetry that
fulfill the modern mathematical criteria for aperiodic
quasicrystallinity with scale-invariant self-similarity. Several contemporary artists and designers working with
Islamic geometric design, including the author, are exploring the application of this ancient design discipline to these
areas of mathematical inquiry.[86] While such innovation is

---

[86] Of particular note are the original recursive designs of Jean-Marc
Castéra and Marc Pelletier.

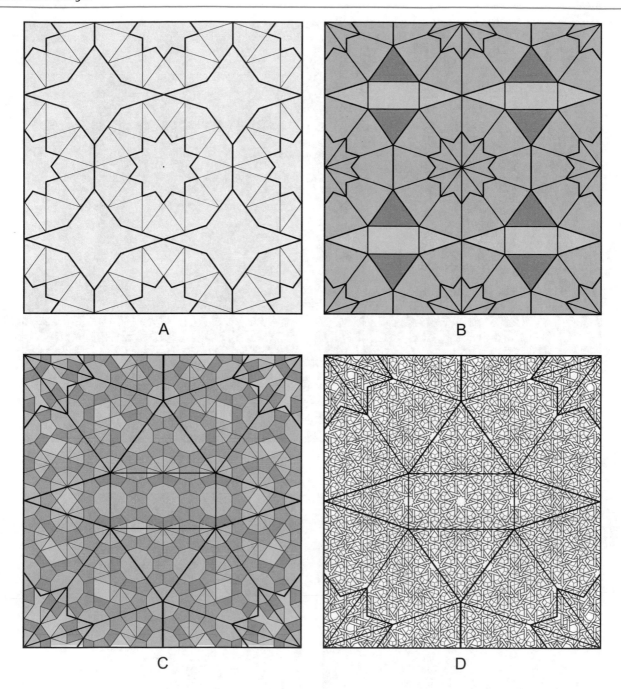

Fig. 477

beyond the scope of this work, the small selection of examples that follow reveal the remarkable similitude between these areas of modern mathematical discovery and historical design systems developed by Muslim artists a thousand years previously.

The *median* pattern in Fig. 480 (by author) has no translation symmetry. This is an aperiodic design that employs the 2 fivefold rhombi with the edge matching rules discovered by Sir Roger Penrose in the 1970s: one with 36° and 144° included angles, and the other with 72° and 108°

included angles. In this instance, the edge matching rules are manifest in the geometric pattern matrix that is applied to each variety of rhombus. As demonstrated previously, Muslim artists used both of these rhombi historically as repeat units. However, by synchronizing the pattern lines upon the relevant edges of each rhombus so that they conform to Penrose matching rules, the fivefold designs created from tessellating with these two rhombi have forced aperiodicity. While this example is aperiodic, it only has a single level of design. However, just as the two Penrose rhombi are well

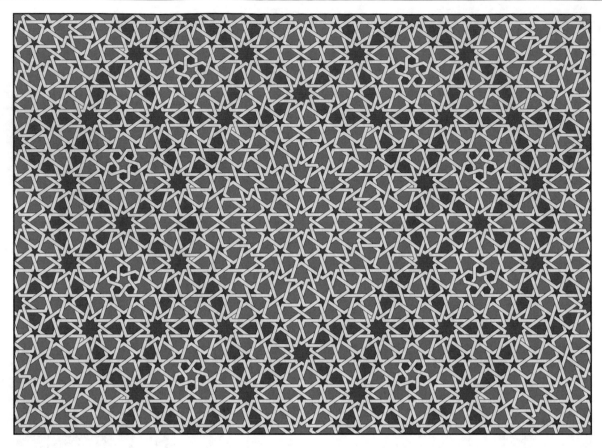

Fig. 478

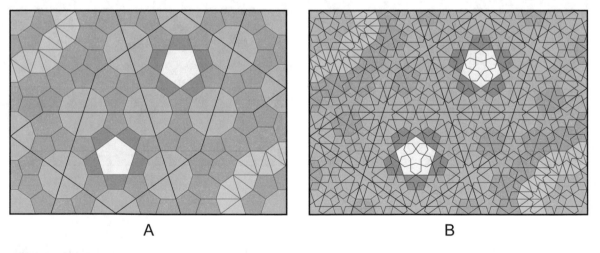

A   B

Fig. 479

known for their ability to infinitely inflate and deflate, so also can the geometric patterns that are applied to these rhombi be provided with inflation and deflation. Such designs have both scale-invariant self-similarity and true quasicrystallinity. Figure 481 shows an example of a three-level quasicrystalline design (by author) created from the *fivefold system*. As with the many historical dual-level designs created from the *fivefold system*, there are multiple regions with

tenfold local rotation symmetry in each level. However, the overall structure of this example is aperiodic, and this same aperiodic pattern is recursively replicated at each successive level, providing this design with scale invariant self-similarity. The pattern in any given level will inflate and deflate infinitely, and the scaling ratio is phi, or 1:1.6180. . .. The aesthetic treatment of this example is not at all traditional, and employs transparency in the overlay of the three

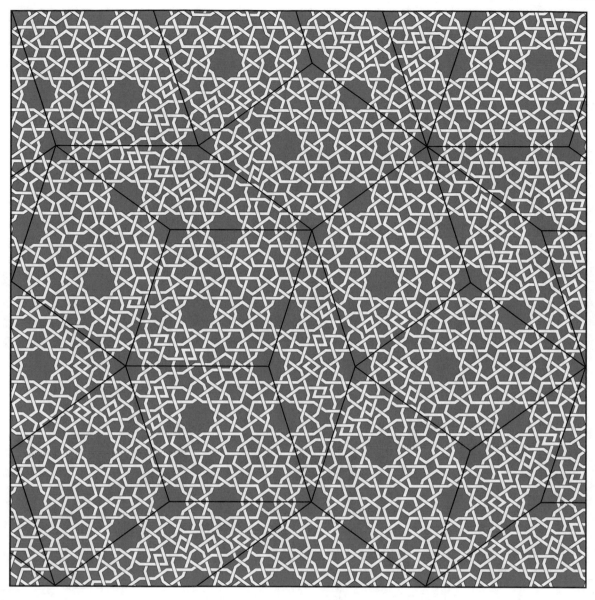

Fig. 480

levels. This visually emphasizes a feature of particular note in this geometric multilevel geometric structure: the frequency at which the intersection points of one level align with key positions in the pattern lines of another level. Figure 482 illustrates the two rhombic repetitive cells that produce the aperiodic designs in Figs. 480 and 481. As determined by both the underlying generative tessellation and the resulting geometric design, the edges of these two rhombi are constrained by Penrose matching rules. The applied geometric pattern lines within each rhombus are governed by lines of reflected symmetry (as per the black dashed lines). These identify Penrose's inflationary and deflationary subdivisions. The Inflation and deflation of the geometric design in Fig. 481 is achieved through recursively applying scaled versions of these two rhombi into the sequential subdivisions at each level; thereby insuring that each level is aperiodic and provided with scale invariant self-similarity. This process of the recursive application of scaled down modules is controlled by the Penrose subdivision rule, and is sometimes referred to as substitution tiling. Figure 483 illustrates this process as applied to the Penrose rhombi. As they inflate and deflate to each successive level the matching rules are maintained, providing each recursive level with self-similar aperiodicity. The inflation and deflation ratio is *phi*, or 1:1.6180.... The three-level design in Fig. 481 applies the geometric designs for both the rhombi in Fig. 482 to each successive level of recursive subdivision. The recursive application of these two decorated rhombi infinitely fills the two-dimensional plane aperiodically, and infinitely inflates and deflates.

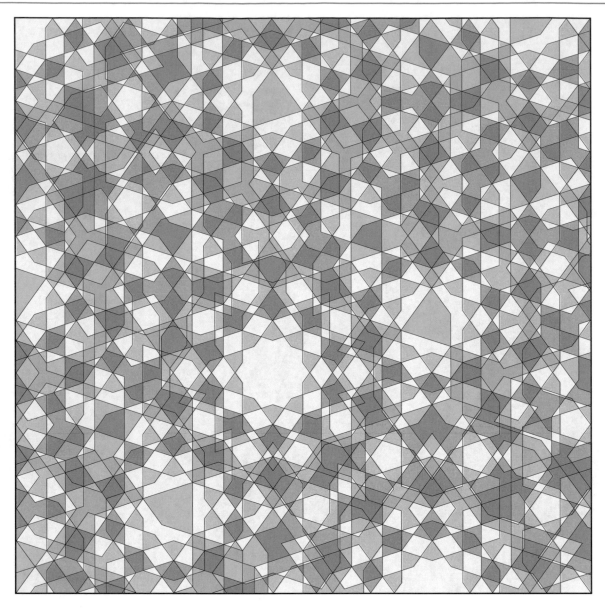

Fig. 481

The *fivefold system* can also be used to create self-similar designs and quasicrystalline designs with radial symmetry. The example in Fig. 484 (by author) is a *two-point* design with tenfold rotational symmetry (only 1/4 shown) that has the added feature of diminishing in scale as the design moves outward from the center. Except for their scale, the primary design (black) and the secondary design (red) are exactly the same, providing this self-similar design with scale invariance. The proportional scale between the two levels is $1 + \varphi$, or 1:2.6180…. As with all scale invariant self-similar designs, this same design has the ability to be recursively applied infinitely. Figure 485 demonstrates how the outward diminution of the primary and secondary patterns in Fig. 484 results from the use of underlying polygonal modules from the *fivefold system* that sequentially reduce in scale as they

move from the center to the periphery. In this illustration the outward expansion has been stopped arbitrarily, but there is no limit to this expansion: all the while with the tessellating polygons becoming smaller and smaller. In fact, the outward expansion of this self-similar structure is an example of Zeno's paradox whereby the ongoing diminishing expansion infinitely approaches, but never arrives at a theoretical limit. Figures 269 and 270 illustrate two of the exceedingly rare historical fivefold patterns that make use of underlying generative tessellations comprised of systematic modules that have more than a single scale. However, the variable scales in these historical examples phase back and forth between just two scales. They do not diminish infinitely outward, nor are they dual level. As mentioned previously, there are two edge lengths in the polygonal modules of the *fivefold system*.

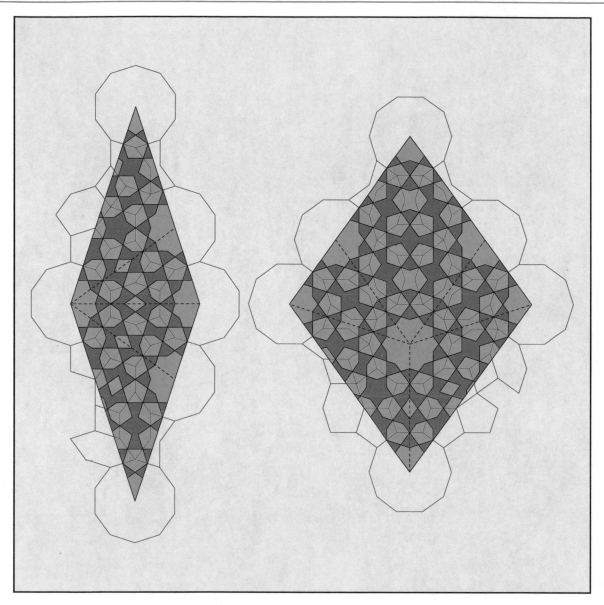

Fig. 482

The diminishing scale of both the primary and secondary underlying tessellations in this design is the result of scaling down the polygons so that their long polygonal edges match the shorter edges of the un-scaled set of polygons, thereby sequentially reducing size as they move outward. Whereas the ratio between the recursive application of the primary and secondary levels is $1 + \varphi$, the ratio between each sequential reduction of the polygonal modules within each tessellation is just *phi*.

The dual-level design in Fig. 486 (by author) is created from the *sevenfold system*. This repeats upon a rhombic grid and is self-similar in that the 14-s4 *median* pattern family is used at both scales, and the application of this variety of

pattern can be recursively applied to each new level infinitely. However, while there are regions of similitude between the levels, this design does not have scale invariance. Also, this design has translation symmetry by virtue of its rhombic repeat unit, and while self-similar, is not aperiodic. Figure 487 illustrates the method for created the design from Fig. 486. This follows the procedure used in historical examples of dual-level designs wherein scaled-down primary polygons—in this case tetradecagons—are applied to the vertices of the primary pattern. As with dual-level designs created from other generative systems, the size of the scaled-down tetradecagons is determined by their application to the vertices of the primary pattern. Figure 487a

Fig. 483

shows three edge configurations: one where tetradecagons are meeting edge-to-edge; one where they overlap and intersect at their corners; and one where they are separated by the convex hexagons from the *sevenfold system*. Figure 487b demonstrates how these scaled tetradecagons are placed on the vertices throughout the primary pattern. Figure 487c fills in the edge lengths of the primary pattern with further polygonal modules from the *sevenfold system*; and Fig. 487d shows the completed application of secondary modules to the total design. *Rho* [ρ] and *delta* [σ] being the two proportions inherent to the heptagon [Fig. 277], the scaling ratio between the primary and secondary levels is 1:8.0978... which can be expressed as [$\rho + \sigma$] $\times$ 2.

The three-level example in Fig. 488 (by author) is also created from the *sevenfold system*. This design is quasicrystalline, with 14-fold rotational symmetry (only 1/4 shown), and its self-similarity is scale invariant. As an alternative to standard historical dual-level methodology, this design places scaled-down 14-pointed stars on the vertices of the *underlying tessellation* rather than at the vertices of the geometric pattern. Each of the three levels of this design uses the crossing pattern lines of the 14-*s4 median* family [Fig. 272]. The recursive iterations of this pattern are scaled down from the center of the design (upper left corner). The primary pattern is blue, the first recursion is green, and the second brown. Each of these is in a tiling treatment, and

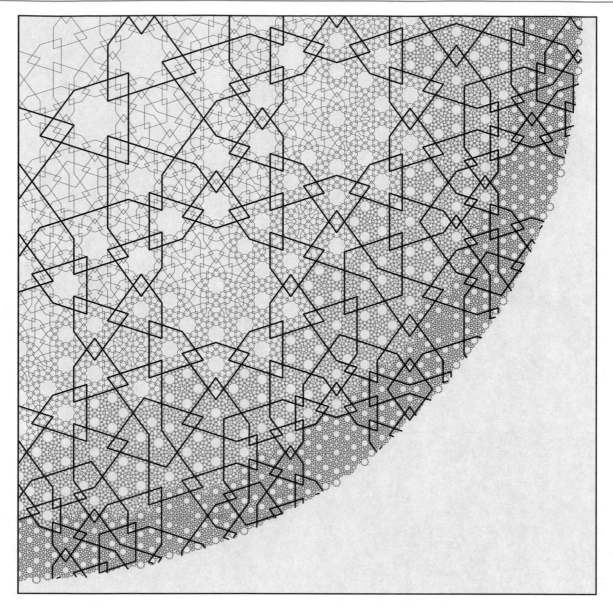

Fig. 484

provided with a level of transparency. Needless to say, such transparency is not a feature of historical practices. Each subsequent iteration can be extended outward from the center infinitely. The scaling ratio between each iteration is $1 + \rho + \sigma$, or 1:5.0489. ... Figure 489 illustrates the three underlying generative polygonal structures for the sevenfold quasicrystalline design from Fig. 488. The first level of pattern is created from the large bounding tetradecagon (black) in Fig. 489a. The second level of pattern is created from the polygonal infill of this bounding tetradecagon in the same illustration. The scale of the secondary tessellation is determined by placing tetradecagons at the vertices of the primary tetradecagon that are separated by the concave

hexagons associated with this system [Fig. 271]. A further secondary tetradecagon of the same size is strategically placed at the center of the primary tetradecagon. Figure 489b scales down the polygonal modules in the secondary tessellation in Fig. 489a so that the same configuration fits within the small secondary tetradecagon at the center of the primary tetradecagon. This scaled-down assembly of polygons is also applied within the secondary (partial) tetradecagons located at the vertices of the primary tetradecagon. Small third-level tetradecagons are then placed at each vertex of the secondary tessellation, and further infill with modules from the *sevenfold system* completes the third level of generative tessellation. This

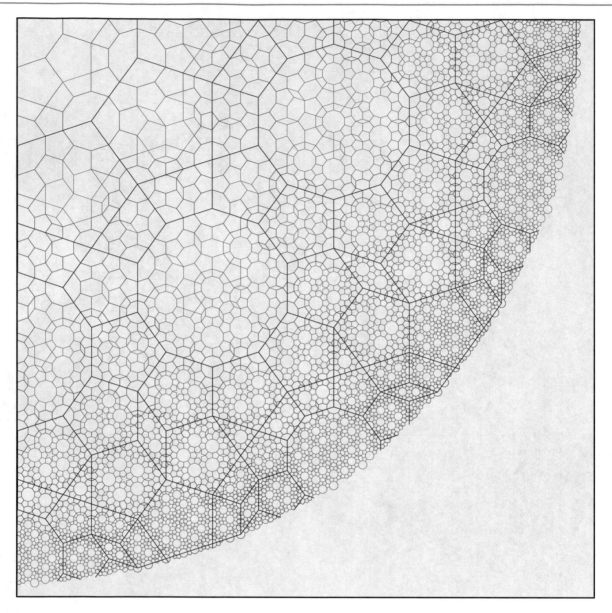

Fig. 485

scale invariant self-similar process can be recursively iterated infinitely, always with the same scaling ratio of $1 + \rho + \sigma$, producing a quasicrystal with 14-fold rotational symmetry.

3.4   Geometric Ornament on Domes: Radial Gore Segments

There are two historical repetitive stratagems for applying geometric designs onto the surfaces of domical structures: polyhedral symmetries, and radial gore segments. Both of these domical methodologies were pioneered by the Seljuks, and both were employed by subsequent Muslim cultures for

applying geometric designs to the interior and exterior surfaces of domes, as well as to the quarter dome hoods of *mihrab* niches. Both of these repetitive methodologies lend themselves to the three-dimensional application of the polygonal technique, and both are aesthetically successful, albeit visually distinct from one another. However, while a large number of examples that employ radial gore segments are found throughout Muslim cultures, only a relatively small number of polyhedral examples are known to the historical record.

The use of radial gore segments in applying geometric designs onto the surfaces of domical structures has the advantage that it will work equally well with both hemispherical domes and domes with a pointed apex. There are

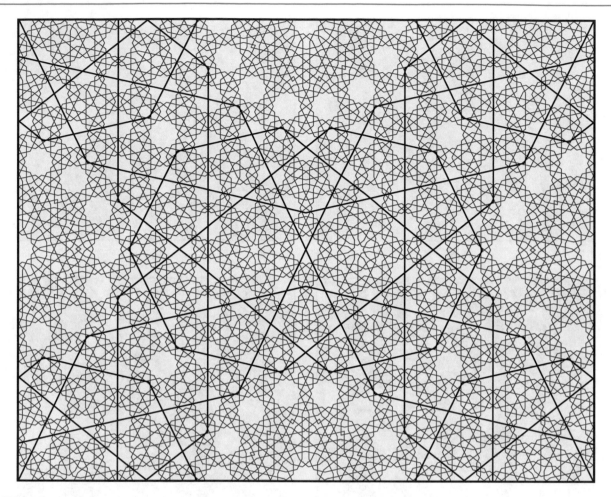

Fig. 486

two convenient methods of deriving profile curvatures for
this latter category of dome: the use of key points associated
with either the orthogonal grid or individual polygons. These
will sometimes have a single center point for each side of the
profile, but are more frequently created from two points of
curvature for each side.[87] The gore segmentation of the
surface of domes historically favored 8-, 12-, and occasionally 16-fold radial divisions, although other divisions, such
as 6-fold and even 24-fold, are also known. As a rule, these
divisions adhere to the symmetry of the supporting chamber,
and with the vast majority of domes being supported by
structures with a square floor plan, the radial divisions are
almost always multiples of four. Figure 490 illustrates the
eightfold radial segmentation of the hemisphere. Figure
490a shows a single gore segment laid flat upon the
two-dimensional plane, Fig. 490b shows the dome in elevation, Fig. 490c shows the dome in plan, and Fig. 490d shows

an array of the eight radial segments laid flat. Of course a
dome is a double-curved surface and will not unfurl onto the
two-dimensional plane without distortion. The following
illustrations of historical geometric gore segments are therefore only representational of the actual geometry, but nonetheless demonstrate the prevalent use of the polygonal
technique in laying out geometric designs on gore segments
with widely diverse proportions.

The earliest extant dome that is ornamented with a geometric design based upon radial gore segments is from the
Friday Mosque at Gulpayegan (1105-18). Figure 491a
illustrates the underlying polygonal tessellation that creates
the design on this Seljuk dome. The ring of eight edge-to-
edge heptagons that create the ring of seven-pointed stars
illustrate a fundamental principle in the placement of geometric designs upon gore segments: the precise curvature of
the dome and the underlying generative tessellation are
intrinsic to one another. Figure 491b represents the unfurled
eightfold segments. At the outer corners of each segment are
1/4 portions of eight-pointed stars. These produce the 8 half
eight-pointed stars located at the base of this dome. It is
interesting to note the similarity between the 2 ten-sided

---

[87] The methodology that governs the design of Islamic arch and dome
profiles has not received the scholarly attention that it deserves, but is
beyond the scope of the current work.

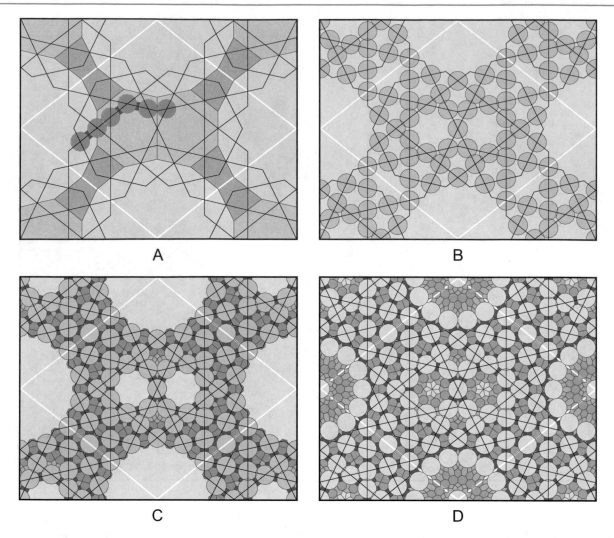

A   B

C   D

Fig. 487

motifs directly below, and one immediately above the seven-pointed stars and those found in two of the patterns from the northeast dome chamber in the Friday Mosque at Isfahan [Figs. 261 and 309b]. This was evidently a popular devise among Seljuk artists during the turn of the twelfth century.

Figures 492 and 493 illustrate the use of the polygonal technique for creating the geometric designs of four Mamluk domes based upon radial gore segmentation.[88] The first three of these four examples are from the notable group of Cairene domes with exterior monochromatic carved stone that achieves design clarity through high relief and resulting shadow. As with the previous Seljuk example from Gulpayegan, the precise curvature and proportions of the gore segments for each of these examples supports the

arrangement of underlying polygons that produce the geometric designs. Figure 492a represents the 1/20 gore segment of the dome covering Sultan Barsbay's mausoleum at the Sultan al-Ashraf Barsbay funerary complex at the northern cemetery in Cairo (1432-33) [Photograph 61]. This dome places sequential rings of 20 half eight-pointed stars at the base, followed by full eight-pointed stars. The pattern in this lower section of the dome is a *median* design with 60° angular openings applied to the 4.8² tessellation of octagons and squares [Fig. 126a]. Immediately above these eight-pointed stars is a ring of seven-pointed stars, and above this is a region of interweaving hexagons and triangles that can be described as a tapered form of the well-known *median* pattern, albeit with pattern lines that continue through the otherwise six-pointed stars. This design is easily created from the 6³ grid of regular hexagons [Fig. 95b]. This section is surmounted by a ring of distorted 7-pointed stars, with a 20-pointed star the apex of the dome that is only

---

[88] The proportional representations of the illustrated gore segments in this section are approximations of the actual proportions of the cited examples.

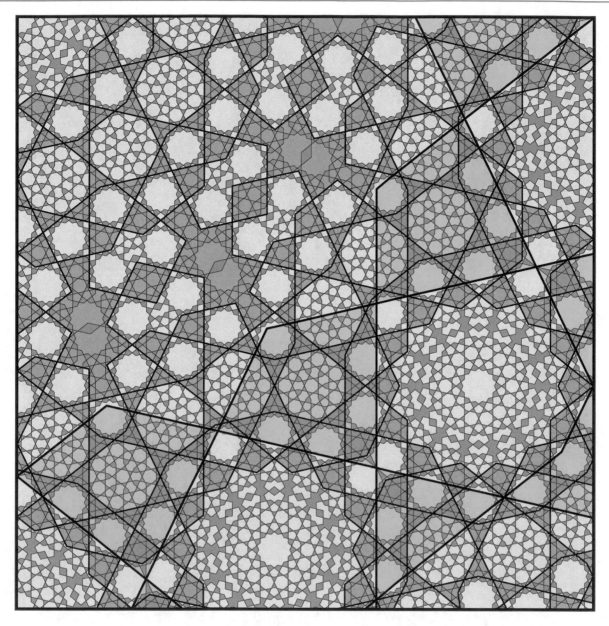

Fig. 488

visible in toto from a bird's eye view. The bold interweaving widened lines that comprise this example are represented in Fig. 493a. The gore segment in Fig. 492b is from a dome at the same Mamluk funerary complex, and covers the tomb of an anonymous Barsbay family member. The radial gore of this domical geometric design employs a sixfold segmentation of the dome. The primary 8- and 12-pointed stars are derived from a tessellation of underlying dodecagons and octagons in the lower 3/4 of the design. It is interesting to note that two-dimensional version of the domical design produced from these underlying dodecagons and octagons was used by Mamluk artists just 10 or 15 years previous at the Sultan al-Mu'ayyad Shaykh complex in Cairo (1415-22)

[Fig. 380e]. The design in this lower region transitions into sequential bands of 5-pointed stars, culminating in a single 12-pointed star at the apex of the dome. Figure 493b illustrates the interweaving treatment of this design. The design in Fig. 492c shows the geometric component from the dome of the Sultan Qaytbay funerary complex in the northern cemetery in Cairo (1472-1474) [Photograph 2]. This exceptional dome incorporates a floral motif (not shown) that meanders beneath the geometric design. The ornament of this dome repeats upon a eightfold radial segmentation, and each gore segment places a ring of half 10-pointed stars at the base of the dome, followed by a ring of 9-pointed stars, followed by a ring of rather distorted 5-pointed stars, and

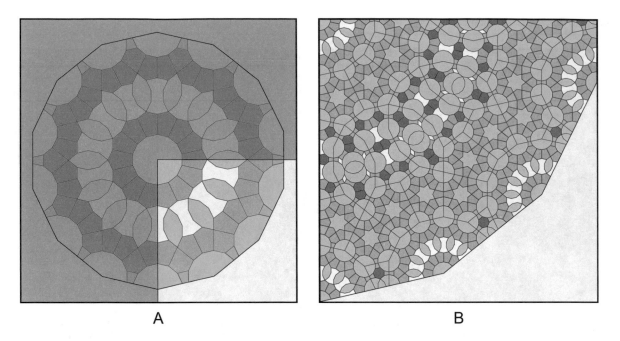

Fig. 489

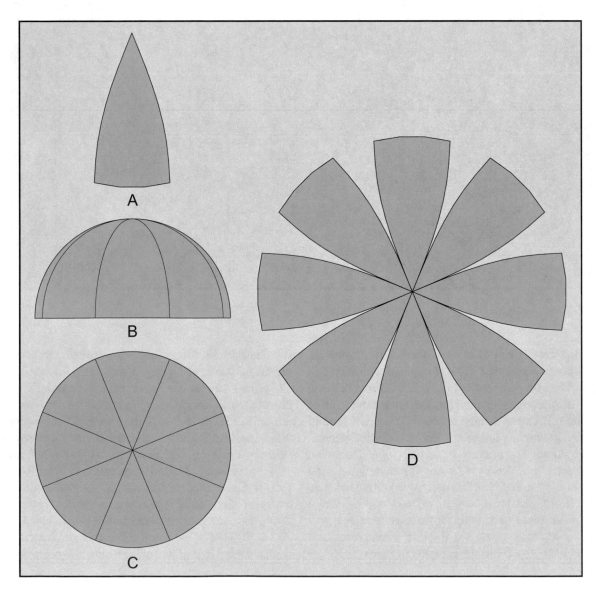

Fig. 490

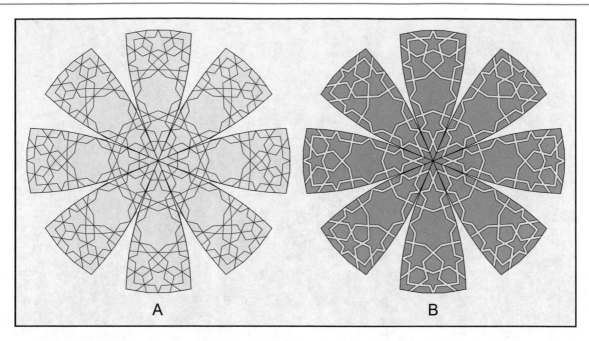

Fig. 491

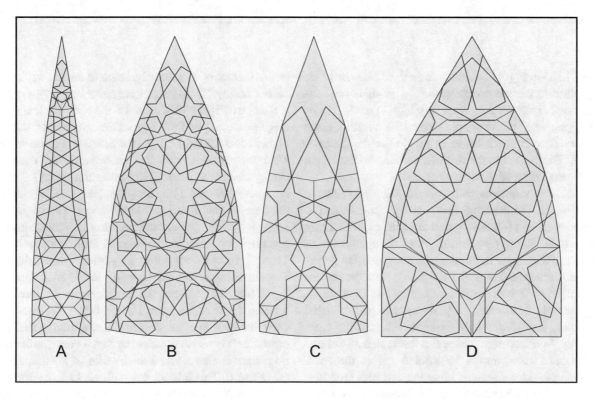

Fig. 492

culminating in a 16-pointed star at the apex. Figure 493c illustrates the widened geometric lines from the 1/8 gore segment of this historical example. Figure 492d shows a design from the mosaic *mihrab* hood at the Amir Qijmas al-Ishaqi mosque in Cairo (1479-81). This is based upon an sixfold radial segmentation, with half 12-pointed stars at the base of the semi-dome, followed by 10-pointed stars at the middle of each radial gore, and half a 12-pointed star at the apex of the dome. Figure 493d illustrates the widened line treatment of the geometric ornament of this example. The very different proportions of this 1/6 gore segment and the previous 1/6 gore segment from the anonymous Barsbay family

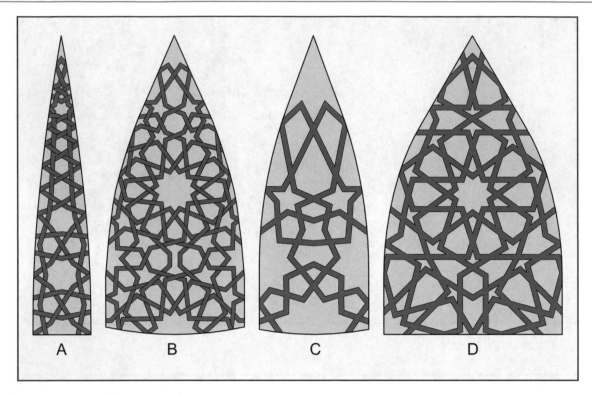

Fig. 493

member's tomb in Fig. 492b results from the difference in the dome profiles of these two examples. The example from the Barsbay family member's tomb is very high in relation to its diameter, whereas the example from the Amir Qijmas al-Ishaqi mosque is much lower in relation to the diameter of its base. This results in the differences between the height and width proportion of these two examples.

The four gore segments with applied geometric patterns in Figs. 494 and 495 are from Persia and Central Asia. The ornament of each of these four domes is polychromatic cut-tile mosaic. Figure 494 demonstrates the application of the polygonal technique to domical gore segments to generate these designs. The design in Fig. 494a is from the interior of a Muzaffarid dome in the Friday Mosque at Yazd (1324) [Photograph 91]. This design repeats upon a 16-fold radial segmentation, and places half 6-pointed stars at the base, followed by 7-pointed stars, more 6-pointed stars, 5-pointed stars, 4-pointed stars, and a 16-pointed star at the apex. Figure 495a provides a widened line version with representative color differentiation within the background regions. The design in Fig. 494b is from the interior dome of the mausoleum of Turabek-Khanym in Konye-Urgench, Turkmenistan (1370) [Photograph 92]. This was produced during the short-lived reign of the Sufi Dynasty, and is an aesthetic precursor to the remarkable architectural ceramics of the Timurids. This dome is divided into 12 radial gore segments, with half ten-pointed stars at the base, followed by ten-pointed stars, and nine-pointed stars. Between these

primary stars is a connective pattern matrix typical of the *obtuse* family. This design transitions into a 24-pointed star at the apex. Figure 495b provides the widened line version of this design as per the historical example, but without the highly ornate floral background mosaics. Figure 494c is the 1/8 gore segment from the Safavid exterior dome of the Friday Mosque at Saveh (late sixteenth century). The underlying tessellation for this design employs half dodecagons at the base of the dome that starts an ascending progressive sequence of octagons, hendecagons, and nonagons that are separated by elongated hexagons and concave hexagons. These produce an *obtuse* pattern that places half 12-pointed stars at the base of the dome, followed by 8-pointed stars, 11-pointed stars, 9-pointed stars, and an 8-pointed star at the apex. Figure 495c represents the widened line version of this design that was used in the historical example. It is worth mentioning that a very similar geometric pattern is also used on the interior of the dome in Saveh (not shown). This interior dome places half ten-pointed stars at the base, followed by nine-pointed stars, ten-pointed stars, seven-pointed stars, and culminating with an eight-pointed star at the apex. The combined use of sequential numbered stars (seven-, eight-, nine- and ten-pointed) is an artistic device that was also used in many two-dimensional designs—especially by artists working during the Seljuk Sultanate of Rum. The design in Fig. 494d is from one of the most well-known exterior geometric domes in Iran: the Safavid shrine of Aramgah-i Ni'mat Allah Vali in Mahan

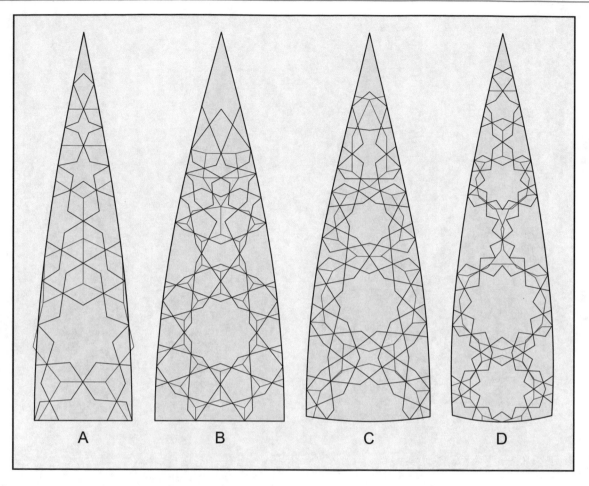

Fig. 494

(1601) [Photograph 93]. The design on this dome is regulated by a 12-fold radial segmentation. The base of the dome has half 8-pointed stars, followed by 10-pointed stars, 9-pointed stars, 11-pointed stars, 12-pointed stars, 9-pointed stars, 7-pointed stars, 5-pointed stars, and culminating in a 12-pointed star at the apex. Here again the artist strove to construct a design with a consecutive numeric sequence (7-, 8-, 9-, 10-, 11-, and 12-pointed stars). This achievement required considerable skill, even if some of the stars are relatively irregular, and not placed in ascending numeric order. Figure 495d provides a representation of the widened line treatment of the pattern on this dome.

3.5 Geometric Ornament on Domes: Platonic and Archimedean Polyhedra

The most geometrically interesting and visually arresting non-Euclidean Islamic geometric designs are the very rare examples that employ polyhedral symmetry as their repetitive schema. In the case of applying patterns to domical gore segments, other than having to balance the underlying generative tessellation with the tapering curvature of the *n*-fold segment, the basic geometric constraints are more or less the same as those governing two-dimensional pattern making. By contrast, applying geometric designs onto the surface of the sphere involves geometric conditions that have no parallel on the two-dimensional plane. This creates interesting geometric challenges and aesthetic opportunities for that artist, and it is surprising that there are so few historical examples of this form of domical ornament.

The earliest known example of polyhedral geometric ornament is from the interior of the magnificent northeast dome in the Friday Mosque at Isfahan (1088-89) [Photograph 30]. This geometric design is created from the pentagonal faces of the dodecahedron projected onto the curvature or the domical surface. Each of these pentagonal faces is used as an underlying generative module in the same way that pentagons are used to create fivefold patterns on the two-dimensional plane, the difference being that on the sphere pentagonal faces can tessellate on their own, whereas on the plane they require at least one other module from the *fivefold system* for complete surface coverage. Figure 496 illustrates the *two-point* pattern constructed

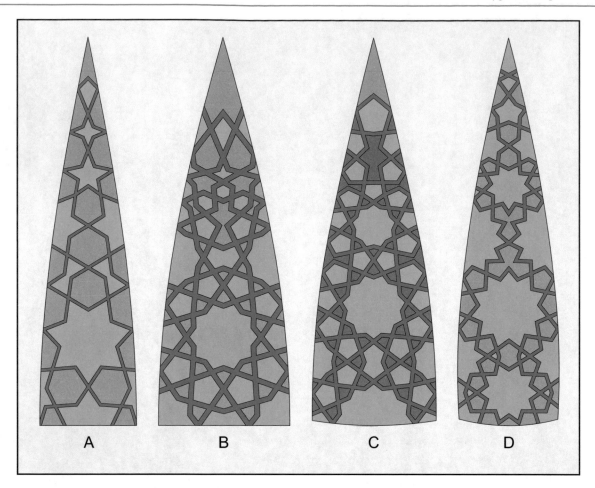

Fig. 495

Fig. 496

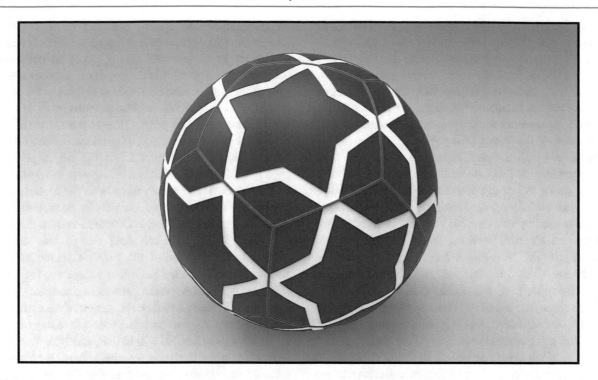

Fig. 497

from the spherical projection of the dodecahedron. This follows the conventions of the polygonal technique by using the projected pentagonal faces (orange lines) as the underlying generative tessellation. This example from Isfahan centers the design at the apex of the dome on one of the pentagonal background elements; thereby creating a fivefold rotational symmetry that descends from the apex to the base of the dome. This is at odds with the ascending 4-, 8-, and 16-fold symmetry within the cubical supporting structure for this dome. One would expect this dome to be a hemisphere that would correspond with half of the spherically projected dodecahedron. However, this remarkable dome is provided with additional loft wherein the apex is raised above the surface of the hemisphere. The artist's decision to break the symmetry of the dodecahedron provides the domed chamber with greater spatial volume, introducing a central point for the eye to fix upon. Considering that this is a mosque, one can assume that this may have also been intended as a way of spatially emphasizing a concept of religious ascendancy. The application of this geometric design therefore required the projection of the original pentagonal faces of the dodecahedron beyond the hemispherical surface onto the elevated surface. The resulting distortion is minimal, and in no way detracts from the beauty of this dome. In fact, this innovative change to the profile of the dome augments the beauty of the geometric design. A more conventional example of this same non-Euclidean design was used on a projecting

hemispherical stone detail in the arch spandrel of the entry portal at the Sahib Ata mosque in Konya (1258). Other than the fact that this example is hemispherical, the main difference between this example from the Seljuk Sultanate of Rum and the earlier example from Isfahan is that the pattern lines of the later example are given a curvilinear treatment. This creates fivefold floriated elements within the regions that are otherwise pentagons.

Another projecting carved stone hemisphere from the Seljuk Sultanate of Rum also makes use of the dodecahedron as the repetitive schema for its geometric design. This is found in the Huand Hatun complex in Kayseri (1237), and like the example from the Sahib Ata mosque, this example is also located within an arch spandrel, although this example is in a *mihrab* rather than an entry portal. Figure 497 illustrates a full spherical representation of the *median* pattern that was used for the high relief stone ornament of the projected hemisphere at the Huand Hatun complex. Like the example from Isfahan, this makes use of the projected pentagonal faces as the underlying tessellation upon which the *median* pattern lines are applied. This pattern is characterized by a simple spherical matrix of five-pointed stars located within each underlying pentagon, and ditrigonal hexagons with threefold symmetry located at each vertex of the underlying tessellation.

The previous domical examples are derived from the dodecahedron. Each of the five Platonic solids has only a single type of polygonal face, and a single variety of vertex condition: the tetrahedron has four triangular faces and four 3^3 vertices; the cube has six square faces and eight 4^3 vertices; the octahedron has eight triangular faces and six 3^4 vertices; the dodecahedron has twelve pentagonal faces and twenty 5^3 vertices; and the icosahedron has twenty triangular faces and twelve 3^5 vertices. Of the five Platonic solids, only the dodecahedron appears to have been used in a fashion whereby the faces of the polyhedra are treated as an underlying generative tessellation. However, out of the 13 Archimedean solids, at least two historical domical examples make use of the projected spherical faces of the polyhedra as underlying tessellations. Archimedean solids are characterized by two or more varieties of regular polygonal face and identical vertices. If the Platonic solids are analogous to the two-dimensional regular grids, the Archimedean solids are analogous to the two-dimensional semi-regular grids [Fig. 89]. Other than the dodecahedron and icosahedron, the use of the Platonic solids as underlying generative tessellations would produce very simplistic patterns, and it is perhaps not surprising that these polyhedra were not used to create spherical designs. However, the greater complexity of the Archimedean solids would have afforded artists significant design potential, and it is surprising that only two of these polyhedra appear to have been used historically as underlying generative tessellations.

Figure 498 illustrates a *median* design applied to a spherical projection of the truncated icosahedron. This Archimedean solid is made up of 12 pentagonal and 20 hexagonal faces, and 60 identical vertices with a 5.6^2 arrangement of pentagons and hexagons. This is commonly recognized as the standard soccer ball. This polyhedra was used to create a geometric design in the hood of the *mihrab* arch at the Lower Maqam Ibrahim in the citadel of Aleppo (1168). This masterpiece of Zangid woodwork is signed by Ma'ali ibn Salam. The pattern lines that make up this spherical design are a combination of great curves and offset great curves. The double curvature of the offset great curves applies to the parallel lines that connect neighboring five-pointed stars. These would have been particularly complex to calculate and construct accurately in wood, and the precision of this spherical woodwork is testament to the genius of this artist. This same spherical design was used many centuries later for one of the purely ornamental Ottoman hollow pierced wooden balls that hang from the ceiling in the mausoleum of Mevlana Jalal al-Din al-Rumi in Konya. The only difference between the design of the earlier Zangid example from Aleppo and that found in Konya is that the six-pointed stars located within the underlying hexagons are open stars in the Ottoman wooden sphere. By contrast, the lines of the six-pointed stars in the Zangid example run through their underlying hexagon, thereby creating a central hexagon surrounded by six triangles.

The spherical design in Fig. 499 can be produced from the truncated cube. This example is from the Seljuk Sultanate of Rum, and is found in another projecting stone hemisphere in the arch spandrel of the portal at the Susuz Han in Susuzköy,

Fig. 498

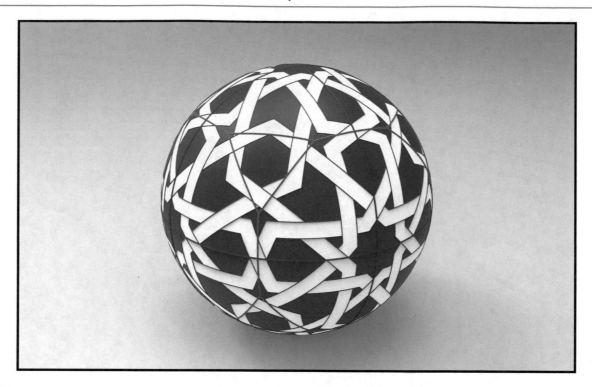

Fig. 499

Turkey (1246). The truncated cube has 14 faces, 8 being triangular and 6 being octagonal. The 24 identical vertices are in a 3.8^2 arrangement. The pattern lines place an eight-pointed star rosette within each underlying octagon. These eight-pointed stars are located upon the vertices of the octahedron. The proportion and placement of the parallel lines of the eight-pointed stars are determined by their point of origin being the midpoints of the adjacent underlying triangles, thereby creating a hexagon within each underlying triangle. It is worth noting that this pattern can also be created from an underlying polygonal network made up of octagons connected with trapezoids and triangles that are located within the eight triangular faces of the octahedron. The generative schema for this spherical design has two analogous two-dimensional examples from Turkey that are roughly of the same time period. The same application of pattern lines was used in the two-dimensional triangular repetitive cells from a panel in the Mengujekid *minbar* at the Great Mosque of Divrigi (1228-29), and in the carved stone ornament at the Çifte Minare *madrasa* in Sivas, Turkey (1271). Other than the difference in spherical versus planar topology, these examples differ in the type of primary stars located at the vertices where the triangular elements come together. In the case of the octahedron, the eight-pointed stars result from there being two points at each corner of the triangular cell, and four projected triangular faces meeting at each vertex. By contrast, in the case of the two-dimensional design, the vertices have six triangular repetitive cells, and the two points

within each corner of the triangle produces the 12-pointed stars [Fig. 320]. The fact that both this two-dimensional design and its analogous spherical variant were being used in the same region and during the same period of time would appear to be more than coincidental.

As referenced in the previous example, while spherical projections of the less complex Platonic solids were not particularly suitable for use as underlying tessellations, they were occasionally used as repetitive devices upon which more complex nonsystematic underlying tessellations could be constructed. Figure 500 illustrates a spherical design that uses the octahedron (black lines) as its governing repetitive structure. This design was used in the Ayyubid *mihrab* hood of the al-Sharafiyah *madrasa* in Aleppo (before 1205). This is produced in low relief carved stone and is signed by 'Abd al-Salâh Abû Bakr. The underlying generative tessellation for this design places octagons at the vertices of the octahedron, and surrounds each of these with a ring of eight pentagons (red lines). Like the spherical design from the Susuz Han, the triangular repeats of this Ayyubid example also have a two-dimensional analogue, and like the previous example, the 8-pointed stars at the vertices of the octahedron become 12-pointed stars on the two-dimensional plane [Fig. 300a—*acute*].

The spherical design in Fig. 501 is also created from a nonsystematic underlying tessellation that repeats upon the projected triangular faces of the octahedron. This *acute* design was used by the Ayyubid artist Abu al-Husayn bin

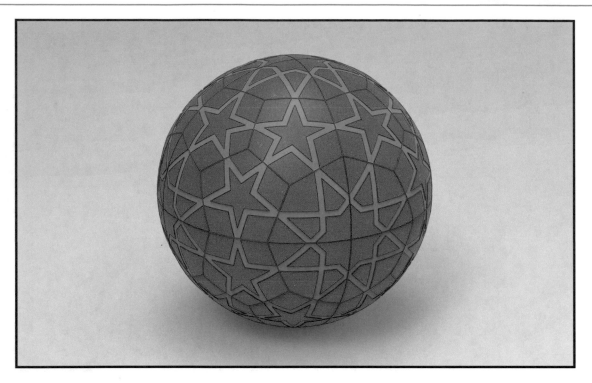

Fig. 500

Fig. 501

Muhammad al-Harrani 'Abd Allah bin Ahmed al-Najjar in the hood of the wooden *mihrab* (1245-46) of the Halawiyya mosque and *madrasa* in Aleppo. This design places underlying octagons at the vertices of the octahedron (black lines), nonagons at the center of each triangular face of the octahedron, and surrounds these with connecting pentagons and barrel hexagons (red lines). Once again, this design is analogous to a well-known two-dimensional pattern that places

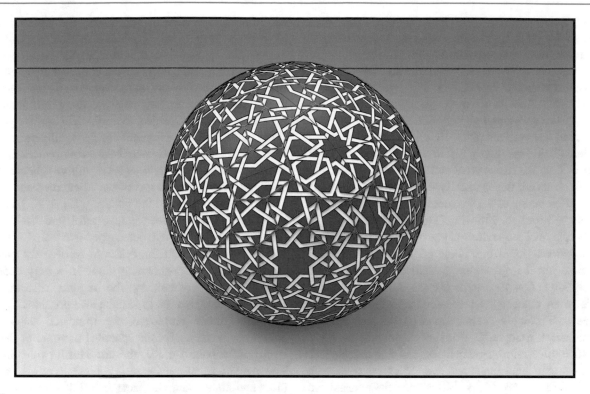

Fig. 502

12-pointed stars at the vertices of the isometric grid and 9-pointed stars at the centers of each repetitive triangle [Fig. 346a]. The octahedral form of this design transforms the 12-pointed stars into 8-pointed stars, creating an immensely successful spherical pattern comprised of 8- and 9-pointed stars connected by a pattern matrix of 5-pointed stars and opposing darts.

The spherical *two-point* design in Fig. 502 is arguably the most complex historical example of an Islamic geometric pattern based on polyhedral geometry. This non-Euclidean design is found in a mosaic arched hood at an anonymous Mamluk mausoleum in the Nouri district of Tripoli, Lebanon. The repetitive structure of this design is the cuboctahedron, comprised of eight triangular faces and six square faces. The 12 identical vertices are in a 3.4.3.4 arrangement. The pattern lines applied to the projected triangular and square faces each have two-dimensional analogues, and each of these analogous two-dimensional patterns was used historically. A Mamluk example of the two-dimensional use of the design contained within just the triangular repetitive cells is found at the Ribat of Ahmad ibn Sulayman al-Rifa'i in Cairo (1291) [Fig. 300b *two-point*]; and a Mamluk example of the analogous two-dimensional square pattern was used in the triangular side panels of the *minbar* at the Princess Asal Bay mosque in Fayyum, Egypt (1497-99) [Fig. 382]. For the patterns in these two repetitive cells to work together they must have identical edge conditions, and in this respect, this domical design shares

the characteristics of two-dimensional hybrid designs. Indeed, the two-dimensional forms of these two repetitive cells will comfortably tessellate the plane in either a $3^2.4.3.4$ or $3^3.4^2$ semi-regular arrangement [Fig. 89].

Other domical geometric designs based on the symmetry of polyhedra include two examples from the Alhambra. The Nasrid barrel vault of the Sala de la Barca (c. 1354-91) is capped at both ends with a quarter spherical rhombicuboctahedron. This polyhedra has 26 spherically projected faces: 8 that are triangles and 18 that are squares. The 24 identical vertices are a 3.4^3 arrangement of these faces. The pattern in the square faces places 12-pointed stars at the center of the repetitive face, 8-pointed stars at the four vertices, and 8-pointed stars at the midpoints of each repetitive edge. The triangular faces also have eight-pointed stars at the midpoints of the repetitive edges and the triangular vertices. The repetitive schema of the pattern in the barrel-vaulted portion of this ceiling is a simple square grid that transforms into the rhombicuboctahedron at each end of the barrel vault. The rings of eight square faces that characterize this polyhedra elegantly allow for this transformation. Unlike the earlier Ayyubid and Zangid wooden polyhedral domical constructions, the geometric design in this Nasrid wooden vault displays significant distortion. For example, the above-referenced placement of eight-pointed stars at the vertices of the repetitive cells works nicely among the square repeat of the barrel-vaulted portion of this ceiling, but is incompatible with the symmetry of the triangle and produces poorly resolved features in the rhombicuboctahedron semi-dome regions of the vault. In essence, the artists responsible for this example were conceptually inspired in their combined use of the barrel vault and rhombicuboctahedron, but underwhelming in their application of the geometric design. Ironically, had the same basic geometric pattern been applied to the polygonal faces with the 12-pointed stars placed upon the vertices rather than at the centers of each face, their 12-fold symmetry would have accommodated the vertices of both the barrel vault and polyhedral regions very acceptably. Indeed, this is exactly the approach to the polyhedral pattern in one of the two domes that grace the nearby Court of the Lions (c. 1354-91). This Nasrid example also employs spherical projections of square and triangular faces. This exceptional polyhedral dome is unusual in that it does not adhere to the geometry of either the Platonic or Archimedean solids. Rather, as discussed in Chap. 1, this is essentially an octacapped truncated octahedron with two varieties of triangle rather than a single type of non-equilateral triangle.[89] This polyhedra has 6 square faces, 40 equilateral triangular faces, and 8 triangular faces with a single 30° acute angle and two 75° obtuse angles. The analysis of Emil Makovicky[90] illustrates the unusual face and vertex configurations of this polyhedral dome, as well as the application of the pattern lines into these repetitive modules. This polyhedron has three varieties of vertex condition. The top half of this hemispherical dome follows the face configuration of the snub cube. Each vertex of this Archimedean solid has a $3^4.4$ arrangement of triangles and square. In applying the pattern lines into this vertex configuration, the placement of three points of a star at each square faced vertex and two points at each triangular face produces 11-pointed stars $[3 + (2 \times 4) = 11]$. This 11-fold symmetry at the polyhedral vertices is distinct from the 12-fold local symmetry that results from the same distribution of points within two-dimensional patterns that employ square and triangular repetitive cells [Fig. 23d]. The second variety of vertex in this unusual polyhedra has five projected equilateral triangular faces combined with the 30° acute angle of the isosceles triangle. This 30° angle provides for one point of a 12-pointed star on the flat plane, and the applied pattern lines in this arrangement thereby also produce an 11-pointed star $[1 + (2 \times 5) = 11]$. The third vertex of this polyhedron combines one square, two equilateral triangles, and two of the 75° obtuse angles from the edge-to-edge isosceles triangles. These back-to-back triangular faces produce a rhombus with a 150° included angle, thereby providing for five points of a 12-pointed star in two dimensions. In this

way, the third variety of vertex also has 11-pointed stars $[3 + (2 \times 2) + 5 = 11]$. This spherical variant of the octacapped truncated octahedron has bilateral symmetry wherein the top and bottom halves of the sphere are mirrored upon the equatorial plane. In this way, the pattern lines at the base of the dome are nicely resolved along the equator. This polyhedron does not project to a spherical surface without a small amount of distortion among the equilateral triangular faces. Had the artist responsible for this remarkable dome chosen to use the snub cube as the repetitive schema, thereby continuing the thematic use of the 11-pointed stars at each vertex, the resolution of the pattern lines at the base of the dome would not have been as successful, and it is likely that the development of the octacapped truncated octahedral variant stemmed from the desire to resolve the pattern at the base of the hemispherical dome in a very acceptable fashion while maintaining the regular distribution of 11-pointed stars that the snub cube also provides.

Nasrid artists employed the truncated cube as the governing geometry for the spherical pommel of the Jineta sword of Muhammad XII, the last Muslim ruler in Spain. This Archimedean polyhedra is comprised of six octagonal faces and eight triangular faces in a 3.8^2 condition at each vertex. This is analogous to the two-dimensional 4.8^2 tessellation of squares and octagons, and the resulting spherical pattern is generated in the same fashion as the well-known star-and-cross design [Fig. 124b] whereby the octagonal faces produce eight-pointed stars. However, the replacement of the squares with triangles means that the eight-pointed stars are separated by trifold elements rather than fourfold crosses. Compared to the other Nasrid polyhedral geometric designs, this is considerably less complex, and its beauty is derived more from the opulence of the cloisonné enamel work than from its geometric ingenuity.

3.6 Conclusion

This survey of the methodological practices encompassed by the polygonal technique reveals the historical development from simplicity to ever-greater complexity. This trajectory built upon the early innovations and experimental approach of artists during the eighth to eleventh centuries to evolve into the highly formalized design conventions of the fully mature tradition during the twelfth to fourteenth centuries. Trained in these formalized methodological practices, Muslim artists expanded this tradition to include highly ingenious nonsystematic patterns with multiple regions of local symmetry, and systematic dual-level designs with varying degrees of self-similarity. Throughout this growth in complexity, nuanced design practices expanded the aesthetic repertoire available to artists. Some of these practices can be regarded as general rules and others as stylistic

---

[89] – Makovicky (2000), 37–41.
   – O'Keeffe and Hyde (1996).
[90] Makovicky (2000).

conventions. Yet even those practices that are constant features of the polygonal technique were occasionally modified or dispensed with, and the one criteria that applies above all others is whether a given design falls comfortably within the prevailing aesthetics of the time and place. The following brief encapsulation of the methodological practices inherent to this tradition provides fundament criteria to the nuanced understanding of this ornamental tradition, as well as valuable methodological praxis for contemporary artists and designers engaged with this geometric art form.

- The polygonal technique is comprised of two methodological categories: systematic and nonsystematic. Both rely on underlying polygonal tessellations upon which pattern lines are distributed. Systematic designs employ a limited set of polygonal modules, each with prescribed pattern lines, which assemble into an unlimited number of combinations. By contrast, the proportions of underlying polygons that comprise nonsystematic designs are specific to each tessellation and do not recombine into other tessellations.
- There are five historical design systems: the *system of regular polygons*, the *fourfold system A*, the *fourfold system B*, the *fivefold system*, and the *sevenfold system*. Different Muslim cultures had greater or lesser affinities with each of these design systems.
- There are four primary historical pattern families common within this design tradition: *acute*, *median*, *obtuse*, and *two-point*. Each underlying tessellation is capable of producing patterns in each of these pattern families, although some generative tessellations are not suited to produce acceptable designs in each family.
- *Acute*, *median*, and *obtuse* crossing pattern lines are applied to the midpoints of the underlying polygonal edges unless the pattern is improved by moving the crossing pattern lines to a nearby point along the polygonal edge. Generally, the placement of the crossing pattern lines within systematic designs maintains their location upon the midpoints of the underlying polygonal edges. In more complex nonsystematic patterns, the crossing pattern lines may be move off of the polygonal edge if this provides a more acceptable visual result.
- The location of the pattern lines in the *two-point* family can vary, and, depending on the resulting aesthetic quality, 1/3 and 1/4 divisions of the polygonal edges are both common.
- As pattern lines cross one another, they should ideally continue in a straight line beyond the intersection. Pattern lines that change direction at the point of intersection, thereby loosing their collinearity, generally appear awkward and are best avoided whenever possible.

- In placing the crossing pattern lines on or near the midpoint of the underlying polygonal edge, the angular opening of the crossing pattern lines determines whether the design will be of the *acute*, *median*, or *obtuse* pattern family. With systematic patterns the precise angle is determined by the inherent geometry of the given system. With nonsystematic patterns the angular openings are ultimately determined through aesthetic judgment, and may vary slightly from location to location. Generally, the inherent angles of the *fivefold system* provide a useful comparative aesthetic when working with nonsystematic designs.
- Other than designs with radial symmetry, historical Islamic geometric designs invariably employ translation symmetry, and will adhere to one of the 17 plane symmetry groups. A diverse range of repeat units were used historically, including; squares, regular hexagons, rectangles, rhombi, and non-regular hexagons. Equilateral triangles were frequently used as repetitive cells, but must be either mirrored to form a rhombus, or rotated to form a hexagon before providing translation symmetry.
- All repetitive geometric patterns have a fundamental domain that contains all of the geometric information necessary to complete the design. The singular or combined application of rotation, reflection, and glide reflection to the fundamental domain fills the repeat unit, allowing for translation symmetry. In some cases, the fundamental domain repeats with translation symmetry alone.
- Hybrid patterns can be constructed from using two or more repetitive cells in combination. This design methodology creates greater complexity within the completed pattern. Typically, each of these repetitive cells will tessellate the plane independently. A criteria of each repetitive cell within a given hybrid design is that all edges of equal length share the same pattern conditions and underlying generative tessellation conditions.
- As a general rule, the *n*-fold symmetry of primary stars will correspond with the symmetry of its location within the repetitive structure. For example, orthogonal patterns will place stars with points that are multiples of four at the vertices of the square grid; isometric patterns will place stars with points that are multiples of six at the vertices of the triangular grid, and regular hexagonal grids will place stars that are multiples of three at each vertex. The same rule applies to the centers of each repetitive cell. The reflection symmetry of stars placed upon an axis of reflection, such as the edges and diagonals of a repeat unit, must align with the axis of reflection, and may be either even or odd numbered.
- The proportions of rectangular repeat units are determined by the *n*-fold local symmetries placed at each

vertex of the rectangular grid, and, where relevant, by the n-fold symmetries of secondary locations, such as the center and diagonals, within the repeat unit.

- The proportions of rhombic and non-regular hexagonal repetitive grids are determined by the internal angles of each repeat unit corresponding with the n-fold symmetry of the primary stars placed at these locations.

- Field patterns have no primary stars. This category of pattern is created from underlying generative tessellations that do not include primary polygons with a larger number of edges.

- The flexibility of oscillating square and rotating kite designs allows for significant geometric manipulation, including the incorporation of unexpected n-fold local symmetries into an orthogonal structure. Such designs generally conform to the *p4g* plane symmetry group.

- Framing rectangles almost always adhere to the geometry of the repetitive grid. In determining the frame for a given pattern, the edges should ideally correspond with lines of symmetry within the overall pattern whenever possible.

- There are many forms of line treatment that can be applied to the basic plain line version of a given design. Each has its own aesthetic quality, and may be more or less appropriate to a given design, the aesthetic sensibilities of an artist, and different materials and techniques of fabrication. Regional styles can also be influenced by line treatment. Such treatments include tiling with two or more colors (as per a chess board), various thicknesses of widened lines with or without interweaving, and various forms of double-line applications.

- The widening of pattern lines can be derived from key points within the polygonal sub-grid, or can be a purely arbitrary decision based upon visual preference. The thickness can also be determined from the geometry of design itself. For example, the pattern lines can be widened to their maximum extent, corresponding to the center of the smallest background element.

- Patterns that are unsuccessful due to large discrepancies in the size of the background shapes can sometimes be corrected by the widening of the pattern lines with a single sided offset. This reduces the size of the larger elements while keeping the smaller elements the same size.

- The primary stars in systematic designs are generally only of one variety: 8-pointed for both the *fourfold system A* and *fourfold system B*, 10-pointed for the *fivefold system*, and 14-pointed for the *sevenfold system*. Occasionally, stars with double the number of points will be incorporated into patterns created from these systems. Patterns created from the *system of regular polygons* will have 6-pointed and/or 12-pointed stars; and the primary stars of patterns created from the 4.8^2 tessellation of squares and octagons will invariably be 8-pointed.

- Patterns can be modified through either additive or subtractive processes. These generally involve adding a secondary network of pattern lines that fit within the original design, but are independent of the underlying tessellation, or the removal of portions of the original set of pattern lines. In either case, these are arbitrary modifications, often subject to cultural predilections, that can substantially change the appearance of a given design.

- The lines of the primary stars in each of the five historical systems can be modified in several fashions, leading to distinctive stylistic variation of a given design. This can include an infill whereby a pattern with primary stars is transformed into a field pattern. The same type of modification to the primary stars can also be applied to nonsystematic designs.

- The primary stars of *median* patterns, *two-point* patterns, and occasionally *acute* patterns, can be modified such that the outer points are replaced with lines that extend into the primary polygons and form a new star rosette. This modification is especially associated with Mamluk aesthetics.

- Another form of pattern modification involves replacing the straight lines of a design with curvilinear lines. This produces a floriated variation that can be comprised of circles, arcs, and s-curves.

- In creating underlying tessellations with the modules from one of the design systems, interstice regions are sometimes produced. These will often produce satisfactory pattern characteristics in some, but not necessarily all of the four pattern families.

- The *system of regular polygons* makes use of underlying generative tessellations comprised of regular polygons, including the triangle, square, hexagon, and dodecagon. Some patterns created from this system also include a hexagonal ditrigon that is derived from six overlapping squares. These underlying generative tessellations correspond with the regular grids, semi-regular grids, *two-uniform* grids, and *three-uniform* grids.

- There is greater historical variability in the angular openings of the crossing pattern lines within the *system of regular polygons*. The *acute* angles within this system have 30° crossing pattern lines; the *median* have either 60° or 90°; and the *obtuse* have either 120° or 135°. Similarly, there are more types of *two-point* pattern line application within this system. The crossing pattern lines within the other four historical systems are generally limited to a single variety for each pattern family.

- In some of the earlier designs created from the *system of regular polygons*, different polygonal cells within the overall underlying tessellation were treated as either active of passive. The active cells are used to generate the pattern lines, whereas the pattern lines extend into the passive cells from their active neighbors.

- The 4.8^2 tessellation of squares and octagons is the generative basis for a very large number of historical designs. This tessellation is one of the semi-regular grids, and as such, this can qualify as part of the *system of regular polygons*. However, this is the only tessellation created from regular polygons that includes the octagon (it will not tessellate with the other regular polygons in any other manner). The octagon and square are also components of the *fourfold system A*. The fact that these two modules are shared by both systems, and that they have been used historically to create so many distinctive patterns, provides them with a stand-alone quality for separate consideration.

- The shape and proportions of the secondary polygonal modules in the *fourfold system A* and *fivefold system* are easily derived from their primary polygon: the octagon and decagon respectively. Within the *fourfold system B* and the *sevenfold system*, the secondary polygonal modules are derived primarily from interstice regions that occur when tessellating with already established modules. Modules can also be created through the truncation of other modules, through overlapping other modules, and through the union of other modules.

- The underlying primary polygons in both systematic and nonsystematic tessellations occasionally overlap with one another, creating a larger conjoined polygon. This creates dual star forms that are often very satisfactory in one or another pattern family.

- In each of the polygonal systems, the dual of some underlying tessellations are also comprised of polygonal modules from the same system. In such incidences, each will create the same geometric pattern. In this way, the same geometric design can often be created from more than a single underlying tessellation.

- Some configurations of polygons work very well with one or two of the pattern families, but very poorly with the others. For example, the arrangement of six pentagons surrounding a thin rhombus works well with the *obtuse* and *two-point* families but not with the *acute* and *median* families. In such cases, the underlying pentagons can be changed into trapezoids to produce successful designs in these latter two families. This same principle applies to nonsystematic design methodology.

- The modules of the *fivefold system* can be used at two different scales to create a single level pattern with variable pattern density. There are two edge lengths in the modules that comprise the *fivefold system*. The ratio of these edge lengths is *phi*: the golden section (1:1.6180...). The diminishing scales of the two

historical examples that employ this device are based upon this proportion. Historical examples of this type of systematic design are very rare, but this scaling feature offers tremendous scope to contemporary artists.

- Nonsystematic design methodology involves three phases: (1) the creation of the radii matrix; (2) the creation of the underlying tessellation; and (3) the creation of the geometric design. Each succeeding phase is directly dependant on its predecessor. Radii matrices are fundamental to the nonsystematic design process and examples of their historical use are found within the Topkapi Scroll.

- In each of the four pattern families, the *fivefold system* provides the aesthetic criterion for achieving success in producing nonsystematic designs. Applying methodological conventions established in the *fivefold system* to nonsystematic pattern making allows for greater design flexibility and diversity. For example, modifications to underlying tessellations that are standard to the *fivefold system* will work analogously in similar nonsystematic situations. In particular, the ring of pentagons that typically surround the decagons of the *fivefold system* is an immensely successful formative device that can also be used to great effect with nonsystematic patterns.

- Nonsystematic patterns can have a single variety of primary star or multiple star forms. Among the most interesting nonsystematic designs are those that employ several star forms that are in numeric sequence, such as 9-, 10-, 11-, and 12-pointed stars. Also of particular interest are patterns the employ the *principle of adjacent numbers*. Such patterns work on the premise that if an individual star with *n*-fold rotation symmetry works particularly well for constructing patterns, then patterns that have both (*n* + 1)-fold symmetry and (*n* − 1)-fold symmetry can also create acceptable patterns. In this way, since eight-pointed stars make very good designs, patterns with seven- and nine-pointed stars should also work well together. Similarly, 10- provides for 9- and 11-pointed stars; and 12- provides for 11- and 13-pointed stars.

- Nonsystematic patterns are particularly flexible in the varieties of repetitive cells that they encompass. Of course the square, equilateral triangle, and regular hexagon are especially common, but rectangular, rhombic, and non-regular hexagons are also represented. A limited number of patterns with radial symmetry are also known to the historical record. The inherent flexibility of this design tradition extends into the realm of repetitive grids, and there are many new approaches that open the door to original designs.

- Each of the five historical design systems lend themselves to the creation of dual-level designs with self-similar characteristics. This methodology involves scaling down the polygonal modules (with their applied pattern lines) at a ratio that allows these smaller modules to seamlessly populate the primary pattern lines and background regions.

- There are four historical varieties of dual-level design. Type A designs have a single line for the primary pattern with full coverage of the secondary pattern. Type B designs have widened lines for the primary pattern that are filled with the secondary pattern. Type C designs also employ widened lines of the primary pattern, but the secondary pattern fills both the widened lines (as per type B) as well as the background regions, thereby providing full surface coverage (as per type A). Type D dual-level designs are native to the Maghreb and differentiate the primary and secondary patterns through color.

- Each variety of dual-level design involves the placement of scaled-down polygonal modules with associated pattern lines from one or another of the five historical design systems onto key locations of the primary pattern or its underlying tessellation. The scaling ratio is determined by the secondary polygonal modules fitting edge-to-edge into the primary pattern matrix, and is always a factor of the proportional relationships inherent within the given design system: for example, *phi* for the fivefold system.

- The self-similarity of historical dual-level designs is rarely scale invariant, but isolated regions within a given design will occasionally have scale invariance. More often, the self-similarity of recursive multilevel designs is a product of their use of the same pattern family, with identical design characteristics at each successive level.

- The dual-level design methodology can, in theory, be applied recursively to multiple levels *ad infinitum*.

- Although no examples of true quasicrystalline designs are known to the historical record, the recursive dual-level design methodology can be used to create designs that meet the criteria of aperiodicity and quasicrystallinity. Such designs can be constructed so that they adhere to the Penrose matching rules and incorporate inflation and deflation.

- There are two historical conventions for applying geometric designs to the surfaces of domes. The most common employs gore segments as the radial repeat unit. The second method is rarely encountered, and projects the polyhedral symmetries of the Platonic and Archimedean solids onto the domical surfaces. The Mughals in India practiced a third technique wherein 2/10 of a two-dimensional tenfold radial pattern were removed from the design and the remaining 8/10 were closed to form a cone. This was then applied, with distortion, to the surface of the dome.

- If there is one overarching principle that is responsible for the longevity and success of this design tradition and is, indeed, fundamental to the revival of this artistic discipline it is innovation. The codified practices as outlined above are a means of working within established aesthetic constraints, but these "rules" are always open to bending and even breaking if the results are beautiful and expand the aesthetic horizons into uncharted territories. The need for innovation is no less relevant today than it was throughout the history of Islamic geometric pattern.

This conclusion is actually just the beginning of an endless quest to better understand and appreciate methodological approaches to creating beauty with geometry. May all who endeavor along this path find creative inspiration from the geometry that permeates our world and universe, from the masterpieces of geometric art of virtually every culture, and from the deep rooted affinity for geometry that is inherent within the human mind and heart.

Computer Algorithms for Star Pattern Construction

4

4.1 Introduction

The primary purpose of this book is to demonstrate the historical development, range, sophistication, and structure of Islamic geometric patterns. Throughout the exposition, it has been taken for granted that these patterns could simply be conjured into existence as computer-based drawings, with no hint at the provenance of those drawings. As it happens, the figures in this book have been produced almost entirely "by hand"; that is, they were created via manual interaction with drawing software. And indeed, a sufficiently proficient computer user should have no difficulty reproducing most of the drawings in this book by hand using software such as Adobe® Illustrator®, the freely available Inkscape, or (in Jay Bonner's case) Rhino3D® and AutoCAD®.

However, the relationship between computers and Islamic patterns need not end there. Given the level of geometric rigor that appears to underlie the construction of these patterns, we might naturally seek to automate some part of that construction by translating it into software algorithms. In any task related to the creation of ornamental designs, there are obvious benefits to computer automation. It becomes fast and painless to explore vast design spaces, without the need to execute each design laboriously by hand; mistakes can trivially be reversed; it becomes possible to generate designs procedurally that might have been difficult or impossible to work out by hand; and computer-generated designs can feed directly into a number of contemporary computer-controlled manufacturing processes to produce real-world artifacts.

As a computer scientist, the study of Islamic geometric patterns has formed one theme of my larger research program throughout my entire career. In this closing chapter, I will develop the mathematical and computational tools needed to construct and render a wide range of patterns. My goal is to present a computational take on the polygonal technique, the style of pattern construction discussed in the rest of the book. This technique was first presented by Hankin (1925a), who referred to it as the "polygons-in-contact technique."

As much as possible, I intend this chapter to be self-contained, so that the exposition will not rely heavily on consultation of other texts as background material. In some cases, this goal had to give way to practicality; for example, the full implementation of a planar map data structure (see Sect. 4.2.3) would likely require a longer explanation than this entire chapter! Naturally, I also assume some minimum level of mathematical sophistication and programming ability. The reader will need an understanding of basic linear algebra (vectors and matrices, systems of linear equations, coordinate systems, and so on), as well as the ability to implement moderate-sized algorithms in their favorite programming language.

I have assumed that the primary goal for a reader of this chapter is to implement a practical software tool, and not to study in complete depth the mathematical ideas that support its creation. To that end, I have made an effort to avoid technical jargon, and to omit complex definitions and proofs when they are not strictly necessary in order to develop software. Of course, my love for the elegance and power of geometry will inevitably reveal itself in my writing; I ask the reader to indulge me, or better yet to join me in my pursuit of mathematical beauty.

4.1.1 Example Construction

In order to get a high-level sense of the mathematical and computational techniques that will play a role in the construction of star patterns, I begin by walking through a single example. The sections that follow in this chapter will then elaborate on the steps presented here.

We begin with a tiling of the plane by regular decagons, regular pentagons, and irregular barrel-shaped hexagons, one that should by now be familiar to readers of this book (see for example Fig. 200c). The tiling is shown in Fig. 503,

J. Bonner, *Islamic Geometric Patterns*, DOI 10.1007/978-1-4419-0217-7_4

Fig. 503 An example of a tiling that will be used as an underlying polygonal tessellation in the construction of a star pattern

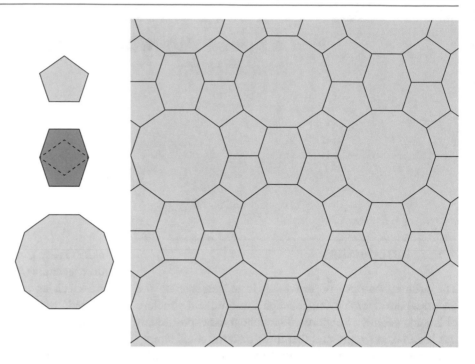

together with isolated copies of each of the three tile shapes. The hexagon's shape is easily derived by superimposing two pentagons, as shown in the figure. Note in particular that most tiles are regular polygons, which the reader can guess will be the source of the stars in the final pattern. This tiling will serve as scaffolding for the construction of a star pattern. I will use the term "template tiling" to refer to such tilings in this chapter, and occasionally also refer to them as "underlying polygonal tessellations," as in the rest of the book.

I have conjured this tiling from thin air, as it were, without offering any hint of its provenance. I am also taking for granted the fact that the tiling "exists" in a mathematical sense. Existence in this context means two things: first, that these shapes really do fit together seamlessly, without subtle gaps, overlaps, or deformations; and second, that the small excerpt shown can, in principle, be extended to cover the entire plane in an obvious way. For now, we can accept this tiling as given, and indeed could simply choose to rely on the vast library of tilings known to be useful for Islamic design, as demonstrated in this book. Later, I will discuss techniques for constructing tilings from scratch.

To construct a star pattern, we develop a motif for each unique tile shape, and copy the motifs to any desired arrangement of those tiles. The motifs must be developed so that they link together to form a seamless design.

Following the methods introduced in this book, we choose a "contact angle" θ, and invent motifs based on rays that form the angle θ with the midpoints of tile edges. The geometric process will involve truncating those rays into line segments where they meet rays coming from

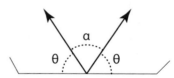

Fig. 504 A depiction of a single edge of an underlying polygonal tessellation, with two rays emanating from its midpoint. The rays are elevated at a "contact angle" of θ; equivalently, they are separated by an "angular opening" of α

other edges. Here, the contact angle represents the elevation of each ray with respect to the tile edge, as shown in Fig. 504. The earlier chapters of this book instead measure the angle formed by the rays themselves, labeled α in Fig. 504. These two measurements are related through the simple equation $2\theta + \alpha = 180°$.[1]

This construction of the rays is illustrated for the pentagon in Fig. 505. In this case, the combination of the tile shape and the contact angle of $72°$ produces a pentacle as the motif.

A similar process yields a motif for the decagon in Fig. 506. Because of the size of the tile, we use a variation of the technique applied to the pentagons. We intersect rays coming not from adjacent decagon edges, but from every second edge. Equivalently, we allow rays to travel until their second intersections, rather than their first. This variation produces a ring of kite shapes as an extra layer of geometry

---

[1] All angles in this chapter are given in degrees. However, many programming languages and libraries expect angles to be given in radians. We can convert between units by noting that $360° = 2\pi$ radians.

Fig. 505 The development of a motif for a pentagonal tile, with a contact angle of $\theta = 72°$. Rays are constructed emanating from every edge midpoint with this angle (*left*). They are cut off where they intersect other rays (*centre*). The remaining line segments define the motif, in this case a pentacle (*right*)

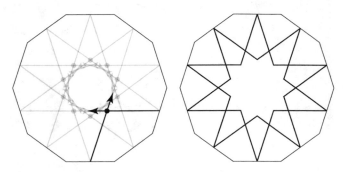

Fig. 506 The development of a motif for a decagonal tile. The rays are permitted to continue to their second intersections, producing an outer layer of kite shapes surrounding a central star

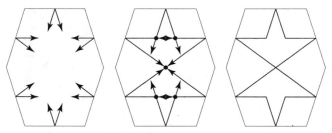

Fig. 507 The development of a motif for the barrel-shaped hexagonal tile. Because the tile is not a regular polygon, we determine all intersections separately

around a central ten-pointed star. Without this extra layer, the motifs inside large regular polygons can appear too sparse.

The situation for the barrel-shaped hexagon is not quite as simple. For tiles that are regular polygons, the intersection points must necessarily be arranged in a symmetric ring around the center of the polygon. Thus, once a single intersection is computed, all others can be constructed via rotations. With more general polygons, we may have to determine each intersection point independently. In this case, we can still intersect each ray with the opposing ray from an adjacent edge to produce the motif shown in Fig. 507. But note that we must be prepared for special cases, such as a point where two intersections coincide, as in the four rays that meet in the center of this hexagon.

As mentioned previously, we can assemble the star pattern by transferring the motifs from the three decorated tile shapes to the instances of those tiles in some final arrangement. The result will resemble the drawing in Fig. 508. Of course, while this design already has a natural geometric elegance, it is more a mathematical schematic than a finished artwork. Part of the beauty of Islamic geometric design comes from the decorative treatments applied to the strands of the design and to the regions they enclose. Two typical examples, including the very important interlaced style, are shown in Fig. 509. The generated geometry can even be used as input for computer-aided manufacturing; Fig. 510 shows an example of a real-world panel cut from Corian® (a synthetic marble-like material) using laser cutting.

To summarize, the major steps in the construction process are the specification of the template tiling, construction of motifs for the individual tile shapes, and decoration of the resulting design. I will discuss each of these steps in the following sections. I will finish the chapter with a brief overview of some new aesthetic possibilities that are made available by applying modern mathematical and computational ideas to the construction of star patterns.

4.2 Basic Building Blocks

It is helpful to begin by reviewing the basic mathematical and computational concepts that make up a toolkit for computer-based geometric art. In this section I give a general overview of the mathematics, algorithms, and data structures that are likely to be found in any two-dimensional design project based on polygons. Readers with experience in computer graphics and geometry will likely be able to skim this section or skip it altogether.

4.2.1 Points, Lines, and Polygons

Every *point* in the plane can be described via a pair of real-valued coordinates; for example, we might define a point $P = (x,y)$. If $P = (x_1,y_1)$ and $Q = (x_2,y_2)$ are distinct points,

Fig. 508 A star pattern
assembled from decorated tiles

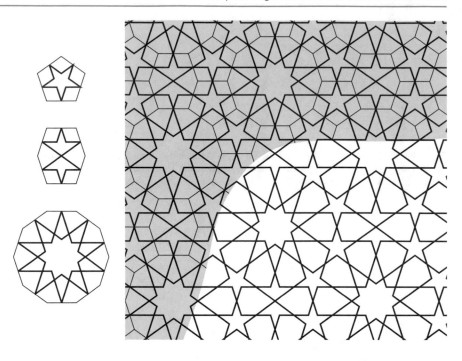

Fig. 509 Two examples of
decorative treatments applied to
star patterns: interlaced strands
(*left*) and filled regions (*right*)

then they define a unique *line l* in the plane. This line can be described in *parametric form* via the function $l(t) = P + t(Q - P) = (x_1 + t(x_2 - x_1), y_1 + t(y_2 - y_1))$. Every distinct real value of t produces a distinct point on the line. If we restrict ourselves to nonnegative values of t we obtain the definition of a *ray* emanating from P, sometimes denoted $\overrightarrow{PQ}$, consisting here of all those points R on the line such that P does not fall between Q and R. If we further restrict t to lie

between 0 and 1 (inclusive), the function $l(t)$ traces out the *line segment* that joins P and Q. A comparison between a line, a ray, and a segment is shown in Fig. 511. Lines, rays, and segments can all be represented in software via pairs of points; the interpretation of those points might therefore depend on the context.

Let A, B, C, and D be points with $A \neq B$ and $C \neq D$. Let A have coordinates (x_A, y_A), and assume that the other points

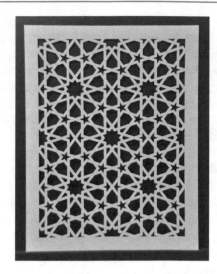

Fig. 510 An example of a star pattern executed in Corian® via laser cutting

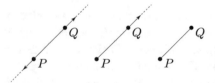

Fig. 511 Examples of a line containing points P and Q (*left*), a ray emanating from P in the direction of Q (*center*), and a line segment joining points P and Q

are defined analogously. These four points define two lines with parametric equations $l_{AB}(t) = A + t(B - A)$ and $l_{CD}(s) = C + s(D - C)$. Assuming that the two lines are not parallel, they must intersect at some unique point P. P can be found by setting $l_{AB}(t) = l_{CD}(s)$, which after some rearrangement can be expressed as the following system of linear equations:

$$\begin{pmatrix} x_B - x_A & x_C - x_D \\ y_B - y_A & y_C - y_D \end{pmatrix} \begin{pmatrix} t \\ s \end{pmatrix} = \begin{pmatrix} x_C - x_A \\ y_C - y_A \end{pmatrix}$$

The intersection can therefore be found by solving the system to obtain t and s, and computing $P = l_{AB}(t)$ (or $P = l_{CD}(s)$). This process can further be refined to intersect any combination of lines, rays, or segments by rejecting intersections that fall outside the implied legal ranges on t and s. When intersecting two rays, for example, if either of the computed values of t and s is negative then we simply decree that the rays do not intersect.

Let $P_1, \ldots, P_n$ be a sequence of n points in the plane, with $P_i = (x_i, y_i)$. (In software, this sequence might be represented using a sequence type such as a list or an array.) We construct a closed path by connecting each P_i to P_{i+1} with a line

segment, finishing by connecting P_n back to P_1. If the resulting path never intersects itself, then the P_i collectively define a *polygon*, which we take to comprise the path together with the region of the plane that it encloses. (This definition ignores more complex objects such as self-intersecting polygons or polygons with internal holes, which are not useful for our purposes.)

One important class of polygons consists of the *regular polygons*, in which all edge lengths and internal angles are identical. We can imagine tracing the outline of a *regular n-gon* by repeatedly taking a single step forward and turning left by $\frac{360}{n}$ degrees. The polygon's internal angles must therefore all be $180 - \frac{360}{n} = 180\frac{n-2}{n}$. Alternatively, some simple trigonometry allows us to inscribe a regular n-gon in a circle of radius r by setting $P_i = \left(r \cos \frac{360i}{n}, r \sin \frac{360i}{n} \right)$. One might therefore represent a regular n-gon of radius 1 in software via the integer n alone, and construct the points later if they are needed.

4.2.2   Transformations

The essence of most mathematical patterns, and of software for drawing them, is the ability to apply a set of transformations to a basic motif. Apart from the extensions to be discussed at the end of this chapter, we will make use of a set of transformations of the plane known as the *similarities*. In the plane, similarities are transformations that affect the position, orientation, and size of an object without changing its underlying shape.

Every similarity in the plane can be encoded in a 3×3 matrix for which the bottom row is fixed as $(0,0,1)$. To transform a given point $P = (x,y)$ by a similarity T in matrix form, we temporarily augment P with an additional coordinate,[2] set to 1:

$$T(P) = \begin{pmatrix} a & b & c \\ d & e & f \\ 0 & 0 & 1 \end{pmatrix} \begin{pmatrix} x \\ y \\ 1 \end{pmatrix} = (ax + by + c, dx + ey + f)$$

In software we do not need to represent the bottom row of the matrix or the extra coordinate attached to points, but it is helpful to include them when presenting the mathematical ideas.

We can immediately single out a few simple similarities, together with their matrix representations:

---

[2] The geometric meaning of this extra "homogeneous coordinate" can be made mathematically rigorous, but the details are not needed in this chapter.

Name	Description	Matrix
Translation	Displace every point by a vector $\mathbf{v} = (t_x, t_y)$; that is, move every point (x,y) to a point $(x + t_x, y + t_y)$	$\begin{pmatrix} 1 & 0 & t_x \\ 0 & 1 & t_y \\ 0 & 0 & 1 \end{pmatrix}$
Rotation	Rotate every point counterclockwise around the origin by an angle θ	$\begin{pmatrix} \cos\theta & -\sin\theta & 0 \\ \sin\theta & \cos\theta & 0 \\ 0 & 0 & 1 \end{pmatrix}$
Horizontal reflection	Reflect in the x-axis: transform every point (x,y) into $(x,-y)$	$\begin{pmatrix} 1 & 0 & 0 \\ 0 & -1 & 0 \\ 0 & 0 & 1 \end{pmatrix}$
Uniform scaling	Grow or shrink the entire plane by a factor α: transform every point (x,y) into (ax, ay)	$\begin{pmatrix} \alpha & 0 & 0 \\ 0 & \alpha & 0 \\ 0 & 0 & 1 \end{pmatrix}$

It suffices to articulate the matrices above, in the sense that any planar similarity may be expressed as a composition of similarities of these four types, and therefore computed as the product of the corresponding matrices. Note also that these 3×3 matrices can encode a broader class of transformations than similarities (a class known as the "affine transformations"); but by working only with known similarities together with their products and inverses, it is easy to avoid producing a more general matrix by accident.

The class of similarities that do not change sizes (i.e., that do not incorporate any nontrivial uniform scaling operations) are known as the *rigid motions*. The rigid motions can be thought of as those transformations that precisely preserve the shapes and sizes of objects in the plane. We can therefore use rigid motions to define congruence: we say that two shapes are *congruent* if there exists a rigid motion that brings the first into exact coincidence with the second. Congruence is the most appropriate mathematical analogue for the informal notion of two shapes being "the same."

4.2.3   Planar Maps

If our goal in constructing a drawing is to decompose the canvas into independent polygonal regions that never interact with one another, it might suffice to represent those regions via a list of polygons. However, we will encounter situations where the relationships between those regions are important as well. For example, we may want to know which other polygons share an edge with some given polygon, or which polygons contain edges that start or end at a given point. When constructing motifs in Sect. 4.4, we will need to compute the intersections of rays, and divide the plane into regions based on those intersections. In general, software that operates based on these geometrical relationships will benefit from a data structure that encodes them explicitly.

Generally, we are interested in representing decompositions of the plane into nonoverlapping *faces*, where each face is a polygon bounded by a sequence of edges, and each edge is a line segment that terminates in two *vertices*. Such geometric configurations are known by various names; I will refer to them as *planar maps*.

The key to an effective planar map implementation is to ensure that every vertex, edge, and face have direct, efficient access to the other geometrical primitives to which they are adjacent. There are a few ways to build a data structure that supports this efficiency. The most robust is the *doubly connected edge list* (DCEL), a complex, pointer-based data structure (de Berg et al., 2008). A complete DCEL implementation can be quite challenging to create. It might be more practical to trade in some of the efficiency for a simpler implementation. It may also be possible to build upon a preexisting library, such as 2D Arrangements in CGAL (www.cgal.org), or the Java Topology Suite (www.vividsolutions.com/jts).

4.3   Tilings

It should be clear by observation that a significant amount of mathematical structure underlies the polygonal tessellations used throughout this book. While some of that structure can be appreciated and applied intuitively (possibly with the aid of basic constructions by compass and straightedge), we can benefit from a more rigorous theory in the development of computational methods for star patterns. The field of mathematics known as *tiling theory* offers many important tools for exploring the range of tilings that might be used productively in Islamic design. The closely related field of *symmetry theory* will help us specify algorithms and data structures for representing simple tilings. In this section I present just enough tiling theory, focusing on tilings by polygons, to support the algorithms to follow. I also discuss a few families of especially relevant tilings.

Tiling theory is a beautiful, rich area of mathematics, full of both unsolved problems and aesthetic opportunities. It is likely to appeal to anyone with a passion for Islamic geometric design. Interested readers are encouraged to seek out Grünbaum and Shephard's seminal work on tiling theory (Grünbaum and Shephard 2016), or my introductory book on tiling theory for computer graphics (Kaplan, 2009a).

4.3.1   Patches and Tilings

In full generality, tilings can exhibit many abstruse mathematical behaviors that are not relevant in the context of Islamic geometric design. In this section, I develop a simplified theory containing only those facts about tilings that are necessary for our purposes. (In the language of Grünbaum and Shephard (2016), these would be *normal, k-hedral* tilings by polygons.)

Let $\mathcal{T} = \{\mathcal{P}_1, \mathcal{P}_2, \ldots\}$ be an infinite collection of polygons. We say that $\mathcal{T}$ is a *tiling of the plane* if the polygons in $\mathcal{T}$ cover the entire plane, leaving no gaps (points that are not covered by any polygon) or overlaps (points that are in the interior of more than one polygon).

One immediate consequence of this definition is that a tiling must necessarily be a mathematical abstraction. The fact that a given set of polygons is a tiling of the plane will always be established through high-level reasoning, rather than by enumerating an infinity of individual tiles. This implicit point of view informs the computational approach to tilings: we regard a program for tiling not as a machine that churns out an infinite list of tiles, but as a procedural rule that, when "challenged" with any finite region of the plane, responds with just enough tiles to cover that region completely. To this end, we define a *patch* to be a finite collection of nonoverlapping polygons, whose union is a single connected region of the plane with no internal holes (see Fig. 512). In light of that definition, a tiling might be viewed computationally as a function that consumes regions and produces patches.

When two polygonal tiles are adjacent in a patch or tiling, it is possible for a vertex of one polygon to lie somewhere along an edge of the other polygon. We refer to such an arrangement as a *T-junction*. A patch or tiling will be called *corner-to-corner* if it contains no T-junctions (see Fig. 513). When constructing motifs with the polygonal technique (see Sect. 4.4), we will need to watch out for tilings that are not corner-to-corner.

4.3.2 Periodic Tilings

The simple definition of a tiling given above permits tilings with no bound on the complexity of the shapes of the tiles or their arrangement. In this section I introduce the mathematical and algorithmic machinery of periodic tilings, which greatly limits both of these forms of complexity. Periodic tilings, and the broader class of periodic designs to which they belong, play a significant role in ornamental design traditions around the world (Washburn and Crowe 1992). Symmetry theory is the standard mathematical tool for characterizing repetition in patterns. A complete discussion of symmetry theory and its relationship to tiling theory would take us too far afield; interested readers should consult Grünbaum and Shephard (2016), the gentler introduction by Farmer (1996), or the more recent topology-based treatise by Conway, Burgiel, and Goodman-Strauss (2008).

A rigid motion in the plane is a *symmetry* of a given tiling if the transformed tiling lines up precisely with the original, in the sense that every transformed tile lies directly atop some untransformed tile (possibly itself). Note that the *identity* transformation, which leaves every point in the plane where it is, is a symmetry of every tiling according to this definition; thus every tiling has some nonempty set of symmetries. A tiling is *periodic* if its symmetries include translations in two nonparallel directions. These translations will necessarily form a two-dimensional family through which the tiling will repeat across the entire plane. For this reason, periodic patterns are also sometimes called *wallpaper patterns* or *all-over patterns*.

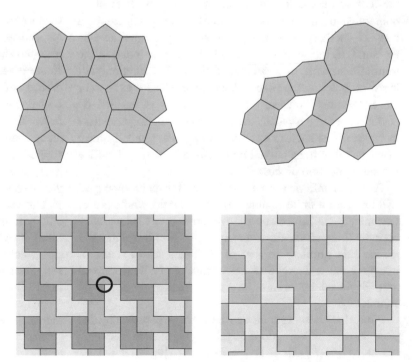

Fig. 512 An example of a patch of tiles (*left*), constructed from the tile shapes used in Sect. 4.1.1. The collection of tiles on the *right* does not constitute a patch, because the tiles are in two disconnected pieces, and one piece has an internal hole

Fig. 513 A demonstration of T-junctions in polygonal tilings. Both tilings use copies of the same L-shaped triomino. The tiling on the *left* contains many T-junctions, one of which is *circled*. The tiling on the *right* has no T-junctions, and is therefore corner-to-corner

Fig. 514 The structure of periodic tilings. The diagram on the *left* shows the tiling of Sect. 4.1.1, with a single translational unit outlined in *bold* and translational symmetry vectors $\mathbf{v_1}$ and $\mathbf{v_2}$ superimposed. On the *right*, the tiling is reconstructed from copies of the translational unit, translated by integer linear combinations of $\mathbf{v_1}$ and $\mathbf{v_2}$

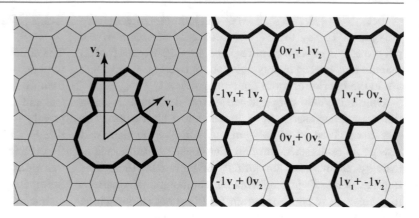

Periodicity implies a significant amount of redundancy in a tiling, a fact that can readily be exploited in creating software for manipulating periodic tilings. In particular, we can "factor out" all the redundancy, reducing the structure of a periodic tiling down to the following pieces of information:

- A finite patch of tiles, called a *translational unit* (or a "repeat unit" in previous chapters)
- Two vectors $\mathbf{v_1} = (x_1, y_1)$ and $\mathbf{v_2} = (x_2, y_2)$

An example of this decomposition is shown in Fig. 514.

The idea is that the entire tiling can be recovered by assembling transformed copies of the translational unit, translated by vectors $a\mathbf{v_1} + b\mathbf{v_2} = (ax_1 + by_1, ax_2 + by_2)$ for all integers a and b. For this construction to work seamlessly, the translational unit must be chosen carefully. It must be nonredundant, in the sense that no tile in the patch can be a copy of another tile translated by $\mathbf{v_1}$ or $\mathbf{v_2}$. It must also be maximal, in the sense that enough tiles are included in the translational unit to guarantee that the construction will leave no holes in the plane. (It is not strictly necessary for the translational unit to be contiguous, as is implied by declaring it to be a patch; but it is usually more convenient to choose it as such.) A periodic tiling will typically have many possible translational units, any of which will suffice for the replication process.

As was mentioned in Sect. 4.3.1, our computational goal will be to fill a finite region with tiles, not the whole plane. Fortunately, the information above can be used to elaborate a periodic tiling efficiently over any region $\mathcal{R}$ of the plane. We reduce this problem to that of finding a minimal set of integer pairs a_i, b_i so that the copies of the translational unit displaced by $a_i\mathbf{v_1} + b_i\mathbf{v_2}$ are sufficient to cover $\mathcal{R}$ completely.

Note that these vectors can be seen as partitioning the plane into an infinite grid of parallelograms; each has vertices at $\{a\mathbf{v_1} + b\mathbf{v_2}, (a + 1)\mathbf{v_1} + b\mathbf{v_2}, (a + 1)\mathbf{v_1} + (b + 1)\mathbf{v_2},$ $\mathbf{v_1} + (b + 1)\mathbf{v_2}\}$ for some integers a and b. These *period parallellograms* correspond one to one with copies of the translational unit. Let us temporarily consider the special case in which the translational units are precisely these parallelograms. We must determine which such parallelograms overlap a given region $\mathcal{R}$.

The solution becomes more obvious if we consider the geometry of the situation in a coordinate system where the coordinate vectors are $\mathbf{v_1}$ and $\mathbf{v_2}$, as shown in Fig. 515. This change of coordinates can be carried out explicitly via a matrix equation; the new representation (x', y') of a point (x, y) in standard Cartesian coordinates is given by

$$\begin{pmatrix} x' \\ y' \\ 1 \end{pmatrix} = \begin{pmatrix} x_1 & x_2 & 0 \\ y_1 & y_2 & 0 \\ 0 & 0 & 1 \end{pmatrix}^{-1} \begin{pmatrix} x \\ y \\ 1 \end{pmatrix}$$

In this new coordinate system, the period parallelograms are mapped to unit squares and the region $\mathcal{R}$ is mapped to a distorted shape $\mathcal{R}'$. The period parallelograms that intersect $\mathcal{R}$ are precisely the unit squares that intersect $\mathcal{R}'$. But the problem of determining which unit squares intersect a shape is analogous to the problem of representing that shape on a raster display by turning on the correct pixels. There are many standard *scan conversion* algorithms that can be applied here to "fill" $\mathcal{R}'$ (Hughes et al., 2013), and the pixel coordinates these algorithms report will correspond to the integer pairs a_i, b_i required above.

This mapping is not perfect—in general, the translational unit will only overlap the period parallelogram without filling it exactly; see Fig. 515f. With a bit more work, one can calculate how much "padding" is needed (in the form of extra integer pairs) to ensure that a given region $\mathcal{R}$ is covered in its entirety.

Frequently, a translational unit will contain additional redundancies that are not accounted for by the translation symmetries. For example, any translational unit for the tiling of Fig. 514 will contain multiple regular pentagons, which

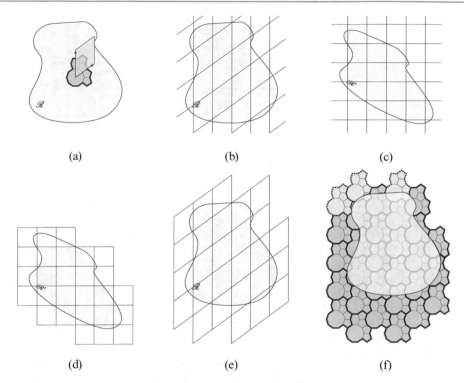

(a) (b) (c)

(d) (e) (f)

Fig. 515 A visualization of region filling for periodic tilings, based on change of coordinate systems. A region $\mathcal{R}$ is shown in (**a**), surrounding a single translational unit; the associated period parallelogram is superimposed on the unit. In (**b**), the region is embedded in a (conceptually) infinite tiling by copies of the period parallelogram. This tiling is transformed via a change of basis into a tiling by unit squares in (**c**); the same transformation produces a distorted region $\mathcal{R}'$.

Standard scan conversion produces the squares in (**d**) that overlap $\mathcal{R}'$. These squares correspond to the needed parallelograms in (**e**), which guide the placement of translational units in (**f**). Three units are drawn with *dashed outlines* in (**f**). These are not accounted for by the process, indicating that the algorithm should conservatively "pad" its answer to guarantee that $\mathcal{R}$ is completely covered

we will almost certainly wish to treat as identical. Furthermore, as indicated in Sect. 4.2.1, special-case tile shapes such as regular polygons are more easily defined in their own "native" coordinates rather than the coordinates describing their positions in a translational unit. In practice, then, we add a layer of indirection to the representation of a translational unit, storing the following information:

- An array of k distinct prototiles, each defined in whatever local coordinate system is most convenient.
- An array of *placed prototiles*: Each placed prototile is a pair consisting of an index into the array of prototiles, together with a 3×3 transformation matrix representing the similarity that transforms the prototile shape to its position in the translational unit.

In this way, whatever procedure we develop for filling tile shapes with motifs, we will need to apply this procedure only once for each unique tile shape. The resulting motifs can be placed in the final pattern using the stored transformation matrices.

4.3.3 Regular and Archimedean Tilings

The simplest tilings of the plane are corner-to-corner tilings by congruent regular polygons, known as *regular tilings*. It is easy to see that in the Euclidean plane there are only three possible regular tilings, constructed from equilateral triangles, squares, and regular hexagons, as shown in Fig. 516. We name each of these tilings using a "word," consisting of a period-delimited list of the sizes of the polygons encountered around each vertex of the tiling, leading to the names (3.3.3.3.3.3), (4.4.4.4), and (6.6.6), respectively. As in previous chapters we use exponentiation for brevity, turning these names into (3^6), (4^4), and (6^3). These tilings appear throughout the world's decorative traditions, both explicitly and as a basis for more complex design. They will be useful in Islamic geometric design both as underlying polygonal tessellations and as a starting point for constructing other tilings to serve in that role.

Let us loosen the constraints above, and consider tilings in which every vertex can be described by a word $a_1.a_2 \ldots a_n$, where each a_i is an integer greater than 2. This generalization

Fig. 516 The three regular tilings of the plane

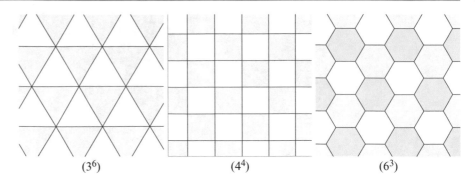

(3^6) (4^4) (6^3)

Fig. 517 The eight non-regular Archimedean tilings

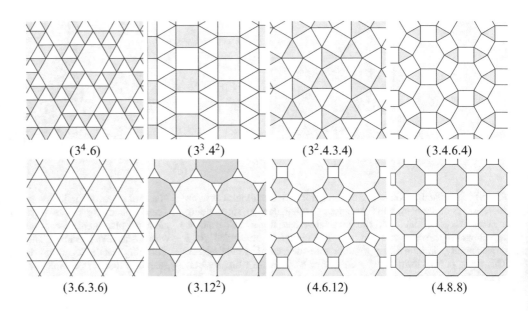

$(3^4.6)$ $(3^3.4^2)$ $(3^2.4.3.4)$ $(3.4.6.4)$

$(3.6.3.6)$ (3.12^2) $(4.6.12)$ $(4.8.8)$

introduces eight additional words that can be used at every vertex of a tiling. Canonical tilings exhibiting those words are shown in Fig. 517. These eight, together with the three regular tilings, are known collectively as the *semi-regular* or *Archimedean* tilings; within the context of Islamic geometric design, they are directly associated with the *system of regular polygons* detailed previously in this book. Many other tilings can be constructed in which every tile is a regular polygon; for example, the regular and Archimedean tilings can be seen as the first step in the enumeration of *k-uniform* tilings for all integers $k > 0$, as discussed in Grünbaum and Shephard (2016), Sect. 2.2.

4.3.4 Axis-Based Construction of Tilings

In this section I introduce a technique that can produce a large family of tilings of particular relevance to Islamic design, because they are guaranteed to contain many regular polygons. Full details on this technique, including its application in non-Euclidean geometry, can be found in an earlier article (Kaplan and Salesin, 2004). The technique is based

on identifying points in the plane where regular polygons can be placed, and scaling those polygons so that they link together to define a tiling.

Consider the regular tiling (6^3), as shown in Fig. 516. The center of every hexagon in this tiling is the focal point of a collection of symmetries of the tiling. We can rotate the tiling around this point by any multiple of $60°$ and bring it into coincidence with itself. We can also reflect the tiling across six distinct lines through this point, which pass through either the corners or the edge midpoints of the surrounding hexagon. Generally, we refer to any point that acts as the center of a $360/n$ degree rotational symmetry as an *n-fold axis*.

The hexagon is only one member of an infinite family of regular polygons that are compatible with all of these symmetries, in the sense that any member of this family can be positioned to map to itself under the given rotations and reflections. If m is any positive integer, then a regular $6m$-gon placed concentrically with the hexagon will share its rotations. Moreover, if we distinguish one ray leaving the hexagon's center along a line of reflection, then there are two distinct orientations for the $6m$-gon that are compatible with

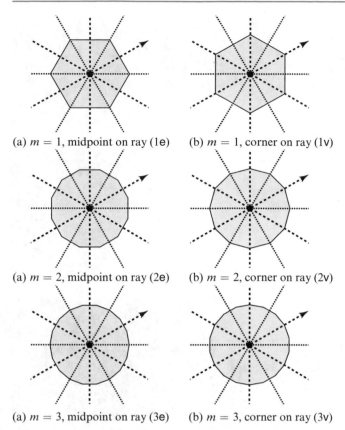

(a) $m = 1$, midpoint on ray (1e) | (b) $m = 1$, corner on ray (1v)

(a) $m = 2$, midpoint on ray (2e) | (b) $m = 2$, corner on ray (2v)

(a) $m = 3$, midpoint on ray (3e) | (b) $m = 3$, corner on ray (3v)

Fig. 518 The first six ways to place a regular $6m$-gon around a sixfold axis. The *dotted* and *dashed lines* indicate lines of reflection of a hexagon. One of the 12 rays emanating from their intersection point is distinguished for the purposes of indicating orientation, and marked here with an *arrowhead*

the lines of reflection: the ray can pass through either a corner or an edge midpoint. We can therefore represent the choice of polygon to place in this hexagon with two values: an integer multiplier m, and a Boolean indicating whether to place a corner or an edge midpoint on the distinguished ray. The first six combinations of m and orientation are visualized in Fig. 518.

In addition to the sixfold axes at hexagon centers, two other families of rotational symmetries lurk in this tiling: threefold axes at hexagon corners, and twofold axes at edge midpoints. As above, we can choose multipliers and orientations for regular polygons to be centered on these axes.

We can visualize this set of choices by constructing a 30–60–90 triangle from a hexagon center, the midpoint of an edge of that hexagon, and a vertex of that edge; we call such a triangle a *flag*. For convenience, we label the 30°, 90°, and 60° corners A, B, and C, respectively. The polygons described above will be centered at A, B, and C, and we can use the rays $\overrightarrow{AB}$, $\overrightarrow{BC}$, and $\overrightarrow{CA}$ as a basis for their orientations.

With these labels in place, let m_A, m_B, and m_C be the integer multipliers for the regular polygons at A, B, and C, and let o_A, o_B, and o_C be the choices of orientation. I use the letters e and v to denote edge and vertex (i.e., corner) orientation; for example, if o_A is e, then the polygon at A should intersect the ray $\overrightarrow{AB}$ at an edge midpoint. The complete arrangement can then be summarized using the notation $[(6^3);m_A o_A, m_B o_B, m_C o_C]$. The initial (6^3) indicates that the construction is based on the regular hexagonal tiling. Any of the multipliers can be zero, indicating that no polygon should be placed on that axis, in which case the orientation can be omitted.

This notation does not quite define a tiling, because it fails to provide (up to) three additional numbers: the sizes of the polygons. Because the ultimate goal is to produce a tiling of the plane, it is likely that we will want to scale the polygons until they come into contact with each other. With that in mind, we scale the polygons according to a portfolio of options:

1. If only one of the multipliers is nonzero, we scale the single polygon until it comes into contact with the opposite edge of the triangle. For example, in $[(6^3);2\mathsf{e},0,0]$, we would scale the 12-gon at A until it touches edge BC.
2. If two of the multipliers are nonzero, we scale the two polygons until they come into contact with each other. In general, this point of contact can take place anywhere along the flag edge between the two polygons. Usually we add the constraint that the two polygons should have the same edge length, in which case there will be unique radii that bring them into contact.

   In the context of Islamic geometric patterns, we would like these two polygons to meet with two corners or two edge midpoints on the flag edge. While we can contrive with this notation to have a corner of one polygon scale to meet the edge of another, the resulting tilings will not generally be compatible with the star pattern construction method to be described later.
3. If all three multipliers are nonzero, we choose one of the two possibilities. The most direct is based on the previous case: we pick two of the polygons and scale them until they meet and have identical edge lengths, and then scale the third until it touches the nearer of the other two. Or we can scale all three polygons simultaneously until they form a three-way contact. This approach requires solving three equations (which express the fact that the polygons touch) with three unknowns (the sizes of the polygons), suggesting that there will be a unique solution. Of course, we cannot also guarantee in this case that the three polygons will have identical edge lengths. We might therefore produce a non-corner-to-corner tiling (as in Fig. 519c), a fact that we will need to take into account later.

Fig. 519 Three examples of the axis-based construction of tilings. Each tiling is shown with the notation described in Sect. 4.3.4. From *left* to *right*, the diagram shows a flag with its vertices labeled, a flag in which regular polygons have been placed according to the notation, a scaled copy of each regular polygon and the irregular polygons created to fill the spaces between them, and the final tiling. A single translational unit is outlined in *bold*. The *bottom* example produces irregular polygons with T-junctions, one of which is *circled* in the *inset* close-up

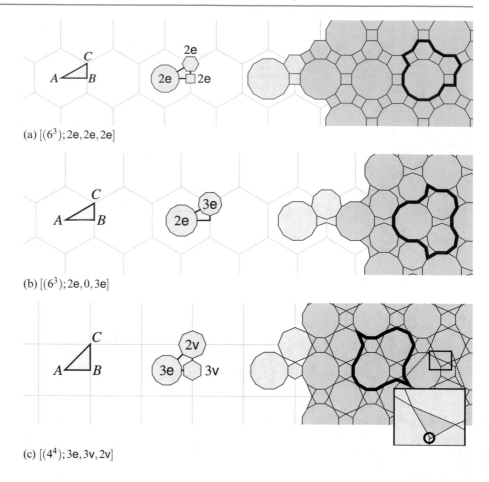

(a) $[(6^3); 2e, 2e, 2e]$

(b) $[(6^3); 2e, 0, 3e]$

(c) $[(4^4); 3e, 3v, 2v]$

We can now begin to assemble a translational unit, as described in Sect. 4.3.2, from translated, rotated, and scaled copies of these regular polygons. The translational unit will contain a single copy of the A polygon if $m_A \neq 0$, three copies of the B polygon if $m_B \neq 0$, and two copies of the C polygon if $m_C \neq 0$.

Except for a small number of fortuitous cases, the regular polygons will not completely cover the flag, implying that the tiles described so far will leave gaps in the plane. We must therefore introduce additional irregular tiles to fill these holes. The construction given above guarantees that all holes will be congruent copies of a single tile. Depending upon the particular disposition of the regular polygons, the hole filler may span more than one flag, and the translational unit may require 1, 2, 3, 6, or 12 copies of it. The hole is most easily found by first subtracting the regular polygons from the flag triangle (a standard operation in constructive planar geometry), and then possibly merging multiple copies of the resulting shape into a single larger tile.

The preceding construction was presented in terms of the regular tiling (6^3), but it applies almost unchanged to (4^4), the regular tiling by squares. The difference is that this tiling has two families of fourfold axes instead of sixfold and threefold axes. The flag is therefore a 45–45–90 triangle,

and some of the numeric constants above must be adapted to this configuration. We could also begin with the regular tiling (3^6), but this case need not be considered explicitly because it has the same axes (and hence yields the same tilings) as (6^3).

Figure 519 illustrates the preceding process in the construction of three different tilings. This notation is rich enough to encompass the three regular tilings, and five of the Archimedean tilings, as follows:

(3^6)	$[(6^3); 0, 0, 1v]$	(4^4)	$[(4^4); 1e, 0, 0]$
(6^3)	$[(6^3); 1e, 0, 0]$	$(3.4.6.4)$	$[(6^3); 1e, 2e, 1v]$
$(3.6.3.6)$	$[(6^3); 1v, 0, 1e]$	(3.12^2)	$[(6^3); 2e, 0, 1e]$
$(4.6.12)$	$[(6^3); 2e, 2e, 2e]$	$(4.8.8)$	$[(4^4); 2e, 2e, 2e]$

Other tilings used as underlying polygonal tessellations can also be interpreted as arising from related constructions. Consider, for example, the tiling shown on the right in Fig. 520, made up of decagons and bowtie-shaped hexagons. The centers of the decagons define a lattice of rhombs with angles of 72° and 108°. Can one-quarter of such a rhomb, namely a 36–54–90 triangle, function as a flag? Note that 36 and 54 are both multiples of 18, and 18° is the angle between adjacent lines of reflection passing through the center of a decagon. In other words, we can place regular $10m$-gons

Fig. 520 A tiling construction inspired by the axis-based method, applied to a tiling by rhombs with interior angles of 72° and 108°

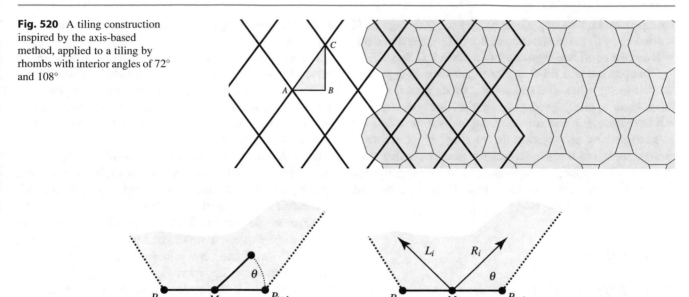

Fig. 521 The construction of the two rays L_i and R_i used to develop motifs within tiles. On the *left*, point P_{i+1} is rotated by angle θ around midpoint M_i to define the ray R_i emanating from M_i. Its counterpart L_i is constructed similarly; the two rays are shown on the *right*

around the 36° and 54° angles of this "flag" and proceed roughly as before. This tiling is closely related to the opening example in Fig. 503; the connection will be made clear in Sect. 4.4.4. In Sect. 4.6.3, we will see that this flag-based construction can be greatly expanded by considering the generalization of regular tilings to non-Euclidean geometry.

4.4 Motifs

Let us assume that we have used the techniques of the previous section to obtain a periodic tiling of the plane based on a small set of regular and irregular polygons. The next step is to produce a line-based motif for each unique tile shape. (We assume for now that we are interested only in the "raw geometry" of the design; the next section will discuss the decoration of Islamic geometric patterns in greater detail.) Once these motifs are constructed, they can be transformed into their final positions to produce a design.

As outlined in Sect. 4.1.1, the general approach is to choose a contact angle θ between 0° and 90°, and to construct two rays at the midpoint of each of a tile's edges. An edge's rays point towards the interior of the tile, forming two angles of size θ with the edge. If the corners of the polygonal tile are labeled $P_1, \ldots, P_n$, then we might denote the two rays originating at the midpoint M_i of P_iP_{i+1} by L_i and R_i; the "left ray" L_i is the one that points to the left of the edge's perpendicular bisector, and the "right ray" R_i points to the right. These rays can easily be constructed in software, as shown in Fig. 521. For example, the origin of R_i is M_i; the point defining the direction of the ray can be computed by transforming P_{i+1} by a rotation about this midpoint by angle θ.

In a computational setting, it is easy to choose any continuous contact angle between 0° and 90°. But as the earlier chapters of this book reveal, the design tradition favors a small set of contact angles based on the geometry of the underlying polygonal tessellation. For example, in the context of the fivefold system described in Chap. 3, the contact angles for the *acute*, *median*, and *obtuse* families would be 72°, 54°, and 36°, respectively.

The key to constructing motifs is to make precise the manner in which these rays encounter rays emanating from other edges, and are truncated into line segments at those meeting points.

Because regular polygons are so simple, I begin by examining them as a special case. I then proceed to develop a general algorithm that can produce motifs for a wide range of polygonal tiles.

4.4.1 Regular Polygons

Let us assume that we are trying to construct a motif for a regular n-sided polygon $P_1, \ldots, P_n$. We can further assume that the polygon's corners lie on a unit circle, as in Sect. 4.2.1. A basic motif can always be formed by finding the intersection, for every i, of rays R_i and L_{i+1} (with indices taken circularly, so that we intersect R_n and L_1). With a bit of trigonometry, we can derive an analytical expression for the locations of these intersections, but in a software implementation it is easier to write a short function that computes the intersection numerically. The rays can be constructed as above, and intersected in the manner discussed in Sect. 4.2.1. We can further simplify this process by computing a single intersection this way, and deriving the other $n - 1$ by rotating it around the origin by multiples of $360/n$ degrees. If we denote by C_i the intersection of R_i and L_{i+1}, then the motif will consist of all line segments of the form R_iC_i and C_iL_{i+1}.

This process will fill a regular n-gon with an n-pointed star. However, when n is large and θ is small, the star will fill the polygon sparsely, producing a design that typically does not have enough visual detail. Traditionally, this deficiency is mitigated by propagating the rays by one extra step, cutting them off at their second intersections rather than their first. This change is equivalent to intersecting R_i with L_{i+2} for all i. An easy heuristic is to add this extra layer of geometry whenever $n \geq 6$.

4.4.2 Other Polygons

When a polygonal tile is not regular, we can no longer rely on the simple expedient of intersecting adjacent left and right rays. Consider, for example, the bowtie-shaped hexagon mentioned at the end of Sect. 4.3.4, with a contact angle of 54°. Naively joining adjacent rays produces the ungainly motif shown on the left in Fig. 522. As seen previously in Fig. 187, the correct motif for this tile and contact angle is a pair of kite-shaped quadrilaterals, in which the long edges of

the bowtie extend rays to opposite edges rather than their neighbors.

The aesthetic of the "correct" solution seems to be derived from a kind of "economy of line," a desire to use as little ink as possible while still linking all rays. With that in mind, we proceed with the heuristic that we should seek to join rays so as to minimize the total length of the lines that make up the resulting motif.

Let us assume that we are given an n-sided polygon, with rays $L_1, R_1, \ldots, L_n, R_n$ constructed as above. We wish to find, for every left ray L_i, a corresponding right ray R_j, from which we compute the intersection C_{ij} of the two rays and insert the two segments L_iC_{ij} and $C_{ij}R_j$ into the motif. This correspondence can be expressed via the pair (i,j). A complete motif can then be found by enumerating a sequence of correspondences of the form $(1,j_1), \ldots, (n,j_n)$ (where the sequence $j_1, \ldots, j_n$ is a permutation of the numbers from 1 to n) so that the total length of the line segments thereby produced is minimized.

We might consider simply generating all permutations of the numbers from 1 to n and testing the sizes of the associated motifs, but this approach quickly becomes inefficient as n grows. In practice, a greedy approximation produces acceptable motifs far more efficiently. First, generate every possible ray correspondence (i,j) and assign that correspondence a weight w_{ij} equal to the sum of the lengths of the two line segments implied by that correspondence. If the rays do not intersect, set w_{ij} to infinity; if they point directly at each other, set w_{ij} to the length of the line segment that joins their origins. As shown in Fig. 523, place the finite-weight triples (i,j,w_{ij}) into a list, sorted by w_{ij}. Next, walk over the triples in sorted order; for every pair of rays with finite weight, add them to the motif if neither ray is yet in use. For reasonable tiles and contact angles, this iterative algorithm will produce a list of correspondences in which

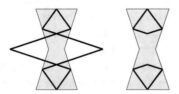

Fig. 522 The motif on the *left* was developed by joining every right ray to its neighboring left ray. On the *right*, the motif is corrected by truncating rays at closer intersection points with rays that are not from adjacent edges in the polygon

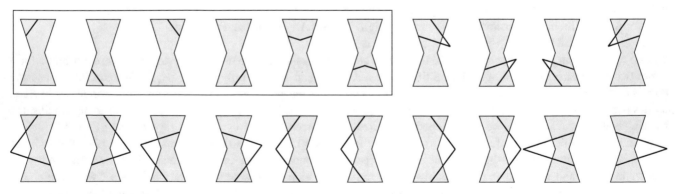

Fig. 523 A demonstration of the greedy method for constructing a motif within the bowtie-shaped irregular polygon of Fig. 522, with a contact angle of 54°. The 20 possible finite-weight ray pairings are shown, sorted by the total length of the line segments that make up the

pairing. By considering the pairings in increasing order we would use the first six (enclosed in a *rectangle*), consume all 12 rays, and produce the motif on the *right* in Fig. 522

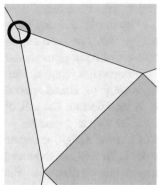

Fig. 524 The non-corner-to-corner tiling $[(6^3);3e,2v,3e]$, an example in which the motif generation algorithm must avoid growing rays out of some edges. In the close-up on the *right*, the *circled* region contains a very short fourth edge of what might appear to be a *triangular* tile. This edge should not be used as a source for rays

Fig. 525 A demonstration of how every contact position must be split into two when constructing a two-point pattern. In the diagram, the contact angle is chosen to be 45°

every ray is used, and from which the motif can be constructed.[3]

In some cases, the axis-based method of constructing tilings, as described in Sect. 4.3.4, can produce tiles that are not corner-to-corner. Consider, for example, the tiling $[(6^3);3e,2v,3e]$, shown in Fig. 524. Although it may not be obvious at a glance, careful inspection reveals that what appears to be a triangular tile is actually a dart-shaped quadrilateral with a very short fourth edge. This fourth edge is a legitimate component of the tiling, but we should not treat it as territory from which rays would emanate, most obviously because the neighboring tile opposite that edge grows no rays at the same location. For this reason, before constructing a tile's rays we consider each of its edge midpoints in turn, and discard midpoints that will not be met by a corresponding midpoint in a neighboring tile.

4.4.3   Two-Point Patterns

As discussed earlier, the *acute*, *median*, and *obtuse* pattern families use rays that emanate from the midpoints of a tessellation's polygonal edges. In *two-point patterns*, the rays grow not from a common origin, but from a pair of points equidistant from the edge's midpoint. The preceding techniques for regular and irregular polygons apply nearly unchanged for two-point patterns. From a given tile edge, construct L_i and R_i as before. Then, given a desired separation $d > 0$, displace L_i and R_i by distances $d/2$ and $-d/2$, respectively, in the direction of the tile edge. That is, compute a unit vector $\vec{u}_i$ that points along the edge and add $(d/2)\vec{u}_i$ to the two points that define the ray L_i (and similarly for R_i), as shown in Fig. 525. The remainder of the matching process can proceed as before, except that we do not permit the motif to make use of the correspondence between L_i and R_i, even though these rays now intersect. We might also need to take this extra intersection into account when decorating the resulting motif (as in Sect. 4.5).

When creating two-point patterns, it is common to set the contact angle to 45°, producing squares centered on edge midpoints. Other contact angles can be used as well, in which case rhombs will be produced instead.

4.4.4   Rosettes

When presented with a complex visual stimulus such as an Islamic geometric pattern, the human eye is predisposed to invent an explanation that accounts for that complexity (Gombrich, 1998). One way in which this predisposition manifests itself is in the tendency to "chunk" the elements

---

[3] A more principled solution is to imagine a weighted bipartite graph in which the $2n$ rays act abstractly as vertices, and L_i is connected to R_j by an edge of weight w_{ij}. The motif with the lowest total length is a minimum-weight bipartite matching of this graph, which can be found using standard flow-based techniques, as described by Cormen et al. (2001), Sect. 26.3.

of an ornamental design into larger "super-elements" that appear together. Examining the design in Fig. 508 and especially the decorated versions in Fig. 509, one such super-element demands attention. It consists of a ten-pointed star surrounded by two layers of geometry: a ring of ten kite-shaped quadrilaterals, and a ring of shield-shaped hexagons. The device is shown in isolation on the left of Fig. 526. Following Lee (1987), I refer to it as a *rosette*.

Rosettes pervade the tradition of Islamic design, as can be seen throughout the rest of this book. Indeed, as shown in Figs. 221–225, there are multiple conventions for the incorporation of rosettes within the Islamic design tradition. And yet, in the pattern of Sect. 4.1.1 they seem to arise by coincidence, as a fortuitous by-product of the arrangement of tiles in the template tiling. In this section I demonstrate two ways in which we can take the geometry of rosettes into account as part of a software system for drawing star patterns. The first extends the motif drawing algorithm to construct rosettes within regular polygons; the second intervenes earlier in the pipeline, modifying tilings so that rosettes will arise as they do in the example.

Lee articulated a simple method for constructing an n-pointed rosette within a regular n-sided polygon (Lee, 1987), illustrated on the right in Fig. 526. The construction hinges on locating what he calls the "shoulder," labeled S in the diagram. We can calculate the location of this point by imposing two aesthetically motivated regularity conditions on the rosette. The first is that the outer edges of the shield hexagons align to form (part of) the outline of a regular n-gon inscribed in the tile. This condition is equivalent to requiring that S lie on the line M_iM_{i+1} joining adjacent edge midpoints. The second constraint is that the four outer shield edges have the same length, which we can fulfill by requiring the shoulder to lie on the angle bisector $\angle OP_{i+1}P_i$ in the diagram. These two constraints are already enough to fix the shoulder position to be the intersection of a line segment and a ray. The remaining constraints on the shape of this rosette follow from requiring the sides of the shield to be parallel to its axis of symmetry.

In practice, rosettes do not always have the canonical shape suggested above (see, for example, Fig. 357). In particular, the first constraint leaves no flexibility in the rosette's contact angle, which creates problems when a template tiling contains regular polygons with different numbers of sides. In previous work I built a parameterized rosette model based on deviations from the canonical construction (Kaplan, 2000). However, it is even better to build the inevitability of rosettes directly into template tilings themselves, in such a way that any compromises that allow them to coexist in a single pattern arise naturally as a consequence of using a single, global contact angle. I present here a transformation I call the "rosette dual," which begins with a tiling containing abutting regular polygons and produces a new tiling that will lead to the formation of rosettes (Kaplan, 2005).

The rosette dual behaves similarly to the construction of motifs in Sect. 4.4, in the sense that the goal is to erect a configuration of line segments within each tile of a template tiling. But instead of declaring these "motifs" to be the endpoint of the construction process, they are stitched together to form a new underlying polygonal tessellation. The main step of the rosette dual is then to explain how to construct the necessary configurations within each tile, a process that depends on the kind of tile under consideration.

When the tile is a regular n-sided polygon with $n \geq 6$, we expect that this polygon will hold the central star within a larger rosette. We fill the tile with a concentric regular n-gon, rotated so that the corners of the internal polygon point to the edge midpoints of the original tile. The "motif" will be made up of the internal polygon, together with a set of n radial line segments joining corners and edge midpoints, as shown in Fig. 527. This process is still governed by a free parameter: the radius of the internal n-gon. We choose this radius so that the internal side length is exactly twice the length of the radial line segments. The rationale is that if a copy of the n-gon is placed adjacent to this one, the adjacent radial edges in their rosette duals will fuse to create tile edges of the same length as the internal edges.

Once again, when the tile is not regular we will face a wider variety of unpredictable situations, and should therefore expect to use a heuristic technique that performs well in practice. As shown in Fig. 528, we proceed in a manner

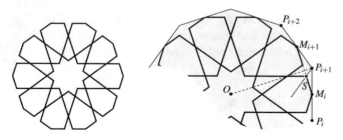

Fig. 526 An example rosette (*left*), together with an illustration of how its geometry is calculated. The angle bisector of $\angle OP_{i+1}P_i$ is shown as a *dotted line*. The intersection of this line with the line segment M_iM_{i+1} defines the "shoulder" S

Fig. 527 An example of computing the rosette dual of a regular polygon, in this case a decagon. We inscribe a smaller concentric decagon within the original, and choose r so that the length of segment AB is precisely twice that of AC

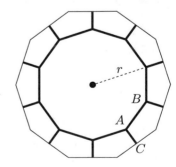

similar to motif inference. Construct a single ray for each tile edge that emanates from its midpoint at an angle of 90° (as might be constructed with a contact angle of 90° in Sect. 4.4). Truncate rays into line segments by cutting them off where they meet each other.

The preceding per-tile transformations generate a set of planar maps, one for each tile shape. We can assemble multiple transformed copies of these planar maps into a single master planar map, from which the tiles of the rosette dual may be extracted, as illustrated in Fig. 529. It is necessary to use at least a 2 × 2 grid of translational units, in order to ensure that planar map edges that form partial tiles along the boundary of the translational unit are completed by edges in neighboring units. When the maps corresponding to neighboring tiles are joined, they will leave behind spurious vertices, which lie on the line joining their two neighbors. All such vertices must be spliced out of the master planar map. The faces of the resulting planar map are all tiles of the rosette dual. With a bit of work, we can extract a subset of those faces that make up a translational unit (with the same translation vectors as the original tiling), and group them by congruence so that multiple copies of a single tile shape are encoded correctly.

The process above can fail when the polygon is sufficiently irregular. As shown in Fig. 530, groups of rays can form families of nearby pairwise intersections without converging on a single point. In practice, we must identify these families and collapse them down to single points. This process can move ray directions away from perpendicular, causing "kinks" when these rays are joined to those from neighboring tiles. The kinks must be identified (and removed) by checking for vertices in the planar map with two neighbors that form an angle close to 180°.

Fig. 528 The construction of the rosette dual of the bowtie polygon in Fig. 520. On the *left*, rays are constructed along the perpendicular bisectors of the tile's edges. The rays are truncated to form the motif on the *right*

The degree to which Islamic geometric artists understood the principle of duality as presented above, or applied duality in the historical canon, cannot be known, but we may still benefit from its explanatory powers and its practical use. The transformation exposes a deep mathematical connection between template tilings that were declared to be intuitively related previously in this book. It conforms to the alternative underlying polygonal tessellations with dual characteristics that were detailed in Chap. 3—see especially Figs. 134, 200, and 354c. It fits well with constructions such as those in Figs. 135, 145, 235, 286, 354, 359, and 377. More immediately, the rosette dual of the tiling used as an example in this section transforms into none other than the example used in Sect. 4.1.1; indeed, constructing explicit rosettes in the decagons of the original tiling yields the same final design as constructing ordinary motifs in the rosette dual.

Moreover, the rosette dual elucidates a seemingly arbitrary adjustment that takes place in several places in this book. Consider the Archimedean tiling (4.8.8), elaborated with explicit rosettes in the octagons. We might expect that building motifs in the rosette dual would produce the same design, but a small discrepancy arises: the shields of the rosettes come out uneven (see Fig. 531). The source of the problem can be seen in the superposition of (4.8.8) with its rosette dual, as in Fig. 532. The radial edge inside the octagon (edge *AB* in the figure) does not have the same length as the four edges that emerge from the center of the square (for example, *CB*), and so where a square meets an octagon they will produce an edge of the rosette dual whose midpoint does not lie on the edge between the two original tiles. However, that original edge must hold the outer tip of the rosette shield if the rosette is to be even. The solution, as demonstrated in Fig. 172, is to adjust the contact position. For tilings that are constructed via the rosette dual, we can now see the source of this adjustment, and calculate exactly how it should be carried out: an edge's contact position should be moved so it lies on the edge's intersection with the corresponding edge from the original tiling. In practice, we modify the representation of a polygonal tiling so that every edge can store an optional explicit contact position if the midpoint does not suffice.

Fig. 529 The construction of a complete rosette dual tiling. The diagrams from *left* to *right* show the original tiling with translational units outlined in *bold*, the same tiles with the motifs of Figs. 527 and 528 inscribed, and the corresponding translational units of the rosette dual tiling

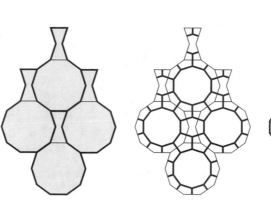

Interestingly, some template tilings presented in this book have the characteristic look of the rosette dual, and produce rosettes when motifs are built within them, but could not have arisen as the duals of simple tilings by regular polygons. Consider the template tiling shown in Fig. 435d, featuring 11- and 13-sided regular polygons surrounded by irregular pentagons. It is tempting to search for a source tiling in which the edges of the regular 11- and 13-gons meet, but as demonstrated in Fig. 435b, such a tiling is impossible. In an ingenious construction, the incompatibility of these two regular polygons is reconciled by distributing the error across the rings of irregular pentagons. In principle, it should be possible to develop an optimization algorithm that can effect similar nonsystematic constructions automatically, searching for a configuration that distributes the error in a manner consistent with the aesthetics of Islamic geometric patterns.

4.5　Decoration

The preceding two sections have provided the fundamental building blocks of Islamic star patterns based on the polygonal technique: the template tilings, and the means of

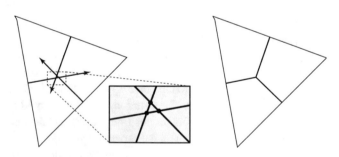

Fig. 530 A difficulty that can arise in computing the rosette dual for irregular polygons, illustrated with the dart-shaped quadrilateral of Fig. 524. On the *left*, the three perpendicular rays do not meet in a single point, but in three pairwise intersections (the short fourth edge does not contribute a ray). We pick a suitable average location and snap the endpoints of the three line segments to that location

elaborating a motif for each different tile shape. These tools are already enough to produce attractive final renderings: take the segments that make up each motif, transform them into position in a translational unit, and stamp out as many units as are needed to fill a region with the pattern.

However, apart from simple schematic drawings, Islamic art is rarely executed in such an abstract manner. Designs are richly decorated in a wide range of styles, media, and colors. Many of those styles are in some sense nonmathematical, being based on floral design or other freeform elements. But there is still a core toolkit of geometric techniques relevant in the production of decorated patterns. In this section I present three such techniques: region filling, thickening, and interlacement.

In what follows, it no longer suffices to consider the line segments that make up a motif in isolation; we must understand the complete disposition of vertices, edges, and faces of a motif within a template tile. To that end, let us assume that we have constructed a "tile planar map" for each tile shape, consisting of the polygonal boundary of the tile together with the inscribed motif. In this planar map we will use the terms "boundary edges" and "motif edges" to distinguish the edges originating, respectively, from these two sources.

4.5.1　Filling

The simplest style of decoration is to assign colors to the different regions in a pattern. In practice, we choose a color for each face in a tile planar map, and draw that face as a polygon filled with its corresponding color. In earlier chapters, this process was referred to as the "tiling treatment." Care should be taken to assign colors to boundary faces so that colors are consistent for the faces that share boundary edges. An example of a filled star pattern was shown in Fig. 509.

There are many traditional coloring schemes for Islamic geometric patterns, which we will not attempt to capture in

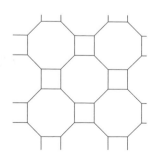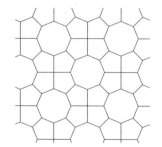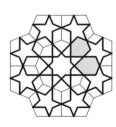

Fig. 531 An illustration of the uneven rosette shields that can arise when naively using motif inference on the rosette dual of a tiling. Unlike the perfect rosette on the *left*, the design on the *right* has shields of two different sizes

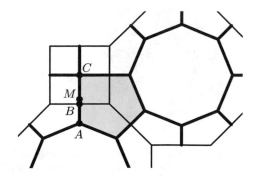

Fig. 532 A visualization of the required adjustment in contact positions in the rosette dual. The shaded pentagon arises naturally in the rosette dual of Archimedean tiling (4.8.8). If we naively use edge midpoint M as a contact position, we obtain uneven rosette shields, as in Fig. 531. The contact position for edge AC should be set to B, its intersection with the edge of the original tiling. That contact position will guarantee congruent rosette shields

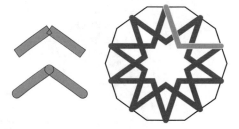

Fig. 533 When drawing thickened bands in an Islamic star pattern, it does not suffice to draw each edge of the design in isolation. Different line cap styles will produce bends that do not close properly, or bends that are rounded on the convex sides (*left*). By joining segments together at bends (*right*), we can simultaneously render sharp corners and support compatibility with motifs in neighboring tiles

algorithmic form. But there is one mathematical feature shared by many of those schemes: frequently, half of the regions are given a single background color, and the other half share a palette of foreground colors. The colors alternate, in the sense no two foreground or background tiles will be adjacent to each other. Mathematically, planar maps for which this sort of coloring is possible are said to be *two-colorable*. Not every planar map is two-colorable, but barring degenerate cases all the maps constructed as motifs in the manner described in Sect. 4.4 will be. In particular, every vertex in a tile planar map will be the meeting point of either two or four edges, which is a sufficient condition to guarantee two-colorability.

This mathematical fact suggests what could be considered a preliminary step to assist in region coloring: apply a two-coloring to the faces of every tile planar map. We can begin by picking any face containing a boundary vertex and coloring it black; we then perform a depth-first search, walking from a face to its neighbors, coloring white the neighbors of a black face, and vice versa. The example on the right in Fig. 509 shows colored tile motifs that might result from such a process, with some additional color changes applied manually for aesthetic purposes. This algorithm guarantees that like-colored map faces will meet across tile edges when tiles are assembled into a pattern.

4.5.2 Thick Bands

Another important visual style is to thicken the lines of each motif so that they are rendered as wide bands (referred to earlier in this book as "widened lines"). Doing so might superficially seem to be a simple matter of altering the desired line thickness in one's rendering library, but some care must be taken for best results. As shown in Fig. 533, we

cannot render each line segment in isolation: where motifs bend, either caps will be obvious or the joint will be rounded on one side. Instead, we join individual segments into paths wherever they meet in degree-2 vertices (bends). Each such path can then be rendered with mitered bends and circular caps (the latter will guarantee that four caps overlap seamlessly at crossings in the motif). We can even render an outline around the bands by rendering slightly thicker bands in a contrasting color in a first pass.

That being said, it is useful in some contexts (particularly interlacement, as discussed below) to compute the complete geometry of the thickened bands from first principles. To this end, we construct a polygon, which I call a "band polygon," corresponding to the thickened version of each individual motif edge in the tile planar map. The corners of the band polygon depend on the behavior of the map vertices at the start and end of the edge. Let w be the desired width of the band, and consider the situation at one of the edge's vertices, at position P (the geometry at the other vertex can be computed analogously). There are several cases to consider:

- If the vertex is a bend in the motif, the band polygon for this edge will contain the inner and outer points of its "mitered join" to the edge on the other side of the bend. If the vertex and its two edges form an angle θ, then these points will lie on the angle's bisector, at distance $w/\left(2\sin\frac{\theta}{2}\right)$ from P.
- If the vertex is the center of a four-way crossing in the motif, we add three corners to the band polygon. The first and third of these corners are the inner points of the mitered joins formed by this edge with the edges to its left and right at the vertex, computed as above; these are placed before and after the second corner, namely P itself.
- The vertex P may also lie on a boundary edge. Usually two motif edges and two boundary edges will be adjacent to P, but in the case of two-point patterns there may be a single motif edge. We determine the geometry by

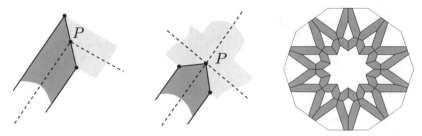

Fig. 534 The computation of the band polygons associated with the rendering of thickened motif edges. The *left* and *center* diagrams show the mitre points that must be added to the edge's polygon at motif vertex *P* in the cases that the vertex is a bend or a crossing. In the case of a crossing, *P* itself must be added as well. In the diagram on the *right*, all of the band polygons in a complete motif are outlined in *black*

imagining copies of the motif edges reflected across the tile boundary. This reflection will give the geometry of a crossing or a bend, from which the location of the polygon corners can be determined as above.

Figure 534 shows sample constructions for a bend and a crossing, together with the result of applying this construction to every motif edge in a tile. Each motif edge will yield a band polygon with four, five, or six points, which can be filled with a desired band color.

4.5.3 Interlacement

A ubiquitous decorative treatment of Islamic geometric patterns involves adding lines, shadows, or other visual cues at crossings to suggest that the crossing is constructed from two ribbons, one passing over the other. These crossings are almost always arranged so that a given ribbon might be traced through a design and be seen to pass alternately over and under the ribbons it encounters in its journey.

A simple means of indicating interlacement is to draw two parallel line segments at every crossing, aligned with the band intended to appear on top. The segments can be oriented in one of two ways at every crossing, and we must choose the orientations consistently. It turns out that there is a close connection between the two-colorability of a planar map and its ability to be represented as an interlacement, and so we can make use of the information prepared in the filled and thickened decoration styles to determine the segments that convey interlacement.

We begin by computing a two-coloring of the faces of the tile planar map, as described above in the context of filling. We then compute and draw the thickened band polygons of the previous decoration style, together with a contrasting border. Now, consider every directed motif edge of the tile planar map in turn. The edges of the planar map have no preferred direction; instead, we treat the edge joining vertices *P* and *Q* as a pair of directed edges, one running from *P* to *Q* and the other from *Q* to *P*. Note that a directed edge

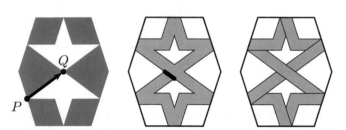

Fig. 535 A demonstration of the placement of line segments to suggest interlacement. On the *left*, we see a two-colored planar map for a tile motif. A directed edge from *P* to *Q* ends in a crossing and has a black face on its left. In the *center* diagram, we join the mitre points at *Q* with a line segment (shown artificially *thickened*). When this process is applied to all directed edges with the same properties, the finished motif on the *right* results

has a definite start and end, as well as unambiguous left and right sides (as one would experience by standing at the start vertex and facing the end vertex). With this information, there is a simple algorithm for placing line segments that suggest interlacement. If the directed edge from *P* to *Q* ends in a degree-4 vertex (a crossing), and has a black face on its left, we draw a line segment joining the mitre points at the end of the thickened edge polygon. In this context, we treat a bend adjacent to a tile boundary as a crossing. This process is illustrated in Fig. 535; a final interlaced design is shown in Fig. 509.

4.6 Extensions

The study of the mathematical structure of Islamic geometric patterns is, to some extent, an exercise in archaeology. We are the benefactors of a vast library of extant patterns, artifacts that have survived or been recreated through more than a thousand years of practice. On the other hand, we have all too little direct documentation on how these patterns were first developed. Even taking into account the evidence for the polygonal technique cited in Chap. 2 (most notably the Topkapı Scroll), the precise manner in which this methodology was put into practice will always remain open to conjecture. We are therefore left with the fascinating puzzle of reconstructing the mindsets and mathematical toolkits of artisans of centuries long past.

But the story need not end there. Regardless of how closely the techniques presented in this chapter might align with their more laborious manual counterparts, they represent a productive source of ideas that can impart momentum to the contemporary practice of Islamic art. The computer is an ideal technology for geometric design, with its capacities for tireless repetition and effortless exploration of large design spaces. Furthermore, we have access to modern mathematical ideas that were beyond the reach of the designers of historical patterns. In my research, I have been fascinated by the search for new kinds of Islamic geometric patterns that would have been inconceivable or impractical without the mathematics and computer science available to us today. I close this chapter by offering a sampling of recent work in this vein, by myself and others. In this instance I will not attempt to explain the required mathematical concepts, but I will provide references with additional information.

4.6.1 Parquet Deformations

The contact angle used to construct motifs can be varied smoothly, producing a continuum of possible designs (though for some template tilings, certain angles are more canonical than others). There is no reason why the contact angle cannot also be varied spatially within a single design, producing a pattern that undergoes a slow, graceful metamorphosis (Kaplan, 2005). I have experimented with these "spatial animations." I generate a long, narrow strip of the template tiling. Then, for each edge midpoint, I choose a contact angle for its rays based on the position of the midpoint as a fraction of the way from the start to the end of the strip. The contact angle might be different for every edge midpoint in a given tile, but the motif generation algorithm can still operate as before. Two examples of this process are shown in Fig. 536. Note that these designs are able to transition gradually between *acute*, *median*, and *obtuse* pattern families.

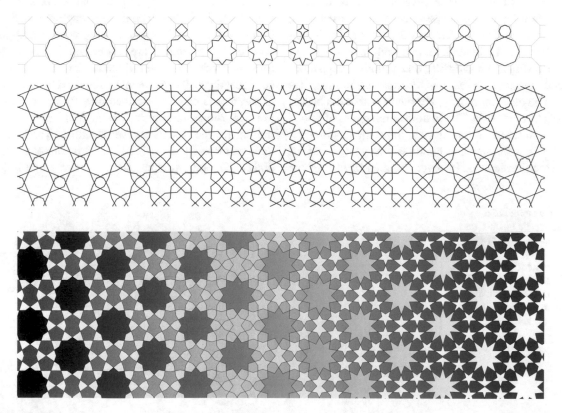

Fig. 536 Two examples of parquet deformations. The *top* diagram shows isolated motifs constructed with continuously varying contact angles, which are then elaborated into a complete design in the *middle*. The *bottom* drawing is a parquet deformation based on the tiling of Fig. 503, with colors added manually in Adobe Illustrator

I name such designs "Islamic Parquet Deformations," after the design style pioneered by Huff and described by Hofstadter (1986). They are also inspired by, and share aesthetic qualities with, Escher's use of metamorphosis (Kaplan, 2008).

4.6.2   Substitution Systems and Non-periodic Tilings

This chapter has focused on the construction of Islamic geometric patterns based on periodic tilings, but many more opportunities await in the use of tilings that are not periodic, or even in patches that cannot obviously be extended to tile the entire plane. There is a clear tradition of historical patterns with radial symmetry that depart from periodicity. For example, Castera chronicles a wide range of tilework fountains in Morocco, most of which feature a complex radial design surrounding a central many-pointed star (Castera, 1999a). The present book demonstrates several contemporary examples of non-periodic Islamic designs, including a multilevel quasicrystalline design based on a substitution tiling and Penrose-like edge matching conditions (Fig. 481); single-level and multilevel sevenfold radial patterns (Figs. 285 and 488); and a single-level pattern with forced aperiodicity as per Penrose matching rules (Fig. 480). The occurrence of aperiodicity within the historical record is problematic. The historical examples of multi-level designs in Chap. 3 have overall translation symmetry, though every translational unit contains a very large amount of geometric information. However, isolated regions of the

top-level tiles can sometimes form a substitution system. Mathematically, such isolated regions can usually be shown to correspond to non-periodic tilings, at least if the underlying tiles can be grouped any number of times into larger units.

As a first example, consider the simple tiling of the plane from rhombs with interior angles of $72°$ and $108°$, radiating outward from a single center of fivefold rotation. This tiling must be non-periodic (any translational symmetry would have to map the fivefold center onto another such point, an impossibility). As discussed at the end of Sect. 4.3.4, we can place regular decagons at every vertex of this tiling. We can then construct the rosette dual of that tiling and place motifs in the tiles to produce a non-periodic star pattern, as shown in Fig. 537. The novelty here is that we cannot rely on the periodic replication algorithm of Sect. 4.3.2; in some sense, the algorithm that fills a region with tiles from this tiling must be developed as a special case.

This example suggests that we might be able to construct star patterns in three phases instead of two, by introducing rhombs as a scaffolding to guide the placement of the template tiles. Each rhomb contains a fragment of the template tiling, with an emphasis on placing regular polygons at the corners. The fragments are chosen so that a consistent set of tiles is produced when rhombs are assembled in legal configurations. The tiles can then be used to construct motifs, or the motifs can be inscribed directly in the rhombs. Figure 538 shows one such construction based on assigning motifs to the two Penrose rhombs, which I previous demonstrated in Sect. 3.10.1 of my dissertation (Kaplan 2002). Several previous authors have proposed such an

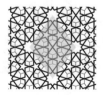

Fig. 537 An example of a simple non-periodic Islamic star pattern, created by extracting a rhombic motif from a simple periodic star pattern and replicating that rhomb in a radial arrangement outward from a fivefold center

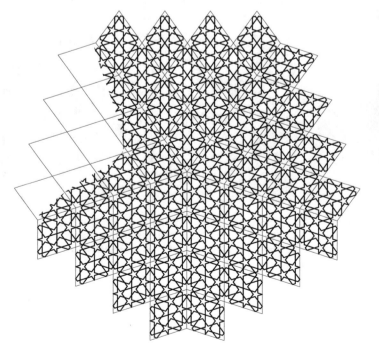

Fig. 538 The development of a non-periodic star pattern based on Penrose rhombs. In (**a**), the two Penrose rhombs are shown with decagons centered at their corners. In (**b**), the resulting template tiles are inscribed with motifs, with some editing needed to handle the two overlapping decagons in the 36° rhomb. Finally, in (**c**), the collected motifs are clipped to the rhombs and assembled into a star pattern using a small fragment of a Penrose tiling

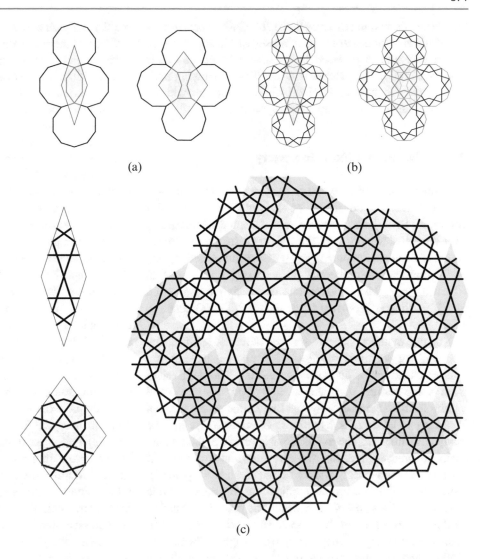

(a)

(b)

(c)

approach for making non-periodic Islamic patterns with the two Penrose rhombs, as well as with the closely related "kite and dart" tiling; as with the example above, some of these decorated tiles can also form periodic patterns (Castera, 1999b; Rigby, 2006). As shown in Fig. 480, it is also possible to construct motifs for the Penrose rhombs that conform to Penrose's original matching conditions, preventing any periodic patterns from being assembled (Makovicky et al., 1998).

The rhombic technique can be further generalized. The Penrose rhombs are an example of a *lattice projection tiling*, an orthogonal projection of a carefully selected set of faces in a high-dimensional cubical lattice (Senechal, 1996). The number of dimensions (five, in the Penrose example) can be chosen arbitrarily, producing tilings with different local orders of symmetry, made from rhombs that can accommodate regular polygons of different orders at their vertices. The main difficulties with these rhombs are choosing additional tiles to fill the spaces between the regular polygons, and dealing with the frequent occurrence of regular polygons that overlap across thin rhombs (Kaplan, 2002).

In a *substitution tiling*, every tile shape is equipped with a rule showing how it can be replaced by a configuration of smaller tiles. See Fig. 483 for an example. More information on substitution tilings can be found in Sect. 6.1 of my book (Kaplan 2009a). By starting with a single tile and iteratively applying the substitution rules, we can produce a patch of tiles of any desired size. Several mathematicians and designers have experimented with substitution as a means of producing template tilings. Research by Makovicky (Makovicky et al., 1998), Bonner (2003), Lu and Steinhardt (2007a), and Cromwell (2009) has speculated on the extent to which historical artisans may have had some intuition for the substitution process, and the extent to which these historical examples might be considered self-similar and/or quasicrystalline. This topic continues to generate considerable debate; Jay Bonner's views appear in previous chapters.

Some past work has also sought to use substitution to produce geometric patterns directly, without the scaffolding of a template tiling. These approaches are sometimes called "modular": they are based on packing the plane with the final elements of a design (Cromwell 2012b). Castera presented a substitution system based on a subset of "Zellij" tiles, in which each tile is replaced by an assembly of the others (Castera, 2003). The patches that result from applying multiple rounds of these sorts of substitutions typically do not offer any suggestion of translation symmetry.

4.6.3 Non-Euclidean Geometry

The foundations of geometry, as practiced by historical Islamic artisans, consisted of Euclid's *Elements*. For centuries, Euclidean geometry was accepted dogmatically as a correct description of the universe. In the nineteenth century, however, mathematics was revolutionized by the realization that formulations of geometry could exist in which the *parallel postulate*, one of the five axioms upon which Euclidean geometry is based, was false (Greenberg, 1993).

Today, mathematicians are comfortable leaving the familiar, flat world of the Euclidean plane and venturing into so-called *non-Euclidean* geometries. One equivalent form of the parallel postulate states that given a line L and a point P, there exists exactly one line L' that passes through P and is parallel to L. In hyperbolic geometry, there may be many such lines L'. The existence of multiple parallels may seem counterintuitive, but we can construct a number of entirely self-consistent *models* of the hyperbolic plane in which they exist. For artistic applications, the most suitable model is the *Poincaré disc*, in which the plane is defined to be the interior of a unit disc and lines are circular arcs that meet the boundary of the disc at right angles (Greenberg, 1993). There is a precedent for the use of the Poincaré disc in art, most notably in M.C. Escher's *Circle Limit* prints (Schattschneider, 2010). From the point of view of Islamic geometric patterns, nearly all the machinery we have developed in this chapter continues to operate. We can still define points, lines, rays, and segments; measure distances and angles; and construct regular and irregular polygons and more general planar maps. Periodicity is no longer a well-defined notion, but the regular and Archimedean tilings generalize into an infinite family of symmetric tilings by

regular polygons (Conway et al., 2008; Dunham et al., 1981), and there are algorithms for filling the Poincaré disc with as many tiles as are desired (Dunham, 1986; Epstein et al., 1992). Even more complex algorithms such as the axis-based construction of tilings and the rosette dual can be adapted to hyperbolic geometry.

With a bit more work, we can produce a self-consistent definition of the geometry of the surface of a sphere (Kay, 1969), where lines are *great circles*, i.e., circles whose radius is the same as that of the sphere. In this world parallel lines do not exist, but what remains is still rich enough to support the construction of Islamic star patterns.

With care, we can even formulate the construction of star patterns in a parallelism-agnostic way, so that the shape of space is merely one more parameter to the algorithm (Kaplan and Salesin, 2004). By fixing other choices (such as the multipliers in the axis-based tiling construction) and varying the choice of geometry, we can then produce conceptually related families of Euclidean and non-Euclidean patterns (see Fig. 539).

It is also possible to begin with an already defined Euclidean design, and translate it into non-Euclidean geometry (accepting some amount of distortion as part of the process). As a simple example, we might consider extracting a suitable square region from a star pattern based on the (4^4) tiling, mapping the square to the faces of a cube, and "inflating" the cube by projecting every point to the surface of a concentric sphere. Mathematicians von Gagern and Richter-Gebert (2009) demonstrated a sophisticated analogue of this process that can translate Euclidean patterns to the hyperbolic plane. As demonstrated in Fig. 540, it is even possible to map Euclidean designs to the surface of a three-dimensional model, as long as the surface can be parameterized in a suitable way (Kaplan, 2009b).

Craig Kaplan is an Associate Professor of Computer Science at the University of Waterloo in Waterloo, Ontario, Canada. He studies the application of computer graphics and mathematics to problems in art, architecture, and design and is an expert on topics such as Islamic geometric patterns and computational applications of tiling theory. He is one of the organizers of the annual Bridges conferences on art and mathematics.

Fig. 539   Three star patterns with similar underlying geometry, rendered in the Euclidean plane, the surface of a sphere, and the hyperbolic plane

Fig. 540   A two-point Islamic star pattern based on the regular tiling by hexagons, mapped onto a 3D Bunny model

Glossary

Design Methodology

Additive pattern A geometric design that is made more complex through the arbitrary addition of secondary pattern elements to an otherwise complete design [Figs. 64–67].

Angular opening In the polygonal technique, the angle created from two rays that emanate from (or occasionally near) the midpoint of a polygonal edge with a bisector that is perpendicular to the polygonal edge [Fig. 504]. The precise angular opening within a given design dictates the geometry of the applied crossing pattern lines that are located at or near the midpoint of each underlying polygonal edge, and thereby determines whether the pattern is of the *acute*, *median*, or *obtuse* pattern family [Fig. 61].

Compass work A type of pattern that is derived by laying out curvilinear design elements with a compass or dividers typically upon the coordinates of either the orthogonal or the isometric grid [Figs. 82 and 83].

Compound pattern A design that has more than a single variety of primary star: for example, 9- and 12-pointed stars. Each of these introduces higher order local symmetry within the overall pattern matrix.

Contact angle In the polygonal technique, the angle between a ray that emanates from (or occasionally near) the midpoint of a polygonal edge and the polygonal edge itself [Fig. 504]. As with the angular opening, this is another way of identifying the variable angular conditions that produce the *acute*, *median*, and *obtuse* pattern families.

Ditrigon A non-regular hexagon with threefold rotation symmetry that is occasionally incorporated into the underlying generative tessellations produced from the *system of regular polygons* [Figs. 116–121], as well as nonsystematic designs [Fig. 309]. Within the *system of regular polygons* this ditrigonal shield-like module is comprised of three 90° and three 150° included angles, with the angular proportions of six overlapping squares in threefold rotation.

Dual grid In the polygonal technique, underlying generative tessellations will often have a reciprocal grid with dual characteristics. As applied to systematic design methodology, the modules of both the initial and dual tessellations will frequently be from the same finite set of modules from a given system. Both self-dualing tessellations will create patterns with equal facility [Figs. 134 and 200]. See also *rosette dual*.

Dual-level design Designs that have primary and secondary components that are distinguished by scale. Muslim artist applied dual-level treatments to both calligraphic and floral ornament, but particular emphasis was given to geometric dual-level design. There are four distinctive fully mature historical styles: type A, type B, type C, and type D [Fig. 442]. Many of the historical examples from each of these have self-similar characteristics. Although not appearing in the historical record, the same recursive design methodology allows for the production of multilevel designs that have scale-invariant self-similarity and quasiperiodicity [Fig. 481].

Field patterns Designs that do not have a primary star form within their pattern matrix, and are consequentially more uniform in their overall density. By way of example within the *fivefold system*, the underlying generative tessellation will forgo the inclusion of decagons, thereby eliminating the presence of ten-pointed stars [Figs. 207–220]. Field patterns are also occasionally produced through the arbitrary addition of pattern elements into the primary stars [Fig. 224].

Grid method A method of generating geometric patterns by connecting specific coordinate points of a grid with line segments. Both the orthogonal and isometric grids can be used to produce patterns of this variety [Figs. 73–76]. Generally speaking, this type of design is less complex and associated with the earlier designs within this tradition. A more complex form of the grid method can be used to create designs in the Maghrebi style of Morocco [Fig. 79].

Hybrid design Designs that incorporate more than a single repetitive cell within their overall repetitive structure, thereby producing greater complexity into the pattern

© Jay Bonner 2017
J. Bonner, *Islamic Geometric Patterns*, DOI 10.1007/978-1-4419-0217-7

matrix. In their least complex form they will have only two repetitive cells [Figs. 182 and 262a], while their more complex expression will incorporate multiple individual repetitive cells [Figs. 263–265]. Contemporary designs that employ the two Penrose rhombi as repetitive cells for non-periodic and aperiodic designs can be considered a variety of hybrid design.

Imposed symmetry A category of design that incorporates local symmetries that are not ordinarily compatible with the overall repetitive structure: for example, the placement of octagons into a threefold or sixfold pattern; or the placement of primary six-pointed stars into an orthogonal repetitive structure [Figs. 51 and 52].

Nonsystematic designs In the polygonal technique, in contrast to systematic patterns, these are created from underlying tessellations comprised of polygons that are specific to the tessellation and will not reassemble into additional arrangements. This variety of design covers the full gamut of repetitive schema [Figs. 309–441].

Oscillating square designs This variety of pattern is characterized by a repetitive square element that oscillates in orientation both vertically and horizontally, and a congruent rhombic element that is placed in an alternating perpendicular layout. These have an overall orthogonal repetitive structure and will occasionally serve as a means of introducing less compatible geometric motifs into a square repeat: for example seven-pointed stars or nonagons [Figs. 23–25]. This type of pattern is closely related to *rotating kite designs*.

Parallel radii extension A relatively rare variety of design that is created from offsetting the lines of a radii matrix in both directions. The lines of the original radii matrix are eliminated, and the parallel offsets are trimmed and/or extended to meet with other offset lines [Figs. 80 and 81].

Pattern family There are four historical pattern families that are ubiquitous to this tradition: the *acute*, *median*, *obtuse*, and *two-point* families. These are intimately associated with the polygonal technique, and each results from the specific angular opening of the crossing pattern lines that are placed upon the polygonal edges of the underlying generative tessellation [Fig. 504]. The *two-point* family places the pattern lines at two points of each polygonal edge. See also *angular opening* and *contact angle*.

Point joining A design methodology that involves the use of a compass and straight edge with which one divides and segments a repeat unit, for example a square, into a formulaic network of dividing lines and circles that provides a matrix of geometric coordinates onto which pattern lines can be established by connecting selected intersection points. Each individual pattern has its own specific step-by-step construction. This methodology is more convenient for reproducing existing patterns than creating original new designs. It is also more applicable to patterns of low and medium complexity, and unsuited to designs of high complexity [Figs. 72 and 77b].

Polygonal technique This design methodology involves the application of pattern lines to designated points located on polygonal edges within a tessellation of polygons. The generative tessellation acts as a temporary scaffolding, and is removed after the design has been created, leaving behind the completed pattern. The polygonal technique has both a *systematic* and *nonsystematic* expression, and is directly responsible for the four pattern families that characterize this tradition. It is the only historical design methodology that provides the means for creating particularly complex geometric patterns with multiple regions of local symmetry. Evidence for the use of the polygonal technique is supported by the historical record. This methodology has been referred to variously as the *Hankin method* and *polygons in contact* (PIC).

Principle of adjacent numbers In the polygonal technique, the principle whereby a pattern that has a singular primary star form that conveniently covers the two-dimensional plane (for example, the 12-pointed star) is an indicator of the potential for a new compound pattern comprised of two alternating star forms that are one numerical step above and below the number of stellate points of the original (keeping our example, 11- and 13-pointed stars). In applying the principle of adjacent numbers the repetitive structure will invariably change [Figs. 429–438].

Radii matrix In the polygonal technique, a network of radii that establish regions of *n*-fold local symmetry at nodal points within a repetitive structure. Radii matrices are employed to create underlying polygonal tessellations that are, in turn, used for generating Islamic geometric patterns. As such, radii matrices are a critical feature of the polygonal technique, and are especially relevant to nonsystematic design methodology.

Ring of pentagons In the polygonal technique, the primary polygons within an underlying generative tessellation will frequently be surrounded by a ring of pentagons that benefits the construction of geometric patterns in several ways: (1) it provides for an easy way of producing the star rosettes that characterize this tradition; (2) in particularly complex patterns it can provide a buffer zone that separates disparate primary polygons, thereby allowing each primary polygon to produce regular *n*-fold stars without the problem of nonaligned connecting radii otherwise disrupting the aesthetic balance within and between these regions; (3) it provides greater potential for design variations within each of the four pattern families.

Rosette dual In the polygonal technique, the algorithmic process of applying interior perpendicular rays that are bisectors of each polygonal tile's edge. Within the primary polygons, these rays are truncated and joined to produce a smaller polygon of the same variety. As applied to each of the polygonal modules within a given template tiling, this produces the rosette dual, or dual grid [Figs. 527–529]. See also *dual-grid*.

Rotating kite designs A variety of pattern that places four quadrilateral kites in rotation around a central square. The rotating kite motif is mirrored into adjacent square cells, creating an overall reciprocating orthogonal structure [Figs. 227–229]. In addition to creating an attractive motif in its own right, the rotating kite structure was occasionally used to introduce regions of local symmetry that are atypical to orthogonal repetitive structures. This variety of pattern is closely related to *oscillating square designs.*

Subtractive pattern A geometric design that is made from removing portions of an otherwise complete pattern [Fig. 126f].

Superimposed pattern A geometric design comprised of two otherwise distinct geometric designs that are superimposed onto one another [Fig. 68].

Systematic designs In the polygonal technique, in contrast to nonsystematic designs, a category of design that employs underlying polygonal modules that can be used interchangeably to create innumerable individual generative tessellations that are thereby systematic. Each of these distinct modules is associated with its own set of applied pattern lines in each of the four pattern families. The rearrangement of these polygonal modules is a very fast and effective method of creating geometric patterns. The five historical design systems are the *system of regular polygons*, the *fourfold system A*, the *fourfold system B*, the *fivefold system*, and the *sevenfold system.*

Template tiling An alternative term for an underlying polygonal tessellation.

Underlying polygonal tessellation In the polygonal technique, a polygonal tessellation that is used for generating geometric patterns. Pattern lines are placed upon key points of the polygonal edges, and after the pattern is completed, the underlying polygonal tessellation is discarded. Sometimes referred to as a *sub-grid.*

Technical Terms

Aperiodic A tessellation of more than one type of repetitive unit cell that can only cover the two-dimensional plane in a non-periodic manner and thereby has forced aperiodicity (as per the Penrose matching rules). Within this work, the term *aperiodic* also applies to geometric patterns that are based upon such tessellations [Fig. 480].

Fundamental domain As pertains to two-dimensional geometric patterns, the fundamental domain is the minimal essential repetitive component of a repeat unit or design. The fundamental domain will sometimes provide translation symmetry on its own, or it can be subject to rotation,

reflection, and/or glide reflection to provide translation symmetry [Figs. 2–4]. It is also referred to as the *fundamental region.*

Golden ratio A proportional ratio that is inherent to the pentagon and decagon and therefore to the *fivefold system* of pattern generation. It is generally denoted by the Greek letter *phi* [φ], and is simply expressed as $(1 + \sqrt{5}) \div 2 = 1.618033987....$

Inflation/deflation The primary property of substitution tiling whereby scaled versions of a geometric figure are hierarchically applied in sequence *ad infinitum*. This can also be expressed as subdivision rules: for example a square is divided into quadrants that are in turn divided into further quadrants, etcetera. Inflation and deflation are applicable to self-similar structures with translation symmetry as well as to non-periodic structures.

Local symmetry A region within a tessellation or geometric design that has *n*-fold rotation symmetry. In geometric patterns, regions of local symmetry are typically expressed as higher order star forms: for example, the ten-pointed stars within fivefold patterns.

Non-Euclidean As pertains to geometric design, the coverage of three-dimensional surfaces including spherical and hyperbolic.

Non-periodic A tessellation of the two-dimensional plane that cannot be constructed from a single type of repetitive cell and is without translation symmetry. However, unlike aperiodic repetitive structures, the repetitive unit cells of such tessellations can also be assembled into periodic structures with translation symmetry. Within this work, the term *non-periodic* also applies to geometric patterns that are based upon such tessellations [Fig. 538].

Penrose matching rules Supplemental edge conditions that are applied to the 2 fivefold Penrose rhombi, or alternatively to the fivefold Penrose kite and dart tiles, that control their edge matching properties and thereby force aperiodicity. In each of these two sets of prototiles there are two types of edge condition [Fig. 483].

Penrose rhombi Two 5-fold rhombic prototiles, one with 36° and 144° included angles, and the other with 72° and 108° included angles. These will cover the plane with translation symmetry either independently or in combination with one another, and will also cover the plane non-periodically. The supplemental inclusion of Penrose matching rules to these rhombi forces aperiodicity. Both of these rhombi were used within the Islamic geometric design tradition, but always periodically [Fig. 5].

Periodic The covering of the two-dimensional plane with a repetitive cell that has translation symmetry. Each of the 17 plane symmetry groups provides periodic surface coverage [Figs. 53–56]. Periodic tiling with regular polygons is of particular relevance to Islamic geometric design, and includes regular, semi-regular (or uniform), 2-uniform, 3-uniform, etc. tilings [Figs. 89–91].

Quasicrystalline Of or relating to a quasicrystal. As pertains to geometric patterns, designs that exhibit a quasiperiodic structure; for example, that of the aluminum-manganese alloy discovered by Dan Shechtman and revealed in electron diffraction patterns that are represented as a two-dimensional quasicrystal diffractograms.

Quasiperiodic A repetitive recurrence of a geometric figure that has irregular periodicity. To date, the terminology associated with non-periodic tilings with hierarchic inflation and deflation substitution rules is still not formally defined. As pertains to the geometric designs within this work, *quasiperiodicity* is interchangeable with *quasicrystallinity*.

Repeat unit Within this work, a repetitive unit cell with translation symmetry. This is distinct from differentiated repetitive cells that can be used in association with one another to create an overall repeat unit (as with hybrid designs), or grouped together to create non-periodic tessellations. See also *translational unit*.

Rotation symmetry The rotation of a figure around a single point. Along with translation, reflection, and glide reflection, rotation is one of the conditions that govern the 17 plane symmetry groups. Within geometric patterns, rotation symmetry provides the regular *n*-fold symmetry of the primary star forms. Also within the tradition of Islamic geometric design, rotation symmetry provides the repetitive structure of radial patterns [Figs. 260, 440 and 441] and domical patterns based upon gore segmentation [Figs. 490–495]. Such radial patterns do not have translation symmetry and are therefore a category of non-period design. As with periodic patterns from this tradition, radial geometric designs can have *scale invariant self-similarity* [Figs. 484, 485, 488, and 489].

Scale invariance In self-similar structures the property of exact duplication at each hierarchic sequence of inflation and deflation *ad infinitum*. With very rare exception, Islamic dual-level designs are not scale invariant.

Self-similarity The property of an object or overall structure to have an identical or analogous scaled-down substructure that, in the abstract, is or can be hierarchically scaled down *ad infinitum*. This recursive process follows a proportional scale with a ratio that is determined by the inherent geometry of the overall structure. Because Islamic dual-level designs are not scale invariant, which is to say that the secondary pattern is not identical to the primary pattern, they are better described as having self-similar characteristics rather than true self-similarity. They are only self-similar by virtue of their primary and secondary patterns being created from modules of the same generative system, and thereby sharing the same pattern characteristics at both levels. However, even with

this looser definition of self-similarity, such designs can in theory have infinite recursion.

Seven frieze groups The seven possible combinations of translation, rotation, reflection, and/or glide reflection that apply to linear bands. All geometric repetitive border designs must adhere to one or another of these seven frieze groups.

Seventeen plane symmetry groups The 17 possible combinations of translation, rotation, reflection, and/or glide reflection that apply to all two-dimensional surface coverage with conventional translation symmetry [Figs. 53–56]. These are also referred to as the *17 wallpaper groups*, and the *17 plane crystallographic groups*.

Subdivision rule For the purposes of this work, the subdivision of the polygons that make up a generative tessellation into smaller polygons. This can be applied with infinite recursions, for example, the subdivisions that provide for the inflation and deflation of the Penrose rhombi [Fig. 483]. See also *substitution tiling*.

Substitution tiling A method of introducing self-similarity by replacing each tile module within a tessellation or pattern with a configuration of smaller tiles [Fig. 483]. Substitution tiling can apply to both periodic and non-periodic structures. See also *subdivision rule* and *inflation/deflation*.

Tessellation A tiling comprised of a single or more than a single closed geometric figure (tile or polygons). A tessellation can have translation symmetry or be non-periodic. Tessellations can employ polygons that are either or both regular and irregular. The classification of tessellations comprised solely of regular polygons includes regular, semi-regular (uniform), 2-uniform, 3-uniform, etc. [Figs. 89–91]. Within the polygonal technique, tessellations provide the underlying structure from which geometric patterns are constructed. See also *underlying polygonal tessellation*.

Tiling The filling of the two-dimensional plane with discrete closed unit cells (tiles or polygons) in a manner whereby they neither overlap nor create interstices. A tiling can have translation symmetry or be non-periodic.

Translation symmetry The copying of a discrete figure by moving it (translating it) to a new vector without changing the figure. In periodic tiling, translation symmetry provides for a repeat unit to cover the two-dimensional plane seamlessly and without gaps. Among the 17 plane symmetry groups, translation symmetry joins with rotation symmetry, reflection, and glide reflection, to provide the geometric conditions for filling the two-dimensional plane.

Translational unit A repetitive unit cell with translation symmetry also referred to as the *repeat unit*.

Non-English Terms

Ablaq A distinctive form of inlayed stone ornament that is typically floral and dichromatic wherein the foreground and background are stylistically the same.

Alif The first letter of the Arabic alphabet that is comprised of a single ascending stroke. The geometry and proportion of the *alif* are fundamental determinants of the cursive calligraphic styles, for example, *Thuluth*.

Allah Arabic word for God used by all Muslims regardless of their parent language. This word provides a frequent motif in *Shatranji Kufi* ornament.

al-Jazirah The region of upper Mesopotamia between the Tigris and Euphrates rivers in what is now southeastern Turkey, northwestern Iraq, and eastern Syria.

Artesonado The Spanish word for a form of wooden vaulted or coffered ceiling with geometric ornament that is typical to Morocco and Muslim Spain, and was maintained by *Mudéjar* artists after the successful *reconquista* in 1492.

Atabeg A Seljuk hereditary title of nobility for a provincial governor who was a subordinate of the sultan.

Ayah A verse from the Quran.

Bandi-rumi Persian for a variety of Anatolian design that is derivative of ornamental knot-work.

Banna'i Persian term for ornamental brickwork ornament. Literally, "of or pertaining to bricklayers" (*banna*).

Beylik A Turkish term for an Anatolian province governed by a local lord, or *Bey*, that was used during the Seljuk Sultanate of Rum and by their successors.

Caliph Title for the religious successor to the Prophet Muhammad and dynastic head of a Caliphate.

Girih Persian term for knot that is used in the textile arts, and is also applied to geometric patterns. Also transliterated as *gereh*, as in *gereh-sazi*, meaning wooden joinery panels with an interlocking geometric design.

Gunbad Middle Persian word for dome and frequently associated with tomb towers. The term *gumbad* is also used.

Hadith Attributed reports of the words and actions of the Prophet Muhammad. Only the Quran is more important among Muslims, and the *Hadith* are fundamental to Islamic jurisprudence (*Sharia*).

Han Turkish word for caravanserai.

Iwan A prominent rectangular façade with a rectangular or vaulted arched opening that serves as the grand entry into a mosque, prayer hall, or mausoleum. The four *iwan* courtyard became a primary feature of many religious buildings throughout Muslim societies, and the *iwans* were sometimes used as open halls for public gatherings.

Jali A variety of pierced stone screen, frequently of geometric design, that originated and was popularized during the Mughal period in the Indian subcontinent.

Kaaba The cubical building in the center of the courtyard of the Al-Masjid Al-Haram (Grand Mosque) in Mecca toward which all Muslims face during their prayers. This houses the black Kaaba stone (al-Hajar al-Aswad). This is the most sacred site for Muslims.

Katshkerim An Armenian carved stone relief panel embedded into the exterior walls of a church or cathedral, often depicting a cross or religious text, and occasionally a geometric pattern.

Khanqah Persian word for a monastery associated with *Sufi* teaching and retreat, and often located adjacent to the tomb of a *Sufi Shaykh*, mosque, or madrasa.

Khatchkar An Armenian carved stone stele that is ornamented with a cross and often includes geometric and floral designs that are stylistically derivative of the stone ornament from the Seljuk Sultanate of Rum.

Khurasan A historical region that includes portions of present-day northeastern Iran, northern Afghanistan, Tajikistan, southern Kyrgyzstan, southeastern Uzbekistan, and eastern Turkmenistan. Principal cities included Nishapur, Tus, Mashhad, Herat, Balkh, Bukhara, Samarkand, and Merv.

Kubachi ware A variety of polychrome underglaze ceramics created in northeastern Persia during the Safavid period, but mistakenly named after the city of Kubachi in Dagestan where a large number of these vessels were found.

Kufi One of the earliest Arabic scripts; named after the city of *Kufa* in central Iraq. This is characterized by angularity and contrasts with the many varieties of Arabic cursive scripts used by calligraphers. *Kufi* was the focus of significant stylistic attention over the centuries, with regional forms and variations associated with artistic media: for example, the *Shatranji Kufi* herringbone brick ornament of Persia and Khurasan.

Lājvard Persian for lapis lazuli, and the deep blue pigment produced from this stone. Mistaken by some as the ceramic colorant responsible for *lājvard* ware in which the closely similar dark blue is produced from cobalt oxide.

Madrasa Arabic word for an educational institution often associated with the teaching of theology and religious law.

Maghreb Arabic word for "sunset," and refers to the region of western North Africa.

Mihrab A niche in the *qibla* wall of a mosque that indicates the direction of Mecca toward which the congregation faces during prayers. The mihrab of a mosque typically receives special ornamental attention.

Minai'i ware A Persian type of on-glaze enameled ceramic ware dating to the twelfth and thirteenth centuries that is associated with the city of Kashan. *Minai'i* ware is polychromatic with highly detailed brushwork that is evocative of the Persian miniature tradition.

Minaret A tower associated with a mosque that is used by the *muezzin* to call Muslims to prayer.

Minbar A typically freestanding stairway placed against the *qibla* wall of a mosque from which the imam addresses his congregation. As with the mihrab, minbars typically receive significant ornamental attention.

Mosque The place of worship for Muslims. In Arabic, *Masjid*; in Persian, *Masjed*; and in Turkish, *Camii* or alternatively *Cami*. A *Jama Masjid*, or *Masjid-i Jami*, is a congregational Mosque where the *Jumu'ah* prayers are performed each Friday; hence the English use of the term Friday Mosque.

Mozarab From the Arabic musta'rib, meaning Arabized. This term was given to Spanish Christians who were greatly influenced by the Muslim culture of al-Andalus, including the adoption of the Arabic language.

Muarak Persian word for cut-tile mosaic.

Mudéjar The term given to the Muslims of al-Andalus who lived under Christian rule, as well as to Muslims who remained in al-Andalus after the final *reconquista* and did not convert to Christianity. The term is also used to describe the Christian architectural style that is heavily derivative of the Islamic architecture and architectural ornament of this region.

Muhandis Arabic word for a person who works with geometry, architecture, engineering, or surveying: basically, anyone who measures—from which the root word derives.

Muqarnas A three-dimensional ornamental device comprised of geometric components that transition between the horizontal and vertical within a building. As such, muqarnas was used extensively in ceiling vaults and domes, in niche hoods, for cornices, and as column capitals.

Mu'tazilite Of or pertaining to the Islamic theological doctrine of *Mu'tazila* that emphasized reason and rationalism and was centered in Iraq during the eighth to tenth centuries.

Pishtaq Persian for a typically arched entry portal set within a projecting rectangular façade frequently found on tomb towers.

Quran The holy scripture of Islam, revealed to the Prophet Muhammad by Allah through the angle Gabriel. The *Quran* is the primary inspiration and subject of the calligraphic arts among Muslim cultures.

Rasmi vault Persian word for a type of vault characterized by a large central star with points that extend down to the periphery. Frequently referred to as *star vaulting* or *squinch net vaulting*.

Reconquista The 770-year-long attempt to rest control of the Iberian peninsula away from Muslim rule and return it purely to Christian control. This was finally achieved in 1492 when Ferdinand and Isabella accepted the surrender of Muhammad XII, the last Nasrid Emir of Granada.

Sabil A fountain or drinking fountain generally established for public use as an act of charity.

Shatranji Kufi A distinctive form of *Kufi* calligraphy that is highly geometric. Also known as "square *Kufi*" or "chessboard *Kufi*" this script universally employs 90° changes of direction, and is consequentially difficult to read. The 90° angles made it very practical as a primary form of brickwork ornament in Persia and Khurasan.

Shaykh Arabic honorific title. In the context of this book, *Shaykh* typically denotes the leader of a Sufi order who is authorized as a teacher.

Shi'a The branch of Islam that considers Ali ibn Abi Talib to be the legitimate successor of the Prophet Muhammad and the first Imam. Ali was married to the Prophet Muhammad's daughter Fatimah.

Sinagoga The Spanish word for Synagogue.

Sufi An adherent of Sufism, the principal vehicle of mysticism and esotericism within Islam. Sufis practice *Tasawwuf*, the inner dimension of Islam, and are inclined toward asceticism. Sufism has a long history, with many Sufi orders (*taruq*) active throughout Muslim societies. The degree to which Sufism has been perceived as heterodox has occasionally put individual Sufi masters, or whole Sufi orders, into conflict with the political and religious authorities. Yet in other times and places, Sufism was firmly rooted within the fabric of society and benefited from the patronage of the nobility. Many founders and prominent *Shaykhs* of these orders were enshrined within tombs that are elaborately ornamented.

Sultan A dynastic ruler with sovereign authority. While not a Caliph, sultans often accepted the religious supremacy of the Caliph.

Sunni The branch of Islam that recognizes Abu Bakr as the first Caliph. He was the father of the Prophet Muhammad's first wife Aisha.

Surah A chapter of the Quran.

Tawhid The fundamental Islamic concept for the indivisible oneness of Allah.

Thuluth One of the six principal historical cursive calligraphic scripts.

Transoxiana The region north of the Oxus River (Amu Darya) that was referred to as Turan by the Persians. It equates with the Central Asian portion of Greater Khurasan and covers the territory between the Amu Darya and the Syr Darya rivers.

Tumar A scroll. In the context of this book, *tumar* were used by artists to record ornamental designs, including geometric patterns, although few examples have survived to the present.

Turkish triangle A Turkish structural and ornamental device for transitioning between the horizontal and vertical that is comprised of multiple flat triangular faces that act as pendentives that support a dome. This same triangular aesthetic was also used for cornices and in column capitals.

Vizir A high-ranked administrative official under an Emir, Sultan, or Caliph.

Yezdi bendi A variety of vault that has affinities with both *muqarnas* and *rasmi* squinch net vaulting primarily associated with the architecture of the Mughal Empire.

Zillij The Moroccan Arabic word for cut-tile mosaic.

References

ABAS, SYED JAN and AMER SHAKER SALMAN. 1995. *Symmetries of Islamic Geometrical Patterns*. Singapore: World Scientific Publishing.

AJLOUNI, RIMA. 2012. The Global Long-Range Order of Quasiperiodic Patterns in Islamic Architecture. *Acta Crystallographica: Section A, Foundations in Crystallography*.

ALLEN, TERRY. 1999. *Ayyubid Architecture*, Occidental, CA: Solipsist Press.

ALLEN, TERRY. 1988. *Five Essays on Islamic Art*. Sebastopol, CA: Solipsist Press.

ALLEN, TERRY. 2003. Portal of the Bimaristan Arghun. *Ayyubid Architecture*. Occidental, CA: Solipsist Press.

ASLAKSEN, HELMER. 2006. In Search of Demiregular Tilings. *Bridges London: Bridges Conference Proceedings*. London.

ATIL, ESIN. 1982. *Art of the Mamluks*. Washington D.C.: Smithsonian Institution Press.

AZARIAN, LEVON. 1973. *Armenian Khatchkars*. Holy See of Etchmiadzin.

BAKIRER, OMUR. 1981. *Selcuklu Oncesi Ve Selcuklu Donemi Anadolu Mimarisinde Tugla Kullanimi*; Ankara: Orta Dogu Teknik Universitesi.

BARLOW, WILLIAM. 1894. "Über die Geometrischen Eigenschaften homogener starrer Strukturen und ihre Anwendung auf Krystalle", *Zeitschrift für Krystallographie und Minerologie*, vol. 23,

BIER, CAROL. 2012. The Decagonal Tomb Tower at Maragha and its Architectural Context: Lines of Mathematical Thought. *Nexus Network Journal* (4) 3. Springer.

BIXLER, HARRY. 1980. *A Group Theoretical Analysis of Symmetry in Two-Dimensional Patterns from Islamic Art*. Ph.D. thesis, New York University.

BLAIR, SHEILA. 1991. *The Monumental Inscriptions from Early Islamic Iran and Transoxiana*. Leiden: Brill.

BLOOM, JONATHAN. 1988. The Introduction of the Muqarnas into Egypt. *Muqarnas V; An Annual on Islamic Art and Architecture*, Grabar, O. [ed.] Leiden: E. J. Brill.

BONNER, JAY F. 2000. *Islamic Geometric Pattern: Their Historical Development and Traditional Methods of Derivation*. Unpublished. (Library of Congress copyright TXu000981014/2000-08-21).

BONNER, JAY F. 2003. Three Traditions of Self-Similarity in Fourteenth and Fifteenth Century Islamic Geometric Ornament. R. Sarhangi, N. Friedman (eds.) *Meeting Alhambra: ISAMA Bridges 2003 Conference Proceedings*. Granada, Spain: University of Granada, Faculty of Sciences, pp. 1–12.

BONNER, JAY F. 2016. The Historical Significance of the Northeast Dome Chamber of the Friday Mosque at Isfahan. *Nexus Network Journal*. Springer

BONNER, JAY F. and MARC PELLETIER. 2012. A 7-Fold System for Creating Islamic Geometric Patterns, Part 1: Historical Antecedents. *Bridges Towson: Mathematics, Music, Art, Architecture, Culture, Annual Conference Proceedings*, 141–148. Maryland: Tessellation Publishing.

BOURGOIN, JOULES. 1879. *Les Eléments de l'art arabe: le trait des enterlacs*. Paris: Librairie de Firmin-Didot et Cie.

BROUG, ERIC. 2008. *Islamic Geometric Patterns*, London: Thames and Hudson.

BROUG, ERIC. 2013. *Islamic Geometric Design*, London: Thames & Hudson.

BULATOV, MITKHAT SAGADATDINOVICH. 1988. *Geometricheskaia garmonizatsiia v arkhtekture Srednei Azii IX-XV VV* [Geometric harmonization in the architecture of Central Asia from the ninth to the fifteenth century]. Rev. ed. Moscow: Izdatel'stvo "Nauka." Original edition, Moscow: Izdatel'stvo "Nauka," 1978.

BURCKHARDT, TITUS. 1976. *Art of Islam: Language and Meaning*, London: World of Islam Festival Publishing Company.

CAIGER-SMITH, ALAN. 1985. *Lustre Pottery: Technique, Tradition and Innovation in Islam and the Western World*. London: Faber and Faber.

CASTÉRA, JEAN-MARC. 1996. *Arabesques; Art Decoratif au Moroc*, ACR Édition Internationale.

CASTÉRA, JEAN-MARC. 1999a, *Muqarnas and Quasicrystals*. N. Friedman, J. Barrallo [eds.], ISAMA 1999.

CASTÉRA, J. M. 1999a. *Arabesques: Decorative Art in Morocco*. ACR Edition.

CASTÉRA, J. M. 1999b. Zellijs, Muqarnas and Quasicrystals. N. Friedman, J. Barrallo (eds.) *ISAMA 99 Proceedings*, pp. 99–104.

CASTERA, J. M. 2003. Play with Infinity. J. Barrallo, N. Friedman, J.A. Maldonado, J. Martínez-Aroza, R. Sarhangi, C. Séquin (eds.) *Meeting Alhambra, ISAMA-BRIDGES Conference Proceedings*. Granada, Spain: University of Granada, pp. 189–196. Available online at http://castera.net/entrelacs/public/articles/granada3.pdf.

CASTÉRA, JEAN-MARC. 2011. Flying Patterns. *Bridges Coimbra: Mathematics, Music, Art, Architecture, Culture, Annual Conference Proceedings*, 263–270.

CASTÉRA, JEAN-MARC. 2016. Persian Variations. *Springer: Nexus Networks Journal*.

CHORBACHI, WASMA'A KHALID. 1989. "In the Tower of Babel: Beyond Symmetry in Islamic Design", *Computers & Mathematics with Applications*, Vol. 17, No. 4–6.

CHORBACHI, WASMA'A KHALID. 1992. "An Islamic Pentagonal Seal (From Scientific Manuscripts of the Geometry of Design)." *Five Fold Symmetry*. Istvan Hargittai [ed.]. Singapore: World Scientific.

CHRISTIE, ARCHIBALD. 1910. *Traditional Methods of Pattern Designing: and Introduction to the Study of Decorative Art*. Oxford: Oxford University Press.

CONWAY, J. H., BURGIEL, H. and GOODMAN-STRAUSS, C. 2008. *The Symmetries of Things*. A. K. Peters

COOMARASWAMY, ANANDA K. 1944. The Iconography of Durer's "Knots" and Leonardo's "Concatenation". *The Art Quarterly*. Detroit, VII. 2.

CORMEN, T. H., LEISERSON, C. E., RIVEST, R. L. and STEIN, C. 2001. *Introduction to Algorithms*, 2 edn. The MIT Press.

CRANE, HOWARD and WILLIAM, TROUSDALE. 1972. Helmand-Sistan Project Carved Decorative and Inscribed Bricks from Bust. *East and West*, Vol. 22, No. 3/4.

CRESWELL, KEPPEL ARCHIBALD CAMERON. 1969. *Early Muslim Architecture: Umayyads, Early Abbasids, and Tulunids*. Revised and published in two parts. Oxford: Clarendon Press.

CRITCHLOW, KEITH. 1976. *Islamic Patterns: an Analytical and Cosmological Approach*. New York: Schocken Books.

CROMWELL, PETER. 2009. The Search for Quasi-Periodicity in Islamic 5-fold Ornament. *The Mathematical Intellegencer* 31(1), 36–56.

CROMWELL, PETER. 2010. Islamic Geometric Design from the Topkapi Scroll I: Unusual Arrangements of Stars. *Journal of Mathematics and the Arts* 4 (2).

CROMWELL, PETER. 2012a. Analysis of a Multilayered Geometric Pattern from the Friday Mosque in Yazd. *Journal of Mathematics and the Arts* 6 (4).

CROMWELL, P. R. 2012b. A Modular Design System Based on the Star and Cross Pattern. *Journal of Mathematics and the Arts* 6(1), 29–42. DOI 10.1080/17513472.2012.678269.

CROMWELL, PETER. 2015. Cognitive Bias and Claims of Quasiperiodicity in Traditional Islamic Patterns. *The Mathematical Intellegencer* 37 (4).

CROMWELL, PETER. 2016. Modularity and Hierarchy in Persian Geometric Ornament. *Nexus Network Journal*. Springer

CROMWELL, PETER and ELISABETH BELTRAMI. 2011. "The Whirling Kites of Isfahan: Geometric Variations on a Theme." *Mathematical Intelligencer* 33.

DE BERG, M., CHEONG, O., VAN KREVELD, M. and OVERMARS, M. 2008. *Computational Geometry: Algorithms and Applications*, third edn. Springer-Verlag.

DODDS, JERRILYNN D. [ed]. 1992. *Andalucía: The Art of Islamic Spain*; New York: The Metropolitan Museum of Art.

DUNHAM, D. 1986. Hyperbolic Symmetry. *Computers and Mathematics with Applications* 12B(1/2), 139–153.

DUNHAM, D., LINDGREN, J. and WITTE, D. 1981. Creating Repeating Hyperbolic Patterns. Computer Graphics (*Proc. SIGGRAPH*), pp. 215–223.

EL-SAID, ISSAM. 1993. *Islamic Art and Architecture: The System of Geometric Design*. Reading, UK: Garnet Publishing. Published posthumously.

EL-SAID, ISSAM and AYSE PARMAN. 1976. *Geometric Concepts in Islamic Art*. London: World of Islam Festival Publishing Company.

EPSTEIN, D. B. A., CANNON, J. W., HOLT, D. F., LEVY, S. V. F., PATERSON, M. S. and THURSTON, W. P. 1992. *Word Processing in Groups*. Jones and Bartlett.

ERNST, CARL. 1997. *The Shambhala Guide to Sufism*. Boston: Shambhala Publications.

ETTINGHAUSEN, RICHARD, OLEG GRABAR and MARILYN JENKINS-MADINA. 2001. *Islamic Art and Architecture 650-1250*. New Haven and London: Yale University Press.

FARMER, DAVID. 1996. *Groups and Symmetry: A Guide to Discovering Mathematics*. Providence, Rhode Island: American Mathematical Society.

FEDOROV, EVGRAF. 1891. Simmetrija na ploskosti [Symmetry in the plane]. *Zapiski Imperatorskogo Sant-Petersburgskogo Mineralogicheskogo Obshchestva* [Proceedings of the Imperial St. Petersburg Mineralogical Society].

FURUZANFAR, BADI' AL-ZAMAN. 1956. *Ahadith-i Mathnawi* (Hadith Sayings of the Mathnawi) Tehran.

GOLOMBEK, LISA and DONALD M. WILBER. 1988. *The Timurid Architecture of Iran and Turan*. 2 vols. Princeton, NJ: Princeton University Press.

GOLOMBEK, LISA, ROBERT MASON, PATRICIA PROCTOR and EILEEN REILLY. 2013. *Persian Pottery in the First Global Age: The Sixteenth and Seventeenth Centuries*. Brill.

GOMBRICH, E. H. 1998. *The Sense of Order: A Study in the Psychology of Decorative Art*, second edn. Phaidon Press Limited.

GRABAR, OLEG. 1990. *The Great Mosque of Isfahan*, New York University Press.

GREENBERG, M. J. 1993. *Euclidean and Non-Euclidean Geometries: Development and History*, third edn. W. H. Freeman and Company.

GRÜNBAUM, B. and SHEPHARD, G.C. 2016. *Tilings and Patterns*, second edn. Dover.

GRÜNBAUM, BRANKO, ZDENKA GRÜNBAUM and G. C. SHEPHERD. 1986. "Symmetry in Moorish Art and other Ornaments." Springer: *Computers & Mathematics with Applications*: 12B, 641–653.

HANKIN, ERNEST HANBURY. 1905. "On Some Discoveries of the Methods of Design Employed in Mohammedan Art". London: *Journal of the Society of Arts*, 53.

HANKIN, ERNEST HANBURY. 1925a. *The Drawing of Geometric Patterns in Saracenic Art*, vol. 15. Calcutta: Memoirs of the Archeological Survey of India, Government of India.

HANKIN, ERNEST HANBURY. 1925b. "Examples of Methods of Drawing Geometrical Arabesque Patterns". London: *Mathematical Gazette* 12: 371–373.

HANKIN, ERNEST HANBURY. 1934. "Some Difficult Saracenic Designs II: A Pattern Containing Seven-Rayed Stars". *Mathematical Gazette* 18: 165–168.

HANKIN, ERNEST HANBURY. 1936. "Some Difficult Saracenic Designs III." *Mathematical Gazette* 20: 318–319.

HARGITTAI, ISTVAN. 1986. *Symmetry: Unifying Human Understanding*. New York: Pergamon Press.

HERZFELD, ERNST. 1922. *Die Gumbadh-i Alawiyyan*. A Volume of Oriental Studies: Presented to Professor Edward E. Brown. Cambridge University Press.

HERZFELD, ERNST. 1943. *Damascus: Studies in Architecture-II*. AI 10.

HERZFELD, ERNST. 1954-56. *MCIA. Troisième partie: Syrie du Nord. Inscription et monuments d'Alep*. Vol. 1, MIFAO.

HILLENBRAND, ROBERT. 1994a. *Islamic Architecture: Form Function and Meaning*. New York: Columbia University Press.

HILLENBRAND, ROBERT. 1994b. The Relationship Between Book Painting and Luxury Ceramics in 13th-Century Iran. *The Art of the Saljūqs in Iran and Anatolia: Proceedings of A Symposium Held in Edinburgh in 1982*, Costa Mesa, CA.

HOAG, JOHN D. 1977. *Islamic Architecture*. New York: Harry N. Abrams.

HOFSTADTER, D. 1986. *Metamagical Themas: Questing for the Essence of Mind and Pattern*. Bantam Books.

HOGENDIJK, JAN. 2012. "Mathematics and Geometric Ornament in the Medieval Islamic World." *Newsletter of the European Mathematical Society*; Issue 86.

HUGHES, J. F., VAN DAM, A., MCGUIRE, M., SKLAR, D. F., FOLEY, J. D., FEINER, S. K. and AKELEY, K. 2013. *Computer Graphics: Principles and Practice*, 3rd ed.. Boston, MA, USA: Addison-Wesley Professional.

HUTT, ANTONY and HARROW, LEONARD. 1979. *Iran 2*. London: Scorpion Publications.

JONES, OWEN. 1856. *The Grammar of Ornament*. London: Day and Son.

KAPLAN, CRAIG. 2000. Computer Generated Islamic Star Patterns. R. Sarhangi (ed.) *Bridges 2000 Proceedings*.

KAPLAN, CRAIG. 2002. *Computer Graphics and Geometric Ornamental Design*. Ph.D. thesis, Department of Computer Science & Engineering, University of Washington.

KAPLAN, CRAIG. 2005. "Islamic Star Patterns from Polygons in Contact." *GI '05: Proceedings of the 2005 Conference of Graphics Interface*, pp. 177–185. Canadian Human-Computer Communications Society.

KAPLAN, CRAIG. 2008. Metamorphosis in Escher's Art. *Bridges 2008: Mathematical Connections in Art, Music and Science*, pp. 39–46.

KAPLAN, CRAIG. 2009a. *Introductory Tiling Theory for Computer Graphics*. Morgan & Claypool.

KAPLAN, CRAIG. 2009b. Semiregular Patterns on Surfaces. *NPAR '09: Proceedings of the 7th international symposium on Non-photorealistic animation and rendering*, New York: ACM Press. pp. 35–39.

KAPLAN, CRAIG and SALESIN, D. H. 2004. Islamic Star Patterns in Absolute Geometry. *ACM Transactions on Graphics* 23(2), 97–119. DOI http://doi.acm.org/10.1145/990002.990003.

KAY, D. C. 1969. *College Geometry*. Holt: Rinehart and Winston, Inc.

KEPLER, JOHANNES. 1619. *Harmonices Mundi* [The Harmony of the World]. Austria.

KITZINGER, ERNST. 1965. "Stylistic Developments in Pavement Mosaics in the Greek East from the Age of Constantine to the Age of Justinian." *La Mosaïque Greco-Romaine*, Colloques Internationaux du Centre National de La Recherche Scientifique. Paris.

KUHNEL, ERNST. 1962. *Islamic Art and Architecture*; Cornell University Press.

LALVANI, HARESH. 1982. *Coding and Generating Islamic Patterns*. Ahmedabad, India: National Institute of Design.

LALVANI, HARESH. 1989. Coding and Generating Complex Periodic Patterns. *The Visual Computer*: Volume 5, Issue 4, pp 180–202.

LEE, ANTHONY J. 1987. "Islamic Star Patterns," *Muqarnas* 4:182–197.

LINGS, MARTIN. 1976. *The Quranic Art of Calligraphy and Illumination*; London: World of Islam Festival Trust.

LOVRIC, MIROSLAV. 2003. "Magic Geometry, Mosaics in the Alhambra, Meeting Alhambra." *Meeting Alhambra: ISAMA Bridges Conference Proceedings*. Granada, Spain: University of Granada, Faculty of Sciences.

LU, PETER and STEINHARDT, PAUL 2007a. Decagonal and Quasicrystalline Tilings in Medieval Islamic Architecture. *Science* 315 (5815), 1106–1110.

LU, PETER and PAUL STEINHARDT. 2007b. "Response to Comment on Decagonal and Quasi-Periodic Tilings in Medieval Islamic Architecture." *Science* 318.

MAHERONNAQSH, MAHMUD. 1976. *Five Volumes on Geometric Patterns (Persian)*. Tehran: Reza Abbasi Museum.

MAKOVICKY, EMIL. 1989. "Ornamental Brickwork, Theoretical and Applied Symmetrology and Classification of Pattern", *Computers and Mathematics with Applications*, vol. 17, nos. 4–6.

MAKOVICKY, EMIL. 1992. 800-Year-Old Pentagonal Tiling from Maragha, Iran, and the New Varieties of Aperiodic Tiling It Inspired. *Fivefold Symmetry* [ed. Hargittai, I.]. Singapore-London: World Scientific.

MAKOVICKY, EMIL. 1994. "*The O-D Pattern from "Puerta del Vino", Alhambra, Granada*". *Boletín de la Sociedad Española de Mineralogía*, vol. 17.

MAKOVICKY, EMIL. 1995. "The Twin Star Mosaic from the Niches of the "Sala de la Barca", "Salon de Comares", Alhambra, Granada." *Boletín de la Sociedad Española de Mineralogía*, vol. 18.

MAKOVICKY, EMIL. 1997. "Brick and Marble Ornamental Patterns from the Great Mosque and the Madinat al-Zahra Palace in Cordoba, Spain: I, Symmetry, Structural Analysis and Classification of Two-Dimensional Ornaments." *Boletín de la Sociedad Española de Mineralogía*, vol. 20.

MAKOVICKY, EMIL. 1998. "Decagonal Patterns in the Islamic Ornamental Art of Spain and Morocco", *Boletín de la Sociedad Española de Mineralogía*, vol. 21 (1998).

MAKOVICKY, EMIL. 1999. "Coloured Symmetry in the Mosaics of the Alhambra, Granada, Spain", *Boletín de la Sociedad Española de Mineralogía*, vol. 22.

MAKOVICKY, EMIL. 2000. *Structure of the domes of pavilion in the Patio del los Leones, the Alhambra: a distorted octacapped truncated octahedron*; Boletín de la Sociedad Española de Mineralogía, 23.

MAKOVICKY, EMIL. 2007. "Comment on "Decagonal and Quasi-Periodic Tilings in Medieval Islamic Architecture"." *Science* 318.

MAKOVICKY, EMIL and P. FGENOLL HACH-ALI. 1996. "Mirador de Lindaraja: Islamic ornamental patterns based on quasi-periodic octagonal lattices in Alhambra, Granada, and Alcazar, Sevilla, Spain." *Boletín Sociedad Espanola Mineralogía* 19.

MAKOVICKY, EMIL and MILOTA MAKOVICKY. 1977. "Arabic Geometric Patterns—A Treasury for Crystallographic Teaching." *Neusis Jahrbook für Mineralogie Monatshefte*, 2.

MAKOVICKY, E., PÉEREZ, F. R. and HACH-ALÉ, P. F. 1998. Decagonal Patterns in the Islamic Ornamental Art of Spain and Morocco. *Boletín Sociedad Española Mineralogía* 21, 107–127.

MAMEDOV, K. H. S. 1986. "Crystallographic Patterns". *Computers & Mathematics with Applications*, Vol. 12B, Nos. 3/4.

MARCHANT, PAUL. 2008. "The Essential Structure of Geometry in Nature." *The Minbar of Saladin; Reconstructing a Jewel of Islamic Art* [ed. Lynette Singer]. New York: Thames & Hudson.

MOLS, LUITGARD E. M. 2006. *Mamluk Metalwork Fittings in their Artistic and Architectural Context*. Delft, Netherlands: Eburon Academic Publishers.

MS PERSIAN 169, fols. 180r-199r. Fi tadakhul al-ashkal al-mutashabiha aw al-mutawafiqa (*On Similar and Complementary Interlocking Figures*). Paris: Bibliothèque Nationale de France [Anonymous and undated Persian treatise].

MÜLLER, EDITH. 1944. *Gruppentheoretische und Strukturanalytische Untersuchungen der Maurischen Ornamente aus der Alhambra in Granada*. Ph.D. diss., University of Zurich, Riischlikon.

NECIPOĞLU, GÜLRU. 1995. *The Topkapi Scroll–Geometry and Ornament in Islamic Architecture*. Santa Monica, CA: The Getty Center for the History of Art and the Humanities.

NECIPOĞLU, GÜLRU ed. Forthcoming. *The Arts of Ornamental Geometry: A Persian Compendium on Similar or Complementary Interlocking Figures (A Volume Commemorating Alpay Özdural)*. Leiden, Boston: Brill.

O'KANE, BERNARD. 1987. *Timurid Architecture in Khurasan*. Costa Mesa, CA: Mazdâ Publishers.

O'KEEFFE, MICHAEL and BRUCE HYDE. 1996. "Crystal Structures: Patterns and Symmetry." *Mineralogical Society of America Monographs*. Washington, D.C.

ÖZDURAL, ALPAY. 1995. "Omar Khayyam, Mathematicians, and *Conversasioni* with Artisans". *Journal of the Society of Architectural Historians*, 54.

ÖZDURAL, ALPAY. 1996. "On Interlocking Similar or Corresponding Figures and Ornamental Patterns of Cubic Equations". *Muqarnas*.

ÖZDURAL, ALPAY. 1998. "A mathematical sonata for architecture". *Technology and Culture*, vol. 39.

PANDER, KLAUS. 1982. *Sowjetischer Orient: Kunst und Kultur, Geschichte und Gagenwart der Völker Mittelasiens*. Köln: DuMont Buchverlag.

PELLETIER, MARC and JAY F. BONNER. 2012. "A 7-Fold System for Creating Islamic Geometric Patterns, Part 2: Contemporary Expression." *Bridges Towson: Mathematics, Music, Art, Architecture, Culture, Annual Conference Proceedings*, 149-156. Maryland: Tessellation Publishing.

PENROSE, ROGER. 1974. "The Role of Aesthetics in Pure and Applied Mathematical Reseach." *Bulletin of the Institute of Mathematics and its Applications*, 10, No. 7/8.

PENROSE, ROGER. 1978. "Pentaplexity," *Eureka* 39: 16-22. Reprinted (1979) as "Pentaplexity: A Class of Non-Periodic Tilings of the Plane," *The Mathematical Intelligencer*, 2: 32-37.

PÉREZ-GÓMEZ, RAFAEL. 1987. "The four Regular Mosaics Missing in the Alhambra". Springer: *Computers & Mathematics with Applications*: 12B, 641-53: Vol. 14, No. 2, pp. 133–137.

PICCARD, ANDRÉ. 1983. *Le Maroc et l'artisanat traditionnel islamique dans l'architecture*. Annecy, France: Editions Atelier 74.

PÓLYA, GEORGE. 1924. Über die Analogie der Kristallsymmetrie in der Ebene. *Zeitschrift für Kristallographie*: Vol. 60.

POPE, ARTHUR U. 1965. *Persian Architecture: the Triumph of Form and Color*. George Braziller.

RABBIT, NASSER. 1996. "Al-Azhar Mosque: An Architectural Chronicle of Cairo's History", *Muqarnas—An Annual on the Visual Culture of the Islamic World*. Brill.

REMPEL', LAZAR I. 1961. *Arkhitekturnyi ornament Uzbekistana [The architectural ornament of Uzbekistan]*. Tashkent: Gos. Izdatel'stvo khudozh. Lit-ry UzSSR.

RIGBY, J. 2006. Creating Penrose-Style Islamic Interlacing Patterns. *Bridges 2006: Mathematical Connections in Art, Music and Science*, pp. 41–48.

ROGERS, J. M. and R. M. WARD. 1988. *Suleyman the Magnificent*. Secaucus, New Jersey: Wellfleet Press.

SALIBA, ROBERT ed. 1994. *Tripoli, the Old City: Monument Survey—Mosques and Madrasas*. Beirut: American University of Beirut, Department of Architecture.

SALTZMAN, PETER. 2008. Qausi-Periodicity in Islamic Design. *Nexus VII: Architecture and Mathematics*, Kim Williams Books.

SCHATTSCHNEIDER, DORIS. 1990. *Visions of Symmetry: Notebooks, Periodic Drawings, and Related Work of M. C. Escher*. New York: W. H. Freeman and Company.

SCHATTSCHNEIDER, D. 2010. The Mathematical Side of M.C. Escher. *Notices of the AMS* 57(6).

SCHIMMEL, ANNEMARIE. 1990. *Calligraphy and Islamic Culture*. London: I.B. Tauris and Co.

SCHNEIDER, GERD. 1980. *Geometrische Bauornamente der Seldschuken in Kleinasien*. Wiesbaden: Reichert.

SCHÖNFLIES, ARTHUR. 1891. *Krystallsysteme und Kristallstruktur*, Leipzig: Teubner.

SENECHAL, M. 1996. *Quasicrystals and Geometry*. Cambridge University Press.

SINGER, LYNETTE ed. 2008. *The Minbar of Saladin*. London: Thames and Hudson.

TABBAA, YASSER. 2001. *The Transformation of Islamic Art during the Sunni Revival*. Seattle and London: University of Washington Press.

VON GAGERN, M. and RICHTER-GEBERT, J. 2009. Hyperbolization of Euclidean Ornaments. *The Electronic Journal of Combinatorics* 16(2).

WADE, DAVID. 1976. *Pattern in Islamic Art*, Woodstock, New York: Overlook Press.

WARD, RACHEL. 1993. *Islamic Metalwork;* London: British Museum Press.

WASHBURN, DOROTHY KOSTER and DONALD CROWE. 1988. *Symmetries of Culture*. Seattle, Washington: University of Washington Press.

WASHBURN, D. K. and CROWE, D. W. 1992. *Symmetries of Culture*. University of Washington Press.

WASHBURN, DOROTHY KOSTER and DONALD CROWE. 2004. *Symmetry Comes of Age: the Role of Pattern in Culture*. Seattle, Washington: University of Washington Press.

WATSON, OLIVER. 1973-75. *Persian Lustre-Painted Pottery: Rayy and Kashan Styles*. Transactions of the Oriental Ceramic Society [40], pp. 1–19.

WATSON, OLIVER. 1985. *Persian Lustre Ware*. London; Boston: Faber & Faber.

WEYL, HERMANN. 1952. *Symmetry*. Princeton, New Jersey: Princeton University Press.

WILBER, DONALD N. 1939. The Development of Mosaic Faience in Islamic Architecture n Iran. *Ars Islamica*, Ann Arbor, Michigan: University of Michigan Press.

WILBER, DONALD. 1955. *The Architecture of Islamic Iran: the Ilkhanid Period*. Princeton, New Jersey: Princeton University Press.

WOLFF, HANS E. 1966. *Traditional Crafts of Persia*. Cambridge, MA: MIT Press.

Index

A

Abbasid
Abbas ibn Abd al-Muttalib, 15
Abbasid Palace of the Qal'a, Baghdad, 45–47, 54, 81, 292,
326, 380, 381, 397, 405, 440
Abu'l-Abbas as-Saffah, 15
Atshan palace, 23
Bab al-'Amma, Samarra, 22
Baghdad Gate, Raqqa, 23
Bulawara Palace, Samarra, 22
caliph al-Ma'mun, 24
caliph al-Musta'sim, 84, 114
caliph al-Mu'tadid, 17
Mujda minaret, 23
Mustansiriyah *madrasa,* Baghdad, 39, 45, 93, 132,
336, 381, 405
No Gumbad mosque, Balkh, 22, 254
Ukhaidir Palace, 23
'Umar al-Suhrawardi mausoleum, Baghdad, 38, 176, 322, 397
'Abd al-Salâh Abû Bakr, 66, 541
Abd er-Raham III, 21
Abd er-Rahman I, 20, 21
Ablaq, 58, 63, 65, 85, 171, 318, 579
Abu al-Husayn bin Muhammad al-Harrani 'Abd Allah, 66, 542
Abu 'Ali Muhammad ibn Muqlah. *See* Calligraphy
Abu al-Wafa al-Buzjani, 16, 49, 203, 207, 212
Abu Bakr al-Khalil, 205, 211
Abyaneh, Iran
Friday Mosque of Abyaneh, 21, 33, 36, 228, 232
Additive patterns, 39, 62, 116, 118, 126, 138, 153, 199, 200,
404, 509, 575
Adud ad-Dawla, 26
Aghlabid
Great Mosque of Kairouan, 18
Great Mosque of Sfax, 441
Agra, India
Agra Fort, 121, 171
I'timad ad-Dawla mausoleum, 71, 89, 119, 121, 123, 125,
129, 189, 234, 276, 352, 467
Taj Mahal, 118, 151, 258
Agzikara, Turkey
Agzikara Han, 75, 79, 83, 151, 180, 192, 326, 389, 457
Ahlat, Turkey
Huseyin Timur tomb, 71, 240
Usta Sagirt tomb, 71, 240
Ahmed ibn Tulun, 5, 18–22, 25, 33, 62, 89, 138, 219, 220, 243,
254, 260, 291
Aksaray, Turkey
Alay Han, 73, 80, 95, 272, 389
Cincikh mosque, 71, 241
Great Mosque at Aksaray, 83, 134, 210

Aksehir, Turkey
Great Mosque of Aksehir, 82
Seyit Mahmut Hayrani tomb, 72, 253
Al-Andalus, 3, 9, 20, 21, 32, 85, 105, 106, 111, 112, 139, 145,
149, 215, 228, 291, 493
Ala'uddin Hussain, 49
Alawid
Bahia Palace, Marrakech, 114
Moulay Ishmail mausoleum, Meknès, 109, 293
Moulay Ismail Palace, Meknès, 107, 244
Albrecht Durer, 150
Aleppo, Syria
'Abd al-Salâh Abû Bakr, 66
Altinbugha mosque, 55, 99, 290
Bab Antakeya, city walls, 66, 491
Bayt Ghazalah private residence, 271
Bimaristan Arghun, 53, 58, 60, 71, 82, 88, 236
citadel of Aleppo, 33, 53, 65, 66, 71, 84, 174, 180, 182,
397, 479, 540
Farafra khanqah (Dayfa Khatun), 65, 67, 285
Firdaws *madrasa,* 64, 255
Halawiyya mosque, 64–67, 285, 405
Khan al-Sabun, 89
Lower Maqam Ibrahim, 33, 57, 60, 61, 66, 174, 180,
290, 479, 540
Maqam Ibrahim at Salihin, 37, 57, 65, 73
Palace of Malik al-Zahir, 65, 66, 89, 182
Sabun Khan, 89
Sharafiyya *madrasa,* 64, 65
Zahiriyya *madrasa,* 33, 60, 64, 104, 285, 410
Alexandria, Egypt, 87, 247
Fort Qaytbey, 87, 247
Alhambra. *See* Granada
'Ali ibn Abi Talib, 22
'Ali ibn Hilal. *See* Calligraphy
'Ali Sina Balkhi (Avicenna), 24
Al-Jazirah, 22, 56, 57, 63, 70, 579
Al-Maghtas, Jordan, 12
Al-Majdal Asqalan, Palestine (Ashkelon, Israel), 372
Mashhad Nabi Hussein, 372
Almaliq, western China, 71, 117, 118, 240
Tughluq Temür mausoleam, 71, 117, 118
Al-Mansur, 15, 21
Almohad, 105, 106, 112, 258, 260
al-Kutubiyya mosque, Marrakech, 258
Almoravid, 105, 106
Great Mosque of Tlemcen, 106
Amasya, Turkey, 52, 71–73, 80, 86, 93, 127, 161, 171, 197,
240, 247, 269, 288, 369, 391
Bayezid Pasa mosque, 93, 369
Bimarhane hospital, 52, 75, 288

J. Bonner, *Islamic Geometric Patterns,* DOI 10.1007/978-1-4419-0217-7